Lecture Notes in Computer Scie

Commenced Publication in 1973
Founding and Former Series Editors:
Gerhard Goos, Juris Hartmanis, and Jan van Leeuwen

More information about this series at http://www.springer.com/series/7407

Alexander Gelbukh (Ed.)

Computational Linguistics and Intelligent Text Processing

17th International Conference, CICLing 2016
Konya, Turkey, April 3–9, 2016
Revised Selected Papers, Part I

 Springer

Editor
Alexander Gelbukh
CIC, Instituto Politécnico Nacional
Mexico City
Mexico

ISSN 0302-9743   ISSN 1611-3349 (electronic)
Lecture Notes in Computer Science
ISBN 978-3-319-75476-5   ISBN 978-3-319-75477-2   (eBook)
https://doi.org/10.1007/978-3-319-75477-2

Library of Congress Control Number: 2018934347

LNCS Sublibrary: SL1 – Theoretical Computer Science and General Issues

Printed on acid-free paper

This Springer imprint is published by the registered company Springer International Publishing AG
part of Springer Nature
The registered company address is: Gewerbestrasse 11, 6330 Cham, Switzerland

In Memoriam of Adam Kilgarriff

Preface

CICLing 2016 was the 17th International Conference on Intelligent Text Processing and Computational Linguistics. The CICLing conferences provide a wide-scope forum for discussion of the art and craft of natural language processing research, as well as the best practices in its applications.

In 2015, Adam Kilgarriff, an influential scientist and a wonderful and brave person, passed away prematurely, at the age of only 55. Adam was a great friend of CICLing, a member of its small informal Steering Committee, one who helped to shape CICLing from its inception. Until his last days, already terminally ill, he volunteered to help us with the reviewing process for CICLing 2015. This CICLing event was dedicated to his bright memory, and its proceedings begin with a paper that attempts to summarize his scientific legacy.

This set of two books contains five invited papers and a selection of regular papers accepted for presentation at the conference. Since 2001, the proceedings of the CICLing conferences have been published in Springer's *Lecture Notes in Computer Science* series as volumes 2004, 2276, 2588, 2945, 3406, 3878, 4394, 4919, 5449, 6008, 6608, 6609, 7181, 7182, 7816, 7817, 8403, 8404, 9041, and 9042.

The set has been structured into 14 sections: an In Memoriam section and 13 sections representative of the current trends in research and applications of natural language processing:

- General formalisms
- Embeddings, language modeling, and sequence labeling
- Lexical resources and terminology extraction
- Morphology and part-of-speech tagging
- Syntax and chunking
- Named entity recognition
- Word sense disambiguation and anaphora resolution
- Semantics, discourse, and dialog
- Machine translation and multilingualism
- Sentiment analysis, opinion mining, subjectivity, and social media
- Text classification and categorization
- Information extraction
- Applications

The 2016 event received submissions from 54 countries. A total of 298 papers by 671 authors were submitted for evaluation by the international Program Committee (see Fig. 1 and Tables 1 and 2). This two-volume set contains revised versions of 89 regular papers selected for presentation, with the acceptance rate of 29.8%.

Table 1. Number of submissions and accepted papers by topic[a]

Submissions	Topic
60	Text mining
58	Information extraction
53	Lexical resources
44	Emotions, sentiment analysis, opinion mining
43	Information retrieval
42	Semantics, pragmatics, discourse
39	Clustering and categorization
36	Under-resourced languages
35	Machine translation and multilingualism
33	Morphology
33	Practical applications
28	Social networks and microblogging
25	Named entity recognition
25	POS tagging
23	Syntax and chunking
21	Noisy text processing and cleaning
20	Formalisms and knowledge representation
16	Plagiarism detection and authorship attribution
16	Question answering
16	Summarization
16	Word sense disambiguation
13	Computational terminology
13	Natural language interfaces
12	Other
11	Natural language generation
10	Speech processing
7	Spelling and grammar check
7	Textual entailment
6	Coreference resolution

[a]As indicated by the authors. A paper may belong to more than one topic.

In addition to regular papers and an In Memoriam paper, the books features papers by the keynote speakers:

- Pascale Fung, Hong Kong University of Science and Technology, Hong Kong
- Tomas Mikolov, Facebook AI Research, USA
- Simone Teufel, University of Cambridge, UK
- Piek Vossen, Vrije Universiteit Amsterdam, The Netherlands

Publication of full-text invited papers in the proceedings is a distinctive feature of the CICLing conferences. In addition to a presentation of their invited papers, the keynote speakers organized separate lively informal events; this is also a special feature of this conference series.

Table 2. Number of submitted and accepted papers by country or region

Country or region	Authors	Submissions[b]	Country or region	Authors	Submissions[b]
Algeria	15	6.5	Libya	2	0.5
Argentina	2	1	Mexico	8	5.17
Australia	5	1.33	Morocco	13	4.81
Austria	1	0.33	Nigeria	3	1
Brazil	8	3.75	Norway	10	3.96
Canada	24	10.78	Pakistan	1	1
China	23	11.25	Peru	3	2
Colombia	7	3	Poland	2	1
Czechia	19	8	Portugal	6	2.05
Egypt	17	6	Qatar	11	2.83
Finland	1	0.25	Romania	12	5.75
France	56	19.52	Russia	12	7.25
Germany	15	5.93	Saudi Arabia	5	1.3
Greece	3	1	Singapore	8	2.5
Hong Kong	2	1	South Africa	3	1
Hungary	6	5	South Korea	7	4
India	73	37.3	Spain	9	3.78
Indonesia	3	1	Sri Lanka	12	5
Iran	5	2.33	Switzerland	4	2
Ireland	3	1.25	The Netherlands	1	0.2
Israel	10	3.24	Tunisia	86	40.35
Italy	5	1	Turkey	53	29.22
Japan	13	3.5	Turkmenistan	1	0.33
Jordan	4	3	UAE	4	1.5
Kazakhstan	19	6.75	UK	18	9.17
Latvia	2	1	USA	30	13.4
Lebanon	2	0.67	Vietnam	4	1.25
			Total:	671	298

[b] By the number of authors: e.g., a paper by two authors from the USA and one from UK is counted as 0.67 for the USA and 0.33 for UK.

With this event, we continued our policy of giving preference to papers with verifiable and reproducible results: In addition to the verbal description of their findings given in the paper, we encouraged the authors to provide a proof of their claims in electronic form. If the paper claimed experimental results, we asked the authors to make available to the community all the input data necessary to verify and reproduce these results; if it claimed to introduce an algorithm, we encourage the authors to make the algorithm itself, in a programming language, available to the public. This additional electronic material will be permanently stored on the CICLing's server, www.CICLing.org, and will be available to the readers of the corresponding paper for download under a license that permits its free use for research purposes.

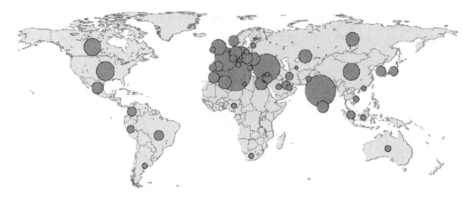

Fig. 1. Submissions by country or region. The area of a circle represents the number of submitted papers

In the long run, we expect that computational linguistics will have verifiability and clarity standards similar to those of mathematics: In mathematics, each claim is accompanied by a complete and verifiable proof, usually much longer than the claim itself; each theorem's complete and precise proof—and not just a description of its general idea—is made available to the reader. Electronic media allow computational linguists to provide material analogous to the proofs and formulas in mathematic in full length—which can amount to megabytes or gigabytes of data—separately from a 12-page description published in the book. More information can be found on http://www.CICLing.org/why_verify.htm.

To encourage providing algorithms and data along with the published papers, we selected three winners of our Verifiability, Reproducibility, and Working Description Award. The main factors in choosing the awarded submission were technical correctness and completeness, readability of the code and documentation, simplicity of installation and use, and exact correspondence to the claims of the paper. Unnecessary sophistication of the user interface was discouraged; novelty and usefulness of the results were not evaluated—instead, they were evaluated for the paper itself and not for the data.

The following papers received the Best Paper Awards, the Best Student Paper Award, as well as the Verifiability, Reproducibility, and Working Description Awards, respectively:

Best Paper 1st Place: "Mining the Web for Collocations: IR Models of Term Associations," by Rakesh Verma, Vasanthi Vuppuluri, An Nguyen, Arjun Mukherjee, Ghita Mammar, Shahryar Baki, and Reed Armstrong, USA
                           "Extracting Aspect Specific Sentiment Expressions Implying Negative Opinions," by Arjun Mukherjee, USA
Best Paper 2nd Place: "Word Sense Disambiguation Using Swarm Intelligence: A Bee Colony Optimization Approach," by Saket Kumar and Omar El Ariss, USA

Best Paper 3rd Place:	"Corpus Frequency and Affix Ordering in Turkish," by Mustafa Aksan, Umut Ufuk Demirhan, and Yeşim Aksan, Turkey
Best Student Paper[1]:	"Generating Bags of Words from the Sums of Their Word Embeddings," by Lyndon White, Roberto Togneri, Wei Liu, and Mohammed Bennamoun, Australia
Verifiability 1st Place:	"Pluralizing Nouns in isiZulu and Related Languages," by Joan Byamugisha, C. Maria Keet, and Langa Khumalo, South Africa
Verifiability 2nd Place:	"A Free/Open-Source Hybrid Morphological Disambiguation Tool for Kazakh,"[2] by Zhenisbek Assylbekov, Jonathan North Washington, Francis Tyers, Assulan Nurkas, Aida Sundetova, Aidana Karibayeva, Balzhan Abduali, and Dina Amirova, Kazakhstan, USA, and Norway
Verifiability 3rd Place:	"An Informativeness Approach to Open IE Evaluation," by William Léchelle and Philippe Langlais, Canada

The authors of the awarded papers (except for the Verifiability Award) were given extended time for their presentations. In addition, the Best Presentation Award and the Best Poster Award winners were selected by a ballot among the attendees of the conference.

Besides its high scientific level, one of the success factors of the CICLing conferences is their excellent cultural program in which all attendees participate. The cultural program is a very important part of the conference, serving its main purpose: personal interaction and making friends and contacts. The attendees of the conference had a chance to visit the Tinaztepe Cavern and nearby attractions; the Underground City and the amazing Cappadocia landscape, probably the most astonishing natural landscape I have seen so far; as well as the attractions and historial monuments of the city of Konya, the ancient capital of the Seljuk Sultanate of Rum and the Karamanids.

The conference was accompanied by two satellite events: the Second International Conference on Arabic Computational Linguistics, ACLing 2016, and the First International Conference on Turkic Computational Linguistics, TurCLing 2016, both founded by CICLing. This is in accordance with CICLing's mission to promote consolidation of emerging NLP communities in countries and regions underrepresented in the mainstream of NLP research and, in particular, in the mainstream publication venues. The Program Committees of ACLing 2016 and TurCLing 2016 have helped the CICLing 2016 committee in evaluation of a number of submissions on the corresponding narrow topics.

---

[1] The best student paper was selected among papers of which the first author was a full-time student, excluding the papers that received Best Paper Awards.

[2] This paper is published in the proceedings of a satellite event of the conference and not in this book set.

I would like to thank all those involved in the organization of this conference. In the first place, the authors of the papers that constitute this book: It is the excellence of their research work that gives value to the book and meaning to the work of all other people. I thank all those who served on the Program Committee of CICLing 2016, the Second International Conference on Arabic Computational Linguistics, ACLing 2016, and the First International Conference on Turkic Computational Linguistics, TurCLing 2016; the Software Reviewing Committee, the Award Selection Committee, as well as additional reviewers, for their hard and very professional work. Special thanks go to Ted Pedersen, Manuel Vilares Ferro, and Soujanya Poria for their invaluable support in the reviewing process.

I also want to cordially thank the conference staff, volunteers, and the members of the local Organizing Committee headed by Hatem Haddad, as well as the organizers of the satellite events, Samhaa R. El-Beltagy of Nile University, Egypt, Khaled Shaalan of the British University in Dubai, UAE, and Bahar Karaoglan of Ege University, Turkey. I am deeply grateful to the administration of the Mevlana University for their helpful support, warm hospitality, and in general for providing this wonderful opportunity of holding CICLing in Turkey. I acknowledge support from the project CON-ACYT Mexico–DST India 122030 "Answer Validation Through Textual Entailment" and SIP-IPN grant 20161958.

The entire submission and reviewing process was supported for free by the Easy-Chair system (www.EasyChair.org). Last but not least, I deeply appreciate the Springer team's patience and help in editing these volumes and getting them printed in very short time—it is always a great pleasure to work with Springer.

December 2016   Alexander Gelbukh

Organization

CICLing 2016 was hosted by the Mevlana University, Turkey, and was organized by the CICLing 2016 Organizing Committee in conjunction with the Mevlana University, the Natural Language and Text Processing Laboratory of the Centro de Investigación en Computación (CIC) of the Instituto Politécnico Nacional (IPN), Mexico, and the Mexican Society of Artificial Intelligence (SMIA).

Organizing Chair

Hatem Haddad   Université Libre de Bruxelles, Belgium

Organizing Committee

Hatem Haddad (Chair)	Université Libre de Bruxelles, Belgium
Armağan Ozkaya	Mevlana University, Turkey
Niyazi Serdar Tunaboylu	Mevlana University, Turkey
Alaa Eleyan	Mevlana University, Turkey
Mustafa Kaiiali	Mevlana University, Turkey
Mohammad Shukri Salman	Mevlana University, Turkey
Gülden Eleyan	Mevlana University, Turkey
Hasan Ucar	Mevlana University, Turkey

Program Chair

Alexander Gelbukh   Instituto Politécnico Nacional, Mexico

Program Committee

Ajith Abraham	Machine Intelligence Research Labs (MIR Labs), USA
Rania Al-Sabbagh	University of Illinois at Urbana-Champaign, USA
Galia Angelova	Bulgarian Academy of Sciences, Bulgaria
Marianna Apidianaki	LIMSI-CNRS, France
Alexandra Balahur	European Commission Joint Research Centre, Italy
Sivaji Bandyopadhyay	Jadavpur University, India
Leslie Barrett	Bloomberg, LP, USA
Roberto Basili	University of Rome Tor Vergata, Italy
Anja Belz	University of Brighton, UK
Christian Boitet	Université Joseph Fourier and LIG, GETALP, France
Igor Bolshakov	Independent Researcher, Russia
Nicoletta Calzolari	Istituto di Linguistica Computazionale – CNR, Italy
Erik Cambria	Nanyang Technological University, Singapore

Nick Campbell	TCD, Ireland
Michael Carl	Copenhagen Business School, Denmark
Niladri Chatterjee	IIT Delhi, India
Dan Cristea	A. I. Cuza University of Iaşi, Romania
Samhaa R. El-Beltagy	Nile University, Egypt
Michael Elhadad	Ben-Gurion University of the Negev, Israel
Anna Feldman	Montclair State University, USA
Alexander Gelbukh	Instituto Politécnico Nacional, Mexico
Dafydd Gibbon	Universität Bielefeld, Germany
Roxana Girju	University of Illinois at Urbana-Champaign, USA
Gregory Grefenstette	IHMC, USA
Hatem Haddad	Université Libre de Bruxelles, Belgium
Eva Hajičová	Charles University, Czechia
Sanda Harabagiu	The University of Texas at Dallas, USA
Yasunari Harada	Waseda University, Japan
Ales Horak	Masaryk University, Czechia
Nancy Ide	Vassar College, USA
Diana Inkpen	University of Ottawa, Canada
Aminul Islam	University of Louisiana at Lafayette, USA
Guillaume Jacquet	JRC, Italy
Miloš Jakubíček	Lexical Computing, Czech Republic
Doug Jones	MIT, USA
Sylvain Kahane	Modyco, Université Paris Ouest Nanterre La Défense and CNRS/Alpage, Inria, France
Dimitar Kazakov	University of York, UK
Alma Kharrat	Microsoft, USA
Philipp Koehn	Johns Hopkins University, USA
Leila Kosseim	Concordia University, Canada
Mathieu Lafourcade	LIRMM, France
Bing Liu	University of Illinois at Chicago, USA
Cerstin Mahlow	University of Stuttgart, Germany
Suresh Manandhar	University of York, UK
Diana Mccarthy	University of Cambridge, UK
Alexander Mehler	Goethe University Frankfurt am Main, Germany
Farid Meziane	University of Salford, UK
Rada Mihalcea	University of North Texas/Oxford University, USA
Evangelos Milios	Dalhousie University, Canada
Ruslan Mitkov	University of Wolverhampton, UK
Dunja Mladenic	Jozef Stefan Institute, Slovenia
Hermann Moisl	Newcastle University, UK
Masaki Murata	Tottori University, Japan
Preslav Nakov	Qatar Computing Research Institute, Qatar Foundation, Qatar
Costanza Navarretta	University of Copenhagen, Denmark
Nicolas Nicolov	J.D. Power and Associates/McGraw-Hill, USA
Joakim Nivre	Uppsala University, Sweden

Kjetil Nørvåg	Norwegian University of Science and Technology, Norway
Attila Novák	Pázmány Péter Catholic University, Hungary
Nir Ofek	Ben-Gurion University, Israel
Partha Pakray	National Institute of Technology Mizoram, India
Ivandre Paraboni	University of São Paulo, Brazil
Patrick Saint-Dizier	IRIT-CNRS, France
Maria Teresa Pazienza	University of Rome Tor Vergata, Italy
Ted Pedersen	University of Minnesota, USA
Viktor Pekar	University of Birmingham, UK
Anselmo Peñas	UNED, Spain
Soujanya Poria	Nanyang Technological University, Singapore
Marta R. Costa-Jussà	Institute for Infocomm Research, Singapore
Fuji Ren	University of Tokushima, Japan
German Rigau	IXA Group, UPV/EHU, Spain
Fabio Rinaldi	University of Zurich, Switzerland
Horacio Rodriguez	Universitat Politècnica de Catalunya, Spain
Paolo Rosso	Universitat Politècnica de València, Spain
Vasile Rus	The University of Memphis, USA
Franco Salvetti	University of Colorado at Boulder/Microsoft, USA
Rajeev Sangal	Language Technologies Research Centre, India
Kepa Sarasola	Euskal Herriko Unibertsitatea, Spain
Fabrizio Sebastiani	Qatar Computing Research Institute, Qatar
Serge Sharoff	UoLeeds, UK
Bernadette Sharp	Staffordshire University, UK
Grigori Sidorov	Instituto Politécnico Nacional, Mexico
Kiril Simov	Linguistic Modelling Laboratory, IICT-BAS, Bulgaria
John Sowa	VivoMind Intelligence, USA
Efstathios Stamatatos	University of the Aegean, Greece
Maosong Sun	Tsinghua University, China
Jun Suzuki	NTT, Japan
Stan Szpakowicz	University of Ottawa, Canada
Hristo Tanev	Independent Researcher, Italy
Juan-Manuel Torres-Moreno	Laboratoire Informatique d'Avignon/UAPV, France
George Tsatsaronis	Technical University of Dresden, Germany
Olga Uryupina	University of Trento, Italy
Manuel Vilares Ferro	University of Vigo, Spain
Aline Villavicencio	Federal University of Rio Grande do Sul, Brazil
Piotr W. Fuglewicz	TiP Sp. z o. o., Poland
Marilyn Walker	UCSC, USA
Andy Way	ADAPT Centre, Dublin City University, Ireland
Bonnie Webber	The University of Edinburgh, UK
Alisa Zhila	IBM, USA

Software Reviewing Committee

Ted Pedersen	University of Minnesota, USA
Florian Holz	University of Leipzig, Germany
Miloš Jakubíček	Masaryk University, Czech Republic
Sergio Jiménez Vargas	Instituto Caro y Cuervo, Colombia
Miikka Silfverberg	University of Colorado, USA
Ronald Winnemöller	University of Hamburg, Germany

Award Committee

Alexander Gelbukh	Instituto Politécnico Nacional, Mexico
Eduard Hovy	Carnegie Mellon University, USA
Rada Mihalcea	University of Michigan, USA
Ted Pedersen	University of Minnesota, USA
Yorick Wilks	University of Sheffield, UK

Additional Reviewers

Abhijit Mishra
Abidalrahman Moh'D
Aitor García
Alisa Zhila
Aljaz Kosmerlj
Amir Hossein Razavi
Avi Hayoun
Benjamin Marie
Bernardo Cabaleiro
Bryan Rink
Carla Parra Escartín
Diana Trandabăț
Egoitz Laparra
Elnaz Davoodi
Enrique Flores
Francisco J. Ribadas-Pena
Hanna Bechara
Helena Gómez Adorno
Hiram Calvo
Ilia Markov
Irazú Hernández Farias
Jan R. Benetka
Jan Rupnik
Janez Starc
Jared Bernstein
Jie Mei
Jumana Nassour-Kassis
Kazuhiro Takeuchi

Kyoko Kanzaki
Leonardo Zilio
Magdalena Jankowska
Mahsa Forati
Maite Giménez
Majid Laali
Marc Franco Salvador
Marcelo Sardelich
Michael Mohler
Miguel Angel Rios Gaona
Olga Kolesnikova
Petya Osenova
Pidong Wang
Pistol Ionut Cristian
Richard Evans
Rodrigo Wilkens
Silvio Ricardo Cordeiro
Sonia Ordoñez Salinas
Svetla Boytcheva
Tal Baumel
Travis Goodwin
Victor Darriba
Victoria Yaneva
Vinita Nahar
Vojtech Kovar
Xin Chen
Zuzana Neverilova

Second Arabic Computational Linguistics Conference

Program Chairs

Samhaa R. El-Beltagy	Nile University, Egypt
Alexander Gelbukh	Instituto Politécnico Nacional, Mexico
Khaled Shaalan	The British University in Dubai, UAE

Program Committee

Bayan Abushawar	Arab Open University, Jordan
Hanady Ahmed	Qatar University, Qatar
Hend Alkhalifa	King Saud University, Saudi Arabia
Mohammed Attia	Al-Azhar University, Egypt
Aladdin Ayesh	De Montfort University, UK
Karim Bouzoubaa	Mohamed V University, Morocco
Violetta Cavalli-Sforza	Al Akhawayn University, Morocco
Khalid Choukri	ELDA, France
Samhaa R. El-Beltagy	Nile University, Egypt
Ossama Emam	ALTEC, Egypt
Aly Fahmy	Cairo University, Egypt
Ahmed Guessoum	University of Science and Technology Houari Boumediene, Algeria
Nizar Habash	New York University Abu Dhabi, UAE
Hatem Haddad	Université Libre de Bruxelles, Belgium
Kais Haddar	MIRACL Laboratory, Faculté des sciences de Sfax, Tunisia
Lamia Hadrich Belguith	MIRACL Laboratory, Tunisia
Imtiaz Khan	King Abdulaziz University, Jeddah, Saudi Arabia
Alma Kharrat	Microsoft, USA
Sattar Izwaini	American University of Sharjah, UAE
Mark Lee	University of Birmingham, UK
Sherif Mahdy Abdou	RDI, Egypt
Farid Meziane	University of Salford, UK
Farhad Oroumchian	University of Wollongong in Dubai, UAE
Hermann Moisl	Newcastle University, UK
Ahmed Rafea	American University in Cairo, Egypt
Allan Ramsay	The University of Manchester, UK
Mohsen Rashwan	Cairo University, Egypt
Horacio Rodriguez	Universitat Politècnica de Catalunya, Spain
Paolo Rosso	Universitat Politècnica de València, Spain
Nasredine Semmar	CEA, France
Khaled Shaalan	The British University in Dubai, UAE
William Teahan	Bangor University, UK
Imed Zitouni	IBM Research, USA

Additional Reviewers

Asma Aouichat
Hind Saddiki
Mohamed Hadj Ameur
Muhammad Shuaib Qureshi

Muhammad Umair
Nora Al-Twairesh
Riadh Belkebir

First International Conference on Turkic Computational Linguistics

Program Chairs

Bahar Karaoglan (Chair)	Ege University, Turkey
Tarık Kışla (Co-chair)	Ege University, Turkey
Senem Kumova (Co-chair)	İzmir Ekonomi University, Turkey
Hatem Haddad (Co-chair)	Université Libre de Bruxelles, Belgium

Program Committee

Yeşim Aksan	Mersin University, Turkey
Adil Alpkocak	Dokuz Eylul University, Turkey
Ildar Batyrshin	Instituto Politécnico Nacional, Mexico
Cem Bozsahin	Middle East Technical University, Turkey
Fazli Can	Bilkent University, Turkey
Ilyas Cicekli	Hacettepe University, Turkey
Gülşen Eryiğit	Istanbul Technical University, Turkey
Alexander Gelbukh	Instituto Politécnico Nacional, Mexico
Tunga Gungor	Boğaziçi University, Turkey
Hatem Haddad	Université Libre de Bruxelles, Belgium
Bahar Karaoglan	Ege University, Turkey
Tarik Kisla	Ege University, Turkey
Senem Kumova Metin	İzmir University of Economics, Turkey
Altynbek Sharipbayev	L. N. Gumilyov Eurasian National University, Kazakhstan
Dzhavdet Suleymanov	Academy of Sciences of Tatarstan, Russia

Additional Reviewers

A. Sumru Özsoy
Ali Erkan
Cem Rıfkı Aydın
Dilyara Yakubova

Ezgi Yıldırım
Nihal Yağmur Aydın
Razieh Ehsani

Website and Contact

The website of the CICLing conference series is http://www.CICLing.org. It provides information about past CICLing conferences and their satellite events, including links to published papers (many of them in open access) or their abstracts, as well as photos and video recordings of keynote talks. In addition, it provides data, algorithms, and open-source software accompanying accepted papers, in accordance with the CICLing verifiability, reproducibility, and working description policy. It also provides information about the forthcoming CICLing events, as well as contact options.

Contents – Part I

Lexical Resources and Terminology Extraction

Best Paper Award, First Place:

Morphology and Part-of-Speech Tagging

Best Paper Award, Third Place:

Syntax and Chunking

Named Entity Recognition

Word Sense Disambiguation and Anaphora Resolution

Best Paper Award, Second Place:

Semantics, Discourse, and Dialog

Contents – Part II

Sentiment Analysis, Opinion Mining, Subjectivity, and Social Media

Invited Paper:

Best Paper Award, First Place:

Applications

Invited Papers:

In Memoriam

Adam Kilgarriff's Legacy
to Computational Linguistics and Beyond

Roger Evans[1], Alexander Gelbukh[2], Gregory Grefenstette[3], Patrick Hanks[4],
Miloš Jakubíček[5,6], Diana McCarthy[7(✉)], Martha Palmer[8], Ted Pedersen[9],
Michael Rundell[10], Pavel Rychlý[5,6], Serge Sharoff[11], and David Tugwell[12]

[1] University of Brighton, Brighton, UK
R.P.Evans@brighton.ac.uk
[2] CIC, Instituto Politécnico Nacional, Mexico City, Mexico
gelbukh@gelbukh.com
[3] IHMC, Ocala, FL, USA
ggrefenstette@ihmc.us
[4] University of Wolverhampton, Wolverhampton, UK
patrick.w.hanks@gmail.com
[5] Lexical Computing, Brighton, UK
milos.jakubicek@sketchengine.co.uk, pary@fi.muni.cz
[6] Masaryk University, Brno, Czech Republic
[7] DTAL University of Cambridge, Cambridge, UK
diana@dianamccarthy.co.uk
[8] University of Colorado, Boulder, USA
martha.palmer@colorado.edu
[9] University of Minnesota, Minneapolis, USA
tpederse@d.umn.edu
[10] Lexicography MasterClass, Brighton, UK
michael.rundell@lexmasterclass.com
[11] University of Leeds, Leeds, UK
s.sharoff@leeds.ac.uk
[12] Independent Researcher, Edinburgh, UK

Abstract. The 2016 CICLing conference was dedicated to the memory of Adam Kilgarriff who died the year before. Adam leaves behind a tremendous scientific legacy and those working in computational linguistics, other fields of linguistics and lexicography are indebted to him. This paper is a summary review of some of Adam's main scientific contributions. It is not and cannot be exhaustive. It is written by only a small selection of his large network of collaborators. Nevertheless we hope this will provide a useful summary for readers wanting to know more about the origins of work, events and software that are so widely relied upon by scientists today, and undoubtedly will continue to be so in the foreseeable future.

© Springer International Publishing AG, part of Springer Nature 2018
A. Gelbukh (Ed.): CICLing 2016, LNCS 9623, pp. 3–25, 2018.
https://doi.org/10.1007/978-3-319-75477-2_1

1   Introduction

The year 2015 was marred by the loss of Adam Kilgarriff who during the last 27 years of his life contributed greatly to the field of computational linguistics[1], as well as to other fields of linguistics and to lexicography. This paper provides a review of some of the key scientific contributions he made. His legacy is impressive, not simply in terms of the numerous academic papers, which are widely cited in many fields, but also the many scientific events and communities he founded and fostered and the commercial **Sketch Engine** software. The Sketch Engine has provided computational linguistics tools and corpora to scientists in other fields, notably lexicography for example [17,50,61], as well as facilitating research in other areas of linguistics [11,12,54,56] and our own subfield of computational linguistics [60,74].

Adam was hugely interested in lexicography from the very inception of his postgraduate career. His DPhil[2] on polysemy and subsequent interest in word sense disambiguation (WSD) and its evaluation was firmly rooted in examining corpus data and dictionary senses with a keen eye on the lexicographic process [20]. After his DPhil, Adam spent several years as a computational linguist advising Longman Dictionaries on the use of language engineering for the development of lexical databases, and he continued this line of knowledge transfer in consultancies with other publishers until realizing the potential of computational linguistics with the development of his commercial software, the Sketch Engine. The origins of this software lay in his earlier ideas of using computational linguistics tools for providing word profiles from corpus data.

For Adam, data was key. He fully appreciated the need for empirical approaches to both computational linguistics and lexicography. In computational linguistics from the 90s onwards there was a huge swing from symbolic to statistical approaches, however the choice of input data, in composition and size, was often overlooked in favor of a focus on algorithms. Furthermore, early on in this statistical tsunami, issues of replicability were not always appreciated. A large portion of his work was devoted to these issues, in his work on WSD evaluation and in his work on building and comparing corpora. His signature company slogan was 'corpora for all'.

This paper has come together from a small sample of the very large pool of Adam's collaborators. The sections have been written by different subsets of the authors and with different perspectives on Adam's work and on his ideas. We hope that this approach will give the reader an overview of some of Adam's main scientific contributions to both academia and the commercial world, while not detracting too greatly from the coherence of the article.

The article is structured as follows. Section 2 outlines Adam's thesis and origins of his thoughts on word senses and lexicography. Section 3 continues with his

---

[1] In this paper, natural language processing (NLP) is used synonymously with computational linguistics.

[2] Like Oxford, the University of Sussex, where Adam undertook his doctoral training, uses **DPhil** rather than **PhD** as the abbreviation for its doctoral degrees.

subsequent work on WSD evaluation in the **Senseval** series as well as discussing his qualms about the adoption of dictionary senses in computational linguistics as an act of faith without a specific purpose in mind. Section 4 summarizes his early work on using corpus data to provide word profiles in a project known as the **WASP-bench**, the precursor to his company's[3] commercial software, the Sketch Engine. Corpus data lay at the very heart of these word profiles, and indeed just about all of Computational Linguistics from the mid 90s on. Section 5 discuss Adam's ideas for building and comparing corpus data, while Sect. 6 describes the Sketch Engine itself. Finally Sect. 7 details some of the impact Adam has had transferring ideas from computational and corpus linguistics to the field of lexicography.

2   Adam's Doctoral Research

To lay the foundation for an understanding of Adam's contribution to our field, an obvious place to start is his DPhil thesis [19]. But let us first sketch out the background in which he undertook his doctoral research. Having obtained first class honours in Philsophy and Engineering at Cambridge in 1982, Adam had spent a few years away from academia before arriving at Sussex in 1987 to undertake the Masters in Intelligent Knowledge-Based Systems, a programme which aimed to give non-computer scientists a grounding in Cognitive Science and Artificial Intelligence. This course introduced him to **Natural Language Processing** (NLP) and lexical semantics, and in 1988 he enrolled on the DPhil program, supervised by Gerald Gazdar and Roger Evans. At that time, NLP had moved away from its roots in Artificial Intelligence towards more formal approaches, with increasing interest in formal lexical issues and more elaborate models of lexical structure, such as Copestake's LKB [6] and Evans and Gazdar's DATR [9]. In addition, the idea of improving lexical coverage by exploiting digitized versions of dictionaries was gaining currency, although the advent of large-scale corpus-based approaches was still some way off. In this context, Adam set out to explore **Polysemy**, or as he put it himself:

> *What does it mean to say a word has several meanings? On what grounds do lexicographers make their judgments about the number of meanings a word has? How do the senses a dictionary lists relate to the full range of ways a word might get used? How might NLP systems deal with multiple meanings?* [19, p. 1]

Two further quotes from Adam's thesis neatly summarize the broad interdisciplinarity which characterized his approach to his thesis, and throughout his research career. The first is from the Preface:

---

[3] The company he founded is **Lexical Computing Ltd.** He was also a partner – with Sue Atkins and Michael Rundell – in another company, **Lexicography Master-Class**, which provides consultancy and training and runs the **Lexicom** workshops in lexicography and lexical computing; http://www.lexmasterclass.com/.

> *There are four kinds of thesis in cognitive science: formal, empirical, program-based and discursive. What sort was mine to be? ... I look round in delight to find [my thesis] does a little bit of all the things a cognitive science thesis might do!* [19, p. 6]

while the second is in the introduction to his discussion of methodology:

> *We take the study of the lexicon to be intimately related to the study of the mind ... . For an understanding of the lexicon, the contributing disciplines are lexicography, psycholinguistics and theoretical, computational and corpus linguistics.* [19, p. 4]

The distinctions made in the first of these quotes provide a neat framework to discuss the content of the thesis in more detail. Adam's own starting point was *empirical*: in two studies he demonstrated first that the range and type of sense distinctions found in a typical dictionary defied any simple systematic classification, and second that the so-called **bank model** of word senses (where senses from dictionaries were considered to be distinct and easy to enumerate and match to textual instances) did not in general reflect actual dictionary sense distinctions (which tend to overlap). A key practical consequence of this is that the then-current NLP WSD systems which assumed the bank model could never achieve the highest levels of performance in sense matching tasks.

From this practical exploration, Adam moved to more *discursive* territory. He explored the basis on which lexicographers decide which sense distinctions appear in dictionaries, and introduced an informal criterion to characterize it – the **Sufficiently Frequent and Insufficiently Predictable** (SFIP) condition, which essentially favors senses which are both common and non-obvious. However he noted that while this criterion had empirical validity as a way of circumscribing polysemy in dictionaries, it did not offer any clear understanding of the nature of polysemy itself. He argued that this is because polysemy is not a 'natural kind' but rather a cover term for several other more specific but distinct phenomena: homonymy (the bank model), alternation (systematic usage differences), collocation (lexically contextualized usage) and analogy.

This characterization led into the *formal/program-based* contribution (which in the spirit of logic-based programming paradigms collapse into one) of his thesis, for which he developed two formal descriptions of lexical alternations using the inheritance-based lexical description language DATR. His aim was to demonstrate that while on first sight much of the evidence surrounding polysemy seemed unruly and arbitrary, it was nevertheless possible, with a sufficiently expressive formal language, to characterize substantial aspects of the problem in a formal, computationally tractable way.

Adam's own summary of the key contributions of his work were typically succinct:

> *The thesis makes three principal claims, one empirical, one theoretical, and one formal and computational. The first is that the Bank Model is fatally flawed. The second is that polysemy is a concept at a crossroads, which*

must be understood in terms of its relation to homonymy, alternations, collocations and analogy. The third is that many of the phenomena falling under the name of polysemy can be given a concise formal description in a manner ... which is well-suited to computational applications. [19, p. 8]

With the benefit of hindsight, we can see in this thesis many of the key ideas which Adam developed over his career. In particular, the beginnings of his empirical, usage-based approach to understanding lexical behaviour, his interest in lexicography and support for the lexicographic process, and his ideas for improving the methodology and development of computational WSD systems probably first came together as the identifiable start of his subsequent journey in [20], and will all feature prominently in the remainder of this review.

What is perhaps more surprising from our present perspective is his advocacy of formal approaches to achieve some of these goals, in particular relating to NLP. Of course, in part this is just a consequence of the times and environment (and supervisory team) of his doctoral study. But while Adam was later in the forefront of lexicographic techniques based on statistical machine learning rather than formal modeling, he still retained an interest in formalizing the structure of lexical knowledge, for example in his contributions to the development of a formal mark-up scheme for dictionary entries as part of the Text Encoding Initiative [8,14,15].

3   Word Sense Disambiguation Evaluation: SENSEVAL

3.1   The Birth of Senseval98

After the years spent studying polysemy, no one understood the complexity and richness of word meanings better than Adam [62]. He looked askance at the NLP community's desire to reduce word meaning to a straight-forward classification problem, as though labeling a word in a sentence with "sense2" offered a complete solution. At the 1997 SIGLEX Workshop organised by Martha Palmer and Mark Light, which used working sessions to focus on determining appropriate evaluation techniques, Adam was a key figure, and strongly influenced the eventual plan for evaluation that gave birth to the **Senseval98** workshop, co-organized by Adam and Martha Palmer. The consensus the workshop participants came to at this meeting were clearly summarized in [24]. During the working session Adam went to great pains to explain to the participants the limitations of dictionary entries and the importance of choosing the right sense inventory, a view for which he was already well known [21,22,25]. This is well in line with the rule of thumb for all supervised machine learning: the better the original labeling, the better the resulting systems. Where word senses were concerned, it had previously not been clearly understood that the sense inventory is the key to the labeling process. This belief also prompted Adam's focus on introducing **Hector** [1] as the sense inventory for Senseval98 [43]. Although Hector covered only a subset of English vocabulary, the entries had been developed by using a corpus-based approach to produce traditional hierarchical dictionary definitions including detailed, informative descriptions of each sense [44].

This focus on high quality annotation extended to Adam's commitment to not just high **inter–tagger agreement** (ITA) but also **replicability**. Replicability measures the agreement rate between two separate teams, each of 3 annotators, who perform double-blind annotation with the third annotator adjudicating. After the tagging for Senseval98 was completed, Adam went back and measured replicability for 4 of the lexical items, achieving a staggering 95.5% agreement rate. Inter-annotator agreement of over 80% for all the tagged training data was also achieved. Adam's approach allowed for discussion and revision of ambiguities in lexical entries before tagging the final test data and calculating the ITA.

Senseval98 demonstrated to the community that there was still substantial room for improvement in the production of annotations for WSD, and spawned a second and then a third Senseval, now known as **Senseval2** [64] and **Senseval3** [59], and Senseval98 is now **Senseval1**. There was a striking difference between the ITA for Senseval98 [26,43], and the ITA for WordNet lexical entries for Senseval2, tagged by Palmer's team at Penn, which was only 71% for verbs. The carefully crafted Hector entries made a substantial difference. With lower ITA, the best system performance on the Senseval2 data was only 64%. When closely related senses were grouped together into more coarse grained senses, the ITA improved to 82%, and the system performance rose a similar amount. By the end of Senseval2 we were all converts to Adam's views on the crucial importance of sense inventories, and especially on full descriptions of each lexical entry. As the community began applying the same methodology to other semantic annotation tasks, the name was changed to **SemEval**, and the series of SemEval workshops for fostering work in semantic representations continues to this day.

3.2   Are Word Senses Real?

Adam was always thought provoking, and relished starting a good debate whenever (and however) he could. He sometimes achieved this by making rather stark and provocative statements which were intended to initiate those discussions, but in the end did not represent his actual position, which was nearly always far more nuanced. Perhaps the best example of this is his article "'I don't believe in word senses",'[4] [25]. Could a title be any more stark? If you stopped reading at the title, you would understand this to mean that Adam did not believe in word senses.[5] But of course it was never nearly that simple.[6]

Adam worked very hard to connect WSD to the art and practice of lexicography. This was important in that it made it clear that WSD really couldn't be

---

[4] This paper is perhaps Adam's most influential piece, having been reprinted in three different collections since its original publication.

[5] The implication that Adam *did* believe in "word senses" is controversial. There are co-authors of this article in disagreement about Adam's beliefs on word senses. Whatever Adam's beliefs were, we are indebted to him for amplifying the debate [13, 30] and for opening our eyes to other possibilities.

[6] In fact, the title is a quote which Adam attributes to Sue Atkins.

treated as yet another classification task. Adam pointed out that our notion of word senses had very much been shaped by the conventions of printed dictionaries, but that dictionary makers are driven by many practical concerns that have little to do with the philosophical and linguistic foundations of meaning and sense. While consumers have come to expect dictionaries to provide a finite list of discrete senses, Adam argued that this model is not only demonstrably false, it is overly limiting to NLP.

In reality then, what Adam did not believe in were word senses *as typically enumerated in dictionaries*. He also did not believe that word senses should be viewed as atomic units of meaning. Rather, it was the multiple occurrences of a word in context that finally revealed the sense of a word. His actual view about word senses is neatly summarized in the article's abstract, where he writes '. . . word senses exist only relative to a task.' Word senses are dynamic, and have to be interpreted with respect to the task at hand.

3.3   Data, Data and More Data

This emphasis on the importance of context guided his vision for leveraging corpora to better inform lexicographers' models of word senses. One of the main advantages of Hector was its close tie to examples from data, and this desire to facilitate data-driven approaches continued to motivate Adam's research. It is currently very effectively embodied in DANTE [35] and in the Sketch Engine [48].[7] This unquenchable thirst for data also led to Adam's participation in the formation of the Special Interest Group on the Web as Corpus (see Sect. 5). Where better to find endless amounts of freely available text than the World Wide Web?

4   Word Sketches

One of the key intellectual legacies of Adam's body of research is the notion that compiling sophisticated statistical profiles of word usage could form the basis of a tractable and useful bridge between corpus data (concrete and available) and linguistic conceptions of word senses (ephemeral and contentious). We refer to such profiles now as **word sketches**, a term which first appeared in papers around 2001 (for example [49,73]), but their roots go back several years earlier.

Following the completion of his doctoral thesis, Adam worked for three years at Longmans dictionary publishers, contributing to the design of their new dictionary database technology. In 1995, he returned to academia, at the University of Brighton, on a project which aimed to develop techniques to enhance (these days we might say 'enrich') automatically-acquired lexical resources. With his thesis research and lexicographic background, Adam quickly identified WSD as a critical key focus for this research, writing:

---

[7] The Sketch Engine, described in Sect. 6, in particular is an incredibly valuable resource that is used regularly at Colorado for revising English VerbNet class memberships and developing PropBank frame files for several languages.

Our hypothesis is that most NLP applications do not need to disambiguate most words that are ambiguous in a general dictionary; for those they do need to disambiguate, it would be foolish to assume that the senses to be disambiguated between correspond to those in any existing resource; and that identifying, and providing the means to resolve, the salient ambiguities will be a large part of the customization effort for any NLP application-building team. Assuming this is confirmed by the preliminary research, the tool we would provide would be for computer-aided computational lexicography. Where the person doing the customization identified a word with a salient sense-distinction, the tool would help him/her elicit (from an application-specific corpus) the contextual clues which would enable the NLP application to identify which sense applied. [Kilgarriff 1995, personal communication to Roger Evans]

This is probably the earliest description of a tool which would eventually become the Sketch Engine (see Sect. 6), some eight years later.

The project followed a line of research in pursuit of this goal, building on Adam's thoughts on usage-based approaches to understanding lexical behavior, methodology for WSD, and his interest in support for the lexicographic process. The idea was to use parsed corpus data to provide profiles of words in terms of their collocational and syntactic behavior, for example predicate argument structure and slot fillers. This would provide a one page summary[8] of a word's behavior for lexicographers making decisions on word entries based on frequency and predictability. Crucially the software would allow users to switch seamlessly between the word sketch summary and the underlying corpus examples [73].

Adam's ideas on WSD methodology were inspired by Yarowky's 'one sense per collocation' [75] and bootstrapping approach [76]. The bootstrapping approach uses a few seed collocations, or manual labels, for sense distinctions that are relevant to the task at hand. Examples from the corpus that can be labeled with these few collocations are used as an initial set of training data. The system iteratively finds and labels more data from which further sense specific collocations are learned, thereby bootstrapping to extend coverage. Full coverage is achieved by additional heuristics such as 'one sense per document' [10]. Adam appreciated that this approach could be used with a standard WSD data set with a fixed sense inventory [49] but importantly also allow one to define the senses pertinent to the task at hand [46].

As well as the core analytic technology at the heart of the creation of word sketches, Adam always had two much more practical concerns in mind in the development of this approach: the need to deliver effective visualisation and manipulation tools for use by lexicographers (and others), and the need to develop technology that was truly scalable to handle very large corpus resources. His earliest experiments focused primarily on the analytic approach; the key deliverable of the follow-on project, WASPS, was the WASP-bench, a tool which combined off-line compilation of word-sketches with a web-based

---

[8] See Fig. 1, below.

interface exploring the sketches and underlying concordance data that supports them [38]; and the practical (and technological) culmination of this project is, of course, the Sketch Engine (see Sect. 6), with its interactive web interface and very large scale corpus resources.

5 Corpus Development

Understanding the data you work with was the key for Adam. As lexicography and NLP became more corpus-based, their appetite for access to more data seemed inexhaustible. Banko and Brill expressed the view that getting more data always leads to improved performance of tasks such as WSD [3]. Adam had a more nuanced view. On the one hand, he was very much in favor of using as much data as possible, hence his interest to using the power of the Web. On the other hand, he also emphasized the importance of understanding what is under the hood of a large corpus: rather than stating bluntly that my corpus is bigger than yours, a more interesting question is *how* my corpus differs from yours.

5.1 Web as Corpus

Adam's work as a lexicographer came from the corpus-based tradition to dictionary building initiated by Sue Atkins and John Sinclair with their COBUILD project, developing large corpora as a basis for lexicographic work for the Collins Dictionary. To support this work, they created progressively larger corpora of English text, culminating in the 100 million word **British National Corpus** (BNC) [53]. This corpus was designed to cover a wide variety of spoken (10%) and written (90%) 20th century British language use. It was composed of 4124 files, each tagged with a domain (informative, 75%; or imaginative, 25%), a medium tag, and a date (mostly post 1975). The philosophy behind this corpus was that it was representative of English language use, and that it could thus be used to illustrate or explain word meanings. McEnery and Wilson [58] said the word 'corpus' for lexicographic use had, at that time, four connotations: 'sampling and representativeness, finite size, machine-readable form, a standard reference.' In other words, a corpus was a disciplined, well understood, and curated source of language use.

Even though 100 million words seemed like an enormous sample, lexicographers found that the BNC missed some of the natural intuitions that they had about language use. We remember Sue Atkins mentioning in a talk that you could not discover that apples were crisp from the BNC, though she felt that *crisp* would be tightly associated with *apple*. Adam realized in the late 1990s that the newly developing World Wide Web gave access to much larger and useful samples of text than any group of people could curate, and argued [39] that we should reclaim the word 'corpus' from McEnery and Wilson's 'connotations', and consider the Web as a corpus even though it is neither finite, in a practical sense, nor a standard reference.

Early web crawlers, such as the now defunct **Altavista** engine, provided exact counts of words and phrases found in their web index. These could be used to predict the individual corpus sizes of given languages, and showed that the Web contained many orders of magnitude more text than the BNC. Language identifiers could distinguish the language of a text with a high rate of accuracy. And the Web was an open source, from which one could crawl and collect seemingly unlimited amount of text. Recognizing the limitations of statistics output by these search engines because the data on which they were based would fluctuate and thus make any experimental conditions impossible to repeat, Adam coined a phrase: 'Googleology is bad science' [31].

So he led the way in exploiting these characteristics of the Web for corpus-related work, gathering together research on recent applications of using the Web as a corpus in a special issue of *Computational Linguistics* [39], and organizing, with Marco Baroni and Silvia Bernardini, a series of **Web as Corpus** (WAC) workshops illustrating this usefulness, starting with workshops in Forli and Birmingham[9] in 2005, and then in Trento in 2006. These workshops presented initial work on building new corpora via web crawling, on creating search engines for corpus building, on detecting genres and types, and annotating web corpora, in other words, recovering some of the connotations of corpora mentioned by McEnery and Wilson. This work is illustrated by such tools at WebBootCat [5], an online tool for bootstrapping a corpus of text, given a set of seed words, part of the Sketch Engine package, of which more is said below.

Initially the multilingual collection of the Sketch Engine was populated with a range of web corpora (I-XX, where XX is AR, FR, RU, etc.), which were produced by making thousands of queries consisting of general words [70]. Later on, this was enriched with crawled corpora in the TenTen family [63], with the aim of exceeding the size of 10^{10} words per corpus crawled for each language.

One of the issues with getting corpora from the Web is the difficulty in separating what is a normal text, which you can expect in a corpus like the BNC, from what is boilerplate, i.e., navigation, layout or informational elements, which do not contribute to running text. This led to another shared task initiated by Adam, namely on evaluation of Web page cleaning. **CleanEval**[10] was a 'competitive evaluation on the topic of cleaning arbitrary web pages, with the goal of preparing web data for use as a corpus, for linguistic and language technology research and development' [55].

Under the impetus of Adam, Marco Baroni and others, Web as Corpus became a special interest group of the ACL, SIGWAC[11] and has continued to organize WAC workshops yearly. The 2016 version, WAC-X, was held in Berlin, in August 2016. Thanks to Adam's vision, lexicography broke away from a limited, curated view of corpus validation of human intuition, and has embraced a computational approach to building corpora from the Web, and using this new corpus evidence as the source for building human and machine-oriented lexicons.

---

[9] Working papers can be found online at http://wackybook.sslmit.unibo.it.
[10] http://cleaneval.sigwac.org.uk/.
[11] https://sigwac.org.uk/.

5.2   Corpus Analysis

The side of Adam's research reported above shows that the Web can be indeed turned into a huge corpus. However, once we started mining corpora from the Web, the next natural question is to assess their similarity to existing resources. Adam was one of the first who addressed this issue by stating 'There is a void in the heart of corpus linguistics' by referring to the lack of measures of corpus **similarity** and **homogeneity** [27]. His answer was to develop methods to show which corpora are closer to each other or which parts of corpora are similar. A frequency list can be produced for each corpus or a part of the corpus, usually in the form of lemmas or lower-cased word forms. Adam suggested two methods, one based on comparing the ranks in those frequency lists using rank statistics, such as Mann-Whitney [27], the other by using **SimpleMaths**, the ratio of frequencies regularized with an indicator of 'commonness' [33].

An extension of this research was his interest in measuring how good a corpus is. We can easily mine multiple corpora from the Web using different methods and for different purposes. Extrinsic evaluation of a corpus might be performed in a number of ways. For example, Adam suggested a lexicographic task: how good a corpus is for extraction of collocates on the grounds that a more homogeneous corpus is likely to produce more useful collocates given the same set of methods [47].

The frequency lists are good not only for the task of comparing corpora. They were recognized early on as one of the useful outputs of corpus linguistics: for pedagogical applications it is important to know which words are more common, so that they can be introduced earlier. Adam's statistical work with the BNC led to the popularity of his BNC frequency list, which has been used in defining the English Language Teaching curriculum in Japan.[12] However, he also realized the limitations of uncritical applications of the frequency lists by formulating the 'whelks' and 'banana' problems [22]. *Whelk* is a relatively infrequent word in English. However, if a text is about whelks, this word is likely to be used in nearly every sentence. In the end, this can considerably elevate its position in a frequency list, even if this word is used only in a small number of texts. Words like *banana* present an opposite problem: no matter how many times in our daily lives we operate with everyday objects and how important they are for the learners, we do not necessarily refer to them in texts we write. Therefore, their position in the frequency lists can become quite low [34], this also explains the *crisp apples* case mentioned above. The 'banana' problem was addressed in the Kelly project by balancing the frequency lists for a language through translation: a word is more important for the learners if it appears as a translation in several frequency lists obtained from comparable corpora in different languages [37].

6   The Sketch Engine

In terms of practical applications – software – Adam' main legacy is undoubtedly the Sketch Engine [36,48]: a corpus management platform hosting by 2016 hun-

---

[12] Personal communication from Adam to Serge Sharoff.

dreds of preloaded corpora in over 80 languages and allowing users to easily build new corpora either from their own texts or using method like the aforementioned WebBootCAT.

Sketch Engine has two fathers: in 2002 Adam' Kilgarriff met at a workshop in Brighton Pavel Rychlý, a computer scientist developing at the time a simple concordancer (Bonito [68]) based on its own database backbone devised solely for the purposes of corpus indexing (Manatee [67]).

This meeting was pivotal: Adam, fascinated by language and the potential of corpus-enabled NLP methods for lexicography and elsewhere, and looking for somebody to implement his word profiling methodology (see Sect. 4) on a large scale; and Pavel, the not-so-fascinated-by-language but eager to find out how to solve all the computationally interesting tasks corpus processing has brought in an effective manner.

resource (noun) British National Corpus freq = 12658 (112.8 per million)

modifier	6477	1.5	object of	3285	2.2	modifies	1906	0.5	subject of	512	0.6
scarce	163	9.53	allocate	194	9.58	allocation	135	9.42	devote	28	7.69
natural	321	8.94	pool	39	8.43	implication	46	7.09	consume	4	5.36
limited	187	8.86	exploit	64	8.23	management	153	6.98	tie	6	4.87
financial	249	8.3	divert	38	7.86	defense	7	6.68	last	4	4.6
mineral	89	8.19	deploy	31	7.67	Stonier	6	6.65	back	5	4.5
additional	107	7.92	devote	44	7.64	utilisation	7	6.63	stretch	4	4.29
valuable	74	7.86	concentrate	62	7.35	committee	132	6.49	result	6	3.93
extra	88	7.53	utilise	22	7.28	centre	158	6.4	depend	6	3.84
human	134	7.38	conserve	17	7.09	allocator	5	6.4	limit	5	3.59
renewable	33	7.31	lack	37	7.0	depletion	6	6.21	match	3	3.58
adequate	49	7.28	reallocate	13	6.98	pack	17	6.2	share	6	3.55
non renewable	25	6.97	mobilise	13	6.83	investigator	8	6.17	earn	3	3.55
existing	53	6.68	mobilize	13	6.79	column	20	6.16	enable	7	3.54
finite	22	6.66	distribute	29	6.73	constraint	14	6.14	remain	12	3.5

Fig. 1. Example word sketch table for the English noun *resource* from the British National Corpus.

A year later the Sketch Engine was born, at the time being pretty much the Bonito concordancer enhanced with word sketches for English. The tool quickly gained a good reputation and was adopted by major British publishing houses, allowing sustainable maintenance and – fortunately for the computational linguistics community – Adam, a researcher dressed as businessman, reinvested all company income always into further development. In a recent survey among European lexicographers the Sketch Engine was their most used corpus query system [52].

Now 12 years later the Sketch Engine offers a wide range of corpus analysis functions on top of billion-word corpora for many language and tries to fulfill Adam's goal of 'Corpora for all' and 'Bringing corpora to the masses'.

Besides lexicographers and linguists (which now implies corpus linguists — almost always) this attracts teachers (not only at universities), students, language learners, and more increasingly translators, terminologists or copywriters.

The name Sketch Engine originates from the system's key function: word sketches, one page summaries of a word's collocational behavior in particular grammatical relations (see Fig. 1). Word sketches are computed by evaluating a set of corpus queries (called the **word sketch grammar**; see [16]) that generate a very large set of headword-collocation candidate pairs together with links to their particular occurrences in the corpus. Next, each collocation candidate is scored using a lexicographic association measure (in this case, the logDice measure [69]) and displayed in the word sketch table sorted by this score (or, alternatively, by raw frequency).

test *(noun)* Alternative PoS: <u>verb</u> (freq: 941,372)
enTenTen [2012] freq = <u>1,915,482</u> (147.70 per million)

Lemma	Score	Freq
testing	0.520	558,727
assessment	0.410	640,347
analysis	0.399	1,196,660
procedure	0.382	1,311,372
study	0.380	3,090,402
method	0.373	2,760,051
application	0.366	3,171,582
program	0.365	6,442,955
datum	0.362	3,165,540
evaluation	0.360	468,130
model	0.357	2,557,538
training	0.354	2,486,409
research	0.354	3,171,715
examination	0.352	375,991
requirement	0.349	1,734,482
exam	0.349	373,769
review	0.348	1,803,362

Fig. 2. Thesaurus entry for the English noun *test* computed from the word sketch database generated from the enTenTen12 corpus.

This hybrid approach – a combination of handcrafted language-specific grammar rules with a simple language independent statistical measure – has proved to be very robust with regard to the noisy web corpora, and very scalable so as to be able to benefit from their large size. Further on, devising a new sketch grammar for another language turned out to be mostly a straightforward task, and usually a matter of a few days of joint work between an informed native speaker and somebody who is familiar with the corpus query language. Sketch grammars can be adapted for many purposes, for example a recent adaptation incorporated automatic semantic annotations of the predicate argument fillers [57]. As of 2016 the Sketch Engine contains word sketch grammars for 26 languages, and new ones are being added regularly. In addition, the same formalism has been successfully used to identify key terms using Adam's SimpleMaths methodology [41].

Two additional features are also provided building on the word sketches: a distributional thesaurus and a word sketch comparison for two headwords called sketch-diff. The distributional thesaurus is computed from the word sketch index by identifying the most common words that co-occur with the same words in the same grammatical relations as a given input headword. Therefore the result is a set of synonyms, antonyms, hyper- and hyponyms — all kinds of semantically related words (see Fig. 2).

The sketch difference identifies the most different collocations for two input headwords (or the same headword in two different subcorpora) by subtracting the word sketch scores for all collocations of both headwords (see Fig. 3).

perceptive	0	34 0.0 6.4	emotionally	0	111 0.0 8.6	being	0	208 0.0 6.1
thought-provoking	0	32 0.0 6.2	artificially	0	57 0.0 7.9	robot	0	77 0.0 6.1
adaptive	0	19 0.0 6.1	fiercely	0	26 0.0 7.0	agent	9	455 0.4 6.0
well-informed	0	24 0.0 6.0	moderately	0	11 0.0 5.7	guess	0	35 0.0 5.5
literate	0	26 0.0 5.9	reasonably	0	54 0.0 5.7	conversation	0	88 0.0 5.1
cultured	0	19 0.0 5.7	culturally	0	12 0.0 5.5	creature	11	137 2.4 5.9
rational	0	clever 6.0	4.0 2.0 0 -2.0 -4.0 -6.0	intelligent	81	80 5.8 5.7
sensitive	8	134 2.0 5.9	wonderfully	20	9 5.4 4.5	fellow	52	14 5.1 3.1
thoughtful	14	121 5.0 7.7	very	1707	596 5.6 4.0	pass	67	9 5.2 2.2
affectionate	6	31 4.5 6.2	too	476	76 5.4 2.8	wordplay	21	0 5.8 0.0
clever	54	30 5.8 4.8	damn	12	0 5.6 0.0	chap	47	0 5.9 0.0
funny	233	103 7.0 5.7	awfully	15	0 6.1 0.0	twist	94	0 6.5 0.0
catchy	19	0 5.8 0.0	terribly	25	0 6.2 0.0	trick	166	0 6.7 0.0

Fig. 3. Sketch-diff table showing the difference in usage of the English adjectives *clever* and *intelligent*.

These core functions (inter-linked with a concordancer) have been subject to continuous development which has in the recent years focused on two major aspects: adaptation for parallel corpora (i.e. bilinguality) and adaptation to multi-word expressions (see [18,45]), so that the Sketch Engine now has both bilingual word sketches for parallel (or comparable) corpora and multi-word sketches showing collocations of arbitrary long headword-collocation combinations like *young man* or *utilize available resource*.

A substantial part of the Sketch Engine deals with corpus building for users. The Sketch Engine integrates dozens of third-party tools that allow researchers to quickly have their text converted into a searchable corpus, for many languages also automatically annotated with lemmas and part-of-speech tags. Underlying processing pipelines used for language-specific sentence segmentation, tokenization, character normalization and tagging or lemmatization represent years of efforts of bringing all of these tools into consistent shape – where the devil is in details which however have huge impact on the final usability of the data.

In this respect Adam's intentions were always to make it as easy as possible for the users to process their data so that they will not need to bother with technical details, but focus on their research. Even close to the end Adam was thinking of ways of facilitating Sketch Engine users. His last revision conducted several months before his death highlights following areas:

- Building Very Large Text Corpora from the Web
- Parallel and Distributed Processing of Very Large Corpora
- Corpus Heterogeneity and Homogeneity
- Corpus Evaluation
- Corpora and Language Teaching
- Language Change over Time
- Corpus Data Visualization
- Terminology Extraction

Lexical Computing Limited is committed to making these latest ideas come to fruition.

7 Lexicography

While collecting data for his DPhil thesis [19] (see Sect. 2), Adam canvassed a number of lexicographers for their views on his developing ideas. Could his theoretical model of how words convey meanings have applications in the practical world of dictionary-making? Thus began Adam's involvement with lexicography, which was to form a major component of his working life for the rest of his career, and which had a transformative impact on the field.

After a spell as resident computational linguist at Longman Dictionaries (1992–1995), Adam returned to academia. Working first with Roger Evans and then with David Tugwell at the University of Brighton, he implemented his ideas for word profiles as the WASP-bench, 'a lexicographer's workbench supporting state-of-the-art word sense disambiguation' [72] (see Sect. 4). The notion of the word sketch first appeared in the WASP-bench, and a prototype version was used in the compilation of the Macmillan English Dictionary [65], a new, from-scratch monolingual learner's dictionary of English. The technology was a huge success. For the publisher, it produced efficiency gains, facilitating faster entry-writing. For the lexicographers, it provided a rapid overview of the salient features of a word's behavior, not only enabling them to disambiguate word senses with greater confidence but also providing immediate access to corpus sentences which instantiated any grammatical relation of interest. And crucially, it made the end-product more systematic and less dependent on the skills and intuitions of lexicographers. The original goal of applying the new WASP-bench technology to entry-writing was to support an improved account of collocation. But the unforeseen consequence was the biggest change in lexicographic methodology since the corpus revolution of the early 1980s. From now on, the word sketch would be the lexicographer's first port of call, complementing and often replacing the use of concordances — a procedure which was becoming increasingly impractical as the corpora used for dictionary-making grew by orders of magnitude.

Lexicography is in a process of transition, as dictionaries migrate from traditional print platforms to electronic media. Most current on-line dictionaries are "horseless carriages" — print books transferred uncomfortably into a new medium — but models are emerging for new electronic artifacts which will show more clearly the relationship between word use and meaning in context, supported by massive corpus evidence. Adam foresaw this and, through his many collaborations with working lexicographers, he not only provided (print) dictionary-makers with powerful tools for lexical analysis, but helped to lay the foundations for new kinds of dictionaries.

During the early noughties, the primitive word sketches used in a dictionary project at the end of the 1990s morphed into the Sketch Engine (see Sect. 6) which added a super-fast concordancer and a distributional thesaurus to the rapidly-improving word sketch tool [48]. Further developments followed as Adam responded to requests from dictionary developers.

In 2007, a lexicographic project which required the collection of many thousands of new corpus example sentences led to the creation of the GDEX tool [40]. The initial goal was to expedite the task of finding appropriate examples, which would meet the needs of language learners, for specific collocational pairings. Traditionally, lexicographers would scan concordances until a suitable example revealed itself, but this is a time-consuming business. GDEX streamlined the process. Using a collection of heuristics (such as sentence length, the number of pronouns and other anaphors in the sentence, and the presence or absence of low-frequency words in the surrounding context), the program identified the "best" candidate examples and presented them to the lexicographer, who then made the final choice. Once again, a CL-based technology delivered efficiency gains (always popular with publishers) while making lexicographers' lives a little easier. There was (and still is) room for improvement in GDEX's performance, but gradually technologies like these are being refined, becoming more reliable and being adapted for different languages [51].

As new components like GDEX were incorporated into the Sketch Engine's generic version, the package as a whole became a de facto standard for the language-analysis stages of dictionary compilation in the English-speaking world. But this was just the beginning. Initially a monolingual resource based around corpora of English, the Sketch Engine gradually added to its inventory dozens, then hundreds, of new corpora for all the world's major languages and many less resourced languages too — greatly expanding its potential for dictionary-making worldwide. This led Adam to explore the possibilities of using the Sketch Engine's querying tools and multilingual corpora to develop tools for translators. He foresaw sooner than most that, of all dictionary products, the conventional bilingual dictionary would be the most vulnerable to the changes then gathering pace in information technology. Bilingual Word Sketches have thus been added to the mix [18].

Adam also took the view that the boundary between lexical and terminological data was unlikely to survive lexicography's incorporation into the general enterprise of Search. In recent years, he became interested in enhancing the Sketch Engine with resources designed to simplify and systematize the work of terminologists. The package already included Marco Baroni's WebBootCat tool for building corpora from data on the web [4]. WebBootCat is especially well adapted to creating corpora for specialized domains and, as a further enhancement, tools have been added for extracting keyword lists and, more recently, key terms (salient 2- or 3-word items characteristic of a domain). In combination, these resources allow a user to build a large and diverse corpus for a specific domain and then identify the terms of art in that field — all at minimal cost. A related, still experimental, resource is a software routine designed to identify, in the texts of a specialized corpus, those sentences where the writer effectively supplies a definition of a term, paving the way (when the technology is more mature) for a configuration of tools which could do most of the work of creating a special-domain dictionary.

Even experiments which didn't work out quite as planned shed valuable light on the language system and its workings. An attempt to provide computational support for the process of selecting a headword list for a new collocation dictionary was only partially successful. But the **collocationality** metric it spawned revealed how some words are more collocational than others — an insight which proved useful as that project unfolded [29].

Adam was almost unique in being equally at home in the NLP and lexicographic communities. A significant part of his life's work involved the application of NLP principles to the practical business of making dictionaries. His vision was for a new way of creating dictionaries in which most of the language analysis was done by machines (which would do the job more reliably than humans). This presupposed a radical shift in the respective roles of the lexicographer and the computer: where formerly the technology simply supported the corpus-analysis process, in the new model it would be more proactive, scouring vast corpus resources to identify a range of lexicographically-relevant facts, which would then be presented to the lexicographer. The lexicographer's role would then be to select, reject, edit and finalize [66]. A prototype version of this approach was the **Tickbox lexicography** [66] model used in the project which produced the DANTE lexical database [2].

Lexicographers would often approach Adam for a computational solution to a specific practical problem, and we have described several such cases here. Almost always, Adam's way of solving the problem brought additional, unforeseen benefits, and collectively these initiatives effected a transformation in the way dictionaries are compiled.

But there is much more. Even while writing his doctoral thesis, Adam perceived the fundamental problem with the way dictionaries accounted for word meanings. Traditional lexicographic practice rests on a view of words as autonomous bearers of meaning (or meanings), and according to this view, the meaning of a sentence is a selective concatenation of the meanings of the words in it. But a radically different understanding of how meanings are created (and understood) has been emerging since at least the 1970s. In this model, meaning is not an inherent property of the individual word, but is to a large degree dependent on context and co-text. As John Sinclair put it,

Many if not most meanings depend for their normal realization on the presence of more than one word [71].

This changes everything — and opens up exciting opportunities for a new generation of dictionaries. Conventional dictionaries identify word senses, but without explaining the complex patterns of co-selection which activate each sense. What is in prospect now is an online inventory of phraseological norms and the meanings associated with them. A "dictionary" which mapped meanings onto the recurrent patterns of usage found in large corpora would in turn make it easier for machines to process natural language. Adam grasped all this at an early stage in his career, and the software he subsequently developed (from the WASP-Bench onwards) provides the tools we will need to realize these ambitious goals.

The cumulative effect of Adam's work with lexicographers over twenty-odd years was not only to reshape the way dictionaries are made, but to make possible the development of radically different lexical resources which will reveal — more accurately, more completely, and more systematically than ever before — how people create and understand meanings when they communicate.

8   Conclusions and Outlook

As this review article highlights, Adam made a huge scientific contribution, not just to the field of computational linguistics but in other areas of linguistics and in lexicography. Adam was a man of conviction. He was eager to hear and take on new ideas but his belief in looking carefully at the data was fundamental. He raised questions over common practice in WSD [23,25], the lack of due care and attention to replicability when obtaining training data [31] as well as assumptions in other areas [28,32]. Though our perspectives of his ideas and work will vary, there is no doubt that our field is the better for his scrutiny and that his ideas have been seminal in many areas.

Adam contributed a great deal more than just ideas and papers. He was responsible, or a catalyst, for the production of a substantial amount of software, evaluation protocols and data (both corpora and annotated data sets). He had a passion for events and loved bringing people together as evidenced by his huge network of collaborators, of which the authors of this article are just a very small part. He founded or co-founded many events including Senseval (now SemEval), the ACL's special interest group on Web as Corpus, and more recently the 'Helping Our Own' [7] exercise which has at its heart the idea of using computational linguistics to help non-native English speakers in their academic writing. This enterprise was typical of Adam's inclusivity.[13] He was exceptional in his enthusiasm for work on languages other than English and fully appreciated the need for data and algorithms for bringing human language technology to the masses of speakers of other languages, as well as enriching the world with access to information regardless of the language in which it was

---

[13] Other examples include his eagerness to encourage participants in evaluations such as Senseval, reminding people to focus on analysis rather than who came top [42] and in his company's aim of 'corpora for all'.

recorded. Adam was willing to go out on a limb for papers for which the standard computational linguistics reviewing response was 'Why didn't you do this in English for comparison?' or 'This is not original since it has already been done in English'. These rather common views mean that those working on other languages, and particularly less resourced languages, have a far higher bar for entry into computational linguistics conferences and Adam championed the idea of leveling this particular playing field.

Right to the very end, Adam thought about the future of language technology and particularly about possibilities for bringing cutting edge resources within the grasp of those with a need for them, but packaged in such a way as to make the technologies practical and straightforward to use. For specific details of the last ideas from Adam see the end of Sect. 6.

The loss of Adam is keenly felt. There are now conference prizes in his name, at eLex[14] and at the ACL *SEM conference. SIGLEX, of which he was president 2000–2004, is coordinating an edited volume of articles *Computational Lexical Semantics and Lexicography Essays: In honor of Adam Kilgarriff* to be published by Springer later this year. This CICLing 2016 conference is dedicated to Adam's memory in recognition for his great contributions to computational linguistics and the many years of service he gave on CICLing's small informal steering committee. There is no doubt that we will continue to benefit from Adam Kilgarriff's scientific heritage and that his ideas and the events, software, data and communities of collaborators that he introduced will continue to influence and enable research in all aspects of language technology in the years to come.

References

1. Atkins, S.: Tools for computer-aided corpus lexicography: the hector project. Acta Linguistica Hungarica **41**, 5–72 (1993)
2. Atkins, S., Rundell, M., Kilgarriff, A.: Database of ANalysed Texts of English (DANTE). In: Proceedings of Euralex (2010)
3. Banko, M., Brill, E.: Scaling to very very large corpora for natural language disambiguation. In: ACL, pp. 26–33 (2001)
4. Baroni, M., Kilgarriff, A., Pomikálek, J., Rychlý, P.: WebBootCat: a web tool for instant corpora. In: Proceedings of Euralex, Torino, Italy, pp. 123–132 (2006)
5. Baroni, M., Kilgarriff, A., Pomikálek, J., Rychlỳ, P.: WebBootCaT: instant domain-specific corpora to support human translators. In: Proceedings of EAMT, pp. 247–252 (2006)
6. Copestake, A.: Implementing Typed Feature Structure Grammars. CSLI Lecture Notes. CSLI Publications, Stanford (2002). http://opac.inria.fr/record=b1098622
7. Dale, R., Kilgarriff, A.: Helping our own: text massaging for computational linguistics as a new shared task. In: Proceedings of the 6th International Natural Language Generation Conference, pp. 263–267. Association for Computational Linguistics (2010)

---

[14] See http://kilgarriff.co.uk/prize/.

8. Erjavec, T., Evans, R., Ide, N., Kilgarriff, A.: The concede model for lexical databases. In: Proceedings of the Second International Conference on Language Resources and Evaluation, pp. 355–362. Athens, Greece (2000)

9. Evans, R., Gazdar, G.: DATR: a language for lexical knowledge representation. Comput. Linguist. **22**(2), 167–216 (1996). http://eprints.brighton.ac.uk/11552/

10. Gale, W., Church, K., Yarowsky, D.: One sense per discourse. In: Proceedings of the 4th DARPA Speech and Natural Language Workshop, pp. 233–237 (1992)

11. Gardner, S., Nesi, H.: A classification of genre families in university student writing. Appl. Linguist. **34**(1), 25–52 (2012). ams024

12. Gilquin, G., Granger, S., Paquot, M.: Learner corpora: the missing link in EAP pedagogy. J. Engl. Acad. Purp. **6**(4), 319–335 (2007)

13. Hanks, P.: Do word meanings exist? Comput. Humanit. **34**(1–2), 205–215 (2000). SENSEVAL Special Issue

14. Ide, N., Kilgarriff, A., Romary, L.: A formal model of dictionary structure and content. In: Heid, U., Evert, S., Lehmann, E., Rohrer, C. (eds.) Proceedings of the 9th EURALEX International Congress. Institut für Maschinelle Sprachverarbeitung, Stuttgart, Germany, pp. 113–126, August 2000

15. Ide, N., Véronis, J.: Encoding dictionaries. Comput. Humanit. **29**(2), 167–179 (1995). http://dx.doi.org/10.1007/BF01830710

16. Jakubíček, M., Rychlý, P., Kilgarriff, A., McCarthy, D.: Fast syntactic searching in very large corpora for many languages. In: PACLIC 24 Proceedings of the 24th Pacific Asia Conference on Language, Information and Computation, Tokyo, pp. 741–747 (2010)

17. Kallas, J., Tuulik, M., Langemets, M.: The basic Estonian dictionary: the first monolingual L2 learner's dictionary of Estonian. In: Proceedings of the XVI Euralex Congress (2014)

18. Kilgarriff, A., Kovar, V., Frankenberg-Garcia, A.: Bilingual word sketches: three flavours. In: Electronic Lexicography in the 21st Century: Thinking outside the Paper (eLex 2013), pp. 17–19 (2013)

19. Kilgarriff, A.: Polysemy. Ph.D. thesis, University of Sussex (1992)

20. Kilgarriff, A.: Dictionary word-sense distinctions: an enquiry into their nature. Comput. Humanities **26**(1–2), 365–387 (1993)

21. Kilgarriff, A.: The hard parts of lexicography. Int. J. Lexicography **11**(1), 51–54 (1997)

22. Kilgarriff, A.: Putting frequencies in the dictionary. Int. J. Lexicography **10**(2), 135–155 (1997)

23. Kilgarriff, A.: What is word sense disambiguation good for? In: Proceedings of Natural Language Processing in the Pacific Rim, pp. 209–214 (1997)

24. Kilgarriff, A.: Gold standard datasets for evaluating word sense disambiguation programs. Comput. Speech Lang. **12**(3), 453–472 (1998)

25. Kilgarriff, A.: I don't believe in word senses. Comput. Humanit. **31**(2), 91–113 (1998). Reprinted in Practical Lexicography: a Reader. Fontenelle (ed.) Oxford University Press (2008). Also reprinted in Polysemy: Flexible patterns of meaning in language and mind Nerlich Todd, Herman and Clarke (eds.) Walter de Gruyter, pp. 361–392. And to be reprinted in Readings in the Lexicon Pustejovsky and Wilks (eds.) MIT Press

26. Kilgarriff, A.: SENSEVAL: an exercise in evaluating word sense disambiguation programs. In: Proceedings of LREC, Granada, pp. 581–588 (1998)

27. Kilgarriff, A.: Comparing corpora. Int. J. Corpus Linguist. **6**(1), 1–37 (2001)

28. Kilgarriff, A.: Language is never ever ever random. Corpus Linguist. Linguist. Theor. **1**(2), 263–276 (2005)

29. Kilgarriff, A.: Collocationality (and how to measure it). In: Proceedings of the 12th EURALEX International Congress, Torino, Italy, September 2006, pp. 997–1004 (2006)

30. Kilgarriff, A.: Word senses. In: Agirre, E., Edmonds, P. (eds.) Word Sense Disambiguation, Algorithms and Applications, pp. 29–46. Springer, Heidelberg (2006). https://doi.org/10.1007/978-1-4020-4809-8

31. Kilgarriff, A.: Googleology is bad science. Comput. Linguist. **33**(1), 147–151 (2007)

32. Kilgarriff, A.: Grammar is to meaning as the law is to good behaviour. Corpus Linguist. Linguist. Theor **3**(2), 195–197 (2007)

33. Kilgarriff, A.: Simple maths for keywords. In: Proceedings of Corpus Linguistics, Liverpool, UK (2009)

34. Kilgarriff, A.: Comparable corpora within and across languages, word frequency lists and the kelly project. In: Procedings of Workshop on Building and Using Comparable Corpora at LREC, Malta (2010)

35. Kilgarriff, A.: A detailed, accurate, extensive, available English lexical database. In: Proceedings of the NAACL HLT 2010 Demonstration Session, pp. 21–24. Association for Computational Linguistics, Los Angeles, June 2010. http://www.aclweb.org/anthology/N10-2006

36. Kilgarriff, A., Baisa, V., Bušta, J., Jakubíček, M., Kovář, V., Michelfeit, J., Rychlý, P., Suchomel, V.: The Sketch Engine: ten years on. Lexicography **1**(1), 7–36 (2014). http://dx.doi.org/10.1007/s40607-014-0009-9

37. Kilgarriff, A., Charalabopoulou, F., Gavrilidou, M., Johannessen, J.B., Khalil, S., Kokkinakis, S.J., Lew, R., Sharoff, S., Vadlapudi, R., Volodina, E.: Corpus-based vocabulary lists for language learners for nine languages. Lang. Resour. Eval. **48**(1), 121–163 (2014)

38. Kilgarriff, A., Evans, R., Koeling, R., Rundell, M., Tugwell, D.: WASPBENCH: a lexicographer's workbench supporting state-of-the-art word sense disambiguation. In: Proceedings of the Tenth Conference on European Chapter of the Association for Computational Linguistics, EACL 2003, vol. 2, pp. 211–214. Association for Computational Linguistics, Stroudsburg (2003). https://doi.org/10.3115/1067737.1067787

39. Kilgarriff, A., Grefenstette, G.: Introduction to the special issue on web as corpus. Comput. Linguist. **29**(3), 333–347 (2003)

40. Kilgarriff, A., Husák, M., McAdam, K., Rundell, M., Rychlý, P.: GDEX: automatically finding good dictionary examples in a corpus. In: Proceedings of the 13th EURALEX International Congress, Barcelona, Spain, July 2008, pp. 425–432 (2008)

41. Kilgarriff, A., Jakubíček, M., Kovář, V., Rychlý, P., Suchomel, V.: Finding terms in corpora for many languages with the Sketch Engine. In: EACL 2014, p. 53 (2014)

42. Kilgarriff, A., Palmer, M.: Introduction to the special issue on SENSEVAL. Comput. Humanit. **34**(1–2), 1–13 (2000). SENSEVAL Special Issue

43. Kilgarriff, A., Palmer, M. (eds.): SENSEVAL98: Evaluating Word Sense Disambiguation Systems, pp. 1–2. Kluwer, Dordrecht (2000)

44. Kilgarriff, A., Rosenzweig, J.: Framework and results for English SENSEVAL. Comput. Humanit. **34**(1–2), 15–48 (2000). SENSEVAL Special Issue

45. Kilgarriff, A., Rychlý, P., Kovář, V., Baisa, V.: Finding multiwords of more than two words. In: Proceedings of EURALEX 2012 (2012)

46. Kilgarriff, A., Rychlý, P.: Semi-automatic dictionary drafting download. In: de Schryver, G.M. (ed.) A Way with Words: Recent Advances in Lexical Theory and Analysis. A Festschrift for Patrick Hanks, Menha (2010)

47. Kilgarriff, A., Rychlỳ, P., Jakubicek, M., Kovár, V., Baisa, V., Kocincová, L.: Extrinsic corpus evaluation with a collocation dictionary task. In: LREC, pp. 545–552 (2014)

48. Kilgarriff, A., Rychlý, P., Smrz, P., Tugwell, D.: The sketch engine. In: Proceedings of Euralex, Lorient, France, pp. 105–116 (2004). Reprinted in Patrick Hanks (ed.) (2007). Lexicology: Critical Concepts in Linguistics. Routledge, London

49. Kilgarriff, A., Tugwell, D.: WASP-Bench: an MT lexicographer's workstation supporting state-of-the-art lexical disambiguation. In: Proceedings of the MT Summit VIII, Santiago de Compostela, Spain, pp. 187–190, September 2001

50. Kosem, I., Gantar, P., Krek, S.: Automation of lexicographic work: an opportunity for both lexicographers and crowd-sourcing. In: Electronic Lexicography in the 21st Century: Thinking Outside the Paper: Proceedings of the eLex 2013 Conference, Tallinn, Estonia, 17–19 October 2013, pp. 32–48 (2013)

51. Kosem, I., Husák, M., McCarthy, D.: GDEX for slovene. In: Proceedings of eLex2011, Bled, Slovenia (2011)

52. Krek, S., Abel, A., Tiberius, C.: ENeL Project: DWS/CQS Survey Analysis (2015). http://www.elexicography.eu/wp-content/uploads/2015/04/ENeL_WG3_Vienna_DWS_CQS_final_web.pdf

53. Leech, G.: 100 million words of English: the British national corpus (BNC). Lang. Res. **28**(1), 1–13 (1992)

54. Louw, B., Chateau, C.: Semantic prosody for the 21st century: are prosodies smoothed in academic contexts? A contextual prosodic theoretical perspective. In: Proceedings of the tenth JADT Conference on Statistical Analysis of Textual Data, pp. 754–764. Citeseer (2010)

55. Baroni, M., Chantree, F., Kilgarriff, A., Sharoff, S.: CleanEval: a competition for cleaning web pages. In: Proceedings of the Sixth International Conference on Language Resources and Evaluation (LREC 2008), Marrakech, Morocco, pp. 638–643 (2008)

56. Mautner, G.: Mining large corpora for social information: the case of elderly. Lang. Soc. **36**(01), 51–72 (2007)

57. McCarthy, D., Kilgarriff, A., Jakubíček, M., Reddy, S.: Semantic word sketches. In: 8th International Corpus Linguistics Conference (CL 2015) (2015)

58. McEnery, T., Wilson, A.: Corpus Linguistics. Edinburgh University Press, Edinburgh (1999)

59. Mihalcea, R., Chklovski, T., Kilgarriff, A.: The SENSEVAL-3 English lexical sample task. In: Mihalcea, R., Edmonds, P. (eds.) Proceedings SENSEVAL-3 Second International Workshop on Evaluating Word Sense Disambiguation Systems, Barcelona, Spain, pp. 25–28 (2004)

60. Nastase, V., Sayyad-Shirabad, J., Sokolova, M., Szpakowicz, S.: Learning noun-modifier semantic relations with corpus-based and WordNet-based features. In: Proceedings of the National Conference on Artificial Intelligence, vol. 21, no. 1, p. 781. AAAI Press/MIT Press, Menlo Park, Cambridge, London 1999 (2006)

61. O'Donovan, R., O'Neill, M.: A systematic approach to the selection of neologisms for inclusion in a large monolingual dictionary. In: Proceedings of the XIII EURALEX International Congress, Barcelona, 15–19 July 2008, pp. 571–579 (2008)

62. Peters, W., Kilgarriff, A.: Discovering semantic regularity in lexical resources. Int. J. Lexicography **13**(4), 287–312 (2000)

63. Pomikálek, J., Rychlỳ, P., Kilgarriff, A., et al.: Scaling to billion-plus word corpora. Adv. Comput. Linguist. **41**, 3–13 (2009)

64. Preiss, J., Yarowsky, D. (eds.): Proceedings of SENSEVAL-2 Second International Workshop on Evaluating Word Sense Disambiguation Systems, Toulouse, France (2001). sIGLEX Workshop Organized by Cotton, S., Edmonds, P., Kilgarriff, A., Palmer, M

65. Rundell, M.: Macmillan English Dictionary. Macmillan, Oxford (2002)

66. Rundell, M., Kilgarriff, A.: Automating the creation of dictionaries: where will it all end? In: Meunier, F. et al. (eds.) A Taste for Corpora. In Honour of Sylviane Granger, pp. 257–281. Benjamins, Amsterdam (2011)

67. Rychlý, P.: Korpusové manažery a jejich efektiví implementace. Ph.D. thesis, Masaryk University, Brno (únor 2000)

68. Rychlý, P.: Manatee/Bonito - a modular corpus manager. In: Proceedings of Recent Advances in Slavonic Natural Language Processing 2007. Masaryk University, Brno (2007)

69. Rychlý, P.: A lexicographer-friendly association score. In: Proceedings of Recent Advances in Slavonic Natural Language Processing, RASLAN 2008, pp. 6–9 (2008)

70. Sharoff, S.: Creating general-purpose corpora using automated search engine queries. In: Baroni, M., Bernardini, S. (eds.) WaCky! Working Papers on the Web as Corpus, Gedit, Bologna (2006)

71. Sinclair, J.: The lexical item. In: Weigand, E. (ed.) Contrastive Lexical Semantics. Benjamins, Amsterdam (1998)

72. Tugwell, D., Kilgarriff, A.: WASP-Bench: a lexicographic tool supporting word-sense disambiguation. In: Preiss, J., Yarowsky, D. (eds.) Proceedings of SENSEVAL-2 Second International Workshop on Evaluating Word Sense Disambiguation Systems, Toulouse, France (2001)

73. Tugwell, D., Kilgarriff, A.: Word sketch: extraction and display of significant collocations for lexicography. In: Proceedings of the ACL Workshop on Collocations, Toulouse, France, pp. 32–28 (2001)

74. Wellner, B., Pustejovsky, J., Havasi, C., Rumshisky, A., Saurí, R.: Classification of discourse coherence relations: an exploratory study using multiple knowledge sources. In: Proceedings of the 7th SIGdial Workshop on Discourse and Dialogue, SigDIAL 2006, pp. 117–125, Association for Computational Linguistics, Stroudsburg (2006). http://dl.acm.org/citation.cfm?id=1654595.1654618

75. Yarowsky, D.: One sense per collocation. In: Proceedings of the ARPA Workshop on Human Language Technology, pp. 266–271. Morgan Kaufman (1993)

76. Yarowsky, D.: Unsupervised word sense disambiguation rivaling supervised methods. In: Proceedings of the 33rd Annual Meeting of the Association for Computational Linguistics, pp. 189–196 (1995)

General Formalisms

A Roadmap Towards Machine Intelligence

Tomas Mikolov[1(✉)], Armand Joulin[1], and Marco Baroni[1,2]

[1] Facebook AI Research, Menlo Park, USA
tmikolov@fb.com
[2] University of Trento, Trento, Italy

Abstract. The development of intelligent machines is one of the biggest unsolved challenges in computer science. In this paper, we propose some fundamental properties these machines should have, focusing in particular on *communication* and *learning*. We discuss a simple environment that could be used to incrementally teach a machine the basics of natural-language-based communication, as a prerequisite to more complex interaction with human users. We also present some conjectures on the sort of algorithms the machine should support in order to profitably learn from the environment.

1 Introduction

A machine capable of performing complex tasks without requiring laborious programming would be tremendously useful in almost any human endeavor, from performing menial jobs for us to helping the advancement of basic and applied research. Given the current availability of powerful hardware and large amounts of machine-readable data, as well as the widespread interest in sophisticated machine learning methods, the times should be ripe for the development of intelligent machines.

Still, since "solving AI" seems too complex a task to be pursued all at once, in the last decades the computational community has preferred to focus on solving relatively narrow empirical problems that are important for specific applications, but do not address the overarching goal of developing general-purpose intelligent machines. In this article, we propose an alternative approach: we first define the general characteristics we think intelligent machines should possess, and then we present a concrete roadmap to develop them in realistic, small steps, that are however incrementally structured in such a way that, jointly, they should lead us close to the ultimate goal of implementing a powerful AI.

The article is organized as follows. In Sect. 2 we specify the two fundamental characteristics that we consider crucial for developing intelligence–at least the sort of intelligence we are interested in–namely *communication* and *learning*. Our goal is to build a machine that can learn new concepts through communication at a similar rate as a human with similar prior knowledge. That is, if one can easily learn how subtraction works after mastering addition, the intelligent machine, after grasping the concept of addition, should not find it difficult to learn subtraction as well. Since, as we said, achieving the long-term goal of building an

© Springer International Publishing AG, part of Springer Nature 2018
A. Gelbukh (Ed.): CICLing 2016, LNCS 9623, pp. 29–61, 2018.
https://doi.org/10.1007/978-3-319-75477-2_2

intelligent machine equipped with the desired features at once seems too difficult, we need to define intermediate targets that can lead us in the right direction. We specify such targets in terms of simplified but self-contained versions of the final machine we want to develop. At any time during its "education", the target machine should act like a stand-alone intelligent system, albeit one that will be initially very limited in what it can do. The bulk of our proposal (Sect. 3) thus consists in the plan for an interactive learning environment fostering the incremental development of progressively more intelligent behavior. Section 4 briefly discusses some of the algorithmic capabilities we think a machine should possess in order to profitably exploit the learning environment. Finally, Sect. 5 situates our proposal in the broader context of past and current attempts to develop intelligent machines. As that review should make clear, our plan encompasses many ideas that have already appeared in different research strands. What we believe to be novel in our approach is the way in which we are combining such ideas into a coherent program.

2   Desiderata for an Intelligent Machine

Rather than attempting to formally characterize intelligence, we propose here a set of desiderata we believe to be crucial for a machine to be able to autonomously make itself helpful to humans in their endeavors. The guiding principles we implicitly considered in formulating the desiderata are to minimize the complexity of the machine, and to maximize interpretability of its behavior by humans.

2.1   Ability to Communicate

Any practical realization of an intelligent machine will have to *communicate* with us. It would be senseless to build a machine that is supposed to perform complex operations if there is no way for us to specify the aims of these operations, or to understand the output of the machine. While other communication means could be entertained, natural language is by far the easiest and most powerful communication device we possess, so it is reasonable to require an intelligent machine to be able to communicate through language. Indeed, the intelligent machine we aim for could be seen as a computer that can be programmed through natural language, or as the interface between natural language and a traditional programming language. Importantly, humans have encoded a very large portion of their knowledge into natural language (ranging from mathematics treatises to cooking books), so a system mastering natural language will have access to most of the knowledge humans have assembled over the course of their history.

Communication is, by its very nature, *interactive*: the possibility to hold a conversation is crucial both to gather new information (asking for explanation, clarification, instructions, feedback, etc.) and to optimize its transmission (compare a good lecture or studying with a group of peers to reading a book alone). Our learning environment will thus emphasize the interactive nature of communication.

Natural language can also channel, to a certain extent, non-linguistic information, because much of the latter can be conveyed through linguistic means. For example, we can use language to talk about what we perceive with our senses, or to give instructions on how to operate in the world (see Louwerse 2011, among others, for evidence that language encodes many perceptual aspects of our knowledge). Analogously, in the simulation we discuss below, a Teacher uses natural language to teach the Learner (the intelligent machine being trained) a more limited and explicit language (not unlike a simple programming language) in which the Learner can issue instructions to its environment through the same communication channels it uses to interact with the Teacher. The intelligent machine can later be instructed to browse the Internet by issuing commands in the appropriate code through its usual communication channels, mastering in this way a powerful tool to interact with the world at large. Language can also serve as an interface to perceptual components, and thus update the machine about its physical surroundings. For example, an object recognition system could transform raw pixel data into object labels, allowing the machine to "see" its real-life environment through a controlled-language modality.

Still, we realize that our focus on the language-mediated side of intelligence may limit the learning machine in the development of skills that we naturally gain by observing the world around us. There seems to be a fundamental difference between the symbolic representations of language and the continuous nature of the world as we perceive it. If this will turn out to be an issue, we can extend the training phase of the machine (its development in a simulated environment such as the one we will sketch below) with tasks that are more perception-oriented. While in the tasks we will describe here the machine will be taught how to use its I/O channels to receive and transmit linguistic symbols, the machine could also be exposed, through the same interface, to simple encodings (bit streams) of continuous input signals, such as images. The machine could thus be trained, first, to understand the basic properties of continuous variables, and then to perform more complex operations in a continuous space, such as identifying shapes in 2D images. Note that including such tasks would not require us to change the design of our learning framework, only to introduce novel scripts.

One big advantage of the single-interface approach we are currently pursuing is that the machine only needs to be equipped with bit-based I/O channels, thus being maximally simple in its interface. The machine can learn an unlimited number of new codes enabling it to interface, through the same channels, with all sorts of interlocutors (people, other machines, perceptual data encoded as described above, etc.). By equipping the machine with only a minimalistic I/O bit-stream interface, we ensure moreover that no prior knowledge about the challenges the machine will encounter is encoded into the structure of the input and output representations, harming the generality of the strategies the machine will learn (compare the difficulty of processing an image when it's already encoded into pixels vs. as raw bits).

Finally, while we propose language as the general *interface* to the machine, we are agnostic about the nature of the internal representations the machine must posit to deal with the challenges it faces. In particular, we are not making claims about the internal representations of the machine being based on an interpretable "language of thought" (Fodor 1975). In other words, we are not claiming that the machine should carry out its internal reasoning in a linguistic form: only that its input and output are linguistic in nature.

To give a few examples of how a communication-based intelligent machine can be useful, consider a machine helping a scientist with research. First of all, the communication-endowed machine does not need to pre-encode a large static database of facts, since it can retrieve the relevant information from the Internet. If the scientist asks a simple question such as: *What is the density of gold?*, the machine can search the Web to answer: $19.3\,\mathrm{g/cm^3}$.

Most questions will however require the machine to put together multiple sources of information. For example, one may ask: *What is a good starting point to study reinforcement learning?* The machine might visit multiple Web sites to search for materials and get an idea of their relative popularity. Moreover, interaction can make even a relatively simple query such as the latter more successful. For example, the machine can ask the user if she prefers videos or articles, what is the mathematical background to be assumed, etc.

However, what we are really interested in is a machine that can significantly speed up research progress by being able to address questions such as: *What is the most promising direction to cure cancer, and where should I start to meaningfully contribute?* This question may be answered after the machine reads a significant number of research articles online, while keeping in mind the perspective of the person asking the question. Interaction will again play a central role, as the best course of action for the intelligent machine might involve entering a conversation with the requester, to understand her motivation, skills, the time she is willing to spend on the topic, etc. Going further, in order to fulfill the request above, the machine might even conduct some independent research by exploiting information available online, possibly consult with experts, and direct the budding researcher, through multiple interactive sessions, towards accomplishing her goal.

2.2   Ability to Learn

Arguably, the main flaw of "good old" symbolic AI research (Haugeland 1985) lied in the assumption that it would be possible to program an intelligent machine largely by hand. We believe it is uncontroversial that a machine supposed to be helping us in a variety of scenarios, many unforeseen by its developers, should be endowed with the capability of *learning*. A machine that does not learn cannot adapt or modify itself based on experience, as it will react in the same way to a given situation for its whole lifetime. However, if the machine makes a mistake that we want to correct, it is necessary for it to change its behavior–thus, learning is a mandatory component.

Together with learning comes *motivation*. Learning allows the machine to adapt itself to the external environment, helping it to produce outputs that maximize the function defined by its motivation. Since we want to develop machines that make themselves useful to humans, the motivation component should be directly controlled by users through the communication channel. By specifying positive and negative rewards, one may shape the behavior of the machine so that it can become useful for concrete tasks (this is very much in the spirit of reinforcement learning, see, e.g., Sutton and Barto 1998, and discussion in Sect. 5 below).

Note that we will often refer to human learning as a source of insight and an ideal benchmark to strive for. This is natural, since we would like our machines to develop human-like intelligence. At the same time, children obviously grow in a very different environment from the one in which we tutor our machines, they soon develop a sophisticated sensorimotor system to interact with the world, and they are innately endowed with many other cognitive capabilities. An intelligent machine, on the other hand, has no senses, and it will start its life as a *tabula rasa*, so that it will have to catch up not only on human ontogeny, but also on their phylogeny (the history of AI indicates that letting a machine learn from data is a more effective strategy than manually pre-encoding "innate" knowledge into it). On the positive side, the machine is not subject to the same biological constraints of children, and we can, for example, expose it to explicit tutoring at a rate that would not be tolerable for children. Thus, while human learning can provide useful inspiration, we are by no means trying to let our machines develop in human-like ways, and we claim no psychological plausibility for the methods we propose.

3   A Simulated Ecosystem to Educate Communication-Based Intelligent Machines

In this section, we describe a simulated environment designed to teach the basics of linguistic interaction to an intelligent machine, and how to use it to learn to operate in the world. The simulated ecosystem should be seen as a "kindergarten" providing basic education to intelligent machines. The machines are trained in this controlled environment to later be connected to the real world in order to learn how to help humans with their various needs.

The ecosystem I/O channels are controlled by an automatic mechanism, avoiding the complications that would arise from letting the machine interact with the "real world" from the very beginning, and allowing us to focus on challenges that should directly probe the effectiveness of new machine learning techniques.

The environment must be challenging enough to force the machine to develop sophisticated learning strategies (essentially, it should need to "learn how to learn"). At the same time, complexity should be manageable, i.e., a human put into a similar environment should not find it unreasonably difficult to learn to communicate and act within it, even if the communication takes place in a language the human is not yet familiar with. After mastering the basic language and concepts of the simulated environment, the machine should be able to interact with and learn from human teachers. This puts several restrictions on the kind of learning the machine must come to be able to perform: most importantly, it will need to be capable to extract the correct generalizations from just a few examples, at a rate comparable to human learners.

Our ecosystem idea goes against received wisdom from the last decades of AI research. This received wisdom suggests that systems should be immediately exposed to real-world problems, so that they don't get stuck into artificial "blocks worlds" (Winograd 1971), whose experimenter-designed properties might differ markedly from those characterizing realistic setups. Our strategy is based on the observation, that we will discuss in Sect. 4, that current machine learning techniques cannot handle the sort of genuinely incremental learning of algorithms that is necessary for the development of intelligent machines, because they lack the ability to store learned skills in long-term memory and compose them. To bring about an advance in such techniques, we have of course many choices. It seems sensible to pick the simplest one. The environment we propose is sufficient to demonstrate the deficiencies of current techniques, yet it is simple enough that we can fully control the structure and nature of the tasks we propose to the machines, make sure they have a solution, and use them to encourage the development of novel techniques. Suppose we were instead to work in a more natural environment from the very beginning, for example from video input. This would impose large infrastructure requirements on the developers, it would make data pre-processing a big challenge in itself, and training even the simplest models would be very time-consuming. Moreover, it would be much more difficult to formulate interrelated tasks in a controlled way, and define the success criterion. Once we have used our ecosystem to develop a system capable of learning compositional skills from extremely sparse reward, it should be simple to plug in more natural signals, e.g., through communication with real humans and Internet access, so that the system would learn how to accomplish the tasks that people really want it to perform.

The fundamental difference between our approach and classic AI blocks worlds is that we do not intend to use our ecosystem to script an exhaustive set of functionalities, but to teach the machine the fundamental ability to *learn how to efficiently learn* by creatively combining already acquired skills. Once such machine gets connected with the real world, it should quickly learn to perform any new task its Teacher will choose. Our environment can be seen as analogous to explicit schooling. Pupils are taught math in primary school through rather artificial problems. However, once they have interiorized basic math skills in this setup, they can quickly adapt them to the problems they encounter in their real life, and rely on them to rapidly acquire more sophisticated mathematical techniques.

3.1   High-Level Description of the Ecosystem

Agents. To develop an artificial system that is able to incrementally acquire new skills through linguistic interaction, we should not look at the training data as a static set of labeled examples, as in common machine learning setups.

We propose instead a dynamic ecosystem akin to that of a computer game. The Learner (the system to be trained) is an actor in this ecosystem.

The second fundamental agent in the ecosystem is the Teacher. The Teacher assigns tasks and rewards the Learner for desirable behaviour, and it also provides helpful information, both spontaneously and in response to Learner's requests. The Teacher's behaviour is entirely scripted by the experimenters. Again, this might be worryingly reminiscent of entirely hand-coded good-old AIs. However, the Teacher need not be a very sophisticated program. In particular, for each task it presents to the learner, it will store a small set of expected responses, and only reward the Learner if its behaviour exactly matches one response. Similarly, when responding to Learner's requests, the Teacher is limited to a fixed list of expressions it knows how to respond to. The reason why this suffices is that the aim of our ecosystem is to kickstart the Learner's efficient learning capabilities, and not to provide enough direct knowledge for it to be self-sufficient in the world. For example, given the limitations of the scripted Teacher, the Learner will only be able to acquire a very impoverished version of natural language in the ecosystem. At the same time, the Learner should acquire powerful learning and generalization strategies. Using the minimal linguistic skills and strong learning abilities it acquired, the Learner should then be able to extend its knowledge of language fast, once it is put in touch with actual human users.

Like in classic text-based adventure games (Wikipedia 2015b), the Environment is entirely linguistically defined, and it is explored by the Learner by giving orders, asking questions and receiving feedback (although graphics does not play an active role in our simulation, it is straightforward to visualize the 2D world in order to better track the Learner's behaviour, as we show through some examples below). The Environment is best seen as the third fundamental agent in the ecosystem. The Environment behaviour is also scripted. However, since interacting with the Environment serves the purpose of observation and navigation of the Learner surroundings ("sensorimotor experience"), the Environment uses a controlled language that, compared to that of the Teacher, is more restricted, more explicit and less ambiguous. One can thus think of the Learner as a higher-level programming language, that accepts instructions from the programmer (the Teacher) in a simple form of natural language, and converts them into the machine code understood by the Environment.

In the examples to follow, we assume the world defined by the Environment to be split into discrete cells that the Learner can traverse horizontally and vertically. The world includes barriers, such as walls and water, and a number of objects the Learner can interact with (a pear, a mug, etc.).

Note that, while we do not explore this possibility here, it might be useful to add other actors to the simulation: for example, training multiple Learners in parallel, encouraging them to teach/communicate with each other, while also interacting with the scripted Teacher.

Interface Channels. The Learner experience is entirely defined by generic *input* and *output* channels. The Teacher, the Environment and any other language-endowed agent write to the input stream. Reward (a scalar value, as discussed next) is also written to the input stream (we assume, however, that the Learner does not need to discover which bits encode reward, as it will need this information to update its objective function). Ambiguities are avoided by prefixing a unique string to the messages produced by each actor (e.g., messages from the Teacher might be prefixed by the string **T:**, as in our examples below). The Learner writes to its output channel, and it is similarly taught to use unambiguous prefixes to address the Teacher, the Environment and any other agent or service it needs to communicate with. Having only generic input and output communication channels should facilitate the seamless addition of new interactive entities, as long as the Learner is able to learn the language they communicate in.

Reward. Reward can be positive or negative $(1/-1)$, the latter to be used to speed up instruction by steering away the Learner from dead ends, or even damaging behaviours. The Teacher, and later human users, control reward in order to train the Learner. We might also let the Environment provide feedback through hard-coded rewards, simulating natural events such as eating or getting hurt. Like in realistic biological scenarios, reward is sparse, mostly being awarded after the Learner has accomplished some task. As intelligence grows, we expect the reward to become *very* sparse, with the Learner able to elaborate complex plans that are only rewarded on successful completion, and even displaying some degree of self-motivation. Indeed, the Learner should be taught that short-term positive reward might lead to loss at a later stage (e.g., hoarding on food with poor nutrition value instead of seeking further away for better food), and that sometimes reward can be maximized by engaging in activities that in the short term provide no benefit (learning to read might be boring and time-consuming, but it can enormously speed up problem solving–and the consequent reward accrual–by making the Learner autonomous in seeking useful information on the Internet). Going even further, during the Learner "adulthood" explicit external reward could stop completely. The Learner will no longer be directly motivated to learn in new ways, but ideally the policies it has already acquired will include strategies such as curiosity (see below) that would lead it to continue to acquire new skills for its own sake. Note that, when we say that reward could stop completely, we mean that users do not need to provide explicit reward, in the form of a scalar value, to the Learner. However, from a human perspective, we can look at this as the stage in which the Learner has interiorized its own sources of reward, and no longer needs external stimuli.

We assume binary reward so that human users need not worry about relative *amounts* of reward to give to the Learner (if they do want to control the amount of reward, they can simply reward the Learner multiple times). The Learner objective should however maximize *average reward over time*, naturally leading to different degrees of cumulative reward for different courses of action (this is analogous to the notion of expected cumulative reward in reinforcement learning, which is a possible way to formalize the concept). Even if two solutions to a task are rewarded equally on its completion, the faster strategy will be favored, as it leaves the Learner more time to accumulate further reward. This automatically ensures that efficient solutions are preferred over wasteful ones. Moreover, by measuring time independently from the number of simulation steps, e.g., using simple wall-clock time, one should penalize inefficient learners spending a long time performing offline computations.

As already mentioned, our approach to reward-based learning shares many properties with reinforcement learning. Indeed, our setup fits into the general formulation of the reinforcement learning problem (Kaelbling et al. 1996; Sutton and Barto 1998) see Sect. 5 for further discussion of this point.

Incremental Structure. In keeping with the game idea, it is useful to think of the Learner as progressing through a series of levels, where skills from earlier levels are required to succeed in later ones. Within a level, there is no need to impose a strict ordering of tasks (even when our intuition suggests a natural incremental progression across them), and we might let the Learner discover its own optimal learning path by cycling multiple times through blocks of them.

At the beginning, the Teacher trains the Learner to perform very simple tasks in order to kick-start linguistic communication and the discovery of very simple algorithms. The Teacher first rewards the Learner when the latter repeats single characters, then words, delimiters and other control strings. The Learner is moreover taught how to repeat and manipulate longer sequences. In a subsequent block of tasks, the Teacher leads the Learner to develop a semantics for linguistic symbols, by encouraging it to associate linguistic expressions with actions. This is achieved through practice sessions in which the Learner is trained to repeat strings that function as Environment commands, and it is rewarded only when it takes notice of the effect the commands have on its state (we present concrete examples below). At this stage, the Learner should become able to associate linguistic strings to primitive moves and actions (*turn left*). Next, the Teacher will assign tasks involving action sequences (*find an apple*), and the Learner should convert them into sets of primitive commands (simple "programs"). The Teacher will, increasingly, limit itself to specify an abstract end goal (*bring back food*), but not recipes to accomplish it, in order to spur creative thinking on behalf of the Learner (e.g., if the Learner gets trapped somewhere while looking for food, it may develop a strategy to go around obstacles). In the process of learning to parse and execute higher-level commands, the Learner should also be trained to ask clarification questions to the Teacher (e.g., by initially granting reward when it spontaneously addresses the Teacher, and by the repetition-based strategy we illustrate in the examples below). With the orders becoming more general and complex, the language of the Teacher will also become (within the limits of what can be reasonably scripted) richer and more ambiguous, challenging the Learner capability to handle restricted specimens of common natural language phenomena such as polysemy, vagueness, anaphora and quantification.

To support user scenarios such as the ones we envisaged in Sect. 2 above and those we will discuss at the end of this section, the Teacher should eventually teach the Learner how to "read" natural text, so that the Learner, given access to the Internet, can autonomously seek for information online. Incidentally, notice that once the machine can read text, it can also exploit distributional learning from large amounts of text (Erk 2012; Mikolov et al. 2013; Turney and Pantel 2010) to induce word and phrase representations addressing some of the challenging natural language phenomena we just mentioned, such as polysemy and vagueness.

The Learner must take its baby steps first, in which it is carefully trained to accomplish simple tasks such as learning to compose basic commands. However, for the Learner to have any hope to develop into a fully-functional intelligent machine, we need to aim for a "snow-balling" effect to soon take place, such that later tasks, despite being inherently more complex, will require a lot less explicit coaching, thanks to a combinatorial explosion in the background abilities the Leaner can creatively compose (like for humans, learning how to surf the Web should take less time than learning how to spell).

Time Off. Throughout the simulation, we foresee phases in which the Learner is free to interact with the Environment and the Teacher without a defined task. Systems should learn to exploit this time off for undirected exploration, that should in turn lead to better performance in active training stages, just like, in the dead phases of a video-game, a player is more likely to try out her options than to just sit waiting for something to happen, or when arriving in a new city we'd rather go sightseeing than staying in the hotel. Since curiosity is beneficial in many situations, such behaviour should naturally lead to higher later rewards, and thus be learnable. Time off can also be used to "think" or "take a nap", in which the Learner can replay recent experiences and possibly update its inner structure based on a more global view of the knowledge it has accumulated, given the extra computational resources that the free time policy offers.

Evaluation. Learners can be quantitatively evaluated and compared in terms of the number of new tasks they accomplish successfully in a fixed amount of time, a measure in line with the reward-maximization-over-time objective we are proposing. Since the interactive, multi-task environment setup does not naturally support a distinction between a training and a test phase, the machine must carefully choose reward-maximizing actions from the very beginning. In contrast, evaluating the machine only on its final behavior would overlook the number of attempts it took to reach the solution. Such alternative evaluation would favor models which are simply able to memorize patterns observed in large amounts of training data. In many practical domains, this approach is fine, but we are interested in machines capable of learning truly general problem-solving strategies. As the tasks become incrementally more difficult, the amount of required computational resources for naive memorization-based approaches scales exponentially, so only a machine that can efficiently generalize can succeed in our environment.

We will discuss the limitations of machines that rely on memorization instead of algorithmic learning further in Sect. 4.3 below.

We would like to foster the development of intelligent machines by employing our ecosystem in a public competition. Given what we just said, the competition would not involve distributing a static set of training/development data similar in nature to the final test set. We foresee instead a setup in which developers have access to the full pre-programmed environment for a fixed amount of time. The Learners are then evaluated on a set of new tasks that are considerably different from the ones exposed in the development phase. Examples of how test tasks might differ from those encountered during development include the Teacher speaking a new language, a different Environment topography, new obstacles and objects with new affordances, and novel domains of endeavor (e.g., test tasks might require selling and buying things, when the Learner was not previously introduced to the rules of commerce).

3.2 Early Stages of the Simulation

Preliminaries. At the very beginning, the Learner has to learn to pay attention to the Teacher, to identify the basic units of language (find regularity in bit patterns, learn characters, then words and so on). It must moreover acquire basic sequence repetition and manipulation skills, and develop skills to form memory and learn efficiently. These very initial stages of learning are extremely important, as we believe they constitute the building blocks of intelligence.

However, as bit sequences do not make for easy readability, we focus here on an immediately following phase, in which the Learner has already learned how to pay attention to the Teacher and manipulate character strings. We show how the Teacher guides the Learner from these basic skills to being able to solve relatively sophisticated Environment navigation problems by exploiting interactive communication. Because of the "fractal-like" structure we envisage in the acquisition of increasingly higher-level skills, these steps will illustrate many of the same points we could have demonstrated through the lower-level initial routines. The tasks we describe are also incrementally structured, starting with the Learner learning to issue Environment commands, then being led to take notice of the effect these commands have, then understanding command structure, in order to generalize across categories of actions and objects, leading it in turn to being able to process higher-level orders. At this point, the Learner is initiated to interactive communication.

Note that we only illustrate here "polite" turn-taking, in which messages do not overlap, and agents start writing to the communication channels only after the end-of-message symbol has been issued. We do not however assume that interaction must be constrained in this way. On the contrary, there are advantages in letting entities write to the communication channels whenever they want: for example, the Teacher might interrupt the Learner to prevent him from completing a command that would have disastrous consequences, or the Learner may interrupt the Teacher as soon as it figured out what to do, in order to speed up reward (a simple priority list can be defined to solve conflicts, e.g., Teacher's voice is "louder" than that of Environment, etc.).

Note also that our examples are meant to illustrate specific instances from a larger set of trials following similar templates, that should involve a variety of objects, obstacles and possible actions. Moreover, the presented examples do not aim to exhaustively cover all learning-fostering strategies that might be implemented in the ecosystem. Finally, we stress again that we are not thinking of a strict ordering of tasks (not least because it would be difficult to fix, *a priori*, an ordering that is based on some objective, learning-model-independent difficulty criterion), but rather about clusters of related tasks organized into levels, such that the Learner could cycle through them and discover its own optimal progression to solve them.

Notation. We present the input to the Learner (messages from Teacher and Environment, as well as Reward, prefixed by **T:**, **E:** and **R:**, respectively) in the left column, and the Learner output on the right (messages addressed to the Teacher are prefixed by **@T:**, those for the Environment are prefixed by **@E:**). We assume different I/O identification prefixes to disambiguate meta-linguistic statements (e.g., "**T:** give order **@E:** I move." in the first exchange below, where it's only the Teacher speaking all the way). The full stop serves as end-of-message delimiter (line breaks are used for formatting purposes, and should be treated as white space; similarly, hyphens are inserted for readability and should be ignored). We use an ellipsis (three dots) to mark a sequence of messages on any channel we do not report (e.g., the Learner exploring a number of solutions before zeroing in on the right one). Finally, we interleave the conversations with comments, shown in grey.

The Learner Learns to Issue Environment Commands. We start with the Teacher teaching the Learner to produce Environment commands, first by rote repetition channeled to the right message recipient (**@E**), then by encouraging it to translate from the Teacher's own natural language to the controlled Environment language. At this stage, reward is not tied to understanding what happened (the "semantics" of the commands). Still, a good Learner might take note of the actions triggered by the commands. The following are example conversations that may happen during the learning process:

Input: *Output:*
T: give order **@E:** I move.

           fdakjf13ktj53u9gs5 5 3#.

character soup without addressee, no reward

           . . .

           @T: dsfjkasjfk;aljsf.

character soup with wrong addressee, no reward

           . . .

           @E: fglk4$3wfgg.

Input: *Output:*
right addressee, character soup, no reward

                                                    . . .
                                    give order **@E:** I move.
E: you moved.
R: 1.
Learner moves and is rewarded, although the solution is inefficient
T: give order **@E:** I move.

                                        . . .
                                    @E: I move.
E: you moved.
R: 1.
Learner figured out minimal string leading to reward

Further skills the Learner will need to develop include segmentation of Teacher's input, and production of separate commands for the Environment. At first, this may be rather trivial, as shown in this example:

Input: *Output:*
T: give orders **@E:** I move and **@E:** I look.
                                    @E: I move and **@E:** I look.
no reward, because Environment does not parse complex commands

                                        . . .
T: give orders **@E:** I move and **@E:** I look.
                                    @E: I move.
E: you moved.
                                    @E: I look.
E: there is an apple.
R: 1.

Eventually, the Learner should learn to switch fluently between Teacher and Environment communication codes, translating the Teacher's linguistic messages into motor commands to the Environment (an example of such exchange is visualized in Fig. 1).

Associating Language to Actions. Next, the Learner is encouraged to take notice of what happens in the Environment and associate the Teacher's language to states of the world, by being rewarded only when its orders result in the corresponding actions being effectively carried out in the world. An example task is illustrated in Fig. 2, where the Learner will not get reward because it is no longer sufficient for it to correctly translate instructions into Environment commands: the commands must also have the intended effect on the world. In this specific case, order execution fails because the Learner encountered a barrier.

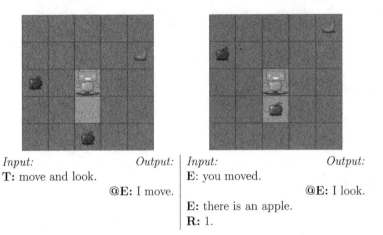

Input: Output: | Input: Output:
T: move and look. | **E:** you moved.
                    @E: I move. | **@E:** I look.
                                         | **E:** there is an apple.
                                         | **R:** 1.

Fig. 1. Example of how the simulation might be visualized to help developers track Learner's behaviour. The left panel represents the Environment at the stage in which Learner issues the move command, the right panel depicts the Environment after this command is executed. A lighter shade of green marks the cell the Learner occupies, and the one it is turned towards. These cells are directly observable. (Best viewed in color.) (Color figure online)

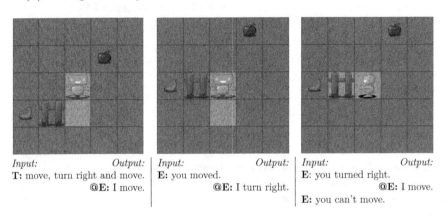

Input: Output: | Input: Output: | Input: Output:
T: move, turn right and move. | **E:** you moved. | **E:** you turned right.
                  @E: I move. | **@E:** I turn right. | **@E:** I move.
                                   | | **E:** you can't move.

Fig. 2. The Teacher gives the order *move, turn right and move*. The Learner is however blocked by a barrier after it turns right, and reward is delayed, because the order is not executed. (Best viewed in color.) (Color figure online)

Learning to Generalize. By adding sufficient variety to the trials above, the Learner should start to get hold of the compositional nature of commands (*turn right* and *turn left* share systematic characteristics, e.g., they both cause a change in what the Learner sees; the Learner can pick X only when X is in the cell in front of him, etc.). The next trials help the Learner further move away from literal command string memorization, and generalize across categories of actions and objects (*turn left/right* are instances of *turning*, *apples* and *pears* are *objects*, etc.). This is demonstrated in the following example:

Input:	*Output:*
T: turn and move.	
	@E: I turn.
E: *(no action)*	
	. . .
	@E: I turn left.
E: you turned left.	
	@E: I move.
E: you can't move.	
there is a barrier in front of Learner	
	. . .
	@E: I turn right.
E: you turned right.	
	@E: I move.
E: you moved.	
R: 1.	

In the next example, the Learner is asked to pick some object that is in front of him, without specifying what the object is (we assume that the simple, unambiguous language of the Environment has no words for general categories such as that of object, so the Learner must eventually find out that it must name specific objects instead):

Input:	*Output:*
T: pick an object.	
	@E: I pick an object.
E: *(no response)*	
	@E: I pick the apple.
E: *(no response)*	
	@E: I pick the pear.
E: you picked the pear.	
R: 1.	

As we just illustrated, initially the Learner will apply an exhaustive search strategy, listing all objects it knows of to find one it can pick. Later, the Teacher should teach the *look* command, and the Learner should be able to discover a faster strategy than exhaustive search:

Input:	*Output:*
T: pick an object.	
	@E: I look.
E: you see a pear.	
	@E: I pick the pear.

Input: *Output:*
E: you picked the pear.
R: 1.

Learning new strategies of course does not imply that the Learner can safely forget the ones it has previously acquired. For example, in some tasks the *look* command might not work (because, say, it is too dark and the Learner cannot see what is in front of him). In such case, an efficient Learner should find it easy to re-use previously learned skills, such as exhaustive search.

Understanding Higher-Level Orders. The next batch of trials aims at developing the ability to decompose higher-level commands into a sequence of basic actions, first fixed (*move twice*), then flexible (*find an apple*). The general teaching strategy is to provide sets of paired trials: In the first trial, the Teacher describes the task explicitly as a sequence of primitive actions, and gives the high-level name of the order. In the second trial, the Teacher issues an equivalent high-level order, and the Learner is rewarded if it goes through the same sequence as in the first trial (we are aware that many complex linguistic expressions are not amenable to a straightforward decompositional analysis, see, e.g., Boleda and Erk 2015: we will use some that are in order to teach the Learner about how to decompose complex concepts into step-by-step plans). The Teacher could start by teaching iteration counters:

Input: *Output:*
T: move and move.
                                      @E: I move.

E: you moved.
                                      @E: I move.

E: you moved.
R: 1.
T: this is called move two times.
                                        . . .

T: move two times.

                                        . . .
                                      @E: I move.

E: you moved.
                                      @E: I move.

E: you moved.
R: 1.

                                        . . .

T: move and move and move.
                                      @E: I move.

E: you moved.
                                      @E: I move.

E: you moved.
                                      @E: I move.

Input:	*Output:*
E: you moved.	
R: 1.	
T: this is called move three times.	
	. . .
T: move three times.	
	. . .
	@E: I move.
E: you moved.	
	@E: I move.
E: you moved.	
	@E: I move.
E: you moved.	
R: 1.	

The Learner should eventually be able to apply modifiers productively to different actions without much further training (e.g., apply *three times* to *turn left*).

Next, the Teacher can show how to segment high-level tasks, such as *finding something*, into atomic action sequences. For example, to find an apple, one can develop a simple strategy of going forward until an apple is found. The Teacher might initiate the Learner to this new skill as follows:

Input:	*Output:*
T: move and look.	
	@E: I move.
E: you moved.	
	@E: I look.
E: you see grass.	
T: move and look.	
	@E: I move.
E: you moved.	
	@E: I look.
E: you see an apple.	
R: 1.	
T: this is called find an apple.	

Note that one such example is not sufficient for the Learner to understand what exactly we call *finding an apple*, as multiple interpretations are valid: maybe we just wanted it to execute the given commands twice. Thus, there should be multiple training sequences of variable length, to clarify that the task is really to apply an `until` loop, i.e., to iterate *move* and *look* commands until the object the Learner is searching for is found.

Further tasks can define composite skills, such as *getting an apple*, which would consist of first finding an apple, and then picking it up. Another generalization would be to include multiple objects the Learner should be searching for. Adding obstacles to the Environment would further complicate the challenge.

The previous strategy of simply going forward until an object is found will not work anymore, as the Learner would stop at the first obstacle. We can either expect the Learner to independently develop better search strategies involving turning, or we can add further examples where the Teacher shows the Learner how to improve through more direct supervision.

Interactive Communication. Tasks such as finding an apple without directions might involve a long random search. Thus, we want to kick-start interactive communication, so that the Learner can be efficiently directed by the Teacher (and eventually by humans) to the right cell. In a first set of trials, the Learner is rewarded for repeating a *how to* request uttered by the Teacher (addressing it back to the teacher via the **@T:** prefix), and following the precise instructions produced by the Teacher in response to the request:

Input:	*Output:*
T: ask me how to find an apple.	
	@E: ask me how to find an apple.
wrong addressee, wrong request prefix	
	. . .
	@T: how to find an apple.
T: turn right and move and move and pick the apple.	
	@E: I turn right.
E: you turned right.	
	@E: I move.
E: you moved.	
	@E: I move.
E: you moved.	
	@E: I pick the apple.
E: you picked the apple.	
R: 1.	

Trials such as this one are later interspersed with trials where the Learner is assigned a task it can in principle accomplish by random search, but taking the initiative by issuing a *how to* request and then following the precise directions provided by the Teacher will considerably speed up reward.

Algorithmic Knowledge. Some tasks illustrated above require understanding basic control flow structures. For example, parsing action modifiers implies a simple form of counting, and in order to find things the Learner must implement an until (equivalently, while not) loop. Similarly, the command *get out of the grass* calls for a while loop. Efficient completion of more advanced tasks, e.g., *return home*, implies development of more complex algorithms, such as pathfinding. After acquiring a bank of such algorithms (which might be encoded in the internal machine representation by something very different from programming language constructs), the Learner should be able, in advanced stages of the simulation, to productively combine them in order to succeed in full-fledged novel missions that involve accomplishing a large number of hierarchically-structured sub-goals (*find somebody who will trade two apples for a banana*).

As we discussed in Sect. 3.1, the Learner's functionality could essentially be interpreted as learning how to compose programs based on the descriptions given in natural language by the Teacher. The programs produce very simple instructions that are understood by the Environment, which can be seen as a sort of CPU. From this point of view, the intelligent system we aim to train is a bridge between the Teacher (later to be replaced by a human operator) and a traditional computer that understands only a limited set of basic commands and needs to be manually programmed for each single task. Thus, we believe that successful construction of intelligent machines could automate computer programming, which will likely be done in the future simply through communication in natural language.

3.3 Interacting with the Trained Intelligent Machine

To conclude the illustration of our plan, we provide a motivating example of how an intelligent machine schooled in our ecosystem could later make itself useful in the real world. We consider a scenario in which the machine works as an assistant to Alice, an elderly person living alone. Bob is Alice's son, and he also interacts with the machine.

We assume that, as part of its training, the machine has been taught how to issue Internet commands and process their outcomes. In the example dialogue, we give a general idea of how the machine would interface to the Internet, without attempting to precisely define the syntax of this interaction. Most importantly, the Internet queries in the example are meant to illustrate how the machine does not need to store all the knowledge it needs to accomplish its duties, as it can retrieve useful information from the Web on demand, and reason about it.

Input:

Bob: I just spoke to the doctor, who said my mother needs to move for at least one hour per day, please make sure she does get enough exercise.

Output:

. . .

following conversation takes place the next day:

@Alice: Are you in the mood for some light physical exercise today?

Alice: Yes, but no more than 15 minutes, please.

@INTERNET: [query search engine for keywords *elderly, light activity, 15 minutes*]

shortly afterwards. . .

Input:

Output:
@Alice: I have downloaded a YouTube video with a 15-minute yoga routine, please watch it whenever you can.

a few hours later...
Alice: I'm going out to buy groceries.

@INTERNET: [query search engine with keywords *average walking speed, elderly person*]
@INTERNET: [search maps app for distance to grocery stores in Alice's neighborhood]
@Alice: Hey, why don't you walk to the Yummy Food Market today? It should take you about 45 minutes to and from, so you'll get the rest of your daily exercise.

@Alice: Thanks for the suggestion.

The communication-based intelligent machine should adapt to a whole range of tasks it was not explicitly programmed for. If necessary, the user can give it further explicit positive and negative reward to motivate it to change its behavior. This may be needed only rarely, as the machine should be shipped to the end users after it already mastered good communication abilities, and further development should mostly occur through language. For example, when the user says *No, don't do this again*, the machine will understand that repeating the same type of behavior might lead to negative reward, and it will change its course of action even when no explicit reward signal is given (again, another way to put this is that the machine should associate similar linguistic strings to an "interiorized" negative reward).

The range of tasks for intelligent machines can be very diverse: besides the everyday-life assistant we just considered, it could explain students how to accomplish homework assignments, gather statistical information from the Internet to help medical researchers (see also the examples in Sect. 2.1 above), find bugs in computer programs, or even write programs on its own. Intelligent machines should extend our intellectual abilities in the same way current computers already function as an extension to our memory. This should enable us to perform intellectual tasks beyond what is possible today.

We realize the intelligent machines we aim to construct could become powerful tools that may be possibly used for dubious purposes (the same could be said about any advanced technology, including airplanes, space rockets and computers). We believe the perception of AI is skewed by popular science fiction movies. Instead of thinking of computers that take over the world for their own reasons, we think AI will be realized as a tool: A machine that will extend our capability to reason and solve complex problems. Further, given the current state of the technology, we believe any discussion on "friendliness" of the AI is at this moment premature. We expect it will take years, if not decades to scale basic intelligent machines to become competitive with humans, giving us enough time to discuss any possible existential threats.

4   Towards the Development of Intelligent Machines

In this section, we will outline some of our ideas about how to build intelligent machines that would benefit from the learning environment we described. While we do not have a concrete proposal yet about how exactly such machines should be implemented, we will discuss some of the properties and components we think are needed to support the desired functionalities. We have no pretense of completeness, we simply want to provide some food for thought. As in the previous sections, we try to keep the complexity of the machine at the minimum, and only consider the properties that seem essential.

4.1   Types of Learning

There are many types of behavior that we collectively call learning, and it is useful to discuss some of them first. Suppose our goal is to build an intelligent machine working as a translator between two languages (we take here a simplified word-based view of the translation task). First, we will teach the machine basic communication skills in our simulated environment so that it can react to requests given by the user. Then, we will start teaching it, by example, how various words are translated.

There are different kinds of learning happening here. To master basic communication skills, the machine will have to understand the concept of positive and negative reward, and develop complex strategies to deal with novel linguistic inputs. This requires discovery of algorithms, and the ability to remember facts, skills and even learning strategies.

Next, in order to translate, the machine needs to store pairs of words. The number of pairs is unknown and a flexible growing mechanism may be required. However, once the machine understands how to populate the dictionary with examples, the learning left to do is of a very simple nature: the machine does not have to update its learning strategy, but only to store and organize the incoming information into long-term memory using previously acquired skills. Finally, once the vocabulary memorization process is finished and the machine starts working as a translator, no further learning might be required, and the functionality of the machine can be fixed.

The more specialized and narrow the functionality of the machine is, the less learning is required. For very specialized forms of behavior, it should be possible to program the solution manually. However, as we move from roles such as a simple translator of words, a calculator, a chess player, etc., to machines with open-ended goals, we need to rely more on general learning from a limited number of examples.

One can see the current state of the art in machine learning as being somewhere in the middle of this hierarchy. Tasks such as automatic speech recognition, classification of objects in images or machine translation are already too hard to be solved purely through manual programming, and the best systems rely on some form of statistical learning, where parameters of hand-coded models are estimated from large datasets of examples. However, the capabilities of state-of-the-art machine learning systems are severely limited, and only allow a small degree of adaptability of the machine's functionality. For example, a speech recognition system will never be able to perform speech translation by simply being instructed to do so–a human programmer is required to implement additional modules manually.

4.2 Long-Term Memory and Compositional Learning Skills

We see a special kind of long-term memory as the key component of the intelligent machine. This long-term memory should be able to store facts and algorithms corresponding to learned skills, making them accessible on demand. In fact, even the ability to learn should be seen as a set of skills that are stored in the memory. When the learning skills are triggered by the current situation, they should compose new persistent structures in the memory from the existing ones. Thus, the machine should have the capacity to extend itself.

Without being able to store previously learned facts and skills, the machine could not deal with rather trivial assignments, such as recalling the solution to a task that has been encountered before. Moreover, it is often the case that the solution to a new task is related to that of earlier tasks. Consider for example the following sequence of tasks in our simulated environment:

- find and pick an apple;
- bring the apple back home;
- find two apples;
- find one apple and two bananas and bring them home.

Skills required to solve these tasks include:

- the ability to search around the current location;
- the ability to pick things;
- the ability to remember the location of home and return to it;
- the ability to understand what *one* and *two* mean;
- the ability to combine the previous skills (and more) to deal with different requests.

The first four abilities correspond to simple facts or skills to be stored in memory: a sequence of symbols denoting something, the steps needed to perform a certain action, etc. The last ability is an example of a compositional *learning skill*, with the capability of producing new structures by composing together known facts and skills. Thanks to such learning skills, the machine will be able to combine several existing abilities to create a new one, often on the fly. In this way, a well-functioning intelligent machine will not need a myriad of training examples whenever it faces a slightly new request, but it could succeed given a single example of the new functionality. For example, when the Teacher asks the Learner to find one apple and two bananas and bring them home, if the Learner already understands all the individual abilities involved, it can retrieve the relevant compositional learning skill to put together a plan and execute it step by step. The Teacher may even call the new skill generated in this way *prepare breakfast*, and refer to it later as such. Understanding this new concept should not require any further training of the Learner, and the latter should simply store the new skill together with its label in its long-term memory.

As we have seen in the previous examples, the Learner can continue extending its knowledge of words, commands and skills in a completely unsupervised way once it manages to acquire skills that allow it to compose structures in its long-term memory. It may be that discovering the basic learning skills, something we usually take for granted, is much more intricate than it seems to us. But once we will be able to build a machine which can effectively construct itself based on the incoming signals –even when no explicit supervision in the form of rewards is given, as discussed above– we should be much closer to the development of intelligent machines.

4.3   Computational Properties of Intelligent Machines

Another aspect of the intelligent machine that deserves discussion is the computational model that the machine will be based on. We are convinced that such model should be unrestricted, that is, able to represent any pattern in the data. Humans can think of and talk about algorithms without obvious limitations (although, to apply them, they might need to rely on external supports, such as paper and pencil). A useful intelligent machine should be able to handle such algorithms as well.

A more precise formulation of our claim in the context of the theory of computation is that the intelligent machine needs to be based on a Turing-complete computational model. That is, it has to be able to represent any algorithm in fixed length, just like the Turing machine (the very fact that humans can describe Turing-complete systems shows that they are, in practical terms, Turing-complete: it is irrelevant, for our purposes, whether human online processing capabilities are strictly Turing-complete–what matters is that their reasoning skills, at least when aided by external supports, are). Note that there are many Turing-complete computational systems, and Turing machines in particular are a lot less efficient than some alternatives, e.g., Random Access Machines. Thus, we are not interested in building the intelligent machine around the concept of the Turing machine; we just aim to use a computational model that does not have obvious limitations in ability to represent patterns.

A system that is weaker than Turing-complete cannot represent certain patterns in the data efficiently, which in turn means it cannot truly learn them in a general sense. However, it is possible to memorize such complex patterns up to some finite level of complexity. Thus, even a computationally restricted system may appear to work as intended up to some level of accuracy, given that a sufficient number of training examples is provided.

For example, we may consider a sequence repetition problem. The machine is supposed to remember a sequence of symbols and reproduce it later. Further, let's assume the machine is based on a model with the representational power of finite state machines. Such system is not capable to represent the concept of storing and reproducing a sequence. However, it may appear to do so if we design our experiment imperfectly. Assume there is a significant overlap between what the machine sees as training data, and the test data we use to evaluate performance of the machine. A trivial machine that can function as a look-up table may appear to work, simply by storing and recalling the training examples. With an infinite number of training examples, a look-up-table-based machine would appear to learn any regularity. It will work indistinguishably from a machine that can truly represent the concept of repetition; however, it will need to have infinite size. Clearly, such memorization-based system will not perform well in our setting, as we aim to test the Learner's ability to generalize from a few examples.

Since there are many Turing-complete computational systems, one may wonder which one should be preferred as the basis for machine intelligence. We cannot answer this question yet, however we hypothesize that the most natural choice would be a system that performs computation in a parallel way, using elementary units that can grow in number based on the task at hand. The growing property is necessary to support the long-term memory, if we assume that the basic units themselves are finite. An example of an existing computational system with many of the desired properties is the cellular automaton of Von Neumann et al. (1966). We might also be inspired by string rewriting systems, for example some versions of the L-systems (Prusinkiewicz and Lindenmayer 2012).

An apparent alternative would be to use a non-growing model with immensely large capacity. There is however an important difference. In a growing model, the new cells can be connected to those that spawned them, so that the model is naturally able to develop a meaningful topological structure based on functional connectivity. We conjecture that such structure would in itself contribute to learning in a crucial way. On the other hand, it is not clear if such topological structure can arise in a large-capacity unstructured model. Interestingly, some of the more effective machine-learning models available today, such as recurrent and convolutional neural networks, are characterized by (manually constrained) network topologies that are well-suited to the domains they are applied to.

5   Related Ideas

We owe, of course, a large debt to the seminal work of Turing (1950). Note that, while Turing's paper is most often cited for the "imitation game", there are other very interesting ideas in it, worthy of more attention from curious readers, especially in the last section on learning machines. Turing thought that a good way to construct a machine capable of passing his famous test would be to develop a *child machine*, and teach it further skills through various communication channels. These would include sparse rewards shaping the behavior of the child machine, and other information-rich channels such as language input from a teacher and sensory information.

We share Turing's goal of developing a child machine capable of independent communication through natural language, and we also stress the importance of sparse rewards. The main distinction between his and our vision is that Turing assumed that the child machine would be largely programmed (he gives an estimate of sixty programmers working on it for fifty years). We rather think of starting with a machine only endowed with very elementary skills, and focus on the capability to learn as the fundamental ability that needs to be developed. This further assumes educating the machine at first in a simulated environment where an artificial teacher will train it, as we outlined in our roadmap. We also diverge with respect to the imitation game, since the purpose of our intelligent machine is not to fool human judges into believing it is actually a real person. Instead, we aim to develop a machine that can perform a similar set of tasks to those a human can do by using a computer, an Internet connection and the ability to communicate.

There has been a recent revival of interest in tasks measuring computational intelligence, spurred by the empirical advances of powerful machine-learning architectures such as multi-layered neural networks (LeCun et al. 2015), and by the patent inadequacy of the classic version of Turing test (Wikipedia 2015c). For example, Levesque et al. (2012) propose to test systems on their ability to resolve coreferential ambiguities (*The trophy would not fit in the brown suitcase because it was too big... What was too big?*). Geman et al. (2015) propose a "visual" Turing test in which a computational system is asked to answer a set of increasingly specific questions about objects, attributes and relations in a picture (*Is there a person in the blue region? Is the person carrying something? Is the person interacting with any other object?*). Similar initiatives differ from ours in that they focus on a specific set of skills (coreference, image parsing) rather than testing if an agent can learn new skills. Moreover, these are traditional evaluation benchmarks, unlike the hybrid learning/evaluation ecosystem we are proposing.

The idea of developing an AI living in a controlled synthetic environment and interacting with other agents through natural language is quite old. The Blocks World of Winograd (1971) is probably the most important example of early research in this vein. The approach was later abandoned, when it became clear that the agents developed within this framework did not scale up to real-world challenges (see, e.g., Morelli et al. 1992). The knowledge encoded in the systems tested by these early simulations was manually programmed by their creators, since they had very limited learning capabilities. Consequently, scaling up to the real world implied manual coding of all the knowledge necessary to cope with it, and this proved infeasible. Our simulation is instead aiming at systems that encode very little prior knowledge and have strong capabilities to learn from data. Importantly, our plan is not to try to manually program all possible scripts our system might encounter later, as in some of the classic AI systems. We plan to program only the initial environment, in order to kickstart the machine's ability to learn and adapt to different problems and scenarios. After the simulated environment is mastered, scaling up the functionality of our Learner will not require further manual work on scripting new situations, but will rather focus on integrating real world inputs, such as those coming from human users. The toy world itself is already designed to feature novel tasks of increasing complexity, explicitly testing the abilities of systems to autonomously scale up.

Still, we should not underestimate the drawbacks of synthetic simulations. The tasks in our environment might directly address some challenging points in the development of AI, such as learning with very weak supervision, being able to form a structured long-term memory, and the ability of the child machine to grow in size and complexity when encountering new problems. However, simulating the real world can only bring us so far, and we might end up overestimating the importance of some arbitrary phenomena at the expense of others, that might turn out to be more common in natural settings. It may be important to bring reality into the picture relatively soon. Our toy world should let the intelligent machine develop to the point at which it is able to learn from and cooperate with actual humans. Interaction with real-life humans will then naturally lead the machine to deal with real-world problems. The issue of when exactly a machine trained in our controlled synthetic environment is ready to go out in the human world is open, and it should be explored empirically. However, at the same time, we believe that having the machine interact with humans before it can deal with basic problems in the controlled environment would be pointless, and possibly even strongly misleading.

Our intelligent machine shares some of its desired functionalities with the current generation of automated personal assistants such as Apple's Siri ad Microsoft's Cortana. However, these are heavily engineered systems that aim to provide a natural language interface for human users to perform a varied but fixed set of tasks (similar considerations also apply to artificial human companions and digital pets such as Tamagotchi, see Wikipedia 2015a). Such systems can be developed by defining the most frequent use cases, choosing those that can be solved with the current technology (e.g., book an air ticket, look at the weather forecast and set the alarm clock for tomorrow's morning), and implementing specific solutions for each such use case. Our intelligent machine is not intended to handle just a fixed set of tasks. As exemplified by the example in Sect. 3.3, the machine should be capable to learn efficiently how to perform tasks such as those currently handled by personal assistants, and more, just from interaction with the human user (without a programmer or machine learning expert in the loop).

Architectures for software agents, and more specifically *intelligent* agents, are widely studied in AI and related fields (Nwana 1996; Russell and Norvig 2009). We cannot review this ample literature here, in order to position our proposal precisely with respect to it. We simply remark that we are not aware of other architectures that are as centered on learning and communication as ours. Interaction plays a central role in the study of multiagent systems (Shoham and Leyton-Brown 2009). However, the emphasis in this research tradition is on how conflict resolution and distributed problem solving evolve in typically large groups of simple, mostly scripted agents. For example, traffic modeling is a classic application scenario for multiagent systems. This is very different from our emphasis on linguistic interaction for the purposes of training a single agent that should become independently capable of very complex behaviours.

Tenenbaum (2015), like us, emphasizes the need to focus on basic abilities that form the core of intelligence. However, he takes naive physics problems as the starting point, and discusses specific classes of probabilistic models, rather than proposing a general learning scenario. There are also some similarities between our proposal and the research program of Luc Steels (e.g., Steels 2003, 2005), who lets robots evolve vocabularies and grammatical constructions through interaction in a situated environment. However, on the one hand his agents are actual robots subject to the practical hardware limitations imposed by the need to navigate a complex natural environment from the start; on the other, the focus of the simulations is narrowly on language acquisition, with no further aim to develop broadly intelligent agents.

We have several points of contact with the semantic parsing literature, such as navigation tasks in an artificial world (MacMahon et al. 2006) and reward-based learning from natural language instructions (Chen and Mooney 2011; Artzi and Zettlemoyer 2013). The agents developed in this area can perform tasks, such as learning to execute instructions in natural environments by interacting with humans (Thomason et al. 2015), or improving performance on real-life videogames by consulting the instruction manual (Branavan et al. 2012), that we would want our intelligent machines to also be able to carry out. However, current semantic-parsing-based systems achieve these impressive feats by exploiting architectures tuned to the specific tasks at hand, and they rely on a fair amount of hard-wired expert knowledge, in particular about language structures (although recent work is moving towards a more knowledge-lean direction, see for example Narasimhan et al. 2015, who train a neural network to play text-based adventure games using only text descriptions as input and game reward as signal). Our framework is meant to encourage the development of systems that should eventually be able to perform similar tasks, but getting there incrementally, starting with almost no prior knowledge and first learning from their environment a set of simpler skills, and how to creatively merge them to tackle more ambitious goals.

The last twenty years have witnessed several related proposals on learning to learn (Thrun and Pratt 1997), lifelong learning (Silver et al. 2013) and continual learning (Ring 1997). Much of this work is theoretical in nature and focuses on algorithms rather than on empirical challenges for the proposed models. Still, the general ideas being pursued are in line with our program. Ring (1997), in particular, defines a continual-learning agent whose experiences "occur sequentially, and what it learns at one time step while solving one task, it can use later, perhaps to solve a completely different task." Ring's desiderata for the continual learner are remarkably in line with ours. It is "an autonomous agent. It senses, takes actions, and responds to the rewards in its environment. It learns behaviors and skills while solving its tasks. It learns incrementally. There is no fixed training set; learning occurs at every time step; and the skills the agent learns now can be used later. It learns hierarchically. Skills it learns now can be built upon and modified later. It is a black box. The internals of the agent need not be understood or manipulated. All of the agent's behaviors are developed through training, not through direct manipulation. Its only interface to the world is through its senses, actions, and rewards. It has no ultimate, final task. What the agent learns now may or may not be useful later, depending on what tasks come next." Our program is definitely in the same spirit, with an extra emphasis on interaction.

Mitchell et al. (2015) discuss NELL, the most fully realized concrete implementation of a lifelong learning architecture. NELL is an agent that has been "reading the Web" for several years to extract a large knowledge base. Emphasis is on the never-ending nature of the involved tasks, on their incremental refinement based on what NELL has learned, and on sharing information across tasks. In this latter respect, this project is close to multi-task learning (Ando and Zhang 2005; Caruana 1997; Collobert et al. 2011), that focuses on the idea of parameter sharing across tasks. It is likely that a successful learner in our framework will exploit similar strategies, but our current focus lies on defining the tasks, rather than on how to pursue them.

Bengio et al. (2009) propose the related idea of curriculum learning, whereby training data for a single task are ordered according to a difficulty criterion, in the hope that this will lead to better learning. This is motivated by the observation that humans learn incrementally when developing complex skills, an idea that has also previously been studied in the context of recurrent neural network training by Elman (1993). The principle of incremental learning is also central to our proposal. However, the fundamental aspect for us is not a strict ordering of the training data for a specific task, but incrementality in the *skills* that the intelligent machine should develop. This sort of incrementality should in turn be boosted by designing separate tasks with a compositional structure, such that the skills acquired from the simpler tasks will help to solve the more advanced ones more efficiently.

The idea of incremental learning, motivated by the same considerations as in the papers we just mentioned, also appears in Solomonoff (2002), a work which has much earlier roots in research on program induction (Solomonoff 1964, 1997; Schmidhuber 2004). Within this tradition, Schmidhuber (2015) reviews a large literature and presents some general ideas on learning that might inspire our search for novel algorithms. Genetic programming (Poli et al. 2008) also focuses on the reuse of previously found sub-solutions, speeding up the search procedure in this way. Our proposal is also related to that of Bottou (2014), in its vision of compositional machine learning, although he only considers composition in limited domains, such as sentence and image processing.

We share many ideas with the reinforcement learning framework (Sutton and Barto 1998). In reinforcement learning, the agent chooses actions in an environment in order to maximize some cumulative reward over time. Reinforcement learning is particularly popular for problems where the agent can collect information only by interacting with the environment. Given how broad this definition is, our framework could be considered as a particular instance of it. Our proposal is however markedly different from standard reinforcement learning work (Kaelbling et al. 1996) in several respects. Specifically, we emphasize language-mediated, interactive communication, we focus on incremental strategies that encourage agents to solve tasks by reusing previously learned knowledge and we aim to limit the number of trials an agent gets in order to accomplish a certain goal.

Mnih et al. (2015) recently presented a single neural network architecture capable of learning a set of classic Atari games using only pixels and game scores as input (see also the related idea of "general game playing", e.g., Genesereth et al. 2005). We pursue a similar goal of learning from a low-level input stream and reward. However, unlike these authors, we do not aim for a single architecture that can, disjointly, learn an array of separate tasks, but for one that can incrementally build on skills learned on previous tasks to perform more complex ones. Moreover, together with reward, we emphasize linguistic interaction as a fundamental mean to foster skill extension. Sukhbaatar et al. (2015) introduce a sandbox to design games with the explicit purpose to train computational agents in planning and reasoning tasks. Moreover, they stress a curriculum strategy to foster learning (making the agent progress through increasingly more difficult versions of the game). Their general program is aligned with ours, and the sandbox might be useful to develop our environment. However, they do not share our emphasis on communication and interaction, and their approach to incremental learning is based on increasingly more difficult versions of the same task (e.g., increasing the number of obstacles), rather than on defining progressively more complex tasks, such that solving the later ones requires composing solutions to earlier ones, as we are proposing. Furthermore, the tasks currently considered within the sandbox do not seem to be challenging enough to require new learning approaches, and may be solvable with current techniques or minor modifications thereof.

Mikolov (2013) originally discussed a preliminary version of the incremental task-based approach we are more fully outlying here. In a similar spirit, Weston et al. (2015) present a set of question answering tasks based on synthetically generated stories. They also want to foster non-incremental progress in AI, but their approach differs from ours in several crucial aspects. Again, there is no notion of interactive, language-mediated learning, a classic train/test split is enforced, and the tasks are not designed to encourage compositional skill learning (although Weston and colleagues do emphasize that the same system should be used for all tasks). Finally, the evaluation metric is notably different from ours - while we aim to minimize the number of trials it takes for the machine to master the tasks, their goal is to have a good performance on held out data. This could be a serious drawback for works that involve artificial tasks, as in our view the goal should be to develop a machine that can learn as fast as possible, to have any hope to scale up and be able to generalize in more complex scenarios.

One could think of solving sequence-manipulation problems such as those constituting the basis of our learning routine with relatively small extensions of established machine learning techniques (Graves et al. 2014; Grefenstette et al. 2015; Joulin and Mikolov 2015). As discussed in the previous section, for simple tasks that involve only a small, finite number of configurations, one could be apparently successful even just by using a look-up table storing all possible combinations of inputs and outputs. The above mentioned works, that aim to learn algorithms from data, also add a long-term memory (e.g., a set of stacks), but they use it to store the data only, not the learned algorithms. Thus, such approaches fail to generalize in environments where solutions to new tasks are composed of already learned algorithms.

Similar criticism holds for approaches that try to learn certain algorithms by using an architecture with a strong prior towards their discovery, but not general enough to represent even small modifications. To give an example from our own work: a recurrent neural network augmented with a stack structure can form a simple kind of long-term memory and learn to memorize and repeat sequences in the reversed order, but not in the original one (Joulin and Mikolov 2015). We expect a valid solution to the algorithmic learning challenge to utilize a small number of training examples, and to learn tasks that are closely related at an increasing speed, i.e., to require less and less examples to master new skills that are related to what is already known. We are not aware of any current technique addressing these issues, which were the very reason why algorithmic tasks were originally proposed by Mikolov (2013). We hope that this paper will motivate the design of the genuinely novel methods we need in order to develop intelligent machines.

6   Conclusion

We defined basic desiderata for an intelligent machine, stressing learning and communication as its fundamental abilities. Contrary to common practice in current machine learning, where the focus is on modeling single skills in isolation, we believe that all aspects of intelligence should be holistically addressed within a single system.

We proposed a simulated environment that requires the intelligent machine to acquire new facts and skills through communication. In this environment, the machine must learn to perform increasingly more ambitious tasks, being naturally induced to develop complex linguistic and reasoning abilities.

We also presented some conjectures on the properties of the computational system that the intelligent machine may be based on. These include learning of algorithmic patterns from a few examples without strong supervision, and development of a long-term memory to store both data and learned skills. We tried to put this in contrast with currently accepted paradigms in machine learning, to show that current methods are far from adequate, and we must strive to develop non-incrementally novel techniques.

This roadmap constitutes only the beginning of a long journey towards AI, and we hope other researchers will be joining it in pursuing the goals it outlined.

Acknowledgments. We thank Léon Bottou, Yann LeCun, Gabriel Synnaeve, Arthur Szlam, Nicolas Usunier, Laurens van der Maaten, Wojciech Zaremba and others from the Facebook AI Research team, as well as Gemma Boleda, Katrin Erk, Germán Kruszewski, Angeliki Lazaridou, Louise McNally, Hinrich Schütze and Roberto Zamparelli for many stimulating discussions. An early version of this proposal has been discussed in several research groups since 2013 under the name *Incremental learning of algorithms* (Mikolov 2013).

References

Ando, R., Zhang, T.: A framework for learning predictive structures from multiple tasks and unlabeled data. J. Mach. Learn. Res. **5**, 1817–1853 (2005)

Artzi, Y., Zettlemoyer, L.: Weakly supervised learning of semantic parsers for mapping instructions to actions. Trans. Assoc. Comput. Linguist. **1**(1), 49–62 (2013)

Bengio, Y., Louradour, J., Collobert, R., Weston, J.: Curriculum learning. In: Proceedings of ICML, Montreal, Canada, pp. 41–48 (2009)

Boleda, G., Erk, K.: Distributional semantic features as semantic primitives-or not. In: Proceedings of the AAAI Spring Symposium on Knowledge Representation and Reasoning: Integrating Symbolic and Neural Approaches, Stanford, CA, pp. 2–5 (2015)

Bottou, L.: From machine learning to machine reasoning: an essay. Mach. Learn. **94**, 133–149 (2014)

Branavan, S., Silver, D., Barzilay, R.: Learning to win by reading manuals in a Monte-Carlo framework. J. Artif. Intell. Res. **43**, 661–704 (2012)

Caruana, R.: Multitask learning. Mach. Learn. **28**, 41–75 (1997)

Chen, D., Mooney, R.: Learning to interpret natural language navigation instructions from observations. In: Proceedings of AAAI, San Francisco, CA, pp. 859–865 (2011)

Collobert, R., Weston, J., Bottou, L., Karlen, M., Kavukcuoglu, K., Kuksa, P.: Natural language processing (almost) from scratch. J. Mach. Learn. Res. **12**, 2493–2537 (2011)

Elman, J.: Learning and development in neural networks: the importance of starting small. Cognition **48**, 71–99 (1993)

Erk, K.: Vector space models of word meaning and phrase meaning: a survey. Lang. Linguist. Compass **6**(10), 635–653 (2012)

Fodor, J.: The Language of Thought. Crowell Press, New York (1975)

Geman, D., Geman, S., Hallonquist, N., Younes, L.: Visual turing test for computer vision systems. Proc. Natl. Acad. Sci. **112**(12), 3618–3623 (2015)

Genesereth, M., Love, N., Pell, B.: General game playing: overview of the AAAI competition. AI Mag. **26**(2), 62–72 (2005)

Graves, A., Wayne, G., Danihelka, I.: Neural turing machines (2014). http://arxiv.org/abs/1410.5401

Grefenstette, E., Hermann, K., Suleyman, M., Blunsom, P.: Learning to transduce with unbounded memory. In: Proceedings of NIPS, Montreal, Canada (2014) (in Press)

Haugeland, J.: Artificial Intelligence: The Very Idea. MIT Press, Cambridge (1985)

Joulin, A., Mikolov, T.: Inferring algorithmic patterns with stack-augmented recurrent nets. In: Proceedings of NIPS, Montreal, Canada (2015) (in Press)

Kaelbling, L.P., Littman, M.L., Moore, A.W.: Reinforcement learning: a survey. J. Artif. Intell. Res. **4**, 237–285 (1996)

LeCun, Y., Bengio, Y., Hinton, G.: Deep learning. Nature **521**, 436–444 (2015)

Levesque, H.J., Davis, E., Morgenstern, L.: The Winograd schema challenge. In: Proceedings of KR, Rome, Italy, pp. 362–372 (2012)

Louwerse, M.: Symbol interdependency in symbolic and embodied cognition. Top. Cogn. Sci. **3**, 273–302 (2011)

MacMahon, M., Stankiewicz, B., Kuipers, B.: Walk the talk: connecting language, knowledge, and action in route instructions. In: Proceedings of AAAI, Boston, MA, pp. 1475–1482 (2006)

Mikolov, T.: Incremental learning of algorithms. Unpublished manuscript (2013)

Mikolov, T., Chen, K., Corrado, G., Dean, J.: Efficient estimation of word representations in vector space (2013). http://arxiv.org/abs/1301.3781/

Mitchell, T., Cohen, W., Hruschka, E., Talukdar, P., Betteridge, J., Carlson, A., Mishra, B., Gardner, M., Kisiel, B., Krishnamurthy, J., Lao, N., Mazaitis, K., Mohamed, T., Nakashole, N., Platanios, E., Ritter, A., Samadi, M., Settles, B., Wang, R., Wijaya, D., Gupta, A., Chen, X., Saparov, A., Greaves, M., Welling, J.: Never-ending learning. In: Proceedings of AAAI, Austin, TX, pp. 2302–2310 (2015)

Mnih, V., Kavukcuoglu, K., Silver, D., Rusu, A., Veness, J., Bellemare, M., Graves, A., Riedmiller, M., Fidjeland, A., Ostrovski, G., Petersen, S., Beattie, C., Sadik, A., Antonoglou, I., King, H., Kumaran, D., Wierstra, D., Legg, S., Hassabis, D.: Human-level control through deep reinforcement learning. Nature **518**, 529–533 (2015)

Morelli, R., Brown, M., Anselmi, D., Haberlandt, K., Lloyd, D. (eds.): Minds, Brains, and Computers: Perspectives in Cognitive Science and Artificial Intelligence. Ablex, Norwood (1992)

Narasimhan, K., Kulkarni, T., Barzilay, R.: Language understanding for text-based games using deep reinforcement learning. In: Proceedings of EMNLP, Lisbon, Portugal, pp. 1–11 (2015)

Nwana, H.: Software agents: an overview. Knowl. Eng. Rev. **11**(2), 1–40 (1996)

Poli, R., Langdon, W., McPhee, N., Koza, J.: A field guide to genetic programming (2008). http://www.gp-field-guide.org.uk

Prusinkiewicz, P., Lindenmayer, A.: The Algorithmic Beauty of Plants. Springer Science & Business Media, New York (2012)

Ring, M.: CHILD: a first step towards continual learning. Mach. Learn. **28**, 77–104 (1997)

Russell, S., Norvig, P.: Artificial Intelligence: A Modern Approach, 3d edn. Pearson Education, New York (2009)

Schmidhuber, J.: Optimal ordered problem solver. Mach. Learn. **54**(3), 211–254 (2004)

Schmidhuber, J.: On learning to think: algorithmic information theory for novel combinations of reinforcement learning controllers and recurrent neural world models (2015). http://arxiv.org/abs/1511.09249

Shoham, Y., Leyton-Brown, K.: Multiagent Systems. Cambridge University Press, Cambridge (2009)

Silver, D., Yang, Q., Li, L.: Lifelong machine learning systems: beyond learning algorithms. In: Proceedings of the AAAI Spring Symposium on Lifelong Machine Learning, Stanford, CA, pp. 49–55 (2013)

Solomonoff, R.J.: A formal theory of inductive inference. Part I. Inf. Control **7**(1), 1–22 (1964)

Solomonoff, R.J.: The discovery of algorithmic probability. J. Comput. Syst. Sci. **55**(1), 73–88 (1997)

Solomonoff, R.J.: Progress in incremental machine learning. In: NIPS Workshop on Universal Learning Algorithms and Optimal Search, Whistler, BC. Citeseer (2002)

Steels, L.: Social language learning. In: Tokoro, M., Steels, L. (eds.) The Future of Learning, pp. 133–162. IOS, Amsterdam (2003)

Steels, L.: What triggers the emergence of grammar? In: Proceedings of EELC, Hatfield, UK, pp. 143–150 (2005)

Sukhbaatar, S., Szlam, A., Synnaeve, G., Chintala, S., Fergus, R.: MazeBase: a sandbox for learning from games (2015). http://arxiv.org/abs/1511.07401

Sutton, R., Barto, A.: Reinforcement Learning: An Introduction. MIT Press, Cambridge (1998)

Tenenbaum, J.: Cognitive foundations for knowledge representation in AI. Presented at the AAAI Spring Symposium on Knowledge Representation and Reasoning (2015)

Thomason, J., Zhang, S., Mooney, R., Stone, P.: Learning to interpret natural language commands through human-robot dialog. Proceedings IJCAI, pp. 1923–1929. Buenos Aires, Argentina (2015)

Thrun, S., Pratt, L. (eds.): Learning to Learn. Kluwer, Dordrecht (1997)

Turing, A.: Computing machinery and intelligence. Mind **59**, 433–460 (1950)

Turney, P., Pantel, P.: From frequency to meaning: vector space models of semantics. J. Artif. Intell. Res. **37**, 141–188 (2010)

Von Neumann, J., Burks, A.W., et al.: Theory of self-reproducing automata. IEEE Trans. Neural Netw. **5**(1), 3–14 (1966)

Weston, J., Bordes, A., Chopra, S., Mikolov, T.: Towards AI-complete question answering: a set of prerequisite toy tasks (2015). http://arxiv.org/abs/1502.05698

Wikipedia: Artificial human companion (2015a). https://en.wikipedia.org/w/index.php?title=Artificial_human_companion&oldid=685507143. Accessed 15 Oct 2015

Wikipedia: Interactive fiction (2015b). https://en.wikipedia.org/w/index.php?title=Interactive_fiction&oldid=693926750. Accessed 19 Dec 2015

Wikipedia: Turing test (2015c). https://en.wikipedia.org/w/index.php?title=Turing_test&oldid=673582926. Accessed 30 July 2015

Winograd, T.: Procedures as a representation for data in a computer program for understanding natural language. Technical report AI 235, Massachusetts Institute of Technology (1971)

Algebraic Specification for Interoperability Between Data Formats: Application on Arabic Lexical Data

Malek Lhioui[1(✉)], Kais Haddar[1], and Laurent Romary[2,3]

[1] Laboratory MIRACL Multimedia Information Systems and Advanced Computing Laboratory, Sfax University, Sfax, Tunisia
ma.lhioui@gmail.com, kais.haddar@yahoo.fr
[2] Inria, Le Chesnay Cedex, France
Laurent.romary@inria.fr
[3] Mark Bloch Center, Berlin, Germany

Abstract. Linguistic data formats (LDF) became, over the years, more and more complex and heterogeneous due to the diversity of linguistic needs. Communication between these linguistic data formats is impossible since they are increasingly multiplatform and multi-providers. LDF suffer from several communication issues. Therefore, they have to face several interoperability issues in order to guarantee consistency and avoid redundancy. In an interoperability resolution context, we establish a method based on algebraic specifications to resolve interoperability among data formats. The proposed categorical method consists in constructing a unified language. In order to compose this unified language, we apply the co-limit algebraic specifications category for each data format. With this method, we establish a complex grid between existing data formats allowing the mapping to the unifier using algebraic specification. Then, we apply our approach on Arabic lexical data. We experiment our approach using Specware software.

Keywords: Data format · Arabic lexicon · Algebraic specifications
Interoperability · Specware

1 Introduction

Algebraic specification is a well-known method belonging to the software engineering area making possible the specification of all object behaviors. It allows strictly defining object structures using logical mathematic concepts. Algebraic specifications exploit equational logic which consists in the substitution of two equal terms (i.e. data formats, languages). Therefore, it seems to be a suitable formalism to factorize involved data formats by means of several operations (i.e. co-limit calculation). In fact, algebraic specifications have a very smart feature permitting researchers to imitate interoperability behavior.

As a matter of fact, interoperability, which is the ability to substitute data and share knowledge among them, is required to carry the disposition of data formats heterogeneity. NLP applications nowadays need such an interoperability notion. In fact, future

© Springer International Publishing AG, part of Springer Nature 2018
A. Gelbukh (Ed.): CICLing 2016, LNCS 9623, pp. 62–74, 2018.
https://doi.org/10.1007/978-3-319-75477-2_3

applications must rely on interoperability in order to be key players; otherwise they would be out of progress. A recent report named TAUS states that: "The lack of interoperability costs the translation industry a fortune". In fact, the highest fortune is compensated primarily for adjusting data formats (TAUS 2011). In the NLP area, each application uses a data format different from the other.

We can have a small idea of existing data formats in the literature; the formalism representation of the linguistic data was standardized in the 1980s using SGML-based markup language. Then, in later in the same century, early in the following century, the TEI suggested a new approach aiming to represent linguistic data. The TEI approach develops multiple markup languages that can be appended to SGML instances. This suggestion can be used for linguistic purposes. For the same approach, two representation methods are possible: using a single DTD or using different TEI DTD (Wörner et al. 2006). In 2003, the community responsible for developing lexicons for Natural Language Processing (NLP) developed the LMF standard (Francopolou 2013). In the speech domain, several representation formalisms are implemented. EXMARaLDA is one of the data models for the representation of spoken interaction based on the annotation graph approach (Bird and Lieberman 1999). Praat, TASX, ELAN and ANVIL are the other data models used for multimodal annotation. Tusnelda is a collection including heterogeneous linguistic data structures. This format was heavily influenced by the work of TEI (Wagner and Zeisler 2004). Another interchange format labeled as PAULA was developed in order to involve various linguistic levels (phonology, syntax, semantic, etc.) (Ide and Romary 2001).

In every NLP application, we have a large overview of heterogeneous data formats. This heterogeneity induces a question of interoperability depending on the diversity of lexicon formats. In order to appease this diversity, we conceive a generic format playing the role of pivot and used as a reference for various NLP applications. The main advantage from using such a format is to avoid having bidirectional translators (i.e. from LMF to TEI, from LMF to HPSG, from TEI to HPSG, etc.). Moreover, using transitivity with the purpose of mapping from one format to another is so hard to achieve; therefore, it will be able to guarantee the interoperability among all existing formats. On the other hand, pivot format makes possible the construction of lexicons subsuming all properties of all involving linguistic markup languages and frameworks. The ensuing pivot format is destined to be a worldwide solution for the ubiquitous dilemma of producing a generic format which subsumes all properties of different markup languages and frameworks. Thus, using this common format, we will be able to provide an enduring archive of varied resources.

In order to develop such a method for interoperability among representation formalisms, we may stand in front of large requirements. The first requirement resides in the ability to represent multi-rooted trees. Therefore, we can effectively model multiple annotation layers. The second requirement concerns the optimal strategy of communication to choose between representation formalisms so that it conserves non-redundancy among them. The pivot format has to allow us to model all structures without restrictions (trees, graphs, etc.).

In this paper, we plan to employ compulsory techniques in order to found a new method for interoperability among representation formalisms. The new method consists in constructing a unified format allowing us to map one format onto another without resorting to bidirectional translation. Thus, we rely on algebraic specifications which permit a rigorous definition of data formats composed of signature (noted $\sum$ composed of a set of sorts S and operations profiles O and axioms E). These specifications show their high potential in areas handling heterogeneous knowledge. Theoretically, algebraic specification allows the abstraction of all problem functionalities. In fact, data formats are relatively succinct, and describing those using algebraic specifications is so reasonable. Since an algebraic specification can be considered as a category in category theory, we can exploit all of its functions. Therefore, using the amalgamated sum of the cited theory, we can build the unified data format. The factorized format will have common specifications of all involving formats without redundancy.

2   Previous Work

Until now, we have had no serious attempt for interoperability between lexicons. However, many works consisting in mapping from one format to another have been done: the mapping process done by (Wilcock 2007) presenting an OWL ontology for HPSG. In the same context, a projecting process from HPSG syntactic lexica towards LMF has been produced by (Haddar et al. 2012). Another mapping process consists in a rule-based system aiming to translate LMF syntactic lexicon into TDL within the LKB platform (Loukil et al. 2010). We can mention also the mapping from LMF input lexicon to OWL-DL (Lhioui et al. 2015). In a further context, normalization is an attempt for interoperability. Standards are essential to exchange, preserve, maintain and integrate data to achieve interoperability. In fact, standards (such as LMF) made serious steps to reach this notion. However, the construction of normalized lexicons has been so far a hard task to achieve because it requires time and human resources. In this paper, we concentrate on algebraic specifications and category theory as a new method to achieve interoperability between lexicons in different formats.

In fact, algebraic specifications allow us to rigorously define data formats. This specification type is inspired from universal algebra areas (Konstantas et al. 2006) of mathematics and abstract types of software engineering. There are several algebraic specification languages including Specware (2009), LARCH (Guttag and Horming 1993), OBJ (Futatsugi et al. 1985), PLUSS (Gaudel 1992). We have chosen Specware language to specify LMF.

Category theory was defined by the two authors Mac Lane and Eilenberg as early as 1940 as a mathematics branch describing abstractly the objects involved and their relations without taking care of their internal properties (Fiadero 2005). From the eighties, this theory has been exploited by computer scientists for a long time. This theory has been explained from two points of view: mathematically and computationally. Mac Lane is one of the authors who explained from a mathematical point of view. However, several works explain category theory such as Mac Lane (1978). Other works are rather computational and tailored especially for software engineering (Barr and Wells works 1990 or Fiadero 2005).

Thus, there are no serious attempts to resolve interoperability among data formats. However, there are methods resolving interoperability among other types of data such as programming languages (Bou Dib 2009). In the data formats area, there are several attempts at resolving mapping from one format to another which are cited previously.

3   Proposed Method

In this paper, we propose a method allowing us to manage interoperability among data formats. Our method builds, by a co-limit calculation on algebraic specifications, a format that unifies concepts formats. The proposed method is characterized by the ability to automatically translate the format to a unified format. The method implementation was equipped using Specware software Kestrel. In order to present our method, we have to explain the two main areas on which it depends: algebraic specifications and theory category. In the following sections, we clarify key notions of algebraic specifications and category theory.

3.1   Algebraic Specifications

We put a specific emphasis on the formal definition of this notion and its application on LMF standard as presented in the following sections.

Algebraic Specification Definition

Formally, an algebraic specification is composed of a signature and axioms.

- We define a signature as a pair named $\sum = (S, O)$:
  - A set of sorts S defining the nomenclature presenting all data types. This set is used to describe the operations set.
  - A set of profiles for operation O are defined; each operation has a domain and co-domain composed of sorts belonging to S.
- The axioms are represented by a set $E: p = (\sum, E)$ that allow us to formalize the behavior of the profiles operations O.

Algebraic signature can be used to represent lexicons by representing the lexical entries as sorts and functionalities as operations. In order to explain the use of algebraic specification, we give in the following section the example of algebraic specification for LMF.

Algebraic Specification for LMF

The specification begins with the word spec and ends with the word endspec. In Specware, sorts are introduced by the word: type. The operation profile is defined by the
word: op. Finally, axioms are defined by the word axiom. In order to apply these theoretical notions, we consider the following LMF lexicon:

```xml
<?xml version="1.0" encoding="UTF-8"?>
  <lexicalRessource dtdVersion="16">
    <globalInformation>
      <feat att="languageCoding" val="ISO 639-3" />
      <feat att="scriptCoding" val="ISO 15 924" />
    </globalInformation>
    <lexicon>
      <feat att="language" val="arab" />
      <lexicalEntry morphologicalPatterns="intransitifVerb">
        <feat att="partOfSpeech" val="verb" />
        <feat att="root" val="س_ل_ج" />
        <feat att="scheme" val="فَعَلَ" />
        <lemma>
          <feat att="writtenForm" val="جَلَسَ" />
          <feat att="writtenForm" val="-" />
          <feat att="type" val="صحيح" />
        </lemma>
        <sens>
          <feat att="id" val="sens0" />
          <definition>
            <feat att="text" val="قام ضد مجرد ثلاثي فعل" />
            <statement>
              <feat att="text" val="جلاس و جلساء ج جليسه فهو معه جلس"
/>
            </statement>
            <statement>
              <feat att="text" val="جلاس و جلوس و ج جالس هو" />
            </statement>
          </definition>
        </sens>
      </lexicalEntry>
    </lexicon>
  </lexicalRessource>
```

In order to apply algebraic specification for LMF, we introduce the following Specware code according to the previous LMF sample:

```
LexicalResource= spec

type dtdVersion
type globalInformation
type lexicon                                                      types
type LexicalResource (dtdVersion, globalInformation, lexicon)

op LexicalResource: dtdVersion × globalInformation × lexicon
 → lexicalResource (dtdVersion × globalInormation × lexicon)
                                                                  operations
op dtdVersion: String → dtdVersion (String)
op globalInformation: feat  → globalInformation(att × val)
op lexicon: feat → lexicon (attx val)
op lexicon: feat × lexicalEntry → lexicon (feat × lexicalEntry)
op lexicalEntry: morphologicalPatterns × feat × lemma × sens
 → lexicalEntry ( morphologicalPatterns × feat × lemma × sens)
op lemma: feat → lemma (att × val)
op sens: feat × definition → sens (feat × definition)
op definition: feat × statement → definition (feat × statement)
op statement: feat → statement (feat)
```

With the previous Specware code, we specify the lexical Resource using algebraic specification. In order to set this specification, we define as necessary types 4 sorts and 10 operations for lexicon representation. Thus, we introduce the type "dtdVersion" in order to type the number of versions of LMF.

3.2 Category Theory

For the first time, mathematicians had difficulties to discover the interest of this theory. Yet, in due time, it has become essential, especially in algebra and topology. In the following subsections, we are going to formally define the new notion in order to illustrate it using existing linguistic data formats.

Category Definition

Formally, a category C is composed of objects O_c and morphisms M_c such as:

1. Each morphism is an application having a source which is a domain and a target called co-domain. Example $f : X \rightarrow Y$ is a morphism. Therefore, $\text{Dom}(f) = X$ and $\text{Cod}(f) = Y$

2. Composition is guaranteed by construction. Example: If we have two morphisms: $f : X \rightarrow Y$ and $g : Y \rightarrow Z$ such as $\text{Cod}(f) = \text{Dom}(g)$, we translate this by $(f; g) : X \rightarrow Z$ with $\text{Dom}(f; g) = \text{Dom}(f)$ and $\text{Cod}(f; g) = \text{Cod}(g)$.

3. The composition rule ';' is associative. Thus, if we have three morphisms such as: $f : X \rightarrow Y$, $g : Y \rightarrow Z$ et $h : Z \rightarrow T$, then $(f; g); h = f; (g; h)$.

4. The identity $\text{idx} : X \rightarrow X$ exists and for each object X satisfying for each morphism $g : Y \rightarrow X$ et $f : X \rightarrow Y$, we have $g; \text{idx} = g$ et $\text{idx}; f = f$.

In order to explain the use of category definition, we give in the following section the example of a category for linguistic data formats.

Linguistic Data Formats Category
In order to present this category, we resort to the graph notion. In order to define a category, we have to extract its components: objects and morphisms. Moreover, we have to ensure the fulfillment of morphisms properties. In our case, we choose linguistic data formats category. In this case, objects are data formats and morphisms are relations between their elements. Figure 1 shows such a category composed of the data formats LMF, TEI and HPSG and the morphism Synonym.

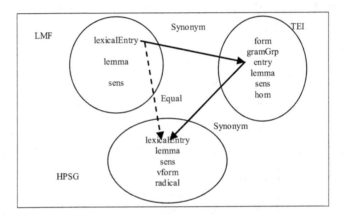

Fig. 1. Category of linguistic data formats

The example of Fig. 1 defines a category. It is composed of the collection of the three objects Oc = {LMF, TEI, HPSG} and the set of morphisms Mc = {Synonym}. The composition rule exists since (lexicalEntry ∈ LMF), (entry ∈ TEI) and (lexicalEntry ∈ HPSG). We have to be able to define a relation Equal such as (lexicalEntry Equal lexicalEntry) if (lexicalEntry Synonym Entry) and (Entry Synonym lexicalEntry).

4   Implementation: Construction of the Unified Format

Our strategy is based on the initiative of formally defining a unified format. This initiative automatically identifies common concepts. Considering formats as objects of a category, we were able to factorize common characteristics between these objects. In fact, the new object is built using the co-limit of the category theory. The originality of our work is to develop this co-limit between two objects. In this case, the co-limit is called pushout or amalgamated sum. The analogy between the pushout and the goal of interoperability seems to be the best answer to the problem.

Using algebraic specifications as categories allows us to apply the categories functions for algebraic specification. For this purpose, we introduce in the following sections algebraic specification category, categorical operations: pushout, pushout priming and morphisms.

4.1   Algebraic Specification Category

The category of algebraic specifications defines an object as an algebraic specification composed of a signature and axioms. The morphisms, in this context, are morphisms of algebraic specifications including a number of features as we will specify them later. Each morphism connects two specifications: source and target. Properties in the source specification are preserved by the target: it is therefore a mapping of signature (sorts and operations) and axioms as shown in Fig. 2.

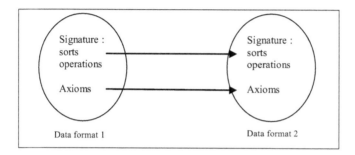

Fig. 2. Morphism of algebraic specifications

The relation of the sorts, in a morphism of algebraic specification, is restricted to a simple renaming. In the case of operations profiles, the morphism is defined so that it keeps the domains and co-domains of each operation without forgetting sorts renaming. In the case of axioms, constraints are stronger. They are such that an axiom of the source is a deductible theorem from the axioms of the target. The morphism in the axioms is much harder to perform. An axiom of the source is a deductible theorem from the axioms of the target.

4.2   Application: Algebraic Morphism of LMF to TEI

In order to deal with more details, we will consider the morphism of LMF to TEI. First of all we can characterize the previous signature of LMF by a sort of lexicalEntry, an operation called lexicalEntry.

```
LMF= spec

type lexicalEntry

op lexicalEntry: morphologicalPatterns × feat × lemma × sens →
lexicalEntry ( morphologicalPatterns × feat × lemma × sens)

axiom a1 is fa (x:lexicalEntry feat(att="scheme" val="فعل" )→ lemma
(feat (att="type" val="ثلاثي مجرد"))

axiom a2 is fa (x:lexicalEntry feat(att="scheme" val="فعّل" )→ lemma
(feat (att="type" val="ثلاثي مزيد"))
```

Then, exploring TEI data format, we illustrate the previous signature using TEI structure by another sort entry, and an operation called entry.

```
TEI= spec

type entry

op entry: form × type → entry ( form × type)

axiom a1 is fa (x:entry form(gramGrp(gram(type="scheme")="فعل" )→
entry (form (gramGrp(pos="ثلاثي مجرد")))

axiom a2 is fa (x:entry form(gramGrp(gram(type="scheme")="فعّل" )→
entry (form (gramGrp(pos="ثلاثي مزيد")))
```

We can establish a morphism of algebraic specifications of LMF to TEI by matching sorts and operators (lexicalEntry and entry) using a simple renaming from "lexicalEntry" to "entry".

4.3 Categorical Operations

In this section, we focus on operations that interest our study; those which help us elaborate a unified language. Categorical operations have two types: universal (applied for all objects) and those having higher order (whose objects are themselves categories). For the first category class, we identify pushout, co-limit and pullback operations. For the other class, we have functor operation. In our study, we focus on universal operations and particularly the pushout.

4.4 Pushout Operation

Pushout is a function having strong potential in the computer science area. It allows a great factorization of common specificities. It is applied in the context of two objects. In this section, we introduce vocabulary required to identify this notion. The first notion to understand is the diagram. The diagram represents objects and their morphisms. In this diagram, objects are represented by nodes and morphisms by arrows.

In category diagram C, commutativity indicates that for each pair of nodes X and Y, all paths from X to Y determine the same and only arrow in the category C. This commutativity constitutes the keystone of our method. Figure 3 represents an example of commutativity.

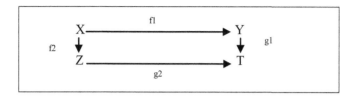

Fig. 3. Commutativity of X-Y-Z-T diagram

4.5 Construction of the Unified Format

In order to obtain a unified format, we need to apply the pushout function on the algebraic specifications. For this purpose, we focus on two key points: the construction of the priming of pushout and the specification of each object.

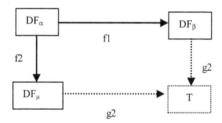

Fig. 4. Operation pushout applied to data formats

DF_β and DF_μ are two algebraic specifications of two data formats. DF_α is the common priming of the two formats DF_β and DF_μ. f1 and f2 are the morphisms defined respectively between the priming-DF_β and the priming-DF_μ. The pushout applied to the two morphisms f1 and f2 produce the format T with its specification and the two morphisms g1 and g2. Therefore, the signature of the T format is composed of common sorts and operations, on the one hand, and those that are specific for DF_β and DF_μ, on the other. In other words, the signature of the priming constitutes the smallest union of DF_β and DF_μ concepts.

4.6 Pushout Priming and Morphism Construction

Priming constitutes the basis for the pushout construction. It allows building an identical correspondence between the concepts of the two objects involved. In the context of algebraic specification, priming contains common sorts and operations. The role of morphisms is to build the correspondence between priming and the other formats concepts. Formally, priming has a specification obtained from the factorization of the concepts of the two objects involved DF_β and DF_μ. Figure 5 shows the construction of priming between LMF and TEI.

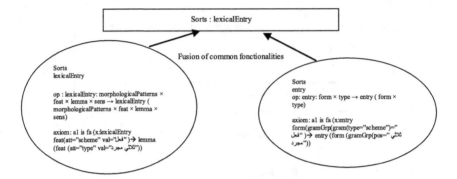

Fig. 5. Construction of the priming.

There is a dependency relation between morphisms and priming. In fact, morphisms are directly deduced from the development of the primer.

4.7 Pushout Object

After the construction of the priming and its morphisms, we have just to apply the pushout operation in order to merge the two objects involved. According to Fig. 4, the priming is DF_α and morphisms are f1 and f2. The pushout operation merges the two formats DF_β and DF_μ. The pushout applied for this example is a kind of minimum union of DF_β and DF_μ specifications: common sorts, operations and axioms (Fig. 6).

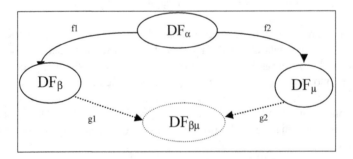

Fig. 6. Construction of pushout result $DF_{\beta\mu}$

The pushout operation has a very attractive feature allowing imitating interoperability behavior between data formats. This operation is based on the construction of the backbone between any objects involved.

5   Evaluation

In order to evaluate our method, experiments have been applied in our case. Our method evaluation has been lead using Specware Software. It simulates a very formal mechanism compared with other simple mapping from one format to another (i.e. from HPSG to OWL) (Wilcock 2007). There are many strong aspects in the method: it is based on theoretic constructions: algebraic specifications and category theory. These constructions make the method very formal. The other important aspect in the built method is that the calculation of the unified format is done automatically using morphisms and the pushout function.

Our method, as it is explained in the previous section, can be applied for a set of arbitrary linguistic data formats. Two cases are possible; if formats belong to the same family, we identify a weak common part; otherwise, if formats belong to the same family, the factorized part is more important than the other case.

The quality of the priming can also be discussed. In the first extreme case, we can envisage empty priming. This is explained by the fact of the two formats involved have no concepts in common. In the other extreme case, if the two formats have strictly the same contents compared with the priming, then the two formats are extremely identical. The interoperability issue was the object of various recent works. Our method is based on theoretical operations such as algebraic specifications and category theory. This guarantees a solid formal basis for the demarche. Moreover, the calculation of the unified format is made automatically due to the pushout function. Another strong point in the method concerns the preservation of properties in the stage of the construction of the priming and the result of the pushout.

6   Conclusion and Perspectives

In this paper, we have proposed a new method for interoperability based on algebraic specifications allowing communication between all existing data formats. We give the application of Arabic lexicons. Our method is efficient, especially on similar semantic data formats and is described as flexible. In fact, the learning of new format or the migration to another is no more required. Our method is validated without resorting to the development of a specific tool. It is based on existing tools: Specware. In the same context, morphisms are generated automatically.

In future works, we have to think of extending our method using the unification of the specifications. In fact, if we combine logics systems and algebraic specification, the institution concept appears to be suitable. Therefore, applying this formalism for this study seems to have promising results.

References

Barr, M., Wells, C.: Category Theory for Computing Science. Prentice-Hall International Series in Computer Science, p. 432 (1990)

Bird, S., Liberman, M.: Annotation graphs as a framework for multidimensional linguistic data analysis. In: Proceedings of the Workshop Towards Standards and Tools for Discourse Tagging. Association for Computational Linguistics (1999)

Bou Dib, A.: Une approche formelle de l'interopérabilité pour une famille de langages dédiés. Informatique. Université Paul Sabatier - Toulouse III (2009)

Fiadeiro, J.L.: Categories for Software Engineering. Springer, Heidelberg (2005)

Francopoulo, G.: Lexical Markup Framework. ISTE Ltd. and Wiley, Great Britain and the United States (2013)

Futatsugi, K., Goguen, J., Jounnaud, J.P., Mesguer, J.: Principles of OBJ2. In: Annual Symposium on Principles of Programming Languages, Proceedings of the 12th ACM SIGACT-SIGPLAN Symposium on Principles of Programming Languages, New Orleans, Louisiana, United States, pp. 52–66 (1985)

Haddar, K., Fehri, H., Romary, L.: A prototype for projecting HPSG syntactic lexica towards LMF. JLCL **27**, 21–46 (2012)

Lhioui, M., Haddar, K., Romary, L.: A prototype for projecting LMF lexica towards OWL. In: CICLING (2015)

Loukil, N., Ktari, R., Haddar, K., Benhamadou, A.: A normalized syntactic lexicon for Arabic verbs and its evaluation within the LKB platform. In: ACSE, Egypt (2010)

Gaudel, M.-C.: Structuring and modularizing algebraic specifications: the PLUSS specification language, evolutions and perspectives. In: Finkel, A., Jantzen, M. (eds.) STACS 1992. LNCS, vol. 577, pp. 1–18. Springer, Heidelberg (1992). https://doi.org/10.1007/3-540-55210-3_169

Guttag, J.V., Horming, J.J.: LARCH: Languages and Tools for Formal Specification, p. 250. Springer, New York (1993)

Ide, N., Romary, L.: Standards for language resources. In: Proceedings of the IRCS Workshop on Linguistic Database, pp. 141–149 (2001)

Kestrel Development corporation and KestreL Technology LLC, Specware to Isabelle in Interface Manual, 2006–2009

Konstantas, D., Bourrières, J.-P., Léonard, M., Boudjlida, N. (eds.): Interoperability of Enterprise Software and Applications. Proceedings of the I-ESA Conferences. Springer, London (2006). https://doi.org/10.1007/1-84628-152-0

Mac Lane, S.: Categories for the Working Mathematician. GTM, vol. 5. Springer, New York (1978). https://doi.org/10.1007/978-1-4757-4721-8

TAUS: Report on a TAUS research about translation interoperability (2011)

Wagner, A., Zeisler, B.: A syntactically annotated corpus of Tibetan. In: Proceedings of LREC, Lisboa, pp. 1141–1144 (2004)

Wörner, K., Witt, A., Rehm, G., Dipper, S. (eds.): Modelling Linguistic Data Structures, Extreme Markup Languages, Montréal, Québec (2006)

Wilcock, G.: An OWL ontology for HPSG. ACL, Finland (2007)

Persianp: A Persian Text Processing Toolbox

Mahdi Mohseni[1(✉)], Javad Ghofrani[2], and Heshaam Faili[1]

[1] University of Tehran, Tehran, Iran
{mahdi.mohseni,hfaili}@ut.ac.ir
[2] Clausthal University of Technology, Clausthal-Zellerfeld, Germany
javad.ghofrani@tu-clausthal.de

Abstract. This paper describes Persianp Toolbox, an integrated Persian text processing system and easily used in other software applications. The toolbox which provides fundamental Persian text processing steps includes several modules. In developing some modules of the toolbox such as normalizer, tokenizer, sentencizer, stop word detector, and Part-Of-Speech tagger previous studies are applied. In other modules i.e. Persian lemmatizer and NP chunker, new ideas in preparing required training data and/or applying new techniques are presented. Experimental results show the strong performance of the toolbox in each part. The accuracies of the tokenizer, the POS tagger, the lemmatizer and the NP chunker are 97%, 95.6%, 97%, 97.2%, respectively.

Keywords: Persianp toolbox · Persian text processing · Tokenizer
POS tagger · Lemmatizer · NP chunker

1 Introduction

The aim of this paper is to describe a Persian text processing toolbox, called Persianp. Persianp is an integrated toolbox and easily usable as a component within larger software applications. Persianp provides several fundamental text processing steps including normalization, tokenization, sentencization, stop word detection, Part-Of-Speech (POS) tagging, lemmatization and noun phrase (NP) chunking. The lemmatizer and the NP chunker of Persianp are developed using novel ideas in (1) automatically generating required training data, and/or (2) processing Persian text efficiently. Other parts of the Persianp are developed based on previous studies such as [4, 8, 10, 14].

So far, several toolkit have been developed to be accessible, easily usable in natural language processing systems such as Stanford CoreNLP [11], UIMA [7] and Gate [5]. The architecture of Stanford CoreNLP has the advantages of simplicity and transparency over systems like UIMA and Gate [11]. In designing Persianp Toolbox, we are inspired by the pipeline architecture of Stanford CoreNLP toolkit.

Researchers active on the field of the Persian language processing are familiar with the diverse challenges of the field. Besides the challenges raised by intrinsic properties of the language, there are two important challenges facing researchers: (1) lack of access to high quality resources and (2) lack of access to fundamental processing tools. Providing available, easily used Persian text processing system is a way to address the latter problem. Although, SteP-1 [20] and ParsiPardaz [19] are two text processing

© Springer International Publishing AG, part of Springer Nature 2018
A. Gelbukh (Ed.): CICLing 2016, LNCS 9623, pp. 75–87, 2018.
https://doi.org/10.1007/978-3-319-75477-2_4

toolbox in Persian, they are not accessible to all researchers. They are only available to organizations through a bureaucratic procedure and only after confirmation. This encouraged us to make Persianp Toolbox freely available to researchers. The first version of the toolbox is already accessible through the links persianp.ir/toolbox.html and persianp-toolbox.sourceforge.net. The next version will be released after integration.

Our innovations and achievements are summarized as follows:

- An integrated and free downloadable toolbox which provides fundamental Persian text processing steps is developed. The first version is now available through our website.
- A context-aware lemmatizer using a hybrid approach of rule-based and machine learning techniques is developed.
- A probabilistic NP chunker by which noun phrases are detected with high accuracy in Persian texts is developed.
- Automatically methods in preparing training data for the lemmatizer and the NP chunker are proposed.
- A 100 k word dataset for evaluating lemmatization in Persian is prepared manually.

In the following sections, at first we describe previous works in developing processing tools for Persian (Sect. 2), then challenges of processing Persian texts are briefly discussed (Sect. 3), after that different parts of Persianp Toolbox and how each part has been developed are explained (Sect. 4), next experiments and evaluations are presented (Sect. 5), and finally the paper is concluded in the last section (Sect. 6).

2 Related Works

High-quality and accessible processing systems can accelerate researches and provide a common bases to compare methods and models. The Persian language not only is a low resource language but it also suffers available high-quality processing tools. A few preprocessing tools has been reported in papers [6, 9, 12–14, 16]. But they have been laboratory single tools which, according to our knowledge, were not available to others nor well-integrated with other processing tools to be a unified package.

STeP-1 [20] is an attempt in developing an integrated software package for processing Persian texts. It normalizes and tokenizes text based on orthographical and inflectional rules of Persian. It also analyzes tokens morphologically and produces all possible combinations of stems and affixes. Wherever there is an ambiguity in morphological analysis of the text all possible results are returned, thus, it is on the user to decide which one is correct. STeP-1 also includes a bigram POS tagger improved by probabilistic morphological analysis which has overall precision 90% on Bijankhan corpus [15]. The online version of this toolbox is available through a web interface but getting access to the package has a bureaucratic process and it is not accessible to all.

ParsiPardaz [19] is a toolkit developed to do fundamental tasks of Persian text processing. Form a long list of abilities that the paper mentions for the toolkit at the beginning, most of them have not been developed or have been reported as under development. ParsiPardaz normalizes text by unifying different Unicode of similar characters and removing diacritics, Tanwin and Hamza. It tokenizes a text according to orthographical and inflectional rules of the language. The POS tagger of the toolkit has been trained using Stanford POS tagger. Similar to SteP-1, stemming is completely independent of the context. The rule-based lemmatizer performs based on inflectional morphology with the help of few databases. Again, the online version of the toolbox is available through a web interface but the toolbox is not available for all. It is only accessible to organizations, though, through a bureaucratic procedure.

3 On Persian Text Processing Challenges

Persian is a morphologically rich language. Inflectional and derivational paradigm of the language is complex. Persian orthography allows diversity in writing words and there is no unified standard to be respected generally. Academy of Persian Language and Literature has published a guideline [2] to be followed in writing. However, the guideline permits more than one form for many words and enumerates several exceptions under each rule. Most important challenges in processing Persian text are as follows:

Detection of word boundaries: whitespace is not always boundaries of text units in Persian. The problem partially returns to morphological properties of the language. For instance, plural morpheme ها /hā/can attached to nouns in more than one forms. For example the plural form of word کتاب /ketāb/'book' has three forms: (کتاب ها, کتابها, کتاب^ها (symbol ^ stands for Zero-Width-Non-Joiner). Multi-unit words are another part of the problem. For example, کتابخانه, کتاب^خانه and کتاب خانه /ketābxāne/'library' are three different written forms of the same word.

Diacritics and rare Arabic characters: writing diacritics are not popular in Persian but their rare appearances can cause problem in processing as they make two similar word to be interpreted differently e.g. in دَر and در /dar/'door, in'. Some rare character such as Tanwin and Hamza which have Arabic origin sometimes attach to a word and creates different forms of the same word, e.g. املاء, املا /emlā/'dictation'.

Irregular plural nouns: Inflection is not only way to make a word plural. Some words, mostly those with Arabic origin, have irregular plural form. For instance, مناطق manātᴄ 'areas' is the irregular plural form of منطقه mantaᴄe 'area'.

Character code ambiguity: There are characters within the Unicode table that have similar appearances but different codes. For example ی /j/and ک /k/each have more than one code in Unicode table. Most problem returns to confusion between codes of similar Persian and Arabic letters in Operation Systems, Editors and different software applications.

4   Persianp Toolbox

Persianp Toolbox consists of several modules which are cohesive with each other. The architecture of Persianp is similar to the pipeline system of Stanford CoreNLP toolkit [11]. Figure 1 depicts different modules of the toolbox and their relations. Each module is dependent on the below one(s), or in the other word each module provides the input of the upper one(s). We have a hybrid approach in development of the lemmatizer and that is why the lemmatizer has two subparts. The modules of the toolbox are describe in the rest of this section. To better show how modules work we pass a sample text (Fig. 2) trough each one and show the result.

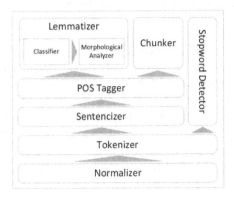

Fig. 1. Modules of Persianp Toolbox and their relations.

Input: از زمان بازگشت به ایران بی صبرانه منتظر اموألم بودم.اما ۳/۵ ماه گذشت و آنها هرگز فرستاده نشدند.
Translation: Since the return to Iran, I was impatiently waiting my belongings. 3.5 months passed and they were never sent.
Transliteration: az zamāne bazgašt be irān bi sabrāne montazere amvālam budam.ammā 3.5 māh gozašt va ānhā hargez ferestāde našodand.

Fig. 2. Sample input text.

4.1   Normalizer

This module is responsible for unification of different Unicode and removing rare and useless characters in the text. For unification, Persian and Arabic letters which are similar in shape but different in Unicode all are converted to the Persian Unicode. For example ی /j/and ک /k/are two letters which have different codes in Persian and Arabic Unicode character set. After Unicode unification step, diacritics, Hamza and Tanwin which are rarely used in Persian texts as well as rare and useless characters are removed from the text. In Fig. 3 the normalizer's output on the sample text is shown. Changed characters are highlighted.

Transliteration: az zamāne bazgašt be irān bi sabrāne montazere amvālam budam.ammā 3.5 māh gozašt va ānhā hargez ferestāde našodand.

In Persian: از زمان بازگشت به ایران بی صبرانه منتظر اموالم بودم.اما 3.5 ماه گذشت و آنها هرگز فرستاده نشدند.

Fig. 3. The normalizer's output on the sample input text.

4.2 Tokenizer

Tokenization is a lexical analysis in which a text is segmented into a sequence of tokens. Whitespaces, punctuation marks and orthographical properties are clues in detection of tokens boundaries. We follow four steps to tokenize a text:

- *Character manipulation*: this step is typical to almost all tokenizers. In this step, we separate number, punctuation marks and non-letter characters from letters.
- *Pattern search*: some tokens comply certain rules, so, they can be detected using regular expressions (regexes). Date, Hour, Number are example of these tokens. We recognize these well-form tokens using a few regexes.
- *Inflected words detection*: as said before Persian has a complex inflectional system and there are many affixes which can attach to words in different ways each with a different written form. In some forms, a whitespace is the between the word and the affix. Thus, splitting text on whitespaces leads to unwanted results for these words. In this step, these affixes are joined to the word which might be the preceding or the following one.
- *Multi-unit words detection*: in Persian a considerable part of words are multi-unit words (MUW). The orthography of Persian not only allows writing a word in some separated units but also it sometimes encourages people to write words separately. Aesthetic and ease of reading factors and have great influence here. In Paykare corpus [3], a 10 million word corpus in Persian, 2.5% of all words are MUWs and from 146 K unique words of the corpus 29 K are MUWs. This has been computed without taking into account verbs and complex predicates. To detect MUWs we perform a vocabulary based approach. The list of MUWs has been extracted from Paykare and a collection of news documents with supervision.

In Fig. 4 the output of the tokenizer on the normalized sample text (Fig. 3) is represented. Symbol ⌣ shows the boundaries of tokens. Number 3.5 has been found using regexes and word بی‌صبرانه /bi sabrāne/'impatiently' has been detected as a multi-unit word. Also, punctuations are separated from letters.

az⌣zamāne⌣bazgašt⌣be⌣irān⌣bi sabrāne⌣montazere⌣amvālam⌣budam⌣.⌣ammā⌣3.5⌣māh⌣ gozašt⌣va⌣ānhā⌣hargez⌣ferestāde⌣našodand⌣.

Fig. 4. The tokenizer's output on the sample input text.

4.3   Sentencizer

Some processing tasks works on sentences, so, after tokenization the sequence of tokens should be sentencized and the boundary of sentences be recognized. In this module, a sequence of tokens, i.e. the output of the tokenizer, is given as input. End of line marks and tokens which are some special punctuations such as single dot and question mark are clues in segmenting the input to a list of sentences. The sample input text is segmented in two sentences (Fig. 5). The single dot token is the position of segmentation.

---

az␣zamāne␣bazgašt␣be␣irān␣bi sabrāne␣montazere␣amvālam␣budam␣.
ammā␣3.5␣māh␣ gozašt␣va␣ānhā␣hargez␣ferestāde␣našodand␣.

Fig. 5. The sentencizer's output on the sample input text.

4.4   Stop Word Detector

Stop words are common words of a language and have no value in many applications. Although, stop words list may change from a domain to another, there are words which are common words in almost all domains. To collect a list of stop words for Persian we used the one published along with Hamshahri corpus [1] as initial list. The list was partially changed to include more general terms. The stop word detector module assigns a label to each word if it is found in the list. The result of applying the stop word detector on the word sequence of the sample input text is shown in Fig. 6. T indicates words that are stop words and F indicates those that have not been found in stop words list.

---

az/T zamāne/T bazgašt/T be/T irān/F bi sabrāne/F montazere/F amvālam/F budam/T ./F
ammā/T 3.5/F māh/F gozašt/F va/T ānhā/T hargez/T ferestāde/F našodand/T ./T

Fig. 6. The output of the stop word detector on the sample input text.

4.5   POS Tagger

Part-Of-Speech (POS) tagging is a fundamental task in many areas of Natural Language Processing (NLP) such as spell checker, text-to-speech, automatic speech recognition systems and machine translation, among others. To develop a POS tagger we relied on past experiences [13, 14]. A first order Hidden Markov Model are trained using Paykare corpus [3]. The tagset contains 14 coarse grain tags in the corpus (Table 1). Two sentences of the sample text tagged by the POS tagger are shown in Fig. 7.

Table 1. POS tags and tag names.

POS tag	Tag name	POS tag	Tag name
N	Noun	NUM	Number
V	Verb	POSTP	Postposition
AJ	Adjective	PRO	Pronoun
ADV	Adverb	PUNC	Punctuation
P	Preposition	CL	Classifier
CONJ	Conjunction	INT	Interjection
DET	Determiner	RES	Residual

az/P zamāne/N bazgašt/N be/P irān/N bi sabrāne/ADV montazere/AJ amvālam/N budam/V ./PUNC
ammā/CONJ 3.5/NUM māh/N gozašt/V va/CONJ ānhā/PRO hargez/ADV ferestāde/AJ našodand/V ./PUNC

Fig. 7. The POS tagger's output on the sample input text

4.6   Lemmatization

Lemmatization is a morphological analysis of a word to remove inflectional affixes and to return the dictionary form of the word. Lemmatization is an important preprocessing task in Persian because the language has a complex inflectional morphology. There are many enclitics with different syntactic functions which can attach to major categories of words such as noun and adjective. Verbs are inflected based on tense, person and number. Even there are cases in which determiners and prepositions are inflected. That is why a large part of words in Persian texts are not in simple from. About 18.5% of Paykare corpus consists of inflected words and 15.6% of the unique words have been experienced inflection at least one time.

Relying on vocabulary and rule-based morphological analysis of the words does not achieve proper results in lemmatization. This is because of various ambiguities in lemmatizing Persian words. For instance, ی /j/is sometimes an inflectional and sometimes derivational suffix. Word دبیری /dabir + i/is constructed by joining ی to دبیر /dabir/ 'teacher'. If ی is an inflectional suffix, the word is a common noun means 'a teacher' and the lemma is دبیر. But if ی is a derivational suffix the word is a proper noun and needs no lemmatization. For another instance, again ی /j/can attach to a common noun like برنامه‌ریز /barnāmeriz/'planner' to create another word برنامه‌ریزی /barnāmeriz + i/. If ی is an inflectional suffix, the new word means 'a planner' and its lemma is برنامه‌ریز. But if ی is a derivational suffix, the new word is another common nouns means 'planning' and it is not an inflected word. A lemmatizer is not able to correctly extract lemmas in these ambiguous cases as long as it is not context-aware and only depends on a vocabulary and morphological rules.

To develop a lemmatizer for Persian, we followed a hybrid approach of rule-based and machine learning techniques. A probabilistic classifier first decides whether a word needs to be lemmatized or not. If the classifier detects the word is an inflected one and needs to be lemmatized the lemma is extracted using rule-based morphological analyses. This procedure avoids egregious errors in Persian word lemmatization.

To train a probabilistic classifier, a dataset of Persian texts in which tokens are labeled with *inflected* or *not-inflected* labels is required. Unfortunately, words in Paykare corpus [3] are not lemmatized, so, Paykare cannot directly be used to train a classifier for our purpose. But words in the corpus have fine-grain tags as well as coarse-grain ones. The combination of Persian inflectional paradigm and fine-grain tags provide us a valuable tool to lemmatized Paykare automatically with a high accuracy. We present a few examples to illustrate the method. The word کتابهایم /ketāb + hā + j +am/'my books' has fine-grain tag N,COM,PL,1 (stands for Noun, Common, Plural, pronominal clitic number 1 i.e. first person). According to information provided by the tag and the inflectional system of Persian, one can interpret that the word construction is کتاب+ی+ها+م because 1 represents م /m/, possessive first person singular pronominal clitic, PL represents ها /hā/which is one of plural suffixes, and ی /j/appears between pronominal clitic and the word when the last letter of the word, here کتابها, is a vowel such as ا /ā/. However, it is not so simple. Ezafe morpheme, an enclitic in Persian, does not always appear in the writing. In دانشجوی /dānešdʒu + je/'student of' with tag N, COM,SING,EZ (stands for Noun, Common, Singular, Ezafe) ی /j/represents Ezafe but in بازی /bāzi + je/'game of' with the same tag ی is a part of the word and Ezafe has no appearance in the writing. These were two example to show the idea and the challenges. For the sake of brevity we don't explain all rules and exceptions.

Using our language knowledge and fine-grain tags of words we are able to lemmatize Paykare corpus automatically. As there are still ambiguities and because of a low error rate in the words tags of Paykare, the automatic lemmatization of the corpus cannot be performed with 100% accuracy but the result is near perfect. In experimental results section we will return to this subject again. A word in the corpus can now be assigned to label *inflected* if the word and its lemma is not the same. Otherwise they are assigned to label *not-inflected*. This is that dataset which will be used to train a probabilistic classifier for automatic lemmatization.

Maximum entropy classifier is used in our lemmatizer to detect whether a word is inflected or not. For each word in a sentence a set of 12 features are used. Features include: the word, the word tag, two ending characters of the word if the word length is greater than two, two preceding words, their POS tags and their already detected inflection label, two succeeding words and their POS tags. We use the Stanford classifier [10] to train our classifier. If the classifier detects a word is not inflected the word itself is returned as the lemma. Otherwise, the word will be passed to the next step i.e. rule-based lemma extraction.

To extract the lemma of a word which has been recognized as inflected word, we exploit the word POS tag, inflectional rules and a vocabulary of simple words. On the one hand, the POS tag which is provided by the POS tagger determines the major categories, such as noun, verb, adjective, etc., of the words. On the other hand, inflectional morphology of the language imposes special rules in each category. So, by following rules specific to the category one can achieve the lemma of the word. There are cases that a rule can applied to a word but it souled not. For example, ت /t/which represent second person singular pronominal clitic can attach to nouns, but not all ت at the end of nouns represent second person singular pronominal clitic. دوست /dust/'friend' has ت in the ending but it is a part of the word. To tackle these ambiguities a vocabulary containing the dictionary form of words are used. No rule applies to the word, if the word is found in the vocabulary. To lemmatized irregular plural nouns another vocabulary containing almost 700 words is used.

Figure 8 represents the result of applying the lemmatizer on the sample text. The lemmas of verbs are in infinitive form such as بودن /budan/'to be'. Word اموالم /amvālam/ 'my belongings' is an irregular plural noun attached to م /m/, first person singular pronominal clitic and its lemma is مال /māl/'belonging'.

az/az zamāne/zamān bazgašt/bazgašt be/be irān/irān bi sabrāne/bi-sabrane
montazere/montazer amvālam/māl budam/budan ./.
ammā/ammā 3.5/3.5 māh/māh gozašt/gozaštan va/va ānhā/ān hargez/hargez
ferestāde/ferestade našodand/šdoan ./.

Fig. 8. The lemmatizer's output on the sample input text.

4.7   NP Chunker

Chunking is the process of identifying and classifying the flat non-overlapping segments of a sentence that constitute the basic non-recursive phrases corresponding to the major parts-of-speech found in most wide-coverage grammars [8]. Our chunker detects noun phrases in a text. To utilize a probabilistic machine learning technique a corpus in which noun phrases are labeled is required. No such corpus is available in Persian. We generate this corpus from Dadegan corpus [18] with some effort. Dadegan is a Persian syntactic dependency treebank consist of approximately 30,000 sentences annotated with syntactic roles. In the treebank each word has one head and might has some dependents, if any. We consider a noun along with its dependents to be a noun phrase if it is not dependent of another noun. However, in the result some undesirable phrases are observed. For example phrases beginning with بر اساس /bar asāse/'according to' or به علت /be ellate/'because of' are not proper. This improper phrases are modified to have a corpus in which noun phrases are marked. Since the POS tags of words are used as feature in training a classifier we finally tag the corpus with our POS tagger.

Noun phrases are represented in the corpus using IOB tagging (Inside, Outside and Beginning of some noun phrase) [17]. To represent each word in the corpus as a set of features a window surrounding the word is considered. The window extends two words before, and two words after the word being classified. Features extracted from this window include: the words themselves, their parts-of-speech, as well as the chunk tags of the preceding inputs in the window [8]. Similar to the lemmatizer, we use Stanford classifier [10] to train a maximum entropy model.

As shown in Fig. 9, the NP chunker finds 3 NPs in the sample text, two of them are single word and the other one is a two-word noun phrase.

az/O zamāne/B bazgašt/O be/O irān/B bi sabrāne/O montazere/O amvālam/B budam/O ./O
ammā/O 3.5/B māh/I gozašt/O va/O ānhā/O hargez/O ferestāde/O našodand/O ./O

Fig. 9. The NP chunker's output on the sample input text.

5   Experimental Results

In this section, the experiments and evaluations of Persianp Toolbox are presented. We could compare Persianp with STeP-1 [20] and ParsiPardaz [19] which are two tool-boxes developed for processing Persian text, if we had access to their package. But their package are not available to us.

In the rest of the section, we firstly talk about datasets and then we present experiments and evaluations of different modules of Persianp Toolbox. Among several modules of Persianp, some modules require no special evaluation. The normalizer performs some character replacement or deletion. The stop word detector is actually a list search. Also, the sentencizer is a simple search on tokens and need no special evaluation. But other modules need to be tested.

5.1   Datasets

Datasets which are used to test different modules of Persianp toolbox are as follows:

- Paykare corpus: Paykare, as mentioned before, is a 10 million word Persian corpus manually annotated by POS tags. This corpus is used for the evaluation of the POS tagger. Moreover, Paykare is automatically lemmatized to train a probabilistic classifier which is a part of the lemmatizer (Sect. 4.6). This new corpus is used to test the classifier performance.
- Dadegan Treebank: As explained in Sect. 4.7, we used Dadegan to generate our training data for NP chunking. This data is used to test the NP chunker.
- A 100 K word dataset: To evaluate tokenization and lemmatization tasks, a dataset have to be created. To do so, a part of Paykare corpus containing 100,000 words is selected and excluded from other parts. This part is lemmatized manually to have a 100 K word test dataset. This dataset is called TD.

5.2   Experiments

We conduct several experiments to test modules of Persianp Toolbox. Test results are as follows:

- *The POS tagger evaluation*: We use 5-fold cross validation over Paykare corpus to test the POS tagger. The overall accuracy is 95.6%.
- *The tokenizer evaluation*: to test the tokenizer the raw text of TD is given to the tokenizer as the input. The accuracy of the tokenizer is 97%.
- *Lemmatization tasks evaluation*: three different experiments are perform to test automatically lemmatization of Paykare, the probabilistic classifier and the lemmatizer module in whole:

- *The evaluation of automatically lemmatization of Paykare*: to prepare a dataset for training a maximum entropy classifier which decides whether a word is inflected or not we require to lemmatize Paykare. This lemmatization is performed automatically using coarse-grain tags of words and morphological rules (Sect. 4.6). Automatically lemmatization cannot be accomplished perfectly because of some ambiguities in applying morphological rules or errors in words tags in Paykare. This process is evaluated on TD. The result shows 98.6% agreement between automatically extracted lemmas and manually ones. This high accuracy allows us to train our probabilistic classifier efficiently.

- *The classifier evaluation*: the maximum entropy classifier of the lemmatizer module is trained on the lemmatized corpus. Before training, the original tags of words are dropped and the corpus is annotated by the developed POS tagger. In this way we have a more fair evaluation. Using 5-fold cross validation the accuracy of 99.1% is achieved. This result shows that using the trained classifier we can almost perfectly detect words which are inflected and need to be lemmatized. This high rate of accuracy avoids egregious errors in Persian word lemmatization.

- *The lemmatizer evaluation*: If a word is detected as inflected word by the classifier, the word lemma is extracted in a rule-based procedure. To evaluate the lemmatizer TD is used. Because in reality texts are not annotated manually with POS tags we drop the original tags of TD and annotate it with the POS tagger to provide a fair test result. The lemmatizer shows accuracy of 97%.

- *The chunker evaluation*: the evaluation is accomplished by 5-fold cross validation on the automatically generated data from Dadegan corpus (Sect. 4.7). The result is 97.2% in accuracy measure.

Table 2 shows all experiments and results together.

Table 2. Evaluation results

Task	Result
POS tagger	95.6%
Tokenizer	97%
Paykare automatic lemmatization	98.6%
Classifier	99.1%
Lemmatizer	97%
Chunker	97.2%

6 Conclusions

Persianp Toolbox has been designed and developed to further enable and accelerate researches on the Persian language. This is a response to an important challenge facing researchers active on the field of the Persian language processing i.e. lack of access to fundamental processing tools. This paper described different modules of Persianp Toolbox and the evaluation of each module was accomplished by experiments.

The results showed the performance and effectiveness of the toolbox. There are still other important modules to be added to the software package. Named entity recognizer and syntactic parser are a priority in our future work.

References

1. AleAhmad, A., Amiri, H., Darrudi, E., Rahgozar, M., Oroumchian, F.: Hamshahri: a standard Persian text collection. Knowl. Based Syst. **22**(5), 382–387 (2009)
2. APLL (Academy of Persian Language and Literate): Persian Writing Style. Asar Press, Iran (2006)
3. Bijankhan, M., Sheykhzadegan, J., Bahrani, M., Ghayoomi, M.: Lessons from building a Persian written corpus: Peykare. Lang. Resour. Eval. **45**(2), 143–164 (2011)
4. Bijankhan, M., Mohseni, M.: Frequency Dictionary: According to a Written Corpus of Today Persian Language. University of Tehran Press, Tehran (2012). (in Persian), ISBN:978-964-03-6296-9
5. Cunningham, H., Maynard, D., Bontcheva, K., Tablan, V.: GATE: an architecture for development of robust HLT applications. In: Proceedings of the 40th Annual Meeting on Association for Computational Linguistics, pp. 168–175 (2002)
6. Dehdari, J., Lonsdale, D.: A Link Grammar Parser for Persian. Aspects of Iranian Linguistics, vol. 1. Cambridge Scholars Press, Cambridge (2008)
7. Ferrucci, D., Lally, A.: UIMA: an architectural approach to unstructured information processing in the corporate research environment. Nat. Lang. Eng. **10**, 327–348 (2004)
8. Jurafsky, D., Martin, J.H.: Speech and Language Processing: An Introduction to Natural Language Processing, Speech Recognition, and Computational Linguistics, 2nd edn. Prentice-Hall, Upper Saddle River (2009)
9. Kiani, S., Akhavan, T., Shamsfard, M.: Developing a Persian Chunker using a hybrid approach. In: International Multiconference on Computer Science and Information Technology, IMCSIT 2009, pp. 227–234 (2009)
10. Manning, C., Klein, D.: Optimization, Maxent models, and conditional estimation without magic. Tutorial at HLT-NAACL. ACL (2003)
11. Manning, C.D., Surdeanu, M., Bauer, J., Finkel, J.R., Bethard, S., McClosky, D.: The Stanford CoreNLP natural language processing toolkit. In: ACL (System Demonstrations), pp. 55–60 (2014)
12. Megerdoomian, K., Zajac, R.: Tokenization in the Shiraz project. Technical report, NMSU, CRL, Memoranda in Computer and Cognitive Science (2000)
13. Mohseni, M., Motalebi, H., Minaei-Bidgoli, B., Shokrollahi-far, M.: A Farsi part-of-speech tagger based on Markov model. In: Proceedings of the 2008 ACM Symposium on Applied Computing, pp. 1588–1589. ACM (2008)
14. Mohseni, M., Minaei-bidgoli, B.: A Persian part-of-speech tagger based on morphological analysis. In: Language Resources and Evaluation Conference (LREC) (2010)
15. Raja, F., Amiri, H., Tasharofi, S., Sarmadi, M., Hojjat, H., Oroumchian, F.: Evaluation of part of speech tagging on Persian text. In: The Second Workshop on Computational Approaches to Arabic Script-Based Languages, Linguistic Institute Stanford University (2007)
16. Rahimtoroghi, E., Hesham F., Shakery, A.: A structural rule-based stemmer for Persian. In: 5th International Symposium on Telecommunications (IST), pp. 574–578 (2010)
17. Ramshaw, L.A., Marcus, M.P.: Text chunking using transformation-based learning. In: Proceedings of the Third Annual Workshop on Very Large Corpora, pp. 82–94 (1995)

18. Rasooli, M.S., Kouhestani, M., Moloodi, A.: Development of a Persian syntactic dependency treebank. In: Proceedings of the 2013 Conference of the North American Chapter of the Association for Computational Linguistics: Human Language Technologies, pp. 306–314 (2013)
19. Sarabi, Z., Mahyar, H., Farhoodi, M.: ParsiPardaz: Persian language processing toolkit. In: Proceedings of the 3rd International eConference on Computer and Knowledge Engineering (ICCKE), pp. 73–79 (2013)
20. Shamsfard, M., Jafari, H.S., Ilbeygi, M.: STeP-1: a set of fundamental tools for Persian text processing. In: Language Resources and Evaluation Conference (LREC) (2010)

Embeddings, Language Modeling, and Sequence Labeling

Generating Bags of Words from the Sums of Their Word Embeddings

Lyndon White[✉], Roberto Togneri, Wei Liu, and Mohammed Bennamoun

The University of Western Australia, 35 Stirling Highway, Crawley, Western Australia
lyndon.white@research.uwa.edu.au,
{roberto.togneri,wei.liu,mohammed.bennamoun}@uwa.edu.au

Abstract. Many methods have been proposed to generate sentence vector representations, such as recursive neural networks, latent distributed memory models, and the simple sum of word embeddings (SOWE). However, very few methods demonstrate the ability to reverse the process – recovering sentences from sentence embeddings. Amongst the many sentence embeddings, SOWE has been shown to maintain semantic meaning, so in this paper we introduce a method for moving from the SOWE representations back to the bag of words (BOW) for the original sentences. This is a part way step towards recovering the whole sentence and has useful theoretical and practical applications of its own. This is done using a greedy algorithm to convert the vector to a bag of words. To our knowledge this is the first such work. It demonstrates qualitatively the ability to recreate the words from a large corpus based on its sentence embeddings.

As well as practical applications for allowing classical information retrieval methods to be combined with more recent methods using the sums of word embeddings, the success of this method has theoretical implications on the degree of information maintained by the sum of embeddings representation. This lends some credence to the consideration of the SOWE as a dimensionality reduced, and meaning enhanced, data manifold for the bag of words.

1 Introduction

The task being tackled here is the *resynthesis* of bags of words (BOW) from sentence embedding representations. In particular the generation of BOW from vectors based on the sum of the sentence's constituent words' embeddings (SOWE). To the knowledge of the authors, this task has not been attempted before.

The motivations for this task are the same as in the related area of sentence generation. Dinu and Baroni (2014) observe that given a sentence has a given meaning, and the vector encodes the same meaning, then it must be possible to translate in both directions between the natural language and the vector representation. A sub-step of this task is the unordered case (BOW), rather than true sentences, which we tackle in this paper. The success of the implementation does indicates the validity of this dual space theory, for the representations considered (where order is neglected). There are also some potential practical

© Springer International Publishing AG, part of Springer Nature 2018
A. Gelbukh (Ed.): CICLing 2016, LNCS 9623, pp. 91–102, 2018.
https://doi.org/10.1007/978-3-319-75477-2_5

applications of such an implementation, often ranging around common vector space representations.

Given suitable bidirectional methods for converting between sentence embeddings and bags of words, the sentence embedding space can be employed as a *lingua franca* for translation between various forms of information – though with loss of word order information. The most obvious of which is literal translation between different natural languages; however the use extends beyond this.

Several approaches have been developed for representing images and sentences in a common vector space. This is then used to select a suitable caption from a list of candidates (Farhadi et al. 2010; Socher et al. 2014). Similar methods, creating a common space between images and SOWE of the keywords describing them, could be used to generate keyword descriptions using BOW resynthesis – without any need for a list. This would allows classical word-based information retrieval and indexing techniques to be applied to images.

A similar use is the replacement of vector based extractive summarisation (Kåagebäck et al. 2014; Yogatama et al. 2015), with keyword based abstractive summarisation, which is the generation of a keyword summary from a document. The promising use of SOWE generation for all these applications is to have a separate model trained to take the source information (e.g. a picture for image description, or a cluster of sentences for abstract summarisation) as its input and train it to output a vector which is close to a target SOWE vector. This output can then be used to generate the sentence.

The method proposed in this paper has an input of a sum of word embeddings (SOWE) as the sentence embedding, and outputs the bag of word (BOW) which it corresponds to. The input is a vector for example $\tilde{s} = [-0.79, 1.27, 0.28, ..., -1.29]$, which approximates a SOWE vector, and outputs a BOW for example $\{$, $:1$, best $:1$, it $:2$, of $:2$, the $:2$, times $:2$, was $:2$, worst $:1\}$ – the BOW for the opening line of Dickens' *Tale of Two Cities*. Our method for BOW generation is shown in Fig. 1, note that it takes as input only a word embedding vocabulary ($\mathcal{V}$) and the vector ($\tilde{s}$) to generate the BOW ($\tilde{c}$).

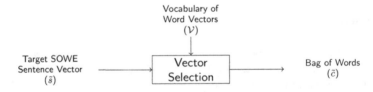

Fig. 1. The process for the regenerating BOW from SOWE sentence embeddings.

The rest of the paper is organized into the following sections. Section 2 introduces the area, discussing in general sentence models, and prior work on generation. Section 3 explains the problem in detail and our algorithm for solving it. Section 4 described the settings used for evaluation. Section 5 discusses the results of this evaluation. The paper presents its conclusions in Sect. 6, including a discussion of future work.

2 Background

The current state of the art for full sentence generation from sentence embeddings are the works of Iyyer et al. (2014) and Bowman et al. (2015). Both these advance beyond the earlier work of Dinu and Baroni (2014) which is only theorised to extend beyond short phrases. Iyyer et al. and Bowman et al. produce full sentences. These sentences are shown by examples to be loosely similar in meaning and structure to the original sentences. Neither works has produced quantitative evaluations, making it hard to determine between them. However, when applied to the various quantitative examples shown in both works neither is able to consistently reproduce exact matches. This motivates investigation on a simpler unordered task, converting a sum of word embeddings to bag of words, as investigated in this paper.

Bag of words is a classical natural language processing method for representing a text, sentence or document, commonly used in information retrieval. The text is represented as a multiset (or bag), this is an unordered count of how often each word occurs.

Word embeddings are vector representations of words. They have been shown to encode important syntactic and semantic properties. There are many different types of word embeddings (Yin and Schtze 2015). Two of the more notable are the SkipGrams of Mikolov et al. (2013a, b) and the Global Vector word representations (GloVe) of Pennington et al. (2014). Beyond word representations are sentence embeddings.

Sentence embeddings represent sentences, which are often derived from word embeddings. Like word embeddings they can capture semantic and syntactic features. Sentence vector creation methods include the works of Le and Mikolov (2014) and Socher (2014). Far simpler than those methods, is the sum of word embeddings (SOWE). SOWE, like BOW, draws significant criticism for not only disregarding sentence structure, but disregarding word order entirely when producing the sentence embedding. However, this weaknesses, may be offset by the improved discrimination allowed through words directly affecting the sentence embedding. It avoids the potential information loss through the indirection of more complex methods. Recent results suggest that this may allow it to be comparable overall to the more linguistically consistent embeddings when it comes to representing meaning.

White et al. (2015) found that when classifying real-world sentences into groups of semantically equivalent paraphrases, that using SOWE as the input resulted in very accurate classifications. In that work White et al. partitioned the sentences into groups of paraphrases, then evaluated how well a linear SVM could classify unseen sentences into the class given by its meaning. They used this to evaluate a variety of different sentence embeddings techniques. They found that the classification accuracy when using SOWE as the input performed very similarly to the best performing methods – less than 0.6% worse on the harder task. From this they concluded that the mapping from the space of sentence meaning to the vector space of the SOWE, resulted in sentences with the same meaning going to distinct areas of the vector space.

Ritter et al. (2015) presented a similar task on spacial-positional meaning, which used carefully constructed artificial data, for which the meanings of the words interacted non-simply – thus theoretically favouring the more complex sentence embeddings. In their evaluation the task was classification with a Naïve Bayes classifier into one of five categories of different spatial relationships. The best of the SOWE models they evaluated, outperformed the next best model by over 5%. These results suggest this simple method is still worth consideration for many sentence embedding representation based tasks. SOWE is therefore the basis of the work presented in this paper.

3   The Vector Selection Problem

At the core of this problem is what we call the Vector Selection Problem, to select word embedding vectors which sum to be closest to the target SOWE (the input). The word embeddings come from a known vector vocabulary, and are to be selected with potential repetition. Selecting the vectors equates to selecting the words, because there is a one to one correspondence between the word embedding vectors and their words. This relies on no two words having exactly the same embeddings – which is true for all current word embedding techniques.

Definition 1. *The Vector Selection Problem is defined on* $(\mathcal{V}, \tilde{s}, d)$ *for a finite vocabulary of vectors* $\mathcal{V}$, $\mathcal{V} \subset \mathbb{R}^n$, *a target sentence embedding* $\tilde{s}$, $\tilde{s} \in \mathbb{R}^n$, *and any distance metric* d, *by:*

$$\operatorname*{argmin}_{\left\{ \forall \tilde{c} \in \mathbb{N}_0^{|\mathcal{V}|} \right\}} \; d\!\left(\tilde{s}, \sum_{\tilde{x}_j \in \mathcal{V}} \tilde{x}_j \, c_j\right)$$

$\tilde{x}_j$ *is the vector embedding for the jth word in the vocabulary* $\tilde{x}_j \in \mathcal{V}$ *and* c_j *is the jth element of the count vector* $\tilde{c}$ *being optimised – it is the count of how many times the* x_j *occurs in approximation to the sum being assessed; and correspondingly it is the count of how many times the jth word from the vocabulary occurs in the bag of words. The selection problem is thus finding the right words with the right multiplicity, such that the sum of their vectors is as close to the input target vector,* $\tilde{s}$, *as possible.*

3.1   NP-Hard Proof

The vector selection problem is NP-Hard. It is possible to reduce from any given instance of a *subset sum problem* to a vector selection problem. The *subset sum problem* is NP-complete (Karp 1972). It is defined: for some set of integers $(\mathcal{S} \subset \mathbb{Z})$, does there exist a subset $(\mathcal{L} \subseteq \mathcal{S})$ which sums to zero $(0 = \sum_{l_i \in \mathcal{L}} l_i)$. A suitable metric, target vector and vocabulary of vectors corresponding to the elements $\mathcal{S}$ can be defined by a bijection; such that solving the vector selection problem will give the subset of vectors corresponding to a subset of $\mathcal{S}$ with the smallest sum; which if zero indicates that the subset sum does exists, and if nonzero indicates that no such subset ($\mathcal{L}$) exists. A fully detailed proof of the reduction from subset sum to the vector selection problem can be found on the first author's website.[1]

3.2   Selection Algorithm

The algorithm proposed here to solve the selection problem is a greedy iterative process. It is a fully deterministic method, requiring no training, beyond having the word embedding mapping provided. In each iteration, first a greedy search (Greedy Addition) for a path to the targeted sum point $\tilde{s}$ is done, followed by correction through substitution (n-Substitution). This process is repeated until no change is made to the path. The majority of the selection is done in the Greedy Addition step, while the n-substitution handles fine tuning.

Greedy Addition. The greedy addition phase is characterised by adding the best vector to the bag at each step (see the pseudo-code in Algorithm 1). At each step, all the vectors in the current bag are summed, and then each vector in the vocabulary is added in turn to evaluate the new distance the new bag would have from the target, the bag which sums to be closest to the target becomes the current solution. This continues until there is no option to add any of the vectors without moving the sum away from the target. There is no bound on the size of the bag of vector (i.e. the length of the sentence) in this process, other than the greedy restriction against adding more vectors that do not get closer to the solution.

Greedy Addition works surprisingly well on its own, but it is enhanced with a fine tuning step, n-substitution, to decrease its greediness.

n-Substitution. We define a new substitution based method for fine tuning solutions called n-substitution. It can be described as considering all subbags containing up to n elements, consider replacing them with a new sub-bag of up that size n from the vocabulary, including none at all, if that would result in the overall bag getting closer to the target $\tilde{s}$.

The reasoning behind performing the n-substitution is to correct for greedy mistakes. Consider the 1 dimensional case where $V = 24, 25, 100$ and $\tilde{s} = 148$, $d(x, y) = |x - y|$. Greedy addition would give $bag_c = [100, 25, 24]$ for a distance of 1, but a perfect solution is $bag_c = [100, 24, 24]$ which is found using 1-substitution. This substitution method can be considered as re-evaluating past decisions in light of the future decisions. In this way it lessens the greed of the addition step.

The n-substitution phase has time complexity of $O\left(\binom{C}{n}V^n\right)$, for $C = \sum \tilde{c}$ i.e. current cardinality of bag_c. With large vocabularies it is only practical to consider 1-substitution. With the Brown Corpus, where $|\mathcal{V}| \approx 40,000$, it was found that 1-substitution provides a significant improvement over greedy addition alone. On a smaller trial corpora, where $|\mathcal{V}| \approx 1,000$, 2-substitution was used and found to give further improvement. In general it is possible to initially use 1-substitution, and if the overall algorithm converges to a poor solution (given the distance to the

---

[1] http://white.ucc.asn.au/publications/White2015BOWgen/.

```
Data: the metric d
the target sum s̃
the vocabulary of vectors 𝒱
the current best bag of vectors bag_c: initially ∅
Result: the modified bag_c which sum to be as close as greedy search can get to
        the target s̃, under the metric d
begin
    t̃ ⟵   ∑   x_i
         x_i∈bag_c
    while true do
        x̃* ⟵ argmin d(s̃, t̃ + x̃_j)        /* exhaustive search of 𝒱        */
              x_j∈𝒱
        if d(s̃, t̃ + x̃*) < d(s̃, t̃) then
        |   t̃ ⟵ t̃ + x̃* bag_c ⟵ bag_c ∪ {x̃*}
        else
        |   return bag_c      /* No further improving step found       */
        end
    end
end
```

Algorithm 1. Greedy Addition. In practical implementation, the bag of vectors can be represented as list of indices into columns of the embedding vocabulary matrix, and efficient matrix summation methods can be used.

target is always known), then the selection algorithm can be retried from the converged solution, using 2-substitution and so forth. As n increases the greed overall decreases; at the limit the selection is not greedy at all, but is rather an exhaustive search.

4 Experimental Setup and Evaluations

4.1 Word Embeddings

GloVe representations of words (Pennington et al. 2014) are used in our evaluations. There are many varieties of word embeddings which work with our algorithm. GloVe was chosen simply because of the availability of a large pre-trained vocabulary of vectors. The representations used for evaluation were pretrained on 2014 Wikipedia and Gigaword 5[2]. Preliminary results with SkipGrams from Mikolov et al. (2013a) suggested similar performance.

4.2 Corpora

The evaluation was performed on the Brown Corpus (Francis and Kucera 1979) and on a subset of the Books Corpus (Zhu et al. 2015). The Brown Corpus was sourced with samples from a 500 fictional and non-fictional works from 1961.

---

[2] Kindly made available online at http://nlp.stanford.edu/projects/glove/.

The Books Corpus was sourced from 11,038 unpublished novels. The Books Corpus is extremely large, containing roughly 74 million sentences. After pre-processing we randomly selected 0.1% of these for evaluation.

For simplicity of evaluation, sentences containing words not found in the pre-trained vector vocabulary are excluded. These were generally rare mis-spellings and unique numbers (such as serial numbers). Similarly, words which are not used in the corpus are excluded from the vector vocabulary.

After the preprocessing the final corpora can be described as follows. The Brown Corpus has 42,004 sentences and a vocabulary of 40,485 words. Where-as, the Books Corpus has 66,464 sentences, and a vocabulary of 178,694 words. The vocabulary sizes are beyond what is suggested as necessary for most uses (Nation 2006). These corpora remain sufficiently large and complex to quantitatively evaluate the algorithm.

4.3 Vector Selection

The Euclidean metric was used to measure how close potential solutions were to the target vector. The choice of distance metric controls the ranking of each vector by how close (or not) it brings the partial sum to the target SOWE during the greedy selection process. Preliminary results on one-tenth of the Books Corpus used in the main evaluation found the Manhattan distance performed marginally worse than the Euclidean metric and took significantly longer to converge.

The commonly used cosine similarity, or the linked angular distance, have an issue of zero distances between distinct points – making them not true distance metrics. For example the SOWE of *"a can can can a can"* has a zero distance under those measures to the SOWE for *"a can can"*.[3] That example is a patholog-ical, though valid sentence fragment. True metrics such as the Euclidean metric do not have this problem. Further investigation may find other better distance metrics for this step.

The Julia programming language (Bezanson et al. 2014), was used to create the implementation of the method, and the evaluation scripts for the results presented in the next section. This implementation, evaluation scripts, and the raw results are available online.[4]. Evaluation was carried out in parallel on a 12 core virtual machine, with 45 Gb of RAM. Sufficient RAM is required to load the entire vector vocabulary in memory.

5 Results and Discussion

Table 1 shows examples of the output. Eight sentences which were used for demon-stration of sentence generation in Bowman et al. (2015) and Iyyer et al. (2014) have the BOW generation results shown. All examples except *(a)* and *(f)* are perfect.

---

[3] The same is true for any number of repetitions of the word *buffalo* – each of which forms a valid sentence as noted in Tymoczko et al. (1995).

[4] http://www.cicling.org/2016/data/97.

Table 1. Examples of the BOW Produced by our method using the Books Corpus vocabulary, compared to the Correct BOW from the reference sentences. The P and C columns show the number of occurrences of each word in the Produced and Correct bags of words, respectively. **Bolded** lines highlight mistakes. Examples a–e were sourced from Iyyer et al. (2014), Examples f–h from Bowman et al. (2015). Note that in example a, the "$\_$_____(n)" represents n repeated underscores (without spaces).

(a) ralph waldo emerson dismissed this poet as the jingle man and james russell lowell called him three-fifths genius and two-fifths sheer fudge

Word	P	C
2008	**1**	**0**
_.....(13)	**1**	**0**
_.....(34)	**1**	**0**
_.....(44)	**1**	**0**
"	**1**	**0**
aldrick	**1**	**0**
and	2	2
as	**0**	**1**
both	**1**	**0**
called	**0**	**1**
dismissed	1	1
emerson	1	1
fudge	1	1
genius	1	1
hapless	**1**	**0**
him	1	1
hirsute	**1**	**0**
james	1	1
jingle	1	1
known	**1**	**0**
lowell	1	1
man	**0**	**1**
poet	1	1
ralph	1	1
russell	1	1
sheer	1	1
the	1	1
this	1	1
three-fifths	1	1
two-fifths	1	1
waldo	1	1
was	**1**	**0**

(b) thus she leaves her husband and child for aleksei vronsky but all ends sadly when she leaps in front of a train

Word	P	C
a	1	1
aleksei	1	1
all	1	1
and	1	1
but	1	1
child	1	1
ends	1	1
for	1	1
front	1	1
her	1	1
husband	1	1
in	1	1
leaps	1	1
leaves	1	1
of	1	1
sadly	1	1
she	2	2
thus	1	1
train	1	1
vronsky	1	1
when	1	1

(c) name this 1922 novel about leopold bloom written by james joyce

Word	P	C
1922	1	1
about	1	1
bloom	1	1
by	1	1
james	1	1
joyce	1	1
leopold	1	1
name	1	1
novel	1	1
this	1	1
written	1	1

(d) this is the basis of a comedy of manners first performed in 1892

Word	P	C
1892	1	1
a	1	1
basis	1	1
comedy	1	1
first	1	1
in	1	1
is	1	1
manners	1	1
of	2	2
performed	1	1
the	1	1
this	1	1

(e) in a third novel a sailor abandons the patna and meets marlow who in another novel meets kurtz in the congo

Word	P	C
a	2	2
abandons	1	1
and	1	1
another	1	1
congo	1	1
in	3	3
kurtz	1	1
marlow	1	1
meets	2	2
novel	2	2
patna	1	1
sailor	1	1
the	2	2
third	1	1
who	1	1

(f) how are you doing ?

Word	P	C
're	**1**	**0**
?	1	1
are	**0**	**1**
do	**1**	**0**
doing	**0**	**1**
how	1	1
well	**1**	**0**
you	**0**	**1**

(g) we looked out at the setting sun .

Word	P	C
.	1	1
at	1	1
looked	1	1
out	1	1
setting	1	1
sun	1	1
the	1	1
we	1	1

(h) i went to the kitchen .

Word	P	C
.	1	1
i	1	1
kitchen	1	1
the	1	1
to	1	1
went	1	1

Table 2. The performance of the BOW generation method. Note the final line is for the Books Corpus, where-as the preceding are or the Brown Corpus.

Corpus	Embedding dimensions	Portion perfect	Mean Jaccard score	Mean precision	Mean recall	Mean F1 score
Brown	50	6.3%	0.175	0.242	0.274	0.265
Brown	100	19.4%	0.374	0.440	0.530	0.477
Brown	200	44.7%	0.639	0.695	0.753	0.720
Brown	300	70.4%	0.831	0.864	0.891	0.876
Books	300	75.6%	0.891	0.912	0.937	0.923

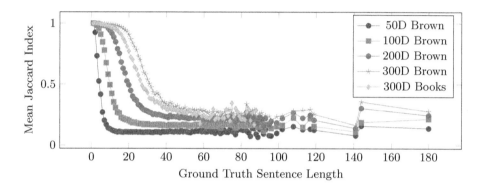

Fig. 2. The mean Jaccard index achieved during the word selection step, shown against the ground truth length of the sentence. Note that the vast majority of sentences are in the far left end of the plot. The diminishing samples are also the cause of the roughness, as the sentence length increases.

Example *(f)* is interesting as it seems that the contraction token *'re* was substituted for *are*, and *do* for *doing*. Inspections of the execution logs for running on the examples show that this was a greedy mistake that would be corrected using 2-substitution. Example *a* has many more mistakes.

The mistakes in Example *(a)* seem to be related to unusual nonword tokens, such as the three tokens with 13, 34, and 44 repetitions of the underscore character. These tokens appear in the very large Books corpus, and in the Wikipedia/Gigaword pretraining data used for word embeddings, but are generally devoid of meaning and are used as structural elements for formatting. We theorise that because of their rarity in the pre-training data they are assigned an unusual word-embedding by GloVE. There occurrence in this example suggests that better results may be obtained by pruning the vocabulary. Either manually, or via a minimum uni-gram frequency requirement. The examples overall highlight the generally high performance of the method, and evaluations on the full corpora confirm this.

Table 2 shows the quantitative performance of our method across both corpora. Five measures are reported. The most clear is the portion of exact matches – this is how often out of all the trials the method produced the exact correct bag of words. The remaining measures are all means across all the values of the measures in each trial. The Jaccard index is the portion of overlap between the reference BOW, and the output BOW – it is the cardinality of the intersection divided by that of the union. The precision is the portion of the output words that were correct; and the recall is the portion of all correct words which were output. For precision and recall word repetitions were treated as distinct. The F_1 score is the harmonic mean of precision and recall. The recall is higher than the precision, indicating that the method is more prone to producing additional incorrect words (lowering the precision), than to missing words out (which would lower the recall).

Initial investigation focused on the relationship between the number of dimensions in the word embedding and the performance. This was carried out on the smaller Brown corpus. Results confirmed the expectation that higher dimensional embeddings allow for better generation of words. The best performing embedding size (i.e. the largest) was then used to evaluate success on the Books Corpus. The increased accuracy when using higher dimensionality embeddings remains true at all sentence lengths.

As can be seen in Fig. 2 sentence length is a very significant factor in the performance of our method. As the sentences increase in length, the number of mistakes increases. However, at higher embedding dimensionality the accuracy for most sentences is high. This is because most sentences are short. The third quartile on sentence length is 25 words for Brown, and 17 for the Books Corpus. This distribution difference is also responsible for the apparent better results on the Books Corpus, than on the Brown corpus.

While the results shown in Table 2 suggest that on the Books corpus the algorithm performs better, this is due to its much shorter average sentence length. When taken as a function of the sentence length, as shown in Fig. 2, performance on the Books Corpus is worse than on the Brown Corpus. It can be concluded from this observation that increasing the size of the vocabulary does decrease success in BOW regeneration. Books Corpus vocabulary being over four times larger, while the other factors remained the same, resulted in lower performance. However, when taking all three factors into account, we note that increasing the *vocabulary size* has significantly less impact than increasing the *sentence length* or the *embedding dimensionality* on the performance.

6   Conclusion

A method was presented for how to regenerate a bag of words, from the sum of a sentence's word embeddings. This problem is NP-Hard. A greedy algorithm was found to perform well at the task, particularly for shorter sentences when high dimensional embeddings are used.

Resynthesis degraded as sentence length increased, but remained strong with higher dimensional models up to reasonable length. It also decreased as the vocabulary size increased, but significantly less so. The BOW generation method is functional with usefully large sentences and vocabulary.

From a theoretical basis the resolvability of the selection problem shows that adding up the word embeddings does preserve the information on which words were used; particularly for higher dimensional embeddings. This shows that collisions do not occur (at least not frequently) such that two unrelated sentences do not end up with the same SOWE representation.

This work did not investigate the performance under noisy input SOWEs – which occur in many potential applications. Noise may cause the input to better align with an unusual sum of word embeddings, than with its true value. For example it may be shifted to be very close a sentence embedding that is the sum of several hundred word embeddings. Investigating, and solving this may be required for applied uses of any technique that solves the vector selection problem.

More generally, future work in this area would be to use a stochastic language model to suggest suitable orderings for the bags of words. While this would not guarantee correct ordering every-time, we speculate that it could be used to find reasonable approximations often. Thus allowing this bag of words generation method to be used for full sentence generation, opening up a much wider range of applications.

Acknowledgements. This research is supported by the Australian Postgraduate Award, and partially funded by Australian Research Council grants DP150102405 and LP110100050. Computational resources were provided by the National eResearch Collaboration Tools and Resources project (Nectar).

References

Bezanson, J., et al.: Julia: a fresh approach to numerical computing (2014). arXiv:1411.1607 [cs.MS]

Bowman, S.R., et al.: Generating sentences from a continuous space (2015). arXiv preprint arXiv:1511.06349

Dinu, G., Baroni, M.: How to make words with vectors: phrase generation in distributional semantics. In: Proceedings of ACL, pp. 624–633 (2014)

Farhadi, A., Hejrati, M., Sadeghi, M.A., Young, P., Rashtchian, C., Hockenmaier, J., Forsyth, D.: Every picture tells a story: generating sentences from images. In: Daniilidis, K., Maragos, P., Paragios, N. (eds.) ECCV 2010. LNCS, vol. 6314, pp. 15–29. Springer, Heidelberg (2010). https://doi.org/10.1007/978-3-642-15561-1_2

Francis, W.N., Kucera, H.: Brown corpus manual. Brown University (1979)

Iyyer, M., Boyd-Graber, J., Daumé III, H.: Generating sentences from semantic vector space representations. In: NIPS Workshop on Learning Semantics (2014)

Kåagebäck, M., et al.: Extractive summarization using continuous vector space models. In: Proceedings of the 2nd Workshop on Continuous Vector Space Models and their Compositionality (CVSC)@ EACL, pp. 31–39 (2014)

Karp, R.M.: Reducibility among combinatorial problems. In: Miller, R.E., Thatcher, J.W., Bohlinger, J.D. (eds.) Complexity of Computer Computations. The IBM Research Symposia Series, pp. 85–103. Springer, Boston (1972). https://doi.org/10.1007/978-1-4684-2001-2_9

Le, Q., Mikolov, T.: Distributed representations of sentences and documents. In: Proceedings of the 31st International Conference on Machine Learning (ICML-2014), pp. 1188–1196 (2014)

Mikolov, T., et al.: Efficient estimation of word representations in vector space (2013a). arXiv preprint arXiv:1301.3781

Mikolov, T., Yih, W.-T., Zweig, G.: Linguistic regularities in continuous space word representations. In: HLT-NAACL, pp. 746–751 (2013b)

Nation, I.: How large a vocabulary is needed for reading and listening? Can. Mod. Lang. Rev. **63**(1), 59–82 (2006)

Pennington, J., Socher, R., Manning, C.D.: GloVe: global vectors for word representation. In: Proceedings of the 2014 Conference on Empirical Methods in Natural Language Processing (EMNLP 2014), pp. 1532–1543 (2014)

Ritter, S., et al.: Leveraging preposition ambiguity to assess compositional distributional models of semantics. In: The Fourth Joint Conference on Lexical and Computational Semantics (2015)

Socher, R.: Recursive deep learning for natural language processing and computer vision. Ph.D. thesis. Stanford University (2014)

Socher, R., et al.: Grounded compositional semantics for finding and describing images with sentences. Trans. Assoc. Comput. Linguist. **2**, 207–218 (2014)

Tymoczko, T., Henle, J., Henle, J.M.: Sweet reason: a field guide to modern logic. In: Textbooks in Mathematical Sciences. Key College (1995). ISBN 9780387989303

White, L., et al.: How well sentence embeddings capture meaning. In: Proceedings of the 20th Australasian Document Computing Symposium (ADCS 2015), Parramatta, pp. 9:1–9:8. ACM (2015). https://doi.org/10.1145/2838931.2838932. ISBN 978-1-4503-4040-3

Yin, W., Schtze, H.: Learning word meta-embeddings by using ensembles of embedding sets (2015). eprint arXiv:1508.04257

Yogatama, D., Liu, F., Smith, N.A.: Extractive summarization by maximizing semantic volume. In: Conference on Empirical Methods in Natural Language Processing (2015)

Zhu, Y., et al.: Aligning books and movies: towards story-like visual explanations by watching movies and reading books (2015). arXiv preprint arXiv:1506.06724

New Word Analogy Corpus for Exploring Embeddings of Czech Words

Lukáš Svoboda[1,2(✉)] and Tomáš Brychcín[1,2]

[1] Department of Computer Science and Engineering, Faculty of Applied Sciences,
University of West Bohemia, Univerzitní 8, 306 14 Plzeň, Czech Republic
{svobikl,brychcin}@kiv.zcu.cz
[2] NTIS—New Technologies for the Information Society, Faculty of Applied Sciences,
University of West Bohemia, Univerzitní 8, 306 14 Plzeň, Czech Republic
http://www.nlp.kiv.zcu.cz

Abstract. The word embedding methods have been proven to be very useful in many tasks of NLP (Natural Language Processing). Much has been investigated about word embeddings of English words and phrases, but only little attention has been dedicated to other languages.

Our goal in this paper is to explore the behavior of state-of-the-art word embedding methods on Czech, the language that is characterized by very rich morphology. We introduce new corpus for word analogy task that inspects syntactic, morphosyntactic and semantic properties of Czech words and phrases. We experiment with Word2Vec and GloVe algorithms and discuss the results on this corpus. The corpus is available for the research community.

1 Introduction

Word embedding is the name for techniques in NLP (Natural Language Processing) where meaning of words or phrases is represented by vectors of real numbers.

It was shown that the word vectors can be used for significant improving and simplifying of many NLP applications [1,2]. There are also NLP applications, where word embeddings does not help much [3].

There has been introduced several methods based on the feed-forward NNLP (Neural Network Language Model) in recent studies. One of the Neural Network based models for word vector representation which outperforms previous methods on word similarity tasks was introduced in [4]. The word representations computed using NNLP are interesting, because trained vectors encode many linguistic properties and those properties can be expressed as linear combinations of such vectors.

Nowadays, word embedding methods Word2Vec [5] and GloVe [6] significantly outperform other methods for word embeddings. Word representations made by these methods have been successfully adapted on variety of core NLP task such as Named Entity Recognition [7,8], Part-of-speech Tagging [9], Sentiment Analysis [10], and others.

© Springer International Publishing AG, part of Springer Nature 2018
A. Gelbukh (Ed.): CICLing 2016, LNCS 9623, pp. 103–114, 2018.
https://doi.org/10.1007/978-3-319-75477-2_6

There are also neural translation-based models for word embeddings [11,12] that generates an appropriate sentence in target language given sentence in source language, while they learn distinct sets of embeddings for the vocabularies in both languages. Comparison between monolingual and translation-based models can be found in [13].

Many researches have investigated the behavior of these methods on English, but only little attention has been dedicated to other languages. In this work we focus on Czech that is a representative of Slavic languages. These languages are highly inflected and have a relatively free word order. Czech has seven cases and three genders. The word order is very variable from the syntactic point of view: words in a sentence can usually be ordered in several ways, each carrying a slightly different meaning. All these properties complicate the learning of word embeddings.

In this article we are exploring whether are word embedding methods as good on highly inflected languages like Czech as they are on English. It has been shown that such word embedding models improve Named Entity Recognition on Czech [8], but we would like to investigate if they capture the semantic and syntactic relationships independently from specific task.

There is a variety of datasets for measuring semantic relatedness between English words, such as *WordSimilarity-353* [14], *Rubenstein and Goodenough (RG)* [15], *Rare-words* [16], *Word pair similarity in context* [4], and many others. To the best of our knowledge, there is only one such corpus for Czech [17], which is essentially only translation of RG corpus into Czech.

Except the similarity between words, we would like to explore other semantic and syntactic properties hidden in word embeddings. A new evaluation scheme based on word analogies were presented in [5]. By examining various dimensions of differences we can achieve interesting results, for example: vector("king") – vector("man") is close to vector("queen") – vector("woman"). Based on this approach and our need to further use and investigate the word embedding methods on Czech, we have decided to build semantic-syntactic word analogy dataset. Especially, we focus on exploring how state-of-the-art word embedding methods carry semantics and syntax of words.

2   Word Embeddings Methods

The backbone principle of word embedding methods is the formulation of *Distributional Hypothesis* in [18] that says *"a word is characterized by the company it keeps"*. The direct implication of this hypothesis is that the word meaning is related to the context where it usually occurs and thus it is possible to compare the meanings of two words by statistical comparisons of their contexts. This implication was confirmed by empirical tests carried out on human groups in [15,19].

The distributional semantics models typically represent the word meaning as a vector, where the vector reflects the contextual information of a word across the training corpus. Each word $w \in W$ (where W denotes the word vocabulary) is associated with a vector of real numbers $w \in \mathbb{R}^k$. Represented geometrically, the word meaning is a point in a high-dimensional space. The words that are closely related in meaning tend to be closer in the space.

In this work we will focus on three monolingual models that produce high quality word embeddings. In general, given a single word in the corpus, these models predict which other words should serve as a substitution for this word.

2.1 CBOW

CBOW (Continuous Bag-of-Words) [5] tries to predict the current word according to the small context window around the word. The architecture is similar to the feed-forward NNLP (Neural Network Language Model) which has been proposed in [20]. The NNLM is computationally expensive between the projection and the hidden layer. Thus, CBOW proposed architecture, where the (non-linear) hidden layer is removed and projection layer is shared between all words. The word order in the context does not influence the projection (see Fig. 1a). This architecture also proved low computational complexity.

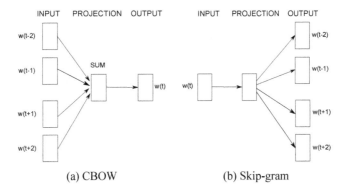

(a) CBOW   (b) Skip-gram

Fig. 1. Neural network models architectures.

2.2 Skip-Gram

Skip-gram architecture is similar to CBOW. Although instead of predicting the current word based on the context, it tries to predict a words context based on the word itself [21]. Thus, intention of the Skip-gram model is to find word patterns that are useful for predicting the surrounding words within a certain range in a sentence (see Fig. 1b). Skip-gram model estimates the syntactic properties of words slightly worse than the CBOW model, but it is much better for modeling the word semantics on English test set [5,21]. Training of the Skipgram model does not involve dense matrix multiplications Fig. 1b and that makes training also extremely efficient [21].

2.3   GloVe

GloVe (Global Vectors) [6] model focuses more on the global statistics of the trained data. This approach analyses log-bilinear regression models that effectively capture global statistics and also captures word analogies. Authors propose a weighted least squares regression model that trains on global word-word co-occurrence counts. The main concept of this model is the observation that ratios of word-word co-occurrence probabilities have the potential for encoding meaning of words.

3   Word Analogy Corpus

In this section we present a new word analogy Czech corpus for testing word embeddings. Inspiration was taken from English corpus revealed in [5]. We follow observation that the state-of-the-art models for word embeddings can capture different types of similarities between words. Given two pairs of words with the same relationship as a question: Which word is related to *export* in the same sense as *minimum* is related to *maximum*? Correct answer should be *import*.

Such a question can be answered with a simple algebraic operation with the vector representation of words:

$$x = \text{vector}(\text{``maximum''}) - \text{vector}(\text{``minimum''}) + \text{vector}(\text{``export''}) \quad (1)$$

Difference between vector("maximum") and vector("minimum") should be similar to difference between vector("export") and vector("import"). For resulting vector x we search in the vector space for the most similar word. When the model works well and is properly trained, we will find that the closest vector representing correct answer for our question is the vector for the word *import*.

If the model has sufficient data, it is able to learn also more complicated semantic relationships between words, such as the main city *Prague* to the state *Czech Republic* is with the similar relation as *Paris* is to *France*, or capturing the presidents of individual states, already mentioned antonyms, plural versus singular words, gradation of adjectives, and other words relationships.

To measure quality of word vectors, we have designed test set containing 8,705 semantic questions and 13,552 syntactic questions. Together, we have 22,257 combinations of questions. Dataset contains only enough frequent words on Czech Wikipedia. We split the dataset into several categories. Each category usually contains about 35–40 pairs of words with same relationship. Question has been built by all combination of word pairs in the same category.

There is a majority of word-to-word relationships, but *Presidents and states* category contains also bigram-to-word (word-to-bigram) relationships such as *Prague* vs. *Czech Republic*.

Semantic questions are represented in categories:

- **Presidents-states-cities:** Consists of 34 pairs of states in Europe and their main cities combining 1,122 questions. There is also 1,122 questions for state with corresponding current president.

- **Antonyms:** This category compounds of three subcategories. In first subcategory we have 38 noun antonym pairs that is resulting in 1406 questions combined. Example of such question is: *anode, cathode* versus *export, import*. Similarly we have 42 adjectives pairs (such as *big, small*) and 34 verb pairs - *buy, sell* versus *give, take*.
- **Family-relations (man-woman):** In this category we have 19 pairs of family representatives with man-woman relation as *brother, sister* versus *husband, wife*.

Syntactic questions are represented in categories:

- **Adjectives-gradation:** In this category we have two antonym pairs with three degrees of adjectives in positive, comparative, and superlative form: *big, bigger* vs *small, smaller*.
- **Nationalities (woman/man):** This category is specific for Czech language, which distinguish between masculine and feminine word relations. Every nationality has its corresponding masculine and feminine word form. For example, English word *Japan* has in Czech masculine form *Japonec* and feminine form *Japonka*. We have 35 such pairs.
- **Nouns-plural:** We find here 37 pairs of nouns and their plural forms.
- **Jobs:** Category with 35 pairs of professions with masculine-feminine word relations.
- **Verb-past:** This category consists of verbs in present form versus verbs in past tense form, such as *play, played* versus *see, saw*.
- **Pronouns:** Last category consists of pairs of pronouns in singular versus plural form.

4 Experiments

In our experiments, we used unsupervised learning of word-level embeddings using Word2Vec [5] and GloVe tool [6]. We used the January 2015 snapshot of the Czech Wikipedia as a source of unlabeled data. The Wikipedia corpus has been preprocessed with the following steps:

1. Removed special characters such as #$&%, HTML tags and others.
2. Filtering XML dumps, removed tables, links converted to normal text. We lowercase all words. We have also removed sentences with less than 5 tokens.

The resulting training corpus contains about 2,6 billion words. For our purpose, it is useful to have vector representation of word phrases, i.e. for bigram representing state *Czech Republic*, it is desirable to have one vector representing those two words. This was achieved by preprocessing the training data set to form the phrases using the *Word2Phrase* tool [21].

We evaluate the word embedding models on our corpus by accuracy that is defined as

$$Acc\% = \frac{NC}{NT}, \tag{2}$$

where NC is the number of correctly answered questions for a category and NT is total number of questions in category.

In our experiments, we use *cosine similarity* as a measure of similarity between two word vectors. Cosine similarity is probably the most used similarity metric for words embedding methods. It characterizes the similarity between two vectors as the cosine of the angle between them

$$S_{\cos}(\boldsymbol{a}, \boldsymbol{b}) = \frac{\boldsymbol{a} \cdot \boldsymbol{b}}{\|\boldsymbol{a}\| \cdot \|\boldsymbol{b}\|} = \frac{\sum a_i b_i}{\sqrt{\sum a_i^2 \sum b_i^2}}, \tag{3}$$

where $\boldsymbol{a}$ and $\boldsymbol{b}$ are two vectors we try to compare. The cosine similarity is used in all cases where we want to find the most similar word (or top n most similar words) for a given type of analogy.

4.1 Models Settings

During the training of Word2Vec (resp. GloVe) models, we limited the size of the vocabulary to 400,000 most frequent single token words and about 800,000 most frequent bigrams. OOV (Out-of-vocabulary) word rate was 6%. It means that out of 22,257 questions was about 1,300 questions not seen in vocabulary.

To train word embedding methods we use context window of size 10. We also explore results with different vector dimension (set to 100, 300, and 500). We choose to compare three training epochs as in [5] for similarly sized training corpus versus ten training epochs for Word2Vec tool. For GloVe tool we choose 10 and 25 iterations, because algorithms cannot be simply compared with the same settings [6]. Other Word2Vec and GloVe settings were on its default values.

4.2 Results on Czech Word Analogy Corpus

In this section we present the accuracies for all tested models (CBOW, Skip-gram, and GloVe) on our word analogy corpus. In all tables below we present results for different vector dimension ranging between 50 and 500, except for Skip-gram model with dimension 500 and 10 training epochs, where the time of computation was much higher than with other methods. Model did not finish after 4 days of training and results of 500 dimension vector does not substantiate such long training time. We use notation n_D in the tables, n means that the correct word must be between n most similar words for a given analogy. D denote the dimension of vectors. Accuracies are expressed in percents.

In Table 1 we present the results for CBOW model. There is a significant improvement between 3 and 10 training epochs. Interesting is also fact that 300-dimensional vectors perform better than 500-dimensional vectors on most categories. Similarly, the results for Skip-gram model are in Table 2. This model performs significantly worse on most categories in comparison with CBOW model. There is also significant overall improvement between 50-dimensional and 100-dimensional vector, but less significant between 100 and 300. Table 3 shows result for GloVe model. This model gives on Czech the worst results compared to both Word2Vec models.

Categories, where the models gives best results are *Verb-past, Noun-plural,* and *State-city*. In general, all models gives better results on tasks exploring syntactic information. Poor accuracy was in categories *State-presidents* and category *Nationality*.

How to achieve better accuracy? It was shown in [21] that sub-sampling of the frequent words and choosing larger Negative Sampling window helps to improve performance. Also, adding much more text with information related to particular categories would help (see [6]), especially for class *State-presidents*.

In this paper we had focused more on how number of training epochs influence overall performance in respect to the reasonable time of training and how vector embeddings holds semantics and syntactic information of individual Czech words. We have relatively large corpus for training so we choose 10 iterations (respectively 25 for GloVe) as maximum to compare. To train such models can take more than 3 days with Core i7-3960X, especially for Skip-gram model and vector dimension set to 500. We also do not expect much improvement with more iterations on our corpus, however, we recommend to do more training epochs than it is set by default.

Table 1. Results for CBOW.

Type	3 training epochs											
	1_50	1_100	1_300	1_500	5_50	5_100	5_300	5_500	10_50	10_100	10_300	10_500
Anton. (nouns)	1.35	4.84	5.55	5.69	3.98	10.88	13.16	10.95	5.69	13.44	16.00	13.30
Anton. (adj.)	4.82	8.86	11.79	13.24	10.63	14.29	18.70	19.16	13.24	17.31	23.64	22.76
Anton. (verbs)	0.20	1.88	2.68	1.25	2.77	3.13	6.25	3.57	2.94	3.84	7.77	4.38
State-president	0.00	0.00	0.18	0.09	0.18	0.00	0.98	0.18	0.45	0.27	1.43	0.71
State-city	14.62	14.8	16.22	8.47	29.77	30.93	32.89	23.26	35.92	39.57	42.96	31.82
Family	6.42	9.01	11.60	9.26	12.10	17.28	21.85	18.64	14.44	21.11	25.80	23.95
Noun-plural	34.46	42.42	41.74	44.60	45.95	53.60	54.35	54.35	50.45	57.43	57.43	57.81
Jobs	2.95	3.87	3.37	2.78	6.57	10.52	10.00	8.92	9.18	14.05	13.80	12.37
Verb-past	14.83	24.29	42.52	34.91	29.94	40.91	60.61	52.00	36.66	48.31	66.50	58.80
Pronouns	1.59	3.84	5.95	3.57	3.97	8.07	12.70	10.05	5.69	9.66	16.00	13.10
Adj.-gradation	12.50	20.00	22.50	15.00	20.00	22.50	22.50	27.50	20.00	27.50	25.00	27.50
Nationality	0.08	0.42	0.33	0.16	0.84	0.92	0.84	1.10	1.26	1.26	1.26	2.01
	10 training epochs											
	1_50	1_100	1_300	1_500	5_50	5_100	5_300	5_500	10_50	10_100	10_300	10_500
Anton. (nouns)	3.84	7.82	8.53	7.40	8.39	15.93	18.49	16.07	10.38	19.42	22.76	20.55
Anton. (adj.)	7.26	11.90	15.45	15.04	13.53	19.63	25.49	23.58	16.49	23.05	30.26	28.92
Anton. (verbs)	0.89	1.88	2.86	3.12	4.01	5.98	6.43	6.07	5.09	6.70	7.59	7.41
State-president	0.18	0.35	0.09	0.09	0.71	0.98	0.62	0.71	1.16	1.60	1.33	1.16
State-city	16.58	27.99	25.94	18.63	37.07	50.62	52.05	39.13	43.49	58.47	61.41	50.71
Family	11.85	15.43	15.68	15.93	19.75	25.55	30.99	29.13	25.56	30.12	38.02	36.42
Noun-plural	50.23	56.68	60.56	57.96	63.21	68.92	70.35	66.52	67.87	72.97	74.02	69.14
Jobs	6.73	10.52	6.82	4.04	14.39	19.78	17.68	13.30	17.59	24.24	23.06	19.36
Verb-past	25.87	38.71	48.53	48.71	46.92	58.95	69.34	68.78	55.10	66.75	76.00	74.94
Pronouns	5.03	6.22	7.80	7.14	10.71	12.17	15.61	15.48	13.76	16.53	19.31	19.84
Adj.-gradation	25.00	25.00	20.00	17.50	25.00	25.00	27.50	25.00	25.00	30.00	32.50	27.50
Nationality	0.67	1.26	0.34	0.42	2.35	2.60	1.68	2.35	3.03	3.19	3.27	2.77

Table 2. Results for Skip-gram.

Type	3 training epochs											
	1_50	1_100	1_300	1_500	5_50	5_100	5_300	5_500	10_50	10_100	10_300	10_500
Anton. (nouns)	0.85	1.71	3.34	5.55	2.20	3.84	8.04	10.74	2.92	5.41	9.67	14.08
Anton. (adj.)	2.26	3.02	5.23	8.48	4.59	5.69	9.00	12.37	6.21	7.14	11.32	14.81
Anton. (verbs)	0.18	0.36	0.36	0.98	0.27	1.61	0.45	2.05	0.89	1.79	0.89	2.68
State-president	0.18	0.18	0.09	0.09	0.53	0.71	0.36	0.62	0.62	1.16	0.71	0.80
State-city	6.60	14.26	8.20	3.48	17.20	27.27	18.89	12.75	22.99	33.69	25.94	21.93
Family	1.98	2.72	2.59	6.79	3.70	6.30	9.01	12.59	6.30	8.52	12.72	16.42
Noun-plural	8.11	14.04	19.14	18.77	15.17	24.62	27.25	36.41	18.17	29.05	31.23	44.59
Jobs	1.77	1.26	1.09	1.01	5.05	3.96	3.45	3.53	6.40	5.81	4.88	5.39
Verb-past	1.72	4.36	4.14	6.08	4.20	8.28	7.67	12.74	6.04	10.62	9.90	19.97
Pronouns	0.79	1.06	0.66	0.40	2.78	2.25	1.72	1.72	3.97	4.23	2.65	2.78
Adj.-gradation	2.50	5.00	5.00	10.00	5.00	7.50	12.50	17.50	5.00	12.50	12.50	25.00
Nationality	0.17	0.08	0.08	0.00	0.84	0.67	0.17	0.42	1.26	1.01	0.25	0.92
	10 training epochs											
	1_50	1_100	1_300	1_500	5_50	5_100	5_300	5_500	10_50	10_100	10_300	10_500
Anton. (nouns)	1.35	2.63	6.19	x	3.27	5.83	10.24	x	4.41	7.25	12.23	x
Anton. (adj.)	1.74	4.82	5.69	x	4.53	9.12	10.05	x	5.57	11.85	12.54	x
Anton. (verbs)	0.36	0.00	0.18	x	0.98	1.96	0.36	x	1.52	2.95	0.62	x
State-president	0.27	0.09	0.27	x	1.07	0.36	0.80	x	1.52	0.62	1.60	x
State-city	4.55	15.15	9.98	x	14.26	31.73	25.85	x	19.88	39.48	35.29	x
Family	3.09	3.70	6.67	x	6.30	9.14	13.46	x	10.37	12.22	16.54	x
Noun-plural	19.22	29.95	23.95	x	31.91	43.92	37.91	x	37.39	47.75	44.59	x
Jobs	2.53	3.03	2.53	x	6.99	7.58	4.88	x	9.93	10.44	7.58	x
Verb-past	2.93	8.25	8.77	x	7.41	15.15	16.69	x	9.73	18.72	20.84	x
Pronouns	0.66	0.66	0.79	x	2.65	2.25	3.44	x	3.84	3.44	4.76	x
Adj.-gradation	2.50	10.00	7.50	x	10.00	15.00	12.50	x	10.00	15.00	15.00	x
Nationality	0.17	0.42	0.08	x	0.50	1.26	0.34	x	0.67	1.60	0.76	x

Our goal was not to achieve maximal overall score, but rather to analyze the behavior of word embedding models on Czech language. In following text, we discuss how well these models hold semantic and syntactic information. From results on semantic versus syntactic accuracy (see Table 4) we can say that for Czech is CBOW approach, which predicts the current word according to the context window better, than predicting a words context based on the word itself as in Skip-gram approach.

Accuracy on category *State-president* is very low with all models. We would assume to achieve similar results as with category State-city. However, such low score was caused by few simple facts. Firstly, we are missing a data, this is supported by argument that this category has 27% OOV of questions, than the probability that resulting word will also be missing in vocabulary is going to be high. Second thing is that even if the correct word for a question is not missing in vocabulary, we have more often different corresponding candidates mentioned as presidents of Czech Republic in training data. For example for a question: *"What is a*

Table 3. Results for GloVe.

Type	3 training epochs											
	1_50	1_100	1_300	1_500	5_50	5_100	5_300	5_500	10_50	10_100	10_300	10_500
Anton. (nouns)	0.36	1.28	0.64	0.81	1.00	2.92	1.99	1.72	1.49	4.27	2.63	2.42
Anton. (adj.)	0.87	0.81	1.34	1.34	2.44	4.01	6.10	5.81	3.60	5.40	8.89	7.62
Anton. (verbs)	0.00	0.00	0.00	0.00	0.36	0.00	0.00	0.00	0.36	0.00	0.18	0.00
State-president	0.00	0.00	0.00	0.00	0.00	0.00	0.00	0.00	0.00	0.00	0.00	0.00
State-city	1.52	0.98	1.16	0.98	3.83	3.21	4.01	2.85	5.17	4.90	6.68	5.81
Family	3.33	4.20	0.99	1.42	6.67	6.42	4.81	3.85	8.52	8.64	7.41	4.35
Noun-plural	14.79	15.32	12.69	5.54	24.47	26.35	25.83	14.30	28.53	31.46	33.03	18.70
Jobs	0.67	0.25	0.00	0.00	1.43	0.76	0.08	0.00	1.68	1.09	0.17	0.00
Verb-past	5.39	6.96	3.15	0.82	11.59	13.71	7.72	2.78	15.11	17.70	10.80	4.71
Pronouns	0.79	0.66	0.00	0.00	1.59	1.32	1.46	0.00	2.12	1.72	2.38	0.00
Adj.-gradation	7.50	7.50	5.00	0.00	10.00	12.50	7.50	7.50	10.00	12.50	10.00	7.50
Nationality	0.00	0.00	0.00	0.00	0.08	0.00	0.00	0.00	0.08	0.17	0.00	0.00
	25 training epochs											
	1_50	1_100	1_300	1_500	5_50	5_100	5_300	5_500	10_50	10_100	10_300	10_500
Anton. (nouns)	0.50	0.85	1.14	1.42	1.28	2.70	4.69	4.05	1.71	4.34	6.33	5.62
Anton. (adj.)	1.68	2.67	1.34	1.34	3.83	6.68	6.56	6.21	5.28	7.96	9.87	8.65
Anton. (verbs)	0.18	0.00	0.00	0.00	0.36	0.18	0.09	0.18	0.89	0.18	0.45	0.36
State-president	0.00	0.00	0.00	0.00	0.00	0.00	0.00	0.00	0.00	0.00	0.00	0.00
State-city	0.98	1.07	0.98	0.45	3.39	4.19	4.01	2.85	4.99	5.97	7.66	6.51
Family	2.35	3.70	2.10	2.22	5.43	5.80	6.05	4.20	7.04	7.65	8.52	5.56
Noun-plural	28.00	30.56	15.32	6.98	39.79	43.84	29.20	18.02	43.47	48.35	38.44	28.23
Jobs	0.17	0.00	0.00	0.00	0.59	0.42	0.00	0.00	0.76	0.76	1.18	0.51
Verb-past	7.86	10.78	3.98	1.13	16.53	19.25	10.07	4.19	20.82	23.64	14.12	6.81
Pronouns	1.32	1.32	0.26	0.00	3.44	2.25	1.06	0.00	4.76	3.57	1.72	0.00
Adj.-gradation	5.00	5.00	5.00	0.00	7.50	10.00	12.50	7.50	15.00	12.50	12.50	7.50
Nationality	0.00	0.00	0.00	0.00	0.00	0.00	0.00	0.00	0.00	0.00	0.00	0.00

Table 4. Accuracy on semantic and syntactic part of corpus.

Type	3 training epochs for CBOW and Skip-gram, 10 training epochs for GloVe											
	1_50	1_100	1_300	1_500	5_50	5_100	5_300	5_500	10_50	10_100	10_300	10_500
CBOW − semantics	4.77	6.57	8.00	6.33	9.90	12.75	15.64	12.63	12.11	15.92	19.6	16.15
Skip-gram − semantics	2.00	4.75	6.66	x	3.71	7.57	9.62	x	3.30	7.62	10.21	x
GloVe − semantics	1.01	1.21	0.69	0.78	2.38	2.76	2.82	2.56	3.19	3.87	4.30	3.63
CBOW − syntactics	11.06	15.81	19.40	16.84	17.85	22.76	26.84	25.65	20.48	26.37	30.00	28.60
Skip-gram − syntactics	2.51	5.51	6.81	x	4.30	7.88	10.54	x	5.02	8.79	10.24	x
GloVe − syntactics	4.86	5.11	3.50	0.98	8.20	9.11	7.10	3.72	9.59	10.77	9.40	5.26
	10 training epochs for CBOW and Skip-gram, 25 training epochs for GloVe											
	1_50	1_100	1_300	1_500	5_50	5_100	5_300	5_500	10_50	10_100	10_300	10_500
CBOW − semantics	6.77	10.90	11.42	10.03	13.91	19.78	22.35	19.12	17.02	23.23	26.90	24.20
Skip-gram − semantics	1.89	4.40	4.83	4.23	5.07	9.69	10.13	8.52	7.21	12.40	13.14	11.79
GloVe − semantics	0.95	1.38	0.93	0.90	2.38	3.26	3.57	2.92	3.32	4.35	5.47	4.45
CBOW − syntax	18.92	23.07	24.01	22.63	27.10	31.24	33.69	31.9	30.34	35.61	38.03	35.59
Skip-gram − syntax	4.67	8.72	7.27	6.04	9.91	14.19	12.63	12.05	11.93	16.16	15.59	15.94
GloVe − syntax	7.06	7.94	4.09	1.35	11.31	12.63	8.81	4.95	14.14	14.80	11.33	7.17

similar word to Czech as is Belarus Alexandr Lukasenko" we are expecting word *Milos Zeman*, who is our current president. However the models tells us that the most similar word is a word *president*, which is good answer, but we would rather like to see actual name. When we explore other most similar word, we will find *Vaclav Klaus*, who was our former president, fourth similar word was the word *Vaclav Havel*, our first and famous president of Czech Republic after 1992. Based on those statements we can say that we had lack of data corresponding to current presidents in our training corpus.

Czech language has a lot of synonyms for every word that is also why there is overall much better improvement in containing more similar words - TOP 10, rather than just comparing again one word with the highest similarity TOP 1. Therefore there is a bigger improvement in TOP 1 versus TOP 10 similar words on semantics over it is on syntactic tasks.

The most interesting results are however for a category Nationality, where we compare nationalities in masculine and feminine form. Complete category is covered in vocabulary. However answers for questions are completely out of topic. For a question which should return feminine form of resident of America, the closest word which model returns is *Oscar Wilde*, respective just his last name, second word is *peacefully philosophy* and another name showing up is *Louise Lasser*. Similar task to category Nationalities with masculine-feminine word form is category Jobs, all models there also perform poorly. This specific task for Czech language seems to be difficult for current state-of-the-art word embeddings methods.

GloVe model seems to give worse results than Word2Vec models, where on English analogy task gives better accuracy [6]. We would probably get better results with tuning the models properties, but that can be achieved with both presented toolkits.

5   Conclusion and Future Work

In this paper we introduced new dataset for measuring syntactic and semantic properties of Czech words. We experimented with three state-of-the-art methods of word embeddings, namely, CBOW, Skip-gram, and GloVe. We achieved almost 27% accuracy on semantic tasks and 38% on syntactic tasks with our best CBOW model with dimension 300 and exploring top 10 most similar words.

Interesting finding is that on Czech, CBOW model performs much better on word semantics rather than Skip-gram, which performs significantly better on English [21].

We made corpus with evaluator script and best trained models publicly available for research purposes at https://github.com/Svobikl/cz_corpus.

For a future work we would like to further investigate properties of other models for word embeddings and try to use external sources of information (such as part-of-speech tags) during training process.

Acknowledgements. This work was supported by the project LO1506 of the Czech Ministry of Education, Youth and Sports and by Grant No. SGS-2016-018 Data and Software Engineering for Advanced Applications. Computational resources were provided by the CESNET LM2015042 and the CERIT Scientific Cloud LM2015085, provided under the programme "Projects of Large Research, Development, and Innovations Infrastructures.

References

1. Collobert, R., Weston, J.: A unified architecture for natural language processing: deep neural networks with multitask learning (2008)
2. Collobert, R., Weston, J., Bottou, L., Karlen, M., Kavukcuoglu, K., Kuksa, P.P.: Natural language processing (almost) from scratch. CoRR abs/1103.0398 (2011)
3. Andreas, J., Klein, D.: How much do word embeddings encode about syntax? In: Proceedings of the 52nd Annual Meeting of the Association for Computational Linguistics (Volume 2: Short Papers), Baltimore. Association for Computational Linguistics, pp. 822–827, June 2014
4. Huang, E.H., Socher, R., Manning, C.D., Ng, A.Y.: Improving word representations via global context and multiple word prototypes. In: Proceedings of the 50th Annual Meeting of the Association for Computational Linguistics (ACL 2012) (Long Papers - Volume 1), Stroudsburg. Association for Computational Linguistics, pp. 873–882 (2012)
5. Mikolov, T., Chen, K., Corrado, G., Dean, J.: Efficient estimation of word representations in vector space (2013)
6. Pennington, J., Socher, R., Manning, C.D.: Glove: global vectors for word representation. In: Proceedings of the 2014 Conference on Empirical Methods in Natural Language Processing (EMNLP), pp. 1532–1543 (2014)
7. Siencnik, S.K.: Adapting word2vec to named entity recognition. In: Proceedings of the 20th Nordic Conference of Computational Linguistics (NODALIDA 2015), pp. 239–243 (2015)
8. Demir, H., Ozgur, A.: Improving named entity recognition for morphologically rich languages using word embeddings. In: 13th International Conference on Machine Learning and Applications (ICMLA), pp. 117–122. IEEE (2014)
9. Al-Rfou, R., Perozzi, B., Skiena, S.: Polyglot: distributed word representations for multilingual NLP (2013)
10. Pontiki, M., Galanis, D., Papageorgiou, H., Manandhar, S., Androutsopoulos, I.: SemEval-2015 task 12: aspect based sentiment analysis. In: Proceedings of the 9th International Workshop on Semantic Evaluation (SemEval 2015), pp. 486–495. Association for Computational Linguistics, Denver (2015)
11. Cho, K., Van Merriënboer, B., Gulcehre, C., Bahdanau, D., Bougares, F., Schwenk, H., Bengio, Y.: Learning phrase representations using rnn encoder-decoder for statistical machine translation. arXiv preprint arXiv:1406.1078 (2014)
12. Bahdanau, D., Cho, K., Bengio, Y.: Neural machine translation by jointly learning to align and translate. arXiv preprint arXiv:1409.0473 (2014)
13. Hill, F., Cho, K., Jean, S., Devin, C., Bengio, Y.: Not all neural embeddings are born equal. CoRR abs/1410.0718 (2014)
14. Finkelstein, L., Gabrilovich, E., Matias, Y., Rivlin, E., Solan, Z., Wolfman, G., Ruppin, E.: Placing search in context: the concept revisited. ACM Trans. Inf. Syst. **20**(1), 116–131 (2002)

15. Rubenstein, H., Goodenough, J.B.: Contextual correlates of synonymy. Commun. ACM **8**(10), 627–633 (1965)
16. Luong, M.T., Socher, R., Manning, C.D.: Better word representations with recursive neural networks for morphology. In: CoNLL, Sofia (2013)
17. Krčmář, L., Konopík, M., Ježek, K.: Exploration of semantic spaces obtained from Czech corpora. In: Proceedings of the DATESO 2011: Annual International Workshop on DAtabases, TExts, Specifications and Objects, Pisek, pp. 97–107, 20 April 2011
18. Firth, J.R.: A synopsis of linguistic theory 1930–1955. Stud. Linguist. Anal. 1–32 (1957)
19. Charles, W.G.: Contextual correlates of meaning. Appl. Psycholinguist. **21**(04), 505–524 (2000)
20. Bengio, Y., Schwenk, H., Senécal, J.S., Morin, F., Gauvain, J.L.: Neural probabilistic language models. In: Holmes, D.E., Jain, L.C. (eds.) Innovations in Machine Learning. Studies in Fuzziness and Soft Computing, vol. 194, pp. 137–186. Springer, Heidelberg (2006). https://doi.org/10.1007/3-540-33486-6_6
21. Mikolov, T., Sutskever, I., Chen, K., Corrado, G.S., Dean, J.: Distributed representations of words and phrases and their compositionality. In: Advances in Neural Information Processing Systems, pp. 3111–3119 (2013)

Using Embedding Models for Lexical Categorization in Morphologically Rich Languages

Borbála Siklósi[✉]

Faculty of Information Technology and Bionics, Pázmány Péter Catholic University,
50/a Práter street, Budapest 1083, Hungary
siklosi.borbala@itk.ppke.hu

Abstract. Neural-network-based semantic embedding models are relatively new but popular tools in the field of natural language processing. It has been shown that continuous embedding vectors assigned to words provide an adequate representation of their meaning in the case of English. However, morphologically rich languages have not yet been the subject of experiments with these embedding models. In this paper, we investigate the performance of embedding models for Hungarian, trained on corpora with different levels of preprocessing. The models are evaluated on various lexical categorization tasks. They are used for enriching the lexical database of a morphological analyzer with semantic features automatically extracted from the corpora.

1 Introduction

Finding a good representation of words and lexemes is a crucial task in the field of natural language processing. The question is what type of representation to use that is able to model the distributional patterns of words including their meaning and their morphosyntactic and syntactic behavior. For English, the use of continuous vector space representations have recently replaced the manual creation of such resources as well as that of sparse discrete representations learned from analyzed or raw texts. The neural-network-based implementations of these continuous representations have proved to be efficient as shown in several publications [2,7,9]. Most studies, however, focus on the application of these models to English, where the moderate number of different word forms and the relatively fixed word order fit well the theory behind these models. The goal of this paper is to investigate the performance of word embedding models applied to Hungarian, an agglutinating language with free word order.

The motivation of our investigation is, however, twofold. First, our goal was to explore the semantic sensibility of embedding methods for a language more complex than English, i.e. whether it is able to locate words consistently in the semantic space when trained on Hungarian texts. On the other hand, the possibility of using the results to augment the stem database of a Hungarian morphological analyzer with semantic features was also investigated [13]. Semantic features

© Springer International Publishing AG, part of Springer Nature 2018
A. Gelbukh (Ed.): CICLing 2016, LNCS 9623, pp. 115–126, 2018.
https://doi.org/10.1007/978-3-319-75477-2_7

may affect the morphological, syntactic and orthographic behavior of words in Hungarian with certain constructions being applicable only to specific classes such as colors, materials, nationalities, languages, occupations, first names etc. and words falling into these categories could be identified and collected from the corpus resulting in exhaustive lists, which could not have been built manually. Moreover, lexical semantic categories extracted by the models can be used to enrich the annotation of argument slots in verbal subcategorization frames like the ones in [5] with further semantic constraints.

The structure of this paper is as follows. First, some of the main characteristics of Hungarian is described, which demonstrates the complexity of the given task. Then a brief summary of related work and continuous embedding models is presented. In the following sections our experiments are described regarding building different models from differently preprocessed corpora, and their use in the task of extracting semantic categories. Finally, both qualitative and quantitative evaluations of the results are presented along with describing some methods created for supporting human evaluation processes.

2   Hungarian

Hungarian is an agglutinating language, and as such, its morphology is rather complex. Words are often composed of long sequences of morphemes, with agglutination and compounding yielding a huge number of different word forms. For example, while the number of different word tokens in a 20-million-word English corpus is generally below 100,000, the number is above 800,000 in the case of Hungarian. However, the 1:8 ratio does not correspond to the ratio of the number of possible word forms between the two languages: while there are about at most 4–5 possible different inflected forms for an English word, there are about a 1000 for Hungarian, which indicates that a corpus of the same size is much less representative for Hungarian than it is for English [14]. These characteristics often make the direct adaptation of NLP methods developed for English unfeasible. The best performing methods for English often perform significantly worse for Hungarian.

For morphologically rich languages, morphological analysis plays a crucial role in most natural language processing tasks, and the quality of the morphological analyzer used is of great importance. In Hungarian, the morphological behavior of words is also affected by certain semantic features. Proper characterization of semantically restricted morphological constructions is only possible if these features are explicitly listed in the stem database of the analyzer.

The aim of this paper is thus twofold. First, to investigate the performance of neural embedding models applied to a morphologically rich language. Second, to provide a methodology for the automatic derivation of semantic categories relevant from the aspect of morphology.

3   Related Work

The main point of distributional semantics is that the meaning of words is closely related to their use in certain contexts [3]. Traditional models of distributional semantics build word representations by counting words occurring in a fixed-size context of the target word [2].

In contrast, a more recent method for building distributional representations of words is using *word embedding models* the most influential implementation of which is presented in Mikolov et al. [7,8]. Different implementations of this technique all build continuous vector representations of word meanings from raw corpora. These vectors point to certain locations in the semantic space consistently so that semantically and/or syntactically related words are close to each other, while unrelated ones are more distant. Moreover, it has been shown that vector operations can also be applied to these representations, thus the semantic relatedness of two words can be quantified as the algebraic difference of the two vectors representing these words. Similarly, the meaning of the composition of two words is generally represented well by the sum of two corresponding meaning vectors [9]. One of the main drawback of this method, however, is that by itself it is not able to handle polysemy and homonymy, since one representational vector is built for one lexical element regardless of the number of its different meanings. There are some studies addressing this issue as well by extending the original implementation of word embedding methods [1,4,18].

When training embedding models, a fixed-size context of the target word is used, similarly to traditional, discrete distributional models. However, this context representation is used as the input of a neural network. This network is used to predict the target word from the context by using back-propagation and adjusting the weights assigned to the connection between the input neurons (each corresponding to an item in the whole vocabulary) and the projection layer of the network. This weight vector can finally be extracted and used as the embedding vector of the target word. Since similar words are used in similar contexts, these vectors optimized for the context will also be similar for such words. There are two types of neural networks used for this task. One of them is the so called CBOW (continuous bag-of-words) model in which the network is used to predict the target word from the context, while the other model, called skip-gram, is used to predict the context from the target word. For both models, the embedding vectors can be extracted from the middle layer of the network and can be used alike in both cases.

4   Experiments

We built two types of models using the word2vec[1] tool, a widely-used framework for creating word embedding representations. This tool implements both models that can be used for building the embedding vectors, however, as the CBOW model has proved to be more efficient for large training corpora, we used this

---

[1] https://code.google.com/p/word2vec/.

model. As a training corpus we used a 3-billion-word raw web-crawled corpus of
Hungarian (applying boilerplate removal). In each experiment, the radius of the
context window was set to 5 and the number of dimensions to 300.

Then we applied different types of preprocessing to the corpus in order to
adapt the method to the agglutinating behavior of Hungarian (or to any other
morphologically rich language having a morphological analyzer/tagger at hand).

4.1 The Model Trained on Raw Text

First, we built a model from the tokenized but otherwise raw corpus (SURF).
This model derived different vectors for the different surface forms of the same
word. Thus, the various suffixed forms of the same lemma were placed at different locations in the semantic space. As a consequence, this model was able to
represent morphological analogies. For example the similarities of the word pairs
jó – rossz 'good – bad' and *jobb – rosszabb* 'better – worse' are much higher in
this model than if we compare the suffixed form and its lemma, i.e. *jó – jobb*
'good – better', and *rossz – rosszabb* 'bad – worse'. Table 1 shows some more
examples for the list of the most similar forms retrieved for some surface word
forms. As it can be seen from the examples, the model represents both semantic
and morphosyntactic similarities. For example the top-n list (containing the n
most similar words for the target word) for the wordform kenyerek 'bread.plur'
has similar pastries listed in their plural form (the $-k$ ending of all of these words

Table 1. Similar words in the model created from a raw corpus. Numbers in parentheses
show corpus frequency.

kenyerek(2270) 'breads'	pirosas(1729) 'reddish'	egerekkel(634) 'with mice'
kiflik(349) 'bagels'	lilás(2476) 'purplish'	patkányokkal(524) 'with rats'
zsemlék(283) 'buns'	rózsaszínes(1638) 'pinkish'	férgekkel(513) 'with worms'
lepények(202) 'pies'	barnás(6463) 'brownish'	majmokkal(606) 'with monkeys'
pogácsák(539) 'scones'	sárgás(7365) 'yellowish'	hangyákkal(343) 'with ants'
pékáruk(771) 'bakery products'	zöldes(5215) 'greenish'	nyulakkal(366) 'with rabbits'
péksütemények(997) 'pastry.pl'	fehéres(2517) 'whitish'	legyekkel(252) 'with flies'
sonkák(613) 'hams'	vöröses(5496) 'reddish'	rágcsálókkal(259) 'with rodents'
tészták(2466) 'pasta.pl'	feketés(1157) 'blackish'	hüllőkkel(241) 'with reptiles'
kalácsok(277) 'cakes'	narancssárgás(429) 'orangish'	pókokkal(436) 'with spiders'
kekszek(1046) 'biscuits'	sárgászöld(723) 'yellowish green'	bogarakkal(425) 'with bugs'
fiaik(1230) 'their sons'	megeszi(7647) 'he eats it'	Vakkalit(5) 'Vakkali.Acc'
lányaik(593) 'their daughters'	eszi(12615) 'he is eating it'	tevedesnek(5) 'as a mistake'
leányaik(251) 'their daughters'	megenné(563) 'he would eat it'	áfa-jának(7) 'of its VAT'
férjeik(759) 'their husbands'	lenyeli(1862) 'he swallows it'	mot-nak(5) 'mot.Dat'
gyermekeik(12028) 'their children'	megeszik(6433) 'they eat it'	Villanysze(5) 'Electrici(an)'
feleségeik(638) 'their wives'	Megeszi(189) 'He eats it'	oktávtól(5) 'from octave'
gyerekeik(5806) 'their children'	megette(7868) 'he ate it'	Isten-imádat(5) 'worship of God'
asszonyaik(458) 'their wives'	megrágja(477) 'he chews it'	Nagycsajszi(5) 'Big Chick'
gyermekei(31241) 'his children'	megeheti(287) 'he may eat it'	-fontosnak(7) '-as important'
fiak(1523) 'sons'	bekapja(977) 'he swallows it'	tárgykörből(5) 'from the subject'

is due to the plural suffix *–k*). The numbers in the lists next to each word are their corpus frequencies.

Even though this model is able to reflect semantic relations to some extent besides morphosyntactic groupings of words, the different surface forms of the same lemma make the model less robust, since the contexts a word is used in are divided between the different surface forms of the same lemma. For example, there are 197 different inflected forms for the lemma kenyér 'bread' in the corpus.

4.2   A Model Built from Annotated Texts

In the other experiment, we used a morphologically annotated version of the corpus. This was done using the PurePos part-of-speech tagger [15] which also performs lemmatization using morphological analyses generated by the Hungarian Humor morphological analyzer [10,11,17]. Each word form in the corpus was represented by two tokens: a lemma token followed by a morphosyntactic tag token ANA. Table 2 shows a sentence preprocessed this way.

Table 2. Analyzed version of the Hungarian sentence *A török megszállás nem feltétlenül jelentette a népesség pusztulását.* 'Turkish occupation did not necessarily lead to the destruction of the population.'

a [Det] török [Adj] megszállás [N] nem [Neg] feltétlenül [Adv] jelent [V.Past.3Sg.Def]
the   Turkish   occupation   not   necessarily   mean
a [Det] népesség [N] pusztulás [N.Poss3Sg.Acc] . [.]
the   population   destruction   .

Since the tags were kept in the actual context of the word they belonged to, the morphosyntactic information carried by the inflections still had a role in determining the embedding vectors. On the other hand, data sparseness was reduced, because the various inflected forms were represented by a single lemma. Table 3 shows some examples of top-n lists generated by this model. While the SURF model is often not capable to capture the semantics of rare word forms reliably (e.g. the most similar entries for the word form *Vakkalit* 'Vakkali.Acc' are completely unrelated forms in Table 1), the ANA model is capable of capturing the semantics of the same lexical items because lemmatization alleviates data sparseness problems and morphosyntactic annotation provides additional grammatical information. The most similar entries of *Vakkali* 'Vakkali' in the ANA model *(Ánanda, Avalokitésvara, Dordzse, Babaji, Bodhidharma, Gautama, Mahakásjapa, Maitreya, Bódhidharma)* clearly indicate that this is the name of a Buddhist personality.

4.3   Spelling Errors and Non-standard Word Forms

Investigating the models also revealed that among the groups of semantically related words, orthographic variations and misspelled forms of these words also

Table 3. Similar words in the model created from a annotated corpus. Numbers in parentheses show corpus lemma frequency.

kenyér	'bread'	eszik	'eat'	csavargó	'vagabond'
hús(136814)	'meat'	iszik(244247)	'drink'	koldus(15793)	'beggar'
kalács(10658)	'milk loaf'	főz(120634)	'cook'	zsivány(3497)	'rogue'
rizs(31678)	'rice'	csinál(1194585)	'make'	haramia(2024)	'ruffian'
zsemle(6690)	'roll'	megeszik(68347)	'eat'	vadember(2497)	'savage'
pogácsa(11066)	'bisquit'	fogyaszt(160724)	'consume'	csirkefogó(2019)	'scoundrel'
sajt(46660)	'cheese'	etet(43539)	'feed'	szatír(1649)	'satyr'
kifli(9715)	'croissant'	zabál(13699)	'gobble'	útonálló(1942)	'highwayman'
krumpli(37271)	'potato '	megiszik(31002)	'drink'	bandita(6334)	'bandit'
búzakenyér(306)	'wheat bread'	eszeget(3928)	'nibble'	suhanc(4144)	'stripling'
tej(113911)	'milk'	alszik(359268)	'sleep'	vándor(14070)	'wanderer'

appear. When initiating the retrieval of top-n lists with such non-standard forms, the resulting lists contained words with the same type of errors as the seed word, but semantic similarity was also represented in the ranking of these words. From the preprocessed model, typical error types of the lemmatizer could also be collected. Misspellings and lexical gaps in the morphological analyzer may lead to cases where the guesser in the tagger erroneously tags and lemmatizes words. Lemmas resulting from similar errors are grouped together by the model. E.g. the 'lemmas' *pufidzsek(i)* 'puffy jacket', *rövidnac(i)* 'shorts', *napszemcs(i)* 'sunglasses', *szemcs(i)* 'glasses', *szmöty(i)* 'gunk' etc. all lack the ending –*i*. They result from the guesser erroneously cutting the ending –*it* from the accusative form of these words. The whole class can be corrected by the same operation, or, as a more permanent solution, all members of the class can easily be added to the lexicon of the morphological analyzer. Similar results can be used to improve the quality of the corpus by correcting these errors in the texts themselves, but also for pinpointing errors in the components of the annotation tool chain (the tokenizer, the lemmatizer or the morphological analyzer) [12]. Another perspective of utilizing this feature of these models is making NLP tools handle OOV items in a more fault tolerant manner by having them annotate unknown words by assigning the annotation of known words that are similar according to the model.

Since the corpus we used was a web-crawled corpus, it also contained a lot of slang and non-standard words coming from user-generated and social media sources. The model works well for these types of texts as well, collecting non-standard words with similar meanings in the top-n lists of such terms. Slang variants *mittomén/mittudomén/mittoménmi/mittudoménmi/nemtommi* 'idunnowhat' are grouped together by the model similarly to representations of laughter *hehehe/hihihi/hahaha/höhö/muhaha/heh/Muhaha/muhahaha/höhöhö*.

4.4 Extracting Semantic Groups

We used the two models to extract coherent semantic groups from the corpus, which could then be used to enrich the lexical categorization system of a morphological analyzer. Since our goal in this task was to organize words along their semantic similarity, rather than their syntactic behavior, we used the ANA model only, i.e. the one trained on the analyzed texts. We created a web application to aid the exploration and visualization of the models and the retrieval of semantically restricted vocabulary. For each category an initial word was selected and the top 200 most similar words were retrieved from the model. Then, the top 200 most similar words were retrieved for items selected by a simple mouse click (taken from the bottom of the previous list). This step was repeated about 10 times. Repeated occurrences were filtered out when retrieving the subsequent lists. The result lists were then merged. Moreover, it was also checked by quick inspection whether the lists did in fact contain mostly relevant items. Those that did not, were deleted by a single click. Throwing these words away, the algorithm was applied again resulting in purer lists. Thus, starting from one word for each category, hundreds or thousands of related words could be retrieved semi-automatically with minimal human interaction that could hardly have been done manually. It was also found that for narrower categories, such as 'materials of clothes', retrieving the top 200 words in each iteration resulted in too much noise, thus in these cases we decreased the size of the top-n list in each iteration to 50.

5 Results

We evaluated the task of semantic categorization by manually counting the number of correct and incorrect words in the given category. However, in order to be able to perform this validation efficiently, the result lists were clustered automatically so that these groups could be reviewed at once. Moreover, the words together with their cluster affiliation were displayed in a two-dimensional plot, providing more visual aid to the human evaluator. Clustering and 2D semantic map visualization was integrated into the web application.

5.1 Clustering

To cluster the lexical elements retrieved from the embedding model, we applied hierarchical clustering. The reason for choosing this type of clustering was based on the argument of [16]. The variety and sophistication of written texts makes the prediction of the number of resulting clusters impossible. However, in a hierarchical clustering, the separation of compact clusters can be performed with regard to the organization of similarities of concept vectors. The input of the clustering algorithm was the set of embedding vectors of candidate words retrieved in the previous step. A complete binary tree was constructed applying Ward's minimum variance method [19] as the clustering criterion, in order to get small, dense subtrees at the bottom of the hierarchy. However, we did not need the whole hierarchy, but separate, compact groups of terms, i.e. well-separated subtrees of the dendrogram. The most intuitive way of defining the cutting points of the tree is to find large jumps in the clustering levels. To put it more formally, the height of each link in the cluster tree is to be compared with the heights of neighboring links below it up to a certain depth. If this difference is larger than a predefined threshold value, then the link is considered inconsistent, and the tree is cut at that point. Cutting the tree at such points resulted in a list of flat clusters containing more closely related words. The density of these clusters can be set by changing the inconsistency value at which point the subtrees are cut dynamically. This clustering of automatically generated word lists effectively grouped items that did not fit the intended semantic category. As a result, instead of checking hundreds of words individually, only the few clusters had to be signed as correct or incorrect, and this judgment could be applied to all words in the cluster. Only in very few cases did we need to break clusters containing both true and false positives. This method decreased the time needed for manual evaluation drastically.

Table 4 shows some examples of the resulting clusters within each category. The closer relations within a cluster can easily be recognized. For example, in the category of occupations, the abbreviated forms of military ranks formed a separate group, or in the case of languages, different dialects of Hungarian were also collected in a single cluster group, and the other clusters are also of similarly good quality if the words really belonged to the target semantic category. Another type of clusters were those which contained words that were semantically relevant from the aspect of the given task, but were not direct members of the category. For example in the case of languages, geographical names which are modifiers of a language or dialect name but are not language names by themselves, (i.e. non-final elements of multiword language names) were grouped together. The third type of clusters were those that contained words definitely not belonging to the given category. These could then easily be identified and removed manually.

5.2   Visualization

Since the embedding vectors place the lexical elements into a semantic space, it is a common practice to visualize this organization. This is done by transforming the high-dimensional vectorspace to two dimensions by applying the t-sne algorithm [6]. The main point of this method is that it places the words in the two-dimensional space so that the distribution of the pairwise distances of elements is preserved. Thus, the organization of the words can easily be reviewed and outstanding groups can easily be recognized.

When applying this visualization to each semantic category, clustering is also represented in the plot by assigning different colors to different clusters. Thus, not only the distance between individual words, but also the distance between clusters can easily be seen in the resulting figure. Figure 1 shows an example of this visualization.

Table 4. Words organized into clusters for four investigated semantic groups

Occupations
költő író drámaszerző prózaíró novellista színműíró regényíró drámaíró
'poet' 'writer' 'drama author' 'prosaist' 'novelist' 'playwright' 'novelist' 'dramatist'
ökológus entomológus zoológus biológus evolúcióbiológus etológus
'ecologist' 'entomologist' 'zoologist' 'biologist' 'evolutionary biologist' 'ethologist'
hidegburkoló tapétázó mázoló szobafestő festő-mázoló szobafestő-mázoló bútorasztalos
'tiler' 'paper hanger' 'painter' 'housepainter' 'painter' 'housepainter' 'cabinetmaker'
tehénpásztor kecskepásztor birkapásztor fejőnő marhahajcsár tehenész marhapásztor
'cowherd' 'goatherd' 'shepherd' 'milkmaid' 'cattleman' 'cowman' 'herdsman'
őrm ftörm zls alezr vőrgy szkv ezds hdgy őrgy szds fhdgy
'Sgt.' 'Sgt. Maj.' 'WO1' 'Lt. Col.' 'Maj. Gen.' 'Corp.' 'Col.' 'Lt.' 'Maj.' 'Capt.' '1Lt.'
Languages
szaúdi kuvaiti szaúd-arábiai jordániai egyiptomi (arab)
'Saudi' 'Kuwaiti' 'Saudi Arabian' 'Jordanian' 'Egyptian (Arabic)'
lengyel cseh bolgár litván román szlovák szlovén horvát
'Polish' 'Czech' 'Bulgarian' 'Lithuanian' 'Romanian' 'Slovak' 'Slovenian' 'Croatian'
osztrák-német német-osztrák elzászi dél-tiroli flamand
'Austrian-German' 'German-Austrian' 'Alsatian' 'South Tyrolean' 'Flemish'
bánsági háromszéki gömöri széki gyimesi felföldi sárközi
Hungarian dialects
Mass nouns
feketeszén kőszén barnaszén lignit feketekőszén barnakőszén
'black coal' 'hard coal' 'brown coal' 'lignite' 'hard coal' 'brown coal'
fluorit rutil apatit aragonit kvarc kalcit földpát magnetit limonit
'fluorite' 'rutile' 'apatite' 'aragonite' 'quartz' 'calcite' 'feldspar' 'magnetite' 'limonite'
konyhasó kálium-klorid nátriumklorid nátrium-klorid
'table salt' 'potassium chloride' 'sodium chloride' 'sodium chloride'
Textiles
selyemszatén bélésselyem düsesz shantung
'silk satin' 'silk lining' 'duchesse' 'shantung'
csipke bársony selyem kelme brokát selyemszövet tafota damaszt batiszt
'lace' 'velvet' 'silk' 'cloth' 'brocade' 'serge' 'taffeta' 'damask' 'batiste'

5.3   Quantitative Evaluation

Due to the clustering and the visualization applied to the sets of words, the validation of the results became very efficient and easy. This was also due to the parameter settings of the clustering, which resulted in smaller but coherent, rather than larger but mixed clusters. The results of the manual evaluation is shown in Table 5.

We evaluated the categorization method for the following semantic categories: languages, occupations, materials and within that textiles, colors, vehicles, greetings and interjections, and units of measure. We categorized the words (or clusters, if they were homogenous) as correct, erroneous or related. The latter category contained words which did not perfectly fit the original category

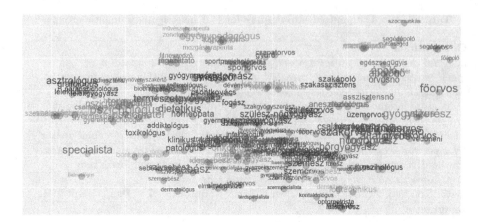

Fig. 1. A fragment of the t-sne visualization of the semantic space of occupations. Distance between words in the diagram is proportional with the semantic similarity of the terms.

Table 5. The results of semantic categorization

	Correct		Errorneous		Related		All
Languages	755	60.69%	98	7.88%	391	31.43%	1244
Occupations	2387	93.32%	134	5.24%	37	1.45%	2558
Materials	1139	84.06%	162	11.96%	54	3.99%	1355
Textiles	120	51.28%	114	48.72%	0	0.00%	234
Colors	870	63.18%	392	28.47%	115	8.35%	1377
Vehicles	1141	72.26%	239	15.14%	199	12.60%	1579
Greetings and interjections	334	24.85%	261	19.42%	749	55.73%	1344
Units of measure	1457	60.08%	909	37.48%	59	2.43%	2425

but were semantically related. In the case of occupations, this category contained words that denote human or humanoid creatures or human roles other than occupation e.g. *srác* 'lad', *öregasszony* 'old woman', *hölgy* 'lady', *albérlö* 'lodger', *élettárs* 'partner', *kobold* 'goblin'. In the case of colors, specific patterns or properties or effects of colors were assigned this category (e.g. *cirmos* 'tabby', *multicolor*, *színehagyott* 'faded'). For vehicles and languages, it contains adjectives which are part of multiword language or vehicle names. The query for greetings resulted in a high number of interjections in addition to greetings, both groups of words exhibiting extreme orthographic variation. In that case, we categorized greetings as correct and interjections as related. The query for units of measures returned also suffixed forms used as heads of adjective phrases, e.g. *5 kilogrammos dinnye* 'a 5 kg melon'. Forms like this are spelled as one word if neither the numeral nor the unit of measure are written using digits or are abbreviated, e.g. *kétórás látogatás* 'a two-hour visit'. Here we categorized forms like this as related. It is evident that, with the exception of textiles, the ratio of correctly categorized items was relatively high. For textiles, on the other hand, it was extremely easy to check the results because all the incorrect words formed a single distinct cluster that contained only articles made of textiles, clothes, foot gear, home textiles.

6   Conclusion

In this paper, it has been shown that the popular method of neural-network-based embedding models can also be applied to morphologically rich languages like Hungarian, especially if the models are generated from an annotated and lemmatized corpus of a reasonable size. In addition to investigating some of the tasks word embedding models are in general applied to, we demonstrated the applicability of the models for a specific application: expanding the lexicon of a morphological analyzer and extending it with semantic features. We have shown that by applying a semi-automatic method for retrieving words of a certain category and providing further aids for the manual evaluation of these, it has become possible and efficient to assign semantic category labels for words that could not have been done manually. The model has been shown to be capable of pinpointing and categorizing corpus annotation errors as well.

Acknowledgment. We thank Márton Bartók for his help in evaluating automatic semantic categorization results.

References

1. Banea, C., Chen, D., Mihalcea, R., Cardie, C., Wiebe, J.: SimCompass: using deep learning word embeddings to assess cross-level similarity. In: Proceedings of the 8th International Workshop on Semantic Evaluation (SemEval 2014), pp. 560–565. Association for Computational Linguistics and Dublin City University, Dublin, Ireland, August 2014
2. Baroni, M., Dinu, G., Kruszewski, G.: Don't count, predict! a systematic comparison of context-counting vs. context-predicting semantic vectors. In: Proceedings of the 52nd Annual Meeting of the Association for Computational Linguistics, Long Papers, vol. 1, pp. 238–247. Association for Computational Linguistics, Baltimore, Maryland, June 2014
3. Firth, J.R.: A synopsis of linguistic theory 1930–1955. In: Studies in Linguistic Analysis, pp. 1–32 (1957)
4. Iacobacci, I., Pilehvar, M.T., Navigli, R.: SensEmbed: Learning sense embeddings for word and relational similarity. In: Proceedings of the 53rd Annual Meeting of the Association for Computational Linguistics and the 7th International Joint Conference on Natural Language Processing, vol. Long Papers, pp. 95–105. Association for Computational Linguistics, Beijing, China, July 2015
5. Indig, B., Miháltz, M., Simonyi, A.: Exploiting linked linguistic resources for semantic role labeling. In: 7th Language and Technology Conference on Human Language Technologies as a Challenge for Computer Science and Linguistics, pp. 140–144, Uniwersytet im. Adama Mickiewicza w Poznaniu, Poznań (2015)
6. van der Maaten, L., Hinton, G.: Visualizing high-dimensional data using t-SNE (2008)

7. Mikolov, T., Chen, K., Corrado, G., Dean, J.: Efficient estimation of word representations in vector space. CoRR abs/1301.3781 (2013)
8. Mikolov, T., Sutskever, I., Chen, K., Corrado, G.S., Dean, J.: Distributed representations of words and phrases and their compositionality. In: Proceedings of a 26 and 27th Annual Conference on Neural Information Processing Systems and Advances in Neural Information Processing Systems 2013, 5–8 December 2013, Lake Tahoe, Nevada, United States, pp. 3111–3119 (2013)
9. Mikolov, T., Yih, W., Zweig, G.: Linguistic regularities in continuous space word representations. In: Proceedings and Conference of the North American Chapter of the Association of Computational Linguistics: Human Language Technologies, 9–14 June 2013, Westin Peachtree Plaza Hotel, Atlanta, Georgia, USA, pp. 746–751 (2013)
10. Novák, A.: A new form of Humor–mapping constraint-based computational morphologies to a finite-state representation. In: Calzolari, N., Choukri, K., Declerck, T., Loftsson, H., Maegaard, B., Mariani, J., Moreno, A., Odijk, J., Piperidis, S. (eds.) Proceedings of the Ninth International Conference on Language Resources and Evaluation (LREC 2014), pp. 1068–1073. European Language Resources Association (ELRA), Reykjavík, Iceland, aCL Anthology Identifier: L14–1207, May 2014
11. Novák, A.: Milyen a jó Humor? [What is good Humor like?]. In: I. Magyar Számítógépes Nyelvészeti Konferencia [First Hungarian Conference on Computational Linguistics], pp. 138–144. SZTE, Szeged (2003)
12. Novák, A.: Improving corpus annotation quality using word embedding models. Polibits 53, pp. 49–53 (2016)
13. Novák, A., Siklósi, B., Oravecz, C.: A new integrated open-source morphological analyzer for Hungarian. In: Calzolari, N., Choukri, K., Declerck, T., Grobelnik, M., Maegaard, B., Mariani, J., Moreno, A., Odijk, J., Piperidis, S. (eds.) Proceedings of the Tenth International Conference on Language Resources and Evaluation (LREC 2016). European Language Resources Association (ELRA), Portorož, Slovenia, May 2016
14. Oravecz, C., Dienes, P.: Efficient stochastic part-of-speech tagging for Hungarian. In: LREC European Language Resources Association (2002)
15. Orosz, Gy., Novák, A.: PurePos 2.0: a hybrid tool for morphological disambiguation. In: Proceedings of the International Conference on Recent Advances in Natural Language Processing (RANLP 2013), pp. 539–545. INCOMA Ltd., Hissar, Bulgaria (2013)
16. Pereira, F., Tishby, N., Lee, L.: Distributional clustering of English words. In: Proceedings of the 31st Annual Meeting on Association for Computational Linguistics, ACL 1993, pp. 183–190, Association for Computational Linguistics, Stroudsburg, PA, USA (1993)
17. Prószéky, G., Kis, B.: A unification-based approach to morpho-syntactic parsing of agglutinative and other (highly) inflectional languages. In: Proceedings of the 37th Annual Meeting of the Association for Computational Linguistics on Computational Linguistics, ACL 1999, pp. 261–268. Association for Computational Linguistics, Stroudsburg, PA, USA (1999)
18. Trask, A., Michalak, P., Liu, J.: sense2vec - A fast and accurate method for word sense disambiguation in neural word embeddings. CoRR abs/1511.06388 (2015)
19. Ward, J.H.: Hierarchical grouping to optimize an objective function. J. Am. Stat. Assoc. 58(301), 236–244 (1963)

A New Language Model Based on Possibility Theory

Mohamed Amine Menacer[1(✉)], Abdelfetah Boumerdas[1], Chahnez Zakaria[1],
and Kamel Smaili[2]

[1] Ecole nationale Supérieure d'Informatique,
BP 68M, 16309 El Harrach, Algiers, Algeria
`am_menacer@esi.dz`
[2] Loria, Campus Scientifique, BP 239, 54506 Vandoeuvre Lès-Nancy, France

Abstract. Language modeling is a very important step in several NLP applications. Most of the current language models are based on probabilistic methods. In this paper, we propose a new language modeling approach based on the possibility theory. Our goal is to suggest a method for estimating the possibility of a word-sequence and to test this new approach in a machine translation system. We propose a word-sequence possibilistic measure, which can be estimated from a corpus. We proceeded in two ways: first, we checked the behavior of the new approach compared with the existing work. Second, we compared the new language model with the probabilistic one used in statistical MT systems. The results, in terms of the METEOR metric, show that the possibilistic-language model is better than the probabilistic one. However, in terms of BLEU and TER scores, the probabilistic model remains better.

Keywords: Machine translation · Probabilistic approach
The possibility theory · Language model

1 Introduction

Language models are essential in many domains such as Statistical Machine Translation (SMT) [2], Automatic Speech Recognition (ASR) [20], Optical Character Recognition (OCR) [23] etc. They are generally used to ensure that a system not only produces the right words but also to output them in a fluent language.

In SMT, the main role of language models is to choose, among the huge number of hypotheses, the right translations. These models are generally based on the probability theory. They proceed by estimating the probability of a linguistic event[1] from a sufficiently large text corpus.

Many language models have been proposed and developed in literature, such as: cache based models [15], trigger based models [16] and multi-gram models [6].

---

[1] A linguistic event is the succession of one or several words.

© Springer International Publishing AG, part of Springer Nature 2018
A. Gelbukh (Ed.): CICLing 2016, LNCS 9623, pp. 127–139, 2018.
https://doi.org/10.1007/978-3-319-75477-2_8

These three models are strongly related with the n-gram model [11], which is the most dominant and the most used model in the language modeling field. Other language models, based on other paradigms than n-gram, have been proposed. For instance, the connexionist language model, proposed by [1], which is very useful when the training data is limited.

There are various limitations of n-gram models. The classical limitations include, for example, the context-dependent issue. In fact, the probability assigned by the model is related to the context in which it occurs. This context is modeled by the size n of the model, which is, in many cases, insufficient to capture the underlying topic information. Moreover, there is another problem related to the probability theory: generally the probability assigned by these language models is significant especially with high and medium frequency events. However, when it comes to low frequency events, the theory breaks down and the model fails to assign reliable probabilities. Therefore, it turns to smoothing techniques to estimate the probability of the corresponding event.

In this paper, we present a new language modeling approach using the possibility theory [17], which is an alternative to the probability theory in order to deal with uncertainty. This paper is organized as follows: the next section briefly presents the probabilistic n-gram language models, followed by a description of the possibility theory and the motivation that encouraged us to use it. Then, the only work, in our knowledge, that has been done to model a possibilistic language model [18] is presented. Finally, we present and describe the language model based on the possibility theory along with some tests to compare its performances with those of the baseline language models.

2   Probabilistic n-gram Language Models

A probabilistic language model could be considered as a function that measures how likely a sentence can be expressed by a speaker in a particular language. In a more formal way, if the following word-sequences: $e = e_1e_2...e_l$ is proposed by a system, then the probability $P(e)$ can be decomposed and calculated as in (1).

$$P(e) = P(e_1e_2...e_l) = \prod_{j=1}^{l} P(e_j|e_1e_2...e_{j-1}) \tag{1}$$

In formula (1), we notice that the probability of the word-sequence e depends on the conditional probability of each word e_j knowing its history $e_1e_2...e_{j-1}$. However, the length of the taken history is a major problem because it is impossible to find all the possible histories that can precede a given word. Consequently, the exact value of the probability $P(e_j|e_1e_2...e_{j-1})$ cannot be calculated. To deal with this issue, the n-grams language model assumes that the word e_j depends only on the $n-1$ words that precede it as it is shown in the Eq. (2).

$$P(e) = \prod_{j=1}^{l} P(e_j|e_{j-n+1}e_{j-n+2}...e_{j-1}) \tag{2}$$

However, according to the previous equation, if the n-gram model fails to find a similar word-sequence in the training data, the probability of the n-gram could be assigned a zero value. This can be misleading because an unseen n-gram doesn't mean that it is wrong.

Generally, some other techniques are used along with the n-gram language model in order to estimate and adjust the value of the probability $P(e)$ and to avoid the problem of assigning zero probability to unseen n-grams; such methods are called the smoothing techniques.

The smoothing techniques can be divided into two groups: the first one concerns the count smoothing methods such as the add-one smoothing, the deleted estimation and Good-Turing smoothing. The second one considers interpolation and back-off based smoothing techniques like: linear interpolation, back-off, Witten-Bell smoothing [4], Kneser-Ney smoothing [12] and Modified Kneser-Ney smoothing [4]. However, there is no technique that performs well in all situations. For all these reasons, we propose to use another theory: the possibility theory.

3   The Possibility Theory

The possibility theory was proposed by Negoita et al. [17] in 1978 as an extension of fuzzy sets theory and fuzzy logic. It is a mathematical framework that deals with the representation of uncertain information resulting from incomplete knowledge [9].

Traditionally, all uncertainties of information are quantified and handled by the probability theory. However, the theory of probability knows some gaps in some cases like the famous Bertrand paradox [21] where one could obtain several different values of probability for the same event only by changing the way of reasoning. The probability theory uses only one measurement to deal with uncertainty, which is the probability measure itself. Therefore, it is limited in certain situations. For example, if we affirm that the sentence *"I eat an apple"* is a well written English sentence with a probability of 0.8, we are faced with two problems. The first one is the percentage of error of this probability, which indicates the inaccuracy of respecting English language usage rules. The second issue concerns the uncertainty, i.e. there is a 80% chance that this sentence faithfully respects all the usage rules of the English language.

To take these phenomena into account, the possibility theory employs two measures: the necessity N and the possibility Π. The first allows to affirm that an event is possible with a certain degree and the second represents the degree of certainty of an event. Consequently, the possibility theory completes the probability theory.

In the same way, as in the probability theory, where the probability P can be obtained from the probability distribution, the possibility and necessity measures can be defined from the possibility distribution π in the possibility theory. But unlike the probability theory, the use of the possibility and the necessity measure in the possibility theory allows us to distinguish what is plausible from what is less plausible, what is normal from what is not, what is surprising from what is expected [8]. These advantages can be very useful to build a possibilistic language model that can reduce some limitations of the probabilistic language models.

The possibility distribution is a function of the universe Ω to the interval unit $[0, 1]$ with the following interpretations: $\pi(s) = 1$ means that the event s is possible and $\pi(s) = 0$ means that the event s is considered as impossible.

In the finite state, the possibility and the necessity measures are defined as described in (3) and (4).

$$\Pi(A) = \max_{s \in A}(\pi(s)) \tag{3}$$

$$N(A) = \min_{s \notin A}(1 - \pi(s)) \tag{4}$$

where A is a subset of the reference set Ω, representing a set of events.

In summary, the uncertainty of a proposition or an event A in the possibilistic model is measured by the couple $(\Pi(A), N(A))$, unlike the probabilistic model where the uncertainty of an event is measured by only the probability P.

4   Related Works

To our knowledge, the only language model based on the possibility theory is the one described in [18]. This work uses the web as an open training corpus to improve language modeling. The reason behind using the web lies in the great and constantly growing volume of its textual documents. The resulting language model was integrated in a speech recognition system.

The basic idea behind the method of Oger and Linarès [18] is to consider the observed events and non-observed ones in the web. Therefore, two theories were employed: probability and possibility. The probability theory is used to take advantage of events observed on the Web, and the possibility theory is used to take advantage of unobserved events on the web.

To benefit from both theories: probability and possibility, Oger and Linarès [18] also proposed a method to combine the probability and the possibility measures.

5   Possibility Estimated on a Text Corpus

The method proposed by Oger and Linarès [18] estimates the possibility of a word-sequence from the web or from a text corpus. The basic idea of this method is the following: for a given word-sequence, the possibility is estimated according to its sub-sequences. Consequently, the more sub-sequences of the original word-sequence exist in the training corpus, the higher this word-sequence is possible. A similar principle is used in the BLEU measure, which is used to evaluate the quality of a translation [19]. Thus, for a higher reliability of results, the size of the sub-sequences is given according to the order of the model. Every time a n-gram sub-sequence does not exist, the existence of all its $(n-1)$-grams sub-sequences is studied until a uni-gram level is reached.

This idea is formulated for a n-gram language model following a recursive equation. This equation determines a set of possibility distributions starting from π_n until reaching π_1:

$$\pi_n(W) = \begin{cases} \frac{|W_n \cap C_n| + \alpha|W_n/C_n|}{|W_n|} & \text{if} \quad n \geq 1 \\ 0 & \text{else} \end{cases} \tag{5}$$

where:

- W: is a sequence of one or more words.
- W_n: is a set of word sequences of size n in W.
- C_n: is the set of word sequences of size n in the corpus C.
- α: is the back-off coefficient with $0 \leq \alpha \leq 1$.
- $/$: is the set subtraction operator.

This formula takes into account, at the same time, the observed and the unobserved events. To calculate the possibility for a word-sequence W, the number of sub-sequences of size n of W present in the training corpus is normalized by the total number of sub-sequences of size n in W. Thus, to propose a precise possibility distribution, a back-off strategy is used recursively to interpolate high order possibilities from its sub-sequence possibilistic scores.

The possibility distribution defined above allows deriving the possibility measure as in Eq. (6).

$$\Pi_n(A) = \max_{W \in A}(\pi_n(W)) \tag{6}$$

where A is a set of sequences of n or more words.

If the whole corpus is considered as a unique element W, then:

$$\Pi_n(W) = \pi_n(W) \tag{7}$$

6   A New Approach Based on the Possibility Theory

The approach proposed by Oger and Linarès [18] has some limits. For instance, when calculating the possibility of a word-sequence in their formula, the sequence is initially divided into a group of n-grams, then each n-gram is checked whether it exists in the training corpus or not. By doing so, all the extracted n-grams will have the same weight. This could be considered as a limit for the language model. In fact, even if a sequence of words belongs to a foreign language[2], the above formulation could give it a high possibility.

To overcome this limit, a novel approach, based on the same idea as that of Oger and Linarès [18] is proposed in our study. In fact, Oger and Linarès [18] define the possibility of a word-sequence based on the existence or the non-existence of this sequence or its sub-sequences of size n in the training corpus. However, in our approach, we consider a weight estimated from a training corpus,

---

[2] This is possible for close languages which share several words such as English and French.

which reinforces the possibility of a word-sequence that contains a sub-sequences often encountered. Furthermore, we interpolate the possibility of a sequence with that of the sequences of lower size. This allows, on one hand, to take into account the case where a sequence does not appear in the training data. On the other hand, it increases the possibility distribution. By considering these assumptions, the possibility-based language model estimates the possibility of a sequence of n words as follows:

$$\pi_n(W) = \sum_{k=1}^{n} \lambda_k \alpha_k \sum_{i=1}^{\alpha} \frac{N(w_i^k)}{\beta_k} \tag{8}$$

where:

- W_i^k: is the i-th sub-sequence of size k in W.
- $N(W_i^k)$: is the number of occurrences of the word w_i of size k in the training corpus C.
- α: is the number of unique sub-sequences w_i of size k in the test corpus W.
- β_k: is the total number of sub-sequences of size k in the training corpus C.
- $\alpha_k = \frac{\alpha'}{\alpha}$: where α' is the number of units of size k in the test corpus that exist in the training corpus where $\alpha' \leq \alpha$.
- λ_k: is the possibility weighting coefficient.
- $\sum_i \lambda_i = 1$.

Suggesting that $\lambda_1 \leq \lambda_2 \leq \ldots \leq \lambda_n$ will ensure that a more important possibility is assigned to longer sequences.

For a word-sequence W, its possibility value is defined according to the sub-sequences composing it. For each sub-sequence of size k, we determine its ponderation estimated from the training corpus. Additionally, we back-off to smaller possibilistic model. This back-off is handled by the two coefficients λ_k and α_k.

To determine the value of the possibility measure for a test corpus C_t made up of several word-sequences of size n or more, we apply the formula (9).

$$\Pi_n(W) = \max_{W \in C_t} (\pi_n(W)) \tag{9}$$

If the test corpus is considered such as a long sentence W, then:

$$\Pi_n(W) = \pi_n(W)) \tag{10}$$

As in the method of Oger and Linarès [18], the calculation of the possibility measure does not require the decomposition of the total possibility in conditional possibilities, which allows us to evaluate a word-sequence totally and not sequence by sequence as done with a probabilistic-based language models.

7   Tests and Results

In order to evaluate the performance of the possibilistic-language model, we consider two types of tests: the goal of the first test is to study the behavior of the new approach compared to that proposed by Oger and Linarès [18]. In the second one, we compare our approach with the probabilistic model used by the Moses decoder during the translation process. This probabilistic language model is calculated using the KenLM language modeling tool [10].

In the following section, the results obtained from the two tests are presented and discussed.

7.1   Experiment on Language Modeling

In this test, we considered the English corpus of EUROPARL [13]. This corpus is composed of the proceedings of the European Parliament since 1996. Table 1 illustrates the number of sentences and the number of words in this corpus.

Table 1. Statistics on the EUROPARL corpus.

Language	$|S|$	$|W|$
English	2 218 201	53 974 751

We subdivided the corpus in three parts as follows:

- 80% used as training data which corresponds to a vocabulary of 111 072 distinct words.
- 10% used for tuning and more specifically for estimating the values of λ_k.
- The last 10% of the corpus used as test data.

We achieved several experiments concerning the possibility-language model on four other corpora. The aim of the first corpus is to study the possibility of a text that is well written in the source language (English). The two other corpora are poorly written in English and the final corpus is written in a foreign language (French). The purpose of these three last tests is to check if the possibilitic-language model estimates correctly their possibility, if so, the possibility of these two corpora should be low.

Statistics on the four test corpora are presented in the Table 2.

Table 2. Statistics on the test corpora.

Corpus	$|S|$	$|W|$
Europarl	35 000	966 990
RandEuroparl	35 000	1 032 197
RandEnglish	35 011	1 031 452
Foreign	34 722	907 024

- Europarl is taken from test data (10% of the Europarl corpus).
- RandEuroparl is randomly generated from the vocabulary wordlist.
- RandEnglish is randomly generated from an English wordlist.
- Foreign is well written in French.

Before testing our possibilistic-language model, we show the influence of the interpolation coefficient λ_k on the calculation of the possibility measure. Initially, we fixed the values of λ_k in order to make the sub-sequence of size n more possible than the one of size $n-1$ and so on. Then, we estimated the values of λ_k using a greed search algorithm on the development corpus. Table 3 illustrates the total number of sentences and words, as well as the size of the vocabulary.

Table 3. Statistics on the development corpora.

Language	$|S|$	$|W|$	$|V|$
English	200 773	5 605 794	45 849

The values of λ_k fixed by hand and evaluated by a greed search algorithm are shown in Table 4. The values of the possibility corresponding to these λ_k and for $n = 4$ are given in Table 5.

Table 4. Values of λ_k fixed by hand and estimated by a greed search algorithm.

n	λ_k	λ^{Hand}	λ^{Calc}
4	λ_1	0.1	0.23
	λ_2	0.2	0.24
	λ_3	0.3	0.26
	λ_4	0.4	0.27
3	λ_1	0.2	0.32
	λ_2	0.3	0.33
	λ_3	0.5	0.35
2	λ_1	0.4	0.49
	λ_2	0.6	0.51

We can notice that the values of the possibility measure obtained after the estimation of the values λ_k are a bit higher compared to the values obtained by fixing λ_k by hand. This is because of the high values of λ_k assigned to the n-grams of lower order (uni-grams and 2-grams), which are generally more frequent than the n-grams of higher order (3 and 4-grams).

Next, we compared the values of the possibility given by our language model and the one proposed by Oger and Linarès [18]. The results of this comparison are illustrated in Table 6.

Table 5. The possibility-language model values with estimates and fixed λ_k.

Corpus	$\Pi_n^{Hand}(w)$	$\Pi_n^{Calc}(w)$
Europarl	0.33	0.46
RandEuroparl	0.01	0.03
RandEnglish	0.03	0.06
Foreign	0.005	0.01

Table 6. Comparison between our language model and the one proposed by Oger and Linarès [18].

Corpus	n	$\Pi_n(W)$	$\Pi_n^{Oger}(W)$
Europarl	4	0.45	0.49
	3	0.6	0.77
	2	0.77	0.95
RandEuroparl	4	0.02	1.52×10^{-7}
	3	0.04	1.52×10^{-5}
	2	0.06	1.52×10^{-3}
RandEnglish	4	0.06	3.56×10^{-7}
	3	0.08	3.56×10^{-5}
	2	0.12	3.56×10^{-3}
Foreign	4	0.012	3.11×10^{-4}
	3	0.017	4.07×10^{-3}
	2	0.026	0.07

We notice that in both approaches the possibility value is low for the test corpora, which are badly written in English (the two corpora RandEuroparl and RandEnglish) or written in a foreign language. For the test corpus that is well written in English (Europarl Corpus), the value of the possibility is high in both approaches. We notice also that for the two approaches, the value of the possibility increases with the reduction of the model size. This is due to the n-grams of the lower order (1 and 2-grams) which are generally frequent in the training corpus.

7.2   Experiment on Machine Translation

In the second test, we achieved a comparison between our possibilistic language model and the probabilistic model within machine translation systems. In order to do that, we train a French to English translation model using Moses decoder [14] and its tools. Some statistics about the data used to train the translation model are illustrated in Table 7.

In order to compare between the two models, a corpus test of 1000 sentences (∼33000 words) has been selected. The translation process uses the possibilistic-

Table 7. EUROPARL French-English bilingual corpus statistics.

Language	$|S|$	$|W|$	$|V|$
French	1 579 312	48 576 991	128 051
English	1 579 312	53 974 751	111 072

language model and then a baseline language model (probabilistic). To achieve this goal, we proceeded as follows:

1. We used Moses decoder and its toolkit to translate the sentences and associate to each one a list of its 1000 best translations proposed by the decoder.
2. For each sentence, we sorted its 1000 best translations, in a decreasing order, using the probabilistic model.
3. For each sentence, we kept only its best translation.
4. Finally, we used the MultEval tool [5] to calculate three scores: BLEU [19], METEOR [7] and TER [22] in order to evaluate the quality of the translations.

The previous steps have been used to evaluate the probabilistic language model. In order to evaluate the possibilistic-language model, the same steps (from 1 to 4) are used except for the second one where we sorted the sentences using our possibilistic model. The results concerning the two models are illustrated in Table 8.

Table 8. Metric scores (BLEU, Meteor and Translation Error Rate -TER-) for all systems.

Metric	System prob[a]	System poss[b]
BLEU ↓	28,6	27,5
METEOR ↑	32,1	33,2
TER ↑	52,9	57,1

[a]The system where probabilistic language model is used.
[b]The system where possibilistic-language model is used.

It is clear that the possibilistic-language model ensures the best values for METEOR compared to the baseline one. However, the two metrics BLEU and TER, confirm that the probabilistic language model achieves better results. This performance is encouraging since in [3], the authors show that the METEOR score correlates most with human judgment when MT is achieved from any language to English (which is the case in our system). However, the BLEU score correlates best when the translation is done from English into another language.

Thus, the advantage of METEOR is that it establishes correspondences between the reference and candidate translation on word matching, synonyms or words with the same root. However, the BLEU and TER scores are based on an explicit word matching between translation and reference.

Some sample translations, using the two language models, compared to the reference translation are given in Table 9. These examples, among many others, show that the approach we propose achieves better translations, which could be considered as new references for the BLEU measure to help improve the quality of the translation even better.

Table 9. Example of a translation using the probabilistic language model and the possibilistic one.

Source	Trans prob LM	Trans poss LM	Reference
nous devons également analyser certains des enjeux créés par le traité d'Amsterdam	We also need to look at some of the issues of the Treaty of Amsterdam	We also have to analyse some of the issues, created by the Treaty of Amsterdam	We also have to look at some of the challenges created by Amsterdam
nous devons conserver ce droit afin de défendre un intérêt national	We should retain the right to defend their national interests	We must keep this law, in order to defend the right to a national interest	We have to retain that right to defend a national interest

8   Conclusion

To conclude, in this paper, we presented a new language model based on the possibility theory while the majority of the language models used in speech recognition and machine translation are based on probabilistic approach. The interest of this new model is that it takes into account the uncertainty. We expect that this new method, which shows the feasibility of the principle, would be an alternative to classical language models. We have tested this method in a real machine translation system, and it achieved promising results. In fact, in terms of METEOR, our model is better than the baseline one while it is less efficient in terms of BLEU. Knowing that METEOR is more correlated with human judgments since this measure has been designed to overcome the weakness of BLEU and NIST. That is why we are optimistic about the interest of this approach in the near future.

References

1. Bengio, Y., Schwenk, H., Senécal, J.S., Morin, F., Gauvain, J.L.: Neural probabilistic language models. In: Holmes, D.E., Jain, L.C. (eds.) Innovations in Machine Learning. Studies in Fuzziness and Soft Computing, vol. 194, pp. 137–186. Springer, Heidelberg (2006). https://doi.org/10.1007/3-540-33486-6_6

2. Brown, P.F., Cocke, J., Pietra, S.A.D., Pietra, V.J.D., Jelinek, F., Lafferty, J.D., Mercer, R.L., Roossin, P.S.: A Statistical Approach to Machine Translation, vol. 16, pp. 79–85. MIT Press, Cambridge (1990)
3. Callison-Burch, C., Fordyce, C., Koehn, P., Monz, C., Schroeder, J.: Further meta-evaluation of machine translation. In: Proceedings of the Third Workshop on Statistical Machine Translation (StatMT 2008), pp. 70–106. Association for Computational Linguistics, Stroudsburg (2008)
4. Chen, S.F., Goodman, J.: An empirical study of smoothing techniques for language modeling. In: Proceedings of the 34th Annual Meeting on Association for Computational Linguistics (ACL 1996), pp. 310–318. Association for Computational Linguistics, Stroudsburg (1996)
5. Clark, J.H., Dyer, C., Lavie, A., Smith, N.A.: Better hypothesis testing for statistical machine translation: controlling for optimizer instability. In: Proceedings of the 49th Annual Meeting of the Association for Computational Linguistics: Human Language Technologies (HLT 2011) (Short Papers - Volume 2), pp. 176–181. Association for Computational Linguistics, Stroudsburg (2011)
6. Deligne, S., Bimbot, F.: Language modeling by variable length sequences: theoretical formulation and evaluation of multigrams. In: International Conference on Acoustics, Speech, and Signal Processing (ICASSP-1995), vol. 1, pp. 169–172. IEEE (1995)
7. Denkowski, M., Lavie, A.: Meteor 1.3: automatic metric for reliable optimization and evaluation of machine translation systems. In: Proceedings of the Sixth Workshop on Statistical Machine Translation, pp. 85–91. Association for Computational Linguistics (2011)
8. Dubois, D., Prade, H.: Possibility theory. Scholarpedia $2(10)$, 2074 (2007)
9. Dubois, D.: Possibility theory and statistical reasoning. Comput. Stat. Data Anal. $51(1)$, 47–69 (2006)
10. Heafield, K.: Kenlm: Faster and smaller language model queries. In: Proceedings of the Sixth Workshop on Statistical Machine Translation (WMT 2011), pp. 187–197. Association for Computational Linguistics, Stroudsburg (2011)
11. Jelinek, F.: Continuous speech recognition by statistical-methods. vol. 64, pp. 532–556. IEEE Institute of Electrical Electronics Engineers, Inc., New York (1976)
12. Kneser, R., Ney, H.: Improved backing-off for m-gram language modeling. In: 1995 International Conference on Acoustics, Speech, and Signal Processing (ICASSP-1995), vol. 1, pp. 181–184. IEEE (1995)
13. Koehn, P.: Europarl: a parallel corpus for statistical machine translation. In: Proceedings of the Conference on Tenth Machine Translation Summit (AAMT), Phuket, pp. 79–86 (2005)
14. Koehn, P., Hoang, H., Birch, A., Callison-Burch, C., Federico, M., Bertoldi, N., Cowan, B., Shen, W., Moran, C., Zens, R., Dyer, C., Bojar, O., Constantin, A., Herbst, E.: Moses: open source toolkit for statistical machine translation. In: Proceedings of the 45th Annual Meeting of the ACL on Interactive Poster and Demonstration Sessions (ACL 2007), pp. 177–180. Association for Computational Linguistics, Stroudsburg (2007)
15. Kuhn, R., De Mori, R.: A cache-based natural language model for speech recognition. IEEE Trans. Pattern Anal. Mach. Intell. $12(6)$, 570–583 (1990)
16. Lau, R., Rosenfeld, R., Roukos, S.: Trigger-based language models: a maximum entropy approach. In: Proceedings of the 1993 IEEE International Conference on Acoustics, Speech, and Signal Processing: Speech Processing (ICASSP 1993), vol. 2, pp. 45–48. IEEE Computer Society, Washington, D.C. (1993)

17. Negoita, C., Zadeh, L., Zimmermann, H.: Fuzzy sets as a basis for a theory of possibility. Fuzzy Sets Syst. **1**, 3–28 (1978)
18. Oger, S., Linarès, G.: Web-based possibilistic language models for automatic speech recognition. Comput. Speech Lang. **28**(4), 923–939 (2014)
19. Papineni, K., Roukos, S., Ward, T., Zhu, W.J.: BLEU: a method for automatic evaluation of machine translation. In: Proceedings of the 40th Annual Meeting on Association for Computational Linguistics (ACL 2002), pp. 311–318. Association for Computational Linguistics, Stroudsburg (2002)
20. Rosenfeld, R.: Two decades of statistical language modeling: where do we go from here? p. 2000 (2000)
21. Saporta, G.: Probabilités, analyse des données et statistique. Editions Technip (2011)
22. Snover, M., Dorr, B., Schwartz, R., Micciulla, L., Makhoul, J.: A study of translation edit rate with targeted human annotation. In: Proceedings of Association for Machine Translation in the Americas, pp. 223–231 (2006)
23. Srihari, R., Baltus, C.: Combining statistical and syntactic methods in recognizing handwritten sentences. In: AAAI Symposium: Probabilistic Approaches to Natural Language, pp. 121–127 (1992)

Combining Discrete and Neural Features for Sequence Labeling

Jie Yang, Zhiyang Teng, Meishan Zhang, and Yue Zhang[✉]

Singapore University of Technology and Design, Singapore, Singapore
{jie_yang,zhiyang_teng}@mymail.sutd.edu.sg,
{meishan_zhang,yue_zhang}@sutd.edu.sg

Abstract. Neural network models have recently received heated research attention in the natural language processing community. Compared with traditional models with discrete features, neural models have two main advantages. First, they take low-dimensional, real-valued embedding vectors as inputs, which can be trained over large raw data, thereby addressing the issue of feature sparsity in discrete models. Second, deep neural networks can be used to automatically combine input features, and including non-local features that capture semantic patterns that cannot be expressed using discrete indicator features. As a result, neural network models have achieved competitive accuracies compared with the best discrete models for a range of NLP tasks.

On the other hand, manual feature templates have been carefully investigated for most NLP tasks over decades and typically cover the most useful indicator pattern for solving the problems. Such information can be complementary the features automatically induced from neural networks, and therefore combining discrete and neural features can potentially lead to better accuracy compared with models that leverage discrete or neural features only.

In this paper, we systematically investigate the effect of discrete and neural feature combination for a range of fundamental NLP tasks based on sequence labeling, including word segmentation, POS tagging and named entity recognition for Chinese and English, respectively. Our results on standard benchmarks show that state-of-the-art neural models can give accuracies comparable to the best discrete models in the literature for most tasks and combing discrete and neural features unanimously yield better results.

Keywords: Discrete features · Neural features · LSTM

1 Introduction

There has been a surge of interest in neural methods for natural language processing over the past few years. Neural models have been explored for a wide range of tasks, including parsing [1–7], machine translation [8–13], sentiment

A. Gelbukh (Ed.): CICLing 2016, LNCS 9623, pp. 140–154, 2018.
https://doi.org/10.1007/978-3-319-75477-2_9

analysis [14–18] and information extraction [19–22], achieving results competitive to the best discrete models.

Compared with discrete models with manual indicator features, the main advantage of neural networks is two-fold. First, neural network models take low-dimensional dense embeddings [23–25] as inputs, which can be trained from large-scale test, thereby overcoming the issue of sparsity. Second, non-linear neural layers can be used for combining features automatically, which saves the expense of feature engineering. The resulting neural features can capture complex non-local syntactic and semantic information, which discrete indicator features can hardly encode.

On the other hand, discrete manual features have been studied over decades for many NLP tasks, and effective feature templates have been well-established for them. This source of information can be complementary to automatic neural features, and therefore a combination of the two feature sources can led to improved accuracies. In fact, some previous work has attempted on the combination. Turian et al. [26] integrated word embedding as real-word features into a discrete Conditional Random Field [27] (CRF) model, finding enhanced results for a number of sequence labeling tasks. Guo et al. [28] show that the integration can be improved if the embedding features are carefully discretized. On the reverse direction, Ma et al. [29] treated a discrete perception model as a neural layer, which is integrated into a neural model. Wang & Manning [30] integrated a discrete CRF model and a neural CRF model by combining their output layers. Greg & Dan [6] and Zhang et al. [18] also followed this method. Zhang & Zhang [31] compared various integration methods for parsing, and found that the second type of integration gives better results.

We follow Zhang & Zhang [31], investigating the effect of feature combination for a range of sequence labeling tasks, including word segmentation, Part-Of-Speech (POS) tagging and named entity recognition (NER) for Chinese and English, respectively. For discrete features, we adopt a CRF model with state-of-the art features for each specific task. For neural features, we adopt a neural CRF model, using a separated Long Short-Term Memory [32] (LSTM) layer to extract input features. We take standard benchmark datasets for each task. For all the tasks, both the discrete model and the neural model give accuracies that are comparable to the state-of-the-art. A combination of discrete and neural feature, gives significantly improved results with no exception.

The main contributions that we make in this investigation include:

– We systematically investigate the effect of discrete and neural feature combination for a range of fundamental NLP tasks, showing that the two types of feature are complimentary.
– We systematically report results of a state-of-the-art neural network sequence labeling model on the NLP tasks, which can be useful as reference to future work.
– We report the best results in the literatures for a number of classic NLP tasks by exploiting neural feature integration.
– The source code of the LSTM and CRF implementations of this paper are released under GPL at https://github.com/SUTDNLP.

2   Related Work

There has been two main kinds of methods for word segmentation. Xue [33] treat it as a sequence labeling task, using B(egin)/I(nternal) /E(nding)/S(ingle-character word) tags on each character in the input to indicate its segmentation status. The method was followed by Peng et al. [34], who use CRF to improve the accuracies. Most subsequent work follows [34–38] and feature engineering has been out of the key research questions. This kind of research is commonly referred to as the character-based method. Recently, neural networks have been applied to character-based segmentation [39–41], giving results comparable to discrete methods. On the other hand, the second kind of work studies word-based segmentation, scoring outputs based on word features directly [42–44]. We focus on the character-based method, which is a typical sequence labeling problem.

POS-tagging has been investigated as a classic sequence labeling problem [45–47], for which a well-established set of features are used. These handcrafted features basically include words, the context of words, word morphologies and word shapes. Various neural network models have also been used for this task. In order to include word morphology and word shape knowledge, a convolutional neural network (CNN) for automatically learning character-level representations is investigated in [48]. Collobert et al. [25] built a CNN neural network for multiple sequence labeling tasks, which gives state-of-the-art POS results. Recurrent neural network models have also been used for this task [49,50]. Huang et al. [50] combines bidirectional LSTM with a CRF layer, their model is robust and has less dependence on word embedding.

Named entity recognition is also a classical sequence labeling task in the NLP community. Similar to other tasks, most works access NER problem through feature engineering. McCallum & Li [51] use CRF model for NER task and exploit Web lexicon as feature enhancement. Chieu & Ng [52], Krishnan & Manning [53] and Che et al. [54] tackle this task through non-local features [55]. Besides, many neural models, which are free from handcrafted features, have been proposed in recent years. In Collobert et al. [25] model we referred before, NER task has also been included. Santos et al. [56] boost the neural model by adding character embedding on Collobert's structure. James et al. [57] take the lead by employing LSTM for NER tasks. Chiu et al. [58] use CNN model to extract character embedding and attach it with word embedding and afterwards feed them into Bi-directional LSTM model. Through adding lexicon features, Chiu's NER system get state-of-the-art performance on both CoNLL2003 and OntoNotes 5.0 NER datasets.

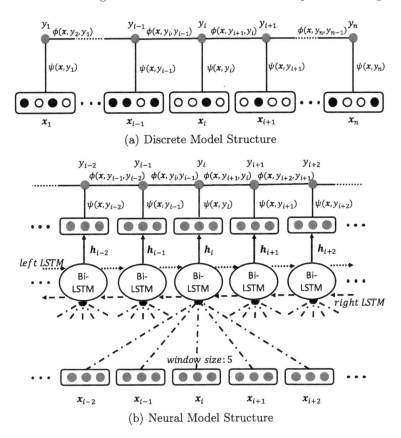

(a) Discrete Model Structure

(b) Neural Model Structure

Fig. 1. Model structures

3   Method

The structures of our discrete and neural models are shown in Fig. 1(a) and (b), respectively, which are used for all the tasks in this paper. Black and white elements represent binary features for discrete model and gray elements are continuous representation for word/character embedding. The only difference between different tasks are the definition of input and out sequences, and the features used.

The discrete model is a standard CRF model. Given a sequence of input $x = x_1, x_2, \ldots, x_n$, it models the output sequence $y = y_1, y_2, \ldots, y_n$ by calculating two potentials. In particular, the *output clique potential* shows the correlation between inputs and output labels,

$$\Psi(x, y_i) = \exp(\theta_o \cdot f_o(x, y_i)) \tag{1}$$

where $f_o(x, y_i)$ is a feature vector extracted from x and y_i, and θ_o is a parameter vector.

The *edge clique potential* shows the correlation between consecutive output labels,

$$\Phi(\boldsymbol{x}, y_i, y_{i-1}) = \exp(\boldsymbol{\theta}_e \cdot \boldsymbol{f}_e(\boldsymbol{x}, y_i, y_{i-1})) \tag{2}$$

where $\boldsymbol{f}_e(\boldsymbol{x}, y_i, y_{i-1})$ is a feature vector extracted from $\boldsymbol{x}$, y_i and y_{i-1}, and $\boldsymbol{\theta}_e$ is a parameter vector.

The final probability of $\boldsymbol{y}$ is estimated as

$$p(\boldsymbol{y}|\boldsymbol{x}) = \frac{\prod_{i=1}^{|\boldsymbol{y}|} \Psi(\boldsymbol{x}, y_i) \prod_{j=1}^{|\boldsymbol{y}|} \Phi(\boldsymbol{x}, y_j, y_{j-1})}{Z(\boldsymbol{x})} \tag{3}$$

where $Z(\boldsymbol{x})$ is the partition function,

$$Z(\boldsymbol{x}) = \sum_{\boldsymbol{y}} \prod_{i=1}^{|\boldsymbol{y}|} \Psi(\boldsymbol{x}, y_i) \prod_{j=1}^{|\boldsymbol{y}|} \Phi(\boldsymbol{x}, y_j, y_{j-1}) \tag{4}$$

The overall features $\{\boldsymbol{f}_o(\boldsymbol{x}, y_i), \boldsymbol{f}_e(\boldsymbol{x}, y_i, y_{i-1})\}$ are extracted at each location i according to a set of feature templates for each task.

The neural model takes the neural CRF structure. Compared with the discrete model, it replaces the output clique features $\boldsymbol{f}_o(\boldsymbol{x}, y_i))$ with a dense neural feature vector $\boldsymbol{h}_i$, which is computed using neural network layer,

$$\boldsymbol{h}_i = BiLSTM((e(x_{i-2}), e(x_{i-1}), e(x_i), e(x_{i+1}), e(x_{i+2})), \boldsymbol{W}, \boldsymbol{b}, \boldsymbol{h}_{i-1})$$
$$\Psi(\boldsymbol{x}, y_i) = \exp(\boldsymbol{\theta}_o \cdot \boldsymbol{h}_i) \tag{5}$$

where $e(x_i)$ is the embedding form of x_i, *BiLSTM* represents bi-directional LSTM structure for calculating hidden state $\boldsymbol{h}_i$ for input x_i, which considers both the left-to-right and right-to-left information flow in a sequence. The *BiLSTM* structure receives input of the embeddings from a window of size 5 as shown in Fig. 1(b).

For the neural model, the edge clique is replaced with a single transition weight $\tau(y_i, y_{i-1})$. The remainder of the model is the same as the discrete model

$$p(\boldsymbol{y}|\boldsymbol{x}) = \frac{\prod_{i=1}^{|\boldsymbol{y}|} \Psi(\boldsymbol{x}, y_i) \prod_{j=1}^{|\boldsymbol{y}|} \Phi(\boldsymbol{x}, y_j, y_{j-1})}{Z(\boldsymbol{x})}$$
$$= \frac{\prod_{i=1}^{|\boldsymbol{y}|} \exp(\boldsymbol{\theta}_o \cdot \boldsymbol{h}_i) \prod_{j=1}^{|\boldsymbol{y}|} \exp(\tau(y_j, y_{j-1}))}{Z(\boldsymbol{x})} \tag{6}$$

Here $\boldsymbol{\theta}_o$ and $\tau(y_i, y_{i-1})$ are model parameters, which are different from the discrete model.

The joint model makes a concatenation of the discrete and neural features at the output cliques and edge cliques,

$$\Psi(\boldsymbol{x}, y_i) = \exp(\boldsymbol{\theta}_o \cdot (\boldsymbol{h}_i \oplus \boldsymbol{f}_o(\boldsymbol{x}, y_i)))$$
$$\Phi(\boldsymbol{x}, y_j, y_{j-1}) = \exp(\boldsymbol{\theta}_e \cdot ([\tau(y_j, y_{j-1})] \oplus \boldsymbol{f}_e(\boldsymbol{x}, y_j, y_{j-1}))) \tag{7}$$

where the $\oplus$ operator is the vector concatenation operation.

The training objective for all the models is to maximize the margin between gold-standard and model prediction scores. Given a set of training examples $\{x_n, y_n\}_{n=1}^{N}$, the objective function is defined as follows

$$L = \frac{1}{N} \sum_{n=1}^{N} loss(x_n, y_n, \Theta) + \frac{\lambda}{2} ||\Theta||^2 \tag{8}$$

Here Θ is the set of model parameters θ_o, θ_e, W, b and τ, and λ is the L_2 regularization parameter.

The loss function is defined as

$$loss(x_n, y_n, \Theta) = \max_{y}(p(y|x_n; \Theta) + \delta(y, y_n)) - p(y_n|x_n, \Theta) \tag{9}$$

where $\delta(y, y_n)$ denotes the hamming distance between y and y_n.

Online Adagrad [59] is used to train the model, with the initial learning rate set to be η. Since the loss function is not differentiable, a subgradient is used, which is estimated as

$$\frac{\partial loss(x_n, y_n, \Theta)}{\partial \Theta} = \frac{\partial p(\widehat{y}|x_n, \Theta)}{\partial \Theta} - \frac{\partial p(y_n|x_n, \Theta)}{\partial \Theta} \tag{10}$$

where $\widehat{y}$ is the predicted label sequence.

Chinese Word Segmentation Features. Table 1 shows the features used in Chinese word segmentation. For "type", each character has five possibilities: 0/Punctuation, 1/Alphabet, 2/Date, 3/Number and 4/others.

Table 1. Feature templates for Chinese word segmentation

1	character unigram: c_i , $-2 \leq i \leq 2$
2	character bigram: $c_{i-1}c_i$, $-1 \leq i \leq 2$; $c_{-1}c_1$ and c_0c_2
3	whether two characters are equal or not: $c_0 == c_{-2}$ and $c_0 == c_1$
4	character trigram: $c_{-1}c_0c_1$
5	character type unigram: $type(c_0)$
6	character types: $type(c_{-1})type(c_0)type(c_1)$
7	character types: $type(c_{-2})type(c_{-1})type(c_0)type(c_1)type(c_2)$

POS Tagging Features. Table 2 lists features for POS tagging task on both English and Chinese datasets. The prefix and suffix include 5 characters for English and 3 characters for Chinese.

NER Features. Table 3 shows the feature template used in English NER task. For "word shape", each character in word is located in one of these four types: number, lower-case English character, upper-case English character and others.

Table 2. Feature templates for POS tagging

1	word unigram: $w_i, -2 \leq i \leq 2$
2	word bigram: $w_{-1}w_0, w_0w_1, w_{-1}w_1$
3	prefix: $Prefix(w_0)$
4	suffix: $Suffix(w_0)$
5	length: $Length(w_0)$ (only for Chinese)

Table 3. Feature templates for English NER

1	word unigram: $w_i, -1 \leq i \leq 1$
2	word bigram: $w_iw_{i+1}, -2 \leq i \leq 1$
3	word shape unigram: $Shape(w_i), -1 \leq i \leq 1$
4	word shape bigram: $Shape(w_i)Shape(w_{i+1}), -1 \leq i \leq 0$
5	word capital unigram: $Capital(w_i), -1 \leq i \leq 1$
6	word capital with word: $Capital(w_i)w_j, -1 \leq i, j \leq 1$
7	connect word unigram: $Connect(w_i), -1 \leq i \leq 1$
8	capital with connect: $Capital(w_i)Connect(w_0), -1 \leq i \leq 1$
9	word cluster unigram: $Cluster(w_i), -1 \leq i \leq 1$
10	word cluster bigram: $Cluster(w_i)Cluster(w_{i+1}, -1 \leq i \leq 0)$
11	word prefix: $Prefix(w_i), 0 \leq i \leq 1$
12	word suffix: $Suffix(w_i), -1 \leq i \leq 0$
13	word POS unigram: $POS(w_0)$
14	word POS bigram: $POS(w_i)POS(w_{i+1}), -1 \leq i \leq 0$
15	word POS trigram: $POS(w_{-1})POS(w_0)POS(w_1)$
16	POS with word: $POS(w_0)w_0$

Table 4. Feature templates for Chinese NER

1	word POS unigram: $POS(w_0)$
2	word POS bigram: $POS(w_i)POS(w_{i+1}), -1 \leq i \leq 0$
3	word POS trigram: $POS(w_{-1})POS(w_0)POS(w_1)$
4	POS with word: $POS(w_0)w_0$
6	word unigram: $w_i, -1 \leq i \leq 1$
7	word bigram: $w_{i-1}w_i, 0 \leq i \leq 1$
8	word prefix: $Prefix(w_i), -1 \leq i \leq 0$
9	word prefix: $Suffix(w_i), -1 \leq i \leq 0$
10	radical: $radical(w_0, k), 0 \leq k \leq 4$; k is character position
11	word cluster unigram: $Cluster(w_0)$

"Connect" word has five categories: "of", "and", "for", "-" and other. Table 4 presents the features used in Chinese NER task. We extend the features used in Che et al. [54] by adding part-of-speech information. Both POS tag on English and Chinese datasets are labeled by ZPar [60], prefix and suffix on two datasets are both including 4 characters. Word clusters in both English and Chinese tasks are same with Che's work [54].

4   Experiments

We conduct our experiments on different sequence labeling tasks, including Chinese word segmentation, Part-of-speech tagging and Named entity recognition.

For these three tasks, their input embeddings are different. For Chinese word segmentation, we take both character embeddings and character bigram embeddings for calculating $e(x_i)$. For POS tagging, $e(x_i)$ consists of word embeddings and character embeddings. For NER, we include word embeddings, character embeddings and POS embeddings for $e(x_i)$. Character embeddings, character bigram embeddings and word embeddings are pretrained separately using word2vec [23]. English word embedding is chosen as SENNA [25]. We make use of Chinese Gigaword Fifth Edition[1] to pretrain necessary embeddings for Chinese words. The Chinese corpus is segmented by ZPar [60]. During training, all these aforementioned embeddings will be fine-tuned. The hyper-parameters in our experiments are shown at Table 5. Dropout [61] technology has been used to suppress over-fitting in the input layer.

Table 5. Hyper parameters

Parameter	Value
dropout probability	0.25
wordHiddensize	100
charHiddensize	60
charEmbSize	30
wordEmbSize	50
wordEmbFineTune	True
charEmbFineTune	True
initial η	0.01
regularization λ	1e-8

4.1   Chinese Word Segmentation

For Chinese word segmentation, we choose PKU, MSR and CTB60 as evaluation datasets. The PKU and MSR dataset are obtained from SIGHAN Bakeoff 2005

---

[1] https://catalog.ldc.upenn.edu/LDC2011T13.

corpus[2]. We split the PKU and MSR datasets in the same way as Chen et al. [62], and the CTB60 set as Zhang et al. [63]. Table 6 shows the statistical results for these three datasets. We evaluate segmentation accuracy by Precision (P), Recall (R) and F-measure (F).

Table 6. Chinese word segmentation datasets statistics

Segmentation datasets (sentences)	PKU	MSR	CTB60
Train	17149	78226	23401
Dev	1905	8692	2078
Test	1944	3985	2795

The experiment results of Chinese word segmentation is shown in Table 7. Our joint models shown comparable results to the state-of-the-art results reported by Zhang & Clark [42], where they adopt a word-based perceptron model with carefully designed discrete features. Compared with both discrete and neural models, our joint models can achieve best results among all three datasets. In particular, the joint model can outperform the baseline discrete model by 0.43, 0.42 and 0.37 on PKU, MSR, CTB60 respectively. In order to investigate whether the discrete model and neural model can benefit from each other, we scatter sentence-level segmentation accuracy of two models for three datasets in Fig. 2. As we can see from Fig. 2, some sentences can obtain higher accuracies in the neural model, while other sentences can win out in the discrete model. This common phenomenon among three datasets suggests that the neural model and the discrete model can be combined together to enjoy the merits from each side.

Table 7. Chinese word segmentation results

Model	PKU			MSR			CTB60		
	P	R	F	P	R	F	P	R	F
Discrete	95.42	94.56	94.99	96.94	96.61	96.78	95.43	95.16	95.29
Neural	94.29	94.56	94.42	96.79	**97.54**	97.17	94.48	95.01	94.75
Joint	**95.74**	**95.12**	**95.42**	97.01	97.39	**97.20**	**95.68**	**95.64**	**95.66**
State-of-the-art	N/A	N/A	94.50	N/A	N/A	**97.20**	N/A	N/A	95.05

---

[2] http://www.sighan.org/bakeoff2005.

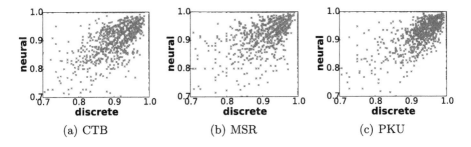

(a) CTB   (b) MSR   (c) PKU

Fig. 2. Chinese word segmentation F-measure comparisons

4.2 POS Tagging

We compare our models on both English and Chinese datasets for the POS
tagging task. The English dataset is chosen following Toutanova et al. [64] and
Chinese dataset by Li et al. [65] on CTB. Statistical results are shown in Table 8.
Toutanova's model [64] exploits bidirectional dependency networks to capture
both preceding and following tag contexts for English POS tagging task. Li et
al. [65] utilize heterogeneous datasets for Chinese POS tagging through bundling
two sets of tags and training in enlarged dataset, their system got state-of-the-art
accuracy on CTB corpus.

Table 8. POS tagging datasets statistics

POS tagging datasets (sentences)	English	Chinese
Train	38219	16091
Dev	5527	803
Test	5462	1910

Both the discrete and neural models get comparable accuracies with state-of-
the-art system on English and Chinese datasets. The joint model has significant
enhancement compared with separated model, especially in Chinese POS tagging task, with 1% accuracy increment. Figure 3 shows the accuracy comparison

Table 9. POS tagging results

Model	English	Chinese
	Acc	Acc
Discrete	97.23	93.97
Neural	97.28	94.02
Joint	**97.47**	**95.07**
State-of-the-art	97.24	94.10

for the discrete and neural models based on each sentence. There are many sentences that are not located at the diagonal line, which indicates the two models gives different results and have the potential for combination. Our joint model outperforms state-of-the-art accuracy with 0.23% and 0.97% on English and Chinese datasets, respectively (Table 9).

4.3   NER

For the NER task, we split Ontonotes 4.0 following Che et al. [54] to get both English and Chinese datasets. Table 10 shows the sentence numbers of train/develop/test datasets.

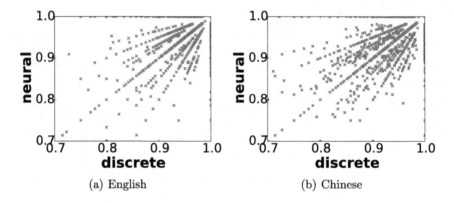

(a) English   (b) Chinese

Fig. 3. POS tagging accuracy comparisons

Table 10. NER datasets statistics

NER datasets (sentences)	English	Chinese
Train	39262	15724
Dev	6249	4301
Test	6452	4346

We follow Che et al. [54] on choosing both the English and the Chinese datasets. Their work induces bilingual constrains from parallel dataset which gives significant enhancement of F-scores on both English and Chinese datasets.

Our discrete and neural models show comparable recall values compared with Che's results [54] on both datasets. Similar with the previous two tasks, the joint model gives significant enhancement compared with separated models (discrete/neural) on all metrics. This shows that discrete and neural model can identify entities using different indicator features, and they can be complementary with each other. The comparison of sentence F-measures in Fig. 4 confirms this observation (Table 11).

Table 11. NER results

Model	English			Chinese		
	P	R	F	P	R	F
Discrete	80.14	79.29	79.71	72.67	73.92	73.29
Neural	77.25	80.19	78.69	65.59	71.84	68.57
Joint	81.90	**83.26**	**82.57**	72.98	**80.15**	**76.40**
State-of-the-art	**81.94**	78.35	80.10	**77.71**	72.51	75.02

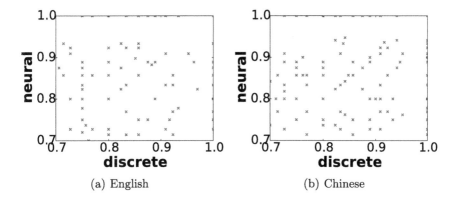

(a) English   (b) Chinese

Fig. 4. NER F-measure comparisons

The joint model outperforms the state-of-the-art on both datasets in the two different languages. The precision of joint models are less than state-of-the-art system. This may be caused by the bilingual constrains in baseline system, which ensures the precision of entity recognition.

5   Conclusion

We proposed a joint sequence labeling model that combines neural features and discrete indicator features which can integrate the advantages of carefully designed feature templates over decades and automatically induced features from neural networks. Through experiments on various sequence labeling tasks, including Chinese word segmentation, POS tagging and named entity recognition for Chinese and English respectively, we demonstrate that our joint model can unanimously outperform models which only contain discrete features or neural features and state-of-the-art systems on all compared tasks. The accuracy/F-measure distribution comparison for discrete and neural model also indicate that discrete and neural model can reveal different related information, this explains why combined model can outperform separate models.

In the future, we will investigate the effect of our joint model on more NLP tasks, such as parsing and machine translation.

Acknowledgments. We would like to thank the anonymous reviewers for their detailed comments. This work is supported by the Singapore Ministry of Education (MOE) AcRF Tier 2 grant T2MOE201301.

References

1. Socher, R., Lin, C.C., Manning, C., Ng, A.Y.: Parsing natural scenes and natural language with recursive neural networks. In: ICML, pp. 129–136 (2011)
2. Chen, D., Manning, C.D.: A fast and accurate dependency parser using neural networks. In: EMNLP, vol. 1, pp. 740–750 (2014)
3. Weiss, D., Alberti, C., Collins, M., Petrov, S.: Structured training for neural network transition-based parsing. In: ACL-IJCNLP, pp. 323–333 (2015)
4. Dyer, C., Ballesteros, M., Ling, W., Matthews, A., Smith, N.A.: Transition-based dependency parsing with stack long short-term memory. In: ACL-IJCNLP, pp. 334–343 (2015)
5. Zhou, H., Zhang, Y., Chen, J.: A neural probabilistic structured-prediction model for transition-based dependency parsing. In: ACL, pp. 1213–1222 (2015)
6. Durrett, G., Klein, D.: Neural CRF parsing. In: ACL-IJCNLP, pp. 302–312 (2015)
7. Ballesteros, M., Carreras, X.: Transition-based spinal parsing. In: CoNLL (2015)
8. Kalchbrenner, N., Blunsom, P.: Recurrent continuous translation models. In: EMNLP, pp. 1700–1709 (2013)
9. Cho, K., Van Merriënboer, B., Gulcehre, C., Bahdanau, D., Bougares, F., Schwenk, H., Bengio, Y.: Learning phrase representations using rnn encoder-decoder for statistical machine translation. arXiv preprint arXiv:1406.1078 (2014)
10. Sutskever, I., Vinyals, O., Le, Q.V.: Sequence to sequence learning with neural networks. In: NIPS, pp. 3104–3112 (2014)
11. Bahdanau, D., Cho, K., Bengio, Y.: Neural machine translation by jointly learning to align and translate. arXiv preprint arXiv:1409.0473 (2015)
12. Ling, W., Trancoso, I., Dyer, C., Black, A.W.: Character-based neural machine translation. arXiv preprint arXiv:1511.04586 (2015)
13. Jean, S., Cho, K., Memisevic, R., Bengio, Y.: On using very large target vocabulary for neural machine translation. In: ACL-IJCNLP, pp. 1–10 (2015)
14. Socher, R., Perelygin, A., Wu, J.Y., Chuang, J., Manning, C.D., Ng, A.Y., Potts, C.: Recursive deep models for semantic compositionality over a sentiment treebank. In: EMNLP, vol. 1631, p. 1642 (2013)
15. Tang, D., Wei, F., Yang, N., Zhou, M., Liu, T., Qin, B.: Learning sentiment-specific word embedding for twitter sentiment classification. In: ACL, vol. 1, pp. 1555–1565 (2014)
16. Nogueira dos Santos, C., Gatti, M.: Deep convolutional neural networks for sentiment analysis of short texts. In: COLING (2014)
17. Vo, D.-T., Zhang, Y.: Target-dependent twitter sentiment classification with rich automatic features. In: IJCAI, pp. 1347–1353 (2015)
18. Zhang, M., Zhang, Y., Vo, D.-T.: Neural networks for open domain targeted sentiment. In: EMNLP (2015)
19. Socher, R., Chen, D., Manning, C.D., Ng, A.: Reasoning with neural tensor networks for knowledge base completion. In: NIPS, pp. 926–934 (2013)
20. Wang, M., Manning, C.D.: Effect of non-linear deep architecture in sequence labeling. In: IJCNLP (2013)
21. Ding, X., Zhang, Y., Liu, T., Duan, J.: Deep learning for event-driven stock prediction. In: ICJAI, pp. 2327–2333 (2015)

22. Mark, G.-W., Stephen, C.: What happens next? event prediction using a compositional neural network. In: AAAI (2016)
23. Mikolov, T., Chen, K., Corrado, G., Dean, J.: Efficient estimation of word representations in vector space. arXiv preprint arXiv:1301.3781 (2013)
24. Pennington, J., Socher, R., Manning, C.D.: Glove: global vectors for word representation. In: EMNLP, vol. 12, pp. 1532–1543 (2014)
25. Collobert, R., Weston, J., Bottou, L., Karlen, M., Kavukcuoglu, K., Kuksa, P.: Natural language processing (almost) from scratch. JMLR **12**, 2493–2537 (2011)
26. Turian, J., Ratinov, L., Bengio, Y.: Word representations: a simple and general method for semi-supervised learning. In: ACL, pp. 384–394 (2010)
27. Lafferty, J., McCallum, A., Pereira, F.C.N.: Probabilistic models for segmenting and labeling sequence data, Conditional random fields (2001)
28. Guo, J., Che, W., Wang, H., Liu, T.: Revisiting embedding features for simple semi-supervised learning. In: EMNLP, pp. 110–120 (2014)
29. Ma, J., Zhang, Y., Zhu, J.: Tagging the web: building a robust web tagger with neural network. In: ACL, vol. 1, pp. 144–154 (2014)
30. Wang, M., Manning, C.D.: Learning a product of experts with elitist lasso. In: IJCNLP (2013)
31. Zhang, M., Zhang, Y.: Combining discrete and continuous features for deterministic transition-based dependency parsing. In: EMNLP, pp. 1316–1321 (2015)
32. Hochreiter, S., Schmidhuber, J.: Long short-term memory. Neural Comput. **9**(8), 1735–1780 (1997)
33. Xue, N., et al.: Chinese word segmentation as character tagging. Comput. Linguist. Chin. Lang. Process. **8**(1), 29–48 (2003)
34. Peng, F., Feng, F., McCallum, A.: Chinese segmentation and new word detection using conditional random fields. In: Coling, p. 562 (2004)
35. Zhao, H.: Character-level dependencies in Chinese: usefulness and learning. In: EACL, pp. 879–887 (2009)
36. Jiang, W., Huang, L., Liu, Q., Lü, Y.: A cascaded linear model for joint chinese word segmentation and part-of-speech tagging. In: ACL (2008)
37. Sun, W.: A stacked sub-word model for joint Chinese word segmentation and part-of-speech tagging. In: HLT-ACL, pp. 1385–1394 (2011)
38. Liu, Y., Zhang, Y., Che, W., Liu, T., Wu, F.: Domain adaptation for CRF-based chinese word segmentation using free annotations. In: EMNLP, pp. 864–874 (2014)
39. Zheng, X., Chen, H., Xu, T.: Deep learning for Chinese word segmentation and pos tagging. In: EMNLP, pp. 647–657 (2013)
40. Pei, W., Ge, T., Baobao, C.: Maxmargin tensor neural network for Chinese word segmentation. In: ACL (2014)
41. Chen, X., Qiu, X., Zhu, C., Huang, X.: Gated recursive neural network for Chinese word segmentation. In: EMNLP (2015)
42. Zhang, Y., Clark, S.: Chinese segmentation with a word-based perceptron algorithm. In: ACL, vol. 45, p. 840 (2007)
43. Sun, W.: Word-based and character-based word segmentation models: comparison and combination. In: Coling, pp. 1211–1219 (2010)
44. Liu, Y., Zhang, Y.: Unsupervised domain adaptation for joint segmentation and pos-tagging. In: COLING (Posters), pp. 745–754 (2012)
45. Ratnaparkhi, A., et al.: A maximum entropy model for part-of-speech tagging. In: EMNLP, vol. 1, pp. 133–142 (1996)
46. Collins, M.: Discriminative training methods for hidden Markov models: theory and experiments with perceptron algorithms. In: EMNLP, pp. 1–8 (2002)

47. Manning, C.D.: Part-of-speech tagging from 97% to 100%: is it time for some linguistics? In: Gelbukh, A.F. (ed.) CICLing 2011. LNCS, vol. 6608, pp. 171–189. Springer, Heidelberg (2011). https://doi.org/10.1007/978-3-642-19400-9_14

48. Santos, C.D., Zadrozny, B.: Learning character-level representations for part-of-speech tagging. In: ICML, pp. 1818–1826 (2014)

49. Perez-Ortiz, J.A., Forcada, M.L.: Part-of-speech tagging with recurrent neural networks. Universitat d'Alacant, Spain (2001)

50. Huang, Z., Xu, W., Yu, K.: Bidirectional LSTM-CRF models for sequence tagging, August 2015

51. McCallum, A., Li, W.: Early results for named entity recognition with conditional random fields, feature induction and web-enhanced lexicons. In: HLT-NAACL, pp. 188–191 (2003)

52. Leong, C.H., Tou, N.H.: Named entity recognition with a maximum entropy approach. In: HLT-NAACL, vol. 4, pp. 160–163 (2003)

53. Krishnan, V., Manning, C.D.: An effective two-stage model for exploiting non-local dependencies in named entity recognition. In: Coling and ACL, pp. 1121–1128 (2006)

54. Che, W., Wang, M., Manning, C.D., Liu, T.: Named entity recognition with bilingual constraints. In: HLT-NAACL, pp. 52–62 (2013)

55. Ratinov, L., Roth, D.: Design challenges and misconceptions in named entity recognition. In: Coling, pp. 147–155 (2009)

56. dos Santos, C., Guimaraes, V., Niterói, R.J., de Janeiro, R.: Boosting named entity recognition with neural character embeddings. In: NEWS (2015)

57. Hammerton, J.: Named entity recognition with long short-term memory. In: Daelemans, W., Osborne, M. (eds.) CoNLL, pp. 172–175 (2003)

58. Chiu, J.P.C., Nichols, E.: Named entity recognition with bidirectional lstm-cnns. arXiv preprint arXiv:1511.08308 (2015)

59. Singer, Y., Duchi, J., Hazan, E.: Adaptive subgradient methods for online learning and stochastic optimization. JMLR **12**, 2121–2159 (2011)

60. Zhang, Y., Clark, S.: Syntactic processing using the generalized perceptron and beam search. Comput. Linguist. **37**(1), 105–151 (2011)

61. Srivastava, N., Hinton, G., Krizhevsky, A., Sutskever, I., Salakhutdinov, R.: Dropout: a simple way to prevent neural networks from overfitting. JMLR **15**(1), 1929–1958 (2014)

62. Chen, X., Qiu, X., Zhu, C., Liu, P., Huang, X.: Long short-term memory neural networks for Chinese word segmentation. In: EMNLP, Lisbon, Portugal, pp. 1197–1206 (2015)

63. Zhang, M., Zhang, Y., Che, W., Liu, T.: Character-level chinese dependency parsing. In: ACL (2014)

64. Toutanova, K., Klein, D., Manning, C.D., Singer, Y.: Feature-rich part-of-speech tagging with a cyclic dependency network. In: NAACL, pp. 173–180 (2003)

65. Li, Z., Chao, J., Zhang, M., Chen, W.: Coupled sequence labeling on heterogeneous annotations: Pos tagging as a case study. In: ACL-IJCNLP, pp. 1783–1792 (2015)

New Recurrent Neural Network Variants
for Sequence Labeling

Marco Dinarelli[1,2,3,4]([✉]) and Isabelle Tellier[1,2,3,4]([✉])

[1] LaTTiCe (UMR 8094), CNRS, ENS Paris,
1 rue Maurice Arnoux, 92120 Montrouge, France
`marco.dinarelli@ens.fr`
[2] Université Sorbonne Nouvelle - Paris 3, Paris, France
`isabelle.tellier@univ-paris3.fr`
[3] PSL Research University, Paris, France
[4] USPC (Université Sorbonne Paris Cité), Paris, France

Abstract. In this paper we study different architectures of Recurrent Neural Networks (RNN) for sequence labeling tasks. We propose two new variants of RNN and we compare them to the more traditional RNN architectures of Elman and Jordan. We explain in details the advantages of these new variants of RNNs with respect to Elman's and Jordan's RNN. We evaluate all models, either new or traditional, on three different tasks: POS-tagging of the French Treebank, and two tasks of Spoken Language Understanding (SLU), namely ATIS and MEDIA. The results we obtain clearly show that the new variants of RNN are more effective than the traditional ones.

1 Introduction

Recurrent Neural Networks [1,2] are effective neural models able to take *context* into account for their decision function. For this reason, they are suitable for NLP tasks, in particular for sequence labeling [3–8], where taking previous labels into account and modeling label dependencies are crucial for the design of effective models. It is indeed not surprising that Conditional Random Fields (CRF) [9] are among the best models for sequence labeling tasks, because their global decision function takes all possible labelings into account to predict the best labeling of a given sequence.

Most of the probabilistic models for sequence labeling (even earlier NLP models) are able to take some context into account for their decision function. In RNN, the contextual information is given by a loop connection in the network architecture. This connection allows to keep, at the current time step, one or more pieces of information predicted at previous time steps. The effectiveness of recurrent architectures for NLP tasks comes from the combination of contextual information and distributional representations, or embeddings.

In the literature about RNNs for NLP tasks, mainly two architectures have been used, also called *simple* RNNs: Elman RNN [2] and Jordan RNN [1]. The

© Springer International Publishing AG, part of Springer Nature 2018
A. Gelbukh (Ed.): CICLing 2016, LNCS 9623, pp. 155–173, 2018.
https://doi.org/10.1007/978-3-319-75477-2_10

difference between these two models lies in the recurrent connection: in Elman RNN the connection is a loop in the hidden layer. This connection allows Elman networks to use, at the current time step, hidden layer's states of previous time steps. In contrast, in Jordan RNN, the recurrent connection is between the output and the hidden layers. In this case the contextual information is made of one on more labels predicted at previous time steps. Since, in this work, we are dealing with sequence labeling tasks, time steps are *positions* in a given sequence. In the last few years, Elman and Jordan RNNs have been very successful in NLP, especially for language modeling [10,11], but also for sequence labeling [3–7,12].

The intuition which originated this paper is that embeddings allow a fine and effective modeling, not only of words, but also of labels and their dependencies, which are very important in sequence labeling tasks. Accordingly, we define two new variants of RNN to achieve this more effective modeling.

In the first variant, the recurrent connection is between the output and the input layers. In other words, this variant just gives labels predicted at previous positions in a sequence as inputs to the network. This contextual information is added to the usual input context made of words, and both are used to predict the label at the current position in the sequence. According to our intuition, and thanks to label embeddings, this variant of RNNs models label dependencies more effectively than in previous simple RNNs. The second variant we propose is a combination of an Elman RNN and of our first variant. This second variant can thus exploit both contextual information provided by the previous states of the hidden layer and the labels predicted at previous positions of a sequence.

A high-level schema of Elman's, Jordan's and our first variant of RNNs are shown in Fig. 1. The schema of the second variant proposed in this paper can be obtained by adding the recursion of our first variant to the Elman architecture. In this figure, w is the input word, y is the predicted label, E, H, O et R are

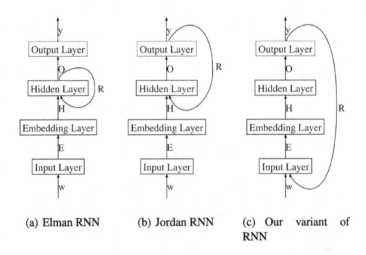

(a) Elman RNN   (b) Jordan RNN   (c) Our variant of RNN

Fig. 1. High level schema of the main RNNs studied in this work.

the parameter matrices between each pair of layers, they will be described in details in the next section. Before this, it is worth discussing the advantages our variants can bring with respect to traditional RNN architectures, and why they are expected to provide better modeling abilities.

First, since the output at previous positions is given as input to the network at the current position, the contextual information flows across the whole network, thus affecting each layer's input and output, and at both forward and backward phases. This allows in return more effective learning. In contrast, in Elman and Jordan RNNs, not all layers are affected at both forward and backward phases.

A second advantage of our variants is given by the use of label embeddings. Indeed, the first layer of our RNNs is just a *look-up* table mapping sparse "one-hot" representations into distributional representations.[1] Since in our variants the output of the network at previous steps is given as input at the current step, the mapping from sparse representations to embeddings involves both words and labels. Label embeddings can be pre-trained from data as it is usually done for words. Pre-trained word embeddings, e.g. with *word2vec*, have already proved their ability to capture very attractive syntactic and semantic properties [13,14]. Using label embeddings, the same properties can also be learned for labels. More importantly, using several predicted labels as embeddings more effectively models label dependencies via the internal state of the network, which is the hidden layer's output.

As a third advantage, label embeddings give the network a stronger "signal" than what is ensured by the recurrent connection of traditional Elman and Jordan RNNs. In the case of Elman RNN, the recurrent connection is a loop in the hidden layer. It thus does not affect the embedding layer in the forward propagation phase. Moreover, Elman RNN does not use previous label predictions as contextual information at all. In contrast, in Jordan RNN, previous labels are used as contextual information since they are given as input to the hidden layer of the network. However, the representation used for labels in Jordan RNN is either a sparse "one-hot" representation, or a probability distribution over labels. It is clear that sparse "one-hot" representations do not provide much information to the network. In contrast, probability distributions over predicted labels allow a softer decision criterion. However, from our analysis, even probability distributions over labels do not provide much more information than sparse "one-hot" representations, as most of the probability mass is concentrated on one or few labels. For instance, this explains why [6] obtains a more effective Jordan RNN when training the network with gold labels than with both a sparse "one-hot" representation and a probability distribution over predicted labels.

Another advantage of having the recurrent connection between output and input layers, and so using a context of label embeddings, is an increased robustness of the model to prediction mistakes. This increased robustness comes from the syntactic and semantic properties embeddings can encode [14]. In contrast, Jordan RNN, using sparse representations or "picked" probability distribution representations of predicted labels, is more affected by prediction error propagation.

---

[1] The "one-hot" representation of an element at position i in a dictionary V is a vector of size $|V|$ where the i-th component has the value 1 while all the others are 0.

All these advantages are also provided by the second variant of RNN proposed in this paper, which uses both a label embeddings context, like the first variant, and the loop connection at the hidden layer, like an Elman RNN.

In order to have a fair and straightforward comparison, together with the results of our new variants of RNNs, we provide the results obtained with our implementation of Elman and Jordan RNNs. Our Elman and Jordan RNNs are very close to the state-of-the-art and not just simple baselines, even if we did not implement every optimization features.

All models are evaluated on the POS-tagging task of the French Treebank [15,16] and on two Spoken Language Understanding (SLU) tasks [17]: ATIS [18] and MEDIA [19], which can be both modeled as sequence labeling tasks. ATIS is a relatively simple task, where effective label dependencies modeling is not really an issue. POS-tagging, given the absolute magnitude of results [16], can be also considered as a relatively simple task in this respect. MEDIA, in contrast, is a more difficult task where the ability of a model to take label dependencies into account is crucial. Evaluating our models in these different conditions allows us to show the general effectiveness of our models, but also their ability to take label dependencies into account.

The results we obtain on these tasks with our implementations, despite they are not always better than the state-of-the-art, provide a stable ranking of the various RNN architectures: at least one of our variants, surprisingly the simpler one, is always better than Jordan and Elman RNNs.

The remainder of the paper is organized as follows: in the next section we describe in details RNNs, starting from traditional Elman and Jordan architectures, to arrive to our variants. In Sect. 3, we describe the corpora used for evaluation, the parameter setting and all the results obtained, in comparison with state-of-the-art models. In Sect. 4 we draw some conclusions.

2   Recurrent Neural Networks

2.1   Elman and Jordan RNNs

The RNNs we consider in this work have the same architecture used also for Feedforward Neural Network Language Models (NNLM) described in [20]. In this architecture we have 4 layers: input, embedding, hidden and output layers. Words are given as input to the network as indexes, corresponding to their position in a dictionary V.

The index of a word is used to select its embedding (or distributional representation) in a real-valued matrix $E \in R^{|V| X N}$, $|V|$ being the size of the dictionary and N being the dimensionality of the embeddings (which is a parameter to be chosen). As shown in Fig. 1, the matrix E is the set of parameters between the input and the embedding layers, and we name $E(v(w_t))$ the embedding of the word w given as input at the position t of a sequence. $v(w_t) = i$ is the index of the word w_t in the dictionary, and it can be seen alternatively as a "one-hot" vector representation (the vector is zero everywhere except at position $v(w)$, where it has value 1).

We name H and O the real-valued matrices of parameters between the embedding and the hidden layers, and between the hidden and output layers, respectively. The output of the network, given the input w_t, is y_t.

The activation function at the hidden layer is a sigmoid.[2] The output layer is a softmax[3], the output of the network y is thus a probability distribution over output labels L.

In contrast to NNLM, RNNs have one more connection, the recursive connection, between two layers depending on the type of RNNs: as mentioned in the previous section, Elman RNNs [2] have a recursion loop at the hidden layer. Since the hidden layer encodes the internal representation of the input to the network, the recurrent connection of an Elman network allows to keep "in memory" words used as input at previous positions in the sequence. Jordan RNNs [1] instead have a recursion between the output and the hidden layer. This means that a Jordan RNN can take previous predicted labels into account to predict the label at current position in a sequence. Whatever the type of RNNs, we name R the matrix of parameters at the recursion connection.

Our implementation of Jordan and Elman RNNs is standard, please see [1,2,21] for more details. Also we find [6] a good reading for understanding RNNs.

2.2 RNN Learning

Learning the RNNs described above consists in learning the parameters $\Theta = (E, H, O, R)$, and the biases, which we omit to keep notation lighter. We use a cross-entropy cost function between the expected label c_t and the predicted label y_t at the position t in the sequence, plus a $L2$ regularization term [22]:

$$C = -c_t \cdot log(y_t) + \frac{\lambda}{2}|\Theta|^2 \tag{1}$$

where λ is an hyper-parameter of the model. Since y_t is a probability distribution over output labels, we can also view the output of a RNN as the probability of the predicted label y_t: $P(y_t|I,\Theta)$. I is the input given to the network plus the contextual information "kept in memory" via the recurrent connection. For Elman RNN $I_{Elman} = w_{t-w}\ldots w_t \ldots w_{t+w}, h_{t-1}$, that is the word input context and the output of the hidden layer at the previous position in the sequence. For Jordan RNN $I_{Jordan} = w_{t-w}\ldots w_t \ldots w_{t+w}, y_{t-1}$, that is the same word input context as Elman RNN, and the label predicted at the previous position. We can thus associate the following decision function in RNNs to predict the label at position t in a sequence:

$$\tilde{l}_t = argmax_{j \in 1,\ldots,|L|} P(y_t^j|I,\Theta) \tag{2}$$

---

[2] Given input x, using a sigmoid the output value is computed as $f(x) = \frac{1}{1+e^{-x}}$.

[3] Using a set of values expressing a numerical preference $l \in L$, the higher the better, the softmax is computed as $g(l) = \frac{e^l}{\sum_{l \in L} e^l}$.

Where $\tilde{l}_t$ is a particular discrete label.

We use the back-propagation and the stochastic gradient descent algorithms for learning the weights Θ. The gradient of the cross-entropy function with respect to the network output, that is the error of the network with respect to its prediction, is:

$$\frac{\delta C}{\delta y_t} = y_t - c_t \tag{3}$$

that is the difference between the output of the network and the "one-hot" representation of the gold output label c_t at position t in the sequence. Using back-propagation algorithm, the error is *back-propagated* in the network to compute the gradient at each layer, with respect to each parameter matrix (more details in [22]). Traditional stochastic gradient descent method can be improved by using momentum [22]: the gradient is accumulated over learning iterations giving some inertia to previous parameters update directions. This can avoid learning to get stuck in local optima. We use this improvement in our networks. Learning is thus performed as:

$$\Theta_{k+1} = \Theta_k + \epsilon \cdot \frac{\Lambda C}{\delta \Theta} \tag{4}$$

where ϵ is the learning rate, an hyper-parameter to be chosen. ΛC is the momentum, which is initialized to zero and then accumulates gradients along training iterations:

$$\frac{\Lambda C}{\delta \Theta} = \mu \cdot \frac{\Lambda C}{\delta \Theta} + \frac{1}{N} \frac{\delta C}{\delta \Theta} \tag{5}$$

μ is another hyper-parameter, giving more or less weight to the momentum with respect to the current gradient, N is the size of the data used at each parameters update. Since we are using Stochastic Gradient Descent, training data are split into mini-batches, parameters update is performed after processing each mini-batch, and N is thus the size of the mini-batch. Please see [22] for more details about hyper-parameters and momentum in neural networks.

2.3   Learning Tricks

In order to improve model learning, we applied some of the recommendations proposed in [22]. In particular we use:

Random training data access: That is, we shuffle training data before each training epoch.
Learning rate decay: The learning rate ϵ is initialized with a given value, and during training epochs it is decreased linearly in the number of epochs.

Another choice for learning models concern the back-propagation algorithm. Indeed, because of the recurrent nature of their architecture, in order to properly learn RNNs the back-propagation through time algorithm should be used [23]. This is supposed to allow RNNs to learn arbitrarily long past context. However, [11] has shown that RNNs for language modeling learn best with just 5 time steps in the past. This may be due to the fact that, at least in NLP tasks, the past information kept "in memory" by the network via the recurrent architecture, actually fades away after some time steps. Since the back-propagation through time algorithm is quite more expensive than the traditional back-propagation algorithm, [6] has preferred to use explicit output context in Jordan RNNs and to learn the model with the traditional back-propagation algorithm, apparently without loosing performance.

In this work we use the same strategy. When using explicitly output labels from previous time steps as context, giving them as input to the hidden layer of a Jordan RNN, the hidden layer activity is computed as:

$$h_t = \Sigma(I_t \cdot H + [y_{t-c+1} y_{t-c+2} \cdots y_{t-1}] \cdot R) \tag{6}$$

where c is the size of the history of previous labels that we want to explicitly use as context to predict next label. $[\cdot]$ indicates the concatenation of vectors.

All the modifications applied to the Jordan RNN so far can be applied in a similar way to the Elman RNN.

2.4   New Recurrent Neural Network Architectures

As mentioned in the introduction, the main difference of the variants of RNN we propose in this work, with respect to traditional RNNs, is to have a recurrent connection between the output and the input layer, as opposed to the Elman and Jordan RNNs which have the recurrent connection at the hidden layer only, and from the output to the hidden layer, respectively.

This simple modification to the architecture of the network has important consequences for the model and, as we will see next, for results. Such consequences have been already mentioned in the introduction, here we describe in more details only the more important ones.

The most interesting consequence is the fact that output labels are mapped into distributional representations, like it is done more usually for input items. Indeed, the first layer of our network, is just a mapping from sparse "one-hot" representations to distributional representations. Such mapping results in very fine features encoding and in attractive syntactic and semantic properties as shown by *word2vec* and related works [13]. Such features and properties can be learnt from data in the same way as we can do for words. In the simplest case, this can be done by using sequences of output labels just like we would use sequences of input items. However it can be done in much more complex ways, using structured information when this is available, like syntactic parse trees or structured semantic labels such as named entities or entity relations. In this work we learn label embeddings using sequences of output labels associated to word sequences in annotated data. It is worth noting that the idea of using label embeddings has been introduced by [24] in the context of dependency parsing. Authors have learned embeddings, but without pre-training, of dependency labels resulting in a very effective parser. In this paper we focus on the use of several label embeddings as context, encoding thus label dependencies, which are very important in sequence labeling tasks. Also, in this work we provide fair comparison of our variants with traditional RNN architectures of Elman and Jordan. As we will see in the evaluation section, despite a local decision function (see Eq. 2), when label dependency modeling is crucial in the task, our variants outperform remarkably Elman and Jordan RNNs.

Using the same notation used for Elman and Jordan RNNs, we name E_w the embedding matrix for words, and E_l the embedding matrix for labels. In the same way as before,

$$I_t = [E_w(v(w_{t-w}))\ldots E_w(v(w_t))\ldots E_w(v(w_{t+w}))]$$

is the concatenation of vectors representing the input words at position t in a sequence, while

$$L_t = [E_l(v(y_{t-c+1}))E_l(v(y_{t-c+2}))\ldots E_l(v(y_{t-1}))]$$

is the concatenation of vectors representing output labels predicted at previous c steps. Then the hidden layer activities are computed as:

$$h_t = \Sigma([I_t L_t] \cdot H) \tag{7}$$

Again, $[\cdot]$ means the concatenation of the two matrices and we omit biases to keep notation lighter. The remainder of the layer activities, as well as the error computation and back-propagation are computed in the same way as before.

Another important consequence of having the recursion between output and input layers is robustness. This is a direct consequence of using embeddings for output labels. Since we use several predicted labels as context at each position t (see L_t above), at least in the later stages of learning when the model is close to the final optimum, it is unlikely to have several mistakes in the same context. Even with mistakes, thanks to the properties of distributed representations [14], mistaken labels have very similar representations to the correct labels. For instance, using an example shown in [14], if we use *Paris* instead of *Rome*, this has no effect in many NLP tasks, e.g. they are both proper nouns in POS-tagging, location in named entity recognition etc. Using distributed representations for labels provides the same robustness on the output side.[4]

Jordan RNNs cannot provide in general the same robustness. This is because the recursion between output and hidden layers of Jordan RNN constrains to use either a "one-hot" representation of labels or the probability distribution given as output by the softmax. In the first case it is clear that a mistake may have more effect than when using a distributed representation, as the only value that is not zero is just at the wrong position in the representation. But also when using the probability distribution, while it may seem a much softer decision than a "one-hot" representation, we have found that most of the probability mass is "picked"

---

[4] Sometimes in POS-tagging, models mistake verbs and nouns, which may seem to have different embeddings. However in these cases, the models make such errors because these particular verbs occur in the same context of nouns (e.g. "the sleep is important"), and so they have similar representations as nouns.

on one or few labels, it doesn't provide thus much more softness than a "one-hot" representation. In any case, as we will see next, distributed representations provide a much more desirable smoothness.

The second variant of RNN that we propose combines together the characteristics of an Elman RNN and our first variant. This allows at the same time to keep in memory the internal state of the network at previous steps and to exploit all advantages coming from the use of label embeddings as context, explained above. In the second variant, the only difference with respect to the first one is the computation of the hidden layer activities, where we use the concatenation of the c previous hidden layer states in addition to the information already used in the first variant:

$$h_t = \Sigma([I_t L_t] \cdot H + [h_{t-c+1} h_{t-c+2} \ldots h_{t-1}] \cdot R) \tag{8}$$

Beyond this difference, the second variant works in the same way as the first variant.

It is important to note that different versions of RNNs exist: *forward, backward* and bidirectional [21]. They allow different modelling of sequence labeling tasks. In this work however we are interested in a comparison of different variants of RNNs than a comparison of different versions of RNNs. While that would provide a wider comparison, it would not change the relative ranking among the different type of RNNs. Taking also space constraints into account, we live a comparison of different RNN versions for future work.

2.5   Recurrent Neural Network Complexity

Before discussing the empirical evaluation of RNN models utilised in this work, we provide an analysis of the complexity in terms of the number of parameters involved in each model.

In Jordan RNN we have:

$$|V| \times D + ((2w + 1)D + c|O|) \times |H| + |H| \times |O| \tag{9}$$

where $|V|$ is the size of the input dictionary, D is the dimensionality of the embeddings, $|H|$ and $|O|$ are the dimensionalities of the hidden and output layers, respectively. w is the size of the window of words used as context on the input side[5]. c is the size of the context of labels, which is multiplied by the dimensionality of the output label dictionary, that is $|O|$.

Using the same symbols, in an Elman RNN and in our first variant we have, respectively:

$$|V| \times D + ((2w + 1)D + c|H|) \times |H| + |H| \times |O| \tag{10}$$

$$|V| \times D + |O| \times D + ((2w + 1)D + cD) \times |H| + |H| \times |O| \tag{11}$$

---

[5] The word input context is thus made of w words on the left and w on the right of the word at a given position t, plus the word at t itself, which gives a total of $2w + 1$ input words.

The only difference between the Jordan and Elman RNNs is in the factors $c|O|$ and $c|H|$. Their difference in complexity depends thus on the size of the output layer (Jordan) with respect to the size of the hidden layer (Elman). Since in sequence labeling tasks the hidden layer is usually bigger, Elman RNN is more complex than Jordan RNN.

The difference between the Jordan RNN and the first variant proposed in this work is in the factors $|O| \times D$ and cD. The first is due to the use of label embeddings[6], the second is due to the use of such embeddings as input to the hidden layer. Since often D and O have sizes in the same order of magnitude, and thanks to the use of vectorized operations on matrices, we didn't found a remarkable difference in terms of training and testing time between the Jordan RNN and our first variant. This simple analysis shows also that our first variant of RNN needs roughly the same number of connection at the hidden layer as a Jordan RNN. Our variant is thus only conceptually simpler, but it is architecturally equivalent to a Jordan RNN.

In contrast, the second variant we propose has the following complexity:

$$|V| \times D + |O| \times D + ((2w+1)D + cD + c|H|) \times |H| + |H| \times |O| \qquad (12)$$

The additional term $c|H|$ with respect to the first variant, is due to the use of the same recurrent connection as an Elman RNN. Even using vectorized operations for matrix calculations, the second variant is more complex and slow in both training and testing time by a factor 1.5 with respect to the other RNNs.

3 Experimental Evaluation

3.1 Corpora

We have evaluated our models on three different corpora:

The Air Travel Information System (ATIS) task [18] has been designed to automatically provide flight information in SLU systems. The semantic representation is frame based and the goal is to find the correct frame and the corresponding semantic slots. As an example, in the sentence *"I want the flights from Boston to Philadelphia today"*, the correct frame is FLIGHT and the words *Boston*, *Philadelphia* and *today* must be annotated with the concepts DEPARTURE.CITY, ARRIVAL.CITY and DEPARTURE.DATE, respectively.

ATIS is a relatively simple task dating from 1993. The training set is made of 4978 sentences taken from the "context independent" data in the ATIS-2 and ATIS-3 corpora. The test set is made of 893 sentences, taken from the ATIS-3 NOV93 and DEC94 datasets. There are not official development data provided with this corpus, we have thus taken a part of the training data at random and used as development data to optimize our model parameters. Please see [18] for more details on the ATIS corpus.

The French corpus MEDIA [19] has been created to evaluate SLU systems in providing tourist information, in particular hotel information in France.

---

[6] We use embeddings of the same size D for words and labels.

It is composed of 1250 dialogues acquired with a Wizard-of-OZ protocol where 250 speakers have applied 5 hotel reservation scenarios. The dialogues have been annotated following a rich semantic ontology. Semantic components can be combined to create complex semantic labels.[7] In addition to the rich annotation used, another difficulty is created by the annotation of coreferences. Some words cannot be correctly annotated without using previous dialog turns information. For example in the sentence *"Yes, the one at less than fifty euros per night"*, the *one* refers to an hotel previously introduced in the dialog. Statistics on training, development and test data from this corpus are shown in Table 1.

Both ATIS and MEDIA can be modelled as sequence labeling tasks using the *BIO* chunking notation [25]. We use these two particular corpora, not just because they provide two different settings for evaluation, but also because several different works compared on ATIS [5–7,26]. [7] is the only work providing results on MEDIA with RNN, it also provides results obtained with CRF, allowing an interesting comparison.

The French Treebank (FTB) corpus is presented in [15]. The version of the corpus we use for POS-tagging is exactly the same used in [16]. This allows a direct comparison of our results with those obtained in that work. In contrast with [16], which obtains the best result using an external lexicon of names with associated POS, we don't use any external resource in this work ([16] provides results also without using the external lexicon). Statistics on the FTB corpus are shown in Table 2.

Table 1. Statistics on the corpus MEDIA

	Training		Development		Test	
# sentences	12,908		1,259		3,005	
	Words	Concepts	Words	Concepts	Words	Concepts
# words	94,466	43,078	10,849	4,705	25,606	11,383
# vocab	2,210	99	838	66	1,276	78
# OOV%	–	–	1.33	0.02	1.39	0.04

Table 2. Statistics on the FTB corpus used for POS-tagging

Section	# sentences	# tokens	# unk. tokens
FTB-train	9881	278083	
FTB-dev	1235	36508	1790 (4,9%)
FTB-test	1235	36340	1701 (4,7%)

---

[7] For example the label `localisation` can be combined with `city`, `relative-distance`, `general-relative-place`, `street` etc.

3.2 Settings

We use the same dimensionality settings used by [5–7] on the ATIS and MEDIA tasks, that is embeddings have 200 dimensions, hidden layer has 100 dimensions. We also use the same context size for words, that is $w = 3$, and we use $c = 6$ as labels context size in our variants and in Jordan RNN. We use the same tokenization, basically consisting of words lower-casing.

In contrast, in our models we use a "more traditional" implementation: the sigmoid activation function at the hidden layer and the $L2$ regularization. While [5–7,26] use the rectified linear activation function and the dropout regularization [22,27].

For the POS-tagging task we have used 200-dimensional embeddings, 300-dimensional hidden layer, again $w = 3$ for the context on the input side, and 6 context labels on the output side. The bigger hidden layer gave better results during validation, due to the larger word dictionary in this task with respect to the others, roughly 25000 for the FTB against 2000 for MEDIA and 1300 for ATIS. In contrast to [16], which has used several features of words (prefixes, suffixes, capitalisation information etc.), we only performed a simple tokenization for reducing the size of the input dictionary: all numbers have been mapped to a

Table 3. Results on the ATIS corpus. Only using words as input (upper part of the table), and using words and classes as input (lower part of the table).

Model	F1 measure
Words	
[6] E-RNN	93.65%
[6] J-RNN	93.77%
[26] E-RNN	**94.98%**
E-RNN	93.41%
J-RNN	93.61%
I-RNN	**93.84%**
I+E-RNN	93.74%
Classes	
[7] E-RNN	96.16%
[7] CRF	95.23%
[5] E-RNN	96.04%
[26] E-RNN	**96.24%**
[26] CRF	95.16%
E-RNN	94.73%
J-RNN	94.94%
I-RNN	**95.21%**
I+E-RNN	94.84%

conventional symbol (NUM), and nouns not corresponding to proper names and starting with a capital letter have been converted to lowercase. We preferred this simple tokenisation, instead of using rich features, because our goal in this work is not obtaining the best results ever, it is to compare Jordan and Elman RNNs with our variants of RNN and show that our variants works better (at least one). Adding many features and/or building sophisticated models would make the message less clear, as results would be probably better but the improvements could be attributed to rich and sophisticated models, instead of to the model itself.

We trained all RNNs with exactly the same protocol: (i) we first train neural language models to obtain word and label embeddings. This language model is like the one in [20], except it uses both words/labels in the past and in the future to predict next word/label. (ii) we train all RNNs using the same embeddings trained at previous steps. We train the RNN for word embeddings for 20 epochs, the RNN for label embeddings for 10 epochs, and we train the RNNs for sequence labeling for 20 epochs. The number of epochs has been roughly optimized on development data. Also we roughly optimized on development data the learning rate and the parameter λ for regularization.

3.3 Evaluation

The evaluation of all models described in this work is shown in Tables 3, 4 and 5, in terms of F1 measure for ATIS and MEDIA, in terms of *Accuracy* on the FTB. In all tables, our implementation of Elman and Jordan RNN are indicated as E-RNN and J-RNN. The new variants of RNN that we propose in this work are indicated as I-RNN and I+E-RNN, the latter being the combination of Elman RNN with the first variant.

The Table 3 shows results on the ATIS corpus. In the higher part of the table we show results obtained using only words as input to the RNNs. In the lower part of the table results are obtained using both words and word-classes available for this task. Such classes concern city names, airports and time expressions like date and time of departures. They allow the models to generalize from specific words triggering concepts.[8]

We would like to note that our results on the ATIS corpus are not always comparable with those published in the literature because of the following reasons: *(i)* Models published in the litterature uses a *rectified linear* activation function at the hidden layer[9] and the *dropout* regularization. Our models use the sigmoid activation function and the $L2$ regularization. These are known to be less effective for learning RNN (see [22] for explanations). *(ii)* For experiments on ATIS we have used roughly 18% of the training data as development

---

[8] For example the cities of *Boston* and *Philadelphia* in the example above are mapped to the class CITY-NAME. If a model has never seen *Boston* during the training phase, but it has seen at least one city name, it will possibly annotate *Boston* as a departure city thanks to some discriminative context, such as the preposition *from*.

[9] $f(x) = max(0, x)$.

Table 4. Results on the MEDIA corpus

Model	F1 measure
[7] E-RNN	81.94%
[7] J-RNN	83.25%
[7] CRF	**86.00%**
E-RNN	82.64%
J-RNN	83.06%
I-RNN	**84.91%**
I+E-RNN	84.58%

Table 5. Results of POS-tagging on the FTB corpus

Model	F1 measure
[16] $MElt_{fr}^{0}$	**97.00%**
E-RNN	96.31%
J-RNN	96.31%
I-RNN	**96.42%**
I+E-RNN	96.28%

data, we thus used a smaller training set for learning our models. *(iii)* The works we compare with, not always give details on how classes available for this task have been integrated into their models. *(iv)* Layers dimensionality and hyper-parameters setting do not always match those of published works. In fact, to avoid running too much experiments, we have fixed our settings based on some works in the litterature, but this doesn't allow a straightforward comparison with other published works.

Despite this, the message we want to pass with this paper is still true for two reasons: *(i)* Some of our results are close, or even better, than state-of-the-art. *(ii)* We provide a fair comparison with our own version of traditional Elman and Jordan RNNs. This shows that our variants of RNNs are improving models that are comparable with state-of-the-art, not simple baselines.

The results in the higher part of Table 3 show that the best model on the ATIS task, with these settings, is the Elman RNN of [26]. We would like to note that it is not clear how the improvements of [26] with respect to [6] (in part due to same authors) have been obtained. Indeed, in [6] authors obtain the best result with a Jordan RNN, while in [26] an Elman RNN obtains the best performance. During our experiments, using exactly the same settings, we have found that improvements of [26] cannot be due to parameters optimization only. We thus conclude that the difference between our results and those in [26] are due to reasons mentioned above.

Beyond this, we note that our Elman and Jordan RNN implementations are roughly equivalent to those of [6]. Also, our first variant of RNN, I-RNN, obtains the second best result (93.84 in bold). Our second variant is still more effective than our Elman and Jordan RNN, but slightly worse than I-RNN.

The results in the lower part of the Table 3 (*Classes*), obtained using both word and classes as input to RNNs, are quite better than those obtained using only words as input. Beyond this, they follow roughly the same behavior, with the difference that in this case our Jordan RNN is slightly better than our second variant. But the first variant I-RNN obtains the best result among our own implementation of RNNs (95.21 in bold). In this case also, we attribute differences with respect to published results to the different settings mentioned above. For comparison, in this part of the table we show also results obtained using CRF.

As a general remark, on the ATIS task, using either words or both words and classes as input for the models, we can see that results are always quite high. This is due to the fact that this task is relatively simple, in particular label dependencies can be easily modeled, as in this task there is basically no segmentation of concepts over different consecutive words (one concept corresponds to one word). In this settings, the potential of the new variant of RNNs to model label dependencies cannot be fully exploited. This limitation is proved also by results obtained by [7, 26] using CRF. Indeed, despite the decision function of RNNs uses several pieces of information to predict the next label, it is still a local decision function. That is, RNNs don't take the whole sequence of labels into account in their decision function. In contrast, CRFs use a global decision function taking all the possible labeling of a given input sequence into account to predict the best sequence of labels. The fact that CRFs are less effective than RNN on ATIS, is a clear sign that label dependency modeling is relatively simple in this task. Despite this, the results in Table 3 shows that the variant I-RNN obtains always the best result among our own implementation of RNNs. This prove the effectiveness of using a context of label embeddings even on a relatively simple task.

In the Table 4 we show results on the corpus MEDIA. As we already mentioned, this task is more difficult because of an annotation semantically richer, but also because of the annotation of coreferences and segmentation of concepts over several words. This last aspect creates relatively long label dependencies, which cannot be taken into account by simple models. The difficulty of this task is confirmed by the absolute magnitude of results in Table 4 (roughly 10 F1 measure points lower than results on the ATIS task).

As we can see in Table 4, CRFs are in general much more effective than RNNs on MEDIA. This outcome can be expected as RNNs use a local decision function not able to keep long label dependencies into account. We can see also that our own implementation of Elman and Jordan RNNs are comparable with those of [7], which are state-of-the-art RNNs using a rectified linear function and dropout regularization. For this reason we can say that our Elman and Jordan RNNs are not simple baselines.

More importantly, results on the MEDIA task shows that in this particular experimental settings where taking label dependencies into account is crucial, the new variant of RNNs proposed in this paper are remarkably more effective than Elman and Jordan RNNs, both our own implementation and the state-of-the-art implementation used in [7]. We attribute this effectiveness to a better modeling of label dependencies, due in turn to the use of a label embeddings context.

In Table 5 we show results obtained on the POS-tagging task of the FTB corpus. On this task we compare our own implementations of all RNNs to the state-of-the-art results obtained in [16] with the model $MElt_{fr}^0$. We would like to underline that the model $MElt_{fr}^0$, while it does not use external resources like the model obtaining the best absolute result in [16], uses several features of words that provide a clear advantage over features used in our RNNs. It can be thus expected that the model $MElt_{fr}^0$ outperforms the RNNs. The message to take from results in Table 5 is that RNNs are all quite close to each other. However the I-RNN variant is slightly more performant, providing once more the best result among RNNs.

We would like to note that we have performed some experiments on ATIS and MEDIA without pre-training label embeddings. Results are not substantially different from those obtained with label embeddings pre-training. This may seem surprising. However on relatively small tasks (see task descriptions above), it is not rare to have similar or even better results without using pre-training. This is due to the fact that learning effective embeddings require a relatively large amount of data. Indeed, [7] shows both results obtained using embeddings pre-trained with *word2vec* and without any embeddings pre-training. Results are indeed very similar. Beyond this, the fact to have roughly the same results on a difficult task like MEDIA without label embeddings pre-training, is a clear sign that our variants of RNNs are superior to traditional RNNs because they use a context made of label embeddings, as the gain with respect to Elman and Jordan RNNs cannot be attributed to the use of pre-trained embeddings. These can encode effectively label dependencies.

It is someway surprising that I-RNN systematically outperforms I+E-RNN, the latter model integrates more information at the hidden layer and thus should be able to take advantage of both Elman and I-RNN characteristics. While an analysis to explain this outcome is not trivial, our interpretation is that using two recursions in a RNN gives actually redundant information to the model. Indeed, the output of the hidden layer keeps the internal state of the network, which is the internal (distributed) representation of the input n-gram of words around position t and the previous c labels. The recursion at the hidden layer allows to keep this information "in memory" and to use it at the next step $t + 1$. However using the recursion of I-RNN the previous c labels are also given explicitly as input to the hidden layer. This may be redundant, and constrain the model to learn an increased amount of noise. A similar idea of hybrid RNN model has been tested in [26] without showing a clear advantage on Elman and Jordan RNNs. Despite this observations, we believe the model I+E-RNN is promising. Using a larger hidden layer or splitting the hidden layer in two parts, one for previous hidden state and one for previous label embeddings, and using an additional hidden layer to merge these two, coupled with the use of *dropout* regularization, can achieve improvements over the I-RNN model. We live this as future work.

What we can say in general on results obtained on all the presented tasks, is that RNN architectures using label embeddings context can model label dependencies in a more effective way, even when such dependencies are relatively simple (like in ATIS and FTB). The two variant of RNNs proposed in this work, but more in particular the I-RNN variant, are for this reason more effective than Elman and Jordan RNNs on sequence labeling tasks.

3.4   Evaluation as Training and Tagging Time

Since our implementations of RNNs are at this moment prototypes[10], it does not make sense to compare them to state-of-the-art in terms of training and tagging time. However it is worth providing training times at least to have an idea and to have a comparison among different RNNs.

As explained in Sect. 2.5, our first variant I-RNN and the Jordan RNN have the same complexity. Also, since the size of the hidden layer is in the same order of magnitude as the size of the output layer (i.e. the number of labels), also Elman RNN has roughly the same complexity as the Jordan RNN. This is reflected in the training time.

The training time for label embeddings is always relatively short, as the size of the output layer, that is the number of labels, is always relatively small. This training time thus can vary from few minutes for ATIS and MEDIA, to less that 1 h for the FTB corpus.

Training word embeddings is also very fast on ATIS and MEDIA, taking less than 30 min for ATIS and roughly 40 min for MEDIA. In contrast training word embeddings on the FTB corpus takes roughly 5 days.

Training the RNN taggers takes roughly the same time as training the word embeddings, as the size of the word dictionary is the dimension that most affects the computational complexity at softmax used to predict next label.

Concerning the second variant of RNN proposed in this work, I+E-RNN, this is slower as it is more complex in terms of number of parameters. Training I+E-RNN on ATIS and MEDIA takes roughly 45 min and 1 h. In contrast training I+E-RNN on the FTB corpus takes roughly 6 days.

We didn't keep track of tagging time, however it is always negligible with respect to training time, and it is always measured in minutes.

All times provided here are with a 1.7 GHz CPU, single process.

4   Conclusions

In this paper we have studied different architectures of Recurrent Neural Networks on sequence labeling tasks. We have proposed two new variants of RNN to better model label dependencies, and we have compared these variants to the traditional architecture of Elman and Jordan RNNs. We explained which are the

---

[10] Our implementations are basically written in Octave https://www.gnu.org/software/octave/.

advantages provided by the proposed variants with respect to previous RNNs. We have evaluated all RNNs, new and traditional architectures, on three different tasks, two of Spoken Language Understanding and one of POS-tagging on the French Treebank. The results show that, despite RNNs don't improve state-of-the-art because of the use of less features, the new variant of RNN I-RNN always outperforms Elman and Jordan RNNs.

References

1. Jordan, M.I.: Serial order: a parallel, distributed processing approach. In: Elman, J.L., Rumelhart, D.E. (eds.) Advances in Connectionist Theory: Speech. Erlbaum, Hillsdale (1989)
2. Elman, J.L.: Finding structure in time. Cogn. Sci. **14**, 179–211 (1990)
3. Collobert, R., Weston, J.: A unified architecture for natural language processing: deep neural networks with multitask learning. In: Proceedings of the 25th International Conference on Machine Learning, ICML 2008, pp. 160–167. ACM, New York (2008)
4. Collobert, R., Weston, J., Bottou, L., Karlen, M., Kavukcuoglu, K., Kuksa, P.: Natural language processing (almost) from scratch. J. Mach. Learn. Res. **12**, 2493–2537 (2011)
5. Yao, K., Zweig, G., Hwang, M.Y., Shi, Y., Yu, D.: Recurrent neural networks for language understanding. In: Interspeech (2013)
6. Mesnil, G., He, X., Deng, L., Bengio, Y.: Investigation of recurrent-neural-network architectures and learning methods for spoken language understanding. In: Interspeech 2013 (2013)
7. Vukotic, V., Raymond, C., Gravier, G.: Is it time to switch to word embedding and recurrent neural networks for spoken language understanding? In: InterSpeech, Dresde, Germany (2015)
8. Xu, W., Auli, M., Clark, S.: CCG supertagging with a recurrent neural network. In: Proceedings of the 53rd Annual Meeting of the Association for Computational Linguistics and the 7th International Joint Conference on Natural Language Processing of the Asian Federation of Natural Language Processing, Short Papers, ACL 2015, Beijing, China, 26–31 July 2015, vol. 2, pp. 250–255 (2015)
9. Lafferty, J., McCallum, A., Pereira, F.: Conditional random fields: probabilistic models for segmenting and labeling sequence data. In: Proceedings of the Eighteenth International Conference on Machine Learning (ICML), Williamstown, MA, USA, pp. 282–289 (2001)
10. Mikolov, T., Karafiát, M., Burget, L., Cernocký, J., Khudanpur, S.: Recurrent neural network based language model. In: 11th Annual Conference of the International Speech Communication Association, INTERSPEECH 2010, Makuhari, Chiba, Japan, 26–30 September 2010, pp. 1045–1048 (2010)
11. Mikolov, T., Kombrink, S., Burget, L., Cernocký, J., Khudanpur, S.: Extensions of recurrent neural network language model. In: ICASSP, pp. 5528–5531. IEEE (2011)
12. Zennaki, O., Semmar, N., Besacier, L.: Unsupervised and lightly supervised part-of-speech tagging using recurrent neural networks. In: Proceedings of the 29th Pacific Asia Conference on Language, Information and Computation, PACLIC 29, Shanghai, China, 30 October–1 November 2015

13. Mikolov, T., Chen, K., Corrado, G., Dean, J.: Efficient estimation of word representations in vector space. CoRR abs/1301.3781 (2013)
14. Mikolov, T., Yih, W., Zweig, G.: Linguistic regularities in continuous space word representations. In: Human Language Technologies: Conference of the North American Chapter of the Association of Computational Linguistics, pp. 746–751 (2013)
15. Abeillé, A., Clément, L., Toussenel, F.: Building a Treebank for French. In: Abeillé, A. (ed.) Treebanks: Building and Using Parsed Corpora, pp. 165–188. Springer, Dordrecht (2003). https://doi.org/10.1007/978-94-010-0201-1_10
16. Denis, P., Sagot, B.: Coupling an annotated corpus and a lexicon for state-of-the-art POS tagging. Lang. Resour. Eval. **46**, 721–736 (2012)
17. De Mori, R., Bechet, F., Hakkani-Tur, D., McTear, M., Riccardi, G., Tur, G.: Spoken language understanding: a survey. IEEE Sig. Process. Mag. **25**, 50–58 (2008)
18. Dahl, D.A., Bates, M., Brown, M., Fisher, W., Hunicke-Smith, K., Pallett, D., Pao, C., Rudnicky, A., Shriberg, E.: Expanding the scope of the ATIS task: the ATIS-3 corpus. In: Proceedings of the Workshop on Human Language Technology, HLT 1994, Stroudsburg, PA, USA, pp. 43–48. Association for Computational Linguistics (1994)
19. Bonneau-Maynard, H., Ayache, C., Bechet, F., Denis, A., Kuhn, A., Lefèvre, F., Mostefa, D., Qugnard, M., Rosset, S., Servan, S., Vilaneau, J.: Results of the French Evalda-Media evaluation campaign for literal understanding. In: LREC, Genoa, Italy, pp. 2054–2059 (2006)
20. Bengio, Y., Ducharme, R., Vincent, P., Jauvin, C.: A neural probabilistic language model. J. Mach. Learn. Res. **3**, 1137–1155 (2003)
21. Schuster, M., Paliwal, K.: Bidirectional recurrent neural networks. Trans. Sig. Proc. **45**, 2673–2681 (1997)
22. Bengio, Y.: Practical recommendations for gradient-based training of deep architectures. CoRR abs/1206.5533 (2012)
23. Werbos, P.: Backpropagation through time: what does it do and how to do it. Proc. IEEE **78**, 1550–1560 (1990)
24. Chen, D., Manning, C.: A fast and accurate dependency parser using neural networks. In: Proceedings of the 2014 Conference on Empirical Methods in Natural Language Processing (EMNLP), Doha, Qatar, pp. 740–750. Association for Computational Linguistics (2014)
25. Ramshaw, L., Marcus, M.: Text chunking using transformation-based learning. In: Proceedings of the 3rd Workshop on Very Large Corpora, Cambridge, MA, USA, pp. 84–94 (1995)
26. Mesnil, G., Dauphin, Y., Yao, K., Bengio, Y., Deng, L., Hakkani-Tur, D., He, X., Heck, L., Tur, G., Yu, D., Zweig, G.: Using recurrent neural networks for slot filling in spoken language understanding. IEEE/ACM Trans. Audio Speech Lang. Process. **23**(3), 530–539 (2015)
27. Srivastava, N., Hinton, G., Krizhevsky, A., Sutskever, I., Salakhutdinov, R.: Dropout: a simple way to prevent neural networks from overfitting. J. Mach. Learn. Res. **15**, 1929–1958 (2014)

Lexical Resources and Terminology Extraction

Mining the Web for Collocations: IR Models of Term Associations

Rakesh Verma(✉), Vasanthi Vuppuluri, An Nguyen, Arjun Mukherjee,
Ghita Mammar, Shahryar Baki, and Reed Armstrong

Computer Science Department, University of Houston, Houston, TX 77204, USA
{rmverma,arjun}@cs.uh.edu, vvuppuluri@uh.edu, anqnguyen@outlook.com,
ghita.mammar@gmail.com, sh.baki@gmail.com, rmarmstr@gmail.com

Abstract. Automatic collocation recognition has attracted considerable
attention of researchers from diverse fields since it is one of the fundamental tasks in NLP, which feeds into several other tasks (e.g., parsing, idioms, summarization, etc.). Despite this attention the problem
has remained a "daunting challenge." As others have observed before,
existing approaches based on frequencies and statistical information have
limitations. An even bigger problem is that they are restricted to bigrams
and as yet there is no consensus on how to extend them to trigrams and
higher-order n-grams. This paper presents encouraging results based on
novel angles of *general* collocation extraction leveraging statistics and
the Web. In contrast to existing work, our algorithms are applicable to
n-grams of arbitrary order, and directional. Experiments across several
datasets, including a gold-standard benchmark dataset that we created,
demonstrate the effectiveness of proposed methods.

1 Introduction

Automatic recognition of semantic associations is a serious challenge and collocations are no different in this regard. Although there is no widely-accepted
definition of the word collocation, Mel'cuk has proposed a characterization and
definition in [28]. We take a relatively broader view of collocations than his proposal, which separates out idioms and quasi-idioms. In this paper, collocations
are arbitrarily restricted lexeme combinations such as *look into* and *fully aware*.[1]
The origin of the word lies in British traditional linguistics. In this paper, we
adopt the notion of collocation in its broadest sense, following Hoey and Colson:
"Collocation has long been the name given to the relationship of a lexical item
with items that appear with greater than random probability in its (textual)
context," [11,21].

As many have observed before, these special lexemes are recognized by native
speakers as belonging together. In [11], the author states: since Hoey's definition is based on a statistical criterion, collocations are likely to correspond to a
broad range of more or less fixed expressions such as compound proper nouns,

---

[1] Our definition includes the semantic phrasemes of [28].

© Springer International Publishing AG, part of Springer Nature 2018
A. Gelbukh (Ed.): CICLing 2016, LNCS 9623, pp. 177–194, 2018.
https://doi.org/10.1007/978-3-319-75477-2_11

compound nouns, compound terms, noun-adjective combinations, idioms, routine formulae, proverbs and sayings, quotations and even well-known song or film titles. We adjust this list as follows: we do not allow well-known song, book or film titles, and we add verb particle constructions (also called phrasal verbs or phrasal-prepositional verbs), compound verbs and light verb constructions. Another important consideration is whether subunits of collocations are considered collocations are not. Although at first sight it seems that (ordered) subunits of collocations should be considered collocations based on the statistical criterion, there could be some difference of opinion on units such as idioms and constructions such as *ad hoc*. Therefore, we report results for both options: unmodified, meaning subunits are not considered collocations and subcollocation, meaning subunits are considered collocations.

Computer recognition of collocations is an important task with many implications. For example, methods that can identify collocations can be appropriately extended to identify multi-word expressions and idioms. Recognition of collocations can significantly improve many important tasks, e.g., summarization [3,42,43], question-answering [4], language translation, topic segmentation [15], authorial style [22], and others.

However, automatic recognition of collocations is a challenging task for many reasons: their rarity even in large or very large corpora, they are often not modelizable by string patterns and the evolution of natural languages with some constructs falling out of favor and new constructs being added with societal changes and advances. Thus, the simple approach of building a large database of collocations and looking up each phrase will over time become obsolete. Statistics, machine learning and data mining techniques can be applied on large corpora for identifying collocations, e.g., [7,17,23,37,39,40,46]. The problems with this approach is finding a good threshold [46] and/or availability of labeled data. We highlight here a few of these and defer the rest for the Related Work Section. Xtract is based on statistical methods for retrieving and identifying collocations from large textual corpora [40] with an estimated precision of 80%. In [16], a semiautomatic method for extracting nested collocations is presented. Parsing and co-occurrences are used in [37,46], but the authors admit that "it is difficult to determine a critical value above which a co-occurrence is a collocation and below which it is not" [46]. Moreover, no results are presented since a collocation reference subset ("gold standard") is not yet constructed.

Another approach would be to search the Web for every phrase in a given text. The Web is huge and contains all types of data, curated and uncurated. There are several hurdles that must be overcome for this approach to be successful: noise, rate limits imposed on queries, sensitivity of search results to small variations in the syntax of query, and results returned are number of page hits rather than number of occurrences of the search query [9]. A full treatment of the many issues involved is beyond the scope of this paper, but see [25], who argue that the advantages often outweigh the disadvantages.

We address automatic recognition of collocations and make the following contributions:

1. We present new collocation extraction algorithms that combine the advantages of the web along multiple dimensions with those of dictionary lookup and minimize their respective disadvantages (Sect. 2). Our algorithms are general, i.e., they work for arbitrary order n-grams and are directional in the sense of Gries [19]. Gries observes that a serious deficiency of many association-based collocation extraction methods is that they use measures that are symmetrical. In other words, the value of the measure is the same regardless of whether the phrase is *look into* or *into look*.
2. We demonstrate the performance of our algorithms on several datasets and compare the performance with that of several baselines including MWE-Toolkit, NSP and Gries's algorithm [19] (Sect. 3).
3. We create a gold-standard dataset derived from the Wiki50 dataset [44] that we will share with the NLP community. We explain the creation of this dataset in detail (Sect. 3.2). The creation of this dataset sheds light on the difficulty of manually annotating corpora for collocations (Sect. 3.4).
4. We present the performance of eight volunteers at the task of collocation extraction. These volunteers were computer science students with some being native speakers of English and some non-native speakers. None of them were experts in linguistics. The volunteers were asked to use dictionaries to look up the phrases as a matter of course. Even when equipped with the Oxford Dictionary of Collocations and Oxford Dictionary of Idioms, performance (F1-score) of the volunteers ranged from 39% to 70%.

2 Collocation Detection Algorithms

In most scenarios collocations tend to have a defined structure. Hence, we design two variants of the methods for extracting collocations, which help us observe the significance of parts-of-speech (POS).

Collocations without POS restrictions. This method ignores the POS of the components in the n-gram to determine whether it is a collocation.

Collocations with POS restrictions. A necessary condition for an n-gram to be a collocation is that the POS of at least one of its components belongs to {Noun, Adjective, Verb, Adverb}.

Although we provide two methods, the steps involved are essentially the same. The first component is splitting the text document into sentences and n-grams. Care must be taken in n-gram extraction to account for punctuation, and, of course, splitting into sentences is itself nontrivial because of abbreviations containing periods. After experiments with off-the-shelf NLP software, we use our own n-gram extractor.

2.1 Dictionary Search

Our first method for collocation recognition is straightforward, viz., lookup. In this work, we used WordNet from NLTK corpus, even though it is small and limited in polylexical expressions, mainly because it is readily available.

Input for algorithms. n-grams extracted from text or given.

Algorithm 1. Collocation Extraction using Lookup

1: **for** each n-gram N **do**
2:      **if** $N \in WordNet$ dictionary **then**
3:           N is a collocation

2.2 Web Search for Title and URL

After searching WordNet, we then explore the largest source of data in determining if an n-gram N is a collocation - the Web. For this we do a phrase query of N using Bing search API[2] and retrieve the top 10 hits of the search. From each result retrieved, the title and URL are extracted. Now, the method checks if any word (substring) that is synonymous to the word 'dictionary', or any dictionary, is present in the title (URL). If the answer is yes, the method then checks if the exact match of N is present in the URL or if the stemmed components of N are present in the stemmed title. This is to avoid missing any component because of different inflectional forms. Snowball stemmer is used to stem the components. If a match is found, N is declared a collocation.

The two steps involved in this method ensure that the n-gram is not a random co-occurrence of lexemes. implying that the n-gram is a collocation. We note that the Bing search API used is not consistent in providing hit counts. Access to a stable web search API will improve this method.

Algorithm 2. Collocation Extraction using Web

1: **for** each n-gram N **do**
2:      Check top 10 search results (Titles/URLs) for words synonymous to 'dictionary'
3:      Titles = search titles that meet the requirement in line 2
4:      URLs = search URLs that meet the requirement in line 2
5:      **if** ($N \in$ Titles) or ($N \in$ URLs) **then**
6:           N is a collocation

We tested this method on six documents selected at random from the Wiki50 dataset using the Wiki50 annotations as gold standard. The F1-scores are in Table 1. We observe that this method alone achieves decent F1-scores, frequently better than 20%. Hence we decided to put this method second in the pipeline when the methods are sequenced.

We noticed that Wikipedia blocks requests after a certain number, and Wikipedia also appears the most in the dictionary websites. The problem is alleviated, however, because most Wikipedia URLs already contain the title.

---

[2] https://datamarket.azure.com/dataset/bing/search.

Table 1. F1 scores for web search on Title and URL

Document	F1-score unmodified	F1-score subcollocation
Bacteriological water analysis	0.2712	0.3552
Bearing an Hourglass	0.1176	0.2026
Budy Caldwell	0.2462	0.3754
Butch Hartman (racer)	0.2394	0.4040
Castlevania chronicles	0.2006	0.2831
Myllarguten	0.1356	0.1875

2.3 Web Search and Substitution

Although the Web search method is often efficient, in some situations the top 10 results may not be sufficient to cover the diversity of myriad collocations. Hence, we use the following technique as a backup to determine if an n-gram N is a collocation. This method uses Bing Search API to obtain hit counts when a phrase query is formed from N. Then each word w in N is replaced by 5 random words that are of the same POS as w. This is done only for words whose POS is from {Noun, Adjective, Verb, Adverb} if we take POS into consideration. After each replacement, the n-gram with one of its words replaced is searched for in the web using Bing search API and the total number of search results returned is obtained. Once all the replacements are done and all search results, $\{S_{11}, S_{12}, S_{13}, S_{14}, S_{15}, S_{21}, ...S_{n5}\}$ are obtained, an average is computed as, S_{avg} of the non-zero S_{ij} values. The final step is to compare S_N against a suitably weighted S_{avg}, where the weight factor is a multiplicative constant c_{sub}. Note that since this method is based on hit counts for a phrase query, it is naturally directional in the sense of Gries.

Algorithm 3. Collocation Extraction using Web and Substitution

---
1: **for** each n-gram N **do**
2:     S_N = Total hit counts for N (phrase query)
3:     N' = new phrase obtained by replacing each word w in N with 5 randomly chosen words of same POS as w
4:     SR = list(Total hit counts returned for each N')
5:     S_{avg} = Average of non-zero values in SR
6:     **if** $S_N > c_{sub}S_{avg}$ **then**
7:         N is a collocation

For this method, we need to find the optimal value of c_{sub}, so we evaluated it on the document "Bearing an Hourglass," from Wiki50 without POS. First, exploring the range $[0, 1]$ with 0.001 increment yielded F1-score lower than 3% for both versions: unmodified and subcollocation. Next we explored the range $[1, 10001]$ with increment of 10. This gave the best F1-score between 6–7% for the subcollocation version and between 5–6% for the unmodified version. Based on the graph of F1-scores, we narrowed the search for c_{sub} to the interval $[1, 2001]$.

Using increment of 1 this interval was explored and the results then narrowed the search to the interval $[1, 101]$. When this interval was explored in increments of 0.001 the best F1-scores of 5.65% and 6.69% were obtained at $c_{sub} = 43.7$ for unmodified version and $c_{sub} = 69.2$ for the subcollocation version.

For validation of these c_{sub} values a different article "Myllarguten" was selected. The optimal values of c_{sub} found above were tried and F1-scores of 1.15% (unmodified) and 4.04% (subcollocation) for c = 43.7, and 1.18% (unmodified) and 3.78% (subcollocation) for c = 69.2 were obtained. These results are significantly worse than the results achieved for "Bearing an Hourglass," so we probed further in the range $[1, 100]$. Again the F1-scores of between 6–7% for subcollocation and between 4–5% for the unmodified versions were observed for c_{sub} in the interval $[1, 20]$. This suggests that: (i) perhaps our sample for tuning c_{sub} values may not be large enough. In other words, combining more articles into the training will give us a c_{sub} value that works better generally. (ii) The "randomness" inherent in the method may be affecting our search for the best c_{sub} value. It could make the best c_{sub} values vastly different for each run, and each article. Therefore, the best c_{sub} value found for only one training run may not be the best for the evaluation. While checking the F1-scores in these experiments, we noticed that the Wiki50 dataset annotations were not consistent with the Oxford Dictionary of Collocations and the Oxford Dictionary of Idioms so we deferred further experiments on the c_{sub} value to post gold-standard dataset creation, which is described below.

2.4   Web Search and Independence

This is another directional approach that does not use as many search queries as the above technique of Sect. 2.3. The idea of this method is to check whether the probability of a phrase exceeds the probability that we would expect if the words are independent. Hit counts are used to estimate these probabilities. There are two variants that differ in Line 8. The steps are described in Algorithm 4 below.

Algorithm 4. Collocation Extraction using Web and Independence - Method I

1: **for** each n-gram N **do**
2:      $T(N) = $ Total hit counts for N
3:      $U_a = $ Universe of web pages containing 'a'
4:      $P(N) = T(N)/U_a$
5:      **for** each word w_i in N **do**
6:          $T(w_i) = $ Total hit counts for w_i
7:          $P(w_i) = T(w_i)/U_a$ /* Prob. of w_i */
8:      **if** $P(N) > f(n)\Pi_{i=1}^{n}P(w_i)$ **then**
9:          N is a collocation

Method-2: The drawback of the first method is that it ignores word repetitions within the phrase, which we fix by modifying Line 8 of Algorithm 4. When the words in the n-gram are repeated, an adjustment is made based on the number of distinct permutations possible from words in the n-gram as follows. When the words in the n-gram are not unique: the n-gram is a collocation if

$$P(N) > f(n)\Pi_{j=1}^{k} n_j! \ \Pi_{i=1}^{n} P(w_i)/n!,$$

where k = number of unique words, n_i = number of occurrences of the ith word in the n-gram.

Optimizing the Independence Methods without POS - function $f(n)$. At first the inequalities for $P(N)$ in the above two algorithms were modified to introduce a similar constant c_{ind}, i.e., $f(n) = c_{ind}$, on the right hand side (RHS) as in the substitution method. However, the results were unsatisfactory. The problem is that the number of hit results for a single word is approximately 10 millions to billions, while the number of hit results for 'a' is 18 billions. The calculated probability of a single word, therefore, can be as low as 10^{-4}. When the length of the N-gram increases one word, the RHS of the inequality decreases by 10^{-4}, while the LHS decreases slightly. Example: Phrase "Bat Durston and the BEMS" had three hits while "Bat Durston and the" had four hits. As the RHS decreases too quickly compared to the LHS, we introduce a balancing function $f(n)$ that grows with the length of the n-gram fast enough to counter this effect. We chose the formula c^{n^p-1} for two reasons: the two parameters c and p give flexibility for optimization and $f(1) = 1$ ensures that unigrams cannot be flagged as collocations. A two-dimensional heat map (Fig. 1) of F1-scores was constructed for c and p ranging in $[1, 3]$ for the unmodified version of Method 1 first. The best F1-score of 11.25% was achieved at $p = 2.93$ and $c = 1.15$. The heat map pattern also shows that close to the highest score is achieved for other combinations of parameters as well.

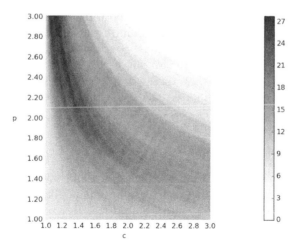

Fig. 1. Heat map for unmodified version

Next the heat map for the subcollocation version of Method 1 was constructed (Fig. 2). The best F1-score of 12% was achieved at surprisingly the same combination of parameter values, $p = 2.93$ and $c = 1.15$. However, the pattern this time shows a narrower band of good parameter choices.

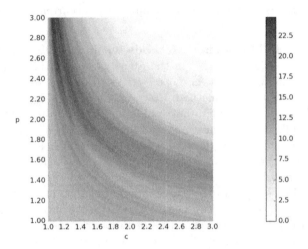

Fig. 2. Heat map for subcollocation version

The heat maps for both versions of Method 2 are similar to those for Method 1 so we omit them here. The highest F-score increases a little bit to 11.30% for the unmodified version at $p = 3$ and $c = 1.21$. For the subcollocation version, the highest F-score increases a little bit to 12.15% at $p = 2.95$ and $c = 1.14$.

Validation on a different article. We ran both Independence methods without POS option using the three sets of p and c values obtained above on the article "Myllarguten." The F1-scores were even better (13.91% to 15.54% for $p = 2.93$ $c = 1.15$), (16.31% to 19.05% for $p = 3$ $c = 1.21$) and (13.79% to 15.31% for $p = 2.95$ $c = 1.14$) than those obtained for "Bearing an Hourglass." Further search for optimal values on the gold-standard dataset we created is described below.

3   Experimental Evaluation

This section details the experiments settings and results. We detail our metrics, the datasets used, comparison between Dataset 3 and Wiki50, the performance of our volunteers, and the performance results.

3.1   Baselines and Overall Methods

We use Mwetoolkit and NSP as our baselines. Our overall pipelines are: **Sub** - which executes Algorithms 1, 2 and 3 in sequence, **T1** - Algorithms 1, 2 and 4 in sequence and **T2** - Algorithms 1, 2 and second variation of 4 that takes into account repetition of words.

3.2   Datasets

For the experiments, we used three different datasets extracted from several sources.

Dataset 1. A set of 400 collocations were extracted from listed websites based on POS structure. This dataset comprises 100 Adjective+Noun collocations from [14,29], 100 Noun+Noun collocations from [45], 100 Verb+Noun collocations from [5,12,24,35,41] and 100 Verb+Preposition collocations from [13,18,30]. Each of these collocations is used as a test to verify the performance of the methods when a complete sentence is not given, and also to compare the performance of our methods with Mwetoolkit, which needs POS patterns of collocations to be extracted.

Dataset 2. This dataset is a collection of idioms obtained from the Oxford Dictionary of Idioms. The text file consisting of 1673 idioms is our input. Since all idioms are essentially MWEs and all MWEs are collocations, idioms would be the perfect choice for testing the software. Also, any non-ASCII characters are ignored while writing the idioms to the text file.

Dataset 3. This is a sentence dataset, which facilitates evaluation of the false positive rates of our methods and the baselines. Its creation was inspired by the discrepancies observed between the Oxford dictionaries and the Wiki50 annotations, while checking the F1-scores of our algorithms during the tuning of the parameters. A set of 100 sentences was selected at random from the Wiki50 dataset [44] and distributed to eight volunteers. Note that if two or more sentences were included in a single quotation, they were counted as a single sentence. Even though the Wiki50 dataset was annotated, we found that it was missing several collocations and a few idioms when we did a spot check with the Oxford Dictionary of Collocations and the Oxford Dictionary of Idioms. So the volunteers were asked to manually annotate all the collocations and the idioms in the 100 sentences using these two resources. Each volunteer was given 25 sentences and each sentence was given to exactly two volunteers.

The volunteers were given instructions on how to annotate since: (i) dictionaries contain abbreviations for generic pronouns such as *sth* for something or *sb* for somebody (ii) the verb forms can be different, e.g., dictionary may contain "make ones way" and sentence may contain "made his way" and (iii) there could be intervening words in the collocations.[3] After the volunteers completed the annotations, one of the co-authors resolved all the conflicts with dictionaries and also checked the phrases on which both volunteers agreed. Then, a different co-author went through all the sentences one more time looking carefully for false negatives and false positives. After this check, each volunteer was given

---

[3] These are also the reasons that automatic creation of gold standard datasets is difficult even if text data is extracted from the Dictionaries mentioned.

feedback on the results and was asked to dispute the findings and find any remaining errors. This process ensured a final check of the results. After this final check, recall, precision and F1-score were calculated for each volunteer and the findings were compared with the Wiki50 dataset. We found that the Wiki50 annotators consistently annotated compound proper nouns and we do include these in our gold-standard collocations as well since we take the most general definition of collocation. Some examples of collocations identified by our process and not identified by the Wiki50 annotators include: "vowing to," "ardent supporter" and "be elected."

3.3   Volunteer Demographics and Performance on Dataset 3

The eight volunteers include two females (25%) and six males (75%). Three are native speakers of English and five speak English as a second language. They range in years from 18 to 25. All are students: one high school senior, one undergraduate senior, two MS, and four PhD students. Their F1-scores range from 39.21% (high-school senior, native speaker) to 70.87% (PhD student, native speaker).

3.4   Comparison with Wiki50 Annotations

A total of 263 collocations were identified in the 100 sentences of which six are idioms. The Wiki50 annotators had identified 159 of these 263 collocations (recall = 60.46%) and missed four out of six idioms we found. Both collocation numbers (ours and Wiki50) include compound proper nouns and compound nouns, except named entities.

3.5   Parameter Value Optimization

Now that we have reasonable confidence[4] in our gold-standard, we undertook an optimization procedure for the parameters on the 100 sentence Dataset 3. The dataset was divided into 60% training, 20% held-out and 20% testing sets. The top three sets of parameter values from the training set for each method and its versions (POS, No POS) and (Unmodified, Subcollocation) were tried on the validation set and the winner proceeded to the test set. No significant difference was observed for POS versus No POS versions. Slight difference was observed for the Unmodified versus Subcollocation versions of the Independence methods and significant difference was observed for the Unmodified (optimal $c_{sub} = 13.1$) and Subcollocation (optimal $c_{sub} = 92.0$) versions of the Substitution method and these values were stable across POS and No POS versions. Here, we present the results on all datasets with $c_{sub} = 13.1$ for all versions of Substitution method and $c = 1.14$, $p = 2.95$ for all versions of Independence method.

---

[4] Note that we do not claim perfection, but we expect mistakes to be rare.

3.6   Results

For datasets 1 and 2, only recall is relevant since the input is the gold standard itself (precision = 100%). Tables 2, 3, 4 and 5 give results for Datasets 1 and 2. Table 6 gives the precision, recall and F1-scores for Dataset 3. The Sub pipeline gives the best recall on all datasets but lower precision. The T1 pipeline gives the best F1-score on Dataset 3. Another interesting trend is that the recall of our methods is usually better for idioms than for collocations overall.

Table 2. Recall on Datasets 1 and 2, $c_{sub} = 13.1$, $c = 1.14$, $p = 2.95$

Recall	Sub		T1		T2	
	No POS	POS	No POS	POS	No POS	POS
Dataset 1	0.744	0.744	0.736	0.736	0.739	0.739
Dataset 2	0.828	0.823	0.826	0.825	0.831	0.819

Table 3. F1-scores on Datasets 1 and 2

F Score	Sub		T1		T2	
	No POS	POS	No POS	POS	No POS	POS
Dataset 1	0.853	0.853	0.848	0.848	0.85	0.85
Dataset 2	0.902	0.899	0.9	0.9	0.903	0.897

Table 4. Recall on Datasets 1 and 2 for subcollocation versions

Recall	Sub		T1		T2	
	No POS	POS	No POS	POS	No POS	POS
Dataset 1	0.652	0.652	0.641	0.641	0.658	0.658
Dataset 2	0.826	0.823	0.824	0.823	0.844	0.82

Table 5. F1-scores on Datasets 1 and 2 for subcollocation versions

F Score	Sub		T1		T2	
	No POS	POS	No POS	POS	No POS	POS
Dataset 1	0.789	0.789	0.781	0.781	0.793	0.793
Dataset 2	0.897	0.895	0.896	0.895	0.905	0.894

Table 6. Percentage precision, Recall and F1-Score for Dataset 3, $c_{sub} = 13.1$, $c = 1.14$ and $p = 2.95$

Corpus	Tech.	Evaluation method	POS			No POS		
			Precision	Recall	F-Score	Precision	Recall	F-Score
Bing	Sub	Unmodified	11.12	58.45	18.6899	10.64	62.46	18.1818
		Subcollocation	13.27	69.21	22.274	13.26	70.79	22.3422
	T1	Unmodified	22.37	51.86	31.2608	19.98	55.59	29.3939
		Subcollocation	34.86	48.03	40.3985	33.22	51.84	40.4933
	T2	Unmodified	19.45	55.01	28.7425	17.76	58.74	27.2788
		Subcollocation	28.19	51.97	36.5571	27.55	55.53	36.8237

3.7 Comparison with Baseline Parameter Values

In the following Tables 7 and 8, we report for comparison the recall on Datasets 1 and 2 that we get with the parameter values chosen so that they do not make any difference $c_{sub} = 1$ and $c = 1$. In Table 9 we present the results for dataset 3 with baseline parameter values. As expected, since parameters were optimized on Dataset 3 where both recall and precision matter, whereas only recall matters for Datasets 1 and 2, the results improve with baseline parameter values for these two datasets. A clear degradation is observable for Dataset 3 with baseline parameter values.

Table 7. Recall on Datasets 1 and 2 for unmodified versions, baseline parameters

Recall	Sub		T1		T2	
	No POS	POS	No POS	POS	No POS	POS
Dataset 1	0.85	0.847	0.779	0.779	0.859	0.859
Dataset 2	0.86	0.865	0.87	0.869	0.925	0.923

Table 8. Recall on Datasets 1 and 2 for subcollocation case, baseline parameters

Recall	Sub		T1		T2	
	No POS	POS	No POS	POS	No POS	POS
Dataset 1	0.828	0.826	0.7	0.7	0.812	0.812
Dataset 2	0.868	0.868	0.87	0.869	0.932	0.931

Comparison: MWEToolkit, NSP and Gries's Delta P. Although Mwe-toolkit is described as MWE extraction software, the definition of an MWE in [33] aligns with our definition of collocation, hence, this is a valid comparison. MWEToolkit needs POS patterns of collocations to be able to extract them.

Table 9. Percentage results on Dataset 3 with baseline parameters

Corpus	Tech.	Evaluation method	POS			No POS		
			Precision	Recall	F-Score	Precision	Recall	F-Score
Bing	Sub	Unmodified	5.4	82.24	10.13	5.4	82.24	10.13
		Subcollocation	8.06	89.61	14.8	8.06	89.61	14.8
	T1	Unmodified	4.76	59.6	8.81	4.84	63.32	9
		Subcollocation	5.7	87.89	10.7	5.7	88.16	10.7
	T2	Unmodified	2.73	66.48	5.24	2.8	70.2	5.38
		Subcollocation	4.82	96.18	9.17	4.82	96.32	9.18

For example, to extract a verb+noun phrase, it requires a pattern 'VN' to be declared prior to execution. So, we use Dataset-1 as input. The recall is 93%, the precision is 10.15% and the F-score is 18.31% even when MWEtoolkit is run on the four different types of collocations separately. The reason for the low precision is that, based on the patterns specified, MWEtoolkit takes the POS tagged words as input and creates phrases. Often, new phrases not part of the input are created and then checked.

NSP provides many association measures for bigrams, a subset of four for trigrams and only log-likelihood (LL) for 4-grams. We used the four trigram measures for both bigrams and trigrams and the best possible thresholds on a development set consisting of 20 sentences of Dataset 3. The highest F-score was 18% for bigrams and 7% for trigrams with PMI, LL and PS. For 4-grams the highest F-score was 4%.

In [19], Gries proposed to use ΔP to differentiate bigram collocations from bigram non-collocations. Although he did not give any thresholds, we took a set of 25 bigram collocations at random from Dataset 3 and 15 bigram idioms at random from Dataset 2 and then found the threshold that gave the best F1-score (32.26%) on Dataset 3, which came out to 1. When this threshold was used for the same set of bigrams but with the British National Corpus corpus supplying the frequencies, the F1-score was 0, since Recall was 0. The threshold that gave the best F1-score of 1.0 for the British National Corpus was -0.4, which when used with the 100 sentences of Dataset 3 for supplying frequencies gave an F1-score of 17.8%.

4   Related Work

Collocation extraction has been well-studied. We include here the closest related work under the following threads:

Statistical measure based approaches: One of the classical methods of discovering collocations is to measure the association strengths of candidate n-grams. The key idea is to ascertain whether appearance of terms in an n-gram is more often than just random chance. In [6], significance of bigrams was computed by measuring the actual frequencies with expected frequencies using a normal approximation to the binomial distribution yielding the z-score. [10] used pointwise mutual information. In [38], an entropy method was proposed relying on the idea that collocations tend to be less noisy than non-collocating n-grams from an information-theoretic perspective. [40] proposed a threshold based approach that first discovers bigrams and then detects collocations based on a threshold and the context of nearby words (or their POS tags) appearing in the sentence. In [8], a technique based on log-odds ratio was proposed. In N-gram statistics package [2], a variety of association measures were implemented. While statistical measures have association evaluation strengths, they are quite dependent on the input corpus and no single method works well in discovering the whole gamut of collocations. In practice, a combination of measures renders better accuracy in collocation discovery [32]. Our proposed methods overcome the major problems with all these approaches, viz., sparsity and lack of directionality [19], by using the Web and devising new directional methods.

Parsing/Multi-lingual/MWEs/Idioms: Since we do not use any information obtained through parsing, these approaches [1,27,31,36,46,47] are not directly comparable to ours. Our algorithms are quite general and suitable for extensions in the *multi-lingual* context,[5] some of which was studied in [17,20,37]; and we note *approaches for MWEs/idioms:* [26,34].

Parsing and dependency based approaches have also been explored. In [27], an information theoretic approach over parse dependency triples were proposed. [31] compares several techniques with his own in which he exploited synonym substitution via WordNet within parse dependency pairs. These could discover collocations such as "emotional baggage" from less frequently occurring phrases "emotional luggage" by substituting luggage by its synonym baggage. [36] explores syntax based approaches for collocation extraction. In [47] shallow syntactic analysis based on compositionality, subsitutability, and modifiability statistics were leveraged to discover collocations. The method was evaluated on a specific cases of German PP-verb combinations. [1], proposed a lexical acquisition technique based on a dependency parser for extracting a verb-particle constructions (e.g., hand in, climb up, drop down, etc.) which are a special case of collocations. Although these methods have made progress, parsing based approaches tend to be sensitive on inherent threshold that need tuning which is often non-trivial and heuristic [46].

Another thread of research exploits aligning multi-lingual corpora for collocation extraction. In [20] lingual collocations were described from sentence-aligned parallel corpora. [17] focused on the special case of verb and its objective noun collocations in bilingual corpora. [37] proposed a framework based on deep syntactic parsing and rule-based machine translation for extracting lexical collocations form multi-lingual corpora. Methods based on syntactic tree-patterns need high quality and large coverage parsers.

---

[5] Languages in which long words can be constructed by glueing together two or more lexemes or languages that have writing systems without word separators are likely to prove much more challenging.

Multiword expressions (MWEs) are a special case of collocations. They range over linguistic constructions such as fixed phrases (per se, by and large), noun compounds (telephone booth, cable car), compound verbs (give a presentation), idioms (a frog in the throat, kill some time), etc. [26] provides a review of linguistic and distributional characteristics of MWEs. [33] developed a system called mwetoolkit and implemented 4 measures (MLE, Dice, t-score, and PMI) for extracting MWEs following certain patterns (e.g., POS sequence patterns). In [34], a distributional similarity of each component word and the overall expression was used in predicting the compositionality of MWEs.

It is difficult to find any free software or research prototype that computes collocations. Xtract is no longer maintained, Collocate http://www.athel.com/colloc.html charges money, and we could not find the system in [37]. We have compared our work with mwetoolkit [33], which is based on user-defined criteria and association measures with counts obtained from Internet search results, and NSP [2]. For MWEtoolkit, the user must first run the Treetagger software on a text file and then process the output with a script in MWEtoolkit to generate an XML file. Then a DTD must be created for the generated XML file and then the XML file can be processed by MWEtoolkit to extract multi-word expressions,[6] which are subsets of collocations. NSP requires preprocessing of text, before constructing n-grams since otherwise it constructs n-grams that span sentences and include punctuation as a separate unit. For instance, "hard" can be a bigram.

In [19], Gries criticized much of the previous work on using association measures for collocation detection because the measures are symmetrical. He then proposed ΔP for differentiating bigram collocations from bigrams that are not collocations. He did not propose any thresholds for his methods, and it is not clear how to extend them from bigrams to higher-order n-grams.

In [11], a web-based search method is proposed that relies on computing the proportion of exact matches of the n-gram in a sample of results that are returned by the API for Yahoo (100 when the paper was written) on a single query. This method requires "subtle manipulation" of the API according to the author. It also requires details of filters for tackling spamdexing and noise (essentially repetitions of lines and paragraphs by the search engine), which are omitted by the author. With these clever techniques in place, the technique yields high recall and precision according to the author when the threshold is chosen by analyzing an unspecified number of collocations selected from a dictionary. The recall is calculated on a set of 3,807 collocations and the precision is calculated on a subset of 5-grams from Google's n-gram collection.

5   Conclusion

We have presented new approaches for detecting collocations that combine the advantages of look-up and Web and minimize their disadvantages. Two other advantages of our approach are that it can be extended to other languages based

---

[6] The term polylexical expressions is preferred by some researchers since it removes reliance on the ill-defined concept of a word.

on availability of a WordNet or dictionary for that language and Web search for it, and, in contrast to approaches such as Mwetoolkit, our approach can be used to directly check phrases without requiring the context. Results of our approach are demonstrated on a variety of test sets including a gold-standard sentence dataset that has been created. We also report on the performance of human volunteers and shed some light on the difficulty of creating collocation datasets manually. Our independence algorithm is within the range of the human volunteers and shows promise for the future. More work is needed to make the Substitution approach robust.

Acknowledgments. We thank the anonymous reviewers for their constructive comments, which have improved the paper. Rakesh's and Vasanthi's research was supported in part by NSF grants DUE 1241772, CNS 1319212, and DGE 1433817. Arjun's research is supported in part by NSF grant CNS 1527364. Reed's research was supported by NSF grant IIS 1359199. Ghita's research was supported by the University of Houston. An Nguyen and Arjun Mukherjee contributed equally to this paper. The authors thank Luis Moraes, Daniel Lee, Avisha Das, Arthur Dunbar, Nirmala Rai and Suvedh Srikanth for participating in the gold-standard creation.

References

1. Baldwin, T.: Deep lexical acquisition of verb-particle constructions. Comput. Speech Lang. **19**(4), 398–414 (2005)
2. Banerjee, S., Pedersen, T.: The design, implementation, and use of the ngram statistics package. In: Gelbukh, A. (ed.) CICLing 2003. LNCS, vol. 2588, pp. 370–381. Springer, Heidelberg (2003). https://doi.org/10.1007/3-540-36456-0_38
3. Barrera, A., Verma, R.: Combining syntax and semantics for automatic extractive single-document summarization. In: Gelbukh, A. (ed.) CICLing 2012. LNCS, vol. 7182, pp. 366–377. Springer, Heidelberg (2012). https://doi.org/10.1007/978-3-642-28601-8_31
4. Barrera, A., Verma, R., Vincent, R.: Semquest: University of Houston's semantics-based question answering system. In: Text Analysis Conference (2011)
5. http://www.bbc.co.uk/worldservice/learningenglish/grammar/learnit/learnitv351.shtml (2015)
6. Berry-Rogghe, G.: The computation of collocations and their relevance in lexical studies. In: The Computer and Literary Studies, pp. 103–112. Edinburgh University Press, Edinburgh (1973)
7. Biber, D.: Co-occurrence patterns among collocations: a tool for corpus-based lexical knowledge acquisition. Comput. Linguist. **19**(3), 531–538 (1994)
8. Blaheta, D., Johnson, M.: Unsupervised learning of multi-word verbs. In: 39th ACL and 10th EACL, pp. 54–60 (2001)
9. Bolshakov, I.A., Bolshakova, E.I., Kotlyarov, A.P., Gelbukh, A.: Various criteria of collocation cohesion in internet: comparison of resolving power. In: Gelbukh, A. (ed.) CICLing 2008. LNCS, vol. 4919, pp. 64–72. Springer, Heidelberg (2008). https://doi.org/10.1007/978-3-540-78135-6_6
10. Church, K.W., Hanks, P.: Word association norms, mutual information and lexicography. In: Proceedings of 27th Annual Meeting of the Association for Computational Linguistics, University of British Columbia, Vancouver, BC, Canada, 26–29 June 1989, pp. 76–83 (1989)

11. Colson, J.-P.: Automatic extraction of collocations: a new web-based method. In: Proceedings of JADT, pp. 397–408 (2010)
12. http://www2.elc.polyu.edu.hk/CILL/eap/2004/u5/verbs&nounspart2.htm (2015)
13. http://www.englishclub.com/vocabulary/ist.htm (2015)
14. http://esl.about.com/od/grammarstructures/a/g_intadj_2.htm (2015)
15. Ferret, O.: Using collocations for topic segmentation and link detection. In: COLING (2002)
16. Frantzi, K.T., Ananiadou, S.: Extracting nested collocations. In: COLING, pp. 41–46 (1996)
17. Fukumoto, F., Suzuki, Y., Yamashita, K.: Retrieving bilingual verb-noun collocations by integrating cross-language category hierarchies. In: COLING, pp. 233–240 (2008)
18. http://grammar.ccc.commnet.edu/grammar/phrasals.htm (23 July 2014)
19. Gries, S.T.: 50-something years of work on collocations: what is or should be. Int. J. Corpus Linguist. **18**(1), 137–166 (2013)
20. Haruno, M., Ikehara, S., Yamazaki, T.: Learning bilingual collocations by word-level sorting. In: Proceedings of the Conference 16th International Conference on Computational Linguistics, COLING 1996, Center for Sprogteknologi, Copenhagen, Denmark, 5–9 August 1996, pp. 525–530 (1996)
21. Hoey, M.: Patterns of Lexis in Text. Oxford University Press, Oxford (1991)
22. Hoover, D.L.: Frequent collocations and authorial style. LLC **18**(3), 261–286 (2003)
23. Ikehara, S., Shirai, S., Uchino, H.: A statistical method for extracting uninterrupted and interrupted collocations from very large corpora. In: COLING, pp. 574–579 (1996)
24. http://www.johnsesl.com/templates/grammar/collocations.php (2015)
25. Kilgarriff, A., Grefenstette, G.: Introduction to the special issue on the web as corpus. Comput. Linguist. **29**(3), 333–347 (2003)
26. Kordoni, V., Egg, M.: Robust automated natural language processing with multiword expressions and collocations. In: ACL (Tutorial Abstracts), pp. 7–8 (2013)
27. Lin, D.: Extracting collocations from text corpora. In: First Workshop on Computational Terminology (1998)
28. Mel'ćuk, I.: Collocations and lexical functions. In: Cowie, A.P. (ed.) Phraseology: Theory, Analysis, and Applications, pp. 23–54. Oxford University Press, Oxford (1998). (2001 [1998])
29. http://www.myenglishteacher.eu/blog/ (2015)
30. http://www.oday.com/erbs/erbs_A.html (2015)
31. Pearce, D.: A comparative evaluation of collocation extraction techniques. In: Proceedings of the Third International Conference on Language Resources and Evaluation, LREC 2002, Las Palmas, Canary Islands, Spain, 29–31 May 2002 (2002)
32. Pecina, P.: An extensive empirical study of collocation extraction methods. In: Proceedings of the Conference ACL 2005, 43rd Annual Meeting of the Association for Computational Linguistics, University of Michigan, USA, 25–30 June 2005 (2005)
33. Ramisch, C., Villavicencio, A., Boitet, C.: Multiword expressions in the wild? The mwetoolkit comes in handy. In: COLING (Demos), pp. 57–60 (2010)
34. Salehi, B., Cook, P., Baldwin, T.: Using distributional similarity of multi-way translations to predict multiword expression compositionality. In: Proceedings of the 14th Conference of the European Chapter of the Association for Computational Linguistics, EACL 2014, Gothenburg, Sweden, 26–30 April 2014, pp. 472–481 (2014)
35. http://www.scielo.cl/pdf/signos/v45n78/a03.pdf (2015)

36. Seretan, V.: Syntax-Based Collocation Extraction. Springer, Dordrecht (2011). https://doi.org/10.1007/978-94-007-0134-2. Syntax-based extraction
37. Seretan, V.: A multilingual integrated framework for processing lexical collocations. In: Przepiórkowski, A., Piasecki, M., Jassem, K., Fuglewicz, P. (eds.) Computational Linguistics. Studies in Computational Intelligence, vol. 458, pp. 87–108. Springer, Heidelberg (2013). https://doi.org/10.1007/978-3-642-34399-5_5
38. Shimohata, S., Sugio, T., Nagata, J.: Retrieving collocations by co-occurrences and word order constraints. In: ACL, pp. 476–481 (1997)
39. Shimohata, S., Sugio, T., Nagata, J.: Retrieving domain-specific collocations by co-occurrences and word order constraints. Comput. Intell. **15**, 92–100 (1999)
40. Smadja, F.A.: Retrieving collocations from text: Xtract. Comput. Linguist. **19**(1), 143–177 (1993)
41. http://www.stgeorges.co.uk/blog/ack/ (2015)
42. Verma, R.M., Chen, P., Lu, W.: A semantic free-text summarization system using ontology knowledge. In: Document Understanding Conference (2007)
43. Verma, R.M., Kent, D., Chen, P.: Ontology driven multi-document summarization. In: Text Analysis Conference (2008)
44. Vincze, V., Nagy, I., Berend, G.: Multiword expressions and named entities in the wiki50 corpus. In: RANLP, pp. 289–295 (2011)
45. http://www.vocabulary.com/lists/201117#view=definitions&word=cross%20hair (2015)
46. Wehrli, E.: Parsing and collocations. In: Natural Language Processing, pp. 272–282 (2000)
47. Wermter, J., Hahn, U.: Collocation extraction based on modifiability statistics. In: Proceedings of the Conference 20th International Conference on Computational Linguistics, COLING 2004, Geneva, Switzerland, 23–27 August 2004 (2004)

A Continuum-Based Model of Lexical Acquisition

Pierre Marchal[1]([✉]) and Thierry Poibeau[2]

[1] ER-TIM, INaLCO, 2 rue de Lille, 75007 Paris, France
pierre.marchal@inalco.fr
[2] LaTTiCe (CNRS, ENS and University Paris 3), PSL Research University and
Univ. Sorbonne Paris Cité,
45 rue d'Ulm, 75005 Paris, France

Abstract. The automatic acquisition of verbal constructions is an important issue for natural language processing. In this paper, we have a closer look at two fundamental aspects of the description of the verb: the notion of lexical item and the distinction between arguments and adjuncts. Following up on studies in natural language processing and linguistics, we embrace the double hypothesis (*i*) of a continuum between ambiguity and vagueness, and (*ii*) of a continuum between arguments and adjuncts. We provide a complete approach to lexical knowledge acquisition of verbal constructions from an untagged news corpus. The approach is evaluated through the analysis of a sample of the 7,000 Japanese verbs automatically described by the system.

1 Introduction

Natural language applications have shown the need for new kinds of lexical resources. Speech transcription or machine translation do not use hand crafted dictionaries as their basic source of knowledge any more, but lexical resources automatically built from the statistical analysis of very large corpora. More precisely, these systems do not usually integrate a component identified as a resource *per se* but make use of very large statistical sources of knowledge (generally called "language models") that incorporate different kinds of linguistic information. Models are generally not readable by humans and are very different from any human readable dictionary.

This does not mean that hand crafted dictionaries are now obsolete, since humans still need practical and usable dictionaries. As a consequence, there seem to be a big divide between these two kinds of lexical resources (those used by computers and those used by humans) although the work of lexicographers relies more and more on the automatic processing of very large corpora. Lexical descriptions produced by lexicographers are now generally established after taking into consideration corpus-based and statistical information.

© Springer International Publishing AG, part of Springer Nature 2018
A. Gelbukh (Ed.): CICLing 2016, LNCS 9623, pp. 195–207, 2018.
https://doi.org/10.1007/978-3-319-75477-2_12

In this paper we propose a model that takes into consideration very large corpora so as to obtain fine-grained information about lexical items. We focus on verbs since this category of words exhibit different features that make their description highly challenging. Like most lexical items, verbs can be ambiguous and one lexical item have most of the time different meanings (*i.e.* different word senses). Describing a verb and determining the different relevant word senses is known to be especially difficult and largely depends on the purpose of the lexical resource (for example, is the resource for a language learner or for a language expert?).

The same problem also arises when it comes to differentiate arguments and adjuncts. As said in [1]: "There are some very clear arguments (normally, subjects and objects), and some very clear adjuncts (of time and 'outer' location), but also a lot of stuff in the middle. Things in this middle ground are often classified back and forth as arguments or adjuncts depending on the theoretical needs and convenience of the author."

Following Manning, we support the idea of gradience in grammar and more generally in languages. Except for practical reasons (*e.g.* in the case of a paper dictionary), there is no reason to determine a fix number of word senses per word or to decide out of context what should be an argument or an adjunct. Of course someone elaborating a paper dictionary has to take this kind of decisions for obvious reasons, but it is not the case of modern dictionaries in electronic format. We believe that lexical descriptions can be more or less fine-grained depending on the goal or the application, and different lexical descriptions of a same lexical item can be equally valid (as long as they are linguistically motivated, of course).

In this paper, we describe a system able to dynamically produce different kinds of dictionaries depending on the user's need. The main source of information are large corpora gathered from the Web. The system collects different kinds of information on verbs and on their complements from these corpora and aims at producing meaningful lexical representations based on an accurate statistical analysis of these data. The end user can then explore more or less fine grained descriptions through different variable settings. Among the parameters that the end user can explore are the number of word senses per verb and the information taken into account to calculate the argumenthood of the different complement.

The system we have developed has been applied to a large corpus of Japanese news stories. Japanese offers specific and interesting challenges since arguments are specified by case particles that are most of the time ambiguous. Various other features (order of the constituents, zero anaphora, etc.) make Japanese a highly challenging language for NLP. In the course of this paper we present a complete system with a very large coverage since information is produce for more than 7,000 Japanese verbs with a high accuracy.

The paper is organized as follows. We first describe the state of the art in lexical acquisition. We then describe our approach to the problem, before giving some details on our experiments on Japanese. We conclude with an evaluation and a discussion on our results.

2   Previous Work

Previous work on the automatic acquisition of lexical data dates back to the early 1990s. The need for precise and comprehensive lexical databases was clearly identified for most NLP tasks (esp. parsing) and automatic acquisition techniques was then seen as a way to solve the resource bottleneck. However, first experiments [2,3] were limited (the acquisition process was dealing with a few verbs only and a limited number of predefined subcategorization frames). The approach was based on local heuristics and did not take into account the wider context.

The approach was then refined so as to take into account all the most frequent verbs and subcategorization frames possible [4–6]. A last step will consist in letting the system infer the subcategorization frames directly from the corpus, without having to predefined the list of possible frames. This approach is supposed to be less precise but most errors are automatically filtered since rare and unreliable patterns can be discovered by a linguistic and statistical analysis. Most experiments have been on verbs that have the most complicated subcategorization frames, but the approach can also be extended to nouns and adjectives [7].

Most developments so far have been done on English, but more and more experiments are now done for other languages as well. See for example, experiments on French [8], German [9] or Chinese [10], among many others. The quality of the result depends of course on the kind of corpus used for acquisition, and even more on the considered language and on the size of the corpus used. Dictionaries obtained with very large corpora form the Web generally give the best performances. The availability of accurate non lexicalized parser is also a key feature for the quality of the acquisition process.

As for Japanese, different experiments have been done in the past, especially by Kawahara and Kurohashi [11,12]. Their approach relies on the idea that the closest case component of a given predicate helps disambiguate its meaning, and thus serves as a clue to merge a set of predicate-argument structures into a case frame. Obtained case frames are further merged based on a similarity measure which combines a thesaurus-based similarity measure between lexical heads and a similarity measure between subcategorization patterns. Their resource has been successfully integrated to a dependency parser; however, we found it failed at describing the continuous aspect of lexical meaning (case frames are organized into a flat structure and no indication on the similarity between them is provided) as well as the continuous aspect of argumenthood (except for the closest case components, no indication on the importance of complements is provided).

3   Description of Our Approach

Although our approach has been applied and evaluated for Japanese, the theoretical framework to calculate the argumenthood of a complement or the structure of lexical entries is partially language independent (although actual case or function markers are of course language dependent and have to be specified for each language considered).

3.1   Calculating the Argumenthood of Complements

We suppose a list of verbs along with their complements that have been auto-matically extracted from a large representative corpus. In our framework, a complement is a phrase directly connected to the verb (or is, in other words, a dependency of the verb), while the verb is the head of the dependents. In what follows we assume that complements are in fact couples made of a head noun and a dependency marker, generally a preposition or a case particle (in the case of Japanese, we will have to deal with case particles but the approach can be generalized to languages marking complement through other means).

Different proposals have been made in the past to model the difference between arguments and adjuncts. For example, [13,14] try to validate linguistic criteria with statistical measures. [1] proposes to estimate the probability of a subcategorization frame associated to a verb. Lastly, [15] following [16] propose to characterize the link between verbs and complements based on productivity measures.

Building on these previous works, we propose a new measure combining the prominent features described in the literature. Our measure is derived from the famous TF-IDF weighting scheme used in information retrieval, with the major difference that we are dealing with complements instead of terms, and with verbs instead of documents. We chose this measure for two main reasons:

1. it is a well documented statistical measure, widely used, and which has already proven effective in numerous information retrieval tasks;
2. it implements common rules of thumb for distinguishing between arguments and adjuncts.

The measure applied to a verb and a complement is thus the following:

$$\text{argumenthood}_{v,c} = (1 + \log \text{count}(v,c)) \log \frac{|V|}{|\{v' \in V : \exists(v',c)\}|} \tag{1}$$

where c is a complement (*i.e.* a tuple made of a lexical head and a case particle); v is a verb; $\text{count}(v,c)$ is the number of cooccurrences of the complement c with the verb v; $|V|$ is the total number of unique verbs; $|\{v' \in V : \exists(v',c)\}|$ is the number of unique verbs cooccurring with this complement.

The first part of the formula, $1 + \log \text{count}(v,c)$, takes into account the cooc-currence frequency of a verb with a given complement (which transposes the idea that arguments are more closely linked to a given verb than a random adjunct). The second part of the formula, $\log \frac{|V|}{|\{v' \in V : \exists(v',c)\}|}$ takes into account the dis-persion of a complement, that is, its tendency to appear with different kinds of verbs. In other words, the more a complement is used with different verbs the more likely it is an adjunct.

The proposed measure assigns a value between 0 and 1 to a complement. 0 corresponds to a prototypical adjunct; 1 corresponds to a prototypical argument.

3.2   Enriching Verb Description Using Shallow Clustering

We introduce a method for merging verbal structures, that is a verb and a set of complements, into minimal predicate-frames using reliable lexical clues. We call this technique *shallow clustering*.

A verbal structure corresponds to a specific sense of a given verb; that is the sense of the verb is given by the complements selected by the verb. Yet a single verbal structure contains a very limited number of complements. So as to obtain a more complete description of the verb sense we propose to merge verbal structures corresponding to the same meaning of a given verb.

Our method relies on two principles:

1. Two verbal structures describing the same verb and having at least one common complement might correspond to the same verb meaning;
2. Some complements are more informative than others for a given verb sense.

As for the second principle, the measure of argumenthood, introduced in the previous section, serves as a tool for identifying the complements which contribute the most to the verb meaning. Our method merges verbal structures in an iterative process; beginning with the most informative complements (*i.e.* complements yielding the highest argumenthood value). Algorithm 1 describes our method for merging verbal structures.

3.3   Modeling Word Senses Through Hierarchical Clustering

We propose to cluster the minimal predicate-frames built during the *shallow clustering* procedure into a dendrogram structure. A dendrogram allows one to define an arbitrary number of classes (using a threshold) and thus fit in with the goal to model a continuum between ambiguity and vagueness. A dendrogram is usually built using a hierarchical clustering algorithm and a distance matrix as the input of the hierarchical clustering algorithm. So as to measure the distance between minimal predicate-frames, we propose to represent minimal predicate-frames as vectors which would serve as the parameters of a similarity function.

We must first define a vector representation for the minimal predicate-frames. Following Partee and Mitchell, we suppose that "the meaning of a whole is a function of the meaning of the parts and of the way they are syntactically combined" [17] as well as all the information involved in the composition process [18]. The following equation summarizes the proposed model of semantic composition:

$$p = f(\mathbf{u}, \mathbf{v}, R, K) \tag{2}$$

where $\mathbf{u}$ and $\mathbf{v}$ are two lexical components; R is the syntactic information associated with $\mathbf{u}$ and $\mathbf{v}$; K is the information involved in the composition process. Following the principles of distributional semantics [19,20] lexical heads can be represented in a vector space model [21]. Case markers (or prepositions) can be used as syntactic information. Finally, we propose to utilize our argumenthood measure to initialize the K parameter as it reflects how important a complement is for a given verb.

Data: A collection **W** of verbal structures $(\mathbf{v}, \mathbf{D})$ with $\mathbf{v}$ a verb and $\mathbf{D}$ a collection of verbal complements

Result: A collection **W'** of minimal predicate-frames

$W' \longleftarrow [\,];$

foreach *verb* $\mathbf{v}$ *such as* $\exists (v, D) \in W$ **do**

    `/* Be C the set of complements c cooccurring with v          */`

    $C \longleftarrow \{c : c \in D \wedge \exists (v, D) \in W\};$

    `/* Be C' the elements of C sorted by decreasing TF-IDF value    */`

    $C' \longleftarrow [c : c \in C \wedge \mathrm{argumenthood}(v, C'[i]) \geqslant \mathrm{argumenthood}(v, C'[i{+}1])];$

    foreach *complement* $\mathbf{c'}$ *of* C' **do**

        `/* Be D' a partial classification of v                         */`

        $D' \longleftarrow [\,];$

        foreach $D : \exists (v, D) \in W$ **do**

            if $c' \in D$ **then**

                add all the complements in D to D';

                remove (v, D) from W;

            end

        end

        foreach $D : \exists (v, D) \in W$ **do**

            if $\forall c \in D \longrightarrow c \in D'$ **then**

                add all the complements in D to D';

                remove (v, D) from W;

            end

        end

        if $|D'| \geqslant 2$ **then**

            add (v, D') to W';

        end

    end

end

Algorithm 1. Shallow clustering of verbal structures

Each minimal predicate-frame is transformed into a vector. The distance between two vectors will represent the dissimilarity between two occurrences of a same verb. Among the very large number of metrics available to calculate the distance between two vectors, we chose the cosinus, since it is (as for the TF-IDF weighting scheme) simple, efficient and perfectly suited to our problem.

The Eq. (3) shows how the cosinus can be calculated for two vectors $\mathbf{x}$ et $\mathbf{y}$ (the cosinus varies between 0 for orthogonal vectors to 1 for identical vectors).

$$cos(\mathbf{x}, \mathbf{y}) = \frac{\mathbf{x} \cdot \mathbf{y}}{|\mathbf{x}||\mathbf{y}|} = \frac{\sum_{i=1}^{n} x_i y_i}{\sqrt{\sum_{i=1}^{n} x_i^2} \sqrt{\sum_{i=1}^{n} y_i^2}} \tag{3}$$

Hierarchical clustering is an iterative process which clusters the two most similar elements of a set into a single element and repeats until there is only one element left. Yet different clustering strategies are possible (*e.g.* single linkage, complete linkage, average linkage). So as to select the best strategy (that is the one which would preserve the most the information from the distance matrix) we propose to apply the cophenetic correlation coefficient as shown in Eq. (4).

$$c = \frac{\sum_{i=1}^{n} \sum_{j=i+1}^{n} (\mathbf{D}_{i,j} - \bar{d})(\mathbf{C}_{i,j} - \bar{c})}{\sqrt{\sum_{i=1}^{n} \sum_{j=i+1}^{n} (\mathbf{D}_{i,j} - \bar{d})^2 \sum_{i=1}^{n} \sum_{j=i+1}^{n} (\mathbf{C}_{i,j} - \bar{c})^2}} \tag{4}$$

Where $\mathbf{D}$ is the initial distance matrix and $\mathbf{C}$ is the cophenetic matrix, that is the inter-cluster distances in the dendrogram. The clustering strategy that maximizes the cophenetic correlation coefficient should be selected.

4 Experiment

4.1 Acquisition and Preprocessing of Textual Data

A large collection of Japanese text is gathered from a selection of news websites using RSS feeds and a set of XPath expressions so as to discard HTML markup and unrelevant content (*e.g.* navigation menu). To comply with external NLP tools (*i.e.* a POS tagger and a parser), specific preprocesses are then applied to the raw textual data: fullwidth form conversion, sentence splitting, etc. In the end, our corpus is made of more than 294 millions characters.

4.2 Verbal Structure Extraction

The next step is to apply a parser to the corpus in order to get a syntactic analysis of the data. The parser must be unlexicalized since our goal is to calculate the argumenthood of the different complement (an unlexicalized parser attaches all the complement to the verb without making any different between arguments and adjuncts). The two most well-known parsers for Japanese are KNP[1] [22] and CaboCha[2] [23] (we are aware other parsers exist as well like EDA[3] [24]). In this work, we have decided to use CaboCha, for efficiency, among other reasons. Since CaboCha is faster than KNP [25], it seems more convenient to process large textual data. We use the default settings.

CaboCha is based on a tagger called MeCab[4] [26] that requires a dictionary of surface forms for tagging. Among the different possible dictionaries, we chose IPAdic [27], that is the recommended dictionary for MeCab.

The next step consists in extracting verbs, along with their complements and case particles. The process is mainly based on the part-of-speech tags from MeCab and on the syntactic links identified by CaboCha. The identification of verbs is not straightforward since some ambiguities or language specificities have to be avoided but we will not detail this part here. As for the particles, nine simple case markers can be identified: が (*ga*), を (*wo*), に (*ni*), へ (*he*), で (*de*), から (*kara*), より (*yori*), まで (*made*), and と (*to*) [28]. As for complex

---

[1] http://nlp.ist.i.kyoto-u.ac.jp/index.php?KNP.
[2] http://taku910.github.io/cabocha/.
[3] http://plata.ar.media.kyoto-u.ac.jp/tool/EDA/home_en.html.
[4] http://taku910.github.io/mecab/.

case markers, the list is not fixed and lots of variation exist among grammars and linguists. In our case we are partly dependent on the list of case markers defined in IPAdic. However, following previous descriptions like [29] or [28], we consider some particles as simple surface variants, like に対して (*ni tai site*), にたいして (*ni tai site*), に対し (*ni tai si*), に対しまして (*ni tai simasite*), and にたいしまして (*ni tai simasite*), that correspond to に対して (*ni tai site*). Last but not least, we consider まで (*made*) as a case particle (and contrary to the choice made by IPAdic). In the end, we have a list of 30 (simple and complex) case particles. Lastly, lexical heads of complement are extracted. When the head can be identified as a named entity, it is replaced by a generic tag; numerical expressions are also replaced by a more generic tag <NUM>.

Finally, verbal structures exhibiting suspicious patterns (*e.g.* two complements marked as direct objects of the verb) are filtered out. In the end we obtain more than 5.5 millions of verbal structure, corresponding to a bit more than 10,000 verbs.

4.3 Measuring the Degree of Argumenthood of Complements

We apply our measure of argumenthood of complements to those obtained during the process of extraction of verbal structures. Here complements are couples made of a lexical head and a case marker. We could assess the suitability of our approach by comparing, for a given verb, complements with the highest degree of argumenthood with complements with the lowest degree of argumenthood. As for the verb 積む (*tumu*, to load, to pill up), the complements with the highest degree of argumenthood all disambiguate the meaning of the verb: 研鑽を[積む] (*kensan wo [tumu]*, to study hard), 修業を[積む] (*syuugyou wo [tumu]*, to train), 経験を[積む] (*keiken wo [tumu]*, to gain experience), etc. On the other hand, none of the complements with the lowest degree of argumenthood help disambiguating the meaning of the verb: 〜氏が[積む] (*si ga [tumu]*, Mr. ... + nominative), <NUM> 人で[積む] (*<NUM>-nin de [tumu]*, <NUM> people + manner), etc.

4.4 Shallow Clustering of the Verbal Structures

We apply our shallow clustering method to the collection of verbal structures. After filtering of the most unfrequent minimal predicate-frames, we obtain a collection of almost 386,000 minimal predicate-frames, associated with 7,116 unique lemmas.

4.5 Hierarchical Clustering

Minimal verbal classes must then be merged gradually through hierarchical clustering, as shown in Sect. 3.3.

Using the cophenetic correlation coefficient we found out that the average linkage was the best clustering strategy. Hierarchical clustering output can be represented as a dendogram, as shown on Fig. 1.

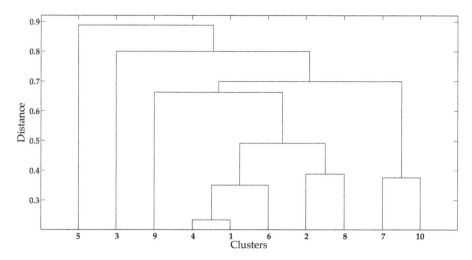

Fig. 1. Dendrogram obtained after the hierarchical clustering of the ten first minimal predicate-frames of the verb 積む (*tumu*).

Each verb is thus described through a variable number of word senses, each word sense being itself defined by the different arguments attached to the verb. It is possible to explore the resource by navigating the hierarchy of word senses, *i.e.* by examining more or less fine-grained description. The interface making it possible to explore the data as well as some comments for the evaluation of the resource are presented in the following section.

5   Results and Discussion

Lexical resource are traditionally evaluated through a comparison with a reference resource [4,6]. Although this approach is intuitive, it is not satisfactory since different lexical descriptions can be valid for a same lexical item, as it has been shown previously. We have nevertheless done a quick comparison with a manually built resource: IPAL [30]. The results show similar results as for other languages, *e.g.* [8]: our system is able to discriminate relevant word senses, but the description is not fully similar to the one obtained with IPAL. Some differences are caused by errors (parsing errors, undetected ambiguities, etc.) but most differences reveal in fact new or interesting word senses that are not described as such in IPAL.

However, the major novelty of our approach is the description of lexical item through a double continuum. In order to make the resource usable by humans, it is necessary to develop a visual interface allowing the end user to navigate the data and explore them in more details. In doing so, it is possible to have a more fine grained comparison with IPAL, which is not only based on a static arbitrary output of the system.

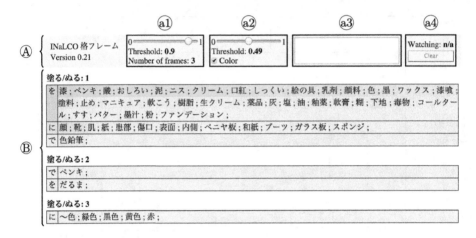

Fig. 2. Screen capture of our graphical interface – Ⓐ control panel: ⓐ1 slider for partitioning of sub-entries; ⓐ2 slider for selection of complements; ⓐ3 notification zone; ⓐ4 sub-entry identifier. – Ⓑ sub-entry panel.

Figure 2 shows the proposed graphical user interface to access our resource. The control panel is the same for all the sub-entries in the resource. It allows the end-user to navigate the data thanks to:

- the slider for the sub-entry partitioning threshold (a1);
- the slider for the complement selection threshold (a2);
- the notification zone (a3);
- the sub-entry identifier (a4).

Beyond the comparison with IPAL, a thorough evaluation of the resource has been done. According to lexicographers the linguistic description is globally accurate, relevant and motivated. The idea of a dynamic description is well received although it increases the complexity of the proposed description.

Some interesting features have also been noted from a linguistic point of view, like the fact that the first divide, for most verbs, is between concrete and abstract usages (and not between more notional word senses). Case particle are more discriminative than head nouns for the definition of word senses at the construction level, which is not too surprising. The most fine grained classes often correspond to very specific word senses that are not always registered in more general resources. A wide variety of idioms and frozen expressions can also be found at this level, making it possible to semi-automatically enriching existing resources.

6   Conclusion

We have shown in this study that it is now possible to develop new kinds of lexical resources based on continuous models and on the automatic analysis of very large corpora. Our system allows one to produce resources that can be finely tuned depending on the task and the expected precision of the foreseen result. Thanks to a relevant user interface, our resource is to the best of our knowledge the first one implementing the idea of a continuum-based representation usable by a professional lexicographer (contrary to most approaches producing a machine-coded language model that is unreadable by humans). The degree of argumenthood and the number of entries per verb can be tuned very easily through the interface and first experiments have proven that valuable, linguistically-motivated distinctions can be observed this way. The full integration of different versions of our resource in natural language processing tools (*e.g.* syntactic parsers) remains to be done. In the near future, different strategies will be explored to determine the best description for a given task, following previous experiments coupling lexical acquisition with practical tasks.

Acknowledgement. Pierre Marchal's research has been partially supported by a national "contrat doctoral" from the ministry of research.

References

1. Manning, C.D.: Probabilistic syntax. In: Bod, R., Hay, J., Jannedy, S. (eds.) Probabilistic Linguistics, pp. 289–341. MIT Press, Cambridge (2003)
2. Manning, C.D.: Automatic acquisition of a large subcategorization dictionary from corpora. In: Proceedings of the Meeting of the Association for Computational Linguistics, pp. 235–242 (1993)
3. Brent, M.R.: From grammar to lexicon: unsupervised learning of lexical syntax. Comput. Linguist. **19**, 203–222 (1993)
4. Briscoe, T., Carroll, J.: Automatic extraction of subcategorization from corpora. In: Proceedings of the 5th ACL Conference on Applied Natural Language Processing, Washington, DC., pp. 356–363 (1997)
5. Korhonen, A.: Subcategorization acquisition. Ph.D. thesis, University of Cambridge (2002)
6. Korhonen, A., Briscoe, T.: Extended lexical-semantic classification of English verbs. In: Moldovan, D., Girju, R. (eds.) Proceedings of the HLT-NAACL 2004: Workshop on Computational Lexical Semantics, Boston, Massachusetts, USA, 2–7 May 2004, pp. 38–45. Association for Computational Linguistics (2004)
7. Preiss, J., Briscoe, T., Korhonen, A.: A system for large-scale acquisition of verbal, nominal and adjectival subcategorization frames from corpora. In: Proceedings of the Meeting of the Association for Computational Linguistics, Prague, pp. 912–918 (2007)
8. Messiant, C., Poibeau, T., Korhonen, A.: LexSchem: a large subcategorization lexicon for French verbs. In: Proceedings of the International Conference on Language Resources and Evaluation, LREC 2008, Marrakech, Morocco, 26 May–1 June 2008 (2008)

9. im Walde, S.S., Müller, S.: Using web corpora for the automatic acquisition of lexical-semantic knowledge. JLCL **28**(2), 85–105 (2013)

10. Han, X., Zhao, T., Qi, H., Yu, H.: Subcategorization acquisition and evaluation for Chinese verbs. In: Proceedings of the 20th International Conference on Computational Linguistics, COLING 2004, Stroudsburg, PA, USA. Association for Computational Linguistics (2004)

11. Kawahara, D., Kurohashi, S.: Case frame compilation from the web using high-performance computing. In: Proceedings of the 5th International Conference on Language Resources and Evaluation, pp. 1344–1347 (2006)

12. Kawahara, D., Kurohashi, S.: A fully-lexicalized probabilistic model for Japanese syntactic and case structure analysis. In: Proceedings of the Human Language Technology Conference of the North American Chapter of the ACL, pp. 176–183 (2006)

13. Merlo, P., Esteve Ferrer, E.: The notion of argument in prepositional phrase attachment. Comput. Linguist. **32**(3), 341–377 (2006)

14. Abend, O., Rappoport, A.: Fully unsupervised core-adjunct argument classification. In: Proceedings of the 48th Annual Meeting of the Association for Computational Linguistics, pp. 226–236 (2010)

15. Fabre, C., Bourigault, D.: Exploiter des corpus annotés syntaxiquement pour observer le continuum entre arguments et circonstants. J. Fr. Lang. Stud. **18**(1), 87–102 (2008)

16. Fabre, C., Frérot, C.: Groupes prépositionnels arguments ou circonstants: vers un repérage automatique en corpus. In: Actes de la 9éme conférence sur le Traitement Automatique des Langues Naturelles (TALN 2002), pp. 215–224 (2002)

17. Partee, B.H.: Lexical semantics and compositionality. In: Gleitman, L.R., Liberman, M. (eds.) An Invitation to Cognitive Science, Second edition, vol. 1: Language, pp. 311–360. MIT Press, Cambridge (1995)

18. Mitchell, J.: Composition in distributional models of semantics. Ph.D. thesis, University of Edinburgh (2011)

19. Firth, J.R.: A synopsis of linguistic theory 1930-1955. In: Studies in Linguistic Analysis, Philological Society, Oxford. Reprinted in F.R. Palmer (ed. 1968), Selected Papers of J.R. Firth 1952-1959, pp. 1–32. Longman, London (1957)

20. Harris, Z.S.: Distributional structure. Word **10**, 146–162 (1954)

21. Salton, G., Wong, A., Yang, C.S.: A vector space model for automatic indexing. Commun. ACM **18**(11), 613–620 (1975)

22. Kurohashi, S., Nagao, M.: KN parser: Japanese dependency/case structure analyzer. In: Proceedings of the Workshop on Sharable Natural Language Resources, pp. 48–55 (1994)

23. Kudo, T., Matsumoto, Y.: Japanese dependency analysis using cascaded chunking. In: The 6th Conference on Natural Language Learning (CoNLL-2002), pp. 63–69 (2002)

24. Flannery, D., Miyao, Y., Neubig, G., Mori, S.: A pointwise approach to training dependency parsers from partially annotated corpora. J. Nat. Lang. Process. **19**(3), 167–191 (2012)

25. Sasano, R., Kawahara, D., Kurohashi, S., Okumura, M.: koubun/zyutugo-kou-kouzou kaiseki sisutemu knp no nagare to tokutyou (2013)

26. Kudo, T., Yamamoto, K., Matsumoto, Y.: Applying conditional random fields to Japanese morphological analysis. In: Proceedings of EMNLP 2004, pp. 230–237 (2004)

27. Asahara, M., Matsumoto, Y.: Ipadic version 2.7.0 users manual (2003)

28. Nihongo Kizyutu Bunpô Kenkyûkai: gendai nihongo bunpou 2: dai-3-bu kaku to koubun; dai-4-bu voisu (2009)
29. Martin, S.E.: A Reference Grammar of Japanese. Yale University Press, New Haven, London (1975)
30. Information-technology Promotion Agency (IPA): IPA lexicon of the Japanese language for computers, basic Japanese verbs (1987)

Description of Turkish Paraphrase Corpus Structure and Generation Method

Bahar Karaoglan[1] , Tarık Kışla[1]([⊠]) , and Senem Kumova Metin[2]

[1] Ege University, İzmir, Turkey
{bahar.karaoglan, tarik.kisla}@ege.edu.tr
[2] Izmir University of Economics, İzmir, Turkey
senem.kumova@ieu.edu.tr

Abstract. Because developing a corpus requires a long time and lots of human effort, it is desirable to make it as resourceful as possible: rich in coverage, flexible, multipurpose and expandable. Here we describe the steps we took in the development of Turkish paraphrase corpus, the factors we considered, problems we faced and how we dealt with them. Currently our corpus contains nearly 4000 sentences with the ratio of 60% paraphrase and 40% non-paraphrase sentence pairs. The sentence pairs are annotated at 5-scale: paraphrase, encapsulating, encapsulated, non-paraphrase and opposite. The corpus is formulated in a database structure integrated with Turkish dictionary. The sources we used till now are news texts from Bilcon 2005 corpus, a set of professionally translated sentence pairs from MSRP corpus, multiple Turkish translations from different languages that are involved in Tatoeba corpus and user generated paraphrases.

Keywords: Turkish · Paraphrase · Corpus generation

1 Introduction

Corpora are the fundamental elements in the development and/or testing of the studies in the fields of Natural Language Processing, Information Retrieval and Computational Linguistics. Building a corpus is much more than putting bunch of texts together. Many things are needed to be decided and done. Some of which are: Sources of the texts and the size of the corpus, the structure to store the texts (e.g. plain texts, html, database), the fields to tag, the metrics to assess the quality of the corpus.

In this paper we present our efforts in developing a Turkish paraphrase corpus, PARDER, hoping to serve for studies in machine translation, summarization, language generation, automatic assessment of answers to essay type questions and plagiarism detection. We first give relevant work done in other languages and Turkish then, describe each of the above points within the context of our studies in the following sections.

© Springer International Publishing AG, part of Springer Nature 2018
A. Gelbukh (Ed.): CICLing 2016, LNCS 9623, pp. 208–217, 2018.
https://doi.org/10.1007/978-3-319-75477-2_13

2 Relevant Work

The literature of paraphrase studies cover numerous corpora that are constructed based on a variety of methods and are holding various features. This variety complicates the classification and the comparison of the paraphrase corpora.

In this study, we will exemplify the notion of paraphrase corpus and paraphrase corpus construction methods in literature by the use of paraphrase corpora listed in Table 1. Henceforth, the corpora in Table 1 will be mentioned by regarding abbreviations given in the second column of the table.

Table 1. Paraphrase Corpora

Corpus	Abbreviation
Microsoft Research Paraphrase [1]	MSRP
User Language Paraphrase [2]	ULP
Question Paraphrase [3]	QP
SIMILAR [4]	SIMILAR
Regneri & Wang [5]	R&W
WiCoPaCo [6, 7]	WiCoPaCo
Question Corpus [8]	QC
FAQFinder [9]	FAQ
Turkish Paraphrase [10]	TP

The paraphrase corpora and different corpus construction methods will be examined and compared based on the following features (if available in the previous studies): text sources of the corpus, pre-processing of the source data, identification of candidate pairs, the annotation of paraphrase/non-paraphrase pairs, the corpus size.

2.1 Text Sources of the Corpus

The paraphrase corpora may be built using different sources such as comparable texts (e.g. similar news texts from different news papers), parallel texts (e.g. answers to the same question) or text corrections (e.g. revisions in Wikipedia articles). The data collected from different sources are parsed into paraphrase units that may be a sentence, paragraph or a collection of words. Following, the candidates are selected from the source data units.

MSRP can be considered as the first major public paraphrase corpus. The candidate sentence pairs are obtained from web-sources. This corpus is not only served for numerous studies but also served as a data source for other paraphrase corpora (e.g. SIMILAR). Sentence is the paraphrase unit in ULP as in the MSRP corpus. ULP corpus involves the sentence pairs that are collected by iSTART system. iSTART system is defined to be an educative support system where the students generate paraphrases to a given set of target sentences to improve their linguistic abilities. The sentence pairs in ULP are the student and target sentence pairs. In QP corpus, question-answer sentence pairs from WikiAnswers are used. The paraphrasing pairs are selected from different questions that are directed to the same answer. R&W corpus is compiled from the subtitles of the TV series House MD. The paraphrasing unit is the sentence, and the corpus involves 14735 sentences from 160 documents. WiCoPaCo corpus is built using the revision logs of Wikipedia. In Wikipedia, the users may add a new content (record) to the system and/or may correct an existing record to improve the quality. In the construction of WiCoPaCo corpus, the sentences that are retrieved from different revisions of the same record have been accepted as paraphrasing candidates. Similar to the WiCoPaCo corpus, the data source of the QC corpus is the log-archive of an online encyclopaedia, Encarta, which includes both the queries and the answers. The paraphrasing unit is again sentence in QC corpus.

A web based-question answering system FAQFinder provides the source data of FAQ corpus. In FAQFinder system, the questions of the users are replied by an answer of a previous similar question. The system involves over 600 files of frequently asked questions.

TP corpus is the first developed Turkish paraphrase corpus that is drawn from four different sources: 1. Two different Turkish translations of Ernest Hemingway's "For Whom the Bell Tolls", 2. Two different subtitles of the film: "The silence of the Lambs", 3. Turkish-English sentence pairs used for machine translation, 4. Paraphrased news sentences. The corpus contains only paraphrased sentences annotated with word and phrase alignments.

2.2   Pre-processing of the Source Data

In paraphrase corpus construction, the source data is commonly pre-processed. The pre-processing involves tasks such as spelling correction, stop word removal and ignoring some parts of source texts. For example, two types of corrections: removal of multiple spaces between tokens and appending the full stop character when there isn't one at the end of the sentence; are performed on the source data of ULP. In QP corpus, following the removal of stop words, TreeTagger and Porter stemmer are used in determination of the roots and the stems of the words in source data, respectively. R&W corpus involves the pre-processing tasks such as spelling correction, POS tagging, named entity recognition, and co-reference resolution. In TP corpus, a tool that is developed by the researchers entails the source texts.

2.3   Identification of Candidate Pairs

The pairs in paraphrase corpora may be determined randomly or using a procedure by researchers. The most comprehensive study on identification of pairs is presented in the construction of MSRP corpus. The identification procedure in MSRP corpus has two steps. Firstly, the source data is filtered by two different criteria sets. Secondly, a support vector machine (SVM) is employed in classification of the filtered pairs. The pairs that are classified as paraphrase by SVM are accepted as candidates of the corpus. This method increases the amount of true paraphrase pairs in MSRP corpus.

In R&W, WiCoPaCo, QC and FAQ corpora, several procedures such as ordering the events in source texts, longest common subsequence filtering, removal of short sentences, are performed. On the other hand, in ULP, QP, SIMILAR and TP corpora, the candidate pairs are selected randomly.

2.4 The Annotation of Paraphrase/Non-paraphrase Pairs

The pairs in a corpus may be annotated in binary mode or within a predefined interval by human annotators. In MSRP corpus, sentence pairs are annotated in binary mode as paraphrase or non-paraphrase as it is common in many other studies (e.g. QP corpus). In ULP corpus, the degree of paraphrasing is annotated within 1–6 interval. Moreover, the quality of user-generated paraphrases is described considering 10 dimensions (garbage, frozen expression, irrelevant, elaboration, writing quality, semantic similarity, lexical similarity, entailment, syntactic similarity, paraphrase quality) of paraphrasing. In R&W corpus, the annotation covers four intervals corresponding to paraphrase, containment, backwards containment, unrelated or invalid tags.

The other important issues in annotation of pairs are the annotation units, the number of annotators and the method to measure annotator agreement. For example, in MSRP, ULP, QP and TP corpus the annotation unit is sentence. Though the annotation unit is considered as the word in SIMILAR corpus. In construction of most of the corpora (e.g. MSRP, R&W, TP), two annotators are employed in classification of pairs as paraphrase or non-paraphrase and one other annotator is employed to resolve the conflictions. In annotation of WiCoPaCo and SIMILAR corpora, the four researchers of the study, a group of six students are assigned respectively. The annotator agreement is given as 63% in SIMILAR corpus.

The average annotator agreement for WiCoPaCo is calculated based on four different evaluation criteria. The highest average agreement value in presented as 0.65 on the semantic differences criterion and the average kappa value f all criteria is given as 0.62. In R&W and TP corpus, the kappa values are reported as 0.55 and 0.416.

2.5 The Size of Corpus

MSRP corpus contains about 5801 sentence pairs where 67% of corpus is annotated as paraphrase. SIMILAR corpus includes 700 pairs from MSRP corpus where the number of paraphrase and non-paraphrase pairs is balanced. The total number of sentence pairs that are annotated in 6 scales is 1998 in ULP corpus.

QP corpus includes 7434 pairs in which there are 1000 question sentences and their corresponding paraphrases. R&W corpus is built from 200 millions of candidate pairs. 1992 of pairs in corpus are annotated as gold standard where 158 pairs are paraphrases, 238 are containments and 194 are tagged as related. 1402 of pairs in R&W corpus is accepted to be unrelated. In WiCoPaCo corpus, 200 pairs are paraphrases and 200 are non-paraphrases. QC corpus contains manually annotated 67379 pairs in which 65750 of pairs are paraphrases and 1629 are non-paraphrases. FAQ and TP corpora involve 679, 1270 paraphrase pairs respectively.

3 PARDER Corpus

In this section, the sources for building the PARDER corpus, the structure of the corpus database, the annotation scheme will be introduced respectively.

3.1 Corpus Sources

The sources for the sentence pairs to be included in the corpus are: Bilcon2005 [11] corpus, translated sentence pairs from MSRP corpus [12, 13], Tatoeba corpus [14], and human generated paraphrase sentences.

Bilcon2005 Turkish news corpus contains 209.305 news, which are collected from five different Turkish news web sources throughout the year 2005. In our study, news texts from Bilcon2005 are parsed into sentences, normalized, and short sentences with less than 3 words and duplicates are removed. For each topic, we then calculated the distance of each sentence to all other sentences in the same topic with 3 different distance metrics: Chebyshev, correlation and Euclid. For each sentence, we selected two sentences with the least distance calculated by each metric as the paraphrase candidates to be marked by the human annotators via a user interface with five marking options: paraphrase, encapsulated, encapsulating, opposite, not-paraphrase. In the user interface, the target statement (sentence) is shown on top of the screen and three annotators labeled each candidate sentence in the list with a label provided via pull down menu.

MSRP corpus is the other source employed in building PARDER. A set of randomly chosen 2000 sentence pairs, which are 60% paraphrase and 40% not paraphrase, from MSRP corpus are translated by a professional. These translated sentences are re-labeled by human annotators considering the fact that translations may not be in parallel with the original labels.

Tatoeba corpus is referred as a multilingual sentence dictionary consisting of cross language translations of sentences between language pairs. In this study, the multiple translations of English sentences to Turkish are utilized. Most of these sentences were very short and different translations of the same English sentence varied by only one word. After eliminating sentences less than 5 words and multiple translations that vary with one word, we obtained 114 paraphrase sentence pairs for PARDER.

Volunteering people, researchers and Turkish Language Education students are provided with a list of sentences from which they can choose to rephrase. Currently there are 2419 Sentence pairs labeled as paraphrase and 1602 labeled as non-paraphrase by the annotators (Table 2).

Table 2. Content Summary of The Corpus

Source	Paraphrase	Non-paraphrase
Bilcon	1.005	802
Tatoeba	114	–
MSRP-translated	1.200	800
User-generated	100	–
Total	2.419	1.602

3.2   Corpus Database Structure

The corpus is created on a database structure to make the labeling as flexible and informative as possible. The database consists of 7 tables: *Documents, Sentences, Similarity, Words, Dictionary, Word-Relation* and *Word-Meaning* table.

Documents Table keeps the physical location (path), type (e.g. news, translation, user generated, etc.), author, source of the document in which the sentence appears.

Similarity Table stores similarity and type information for each sentence pair in the corpus keeping the overall similarity score of the sentence pairs. Similarity column contains a code that represents the number of similarity values assigned by the annotators. For example, the value 21000 of similarity column, as given in Fig. 1, means that two annotators marked the regarding sentence pair as paraphrase; one has marked them as encapsulated. Type column stores the binary decision that denotes if the sentence pair is paraphrase (1) or not (0).

2	1	0	0	0
#of paraphrase judgement	#of encapsulated judgement	#of encapsulating judgement	#of opposite judgement	#of non-paraphrase judgement

Fig. 1. The structure of similarity value (in similarity table).

Words Table has an entry for each word in each sentence holding part of speech tag (POS), morphological analysis, named entity (NE) tag and its position within the sentence. This table is related to the *Word-Meaning Table* enabling the detection of synonym, antonym and meaning relations between the words.

Dictionary Table keeps the words as they appear in the dictionary and their ids. For the fact that words may have more than one meaning another table, *Word-Meaning Table*, is held to keep the meanings attached to part of speech.

Word-Relation Table has an entry for each word in the database to keep synonyms, antonyms and etc. Figure 2 shows snapshot of the database for two sentences drawn from Document #1 of Milliyet newspaper with ids #5 and #121 that are tagged as paraphrases. Sentences #5 and #121 together with their English translations are:

#5: Stabilize yolda aşırı hız nedeniyle sürücünün kontolünden çıkan otobüs, yol kenarında bulunan Aras Nehri'ne uçtu. (Eng: Due to excessive speed on the stabilized road, getting out of control of the driver, the bus flew into Aras River, which is running on the side of the road.)

#121: Ancak, stabilize yolda aşırı hızla ilerleyen otobüs, sürücünün direksiyon hakimiyetin kaybetmesi sonucunda nehre uçtu. (Eng: However, over speeding bus on the stabilized road, as a result of loosing control of the wheel by the driver flew into the river.)

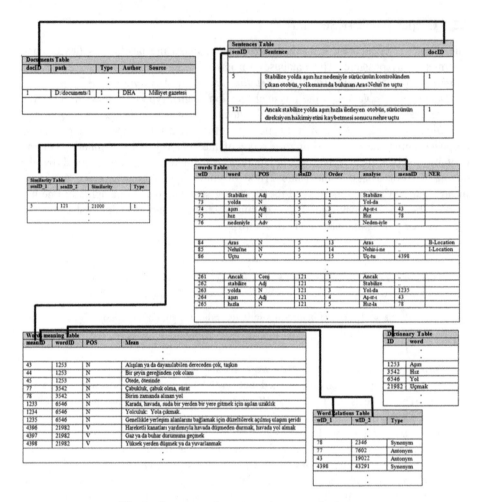

Fig. 2. Sentence pair structure in corpus database.

3.3   Annotation Scheme

The sentence pairs drawn from different sources are all normalized and transferred to the database of the annotation software developed by the researchers. Three human annotators tagged the sentence pairs on pentad scale (paraphrase, encapsulating, encapsulated, non-paraphrase and opposite). This analysis is also done on binary-scaled (paraphrase/non-paraphrase) judgment. For binary scaled judgment, those sentence pairs with similarity score in the similarity table, greater than a predefined threshold score 12 are considered as paraphrase and the rest as non-paraphrase. Similarity score is calculated using similarity value. For example, if we assume that similarity value is 21000, similarity score will be 14 ($2 \times 5 + 1 \times 4 + 0 \times 3 + 0 \times 2 + 0 \times 1$). The reliability of the agreement between the annotators is assessed by Fleiss Kappa values [15] that are given in Table 3.

Table 3. The results of the Fleiss Kappa Analysis

Scale	kappa	SE_{fleiss}*	z	CI_{lower}	CI_{upper}	p
Binary Scaled	0.634	0.004	148.11	0.626	0.642	0.00
Pentad Scaled	0.671	0.003	228.30	0.665	0.667	0.00

*Standard error (SE) values are calculated using formula given in [16].

4   The Analysis of PARDER

PARDER corpus is analyzed considering the averaged values of sentence-pair based attributes: number of the words in sentence (*SL*), the ratio of common words (*MW*), the ratio of sentence lengths (*LS*), the ratio of common consequent sets (*MB*) [17] and the ratio of sequencing (*OW*) [17]. Table 4 gives the overall analysis results of PARDER corpus (P: Paraphrase NP: non-Paraphrase). In the analysis, the attribute values are obtained individually for each set of sentence pairs that are extracted from three different sources.

Table 4. The analysis results of PARDER corpus using syntactic features

Attributes	Sources					
	Bilcon2005		MSRP		Tatoeba	User-generated
	P	NP	P	NP	P	P
SL	18.8	17.5	17	15.5	7	13.9
LS	0.72	0.62	0.85	0.79	0,88	0,83
MW	0.34	0.15	0.55	0.33	0.36	0,31
MB	0.15	0.03	0.31	0.14	0.13	0.24
OW	0.29	0.11	0.50	0.29	0.32	0.37

The semantic attributes that we consider are the difference in the polarity (*DP*) (positive/negative) and tenses between the sentences (*DT*). Table 5 shows the statistics related to the mentioned attributes.

Table 5. The analysis results of PARDER corpus using semantic features

Attributes	Sources					
	Bilcon2005		MSRP		Tatoeba	User-generated
	P	NP	P	NP	P	P
DP	34%	80%	20%	60%	21%	12%
DT	7%	14%	6%	11%	5%	8%

5 Conclusion

In this paper, we have presented Turkish paraphrase corpus, PARDER, and described the corpus construction steps in our on-going project. The data sources, candidate selection procedure, data structure and annotation scheme of the PARDER corpus are introduced. The corpus currently contains nearly 4000 sentences with the ratio of 60% paraphrase and 40% non-paraphrase sentence pairs.

PARDER corpus is built to serve for many purposes in the field of language processing. We aim to increase the corpus size to 6000 pairs and make it accessible on web for researchers in future.

Acknowledgement. This work is carried under the grant of TÜBİTAK – The Scientific and Technological Research Council of Turkey to Project No: 114E126, Using Certainty Factor Approach and Creating Paraphrase Corpus for Measuring Similarity of Short Turkish Texts and Ege University Scientific Research Council Project No 2015/BİL/034, Developing a Paraphrase Corpus for Turkish Short Text Similarity Studies.

References

1. Dolan, B., Quirk C., and Brockett C.: Unsupervised construction of large paraphrase corpora: exploiting massively parallel news sources. In: Proceedings of the 20th International Conference on Computational Linguistics. Association for Computational Linguistics (2004)
2. McCarthy, P.M., McNamara, D.: The user-language paraphrase challenge. In: Special ANLP Topic of the 22nd International Florida Artificial Intelligence Research Society Conference, Florida (2008)
3. Bernhard, D., Gurevych, I.: Answering learners' questions by retrieving question paraphrases from social Q&A sites. In Proceedings of the Third ACL Workshop on Innovative Use of NLP for Building Educational Applications, pp. 44–52. Association for Computational Linguistics, Stroudsburg (2009)
4. Rus, V., Lintean, M., Moldovan, C., Baggett, W., Niraula, N., Morgan, B.: SIMILAR Corpus: a resource to foster the qualitative understanding of semantic similarity of texts. In: LREC, pp. 50–59 (2012)
5. Regneri, M., Wang, R.: Using discourse information for paraphrase extraction. In: Proceedings of the 2012 Joint Conference on Empirical Methods in Natural Language Processing and Computational Natural Language Learning, pp. 916–927 (2012)
6. Max, A., Wisniewski, G.: Mining naturally-occurring corrections and paraphrases from Wikipedia's Revision History. In: LREC (2010)
7. Dutrey, C., Bouamor, H., Bernhard, D., Max, A.: Local modifications and paraphrases in Wikipedia's revision history. Procesamiento del Lenguaje Natural **46**, 51–58 (2010)
8. Zhao, S., Zhou, M., Liu, T.: Learning question paraphrases for QA from encarta logs. In: IJCAI (2007)
9. Lytinen, S., Tomuro, N.: The use of question types to match questions in FAQFinder. AAAI Spring Symposium on Mining Answers from Texts and Knowledge Bases (2002)
10. Demir, S., El-Kahlout, I.D., Unal, E., Kaya, H.: Turkish paraphrase corpus. In: LREC, pp. 4087–4091 (2012)

11. Can, F., Kocberber, S., Baglioglu, O., Kardas, S., Ocalan, H.C., Uyar, E.: New event detection and topic tracking in Turkish. J. Am. Soc. Inform. Sci. Technol. **61**(4), 802–819 (2010)
12. Dolan, W., Brockett, C.: Automatically Constructing a Corpus of Sentential Paraphrases. In Third International Workshop on Paraphrasing (2005)
13. Brockett, C., Dolan, W.: Support vector machines for paraphrase identification and corpus construction. In: Third International Workshop on Paraphrasing (IWP2005) (2005)
14. Tiedemann J.: Parallel data, tools and interfaces in OPUS. In Proceedings of the 8th International Conference on Language Resources and Evaluation (LREC) (2012)
15. Fleiss, J.L.: Measuring nominal scale agreement among many raters. Psychological Bull. **76**(5), 378–382 (1971)
16. Fleiss, J.L., Nee, J.C., Landis, J.R.: Large sample variance of kappa in the case of different sets of raters. Psychological Bull. **86**(5), 974–977 (1979)
17. Islam, A., Inkpen, D.: Semantic text similarity using corpus-based word similarity and string similarity. ACM Trans. Knowl. Discov. Data (TKDD), **2**(2), Article 10, 25 pages (2008)

Extracting Terminological Relationships
from Historical Patterns of Social Media Terms

Daoud Daoud[1(✉)] and Mohammad Daoud[2]

[1] Department of Computer Science, Princess Sumaya University
for Technology, Amman, Jordan
d.daoud@psut.edu.jo
[2] Department of Computer Science,
American University of Madaba, Madaba, Jordan
m.daoud@aum.edu.jo

Abstract. In this article we propose and evaluate a method to extract terminological relationships from microblogs. The idea is to analyze archived
microblogs (tweets for example) and then to trace the history of each term.
Similar history indicates a relationship between terms. This indication can be
validated using further processing. For example, if the term t1 and t2 were
frequently used in Twitter at certain days, and there is a match in the frequency
patterns over a period of time, then t1 and t2 can be related. Extracting standard
terminological relationships can be difficult; especially in a dynamic context
such as social media, where millions of microblogs (short textual messages) are
published, and thousands of new terms are coined every day. So we are
proposing to compile nonstandard raw repository of lexical units with unconfirmed relationships. This paper shows a method to draw relationships between
time-sensitive Arabic terms by matching similar timelines of these terms. We
use dynamic time warping to align the timelines. To evaluate our approach we
elected 430 terms and we matched the similarity between the frequency patterns
of these terms over a period of 30 days. Around 250 correct relationships were
extracted with a precision of 0.65. These relationships were drawn without using
any parallel text, nor analyzing the textual context of the term. Taking into
consideration that the studied terms can be newly coined by microbloggers and
their availability in standard repositories is limited.

Keywords: Arabic terminology · Time-sensitive terminology
Social media analysis · DTW · Preterminology

1 Introduction

Internet users are producing 10,000 Microposts on average every second [1]. Microposts are short messages containing few sentences written in several languages; often,
publishers use concise and informal style. These messages tend to talk about time
sensitive topics [2, 3]. Microposts are rich with terminology [4], not only old and well
defined terminology but also newly coined terms [5]. These new terms are created to
represent a new concept (an event - related concept). The community of internet users
and microbloggers observe a concept and then they will suggest lexical units to

© Springer International Publishing AG, part of Springer Nature 2018
A. Gelbukh (Ed.): CICLing 2016, LNCS 9623, pp. 218–229, 2018.
https://doi.org/10.1007/978-3-319-75477-2_14

describe this concept. Within a short period of time a most popular (frequent) term will be attached to that concept. So the community will produce a suitable lexical unit (term) to a new concept in a natural, dynamic and online fashion. In this paper we are interested in finding the relationships of these terms to position them correctly in a wider terminological network.

Building and maintaining an up-to-date terminological repository is very important for several applications [6], like, machine translation, information retrieval, publishing... However, finding terminology (terms and relationships) is a very difficult task [7], especially for poorly equipped languages, and when the domain is active and changing everyday (new concepts appear every day). Classical approaches in building terminology depend heavily on terminologists and subject-matter experts [8, 9]. This approach is very expensive, and it achieves poor coverage because terminologists have limited capability and subject matter experts are rare for lively domains. Statistical approaches on the other hand are less expensive, but they need large and processed corpus/corpora. Besides, statistical methods might find a list of candidate terms without relationships, so mapping these terms into a lexical network can be difficult. Microblogs are massive and can solve the problem of the availability of a large textual corpus, however, these microblogs have little textual context (A micropost in Twitter is 140 characters only) and they are usually poorly written.

We are working on analyzing terms that appear on microblogs over a period of time to monitor their evolutions. Our idea is that terms with similar histories (frequency patterns over a period of time) are probably similar. For example, if two terms are peaking at the same dates then there is a chance that these terms are used by the internet users synonymously. That way rather than using textual context (which is almost nonexistent in microblogs), we are using historical context to relate between terms. And that will make social media a legitimate source of terminology (terms and relationships). However, building a terminological database is still challenging, because terminology must be standardized and must have a formal body to approve it. We are proposing to extract unconfirmed terminological relationships (preterminology relationships) [10–12] rather than standard terminology, preterminology is considered as raw material for terminology that can be refined to produce standard terminology. Matching timelines for terms is a classical time series problem, where time series are searched for similarities. There are several approaches to search time series. The performance of these approaches depends on the application [13]. We use an algorithm originally used for speech recognition called Dynamic Time Warping algorithm [14] with a normalized Euclidean distance function. This approach will not only measure the distance between timelines, but it will consider the slight shifts in the timelines. And this is very suitable for our application because related terms might not peak on the exact same days.

This article is organized as follows; the following section introduces terminology evolution in big data. The third section presents our approach in finding historical similarity between terms. The fourth section shows our data collection method. The fifth section shows the experimental results and evaluation, and finally we will draw some conclusions.

2 Terminology in Big Data

Extracting knowledge from big data, such as social media generated content, is attracting more and more researchers. Data provided by internet users can be used to find new trends, prevent diseases, combat crimes, and predict future events. Luckily, big data holders may provide free Application Programming Interfaces (APIs) to ease the process of capturing raw data. Twitter.com [15] provides an easy to use API that provides programmatic access to read and write Twitter data, author a new Tweet, read author profile and follower data. Twitter is a successful online social networking service that enables users to send and read short 140-character messages called "tweets". The API makes it possible to read and save the tweet with its metadata (publisher, time, date, geographical location…). Many applications have succeeded in collecting and analyzing twitter data. Because of its popularity and the diversity of its users, Twitter data can be used to study comprehensive and global trends and users' behaviors, as it represents a good share of social media. Extracting terminology or other lexical semantic information from Twitter is an ambitious task. Many succeeded in extracting trending lexical units, finding collocations, classifying tweets, and analyzing positivity/negativity of terms and tweets [16–18]. These attempts consider the textual context of lexical units. However, there is a limitation in using Twitter's textual context as natural language processing of tweets is difficult, because of the following factors:

- Tweets are short messages with little context
- Grammar is not properly followed
- Spelling variations of words
- Tweets contain many spoken words (dialectic terminology)
- A tweet may contain a mix of terms in several languages

Therefore, while there is a need and a possibility to extract real-time terminology from tweets, attempts are faced with challenges. And one can only extract frequent lexical units without actual terminological information and relationships. The difficulty increases when dealing with a poorly equipped language such as Arabic, where there is a gap between the Arabic resources and the language used by Twitter's users.

Traditional terminology has a specific definition that disallows the integration of unconventional resources and unstructured raw lexical units. That is why a classical standard terminological repository suffers from a lack of linguistic and informational coverage, and it cannot deal flexibly with hidden or absent terminology. The problem becomes even harder when the community does not have classical linguistic resources. We suggest extracting unconfirmed terminological relationship between terms (preterms). These possible relationships will have a similarity weight indicating a possible relationship (translation, synonymy, acronym, etc.). Next section describes a method to analyze the history of each preterm to find possible similarities.

3   Timeline Similarity

We monitor the frequencies of terms each day (how many times a particular term appeared in Twitter in a particular day). We create a timeline for each term. The timeline shows the daily frequencies of the term. These timelines illustrate the peaks, bottoms, and possibly the coining date of a term.

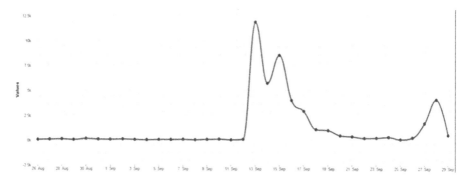

Fig. 1. Timeline example for the term "اقتحام لاقصى" (Al-Aqsa raid)

Figure 1 shows the timeline for the term "اقتحام لاقصى" (Al-Aqsa raid). We can see that the term has peaked on 13 September 2015 with 11,800 frequencies. The tool used to produce the figure is an online Arabic social media monitoring platform built by the second author. We studied a small set of Arabic terms and we observed similarities between the timelines of related terms. Figure 2 shows the timelines of "اسعار النفط, اوبك" (OPEC, oil prices). We can see similarity in the frequencies during the period from 25 August 2015 to 29 September 2015.

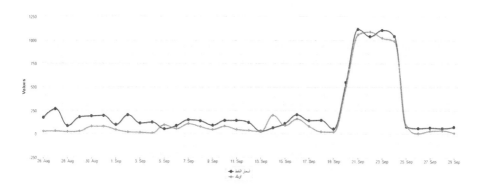

Fig. 2. Timelines of "اسعار النفط, اوبك"

The similarity between terms can occur due to one of the following reasons:

1. Term collocation: terms that co-occur to convey certain meaning, Fig. 3 shows an example.
2. Event co-occurrence: separate events happened at the same time. Each event has related terms that might produce similar timelines.
3. Same event with different concepts (related terms); Fig. 4 shows an example.
4. Same or similar concept with different lexical units (translation, synonymy, Acronym, Hyponymy, Antinomy, Hypernymy). Figure 5 shows an example.

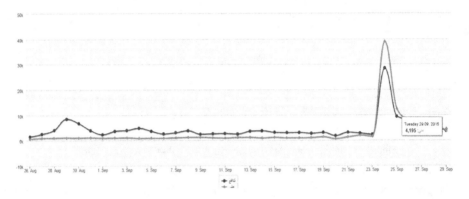

Fig. 3. Timeline example (Term collocation)

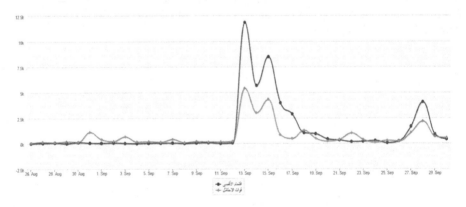

Fig. 4. Event co-occurrence

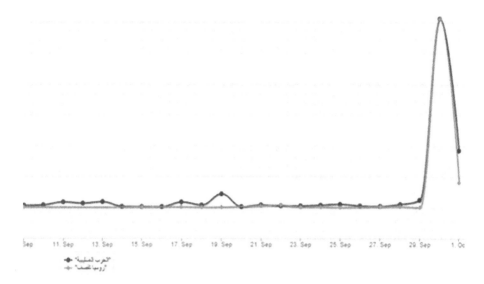

Fig. 5. Community generated synonym

Our objective based on these observations is to search for similar timelines to build a candidate set of relationships between new terms extracted from the community of Arabic social media.

3.1   Time-Series Similarity Search

Similarity search in timelines (time-series) is an interesting research direction to analyze stock prices data, weather forecast, biomedical measurements, etc. While there are several methods to find similarity between time series, the choice of a particular method is an application-dependent. Therefore, we are testing our hypothesis with a standard Dynamic Time Warping algorithm to measure the similarity between terms.

There are several approaches that depend on the application. In our case the approach we need to use must consider the following assumptions:

1. Suppose that t1 and t2 are two timelines for two terms. t1 and t2 are similar if they have similar shapes, regardless of the magnitudes. And even if the frequencies did not match exactly. For example, Fig. 4, from 12/9 to 17/9 shows different frequencies between the two timelines. However, the shapes are similar.
2. Similar terms might not peak in the exact same day. t1 could peak in a particular day and the other t2 might peek in the next day. t1 and t2 are considered similar if they have similar peaking patterns.
3. The presence of the peaks is more important that their magnitudes.

Dynamic time warping (DTW) is a technique that aligns two time series in which one time series may be "warped" by stretching or shrinking its time axis. This alignment can be used to find corresponding regions or to determine the similarity between the two time series. DTW focuses on aligning the peaks of the time lines without

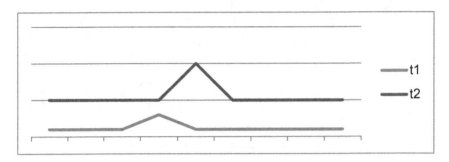

Fig. 6. Two similar time series

focusing on their magnitudes and it matches peaks even if they did not appear at the exact same time. This satisfies the assumptions mentioned above. DTW would consider t1 and t2 in Fig. 6 to be similar.

3.2   DTW Algorithm

DTW is a time series alignment algorithm that was originally used in voice recognition [14] It relates two time series of feature vectors by warping the time axis of one series onto another.

Given two time series X and Y, Where:

$$X = x_1 + x_2 + x_3 + \ldots + x_i + \ldots + x_n$$
$$Y = y_1 + y_2 + y_3 + \ldots + y_i + \ldots + y_n$$

Algorithm 1 will produce the cost of aligning X and Y (warping them) the cost will be low if the two time series are similar.

```
int standardDWT(X, Y) {
    // Where X = x₁ + x₂ + x₃ + ... + xᵢ + ... + xₙ and Y = y₁ + y₂ + y₃ + ... + yᵢ
    + ... + yₙ
        Create DTW[0..n, 0..m]
        Set the first row and column of DTW to infinity
        DTW[0, 0] = 0
        for i = 1 to n
            for j = 1 to m
                DTW[i, j] = d(X[i], Y[j])+ minimum(DTW[i-1, j] ,
                                                   DTW[i , j-1],
                                                   DTW[i-1, j-1])
        return DTW[n, m]
    }
```

Algorithm 1. Standard DWT

We start by filling a distance matrix DTW which has n × m elements; each element represents the warping distance between every two points in the time series. The warping distance between x_i and y_j is measured according to the following equation:

$$DTW(x_i, y_i) = d(x_i, y_i) + minimum(DTW(x_{i-1}, y_j),$$
$$DTW(x_i, y_{j-1}),$$
$$DTW(x_{i-1}, y_{j-1}))$$

Where $d(x_i, y_i)$ is a distance function to calculate the distance between x_i and y_i. This version of DTW satisfies the monotonicity, continuity, boundary constrains demonstrated by [14, 19, 20].

We use the Euclidian distance as a distance function between x_i, y_i. So the distance will be calculated as follows:

$$d(x_i, y_i) = | x_i - y_i |$$

Frequency reading must be normalized to achieve meaningful results and to give more importance to peaks in relation to the average readings of a particular timeline. A frequency reading f is measured according to this equation

$$Norm(f) = f - m$$

Where m is the average of frequencies for that term. The returned value from the algorithm indicates the cost of aligning the two normalized timelines. The similarity score described below indicates the possible similarity between the two timeline:

$$Similarity(X, Y) = 1 - cost/max(n, m)$$

Where cost is the returned value from the algorithm, n and m are the lengths of X and Y respectively. High similarity score means the probability that the two terms are related is high.

4   Data Collection

We are testing our approach with timelines collected by an online platform that addresses Arabic social media content and provides a powerful platform to collect, search, monitor and analyze social media content. Users can see the latest on the ground media from news events, as they happen. It enables users to find trends related to a certain event, person or brand. Top video, images, news stories are also displayed and linked to a specified search term, time frame or any customized filter.

The platform has many functions. However, we are interested in the production of timelines which are archived through the following steps:

1. Data collection: Arabic tweets are collected using Twitter API. The online platform receives live feed from Twitter. Any non-Arabic tweets will be filtered.

2. Indexing: tweets are analyzed and indexed according to the terms they carry. Arabic analysis component is used for stemming and tokenization.
3. Reporting: the platform reports the frequencies for each term per time interval. Thus, we can build a timeline for each term.

The online system is available currently at "http://45.33.23.107". We are using its produced timelines and terms for our experiment.

5   Experimentation and Evaluation

Arabic tweets collected by the online platform during the month of October 2015 were analyzed. We selected 430 timelines for the most popular terms in that month. Then we searched for similarities between these terms. The produced relationships were evaluated based on precision and recall.

5.1   Precision

We are trying to evaluate the precision of the similarity score according to this equation

$$\text{Precision} = C_{th}/T_{th}$$

Where C_{th} is number of correct relationships with a score greater than the threshold *th*. Relationship between t1 and t2 is considered correct if t1 and t2 are event related or if there is a terminological relationship (synonymy, acronym, hyponymy, antonymy, hypernymy) between them.

T_{th} is total number of produced relationships with a score that is greater that *th*.

When *th* is small the produced set of relationships increases but precision might decrease. When *th* = 0.85 the precision is 0.95. Figure 7 shows the precision in relation to the threshold.

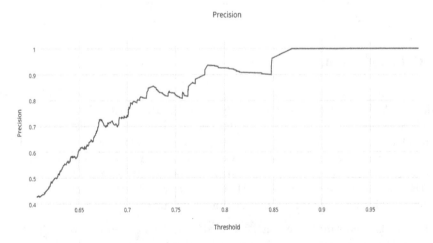

Fig. 7. Precision

As you can see the precision starts to decline when *th* is below 0.7. The similarity score proved to be a good indicator of a relationship between terms.

5.2 Recall

Recall is measured in terms of number of correct relationships extracted by our approach. When the threshold is 0.85 number of correct relationships is 250. Figure 8 shows the recall in relation to the threshold.

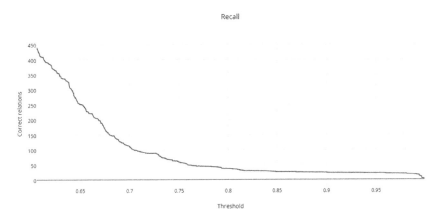

Fig. 8. Recall

When the threshold is 0.65 the precision is 0.62 and 250 correct relation were extracted from 430 terms.

5.3 Assessment and Sample Results

Our approach has correctly identified terminological relationships between time sensitive terms without analyzing the textual context; Table 1 shows sample results. Extracting relationships between terms is a challenging task that needs large corpora, and specialists. The challenge increases when the terms are time sensitive Arabic terms. Our approach extracted 400 of relationships from 430 terms with high precision; these relationships can be used in many applications, such as:

1. Extracted relationships can be post edited by specialists to enrich Arabic terminological databases.
2. Lexicon for social media analysis: auto microblogs classifications, auto tagging, sentiment analysis. In fact, we intend to use these relationships to dynamically extend a polarized lexicon for Arabic sentiment analysis.
3. These relationships can locate newly coined terms on the ontological map.

The approach will be used on a larger scale to automatically discover related terms on-the-fly by analyzing online microblog feeds. The importance of this approach is that

Table 1. Sample results

Term 1	Term 2	Similarity	Note
سقوط الطائرة	الطائرة الروسية	0.9004803	Correct (same event different concepts-related terms)
احتلال	الفلسطينية	0.868922991	Correct (same event different concepts-related terms)
الملك سلمان	خادم الحرمين	0.8493	Correct (synonym)
الكويت	تربية	0.847755429	Incorrect
الصهاينة	اسرائيل	0.842341	Correct (synonym)
احتلال	الصهاينة	0.809440709	Correct (synonym derived by community)
اقتحام	تدنيس	0.798213	Correct (synonym)
مستوطنين	احتلال	0.78232	Correct (synonym)
القدس	فلسطين	0.737100147	Correct (part-of)
موسكو	بشار الاسد	0.727927333	Correct (same event different concepts-related terms)
عدن	القدس	0.726857411	Incorrect
هدايا	الأردن	0.705413478	Incorrect
قصف النظام	القصف الروسي	0.688777279	Correct (same event different concepts-related terms)

it does not rely on textual context; in fact many extracted relations were between terms that did not appear in the same tweet. Most of the wrongly extracted relationships were between key terms describing two separate events that took place at the same time. These errors can be reduced when the timeline is longer (than 30 days).

6 Conclusions

We have presented an approach to extract terminological relationships between time-sensitive Arabic terms. Our hypothesis is that terms that have similar history (timeline) are similar or related. We used Dynamic Time Warping algorithm to measure the similarity between terms. Our experiment produced 250 correct relationships out of 430 terms with a precision of 0.65. The extracted information is crucial because it maps time-sensitive terms into a wider terminological map. The approach can be used to identify and connect terminology on-the-fly by analyzing microblogs feeds online.

References

1. internetlivestats. Internet Live Stats - Internet Usage & Social Media Statistics (2015). http://www.internetlivestats.com/one-second/. Accessed 1 Feb 2015
2. Grinev, M., et al.: Analytics for the realtime web. Proc. VLDB Endow. **4**, 1391–1394 (2011)
3. Kwak, H., et al.: What is Twitter, a social network or a news media? In: The 19th International World Wide Web (WWW) Conference. Raleigh, NC, USA (2010)

4. Uherčík, T., Šimko, M., Bieliková, M.: Utilizing microblogs for web page relevant term acquisition. In: van Emde Boas, P., Groen, F.C.A., Italiano, G.F., Nawrocki, J., Sack, H. (eds.) SOFSEM 2013. LNCS, vol. 7741, pp. 457–468. Springer, Heidelberg (2013). https://doi.org/10.1007/978-3-642-35843-2_39

5. Becker, H., Naaman, M., Gravano, L.: Beyond trending topics: real-world event identification on Twitter. In: Proceedings of the Fifth International Conference on Weblogs and Social Media (2011)

6. Daoud, M., Boitet, C., Kageura, K., Kitamoto, A., Mangeot, M., Daoud, D.: Building specialized multilingual lexical graphs using community resources. In: Lacroix, Z. (ed.) RED 2009. LNCS, vol. 6162, pp. 94–109. Springer, Heidelberg (2010). https://doi.org/10.1007/978-3-642-14415-8_7

7. Cabre, M.T., Sager, J.C.: Terminology: Theory, Methods, and Applications, vol. xii, 247 p. John Benjamins Publishing (1999)

8. Kim, Y.G., et al.: Terminology Construction Workflow for Korean-English Patent MT, vol. X, 5 p. MT Summit, Thailand (2005)

9. Hartley, A., Paris, C.: Multilingual document production: from support for translating to support for authoring. Mach. Transl. 12(1–2), 109–129 (1997)

10. Daoud, M., et al.: Constructing multilingual preterminological graphs using various online-community resources. In: The Eighth International Symposium on Natural Language Processing, SNLP 2009, Thailand, Bangkok. IEEE (2009)

11. Daoud, M., et al.: Constructing multilingual preterminological graphs using various online-community resources. In: the Eighth International Symposium on Natural Language Processing, SNLP 2009, Thailand, pp. 116–121 (2009)

12. Daoud, M., et al.: Passive and active contribution to multilingual lexical resources through online cultural activities. In: NLPKE 2010, Beijing, China, 4 p. (2010)

13. Agrawal, R., Faloutsos, C., Swami, A.: Efficient similarity search in sequence databases. In: Lomet, D.B. (ed.) FODO 1993. LNCS, vol. 730, pp. 69–84. Springer, Heidelberg (1993). https://doi.org/10.1007/3-540-57301-1_5

14. Sakoe, H., Chiba, S.: Dynamic programming algorithm optimization for spoken word recognition. IEEE Trans. Acoust. Speech Signal Process. 26, 43–49 (1978)

15. Twitter. Twitter (2015). twitter.com. Accessed 1 Feb 2015

16. Speriosu, M., et al.: Twitter polarity classification with label propagation over lexical links and the follower graph. In: Proceedings of the First Workshop on Unsupervised Learning in NLP, EMNLP 2011, pp. 53–63 (2011)

17. Zhao, W.X., et al.: Topical keyphrase extraction from Twitter. In: HLT 2011 Proceedings of the 49th Annual Meeting of the Association for Computational Linguistics: Human Language Technologies, vol. 1, pp. 379–388 (2011)

18. Daoud, D., Alkouz, A., Daoud, M.: Time-sensitive Arabic multiword expressions extraction from social networks. Int. J. Speech Technol. 19, 249–258 (2015)

19. Keogh, E., Ratanamahatana, C.: Exact indexing of dynamic time warping. Knowl. Inf. Syst. 7, 358–386 (2004)

20. Salvador, S., Chan, P.: Toward accurate dynamic time warping in linear time and space. Intell. Data Anal. 11, 561–580 (2007)

Adaptation of Cross-Lingual Transfer Methods for the Building of Medical Terminology in Ukrainian

Thierry Hamon[1,2(✉)] ⬡ and Natalia Grabar[3,4]

[1] LIMSI, CNRS, Université Paris-Saclay, 91405 Orsay, France
hamon@limsi.fr
[2] Université Paris 13, Sorbonne Paris Cité, 93430 Villetaneuse, France
[3] CNRS, UMR 8163, 59000 Lille, France
natalia.grabar@univ-lille3.fr
[4] Univ. Lille, UMR 8163 - STL - Savoirs Textes Langage, 59000 Lille, France
https://perso.limsi.fr/hamon/
http://natalia.grabar.perso.sfr.fr/

Abstract. An increasing availability of parallel bilingual corpora and of automatic methods and tools makes it possible to build linguistic and terminological resources for low-resourced languages. We propose to exploit corpora available in several languages for building bilingual and trilingual terminologies. Typically, terminology information extracted in better resourced languages is associated with the corresponding units in lower-resourced languages thanks to the multilingual transfer. The method is applied on corpora involving Ukrainian language. According to the experiments, precision of term extraction varies between 0.454 and 0.966, while the quality of the interlingual relations varies between 0.309 and 0.965. The resource built contains 4,588 medical terms in Ukrainian and their 34,267 relations with French and English terms.

Keywords: Cross-Lingual Transfer · Parallel Corpora · Terminology
Ukrainian

1 Introduction

Automatic acquisition of terminological resources has gone through a very active period and provides nowadays several automatic tools and methods [5,10,21] for several European languages and for Japanese. Nevertheless, other languages remain low-resourced and may require specific Natural Language Processing (NLP) developments, which must take into account morphological specificities of such languages [13,22], for instance. In the existing studies, statistical methods for extracting collocations and repeated segments are often exploited [6,9,22,26] and allow to extract results with reliable recall but low precision.

We propose to take advantage of the advanced research work done in languages like English or French, and to transpose it on low-resourced languages.

© Springer International Publishing AG, part of Springer Nature 2018
A. Gelbukh (Ed.): CICLing 2016, LNCS 9623, pp. 230–241, 2018.
https://doi.org/10.1007/978-3-319-75477-2_15

For such objectives, we propose to exploit the transfer methodology together with parallel and aligned corpora. We consider that the transfer methodology can be suitable for the objectives related to terminology extraction. The principle is the following. Suppose we have parallel and aligned corpora with two languages $L1$ and $L2$, and we have several types of syntactic or semantic annotations and information associated to $L1$. The transfer approach permits to transpose these annotations or information from $L1$ to $L2$, and to obtain in this way the corresponding annotations and information in the $L2$ text. From this point of view, $L1$ is considered as the source language while $L2$ is considered as the target language. This kind of approach is particularly interesting when working with low-resourced languages for which less tools and semantic resources are available. An increasing availability of parallel bilingual corpora, and of automatic methods and tools for their processing makes it possible to build linguistic and terminological resources using the transfer methodology [15,29]. Very few works have been done in this direction, and we assume they open novel and efficient ways for the processing of multilingual texts in particular from low-resourced languages [16,30]. Notice that the modeling of cross-language features aims at using language-independent features to create various types of annotations. Among such features, we can mention part-of-speech, semantic categories or even acoustic and prosodic features.

In the following of this paper, we start with the presentation of the motivation for this study (Sect. 2). We then present the material used for the acquisition of terminology (Sect. 3), and the methods designed for achieving this objective (Sect. 4). We discuss the results we obtain (Sect. 5), and conclude with directions for future work (Sect. 6).

2 Motivation and Rationale

The motivation of our work is double. We want (1) to design specific methods for the acquisition of terminological resources for low-resourced languages such as Ukrainian, and (2) to automatically build medical terminology for Ukrainian.

If little digitized resources are currently available for Ukrainian, terminological work is an active research area there, although it mainly is concerned with theoretical and linguistic issues. For instance, terminological descriptions are available for several specialized domains and languages: physics [35]; law [40]; computer science [7,34]; religion [39]; literature [24]; Crimean Tatar language [1,17]. Then, following recent research orientations, work on construction of electronic corpora [11,33], on their use for building of terminologies and dictionaries [31,32,38], and on transformation of traditional dictionaries in electronic format [37] appear. Still, little work is oriented on automatic building and utilization of terminologies. Since terminology extraction tools require Part-of-Speech (POS) tagging of texts and the two POS taggers developed for Ukrainian are not easily available or usable (the UGtag tagger [12] does not perform the syntactic and morphological disambiguation, and a module for the TNT POS tagger [3] is difficult to obtain), it remains difficult to use such tools for the pre-processing of corpora in order to prepare traditional terminology acquisition process. As for the utilization of terminologies, we can cite for instance localization of tools [25] and indexing of a language therapy terminology [36].

As we indicated above, we propose to take advantage of the advanced research work done in languages such as English or French, and to transpose it on lower-resourced languages using cross-lingual transfer methodology. We work with medical data and in three languages: Ukrainian, French, and English.

The general rationale of our approach is the following. Our work is based on exploitation of two kinds of corpora: Wikipedia in Ukrainian which provides several useful kinds of information (such as term labels and their codes) with a high level of quality, and parallel corpus *MedlinePlus*. Each corpus is exploited through dedicated methods. The *MedlinePlus* corpus provides the basis for the building of the terminology, while the *Wikipedia* corpus permits to enrich this information and helps the word-level alignment of the *MedlinePlus* corpus.

3   Material

3.1   Corpora

We use two kinds of corpora:

- *MedlinePlus:* parallel medical corpus from MedlinePlus[1]. The source data are built by MedlinePlus from the National Library of Medicine. They contain patient-oriented brochures on several medical topics (body systems, disorders and conditions, diagnosis and therapy, demographic groups, health and wellness). These brochures have been created in English and then translated in several other languages, among which French and Ukrainian. In Fig. 2, we present an excerpt from this corpus;

Fig. 1. Example of Ukrainian Wikipedia source page (for *Dwarfism*). The infobox with the coding is on the right.

---

[1] http://www.nlm.nih.gov/medlineplus/healthtopics.html.

– *Wikipedia:* medicine-related articles from Wikipedia. This corpus is extracted from the Ukrainian part of Wikipedia using medicine-related categories, such as *Медицина* (*medicine*) or *Захворювання* (*disorders*). The corpus potentially covers a wide range of medical notions. In Fig. 1, we indicate an example of source page composed of three parts: the navigation frame on the left, the text and explanations in the center, and the infobox with illustrations and codings on the right.

In Table 1, we indicate the size of the corpora. Not surprisingly, the *Wikipedia* corpus is much larger although only part of its information is exploited, as we explain it in Sect. 4.

Table 1. Size of the exploited corpora (Ukrainian = UK, French = FR, English = EN)

Corpus	Size (occ. of words)
Wikipedia/UKmed	246,368,411
MedlinePlus/UK	43,184
MedlinePlus/FR	53,067
MedlinePlus/EN	46,544

3.2 UMLS: Unified Medical Language System

The UMLS (Unified Medical Language System) [14] merges over 100 biomedical terminologies, such as international terminologies MeSH [19] and ICD [4]. Such terminologies may exist in several languages. For instance, French and English versions of MeSH are included in the UMLS. No terminologies in Ukrainian are part of the UMLS. Each UMLS term is provided with unique identifier, which allows finding the corresponding terms in other terminologies or languages.

3.3 Stopwords in Ukrainian

We use a list with 385 stopwords in Ukrainian (*на(on)*, *або(or)*, etc.) issued from an existing resource dedicated to the localization of graphical interfaces[2].

4 Methods

The methods we propose for the extraction of bilingual terminology are adapted to each kind of corpora and of data they contain: the *MedlinePlus* corpus (Sect. 4.1) and the *Wikipedia* corpus (Sect. 4.2). We then present their cross-fertilization (Sect. 4.3), and evaluation of the results (Sect. 4.4).

---

[2] https://github.com/fluxbb/langs/blob/master/Ukrainian/stopwords.txt.

4.1   Extraction of Bilingual Terminology from the *MedlinePlus* Corpus

Prior to the exploitation of the *MedlinePlus* data, the documents are first transformed in a suitable format: (1) the source pdf documents are converted in text format; (2) in each language, documents are segmented in paragraphs; (3) French/Ukrainian and English/Ukrainian alignments are generated, in which n_{th} paragraph from one language is put in front of the n_{th} paragraph from the other language; (4) alignments within pairs of languages are then verified manually. In Fig. 2, we present an excerpt from the English/Ukrainian aligned corpus.

English	*Ukrainian*
Cancer cells grow and divide more quickly than **healthy cells**. **Cancer treatments** are made to work on these **fast growing cells**.	**Ракові клітини** ростуть і діляться швидше, ніж **здорові клітини**. При лікуванні раку здійснюється вплив на ці **клітини, що швидко ростуть**.
- **Tiredness**	- **Втома**
- **Nausea** or **vomiting**	- **Нудота** або **блювота**
- **Pain**	- **Біль**
- **Hair loss** called **alopecia**	- **Втрата волосся**, що називається алопецією

Fig. 2. Example of the paragraph-aligned MedlinePlus corpus (English/Ukrainian)

In French and English, we can use terminology extraction tools for bootstrapping the acquisition of terminology. Hence, we use the YaTeA term extractor [2], that is applied to POS-tagged documents (TreeTagger [23], accompanied by Flemm morphological analyzer [18] in French). On the left of Fig. 2, we show in bold the extracted candidate terms in English. Then, exploitation of the *MedlinePlus* parallel and aligned corpus is performed in several ways (Fig. 3).

Transfer 1. The simplest situation is when the two aligned lines contain one candidate term in each language: these terms are recorded as candidates for the alignment. For instance, in Fig. 2, the pairs {*Tiredness, Втома*} and {*Pain, Біль*} are issued from this kind of alignment.

Transfer 2. When paragraphs contain complex expressions or sentences, the processing is done as follows (Fig. 3):

1. paragraph-aligned corpora are aligned at the word level using GIZA++ [20],
2. in each paragraph pair (French/Ukrainian and English/Ukrainian) of the word-aligned corpora, terms recognized in French and English are transferred on Ukrainian paragraph;
3. extracted alignments are recorded as candidates for building the terminology.

For instance, in Fig. 2, the term *Cancer cells* is automatically extracted from the English corpus. GIZA++ proposes that *Cancer cells* is aligned with *Ракові клітини*. Thus, through the word-aligned text, we can propose that *Cancer cells*

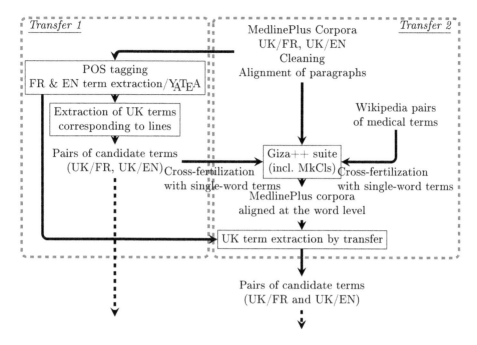

Fig. 3. Extraction of medical terms from the *MedlinePlus* corpora (Ukrainian = UK, French = FR, English = EN)

is the translation of *Ракові клітини*. This processing is performed on two pairs of languages (French/Ukrainian and English/Ukrainian).

As indicated in Table 1, our corpora are rather small for statistical alignment performed by GIZA++. For this reason, we provide GIZA++ with a bilingual dictionary in order to help alignment at the word level. Besides, we remove term pairs in which Ukrainian unit, corresponding to stopwords, is aligned as candidate term with French or English terms.

4.2 Extraction of Bilingual Terminology from the *Wikipedia* Corpus

The *Wikipedia* corpus is used to complete and to help the method applied to the *MedlinePlus* corpus. The exploited content is extracted from infoboxes (on the right in Fig. 1) and is reachable through the MediaWiki source code of Wikipedia. This provides labels of medical terms in Ukrainian and their MeSH codes. The process is the following (Fig. 4):

1. the infobox content is parsed with the Perl module `Text::MediawikiFormat`[3] which provides label and MeSH code of terms,
2. the MeSH code is used to query the UMLS, and to get the corresponding French and English terms,
3. the term pairs French/Ukrainian and English/Ukrainian are associated and provide good candidates for the bilingual terminology building.

---

[3] http://search.cpan.org/~szabgab/Text-MediawikiFormat.

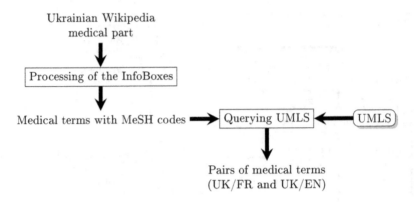

Fig. 4. Extraction of medical terms from Wikipedia (Ukrainian = UK, French = FR, English = EN)

For instance, in Fig. 1, the term *нанізм* is extracted, as well as its MeSH code D004392. In the UMLS, the corresponding English terms are *dwarfism* and *nanism*, while the corresponding French term is *nanisme*. Notice that similar method has been used for building of medical terminology in the Arabic language [28]. This part of the method exploits specific and intentionally created content and provides reliable information.

4.3 Cross-Fertilization

Two kinds of cross-fertilization between the two methods (Sects. 4.1 and 4.2) are performed:

- Wikipedia terms are used to enrich the extracted terminology,
- single-word terms extracted from *Wikipedia* and/or the Transfer 1 method in *MedlinePlus* are provided to GIZA++, as additional bilingual dictionary, in order to help the alignment of *MedlinePlus* at the word level.
  In our best setting, both resources of single-word term pairs are used.

4.4 Evaluation

Evaluation is handled manually in order to check whether candidate terms extracted for building bilingual terminologies are correct. The evaluation is performed by a native Ukrainian speaker, non expert in medicine but with several years experience in medical informatics. Terms are validated independently in each language. Besides, bilingual and trilingual relations between the Ukrainian, English and French terms are also evaluated. Thus, precision of results can be computed, *i.e.* the ratio between correct answers and all the answers.

5   Results and Discussion

Table 2 presents results and precision of the terms extracted by the three methods. Table 3 presents results and precision concerning the pairs and triples of terms.

Table 2. Evaluation of terms extracted (Ukrainian = UK, French = FR, English = EN)

Source	UK		FR		EN	
	#terms	Prec.	#terms	Prec.	#terms	Prec.
Wikipedia	357	1	1,428	1	3,625	1
MedlinePlus$_{Transfer1}$	436	0.966	316	0.971	354	0.989
inexact match		0.998		0.987		0.997
MedlinePlus$_{Transfer2}$	9,040	0.454	3,671	0.674	3,597	0.761
inexact match		0.840		0.726		0.799
Total	9,529	0.481	5,200	0.769	7,335	0.883
Total of correct terms	4,588		3,998		6,476	

Table 3. Evaluation of pairs and triples (Ukrainian = UK, French = FR, English = EN)

Source	UK/FR		UK/EN		UK/FR/EN		Total	
	#rel.	Prec.	#rel.	Prec.	#trpl.	Prec.	#trpl.	Prec.
Wikipedia	1,515	1	3,789	1	28,840	1	28,840	1
MedlinePlus$_{Transfer1}$	63	0.937	115	0.965	282	0.954	460	0.954
inexact match		0.984		1		0.982		0.987
MedlinePlus$_{Transfer2}$	3,724	0.309	4,745	0.401	4,724	0.419	13,218	0.381
inexact match		0.751		0.840		0.586		0.724
Total	3,798	0.318	4,819	0.41	33,845	0.918	42,462	0.807
Total of correct relations	1,207		1,974		31,086		34,267	

5.1   Extraction of Bilingual Terminology from the *Wikipedia* Corpus

Exploitation of Wikipedia infoboxes allows to collect 357 Ukrainian medical terms among which 177 are single-word terms. By querying the UMLS with MeSH codes, those terms are associated with 1,428 French terms (including 339 single-word terms) and 3,625 English terms (including 448 single-word terms). Higher number of French and English terms, compared to the number of Ukrainian terms, is due to synonyms proposed by MeSH. As for the pairs of terms, we obtain 1,515 Ukrainian/French and 3,789 Ukrainian/English term pairs, including, 270 and 405 pairs between single-word terms, respectively. Since each Ukrainian term is associated with at least one French and English term, it allows to build 28,840 triples. We assume that precision of this terminology is 1 because the source information is created manually and is highly reliable.

5.2 Extraction of Bilingual Terminology from the *MedlinePlus* Corpus

Use of the Transfer 1 method allows to extract 436 Ukrainian terms with a high precision (0.966). These terms are associated with 316 French and 354 English terms, within 282 triples between Ukrainian, French and English terms, with 0.954, 0.937 and 0.965 precision, respectively. Within this set, 63 pairs exist only between Ukrainian and French terms, and 115 pairs only between Ukrainian and English terms. Then, the Transfer 2 method allows to collect 334 Ukrainian/French term pairs (including 108 pairs with single-word terms) and 380 Ukrainian/English term pairs (including 135 pairs with single-word terms). We observe that these relations can involve synonyms in either language: {*фаллопієва труба, trompes de fallope/trompe utŕine*} (*fallopian tube*), {*втрата слуху/втрачається слух, hearing loss*}, {*втома, fatigue/tiredness*}. Besides, in Ukrainian, several inflectional forms can be associated to a given English or French form: {*вагітність, pregnancy*} and {*вагітності, pregnancy*}.

As the precision values suggest, the Transfer 1 method generates few errors. Their analysis indicates that they are mainly due to partial or non-literal translations: {*появу виразок у роті, mouth sores*} – lit. *(appearance of) mouth sores*, – lit. *you can sleep*. The silence of the method can be explained by two reasons: (1) Again translation can prevent the Transfer 1 method from extracting terms in French or English. For instance, *Soins* is the French translation of *Your care*, which gives simple term in one language and complex term in another language and makes impossible their extraction by the Transfer 1 method. The Transfer 2 method will solve this problem; (2) However, the main reason of the silence is the incapacity of the term extractor to identify French or English terms because of errors in the POS tagging.

As for the Transfer 2 method, we present the results obtained when the pairs of single-word terms from the *MedlinePlus* corpus and *Wikipedia* are used to help the GIZA++ alignment. Then, the Transfer 2 method allows to extract 9,040 Ukrainian terms with 0.454 precision (exact match). Precision of French and English terms is higher: 0.674 and 0.761 respectively (exact match). Moreover, number of French and English terms is dramatically lower (about -45% and -40%) than in Ukrainian: the rich morphology of Ukrainian provides indeed several inflected forms for a given term ({*напад, нападу*} – *attack*, {*припадків, припадки*} – *seizure*, – *bones*). Besides, we can also extract synonymous terms ({*приступам, припадків*} – *attacks/seizures*, {*биття, удару*} – *beats*). Precision of inexact match, when the correct answer is included or includes the candidate term, is much higher and gains up to 0.40 points in Ukrainian and 0.05 in French and English. We assume this difference on Ukrainian is mainly due to the improvement of the alignment quality. As for the interlingual relations, the Transfer 2 method collects 3,724 pairs of Ukrainian/French terms with 0.309 precision, 4,745 pairs of Ukrainian/English terms with 0.401 precision and 4,724 triples with 0.419 precision.

An analysis of the results shows that most of the errors are due to the alignment problems. Indeed, we observe that when the alignment is correct, the Ukrainian terms are correctly extracted by the transfer methods. Otherwise, the errors occur.

As we indicated, the *MedlinePlus* corpus contains patient-oriented brochures, which are not highly specialized. Yet, most of the extracted terms are specific to the medical domain ({*трахеотомiею, tracheotomy*}), {*фактори ризику, risk factors*}, {*шприца, syringe*}, {*холестерину, cholesterol*}). The extracted terms can also refer to close and approximating notions which reflect this type of documents: interesting observation is that some French and English terms correspond to propositions in Ukrainian: pain).

Finally, all the methods combined allow to build a terminological resource containing 4,588 Ukrainian medical terms and their 34,267 relations with French and English terms.

6   Conclusion and Future Work

In this work, we propose to exploit two kinds of freely available multilingual corpora in French, English and Ukrainian. Each corpus is exploited with appropriate methods which allow extracting candidate terms and creating Ukrainian/French and Ukrainian/English term pairs. In particularly, French and English corpora are processed with NLP and term extraction tools. Then, thanks to the transfer methods these terms are transposed on Ukrainian. We also propose to use existing terminologies and to exploit simple terms for improving the alignment performed at the word level with GIZA++.

Our future work will address enrichment of the created resource with terms from other corpora. Besides, in the *Wikipedia* corpus, we can use other codes, such as those from MKX-10 (ICD10) or MedlinePlus. This will also augment the coverage of the term pairs extracted in the current work. Another perspective of this work is the improvement of the bilingual alignment of documents at the word level, which is currently a major source of errors. In that respect, we plan to investigate the use of other alignment algorithms, such as FastAlign [8] or the Lingua::Align toolbox [27].

Acknowledgments. This work is funded by the LIMSI-CNRS AI project *Outiller l'Ukranien*.

References

1. Alieva, V.: Onomatopoeic words in the Crimean Tatar language. Uchenye zapiski **18**(57), 8–11 (2005)
2. Aubin, S., Hamon, T.: Improving term extraction with terminological resources. In: Salakoski, T., Ginter, F., Pyysalo, S., Pahikkala, T. (eds.) FinTAL 2006. LNCS (LNAI), vol. 4139, pp. 380–387. Springer, Heidelberg (2006). https://doi.org/10.1007/11816508_39

3. Babych, S., Eberle, K., Babych, B.: Development of hybrid machine translation systems for under-resourced languages: automated creation of lexical and morphological resources for MT. In: Applied and Literary Translation and Interpreting: Theory, Methodology, Practice. p. 5. Kyiv, Ukraine, April 2013

4. Brämer, G.: International statistical classification of diseases and related health problems. Tenth revision. World Health Stat. Q. **41**(1), 32–6 (1988)

5. Cabré, M., Estopà, R., Vivaldi, J.: Automatic Term Detection: A Review of Current Systems, pp. 53–88. John Benjamins, Amsterdam (2001)

6. Delač, D., Krleža, Z., Šnajder, J., Dalbelo Bašić, B., Šarić, F.: *TermeX*: a tool for collocation extraction. In: Gelbukh, A. (ed.) CICLing 2009. LNCS, vol. 5449, pp. 149–157. Springer, Heidelberg (2009). https://doi.org/10.1007/978-3-642-00382-0_12

7. Dmytruk, V.: Typological features of word-formation in computing, the internet and programming in the first decade of the XXI century. In: УДК, pp. 1–11 (2009)

8. Dyer, C., Chahuneau, V., Smith, N.A.: A simple, fast, and effective reparameterization of IBM model 2. In: NAACL/HLT, pp. 644–648 (2013)

9. Grigonyte, G., Rimkute, E., Utka, A., Boizou, L.: Experiments on Lithuanian term extraction. In: NODALIDA, vol. 2011, pp. 82–89 (2011)

10. Kageura, K., Umino, B.: Methods of automatic term recognition. In: National Center for Science Information Systems, pp. 1–22 (1996)

11. Kelih, E., Buk, S., Grzybek, P., Rovenchak, A.: Project description: designing and constructing a typologically balanced ukrainian text database. In: Методианалізутексту, pp. 125–132 (2009)

12. Kotsyba, N., Mykulyak, A., Shevchenko, I.V.: UGTag: morphological analyzer and tagger for the Ukrainian language. In: Proceedings of the International Conference Practical Applications in Language and Computers, PALC 2009 (2009)

13. Kruglevskis, V., Vancane, I.: Term extraction from legal texts in Latvian. In: Second Baltic Conference on Human Language Technologies (2005)

14. Lindberg, D., Humphreys, B., McCray, A.: The unified medical language system. Methods Inf. Med. **32**(4), 281–291 (1993)

15. Lopez, A., Nossal, M., Hwa, R., Resnik, P.: Word-level alignment for multilingual resource acquisition. In: LREC Workshop on Linguistic Knowledge Acquisition and Representation: Bootstrapping Annotated Data, Las Palmas, Spain (2002)

16. McDonald, R., Petrov, S., Hall, K.: Multi-source transfer of delexicalized dependency parsers. In: EMNLP (2011)

17. Memetova, E.: Lexicophraseological expressive means of the Crimean Tatar language. Uchenye zapiski **18**(57), 37–39 (2007)

18. Namer, F.: FLEMM: un analyseur flexionnel du français à base de règles. Traitement automatique des langues (TAL) **41**(2), 523–547 (2000)

19. National Library of Medicine, Bethesda, Maryland: Medical Subject Headings (2001). www.nlm.nih.gov/mesh/meshhome.html

20. Och, F., Ney, H.: Improved statistical alignment models. In: ACL, pp. 440–447 (2000)

21. Pazienza, M.T., Pennacchiotti, M., Zanzotto, F.: Terminology extraction: an analysis of linguistic and statistical approaches. In: Sirmakessis, S. (ed.) Knowledge Mining, Studies in Fuzziness and Soft Computing, vol. 185, pp. 255–279. Springer, Berlin Heidelberg (2005). https://doi.org/10.1007/3-540-32394-5_20

22. Pinnis, M., Ljubešić, N.,Ştefănescu, D., Skadiņa, I., Tadić, M., Gornostay, T.: Term extraction, tagging, and mapping tools for under-resourced languages. In: TKE 2012, pp. 193–208 (2012)

23. Schmid, H.: Probabilistic part-of-speech tagging using decision trees. In: International Conference on New Methods in Language Processing, pp. 44–49 (1994)
24. Shatalina, O.: Literature terminology of the old Ukrainian literature of the 18th century. Uchenye zapiski **18**(57), 5–7 (2005)
25. Shyshkina, N., Zorko, G., Lesko, L.: Terminology work and software localization in Ukraine. In: Problems of Cybernetics and Informatics, pp. 17–20 (2010)
26. Tadić, M., Šojat, K.: Finding multiword term candidates in Croatian. In: IESL Workshop, RANLP Conference, pp. 102–107 (2003)
27. Tiedemann, J., Kotzé, G.: A discriminative approach to tree alignment. In: Ilisei, I., Pekar, V., Bernardini, S. (eds.) International Workshop on Natural Language Processing Methods and Corpora in Translation, Lexicography and Language Learning, pp. 33–39 (2009)
28. Vivaldi, J., Rodríguez, H.: Arabic medical term compilation from Wikipedia. In: Proceedings of the CIST 2014 (2014)
29. Yarowsky, D., Ngai, G., Wicentowski, R.: Inducing multilingual text analysis tools via robust projection across aligned corpora. In: HLT (2001)
30. Zeman, D., Resnik, P.: Cross-language parser adaptation between related languages. In: NLP for Less Privileged Languages (2008)
31. Bulgakov, O.: Building a semantic dictionary of prepositional constructions based on the Ukrainian national linguistic corpus. Technical report, Ukrainian language-information fund of NAS of Ukraine, Kiev, Ukraine (2006). (In Russian)
32. Glibovets, A., Reshetnev, I.: Iterative method for the construction of terminology from scientific corpora in Ukrainian. Cybern. Syst. Anal. **50**(6), 53–62 (2014). (In Russian)
33. Demska, O.: Textual corpus: idea of another form. VPC NaUKMA, Kyiv (2011). (In Ukrainian)
34. Kossak, O.: Ukrainian computational terminology. In: Modern Problems of Computer Science, pp. 39–42 (2000). (In Ukrainian)
35. Kocherha, O., Meinarovych, E.: Scientific English-Ukrainian-English dictionary. Physics and close areas. Nova knyha, Vinnytsia (2010). (In Ukrainian)
36. Lalaieva, R., Surovanets, I., Tychtchenko, O.: Indexing of Polish, Russian and Ukrainian speech therapy terminology. Lexicographical J. **10**, 29–36 (2004). (In Ukrainian)
37. Levchenko, O., Kulchytsky, I.: Technology for transforming a five-language comparative dictionary in digital format. In: Information Systems and Networks, pp. 129–138 (2013). (In Ukrainian)
38. Monakhova, T.: Exploitation of corpus linguistics methods in lexicography. Sci. Works **98**(85), 55–60 (2009). (In Ukrainian)
39. Puriaeva, N.: Analysis of religious language in general and of the religious dictionary in particular. Lexicographical J. **10**, 36–42 (2004). (In Ukrainian)
40. Tymenko, L.: Lexical and thematic clusters of Ukrainian law terminology at the beginning of the XX century. Lexicographical J. **10**, 65–70 (2004). (In Ukrainian)

Adaptation of a Term Extractor to Arabic Specialised Texts: First Experiments and Limits

Wafa Neifar[1,2], Thierry Hamon[1,3(✉)] [iD], Pierre Zweigenbaum[1],
Mariem Ellouze Khemakhem[2], and Lamia Hadrich Belguith[2]

[1] LIMSI, CNRS, Université Paris-Saclay, 91405 Orsay, France
{neifar,hamon,pz}@limsi.fr
[2] MIRACL Laboratory, Sfax University, B.P-3018, Sfax, Tunisia
mariem.ellouze@planet.tn, l.Belguith@fsegs.rnu.tn
[3] Université Paris 13, Sorbonne Paris Cité, 93430 Villetaneuse, France

Abstract. In this paper, we present an adaptation to Modern Standard Arabic of a French and English term extractor. The goal of this work is to reduce the lack of resources and NLP tools for Arabic language in specialised domains. The adaptation firstly focuses on the description of extraction processes similar to those already defined for French and English while considering the morpho-syntactic specificity of Arabic. Agglutination phenomena are further taken into account in the term extraction process. The current state of the adapted system was evaluated on a medical text corpus. 400 maximal candidate terms were examined, among which 288 were correct (72% precision). An error analysis shows that term extraction errors are first due to Part-of-Speech tagging errors and the difficulties induced by non-diacritised texts, then to remaining agglutination phenomena.

1 Introduction

For several decades, several approaches for terminology acquisition have been proposed to extract terms from texts in order to assist terminologists when creating terminological resources [1–4]. Assisted terminology acquisition fosters the use of terminologies in applications and therefore facilitates access to information contained in specialised texts [3,5].

However, all the languages are not resourced similarly, and approaches have been proposed to identify and extract terms and acquire relations mainly from European and Japanese texts [1]. In contrast, languages such as Modern Standard Arabic (MSA) lack linguistic resources and dedicated approaches for defining such resources from texts. Moreover, the conception of Natural Language Processing (NLP) methods faces difficulties which are inherent to the intrinsic characteristics of Arabic. These obstacles restrain the development of approaches for terminology acquisition. In this context, we aim at reducing the lack of available resources and NLP tools for Arabic language in specialised domains.

© Springer International Publishing AG, part of Springer Nature 2018
A. Gelbukh (Ed.): CICLing 2016, LNCS 9623, pp. 242–253, 2018.
https://doi.org/10.1007/978-3-319-75477-2_16

To address this problem, we propose to adapt an existing term extractor, Y$_{A}$T$_{E}$A [6], to the Arabic language. This system was initially developed for English and French texts. It relies on a linguistic description of the terms to extract, and offers the possibility to consider specific terminological phenomena and to be configured for another language. We present the methodology we followed to adapt it to some morphological and morpho-syntactic specificities of MSA.

The paper is organized as follows. We present the state of the art on MSA terminology acquisition in Sect. 2. Then, we describe the process for the adaptation of Y$_{A}$T$_{E}$A to specialised MSA texts (Sect. 3). First experiments on a medical corpus are presented and discussed in Sect. 4, then we conclude.

2   Terminology Acquisition in Specialised MSA Texts

Despite the increasing research on Arabic Natural Language Processing, few authors address terminological acquisition from specialised Arabic texts. The main cause is the linguistic complexity of this language which impedes the development of NLP and term extraction approaches [7]: numerous linguistic phenomena, such as non-diacritisation, agglutination, and morphological and syntactic ambiguity, have to be considered. Another identified cause is the linguistic variation in the Arabic language family which includes MSA, Classical Arabic (CA), and many dialects [8].

Terminology planning strategy is required in many domains such as agriculture, geology, environmental protection or law, and may depend on the country [9]. Even though specialists of some specialised domains, such as medicine, use French and English in their practice or for teaching [10], an understanding of specialised notions by lay people (e.g. the patients) in their own language is required. Moreover, for the medical domain, Arabic is one of the official languages of the World Health Organisation. It is thus all the more crucial to endow it with terminological acquisition and management systems.

[7] consider that the particularities of Arabic impede the use of statistical approaches and argue that terminological extraction on Arabic texts can only be achieved with linguistic methods. The approach they propose therefore mainly relies on a precise description of term formation and variation, similarly to previous work on English and French [11]. Actually, most of the approaches used for Arabic terminological acquisition are usually similar to those proposed for specialised texts in other languages [1,2]. [12] propose to extract complex candidate terms using a two-step hybrid approach. First, a linguistic filter based on morphological analysis and Part-of-Speech tagging identifies and parses sequences of tokens. Secondly, statistical filtering based on the Log-Likelihood Ratio (LLR) [13] is applied to select the best solution in case of ambiguity. [14] propose a similar hybrid approach but use two statistical measures to measure *unithood*, i.e. the stability of syntagmatic combinations, and *termhood*, i.e. the degree of association of a terminological unit with the specific domain, of the candidate sequence [15]. They rely on the LLR in the former case and on the C-Value [16] in the latter case. Both approaches are evaluated using an Arabic corpus in the

environmental domain. Other approaches for extracting terms from MSA follow a similar framework. [17] adapt general-language methods to parse domain-specific texts (in the religious domain) and to extract single- and multi-word terms automatically: a corpus of CA and MSA texts is collected from archives of Islamic newspapers and Islamic sites, and used to evaluate the approach. The termhood of single-word terms is measured with the TF-IDF score, while several other statistical measures (PMI, Kappa, Chi-square, T-test, Shapiro-Piatersky and Rank Aggregation) are used to capture the degree of association of term components.

Currently, the approaches used are very similar to hybrid methods already proposed for English or French. Moreover, except [18], those methods do not take into account some linguistic particularities of Arabic term formation, such as its morphological and syntactic ambiguities, non-diacritisation and agglutination. Besides, as pointed out by [19], the evaluation of Arabic term extraction methods is often questionable: only a few top hundred words out of thousands are evaluated manually, whereas it is well-known that the quality of the very first extracted candidate terms is generally high regardless of the approach [20,21]. This can be explained by the lack of gold standards for the evaluation of terminological acquisition methods, as previously observed in many other languages, including French, but also by the scarcity of available domain-specific Arabic text collections.

In this work, we intend to go beyond these difficulties. We propose to adapt a freely available term extractor which offers the possibility to integrate particularities of the Arabic language in the term extraction process. Furthermore, we perform a full evaluation of the candidate terms extracted from medical texts.

3   Adaption of Y$_{A}$T$_{E}$A to MSA

3.1   Term Extraction for French

As mentioned above, we aim at proposing an approach to identify Arabic single and multi-word terms in domain-specific texts. To achieve this goal, we adapt the freely available term extractor Y$_{A}$T$_{E}$A[1] [6] to MSA.

This term extractor offers the possibility to take into account the linguistic specificities of a language, in our case MSA, with the definition of linguistic rules and resources describing the term extraction process. Since Y$_{A}$T$_{E}$A was first developed to extract terms from French and English texts, the adaptation of the approach to a Semitic language is a challenge. This may also allow us to discover some limits of the methods implemented in Y$_{A}$T$_{E}$A.

The Y$_{A}$T$_{E}$A term extractor performs shallow parsing of POS tagged and lemmatized text:

- **Step 1**: the text is chunked using positive and negative syntactic boundaries. These boundaries can be POS tags (pronouns, conjugated verbs, typographical marks, etc.) but also words or phrases. This first step provides maximal

---

[1] http://search.cpan.org/~thhamon/Lingua-YaTeA/.

noun phrases that can be candidate terms or contain candidate terms. For instance, the use of syntactic boundaries on Example 1(a) identifies the maximal noun phrases *oxygen transported* and *blood* (Fig. 1(a)).

– **Step 2**: Syntactic analysis patterns take into account morpho-syntactic term variation and are recursively applied to the previously identified noun phrases to propose parsed candidate terms. Each candidate term is represented as a binary syntactic tree where nodes are term constituents (we refer to them as 'components') with a head (T) or modifier (M) syntactic role. Components are also considered as candidate terms. Phrases which cannot be parsed are considered irrelevant and rejected. This step produces single-word terms and multi-word terms parsed into their head and modifier components.

For instance, the maximal noun phrase *oxygen transported* is parsed into its head component *oxygen* and its modifier component *transported* (Fig. 1(c)).

– **Step 3**: Statistical measures such as frequency and C-Value [16] are then associated with these candidate terms [21].

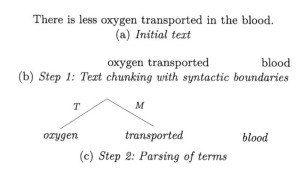

There is less oxygen transported in the blood.
(a) *Initial text*

oxygen transported   blood
(b) *Step 1: Text chunking with syntactic boundaries*

oxygen   *transported*   *blood*
(c) *Step 2: Parsing of terms*

Fig. 1. Steps for term extraction from an English sentence

3.2   Linguistic Description of Terms

The proposed adaptation of YₐTₑA to MSA involves Step 1 (Sect. 3.2) and Step 2 (Sect. 3.3). Figure 2 summarises the process of term extraction from texts in MSA. We first assume that the texts have been morphologically analysed and POS-tagged. Morphological analysis and POS tagging are performed with the MADA+TOKAN analyser [22]. In the following, we define POS tags for each word as the concatenation of grammatical category, gender, number, case and state when it is relevant (e.g. `noun-m-s-g-d`, `verb-m-s-na-na`).

Our adaptation of YₐTₑA to domain-specific MSA texts follows traditional terminology best practices. Thus, in Step 1 (text chunking), we define the syntactic boundaries specific to MSA given observations on MSA texts. Similarly to other languages, pronouns, punctuation, conjugated verbs are considered as textual elements which cannot occur in a term. We also take into account some specific lexical elements of the Arabic language, such as pseudo-verbs (كان, تكون, إن) (*it was,*

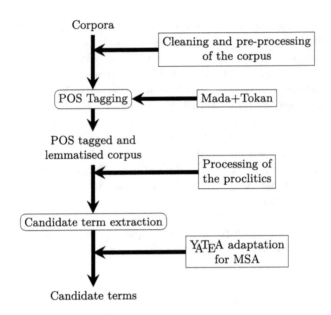

Corpora

Cleaning and pre-processing
of the corpus

POS Tagging ← Mada+Tokan

POS tagged and
lemmatised corpus

Processing of
the proclitics

Candidate term extraction

Y$_A$T$_E$A adaptation
for MSA

Candidate terms

Fig. 2. MSA term extraction process

she is, certainly), adverbs (فقط, هنا, ربما) (*maybe, here, only*) or lexical phrases or
patterns (في بعض الأوقات) (*sometimes*) that cannot occur in terms either. These
syntactic boundaries gather 39 POS tags, 28 words in their inflected forms, 20 lemmas, and 6 exceptions, *i.e.* inflected forms of syntactic boundaries which may occur
in the candidate terms. A filter is applied to the chunks to consider maximal noun
phrases that contain at least one noun. Finally, in order to ensure that candidate
terms are linguistically well-formed, we determine which POS tags cannot appear
at the beginning or end of a term. This is mainly prepositions (من (*from*), إلى (*to*),
بَيْنَ (*between*), عند (*when*)) to which MADA+TOKAN attributes incorrect morphosyntactic categories.

For example, in the sentence shown in Fig. 3(a), syntactic boundaries تكون
(pseudo-verb), التي (*that/which*), يحملها (*transports it*) and أقل (*less*) identify the
maximal noun phrases كمية الأكسجين (*quantity of oxygen*) and الدم (*the blood*).

In Step 2, we define specific patterns for the parsing of the previously identified maximal noun phrases. These patterns are applied to maximal noun phrases
to identify the syntactic roles (head and modifier) of noun phrases components.
At the same time they filter out word sequences which cannot be analysed with
the defined patterns and are thus hypothesized to be useless. Patterns are minimal
sequences of POS tags and prepositions and can specify morphological features
such as gender and number, as well as the case of the constituents. Specifically,
the genitive is very useful in the analysis of noun phrases. We identified 696 patterns for the analysis of expressions containing two content words (e.g. درجة الالم –
degree of pain), and 21 patterns for the analysis of three-content-word expressions

(e.g. ارتفاع معدلات الكولسترول – *elevation of cholesterol levels*). All these patterns are applied recursively to the maximum noun phrases and their components.

For instance, the pattern `noun-m-s-g-d(modifier) noun-f-s-n-c(head)`[2] allows us to analyse the maximal noun phrase كمية الأكسجين (*quantity of oxygen*) as presented in Fig. 3(c).

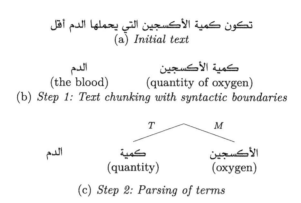

تكون كمية الأكسجين التي يحملها الدم أقل
(a) *Initial text*

الدم   كمية الأكسجين
(the blood)   (quantity of oxygen)
(b) *Step 1: Text chunking with syntactic boundaries*

$$T \qquad M$$

الدم   كمية   الأكسجين
   (quantity)   (oxygen)
(c) *Step 2: Parsing of terms*

Fig. 3. Steps for term extraction from an Arabic sentence. (literally: *The quantity of oxygen in the blood is lesser*)

3.3 Taking into Account Arabic-Specific Phenomena

Diacritisation. Arabic writing is characterised by the use of diacritics, non-alphabetic signs added to a letter, which are generally encoded by characters added after that letter. Words, and by extension texts, can occur in two forms: a non-diacritised form فصل and a diacritised form فَصَل (*He dismissed*) or فَصْل (*chapter/season*). Non-diacritised writing is the cause of many ambiguities and it is sometimes difficult or impossible to infer the meaning of some non-diacritised words if the context is unknown [23].

As most of Arabic documents, our corpus is a collection of non-diacritised texts. The absence of diacritisation in texts leads to many POS tagging errors because of the ambiguities of non-diacritised forms, and consequently, to a degradation in term extraction results. Thus, since MADA+TOKAN associates a lemma in diacritised form to each word of the corpus, we address this phenomenon from the very beginning of pattern definition.

Agglutination and Clitics. Agglutination is a basic characteristic of the Arabic language. It consists in associating particular lexicon elements called clitics, with the base word to which they relate. Clitics are usually pronouns, articles, prepositions or coordinating conjunctions. Since they are halfway between affixes and words,

---

[2] `noun-m-s-g-d`: defined singular masculine noun in genitive case. `noun-f-s-n-c`: constructed singular feminine noun in nominative case.

they are morphologically related to another word, even if they are grammatically independent.

There are two kinds of clitics according to their position relative to the base word: proclitics and enclitics. Proclitics are lexical items which precede and act as a prefix of the word. They may represent an article (ال) (*the*), a preposition (ل, ب) (*for, by*), a conjunction (و) (*and*), etc. Enclitics are elements following the word, similarly to suffixes, such as possessive pronouns (ها, هم) (*her, their*) [8]. For instance, in the word معصميك (*your wrists* or more precisely *both your wrists*), we distinguish the word معصمي (whose nominative form is معصمان – *two wrists*) followed by the enclitic ك which is a possessive pronoun (*your*). Similarly, in the word والأسنان (*and the teeth*), we distinguish the following elements: و (*and*) a proclitic representing a coordinating conjunction, ال (*the*) a proclitic representing a article, and the word أسنان (*teeth*).

From an Arabic language processing perspective, it is sometimes difficult to distinguish a proclitic or an enclitic from a character of the base word. In the following example, وسائل two analyses are possible: (i) it can be only one word وسائل (*methods*), (ii) it can also be decomposed as a proclitic و which is a coordinating conjunction (*and*) followed by a name سائل (*liquid*). Consequently, this characteristic leads to morphological ambiguity in the analysis. Thus, in our case, if the agglutination is not taken into account, it makes more complex the definition of syntactic patterns and causes errors when extracting terms. Currently, we chose to focus on proclitics and to separate the words with which they are associated. We use the morphological analysis done by MADA+TOKAN which provides decomposition of words in the corpus by separating proclitics and enclitics. Considering the proclitics in the term extraction process leads to the definition of 21 additional syntactic analysis patterns such as: noun-m-s-a-c(Head) و noun-m-s-n-c(Modifier) or noun-f-s-n-d(Head) (adj-f-s-g-i(Modifier) noun-m-s-g-d(Head))(Modifier).

Case Marker and Nunation. All Arabic nouns bear inflection according to case. Case has three value in Arabic, marked by the three short vowels or the corresponding nunated forms for these three vowels, marked by diacritics [8]:

- a noun in the nominative form is marked by the short vowel *damma*: ُ or the nunated form *dammatan* ٌ ;
- a noun in the accusative form is marked by the short vowel *fatha*: َ or the nunated form *fathatan* ً ;
- a noun in the genitive form is marked by the short vowel *kasra*: ِ or the nunated form *kasratan* ٍ ;

Arabic nouns also bear inflection according to state, which has three values:

- the *definite* state corresponds to the nominal form most typically present with an article;
- the *indefinite* state designates a non-specific instance of a noun;
- the *construct* state, also called *al- 'idāfah* (possession), indicates that the noun is the head of an *Idafa* (possessor), *i.e.* a noun is added to form a possession relation.

For our purposes, state is useful to identify the syntactic role of a noun as component of a term. It is important to consider this feature when parsing phrases extracted from Arabic texts. Thus, we have taken into account these specificities when defining syntactic patterns. Currently, we use the POS tags and information provided by the morphological analysis of MADA+TOKAN, but since this includes some incorrect information which has a strong impact on term extraction quality, we plan to define specific treatments.

4   Evaluation of the Adapted Term Extractor

The adapted term extractor has been evaluated on 30 MSA medical texts (see Sect. 4.1). Two experiments were conducted to evaluate the contribution of the agglutination treatment to the term extraction process. Results are presented and discussed in Sect. 4.2. Term extraction errors are analysed in Sect. 4.3.

4.1   Medical Corpus

As stated in [24], very few domain-specific corpora are freely available for Arabic. We therefore built our own domain-specific corpus by collecting medical texts from MedlinePlus[3]. These texts are dedicated to patients and explain medical procedures and situations. Therefore, even though they are not highly specialised, this text collection is consistent with the situation of Arabic language in the medical and public health domains described in Sect. 2. Currently, our corpus contains 504 Arabic-French-English parallel texts available in PDF. These documents need to be converted into plain text and to be pre-processed in order to standardise their character set and to fix their spelling [8]. This cleaning and normalisation step also requires a time-consuming manual checking given the inconsistency of these encountered problems and errors. Currently, we completed these steps for 30 Arabic medical texts (15,532 words) that we use in our experiments.

For the morpho-syntactic analysis of these texts, we studied the results of two tools: the Stanford Tagger [25] and MADA+TOKAN [26]. Given the information provided by these two Arabic POS taggers we chose MADA+TOKAN: besides good results and POS tags and lemma, MADA+TOKAN provides a morphological analysis which we also use in the term extraction process.

4.2   Experiments and Results

We ran the adapted term extractor and evaluated the impact of agglutination processing on term extraction. The results of the experiments are shown in Table 1 and commented below.

First, we applied the term extractor adapted to MSA without considering the agglutination phenomenon: enclitics and proclitics were considered as parts of the words. Thanks to the syntactic boundaries defined in Step 1, we identified 1,972

---

[3] http://www.nlm.nih.gov/medlineplus/languages/all_healthtopics.html.

Table 1. Results of term extraction on Arabic medical texts (MNP: maximal noun phrases, ST: simple candidate terms, CT: complex candidate terms, among which CTmax: complex candidate terms which are maximal noun phrases, (P %): Precision as a percentage)

	Step 1		Step 2			
	MNP	ST	ST	CT	CTmax (P %)	Total
Without agglutination processing	1,972	262	262	1,133	590 (65.7)	1,395
With agglutination processing	1,916	298	298	824	400 (72.0)	1,122

maximal noun phrases (MNP) and 262 nouns (ST) that are considered as simple candidate terms. Parsing the maximal noun phrases (Step 2) led to keeping 590 maximal noun phrases (CTmax) as complex candidate terms. The components of complex terms are also considered as candidate terms. For example, we consider as terms at the same time ارتفاع معدلات الكولسترول (*elevation of cholesterol levels*), as well as معدلات الكولسترول (*cholesterol levels*), ارتفاع (*elevation*), معدلات (*levels*) and الكولسترول (*cholesterol*). Therefore, at the end of the term extraction process, we obtain a set of 1,395 candidate terms, among which 1,133 are complex candidate terms (CT).

We performed a manual analysis of the 262 simple terms (ST) and 590 candidate terms corresponding to maximal noun phrases (CTmax). We evaluated parsing quality as well as the relevance of the extracted candidate terms regarding the medical domain.

Among the 590 maximal noun phrases, 388 candidate terms (65.7%) were correctly analysed and relevant to the medical domain. As expected, it appears that agglutination is the main source of errors during text chunking and noun phrase parsing. Other errors come from candidate terms which are not relevant to the domain and from chunking errors (see Sect. 4.3 for details).

The second experiment applies the specific treatments which separate proclitics from their base words and evaluates their contribution. In this setting, Step 1 identifies 298 simple candidate terms (ST) and 1,916 maximal noun phrases (MNP). Among them, 400 are kept at the end of Step 2, and result in proposing 824 complex candidate terms. We also analysed these 400 maximal noun phrases: 288 (72.0%) were judged relevant to the domain. Figure 4 presents terms extracted from medical texts. The handling of agglutination, specifically for proclitics, in the term extraction process, leads to a reduction in the number of complex candidate terms produced and to an increase in the number of simple terms. This may reflect the fact that initially agglutinated elements are then considered as syntactic boundaries. We also observe that the number of maximal noun phrases which are not analysed during Step 2 increases. We assume that the words whose proclitics are removed are grouped more or less fortuitously into larger maximal noun phrases which become more difficult to analyse. A thorough analysis of these maximal noun phrases may confirm this hypothesis and lead us to propose new syntactic patterns.

سرطان الثدي (breast cancer) | سرعة ضربات القلب (rapid heartbeat)
الاوعية الدموية (the blood vessels) | الذراع الايمن (the right arm)
تمارين الكاحلين (ankle exercises) | درجة الالم (degree of pain)
الزائدة الدودية (appendix) | العلاج الكيماوي (chemotherapy)
العلاج الاشعاعي (the radiotherapy) | اخذ عينات انسجة الثدي (breast biopsies)
مرض السكر (the diabetes) | الرئتين (both lungs)
التهاب شعب هوائية (bronchitis) | آلام (pains)
العلاج (the treatment) | الرأس (the head)

Fig. 4. Example terms extracted from Arabic medical texts

4.3   Error Analysis

We have analysed the results in order to characterise the reasons that impede term extraction during Steps 1 and 2. First, erroneous POS tagging leads to incorrect chunking and then to semantically irrelevant candidate terms (درجة حرارة لضمان سلامة – *Temperature to ensure the safety*), but also to incorrect term analysis or extraction because of erroneous morpho-syntactic categories and morphological features. For instance, in the noun phrase العضلة رباعية الرؤوس (quadriceps), the word العضلة (muscle) is considered as an adjective instead of a noun, and the word الرؤوس (*the heads*) is identified as a masculine singular noun instead of a plural noun. Besides, similarly to other languages, POS tagging errors may be due to unknown words or to foreign words translated into Arabic: for instance, the word الفولات (*folic acid*) is tagged as a feminine plural noun and lemmatized as the noun *the broad beans*. The quality of POS tagging can also be explained by the non-diacritised texts: the non-diacritised تناول is either a noun (*the intake*) or a verb (*to take*). Its categorisation as a verb prevents from identifying the term تناول الفيتامينات (*vitamin intake*).

Agglutination is also an important source of silence for term extraction. The incorrect identification and analysis of agglutinated words may cause the assignment of wrong POS tags. For instance, the word بثني which can be contextually translated as *bending* is considered as a verb phrase (*he broadcasted to me*), whereas it should be decomposed into a noun (ثني – bending) and the proclitic ب (with). These errors are the main explanations for the silence of the term extraction process.

Finally, errors can be due to the extraction of candidate terms which are judged as irrelevant in the medical domain, *e.g.* مالك الحيوان (*owner of the animal*).

5   Conclusion and Perspectives

We have presented the adaptation of a state-of-the-art term extractor, YATEA, for analysing specialised MSA texts. The term extraction process relies on the linguistic description of terms in Arabic, and takes into account morphological features of this language, with a focus on agglutination. We thus considered information on proclitics in term extraction. To achieve this goal, we inserted the morphological analysis performed by MADA+TOKAN into the term extraction process and defined additional rules for the shallow parsing of Arabic terms containing proclitics.

Evaluation was carried out on a 15,532 word corpus made of 30 medical texts. The precision of results improves when proclitics are taken into account. Even if the number of correctly extracted terms decreases, the precision of the maximal complex terms increases from 65.7 to 72.0%.

In future work, we plan to improve the use of agglutination and diacritisation in the term extraction process. Specifically, we will take into account enclitics. Also, short vowels used to inflect nouns and the *masdar* will be useful to correct POS tagging in specific cases, as well as to guide the shallow parsing of candidate terms. Finally, we intend to confirm our results on other specialised texts in MSA according to their availability.

References

1. Cabré, M.T., Estopà, R., Vivaldi, J.: Automatic term detection: a review of current systems. In: Bourigault, D., Jacquemin, C., L'Homme, M. (eds.) Recent Advances in Computational Terminology. John Benjamins, Amsterdam (2001)
2. Pazienza, M.T., Pennacchiotti, M., Zanzotto, F.: Terminology extraction: an analysis of linguistic and statistical approaches. In: Sirmakessis, S. (ed.) STUDFUZZ. STUDFUZZ, vol. 185, pp. 255–279. Springer, Heidelberg (2005). https://doi.org/10.1007/3-540-32394-5_20
3. Marshman, E., Gariépy, J.L., Harms, C.: Helping language professionals relate to terms: terminological relations and termbases. J. Spec. Transl. **18**, 45–71 (2012)
4. Q. Zadeh, B., Handschuh, S.: The ACL RD-TEC: a dataset for benchmarking terminology extraction and classification in computational linguistics. In: Proceedings of the 4th International Workshop on Computational Terminology (Computerm), Dublin, Ireland, pp. 52–63 (2014)
5. Cohen, K.B., Demner-Fushman, D.: Biomedical Natural Language Processing. John Benjamins Publishing Company, Philadelphia (2013)
6. Aubin, S., Hamon, T.: Improving term extraction with terminological resources. In: Salakoski, T., Ginter, F., Pyysalo, S., Pahikkala, T. (eds.) FinTAL 2006. LNCS (LNAI), vol. 4139, pp. 380–387. Springer, Heidelberg (2006). https://doi.org/10.1007/11816508_39
7. Boulaknadel, S., Daille, B., Aboutajdine, D.: A multi-word term extraction program for arabic language. In: Chair, N.C.C., Choukri, K., Maegaard, B., Mariani, J., Odijk, J., Piperidis, S., Tapias, D. (eds.) Proceedings of the LREC 2008 (2008)
8. Habash, N.: Introduction to Arabic Natural Language Processing. Synthesis Lectures on Human Language Technologies. Morgan & Claypool Publishers, San Raphael (2010)
9. Massoud, R.: La terminologie au liban : réalités et défis. Annales de l'Institut de langues et de traduction (ILT) 10 (2003)
10. Samy, D., Moreno-Sandoval, A., Bueno-Díaz, C., Garrote-Salazar, M., Guirao, J.M.: Medical term extraction in an arabic medical corpus. In: Proceedings of LREC 2012 (2012)
11. Daille, B.: Conceptual structuring through term variations. In: Bond, F., Kohonen, A., Carthy, D.M., Villaciencio, A. (eds.) Proceedings of the ACL 2003 Workshop on Multiword Expressions: Analysis, Acquisition, and Treatment, pp. 9–16 (2003)

12. Bounhas, I., Slimani, Y.: A hybrid approach for arabic multi-word term extraction. In: IEEE International Conference on Natural Language Processing and Knowledge Engineering, NLP-KE 2009, pp. 1–8. IEEE (2009)
13. Dunning, T.: Accurate methods for the statistics of suprise and coincidence. Comput. Linguist. **19**, 61–74 (1993). Special Issue on Using Large Corpora: I
14. AlKhatib, K., Badarneh, A.: Automatic extraction of arabic multi-word terms. In: IMCSIT, pp. 411–418 (2010)
15. Kageura, K., Umino, B.: Methods of automatic term recognition - a review. Terminology **3**, 259–289 (1996)
16. Maynard, D., Ananiadou, S.: Identifying terms by their family and friends. In: Proceedings of COLING 2000, Saarbrucken, Germany, pp. 530–536 (2000)
17. Abed, A.M., Tiun, S., Albared, M.: Arabic term extraction using combined approach on islamic document. J. Theor. Appl. Inf. Technol. **58**, 601–608 (2013)
18. Bounhas, I., Elayeb, B., Evrard, F., Slimani, Y.: Organizing contextual knowledge for arabic text disambiguation and terminology extraction. Knowl. Org. J. **38**, 473–490 (2011)
19. Bounhas, I., Lahbib, W., Elayeb, B.: Arabic domain terminology extraction: a literature review. In: Meersman, R., Panetto, H., Dillon, T., Missikoff, M., Liu, L., Pastor, O., Cuzzocrea, A., Sellis, T. (eds.) OTM 2014. LNCS, vol. 8841, pp. 792–799. Springer, Heidelberg (2014). https://doi.org/10.1007/978-3-662-45563-0_51
20. Korkontzelos, I., Klapaftis, I.P., Manandhar, S.: Reviewing and evaluating automatic term recognition techniques. In: Nordström, B., Ranta, A. (eds.) GoTAL 2008. LNCS (LNAI), vol. 5221, pp. 248–259. Springer, Heidelberg (2008). https://doi.org/10.1007/978-3-540-85287-2_24
21. Hamon, T., Engström, C., Silvestrov, S.: Term ranking adaptation to the domain: genetic algorithm-based optimisation of the *C-Value*. In: Przepiórkowski, A., Ogrodniczuk, M. (eds.) NLP 2014. LNCS (LNAI), vol. 8686, pp. 71–83. Springer, Cham (2014). https://doi.org/10.1007/978-3-319-10888-9_8
22. Roth, R., Rambow, O., Habash, N., Diab, M., Rudin, C.: Arabic morphological tagging, diacritization, and lemmatization using lexeme models and feature ranking. In: Proceedings of ACL-08: HLT, Short Papers, Columbus, Ohio, pp. 117–120 (2008)
23. Hadrich, L.B., Chaaben, N.: Analyse et désambiguïsation morphologiques de textes arabes non voyellés. In: Actes de TALN'06, Leuven, Belgique, pp. 493–501 (2006)
24. Al-Sulaiti, L., Atwell, E.: The design of a corpus of contemporary arabic. Int. J. Corpus Linguist. **11**, 1–36 (2006)
25. Toutanova, K., Klein, D., Manning, C., Singer, Y.: Feature-rich part-of-speech tagging with a cyclic dependency network. In: Proceedings of HLT-NAACL 2003, pp. 252–259 (2003)
26. Habash, N., Rambow, O., Roth, R.: MADA+TOKAN Manual. CCLS-10-01 (2010)

Morphology and Part-of-Speech Tagging

Corpus Frequency and Affix
Ordering in Turkish

Mustafa Aksan🆔, Umut Ufuk Demirhan⁽⊠⁾🆔, and Yeşim Aksan🆔

English Language and Literature Department, Faculty of Science and Letters,
Mersin University, Mersin, Turkey
mustaksan@gmail.com, umutufuk@gmail.com,
yesim.aksan@gmail.com

Abstract. Suffix sequences in agglutinative languages derive complex structures. Based on frequency information from a corpus data, this study will present emerging multi-morpheme sequences in Turkish. Morphgrams formed by combination of voice suffixes with other verbal suffixes from finite and non-finite templates are identified in the corpus. Statistical analyses are conducted on permissible combinations of these suffixes. The findings of the study have implications for further studies on morphological processing in agglutinative languages.

Keywords: Frequency · Affix order · Verbal synthesis
Turkish National Corpus

1 Introduction

Languages exploit their morphological resources in creating new lexemes and in expression of various grammatical categories. The structures resulting from affixation are rich and complex and follow strict ordering restrictions. Despite their significant use, there are relatively few studies on affix ordering [9].

The present study aims to address issues of (i) the order of affixes that are concatenated on a verb root, and (ii) observed frequencies of these verbal affixes. Previous works on affixation have uncovered grammatical aspects of forms and their orderings. As for the frequency of forms, there are only a handful studies since studies on frequencies require large scale language corpora and their construction is a recent undertaking in Turkey.

A balanced and a representative corpus of contemporary Turkish (Turkish National Corpus) provides the data of the study. Corpus data in morphological analysis provides information for researchers on varieties of language use across different domains, time period and medium that are inaccessible otherwise. A corpus-based study reveals quantificational aspects of language structure that help determine fundamental properties of units and patterns.

This paper is organized as follows: in the first section, we will introduce the corpus of the study and our extraction methodology. In the second part, we will present observed frequencies of verbal affixes and emerging "morphgrams" (or morpheme

© Springer International Publishing AG, part of Springer Nature 2018
A. Gelbukh (Ed.): CICLing 2016, LNCS 9623, pp. 257–270, 2018.
https://doi.org/10.1007/978-3-319-75477-2_17

bundles). Finally, we will discuss the findings of statistical analyses for ordering of affixes from different positions in the template.

2 Methodology

2.1 The Corpus

Frequency information of suffixes is obtained from the written part of the Turkish National Corpus (TNC). TNC is constructed following the principles used to construct the British National Corpus (see http://www.natcorp.ox.ac.uk). The size of TNC is 50,997,016 running words, representing a wide range of text categories spanning a period of 23 years (1990–2013). It consists of samples from textual data (98%) and transcribed spoken data (2%).

The distribution of samples in the corpus is determined for each text domain, time, and medium [1]. Tables 1 and 2 shows the distribution of texts in the written part of the TNC across domain and medium, respectively.

Table 1. The distribution of texts according to domains in the TNC

Domain	No. of words	% of words
Imaginative: Prose	9.365.775	18.74%
Informative: Natural and pure sciences	1.367.213	2.74%
Informative: Applied science	3.464.557	6.93%
Informative: Social science	7.151.622	14.31%
Informative: World affairs	9.840.241	19.69%
Informative: Commerce and finance	4.513.233	9.03%
Informative: Arts	3.659.025	7.32%
Informative: Belief and thought	2.200.019	4.4%
Informative: Leisure	8.421.603	16.85%
Total	49.983.288	100.00%

Table 2. The distribution of texts across mediums in the TNC

Medium	No. of words	% of words
Unspecified	10.541	0.02%
Book	31.456.426	62.93%
Periodical	15.968.240	31.95%
Miscellaneous: Published	958.999	1.92%
Miscellaneous: Unpublished	1.589.082	3.18%
Total	49.983.288	100.00%

The representativeness of the TNC is secured through balance and sampling of varieties of contemporary language use. The selection of written texts is done via the criteria of text domain, medium, and time. The criterion of domain means that texts are distributed along imaginative and informative. While the imaginative domain is represented by texts of fiction, the informative domain is represented by texts from the social sciences, arts, commerce-finance, belief-thought, world affairs, applied sciences, natural-pure sciences, and leisure. The criterion of medium refers to text production. The texts collected to represent the written medium are selected from books, periodicals, published or unpublished documents and texts written-to-be-spoken such as news broadcasts and screenplays, among others. The criterion of time defines the period of text production.

2.2   Annotation and Data Processing

Frequency lists of verbal suffixes in Turkish are based on lemmas and morphological tags, analyzed and tagged by the TNC-tagger. Part-of-speech annotation, morphological tagging, and lemmatization of the TNC are done by developing a natural language-processing (NLP) dictionary based on the NooJ_TR module [10]. Semi-automatic process of developing NLP dictionary includes the following steps: (i) automatically annotating the type list with the NooJ_TR module, which follows a root-driven, non-stochastic, rule-based approach to annotate the morphemes of the given types using a graph-based, finite-state transducer; (ii) manually checking and revising the output and eliminating artificial/non-occurring ambiguities and theoretically possible multi-tags. Standardization of non-canonical spellings, tagging of low-frequency entries not processed by the module, and revising the tagged entries are the other tasks carried out at this stage. Finally, the entries of the NLP dictionary and actual running words of the corpus are matched via the PHP and MySQL-based interface of the corpus.

In Turkish, the order of suffixes is relatively fixed, and allomorphy is quite regular. While the order is mostly predictable and easy to process with clear-cut morpheme boundaries, in many cases, the existence of a number of homographic morphemes, lemma+suffix combinations, suffix+suffix combinations as well as homographic lemmas present specific challenges. The challenges constitute the ambiguities that 15% of the TNC tokens contain. In other words, the TNC is only disambiguated at the morphological level, and excludes all beyond-word, contextual resolution of the ambiguities in its NLP dictionary. Hence, these ambiguities are kept in the TNC dictionary. If a given type is ambiguous because of a homophonous suffix, for instance, the frequency of that type is added to the frequencies of all – two in most cases - possible suffixes that can be assigned. This approach minimizes the effect of ambiguities – 15% of all TNC texts - to the overall rankings of the frequency lists.

3   Verbal Templates in Turkish

Suffixes in the verbal domain are organized into two major templates.[1] In both finite and non-finite templates suffixes are assigned to their respective positions [4, 6, 11, 12] (Table 3):

---

[1] We use the term "template" to refer to cited order of affixes with no theoretical implications.

Table 3. Finite verb template

(1)-(y)A	(2)-(y)Abil	(3)-DI	(4)-(y)DI	(5)-DIr
	-(y)Iver	-mIş	-(y)mIş	
	-(y)Agel	-(A/I)r/-z	-(y)sA	
	-(y)Akal	-(y)AcAK		
	-(y)Adur	-(I)yor		
		-mAlI		
		-mAktA		
		-(y)A		

Table 4. Non-finite verb template

Root	Voice	Neg	Subordinator	Agr
V	Causative	-mA	-DIK	Agr
	Passive		-AcAk	
	Reflexive		-Iş	
	Reciprocal			

There are four different groups or "paradigms" of agreement suffixes in Turkish [13]. The attachment of particular agreement marker from any of these paradigms is determined by which of the tense/aspect/modality position it will follow (Table 4).[2]

The positions available preceding the inflectional categories listed above are filled by the voice categories, negation, subordinators and agreement suffixes in this order. All of the four voice categories in Turkish, Passive, Causative, Reciprocal and Reflexive, occupy the position immediately after the root. Their productivity and combination restrictions are evident from their frequencies.

The earliest study on frequencies of Turkish suffixes and their combinations is done by Pierce in 1960s. Based on a very small-sized corpus of written and spoken Turkish, Pierce presents raw frequencies of Turkish derivational and inflectional suffixes [14], as well as top 20 most frequent lexemes [15]. Until the work of [16][3] we find no other principled account of suffix frequencies in Turkish. We may conclude that quantificational aspects of suffixes and also suffix combinations (i.e., morphgrams) are largely set aside mainly due to lack of corpora.

---

[2] Hakkani-Tür et al. [7] propose a morphological disambiguation procedure for agglutinative languages based on inflection groups.

[3] Hankamer [8] presents a rough quantificational description of Turkish affixes. Based on a count of newspaper articles, he reports that average number of affixes is 3.06 and proportions of words with five or more suffixes is 19.8. Güngör [16] working on a 2,200,000-word corpus of mainly newspapers and periodicals finds the maximum number of suffixes as 8. In his calculations, the average number of suffix length is 2.4.

4 Voice Suffixes and Frequency

The following is the list of five most frequent suffixes in Turkish extracted from the written component of TNC, including usages from both nominal and verbal domains:

The nominal inflection markers outnumber the verbal inflectional markers as expected where nominal roots also outnumber the verbals. The Table 5 above includes no voice categories.

Table 5. Most frequent suffixes in the TNC

Rank	Suffix	%	Frequency
1	bare	16,45	13,027,015
2	nominative (nom)	10,25	8,120,248
3	possessive (p3s)	7,41	5,869,510
4	accusative (acc)	5,69	4,503,459
5	person (3s)	5,27	4,176,400

The passive attaches to transitive and intransitive root verbs and functions as a detransitivizer. Second in rank is the causative that attaches both transitive and intransitive root forms and increases transitivity of the clause. Reciprocity of the action expressed in a clause is marked by the reciprocal suffix that attaches to very small number of roots hence its frequency of occurrence in the corpus is relatively small. The reflexive is even more limited in terms of available roots to attach, the least productive among voice categories (Table 6).

Table 6. Frequencies of voice categories

Rank	2-morphgrams	%	Frequency
11	passive (pasv)	2,50	1,976,830
21	causative (caus)	1,35	1,071,278
41	reciprocal (recp)	0,33	262,302
46	reflexive (refl)	0,14	108,156

The templates above predict that a voice suffix may be followed by negative, subordinator and agreement marker individually or in combination. The verb formed by attachment of a voice affix with a verb root may also be attached to any of the affixes from positions ordered in the finite template. The extracted morpheme sequences or morphgrams that include a voice suffix and its observed frequencies of combinations with suffixes from both templates is given Table 7 below:

Table 7. Top five 2-morphgrams

Rank	2-morphgrams	Frequency	Sample
1	past+3s	1,354,449	*ağladı*
2	p3s+loc	1,021,648	*içinde*
3	aor+3s	864,146	*bakar*
4	vi+past	838,367	*vardı*
5	p3s+acc	797,890	*kendisini*

The three verbal suffix sequences in the rank frequency list above are composed of position 2 and position 6 suffixes in the prescribed order, representing the minimal requirement in Turkish, namely, obligatory marking of tense and agreement. We do not find any occurrence of voice categories.

The voice suffixes and their combinations in 2-morphgrams appear in the following Table 8 with their respective observed frequency ranks:

Table 8. Rank frequencies of 2-morphgrams including voice categories

Rank	2-morphgrams	Frequency	Sample
22	pasv+pcan	313,078	*yapılan*
25	pasv+nzma	296,006	*ezilme*
34	caus+pasv	220,228	*yaptırılacak*
48	caus+nzma	157,896	*çıkarma*
84	recp+caus	68,683	*tanıştır*
134	recp+pasv	27,889	*anlaşılan*
157	pasv+pasv	19,251	*bulunulacak*
173	refl+nzma	16,391	*övünme*
179	refl+neg	14,217	*gezinme*

The frequencies of 2-morphgrams indicate order of suffixes from different positions to combine with the voice categories. The most productive orderings of passives are those with the nominalizers -*an* and -*ma*. The passive combines with perfective from the 3rd position in the finite template and followed by modality suffix -*ebil* and negative -*ma* when ranked with respect to their relative frequencies. The second most productive voice category *causative*, first combines with the passive from the same position in non-finite template and then with nominalizer -*ma* in the same template from position 3. Causative combining with a suffix from the finite template appears 3rd in rank is the aorist from position 3 which is followed by negative -*ma* in the frequency ranking. The first citation of reciprocal in 2-morphgrams list is where it is followed by another voice category, the causative and by the passive in its third most productive combination. In between reciprocal combines with nominalizer -*ma*, and finally with the negative. Thus, in the list of frequent 2-morphgrams, all citations of reciprocal are with combinations from the same template. The least productive reflexive combines only with nominalizer -*ma* and the negative, again from its own template.

In 3-morphgrams, we observe exponential increase in the citations of voice categories. 3-morphgrams including a passive appears in the top ten list and five more other combinations of suffixes with categories are cited in the top 20 list. The most frequent passive 3-morphgrams are again including nominalizers from position 2 in non-finite template, both followed by the same nominal agreement marker. In other productive combinations, passive combines with position 3 suffixes from the finite template (Tables 9 and 10).

Table 9. Most frequent 3-morphgrams in the TNC

Rank	3-morphgrams	Frequency	Sample
1	vi+past+3s	646,729	*okudu*
2	pasv+nzma+p3s	227,646	*edilmesi*
3	pcdk+p3s+acc	194,116	*olduğunu*
4	pcdk+p2s+acc	192,053	*olduğunu*
5	imprf+vi+past	186,393	*ediyordu*

Table 10. 3-morphgrams with voice in the TNC

Rank	3-morphgrams	Frequency	Sample
2	pasv+nzma+p3s	227,646	*edilmesi*
12	pasv+perf+3s	113,122	*edilmiş*
29	caus+pasv+nzma	55,745	*geliştirilmesi*
45	caus+past+3s	39,557	*belirtti*
57	recp+caus+nzma	28,229	*bölüştürme*

The list of 3-morphgrams that includes a voice category identifies passive as the most productive. The top two passive combinations, as in the case of 2-morphgrams are formed by attachment of nominalizers from non-finite template. The third affix to be attached is the agreement marker.

The top ten list of 4-morphgrams include more voice categories. Here, we have the first citation of a voice combination where passive outnumbers the other voice categories. In all three citations passive is followed by position 2 and position 3 suffixes, which themselves are followed by position 4 suffixes and then by agreement (Table 11).

As the number of suffixes increases in a morphological complex so the number of voice categories that appear in these complexes (Table 12).

The above list of 4-morphgrams displays that when a voice suffix selects a nominalizer, it is most likely to be attached an agreement marker from the same template which itself precedes the case marker. This is expected since nominalized clauses in Turkish are not differentiated from simple nominals in affixation of nominal markers.

We observe doubling of morphgrams with voice categories in the list of 5-morphgrams. Increase in number of suffixes here increases the number voice marked

Table 11. Most frequent 4-morphgrams in the TNC

Rank	4-morphgrams	Frequency	Sample
1	imprf+vi+past+3s	149,758	*ediyordu*
2	perf+vi+past+3s	127,325	*olmuştu*
3	pasv+perf+cop+3s	71,031	*edilmiştir*
4	aor+vi+past+3s	58,818	*olurdu*
5	pasv+cont+cop+3s	55,648	*görülmektedir*
6	caus+pasv+nzma+p3s	50,398	*geliştirilmesi*
7	aor+vi+avsa+3s	46,222	*olursa*
8	pasv+va1+aor+3s	40,355	*edilebilir*
9	imprf+vi+past+1s	25,903	*biliyordum*
10	va1+neg+aor+3s	25,437	*olamaz*

Table 12. 4-morphgrams with voice in the TNC

Rank	4-morphgrams	Frequency	Sample
3	pasv+perf+cop+3s	71,031	*edilmiştir*
5	pasv+cont+cop+3s	55,648	*görülmektedir*
6	caus+pasv+nzma+p3s	50,398	*geliştirilmesi*
15	pasv+nzma+p3s+acc	23,260	*yapılmasını*
18	pasv+pcdk+p2s+acc	21,975	*bulunduğunu*
20	pasv+neg+aor+3s	21,182	*vazgeçilmez*
28	caus+pasv+perf+3s	16,530	*geliştirilmiş*
35	recp+caus+neg+imp2	13,672	*eşleştirme*
41	caus+cont+cop+3s	11,554	*oluşturmaktadır*
161	refl+pasv+neg+aor	2,713	*kaçınılmaz*
162	refl+imprf+vi+past	2,707	*görünüyordu*

sequences to eight while it does not increase voice combinations. The most productive morphgrams include voice selecting position 3 suffix to follow which is also followed by position 4 suffix attached to the copula. The final in the sequence is the same agreement marker in all top 10 most frequent 5-morphgrams. It is interesting to note that there occur no nominalizers from non-finite template in these sequences. Thus, all top ten 5-morphgrams are composed of suffixes selected from the finite template (Tables 13 and 14).

In 6-morphgrams, the tendency in increasing the number of voice categories entering into the combinations does not increase but it decreases. Furthermore, apart from one citation of causative, there are no occurrences of other voice categories other than the passive. The main contribution to 6-morphgrams comes from the introduction of modality suffixes from position 1 and position 2 and eight citations of negative in the sequences. Eight of the finite template sequences end with 3rd person singular agreement suffix contributing to their relative high frequency. The nominalizer *-ecek* appears for the first time in a productive sequence.

Table 13. Most frequent 5-morphgrams in the TNC

Rank	5-morphgrams	Frequency	Sample
1	pasv+perf+vi+past+3s	19,049	*belirtilmişti*
2	neg+imprf+vi+past+3s	17,498	*istemiyordu*
3	pasv+imprf+vi+past+3s	16,130	*bulunuyordu*
4	neg+perf+vi+past+3s	12,025	*kalmamıştı*
5	neg+aor+vi+past+3s	10,784	*olmazdı*
6	pasv+val+neg+aor+3s	10,238	*edilemez*
7	caus+pasv+perf+cop+3s	9,783	*belirtilmiştir*
8	pasv+aor+vi+avsa+3s	9,322	*bakılırsa*
9	caus+imprf+vi+past+3s	8,832	*sarkıtıyordu*
10	caus+perf+vi+past+3s	7,881	*taşıtmıştı*

Table 14. Most frequent 6-morphgrams in the TNC

Rank	6-morphgrams	Frequency	Sample
1	val+neg+imprf+vi+past+3s	3,910	*bilemiyordu*
2	val+neg+aor+vi+past+3s	3,768	*olamazdı*
3	val+neg+perf+vi+past+3s	2,206	*bulamamıştı*
4	caus+pasv+perf+vi+past+3s	1,965	*belirtilmişti*
5	val+neg+imprf+vi+past+1s	1,879	*bilemiyordum*
6	pasv+val+neg+perf+cop+3s	1,471	*bulunamamıştır*
7	pasv+val+aor+vi+past+3s	1,400	*edilebilirdi*
8	pasv+neg+imprf+vi+past+3s	1,351	*bulunmuyordu*
9	pasv+val+neg+pcck+p3s+acc	1,349	*edilemeyeceğini*
10	pasv+val+neg+pcck+p2s+acc	1,318	*edilemeyeceğini*

In the remaining extracted well-formed morphgrams composed of seven, eight and nine suffixes, all of the most frequent top ten sequences include voice suffixes. In other words, it is almost always the existence of voice categories or their combinations that produce these complex morpheme bundles (Table 15).

Table 15. Most frequent 7-morphgrams in the TNC

Rank	7-morphgrams	Frequency	Sample
1	pasv+val+neg+aor+vi+past+3s	934	*söylenemezdi*
2	pasv+val+neg+imprf+vi+past+3s	394	*dayanamıyordu*
3	caus+val+neg+imprf+vi+past+3s	323	*kestiremiyordu*
4	pasv+val+neg+perf+vi+past+3s	313	*bulunamamıştı*
5	pasv+val+neg+aor+vi+avsa+3s	276	*sağlanamazsa*
6	caus+pasv+val+neg+perf+cop+3s	239	*gerçekleştirilememiştir*

(continued)

Table 15. (*continued*)

Rank	7-morphgrams	Frequency	Sample
7	caus+va1+neg+aor+vi+past+3s	196	*anlatamazdı*
8	pasv+va1+neg+imprf+vi+avsa+3s	188	*yapılamıyorsa*
9	pasv+pasv+neg+aor+vi+past+3s	186	*kaçınılmazdı*
10	caus+va1+neg+perf+vi+past+3s	181	*çıkaramamıştı*

There is no citation of nominalizers from non-finite template, while in all sequences we find the negative. The position of the negative imposes a grammatical constraint that it should always be followed by position 3 and 4 suffixes that are followed by agreement markers. The higher frequency of these morpheme bundles also correlates with the 3rd person singular marking serving as a closing suffix of the sequence (Table 16).

Table 16. Most frequent 8-morphgrams in the TNC

Rank	8-morphgrams	Frequency	Sample
1	caus+pasv+va1+neg+aor+vi+past+3s	54	*anlatılamazdı*
2	caus+pasv+va1+neg+aor+vi+avsa+3s	47	*oluşturulamazsa*
3	caus+pasv+va1+neg+imprf+vi+past+3s	41	*kestirilemiyordu*
4	caus+pasv+va1+neg+perf+vi+past+3s	37	*gerçekleştirilememişti*
5	caus+pasv+va1+neg+imprf+vi+avsa+3s	18	*düşürülemiyorsa*
6	pasv+va1+neg+nzma+p3s+vi+past+3s	18	*alınamamasıyla*
7	recp+pasv+va1+neg+perf+vi+past+3s	15	*anlaşılamamıştı*
8	recp+pasv+va1+neg+imprf+vi+past+3s	13	*anlaşılamıyordu*
9	recp+pasv+va1+neg+aor+vi+past+3s	12	*anlaşılamazdı*
10	pasv+pasv+va1+neg+aor+vi+past+3s	12	*denilemezdi*

The 8-morphgram sequences owe their expansion to the combination of different voice categories. While we find couples in 8-morphgrams, we find triples in 9-morphgrams listed below. In all combinations, passive is the most productive of voice categories. The reciprocal-passive combinations take the same lexical verb root *anla-* 'to comprehend' also sharing the negative, the modality suffix from position 1 and past marking followed the 3rd person singular.

The remaining non-voice 8-morphgrams are nominalized forms. The modality suffix from position 1 of the finite template is followed by negative marker (which is obligatory in case of the modality suffix in question) with further attachment of a nominalizer from non-finite template. Nominal agreement markers in the sequence are then followed by copula which combines with position 4 suffixes from the finite template (Table 17).

Table 17. Non-voice 8-morphgrams in the TNC

Rank	Non-voice 8-morphgrams	Frequency	Sample
1	va1+neg+pcan+pl+abl+vi+past+3s	5	*gidemeyenlerdendi*
2	va1+neg+pcdk+pl+p1p+vi+past+3s	3	*anlayamadıklarımızdı*
3	va1+neg+nzma+p3p+abl+vi+past+3s	2	*bakamamalarındandı*
4	va1+neg+nzma+p3s+abl+vi+past+3s	2	*kurtulamamasındandı*
5	va1+neg+pcdk+pl+p1s+vi+past+3p	2	*geçemediklerimdi*

Table 18. The most frequent 9-morphgrams in the TNC

Rank	9-morphgrams	Freq.	Sample
1	recp+caus+pasv+va1+neg+aor+vi+past+3s	5	*karşılaştırılamazdı*
2	recp+pasv+va1+neg+nzma+p3s+vi+past+3s	2	*anlaşılamamasıydı*
3	caus+caus+pasv+va1+neg+aor+vi+past+3s	2	*çıkartılamazdı*
4	caus+caus+pasv+neg+nzma+p3s+vi+past+3s	1	*çıkartılmamasıydı*
5	caus+caus+pasv+va1+neg+imprf+vi+past+3s	1	*çıkartılamıyordu*
6	recp+caus+pasv+va2+neg+perf+vi+past+3s	1	*geçiştirilivermemişti*
7	recp+caus+pasv+va1+neg+nzma+p3s+cop+3s	1	*ayrıştırılamamasıdır*
8	caus+caus+va1+va1+neg+aor+vi+perf+3s	1	*düşürtebilemezmiş*
9	pasv+va1+neg+pcan+pl+abl+vi+past+3s	1	*dayanamayanlardandı*
10	caus+caus+pasv+va1+neg+imprf+vi+avsa+3s	1	*çıkartılamıyorsa*

Morpheme bundles with 9 suffixes are expansions of voice suffix sequences in general. In the following Table 18 other than three morphgrams, all occurrences are voice triplets. The most productive *caus-caus-pasv* sequences are attached to the same verb root, *çık* 'to go out'. The number of components in these morphgrams increases by the affixation of negative and the modality suffix from position 1 in the finite template.

The gradual increase in the number of suffixes and their combinations and the role of voice suffixes in increase is summarized Table 19 below.

Table 19. Voice affixes in morphgrams

2-morphgrams	0	5-morphgrams	7	8-morphgrams	10
3-morphgrams	1	6-morphgrams	6	9-morphgrams	10
4-morphgrams	4	7-morphgrams	10		

The voice affixes may also select suffixes from their own position, in other words may derive combinations of suffixes of the same type. Voice combinations in Turkish appear to follow predictions of Mirror Principle which suggests that morpheme order mirrors the syntactic operations [2, 5]. The grammatical combinations of voice categories identify the following permissible combinations [6].

1. V-REC-CAUS-PASS   4. V-CAUS-PASS   7. V-PASS-PASS

2. V-REC-CAUS   5. V-CAUS-CAUS

3. V-REC-PASS   6. V-REF-PASS

The written component of TNC cites samples for each of these combinations. The corpus frequencies of these combinations and their cross tabulations are given in Table 20 below.

Table 20. Cross tabulation of voice suffixes

1st slot * 2nd slot cross tabulation						
			2nd slot			
			caus	pasv	recp	Total
1st slot	caus	Observed frequency	17399	220228	0	237627
		Expected frequency	56415,0	179755,1	1456,9	237627,0
	pasv	Observed frequency	0	19251	2223	21474
		Expected frequency	5098,1	16244,2	131,7	21474,0
	recp	Observed frequency	68683	27889	0	96572
		Expected frequency	22927,2	73052,8	592,1	96572,0
	refl	Observed frequency	0	6915	0	6915
		Expected frequency	1641,7	5230,9	42,4	6915,0
Total		Observed frequency	86082	274283	2223	362588
		Expected frequency	86082,0	274283,0	2223,0	362588,0

5 Statistical Tests

The significance of voice+voice combinations is also checked. To determine whether the combination of the two variables - the first and the second slot entries - are statistically significant or not, the chi-square test is applied to the categorical variables.

As is seen in Table 20, since zero cells (0%) have the expected frequency count less than 5, the chi-square result displayed on Table 21 can be interpreted. It should be noted that the values of reflexive suffix were excluded from the chi-square analysis since they do not occur in the second slot.

According to Table 21, there is a statistically significant relation between the first and the second suffixes observed on verbs ($\chi^2 = 198483,646$; $p < 0,05$).

Table 21. Chi-square test

Chi-square tests	Value	df	Asymp. Sig. (2-sided)
Pearson chi-square	198483,646[a]	6	,000
Likelihood ratio	168477,606	6	,000
Linear-by-linear association	108046,078	1	,000
N of valid cases	362588		

[a]0 cells (,0%) have expected count less than 5. The minimum expected count is 42,40.

The results of the 1st chi-square test is as follows:

- 1st slot is filled by **causative** → $\chi^2 = 37552,63$

$$\frac{(G-B)^2}{B} = \frac{(17399 - 56415)^2}{56415} + \frac{(220228 - 179755,1)^2}{179755,1} + \frac{(0 - 1456,9)^2}{1456,9}$$

- 1st slot is filled by **passive** → $\chi^2 = 38875,66$
- 1st slot is filled by **reciprocal** → $\chi^2 = 119829,1$
- 1st slot is filled by **reflexive** → $\chi^2 = 2226,274$

Above, one can see the results of the first chi-square analysis. This shows that passive suffix has the highest chi-square value. In the second stage of the analysis, the passive suffixes were excluded, and the chi-square test was applied once more. The results of the second analysis show that the causative suffix has the highest chi-square value. For the third analysis, upon excluding the most frequent voice suffix namely the causative, the chi-square test was implemented once again. This time the first slot is filled by reciprocal and reflexive. The results of the chi-square test are again significant for both of these variables. Taken together, the results of chi-square test prove that the first and the second slot entries observed on verbs of written part of the TNC are in statistically significant relationship.

The results of the 2nd and the 3rd chi-square tests are as follows:

- ($\chi^2 = 150518,933$; $p < 0,05$) → The chi-square result of the 2nd analysis.
- 1st slot is filled by **causative** → $\chi^2 = 40415,9$
- 1st slot is filled by **reciprocal**: $\chi^2 = 107769$
- 1st slot is filled by **reflexive**: $\chi^2 = 2334,048$
- ($\chi^2 = 14623,349$; $p < 0,05$) → The chi-square result of the 3rd analysis.

6   Conclusion

Corpus data suggests that productive multi-word units in many languages may correspond to multi-morpheme units in agglutinative languages [3]. This study provided a partial answer to identifying contexts use and relevant frequencies of voice categories that are formalized as suffixes attached productively to verb roots. Extracted from a balanced and a representative corpus of contemporary Turkish, we have presented basic quantificational distributions of suffixes from finite and non-finite templates.

A future research will include other suffixes from both templates and derive combinatorial potentials of verbal suffixes in more principled way. The frequency information coupled with rules constraining suffix order may contribute to morphological processing of natural language data.

Acknowledgements. This study is supported by a research grant from the Scientific and Technological Research Council of Turkey (TÜBİTAK, Grant No: 115K135, 113K039).

References

1. Aksan, Y., Aksan, M., Koltuksuz, A., Sezer, T., Mersinli, Ü., Demirhan, U.U., Yılmazer, H., Kurtoğlu, Ö., Atasoy, G., Öz, S., Yıldız, İ.: Construction of the Turkish National Corpus (TNC). In: Proceedings of the 12th International Conference on Language Resources and Evaluation (LREC), pp. 3223–3227 (2012)
2. Baker, M.: The mirror principle and morphosyntactic explanation. Linguist. Inquiry **16**(3), 373–415 (1985)
3. Durrant, P.: Formulaicity in an agglutinative language. Corpus Linguist. Linguist. Theor. **9**(1), 1–38 (2013)
4. Enç, M.: Functional categories in Turkish. In: Proceedings of WAFL 1, MIT Working Papers in Linguistics, vol. 46, pp. 208–225 (2004)
5. Göksel, A.: Levels of representation and argument structure in Turkish. Ph.D. thesis, University of London (1993)
6. Göksel, A., Kerslake, C.: Turkish: A Comprehensive Grammar. Routledge, London (2005)
7. Hakkani-Tür, D.Z., Oflazer, K., Tür, G.: Statistical morphological disambiguation for agglutinative languages. Comput. Humanit. **36**, 381–410 (2002)
8. Hankamer, J.: Morphological parsing and lexicon. In: Marslen-Wilson, W. (ed.) Lexical Representation and Process. MIT Press, Cambridge (1989)
9. Manova, S. (ed.): Affix Ordering Across Languages and Frameworks. OUP, New York (2015)
10. Aksan, M., Mersinli, Ü.: A corpus based NooJ module for Turkish. In: Proceedings of the NooJ 2010 International Conference and Workshop, pp. 29–39. Democritus University Press, Komotini (2011)
11. Sezer, E.: Finite inflection in Turkish. In: Erguvanlı, E. (ed.) The Verb in Turkish. John Benjamins, Amsterdam (2001)
12. Göksel, A.: The auxiliary verb *ol* at the morphology-syntax interface. In: Erguvanlı, E. (ed.) The Verb in Turkish. John Benjamins, Amsterdam (2001)
13. Kornfilt, J.: Turkish. Routledge, London (1997)
14. Pierce, J.E.: A frequency count of Turkish affixes. Anthropol. Linguist. **3**, 31–42 (1961)
15. Pierce, J.E.: Study of grammar and lexicon in Turkish and Sahaptin (Klikitat). Int. J. Am. Linguist. **29**(2), 96–106 (1963)
16. Güngör, T.: Lexical and morphological statistics for Turkish. In: Proceedings of International XII. Turkish Symposium on Artificial Intelligence and Neural Networks (2003)

Pluralising Nouns in isiZulu and Related Languages

Joan Byamugisha[1], C. Maria Keet[1(✉)], and Langa Khumalo[2]

[1] Department of Computer Science, University of Cape Town,
Cape Town, South Africa
{jbyamugisha,mkeet}@cs.uct.ac.za

[2] Linguistics Program, School of Arts, University of KwaZulu-Natal,
Durban, South Africa
Khumalol@ukzn.ac.za

Abstract. There are compelling reasons for a Controlled Natural Language of isiZulu in software applications, which requires pluralising nouns. Only 'canonical' singular/plural pairs exist, however, which are insufficient for computational use of isiZulu. Starting from these rules, we take an experimental approach as virtuous spiral to refine the rules by repeatedly testing two test sets against successive versions of refined rules for pluralisation. This resulted in the elucidation of additional pluralisation rules not included in typical isiZulu textbooks and grammar resources and motivated design choices for algorithm development. We assessed the potential for reuse of the approach and the type of deviations with Runyankore, which demonstrated encouraging results.

1 Introduction

Although the imperative for Human Language Technologies in isiZulu—and, in fact, most Bantu languages—exists, the language is still under-resourced, especially for computational information and knowledge processing [19]. While some results have been obtained in natural language understanding, such as morphological analysers, there are scant results for *generating* isiZulu [11,12]. We take here one aspect of the generation that, perhaps, seems rather basic: pluralising a noun. Automation of pluralisation is useful for software interfaces that need a controlled natural language (CNL) [11], technology-assisted learning, and other uses. A (very) simple example is one that in the calendar or weather forecast it should be '1 day' and '2 days', not '1 days' or '2 day', and it has been noted as a requirement for collecting and displaying information for South Africa's National Indigenous Knowledge Management System [1].

There is no computational approach for pluralising isiZulu nouns yet and, to the best of our knowledge, not for any language in the Bantu language family (of which isiZulu is a member). Looking at better resourced languages, the bulk of automation of pluralisation has been carried out in the 1990s. For instance, Conway's English pluralisation algorithm [7] is based on extensive linguistic resources

© Springer International Publishing AG, part of Springer Nature 2018
A. Gelbukh (Ed.): CICLing 2016, LNCS 9623, pp. 271–283, 2018.
https://doi.org/10.1007/978-3-319-75477-2_18

that easily could be used for specification of the rules computationally. Likewise, rules for German pluralisation and their occurrences are well-known and have been experimentally assessed thanks to accessible databases [15]. Both rely on regular expressions of the ending of the nouns. NLP research for Arabic, a more complex language, took up in the 2000s and, regarding pluralisation, now focuses more on comparing techniques [3] and refinements [2], noting that there are several usable computational resources for Arabic already.

The state-of-the-art for nouns from the linguistics viewpoint covers some theory and experiments on morphological analysers without online resources [5,18], and a basic POS tagger and corpus exist [20]. Some recent advances in linguistics describe what goes in which noun class (NC) with which prefix [16,17,21], and all textbook resources have the 'standard table' of prefixes, as included in Table 1, but this does not state *when* or *why* which prefix is used in the plural, and whether this is the only thing to consider in pluralising nouns. Some of its limitations are known with anecdotical evidence: (i) there are variations for some NCs even in the standard table, but no indication when to use which one; (ii) some are known to be phonologically conditioned; (iii) some do not have plurals; and (iv) for some nouns the categorisation into the NC is not well-established and thus may generate exceptions.

This raises the following three questions:

Q1: How well does the 'standard' table of prefixes work for pluralising nouns?
Q2: What are the exceptions, and are there any rules among the exceptions?
Q3: Do the types of exceptions appear in Bantu languages other than isiZulu?

One option to answer this is to try to avail of corpora; however, the only available corpus consists of the bible and a few fiction novels [20], and, in taking a corpus-based approach, one needs the rules we aim to find. Practically, this is a chicken-and-egg problem. We propose here to take a combined approach to answer the questions, using rules with experimental evaluation to iteratively improve the rules and results of the pluraliser. We achieved over 90% correct pluralisation on the test data eventually, identified types of exceptions, and elucidated eight pluralisation rules beyond the 'standard table'. Question 3 was evaluated with Runyankore, a Bantu language spoken in Uganda and orthographically similar to isiZulu, which achieved higher initial correctness with the first test set, but it also had similar types of exceptions, suggesting that a bootstrapping approach—reusing the approach and similar rules presented here—for orthographically similar Bantu languages would be feasible and reduce development time.

In the remainder of the paper we first provide the basic background to isiZulu (and, by extension, Bantu) nouns and verbalisation of structured knowledge in Sect. 2. Section 3 discusses design considerations for developing the pluraliser and Sect. 4 presents the main results of the pluraliser. We discuss in Sect. 5 and conclude in Sect. 6.

2   Background

It must be accentuated that isiZulu is a computationally under-resourced language. It is a Bantu language that belongs to the Nguni group of languages, which include isiXhosa, isiNdebele and siSwati. It is the most popular language in South Africa spoken as a first (home) language by about 23% of South Africa's over 50 million people. Bantu languages have a characteristically agglutinating morphology, which makes it a challenge to develop computational technologies for them. One of the salient features of isiZulu, which is also true for other Bantu language like Runyankore, is the unique system of noun classes. This will be elaborated on in the next section, after which we introduce a motivating example use case.

2.1   isiZulu Nouns

Each noun is allocated a specific noun class. The canonical NC list is Table 1. The noun comprises of two formatives, the prefix and the stem. The prefix can be identified as a full prefix or an incomplete prefix. It is a full prefix when the augment (pre-prefix) is followed by a prefix proper. It is an incomplete prefix when it only has the augment. An example of a full prefix is *isihlalo* 'chair', *i-* (augment) *si-* (prefix proper) *-hlalo* (stem). An example of an incomplete prefix is *ubaba* 'father', *u-* (augment) *-baba* (stem). Because of the agglutinating nature of isiZulu coupled with a conjunctive writing system, which glues together elements of an isiZulu word, a number of NC prefixes in isiZulu are phonologically conditioned and yet others are homographs.

The NC has received considerable attention recently [17,21,22] in an effort to explicate the generation of, and semantic motivation for, the various NC assignments. Most NCs are paired such that there is a distinctive pattern of a singular form in one class and a plural form in another, and yet other classes are latent. It is notable in isiZulu that the NC prefix of class 1 and 3 is conditioned by the morphology of the stem it takes: *-mu-* before monosyllabic stems and *-m-* for other stems. Similarly, NC 5 and 11 are conditioned: the full prefixes only take monosyllabic stems [21], e.g., nc5 *ili-+-hlo* 'eye' nc11 *ulu-+-thi* 'stick'. Interestingly, the *n* of NCs 9 and 10 merges with the following consonant forming prenasalized consonants.

There are other known deviations with loanwords, vowel-commencing roots, and concordance in the second term of a compound noun, and anecdotal evidence suggests these are not catered for with the 'standard' table of singular-plural pairings. Pluralisation in Runyankore seems to have similar exceptions as isiZulu.

2.2   Use Case: Toward an isiZulu Controlled Natural Language

Controlled Natural Languages (CNLs) are a fragment of the full natural language that are used in applications with contextual text, such as automated generation of weather reports and medical apps, eLearning, and in prescription notes and instruction manuals. There are few results for an isiZulu CNL

Table 1. Zulu noun classes, with the 'canonical' list of prefixes for isiZulu and Runyankore; NC: Noun class, AU: augment, PRE: prefix, n/a: class not used.

NC	isiZulu		Runyankore		NC	isiZulu		Runyankore	
	AU	PRE	AU	PRE		AU	PRE	AU	PRE
1	u-	m(u)-	o-	mu-	9	i(n)-	-	e-	n-, m-
2	a-	ba-	a	ba-	10	i-	zi(n)-	e-	n-
1a	u-	-	n/a		11	u-	(lu)-	o-	ru-
2a	o-	-	n/a		(10)	i-	zi(n)-	n/a	
3a	u-	-	n/a		12	n/a		a-	ka-
(2a)	o-	-	n/a		13	n/a		o-	tu-
3	u-	m(u)-	o-	mu-	14	u-	bu-	o-	bu-
4	i-	mi-	e-	mi-	15	u-	ku-	o-	ku-
5	i-	(li)-	e-	i-, ri-	6	n/a		a-	ma-
6	a-	ma-	a-	ma-	16	n/a		a-	ha-
7	i-	si-	e-	ki-	17	-	ku-	-	ku-
8	i-	zi-	e-	bi-	18	n/a		o-	mu-
9a	i-	-	n/a		20	n/a		o-	gu-
(6)	a-	ma-	n/a		21	n/a		a-	ga-

covering verbalisation patterns [12] and algorithms for subsumption, disjointness, conjunction, existential quantification and its negation [11]. This revealed that a template-based approach is not feasible due to isiZulu being a highly agglutinating language and the many NCs and the concordances it requires. To illustrate this, let us take the common Description Logic (DL) language $\mathcal{ALC}$ [4] as knowledge representation language, which is becoming increasingly popular for CNLs [6], especially in its serialised form, OWL [14]. The straightforward 'all some' pattern ($C \sqsubseteq \exists R.D$ 'each C R at least one D') with for relationship R a present tense verb already generates interesting issues. A few corresponding pretty-printing examples are shown in Fig. 1. While the universal quantification

(1) **Grandmother** $\sqsubseteq$ $\exists$**eats.Apple**
    bonke <u>ogogo</u> badla i-aphula elilodwa
                        ('All <u>grandmothers</u> eat at least one apple')
(2) **Nurse** $\sqsubseteq$ $\exists$**eats.Apple**
    bonke <u>abongi</u> badla i-aphula elilodwa
                        ('All <u>nurses</u> eat at least one apple')
(3) **Leopard** $\sqsubseteq$ $\exists$**eats.Apple**
    zonke <u>izingwe</u> zidla i-aphula elilodwa
                        ('All <u>leopards</u> eat at least one apple')

Fig. 1. Pretty-printing of the verbalisation of 'all x eat at least one apple', with the plural underlined; their respective singulars are *ugogo* 'grandmother' (nc1a), *umongi* 'nurse' (nc1), and *ingwe* 'leopard' (nc9).

∀ (silent on the left-hand side of the subsumption "⊑") is just "each" or "for all" in English, in isiZulu it depends on the NC of the first noun, resulting in *bonke*, *zonke* etc., and, moreover, this first noun has to be in the plural according to [12]. However, although such knowledge is normally represented in the singular and the algorithms in [11] do not describe the pluralisation step, yet three patterns rely on pluralisation to generate a grammatically correct sentence (disjointness and existential quantification). The success of these algorithms hinge on being able to find the correct plural for the given singular.

3   Design Considerations

For the design of the pluralisation rules and algorithms, there are two distinct approaches from a linguistic viewpoint, and several representation options for each. Concerning the language, one can

(1) base it on a purely syntactic/orthographic analysis, relying on the (patterns of the) characters of the string of text;
(2) base it on a semantic analysis using the meaning of the nouns and, implicitly with that, their respective NC.

The consequence for a computational approach using the first option is that it will require some regular expressions with as formal foundation a Finite State Automaton for the first part of the process, being to analyse the word, or more advanced FSMs that also will handle the change in prefix. On cursory glance, it would seem similar to a pluraliser for English that processes a noun's final character(s), as with the design and Perl implementation by [7], but then applied to the beginning character(s) instead. This is not exactly the case, as we shall see below. The second one entails more investment upfront through either adding some encoding of the meaning or, in lieu of that, that at least the NC of the noun is stored with the noun.

We illustrate consequences of these first, basic, design choices through an example with aiming to pluralise a noun in the singular in nc1 into a noun in the plural in nc2; the same considerations hold likewise for other clusters of NCs with the same prefixes (nc1a, nc2a, nc3a, and nc11; and nc5, nc9a, and nc9). To abstract away from actual code, we use an automata-based programming notation, which is like the usual automata as used for morphological analysers, but the transitions/steps are code sections, like a condition, function, or some other routine, rather than consuming characters of the word inputted to the FSM. Three options are shown in Fig. 2, where, e.g., the "*noun*[0 : 1] == um" is to be understood as 'if the first and second character of the variable *noun* equals the character string um' [then transition state from state p to state q] and "$\frac{um}{aba}$" 'replace the first characters um with aba'. Returning the plural is the same for all and omitted for being trivial at this stage for the current argument.

The first option is a pure string-based approach, and systematically going through the NC pairs of Table 1, as depicted in the top-most automaton: check whether the word starts with umu or um, and replace the 2- or 3-character prefix

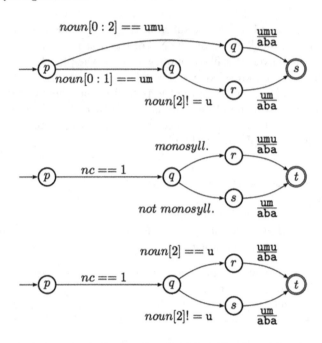

Fig. 2. Three variations in designing the algorithm illustrated for the case for noun classes 1 and 2, using an automata-based programming notation.

with the prefix for the plural, aba; e.g., _umuntu_ (singular 'human') $\mapsto$ _abantu_ (plural). When we arrive at setting up the conditions for nc3, however, the conditions are the same, but the 2- or 3-character prefix is to be replaced by imi instead (e.g., _umumba_ $\mapsto$ _imimba_ 'stuff box containers'). In addition, a mass noun like _ummbila_ 'maize', in nc3, does not have a plural. Thus is, using a string-based approach only is expected to lead to a considerable error percentage in pluralisation.

To distinguish between when to swap um for aba and umu for aba, one either uses characters again, or checks whether the stem is monosyllabic. There are, however, no documented string-based rules for recognising monosyllabic stems, for it depends on the correct identification of the prefix, which is non-deterministic and thus always will have a certain error rate. That is, this is not an option at present, unless a list of monosyllabic stems is constructed upfront and consulted during pluralisation. The third strategy is to use both the NC and the prefix analysis, provided that it is annotated somewhere in the input (bottom figure in Fig. 2). This design in particular would cause a problem only if there were u-commencing stems in NC1. There are none, though there are a few that commence with an a, e, i, or o; e.g. _umongi_ 'nurse'. This strategy, therefore, seems to have the most potential, hence probably results in the least amount of errors.

Combining the separate automata for nc1 and nc3 and with the additional rule for vowel-commencing stems (elaborated on Sect. 4), i.e., commencing with NCs, results in 15 states, as does the one that commences with the string processing and then NC checking; the latter is included in Fig. 3. For *umumba*, the states visited in the execution trace are: $\{q, x\} \rightarrow \{r, y\} \rightarrow \{a\} \rightarrow \{c\} \rightarrow \{d\}$.

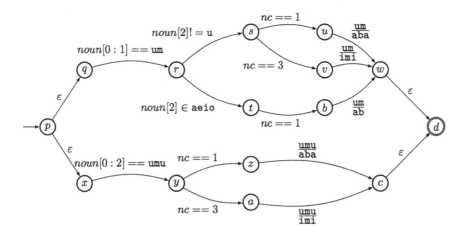

Fig. 3. The automata-based programming notation for words in nc1 and nc3 combined, using NC information of the noun.

4   Evaluation of the Pluraliser

The aim of the evaluation is to examine the rules for pluralisation of isiZulu nouns, to iteratively improve on the encoded rules, and to consider an orthographically similar language, Runyankore, primarily used in Uganda. We first describe the materials & methods, followed by the results for the isiZulu pluraliser and then the results for Runyankore.

4.1   Materials and Methods

We encoded a basic pluraliser based on the standard isiZulu noun prefix table (Table 1), and tested it against the data set (see below) of whole words with(out) their respective NC. Informed by the errors, new rules were repeatedly added to the pluraliser and the set re-tested. The evaluation of the correctness of the computed plurals is carried out with an isiZulu speaker. Accuracy is calculated as $\frac{|\text{correct plural}|}{|\text{nouns}|} * 100$.

The materials consist of the program code with its related input files (in txt). Due to lack of shared resources in isiZulu, we manually compiled two wordlists. The first set of nouns, *Set1*, is a 'random' list of 101 words informed by content of multiple ontologies (african wildlife, building, tourism, wine, and other domain ontologies) and general domain words covering categories of entities as in the DOLCE foundational ontology [13]. This list was created in English so as to ensure that the words were not cherry-picked for their linguistic features in isiZulu, but only on domain coverage. This list was manually translated into isiZulu and the NC manually added, which appeared to cover all isiZulu NCs except 9a and the locative nc17. The second set of 117 words, *Set2*, was compiled by taking the first-listed noun on every left-hand page of the Shuter & Shooter isiZulu Scholar's dictionary's isiZulu-English section [8]; hence, this set is an alphabetically balanced random sample, also with a near-full NC coverage (missing nouns in nc15 and nc17). For both sets, about 2/3 of the nouns are in either nc5, nc7, or nc9.

The same set of 101 English words of *Set1* were manually translated to Runyankore. 13 had no direct translations and four were added to cover NCs 20 and 21, as well as special orthographical cases; this resulted in a set of 92 words, *Set1r*. A second set was obtained from an attempt to extract every singular noun from the Runyankore dictionary [23]; 2542 words were extracted, *Set2r*, which represented all NCs except 20 and 21, mainly because they are considered derogatory and as such are common in speech but not writing.

The materials—code, output, and analysis—are online available at http://www.meteck.org/files/geni/ and http://www.CICLing.org/2016/data/49.

4.2   Results

isiZulu Pluraliser. The accuracy of the pluraliser versions is included in Table 2. The iterations could have been done in any order, and are listed in the order in which they have been carried out.

Table 2. Accuracy of the different versions of the pluraliser on the test sets.

Pluraliser version	Set1	Set2
Whole words	53	45
Words + nc	92	78
Words + nc + compounds	96	78
Words + nc + compounds + mass	99	87
Words + nc + compounds + mass + pl. exceptions	99	87
Words + nc + compounds + mass + pl. exceptions + prefix exceptions	100	91
Words + nc + compounds + mass + pl. exceptions + prefix exceptions + pl. only	100	92

Test "0": just words, and words with NC. The main contributor to the errors is sameness in prefixes for some NCs, as described in Sect. 3, so it pluralises on the first if-statement that evaluates to true, rather than the NC later in the code. Adding the NC explicitly in the input had the greatest reduction in error rate for both sets, as can be seen in Table 2.

Table 3. Additional pluralisation rules, denoted in easily readable pseudo-code (e.g. secondWordMinus1stLetter is in the code `second[1:]`).

No.	Rule
Compound nouns	
1.	**if** nc = 1 **and** 1stLetterOfTheSecondWord != vowel, **then** return 'b' + secondWordMinus1stLetter
2	**if** nc = 3a, **then** return 'aba' + secondWordMinus1stLetter
3	**if** (nc = 9 **or** nc = 7) **and** 1stLetterOfSecondWord != vowel, **then** return 'z' + secondWordMinus1stLetter
4.	**if** (nc = 9 **or** nc = 7) **and** 1stLetterOfSecondWord = vowel), **then** return secondWord1stLetter+'zi'+secondWordFrom4thLetterOnwards
Mass nouns	
5	**if** 'm' in nc, **then** return word
Prefix exceptions/vowel-commencing stems	
6.	**if** nc = 1 **and** thirdLetterOfWord in 'aeio', **then** return 'ab' + wordMinusFirstTwoLetters
7.	**if** nc = 7 **and** thirdLetterOfWord in 'aeou', **then** return 'iz' + wordMinusFirstTwoLetters
Noun in plural only	
8	**if** nc in '24682a' **or** nc = 10, **then** return word

First iteration: addressing compound nouns. Compound nouns in isiZulu typically have a nominal lexeme plus an adjective (e.g., *ukhilimu oyiqhwa* 'ice cream'). The initial morpheme of the adjective must be in agreement with the prefix of the first nominal lexeme, which means that the NC prefix must be in agreement with the possessive concord (PC) (e.g., *u-* must agree with *o-*). The first iteration included devising rules that correctly adds the applicable concord; e.g., *indawo yokubhukuda* (nc9) ↦ *izindawo zokubhukuda* 'swimming pools', where *in-* agrees with *yo-* (from the nc9 PC *ya-+u-* of *ukubhukuda* 'to swim'/'swimming') and *izin-* with *zo-* (from the nc10 PC *za-+u-*). There are some variations to this, such as the extra *e-* in *isilwane esifuyiweyo* (nc7) ↦ *izilwane ezifuyiweyo* 'pets', which is addressed by rules 1–4 in Table 3.

Second iteration: addressing the mass nouns. Mass nouns refer to those entities that are not countable on their own but only in certain quantities—in ontology also called amount of matter or stuff—such as 'water' and 'wine'. The term does not change in isiZulu, like in other languages, and they can be in any NC; e.g., *amanzi* 'water' is in nc6 and *iwayini* 'wine' is in nc5. No known marker exist to determine a noun is a mass noun, and a cursory evaluation on types of mass nouns (such as pure vs mixed stuffs [10]) did not reveal a pattern either. To be able to handle it computationally, we therefore append the NC with an 'm' and add rule 5 (Table 3), i.e., the "word" that will be returned is the same as the original noun.

Third iteration: addressing exceptions to plural classes. Exceptions to the plural classes given a NC for the singular do exist, but are very rare; e.g., *indoda* ↦ *amadoda* 'men'. They are true exceptions without rules and therefore put in a separate look-up file of exceptions for reusability.

Fourth iteration: addressing exceptions to the prefixes. There are (at least) two types of exceptions to the 'standard' prefixes. One is due to the stem beginning with a vowel rather than a consonant, such as *-akhiwo* as stem of *isakhiwo* ↦ *izakhiwo* 'buildings', and other phonologically conditioned plurals, such as *ucingo* ↦ *izingcingo* 'wires'/'telephones' (an additional *-g-*). The former is detectable by checking the third character of the noun for the vowel; e.g., it is *is-/iz-* rather than *isi-/izi-* for nc7, and *ab-* for nc2. This is captured in rules 6 and 7 (Table 3).

Fifth iteration: nouns in plural only. This issue came afore only with *Set2*, although it essentially also existed with *Set1*. Some nouns in both sets exist in the plural only, but were also mass nouns (*imicikilisho* 'slow work' and *amanzi* 'water'), so were addressed by that rule already. However, *amanqamu* 'final act' is not a mass noun, yet also in a plural NC (nc6). To cover any such instances, rule 8 is added (see Table 3).

Remaining errors. Generally, loanwords that have not been assimilated yet are assigned to nc5, such as *iradio* 'radio' and *i-okhestra* 'orchestra'. *Set2* contains the loanword *iaphula* 'apple', which also can be written as *i-aphula* so as to agree with the rule that isiZulu does not allow vowel sequencing. The latter writing would also simplify the rules for pluralisation, for then it is straightforwardly *i-aphula* ↦ *ama-aphula* rather than *iaphula* ↦ *ama-aphula*, and therewith increasing the accuracy of the pluraliser to 93%. The other errors have no obvious possible solution: the name of a disease remains in the singular, e.g., *isichenene* 'bilharzia', and when to use *izin-* or *izim-* and not *izi-* for nc9 and nc11, i.e., the plural class nc10. It has been suggested that *im-/izim-* applies to noun stems commencing with a labial sound (b, f, p, v) [9, p. 43], which is yet to be evaluated.

4.3   Generalisability: Runyankore

The pluraliser had 88% accuracy on *Set1r* without making the NC explicit. This is because classes 1, 3, and 18, all have the prefix *omu* and nc15 and nc17 both have the prefix *oku*. Making the NC explicit increased the accuracy to 92%, with the remaining errors mainly resulting from loanwords and compound nouns.

With further clarification that loan words, such as *univasite*, *guriini*, and *kompuyuta*, belong to nc9/10, and the inclusion of this in the pluraliser, the accuracy further improved to 97%. The mass nouns in *Set1r* belong to the plural class nc6, so are not pluralised. Compound nouns in Runyankore have the genitive of the main noun associated with the second noun: for example *ekyokurya ky'enjangu* 'cat food'. This is pluralised by pluralising the main noun, getting the genitive of this plural, and associating that with the second noun. Thus, *ekyokurya* ↦ *ebyokurya*, its genitive *ebya* is obtained, and this drops its initial vowel, and if the second noun starts with a vowel too, the genitive's ending vowel is replaced with an apostrophe and associated with the second noun. This results in *ebyokurya by'enjangu*. This final step resulted in 100% accuracy.

When this pluraliser was used on the larger *Set2r*, there was an initial accuracy of 67%. Two reasons for this have so far been identified. First, according to the NC table, nc12 should be pluralised as nc13. However, among Runyankore speakers, nc12 is instead pluralised as nc14. This also seems to be in agreement with the Runyankore dictionary, and when applied improved the accuracy to 71%. Second, there are two types of exceptions: (1) those words which do not get pluralised, even if they belong to a singular NC; such words include *eihangwe* 'day light', *eiriho* 'thirst', *orunyankore* (the language Runyankore), *orwakabiri* 'Tuesday', *omururu* 'greed', etc.; and (2) those words whose pluralisation does not follow the standard NC table such as *eka*, which should be in nc9/10 but is pluralised as *amaka* (nc6), and *orutunguru* (nc11) but is pluralized as *obutunguru* (nc14 instead of nc10). When these nouns were placed in a separate look-up file of exceptions, as was proposed for isiZulu, the accuracy improved to 74%.

Finally, some words for different nouns are written the same but can be differentiated when spoken; e.g., *omubazi* in writing can be either 'medicine' (nc3) or 'accountant' (nc1), but their respective plurals are *emibazi* 'medicines' and *ababazi* 'accountants'. These issues appear in isiZulu as well, because the tone is not indicated in the orthography (e.g., *umfundisi* (nc1/- or nc1/2)), but such instances were coincidentally not in the set of test words.

5   Discussion

For both isiZulu and Runyankore, adding the NC was useful, though for isiZulu that effect was more pronounced due to more sameness of prefixes compared to the Runyankore prefixes. Thus, the results show that pluraliser algorithms for Bantu languages, with their emblematic noun class system, will need to take a mixed syntax (word string combinations) and semantics (word meaning; NC at least) approach in order to achieve acceptable results in automated pluralisation. Answering research question Q1 stated in Sect. 1, it depended on the test set how well the 'standard' table works for pluralising nouns (78% and 92% accuracy without the extra rules), thereby deserving further investigation to discover and record additional pluralisation rules. The canonical structures of the nominal class system and the agreement morphology as well as those underlying patterns of the additional rules for isiZulu and Runyankore that were described are remarkably similar, making a compelling case for a bootstrapping approach, answering question Q3 in the affirmative.

The main outcome from a linguistic viewpoint, is that the 'standard' prefix table for each NC is definitely *not* comprehensive for isiZulu plurals. While we knew upfront that just words would return a substantial number of errors, the other assumption was that using the NC as well would be most effective and also suffice for all cases but a few exceptions to the rule. There are more 'exceptions' than anticipated, both true exceptions and regular ones. The combined rules-based and (limited) experiment with manual word analysis made it possible to make explicit, and thus start documenting, those rules that apply in pluralisation of isiZulu nouns beyond the 'standard' ones, contributing to answering question Q2. They may not yet cover all regular exceptions, but is expected to suffice within the CNL use case.

While this set did not show any issues with phonologically conditioning other than concerning vowel-commencing stems, they do exist (e.g., *ucingo* ↦ *izingcingo*). It is not known how they are conditioned, so this is a point of further investigation for linguists.

6   Conclusions

Automation and experimentally evaluating pluralising nouns from their singular to their respective plural showed that including the noun class of the noun is essential to obtaining reasonable success rates for isiZulu. This shows the need for a combination of syntax and semantics in the pluralisation algorithm. Resolving the remaining issues revealed several regular exceptions, which resulted in new pluralisation rules. They concern compound nouns, mass nouns, exceptions to the prefix due to the stem commencing with a vowel, and true exceptions. Some of them were also compared with another Bantu language, Runyankore, suggesting a generalisability of the types of exceptions across the Bantu language family.

Acknowledgements. This work is based on the research supported in part by the National Research Foundation of South Africa (CMK: Grant Number 93397), and by the Hasso Plattner Institute (HPI) Research School in CS4A at the University of Cape Town (JB).

References

1. Alberts, R., Fogwill, T., Keet, C.M.: Several required OWL features for indigenous knowledge management systems. In: Proceedings of the OWLED 2012. CEUR-WS, Crete, Greece, 27–28 May, vol. 849 (2012)
2. Alkuhlani, S., Habash, N.: Identifying broken plurals, irregular gender, and rationality in Arabic text. In: Proceedings of the 13th Conference of the European Chapter of the Association for Computational Linguistics, Avignon, France, 23–27 April 2012, pp. 675–685. ACL (2012)
3. Altantawy, M., Habash, N., Rambow, O., Saleh, I.: Morphological analysis and generation of Arabic nouns: a morphemic functional approach. In: Chair, N.C.C., Choukri, K., Maegaard, B., Mariani, J., Odijk, J., Piperidis, S., Rosner, M., Tapias, D. (eds.) Proceedings of the Seventh International Conference on Language Resources and Evaluation (LREC 2010). European Language Resources Association (ELRA), Valletta, May 2010
4. Baader, F., Calvanese, D., McGuinness, D.L., Nardi, D., Patel-Schneider, P.F. (eds.): The Description Logics Handbook - Theory and Applications, 2nd edn. Cambridge University Press, Cambridge (2008)

5. Baumann, P., Pierrehumbert, J.: Using resource-rich languages to improve morphological analysis of under-resourced languages. In: Maegaard, B., Mariani, J., Moreno, A., Odijk, J., Piperidis, S. (eds.) Proceedings of 9th International Conference on Language Resources and Evaluation (LREC 2014), Reykjavik, Iceland, 26–31 May 2014. European Language Resources Association (ELRA) (2014)
6. Bouayad-Agha, N., Casamayor, G., Wanner, L.: Natural language generation in the context of the semantic web. Semant. Web J. **5**(6), 493–513 (2014)
7. Conway, D.M.: An algorithmic approach to English pluralization. In: Salzenberg, C. (ed.) Proceedings of the Second Annual Perl Conference, San Jose, USA, 17–20 August, 1998. O'Reilly (1998)
8. Dent, G.R., Nyembezi, C.L.S.: Scholar's Zulu Dictionary, 4th edn. Shuter & Shooter Publishers, Songololo (2009)
9. Groenewald, H.C., Turner, N.S.: Siyasiqonda! Basic Zulu for the Health Sciences. 144 p. University of KwaZulu-Natal, South Africa
10. Keet, C.M.: A core ontology of macroscopic stuff. In: Janowicz, K., Schlobach, S., Lambrix, P., Hyvönen, E. (eds.) EKAW 2014. LNCS (LNAI), vol. 8876, pp. 209–224. Springer, Cham (2014). https://doi.org/10.1007/978-3-319-13704-9_17
11. Keet, C.M., Khumalo, L.: Basics for a Grammar engine to verbalize logical theories in isiZulu. In: Bikakis, A., Fodor, P., Roman, D. (eds.) RuleML 2014. LNCS, vol. 8620, pp. 216–225. Springer, Cham (2014). https://doi.org/10.1007/978-3-319-09870-8_16
12. Keet, C.M., Khumalo, L.: Toward verbalizing ontologies in isiZulu. In: Davis, B., Kaljurand, K., Kuhn, T. (eds.) CNL 2014. LNCS (LNAI), vol. 8625, pp. 78–89. Springer, Cham (2014). https://doi.org/10.1007/978-3-319-10223-8_8
13. Masolo, C., Borgo, S., Gangemi, A., Guarino, N., Oltramari, A.: Ontology library. WonderWeb Deliverable D18 (ver. 1.0, 31-12-2003) (2003). http://wonderweb.semanticweb.org
14. Motik, B., Patel-Schneider, P.F., Parsia, B.: OWL 2 web ontology language structural specification and functional-style syntax. W3c recommendation, W3C, 27 October 2009. http://www.w3.org/TR/owl2-syntax/
15. Nakisa, R.C., Hahn, U.: Where defaults don't help: the case of the German plural system. In: Proceedings of the 18th Annual Conference of the Cognitive Science Society, San Diego, USA, 12–15 July 1996, pp. 177–182 (1996)
16. Ngcobo, M.: Loan words classification in isiZulu: the need for a sociolinguistic approach. Lang. Matters: Stud. Lang. Afr. **44**(1), 21–38 (2013)
17. Ngcobo, M.N.: Zulu noun classes revisited: a spoken corpus-based approach. S. Afr. J. Afr. Lang. **1**, 11–21 (2010)
18. Pretorius, L., Bosch, S.E.: Computational aids for Zulu natural language processing. South. Afr. Linguist. Appl. Lang. Stud. **21**(4), 267–282 (2003)
19. Sharma Grover, A., Van Huyssteen, G., Pretorius, M.: The South African human language technology audit. Lang. Resour. Eval. **45**, 271–288 (2011)
20. Spiegler, S., van der Spuy, A., Flach, P.A.: Ukwabelana - an open-source morphological Zulu corpus. In: Proceedings of the COLING 2010, pp. 1020–1028. ACL (2010)
21. van der Spuy, A.: Zulu noun affixes: a generative account. S. Afr. J. Afr. Lang. **29**(2), 195–215 (2009)
22. van der Spuy, A.: Generating Zulu noun class morphology. Lang. Matters **41**(2), 294–314 (2010)
23. Taylor, C.: A Simplified Runyankore-Rukiga-English Dictionary. Fountain Publishers, Kampala (2009)

Morphological Analysis of Urdu Verbs

Aneeta Niazi[(⊠)]

Center for Language Engineering (CLE), KICS, UET Lahore, Lahore, Pakistan
aneeta.niazi@kics.edu.pk

Abstract. The acquisition of knowledge about word characteristics is a basic requirement for developing natural language processing applications of a particular language. In this paper, we present a detailed analysis for the morphology of Urdu verbs. During our analysis, we have observed that Urdu verbs can have 47 different types of inflections. The different inflected forms of 975 Urdu verbs have been analyzed and the details of the analysis have been presented. We propose a new classification scheme for Urdu verbs, based on morphology. The morphological rules proposed for each class have been tested by simulating with a 2-layer morphological analyzer, based on finite state transducers. The analysis and generation of surface forms have been successfully carried out, indicating the robustness of proposed methodology.

Keywords: Morphology · Urdu verbs · Inflection · Classification
Morphological rules · Finite state transducers · Inflected forms

1   Introduction

Urdu is the national language of Pakistan and one of the twenty-three official languages of India [7]. It has around 11 million native speakers, and more than 105 million speakers who speak it as a second language [6]. Urdu and Hindi are considered the different dialects of the same language. There are many similarities in the vocabulary of Urdu and Hindi. Both Urdu and Hindi nouns are either masculine or feminine. In both the languages, verbs change to indicate the gender of their subjects. However, many words in Urdu are borrowed from Arabic and Persian, which makes Urdu vocabulary slightly different from Hindi [11]. Urdu is written in Arabic script, whereas Hindi is written in Dev Naagari script. Another major difference in the writing style of Urdu and Hindi is that Urdu is written from right to left, where as Hindi is written from left to right. In addition to vowels and consonants, Urdu language also contains diacritics for pronunciation (such as zair(ؠ), zabar(ؘ), paesh(ؙ) etc.), but Urdu text can be written and understood without these additional diacritics.

Urdu is a comparatively under-resourced language [7], and for the development of information retrieval systems, data mining, electronic dictionaries, tagging systems, language models and other Natural Language Processing (NLP) applications of any language, morphological analysis and processing are of particular importance [8].

Morphology is a branch of computational linguistics, that deals with the study of the internal structure of words [1]. In order to develop different natural language processing applications for a particular language, the knowledge about the information

© Springer International Publishing AG, part of Springer Nature 2018
A. Gelbukh (Ed.): CICLing 2016, LNCS 9623, pp. 284–293, 2018.
https://doi.org/10.1007/978-3-319-75477-2_19

that each word carries within its structure serves as a pre requisite. A morphological analyzer aims to provide the structured representation of a word, by breaking its inner structure into smallest grammatical units, known as morphemes [9].

While carrying out morphological analysis, a distinction is usually made between inflection and derivation, based on the affixations [10]. Inflection is the process where the addition of an affix does not change the category of a word, for example, "introducing" is an inflected form of the word "introduce" [10]. Derivation, on the other hand, is the process of forming new words from a base word, for example "introduction" is a derivation of the word "introduce" [10].

Urdu is a morphologically rich language, as it contains a wide range of inflected forms. In Urdu, verbs are formed from the inflection of a base verb. It has been reported that Urdu verbs can take up to 57 inflected forms [1]. An inflected form of a verb in Urdu can possibly contain information about person, gender, respect, number (i.e. singular or plural), tense etc. Therefore, a detailed morphological analysis of all inflected forms of Urdu verbs is mandatory for the practical implementation of Urdu natural language processing techniques.

For example, the word "پڑھیں"("*parrhain*", meaning: "read") is a verb of the root "پڑھ"("*parrh*", meaning: "read"), which can be used as follows:

1. Third level of respect for singular second person in imperative form. e.g. "آپ کتاب پڑھیں"("*aap kitaab parrhain*", meaning:: "Read the book.")
2. Second level of respect for plural second person in imperative form. e.g. "آپ سب لوگ کتاب پڑھیں۔"("*aap sab log kitaab parrhain*", meaning: "All of you, read the book.")
3. Second level of respect for singular third person in subjunctive form e.g. "ابو کتاب پڑھیں گے۔"("*abbu kitaab parrhain ge*", meaning:: "Father will read the book.")
4. First level of respect for plural third person in subjunctive form e.g. "وہ سب لوگ۔ کتاب پڑھیں گے"("*wo sab log kitaab parrhain ge*", meaning: "All of them will read the book.")
5. First level of respect for plural first person in subjunctive form e.g. "ہم کتاب پڑھیں گے۔"("*hum kitaab parrhain ge*", meaning: "We will read the book.")
6. Feminine plural perfective form e.g. "انہوں نے کتابیں پڑھیں۔"("*unhaun ne kitaabein parrhein*" meaning: "They read the books.")

From the above mentioned example, it can be observed that a single word can have multiple definitions. The process of generating a word from its given information is known as morphological generation. For example, the word "پڑھیں"("*parrhain*" meaning: "read") should be generated when any of the above definitions is given as input to a morphological generator.

For Urdu language, morphological rules can be simulated with the help of finite state transducers. A finite state transducer can either have a three layered structure (i.e. morphological layer, intermediate layer and phonological layer), or it can have a two layered structure (i.e. morphological layer and intermediate layer). In a two layered structure of finite state transducers, the morphological layer is treated as a morpho-phonological layer.

In this paper, we have discussed the analysis of different inflected forms of Urdu verbs in detail, and we present a classification scheme for Urdu verbs, based on their structural similarities.

2 Literature Review

Since Urdu is an under-resourced language, a limited research has been carried out on the morphology of Urdu verbs. An overview of Urdu language morphology has been presented in [2]. Urdu word formation processes are briefly discussed. It is stated that inflection in Urdu is based on suffixation. The suffixes in Urdu may comprise of a single syllable or a single character, containing information about multiple features. Different inflected forms can have identical orthography.

A two-step process is involved in Urdu morphology development [3]. The first step is to identify the classes of words and their subtypes i.e. the identification of features that are associated with each surface form. The second step is to incorporate the information about word classes and their sub types in the language.

Morphological and inflectional analysis of the Urdu language has been carried out [4]. The presented morphological analysis has been divided into two categories i.e. single word analysis and combination analysis. Single word analysis has been conducted for all space separated parts of a sentence. Combination analysis has been done by combining a word with its post positions and auxiliaries, that is only applicable to noun-clitic, verb-auxiliaries and adjective-clitic compositions. It has been stated that the combination analysis should be carried out at syntax layer, instead of morphological layer, as it can be handled more easily at syntax level.

An analysis of the morphology of Urdu verbs and nouns has been carried out [1]. Verbs of different transitivity and forms have been analyzed on the basis of their common suffixes. A detailed analysis, along with morphological rules for the following verb forms, has been presented:

1. Derived Causative Form e.g. "ہٹانا"("*hatana*", meaning: "remove") from intransitive verb "ہٹنا"("*hatna*", meaning: "retreat").
2. Irregularly Derived Causative Form e.g. "بلانا"("*bulana*", meaning: "call") from intransitive verb "بولنا"("*bolna*", meaning: "speak").
3. Irregular Causative Form e.g. "بلوانا"("*bulwana*", meaning: "making someone speak") from intransitive verb "بولنا"("*bolna*", meaning: "speak").
4. Infinitive Form e.g. "لینا"("*laina*", meaning: "take") from the root "لے"("*le*", meaning: "take").
5. Imperfective Form e.g. "لیتا"("*laita*", meaning: "takes") from the root "لے"("*le*", meaning: "take").
6. Perfective Regular Form e.g. "لکھا"("*likha*", meaning: "wrote") from the root "لکھ"("*likh*", meaning: "write").
7. Perfective Irregular Form e.g. "لیا"("*lia*", meaning: "took") from the root "لے"("*le*", meaning: "take").
8. Subjunctive Form e.g. "لکھو"("*likho*", meaning: "write") from the root "لکھ"("*likh*", meaning: "write").

A detailed linguistic analysis of Urdu verbs and nouns has been presented [5]. The morphological rules developed for verbs and nouns have been tested by successfully stimulating two-level morphology, using finite state transducers. Based on structural similarities for 21 different inflected forms, containing information about tenses, person, gender, number(singular/plural) and respect, Urdu verbs have been classified as follows:

1. Verbs ending with consonants (i.e. all letters except vowels "ے ، ی ، و ، ا").
2. Verbs ending with alif (ا).
3. Verbs ending with wao (و).
4. Verbs ending with choti ye (ی).
5. Verbs ending with barri ye (ے).
6. Irregular Verbs (i.e. verbs that show an exceptional structural behaviour).

In addition, a classification for those Urdu verbs has been presented, that are inflected by the semantic functions of Transitive affixes (e.g. "ابال"(*ubaal*) from the root verb "ابل"(*ubal*)), Direct causative affixes (e.g. "اترا"(*utra*) from the root verb "اتر"(*utar*)) and Indirect causative affixes (e.g. "اتروا"(*utarwa*) from the root verb "اتر"(*utar*)). Such verbs have been divided into two classes:

1. Verbs ending with consonants (i.e. all letters except vowels "ے ، ی ، و ، ا").
2. Verbs ending with vowels (ا ، و ، ی ، ے).

3 Methodology

For the morphological analysis of Urdu verbs, a list of 975 verbal roots, without diacritics, have been extracted from the CLE Urdu Verb List[1]. These verb roots have been divided into 5 different classes, based on the ending letter, as stated by [5].

During the analysis, it has been observed that the 21 inflected forms that have been made the basis of analysis by [5] do not cover all inflectional cases. This is because of the fact that in Urdu, respect levels also exist for singular third person in addition to plural third person. For example, "احمد آ گیا ہے"(*"Ahmed aa gaya hay"*, meaning: "Ahmed has come.") and "احمد آ گئے ہیں"(*"Ahmed aa gaye hain"*, meaning: "Ahmed has come.") exhibit two different levels of respect for a singular third person. Moreover, in Urdu respect levels also exist for singular second person, in addition to plural second person. For example, "تو پڑھ"(*"tu parrh"*, meaning: "You read."), "تم پڑھو"(*"tum parrho"*, meaning: "You read.")"آپ پڑھیں"(*"aap parrhain"*, meaning: "You read.") and "آپ پڑھیے"(*"aap parrhiye"*, meaning: "You read.") exhibit four different levels of respect for a singular second person. Urdu also has levels of respect in past tense, in addition to non past tenses. For example, "وہ آیا"(*"woh aaya"*, meaning: "He came.") and "وہ آئے"(*"woh aaye"*, meaning: "He came.") exhibit two different levels of respect. The number of respect levels in Urdu differ for persons. For example, first

---

[1] http://www.cle.org.pk/software/ling_resources/urduverblist.htm.

persons always have a single level of respect, whereas second persons can have up to four levels and third persons can have up to three levels of respect. The number of respect levels varies for second and third person for different inflected forms. For example, in habitual forms, the second person has only two levels of respect. For example, in "تو کھانا کھاتا ہے"("*tu khaana khaata hay*", meaning: "You eat food."), "کھاتے ہو ۔ تم کھانا"("*tum khaana khaatay ho*", meaning: "You eat food.") and "کھاتے ہیں ۔ آپ کھانا"("*aap khaana khaatay hain*", meaning: "You eat food."), the root verb "کھا"("*kha*", meaning: "eat") has only two different inflected forms, "کھاتا"(*khaata*) and "کھاتے"(*khaatay*).

Therefore, the inflectional forms that should be used for analyzing verbs are given as follows:

1. Root
2. Infinitive Singular
3. Infinitive Plural
4. Infinitive Feminine
5. Past First Person Masculine Singular Honour Level 1
6. Past Second Person Masculine Singular Honour Level 1
7. Past Third Person Masculine Singular Honour Level 1
8. Past First Person Feminine Singular Honour Level 1
9. Past Second Person Feminine Singular Honour Level 1
10. Past Third Person Feminine Singular Honour Level 1
11. Past First Person Masculine Plural Honour Level 1
12. Past Second Person Masculine Plural Honour Level 1
13. Past Third Person Masculine Plural Honour Level 1
14. Past First Person Feminine Plural Honour Level 1
15. Past Second Person Feminine Plural Honour Level 1
16. Past Third Person Feminine Plural Honour Level 1
17. Habitual First Person Masculine Singular Honour Level 1
18. Habitual First Person Masculine Plural Honour Level 1
19. Habitual First Person Feminine Singular Honour Level 1
20. Habitual First Person Feminine Plural Honour Level 1
21. Habitual Second Person Masculine Singular Honour Level 1
22. Habitual Second Person Masculine Plural Honour Level 1
23. Habitual Second Person Feminine Singular Honour Level 1
24. Habitual Second Person Feminine Plural Honour Level 1
25. Habitual Second Person Masculine Singular Honour Level 2
26. Habitual Third Person Masculine Singular Honour Level 1
27. Habitual Third Person Masculine Plural Honour Level 1
28. Habitual Third Person Feminine Singular Honour Level 1
29. Habitual Third Person Feminine Plural Honour Level 1
30. Habitual Third Person Masculine Singular Honour Level 2
31. Non Past First Person Singular Honour Level 1
32. Non Past Second Person Singular Honour Level 1
33. Non Past Second Person Singular Honour Level 2
34. Non Past Second Person Singular Honour Level 3

35. Non Past Third Person Singular Honour Level 1
36. Non Past Third Person Singular Honour Level 2
37. Non Past First Person Plural Honour Level 1
38. Non Past Second Person Plural Honour Level 1
39. Non Past Second Person Plural Honour Level 2
40. Non Past Third Person Plural Honour Level 1
41. Command Singular Honour Level 1
42. Command Singular Honour Level 2
43. Command Singular Honour Level 3
44. Command Singular Honour Level 4
45. Command Plural Honour Level 1
46. Command Plural Honour Level 2
47. Command Plural Honour Level 3

We have carried out a detailed comparison of inflectional behaviour of the verbs for the given 47 inflected forms[2]. It has been observed that the inflectional behaviour of words ending with the letters alif(ﺍ) and wao(ﻭ) is quite similar to that of the verbs ending with consonants.

From the comparison, we have also observed that the verbs ending with alif(ﺍ) and wao(ﻭ) show exactly same behaviour for all inflected forms, i.e. the addition of suffixes for these two classes is the same for all inflected forms. Therefore, these two classes can be merged to form a single class, as stated by [5]. The verbs ending with alif(ﺍ) and wao(ﻭ) have the following differences, when compared with the verbs ending with consonants:

1. Addition of "ﻱ" before the suffix "ﺍ".
2. Addition of "ﻱ" before the suffix "ﻱ".
3. Addition of "ﻱ" before the suffix "ﮮ".
4. Addition of "ﻱ" before the suffix "ﯿﮟ".
5. Addition of diacritic "ٔ" before the suffix "ﻭ".

It can be seen that there are only single character differences between the inflected forms of the verbs ending with consonants and the verbs ending with alif-wao. There exists a difference of a single character in 28 out of the 47 verb forms, while the behaviour of 19 inflectional verb forms is exactly the same.

The single character differences can be considered as phonological differences, and can be handled by applying special rules. Therefore, there remains no need of a separate class for handling verbs ending with alif(ﺍ) and wao(ﻭ), and this class can be merged with the verbs ending with consonants.

The verbs ending with choti ye(ﻱ) have 32 inflected forms that follow the exact same pattern as that of the verbs ending with consonants. The differences observed in the remaining 15 inflected forms are given as follows:

1. Addition of suffix "ﻱ" is not made in the past feminine singular forms, and the root form is used as it is.

---

[2] http://www.CICLing.org/2016/data/26.

2. Addition of suffix "تی" in place of suffix "ہی".
3. For imperative singular and plural forms, the number of respect levels is 3 and 2 respectively, instead of 4 and 3.

It can be observed that although there are only three differences between the infected forms of verbs ending with choti ye(ی) and those ending with consonants, these differences do not involve a simple character addition. Therefore, these two classes cannot be merged to give a single class.

The verbs ending with barri ye(ے) show a very different inflectional behaviour compared with the verbs ending with consonants. There are only 2 inflected forms having the exact same behaviour between the two classes. The differences found between the inflected forms of two classes are given as follows:

1. Replacement of "ے" with "ی" before the addition of suffixes "تا", "تے", "تی", "تیں", "نا", "نے", "نی", "ں", "ا" and "ے".
2. Replacement of "ے" with "ی" instead of addition of suffix "ی" in past feminine singular forms.
3. Replacement of "ے" with "ی" and addition of "ح" before the addition of suffix "چے".
4. Addition of suffix "ے" is not made in the non-past singular second and third person honour level 1 forms and the root form is used as it is.
5. Deletion of "ے" before addition of suffixes "و" and "وں".

Based on the differences listed above, a separate class has been maintained for handling verbs ending with barri ye(ے).

From the analysis, it has been observed that the verbal roots "کر"(kar), "جا"(ja) and "ہو"(ho) exhibit unique inflectional behaviour, that is very different from other roots. These root verbs need to be handled as exceptional cases, that are resolved by independent, specified morphological rules. This can be done by addition of a class for exceptions in the classification scheme.

For the coverage of transitive Urdu verbs (e.g. "اچھال"(uchaal) from the verbal root "اچھل"(uchal)), a class for irregular consonant roots has been created. For transitive verbs, direct causative verbs ("اٹھا"(uttha) from the verbal root "اٹھ"(utth)) and indirect causative verbs ("اٹھوا"(utthwa) from the root verb "اٹھ"(utth)), the same classification patterns have been used as those stated by [5].

The new classification scheme for Urdu verbs consists of the following classes:

1. Verbs ending with consonants (i.e. all letters except "ے ،ی").
2. Verbs ending with choti ye (ی).
3. Verbs ending with barri ye (ے).
4. Exceptional verbs.
5. Irregular consonant verbs.

The proposed classification scheme can be tested with the help of a two-layer morphological analyzer, that considers morphological and phonological layers as a single morpho-phonological layer.

4   Testing and Results

For the morphological analysis of Urdu verbs, a list of 975 verb roots have been used. The proposed classification scheme has been applied on these verb roots, and 5 classes of verbs have been obtained accordingly.

In order to check the validity of the morphological rules made for each class, the rules have been simulated for both analysis and generation with a two-layer Urdu Morphological Analyzer[3]. The analyzer generated 49879 morphological rules, and the generator successfully returned corresponding surface forms.

For example, the analyzer output for the verbs "لکھو"("*likho*", meaning: "write") and "کیا"("*kea*", meaning: "did") is given as follows:

لکھو

analyzer>> لکھ(*likh*)+Verb+NonPast+Second Person+Singular+Honour Two

analyzer>> لکھ(*likh*)+Verb+Non Past+Second Person+Pl+Honour One

کیا

analyzer>> کر(*kar*)+Verb+Past+Masculine+First Person+Singular+Honour One

analyzer>> کر(*kar*)+Verb+Past+Masculine+Second Person+Singular+Honour One

analyzer>> کر(*kar*)+Verb+Past+Masculine+Third Person+Singular+Honour One

When the above mentioned rules obtained from analyzer are given as input to the generator, the following output is obtained:

لکھ(*likh*)+Verb+NonPast+Second Person+Singular+Honour Two

generator>> لکھو

لکھ(*likh*)+Verb+Non Past+Second Person+Pl+Honour One

generator>> لکھو

کر(*kar*)+Verb+Past+Masculine+First Person+Singular+Honour One

generator>> کیا

کر(*kar*)+Verb+Past+Masculine+Second Person+Singular+Honour One

generator>> کیا

کر(*kar*)+Verb+Past+Masculine+Third Person+Singular+Honour One

generator>> کیا

The following table shows the results obtained after the implementation of the proposed methodology (Table 1).

A detailed analysis of the analyzer output has shown that the maximum number of inflected forms obtained for a verbal root is 166. We have observed that on average, verbal roots have 51.16 inflected forms.

From the results, it is quite evident that Urdu is morphologically a very rich language. The successful analysis and generation of morphological rules show that the new classification scheme has been successfully implemented for Urdu verbs.

---

[3] http://www.cle.org.pk/software/langproc/MorphologicalAnalyzer.htm.

Table 1. Number of root verbs, total number of inflected forms, number of orthographically unique inflected forms, number of causative forms and number of indirect causative forms that have been obtained by the implementation of proposed classification scheme and the morphological rules, made for 47 inflected forms.

Verb class no.	Number of roots	Inflected forms	Orthographically unique inflected forms	Causative forms	Indirect causative forms
1	958	48475	16782	156	49
2	2	110	38	1	1
3	2	94	32	–	–
4	3	140	48	–	–
5	10	1060	336	–	–
Total	975	49879	17236	157	50

5   Conclusion

In this paper, a detailed morphological analysis of Urdu verbs has been carried out and a new classification scheme has been proposed. The analysis and generation output of 975 verbs from a two-layer morphological analyzer have shown successful implementation of the presented classification scheme.

6   Future Work

In future, the presented classification scheme could be further extended by implementing it on a larger set of Urdu verbs. The morphological analysis of other grammatical entities of Urdu such as nouns, adjectives etc. can be carried out by using a similar methodology, which can in turn be implemented in the development of Urdu language processing applications.

References

1. Rizvi, S.J., Hussain, D.M.: Analysis, design and implementation of urdu morphological analyzer. In: Student Conference on Engineering Sciences and Technology, Karachi (2005)
2. Qureshi, A.H., Anwar, D.B., Awan, M.: Morphology of the Urdu language. Int. J. Res. Linguist. Soc. Appl. Sci. 1(3), 20–25 (2012)
3. Butt, M., King, T.H.: Urdu and the parallel grammar project. In: Proceedings of the 3rd Workshop on Asian Language Resources and International Standardization (2002)
4. Humayoun, M.: Urdu morphology, orthography and lexicon extraction. MS thesis (2006)
5. Hussain, S.: Finite-state morphological analyzer for Urdu. MS thesis (2004)
6. Grimes, B.F.: Pakistan. In: Ethnologue: Languages of the World, 14th edn. Summer Institute of Linguistics, Dallas (2000)
7. Khan, S.A., Anwar, W., Bajwa, U.I., Wang, X.: A light weight stemmer for Urdu language: a scarce resourced language. In: Proceedings of 24th International Conference on Computational Linguistics, p. 69 (2012)

8. Yousfi, A.: The morphological analysis of arabic verbs by using the surface patterns. Int. J. Comput. Sci. Issues **7**(11), 33–36 (2010)
9. Aronoff, M., Fudeman, K.: What is Morphology?. Blackwell publishing, Oxford (2005)
10. Islam, R.A.: The morphology of loanwords in Urdu: the Persian, Arabic and English strands. Ph.D. thesis (2011)
11. The Hindi Language. http://www.kwintessential.co.uk/language/about/hindi.html

Stemming and Segmentation for Classical Tibetan

Orna Almogi[1,2], Lena Dankin[3], Nachum Dershowitz[3,4(✉)], Yair Hoffman[3],
Dimitri Pauls[1,2], Dorji Wangchuk[1,2], and Lior Wolf[3]

[1] Department of Indian and Tibetan Studies,
Universität Hamburg, Hamburg, Germany
[2] Khyentse Center for Tibetan Buddhist Textual Scholarship,
Universität Hamburg, Hamburg, Germany
[3] School of Computer Science, Tel Aviv University,
Ramat Aviv, Tel Aviv, Israel
nachum@tau.ac.il
[4] Institut d'Études Avancées de Paris, Paris, France

Abstract. Tibetan is a monosyllabic language for which computerized language tools are largely lacking. We describe the development of a syllable stemmer for Tibetan. The stemmer is based on a set of rules that strive to identify the vowel, the core letter of the syllable, and then the other parts. We demonstrate the value of the stemmer with two applications: determining stem similarity of two syllables and word segmentation. Our stemmer is being made available as an open-source tool and word segmentation as a freely-available online tool.

It is worthy of remark that a tongue which in its nature was monosyllabic, when written in the characters of a polysyllabic language like the Sanskrit, had necessarily to undergo some modification.

Sarat Chandra Das, "Life of Sum-pa mkhan-po, also styled Ye-śes dpal-'byor, the author of Rehumig (Chronological Table)", *Journal of the Asiatic Society of Bengal (1889)*

1 Introduction

The Tibetan language belongs to the Tibeto-Burman branch of the Sino-Tibetan family. The language is ergative, with a plethora of (usually) monosyllabic grammatical particles, which are often omitted. Occasionally, the same syllable can be written using one of several orthographic variations, for example, *sogs* and *stsogs*. In the case of verbs, the syllable has various inflectional forms that are often homophones, a fact that can result in variants in the reading due to scribal errors. Examples of such inflectional forms are *sgrub*, *bsgrubs*, *bsgrub*, *sgrubs* (present, past, future and imperative, respectively), all of which are homophones with "stemmic identity". It should be noted that the notion of "stem" does not exist in traditional Tibetan grammar, but has been introduced here for the purpose of identifying virtually identical syllables despite varying orthographies or

© Springer International Publishing AG, part of Springer Nature 2018
A. Gelbukh (Ed.): CICLing 2016, LNCS 9623, pp. 294–306, 2018.
https://doi.org/10.1007/978-3-319-75477-2_20

inflectional forms. In some texts, orthographic abbreviations are very common: letters (consonants or vowels) are omitted within a syllable, while contracting two or more syllables into one; examples are: *bkra shis* → *bkris* and *ye shes* → *yees*. The language is also abundant in homophones without stemmic identity, which lead to scribal errors and to drift in textual content. The latter two phenomena will be disregarded here since they are irrelevant for stemming.

The Tibetan writing system is reported by tradition to have been developed circa the 7th century. It is based on the Indian Brāhmī/Gupta script with adaptations for Tibetan. Tibetologists commonly employ (as we do here) the Wylie [10] system for transliteration into Latin characters, in which no diacritics are used and thus various letters are represented by two or three consonants.

Many of the Mahāyāna Indic Buddhist texts are extant in Tibetan, and sometimes only in Tibetan. The Tibetan Buddhist canon consists of two parts: the Kangyur, which commonly comprises 108 volumes containing what is believed by tradition to be the Word of the Buddha, texts that were mostly translated directly from the Sanskrit original (with some from other languages and others indirectly via Chinese); and the Tengyur, commonly comprising about 210 volumes consisting of canonical commentaries, treatises, and various kinds of manuals that were likewise mostly translated from Sanskrit (with some from other languages and a few originally written in Tibetan).

After a brief introduction to the Tibetan syllable (Sect. 2), we describe the stemming algorithm (Sect. 3), followed by two applications: stem similarity (Sect. 4) and word segmentation (Sect. 5).

2   The Tibetan Syllable

For our purposes, we consider the Tibetan alphabet to consist of 29 consonants and 5 vowels. (Traditionally, *a* is including among a list of 30 consonants, rather than amongst the vowels.) Table 1 gives the distribution of consonants in the Tibetan Buddhist canon. (Digital data provided by Paul Hacket of Columbia University.)

The Tibetan language is monosyllabic (morphemes normally consist of one syllable): each syllable is written separately; but word boundaries – a word consists of one or more syllables – are not indicated in a text, similarly to Chinese. Table 2 presents the most frequent unigrams, bigrams, and trigrams in the corpus, along with their distributions. Notice that the Zipf's law [11], that word frequency is inversely proportional to rank in a frequency table, does not hold at the level of Tibetan syllables; see [9].

Figure 1 shows a syllable with its parts in Tibetan script (left) and a syllable appended with the grammatical particle following it (right). The script is transliterated in Wylie from left to right, with stacked letters transliterated from top to bottom.

Ligatures are standard. Three-letter ligatures are these: *rky, rgy, rmy sky, sgy, spy, sby, smy, skr, sgr, spr, sbr, smr.*

A unique feature of the language is the fact that particles can be added/omitted while the text still retains its meaning and is in fact considered

stacks: A B C D   B E

Fig. 1. left: Tibetan syllable (exemplified by *bsgrup*); right: disyllabic contraction (exemplified by *sgra'ang*)

to be the *same* text (the omission of particles, however, might result in ambiguity). Thus the presence or absence of particles can in most cases be considered inconsequential, and, therefore – at least in the context here – that component may be disregarded for most intents and purposes.

As suggested in [6], every syllable can be decomposed – in a deterministic fashion – into consonants and vowels located in specified positions. In some cases a particle is appended to the syllable preceding it, resulting in a disyllabic (or, rarely, trisyllabic) contraction. To accommodate these cases, we use an octuple (8-tuple), consisting of the following components (somewhat different from the decomposition used in [6] for standardizing the lexicographic ordering of syllables):

$$\langle \quad \text{prescript, superscript, core, subscript,}$$
$$\text{vowel, coda, postscript, appended particle} \rangle$$

Some of the positions may remain empty or contain a special value to indicate their absence.

In Fig. 1, stack A holds the prescript component, stack B holds the superscript, core letter, subscript, and vowel components. Stack C holds the coda (final letter), and D, the postscript. An additional position E holds the appended particle(s).

Thus, for example, the future tense *bsgrub* and the imperative *sgrubs* of the verb *sgrub* (to perform) would take the following forms:

$$bsgrub = \langle \, b, \, s, \, g, \, r, \, u, \, b, \, -, \, - \, \rangle$$
$$sgrubs = \langle \, -, \, s, \, g, \, r, \, u, \, b, \, s, \, - \, \rangle$$

The disyllabic contraction *sgra'ang*, to give another example, would take the form:

$$sgra'ang = \langle \, -, \, s, \, g, \, r, \, a, \, -, \, -, \, 'ang \, \rangle$$

Each location in the tuple is governed by a different set of rules. Some combinations are possible, while other combinations never occur and their appearance would suggest either a transliteration error, scribal error (or a damaged woodblock), or the presence of a non-Tibetan word, such as a Sanskrit word transliterated in Tibetan script.

Table 1. Consonant distribution within the Tibetan Buddhist canon. (The remaining consonants each appear less than 3% of the time.)

s	d	g	b	r	n	y	p	m	ng	l
10%	8%	6.5%	6.5%	6.5%	5.6%	5.5%	5.1%	4.4%	4.4%	3.4%

Table 2. Top 5 unigrams, bigrams, and trigrams in the Tibetan canon.

Unigrams		Bigrams		Trigrams	
n-gram	%	n-gram	%	n-gram	%
pa	5.6	pa dang	0.0071	zhes bya ba	0.0050
dang	2.7	bya ba	0.0068	la sogs pa	0.0028
ba	2.4	zhes bya	0.0067	bya ba ni	0.0025
par	2.3	la sogs	0.0045	bcom ldan'das	0.0020
ni	2.0	thams cad	0.0041	byang chub sems	0.0010

3 Stemming

Since syllables having the same stem may take many different forms, stemming is a crucial stage in almost every text-processing task one would like to perform in Tibetan. Usually, in Indo-European and Semitic languages, stemming is performed on the word level. However, in Tibetan, in which words are not separated by spaces or other marks, a syllable-based stemming mechanism is required even in order to segment the text into lexical items. We should point out that (heuristic) stemming does not mean the same thing as (grammatical) lemmatization, and the stemming process can result in a stem that is not a lexical entry in a dictionary. Moreover, unlike other Indo-European languages, stemming of Tibetan is mostly relevant to verbs and verbal nouns (which are common in the language). Despite being inaccurate in some cases, stemming (for Tibetan as for other languages) can improve tasks such as word segmentation and intertextual parallel detection [8]. Moreover, even for Tibetan words consisting of more than one syllable, stemming each syllable makes sense since all the inflections are embedded at the syllable level. For instance, the words *brtag dbyad* (analysis) and *brtags dpyad* (analyzed) are stemmed to *rtog dpyod* (to analyze, analysis).

The following are the main rules that govern the structure of the syllable [4].

- There are 30 possibilities for the core letter: any of the 29 consonants or the core letter *a* qua consonant.
- There are 5 vowels, one of which must be present: *a, i, u, e, o.*
- There are 3 possible superscripts: *r* (with core *k, g, ng, j, ny, t, d, n, b, m, ts, dz*); *l* (with *k, g, ng, c, j, t, d, p, b, h*); *s* (with *k, g, ng, ny, t, d, n, p, b, m, ts*).
- There are 4 subscripts: *y* (with *k, kh, g, p, ph, b, m*); *r* (with *k, kh, g, t, th, d, p, ph, b, m, sh, s, h*); *l* (with *k, g, b, z, r, s*); *w* (with *k, kh, g, c, ny, t, d,*

ts, tsh, zh, z, r, l, sh, s, h). In rare cases, the combinations *rw* and *yw* may also appear as subscripts, e.g. in the syllables *grwa* and *phywa*.

- There are 10 possible codas (final letters): *g, ng, d, n, b, m, ', r, l, s*.
- There are 5 possible prescripts: the letters *g* (with *c, ny, t, d, n, zh, z, y*,[1] *sh, s, ts*), *d* (with *k, g, ng, p, b, m, ky, gy, py, by, my, kr, gr, pr, br*), *b* (with *k, g, c, t, d, zh, z, sh, s, ky, gy, kr, gr, kl, zl, rl, sl, rk, rg, rng, rj, rny, rt, rd, rn, rts, rdz, lt, sk, sg, sng, sny, st, sd, sn, sts, rky, rgy, sky, sgy, skr, sgr*), *m* (with *kh, g, ng, ch, j, ny, th, d, n, tsh, dz, ky, gy, khr, gr*), *'* (with *kh, g, ch, j, th, d, ph, b, tsh, dz, khy, gy, phy, by, khr, gr, thr, dr, phr, br*).
- There are 2 possible postscripts, which come after the coda: *s, d* (the suffix *d* is archaic and seldom found).
- There are 6 particles that are appended at the end of syllables: *'am, 'ang, 'i, 'is, 'o, 'u*. This is only possible with syllables ending with a vowel (i.e. lacking a final letter, and thus by definition also a postscript), or with the final letter *'*. The appending of the particle results in a disyllabic contraction (while the two vowels are often pronounced as diphthongs). Rarely, two particles can also be appended (e.g. *phre'u'i*). However, since for the stemming we regard the appended particle(s) as a single unit, which is not stemmed, these cases of doubled-appended syllables do not affect stemming and thus are disregarded. There are two additional possible particles that can be appended at the end of a syllable: *s* and *r*. Since both *s* and *r* are also valid codas, this may cause ambiguity. (The potential problem is partially solved for the letter *s* in the normalization stage, but a full solution is difficult to achieve.)

This gives an upper bound of ($30 \times 6 \times 4 \times 5 \times 11 =$) 46,200 for the number of stems (ignoring the rare doubled subscripts). But because not all consonants take all subscripts and superscripts (the ample restrictions concerning the possible combinations of the prescripts remain disregarded), the actual bound is a fraction thereof: 9075 ($= 165 \times 5 \times 11$). See Table 3.

Using the previously described tuple-representation of the Tibetan syllable, we define the stem of a syllable to be the quintuple consisting of the eight original parts minus prescript, postscript, and appended particle:

$$\langle \text{superscript, core, subscript, vowel, coda} \rangle$$

The stem can be written in the following format:

$$\begin{array}{l} \text{superscript} \\ \text{core+vowel -coda} \\ \text{subscript} \end{array}$$

The deleted parts do not change the basic underlying semantics of the syllable.

The stemmer works in the following manner: first, we break the syllable into a list of Tibetan letters. This stage is required because Wylie transliteration represents some Tibetan letters by more than one character (e.g. *zh, tsh*). There is, fortunately, no ambiguity in the process of letter recognition. By design,

---

[1] Transliterated *g.y* to differentiate from core *g* with subscript *y*.

the transliteration scheme ensures that whenever a sequence of two or three characters represents a single letter, it cannot also be interpreted in context as a sequence of distinct Tibetan letters.

Each Tibetan syllable should contain one core letter and one vowel. Other positions (subscript, etc.) are not obligatory; there should be only one letter that

Table 3. Possible superscripts and subscripts.

Core	Superscript			Subscript				Total
k	r	l	s	y	r	l	w	20
kh				y	r		w	4
g	r	l	s	y	r	l	w	20
ng	r	l	s					4
c		l					w	4
ch								1
j	r	l						3
ny	r		s				w	6
t	r	l	s		r		w	12
th					r			2
d	r	l	s		r		w	12
n	r		s					3
p		l	s	y	r			9
ph				y	r			3
b	r	l	s	y	r	l		16
m	r		s	y	r			9
ts	r		s				w	6
tsh							w	2
dz	r							2
w								1
zh							w	2
z						l	w	3
'								1
y								1
r						l	w	3
l							w	2
sh					r		w	3
s					r	l	w	4
h	l				r		w	6
a								1
								165

fits each of the seven places in the tuple, while the eighth place accommodates a syllable, one of 6 possible appended particles. We therefore start with the detection of all the vowels (by definition, each syllable contains one vowel). A contraction consisting of an appended syllable commonly contains two vowels. (As noted earlier, the rare case of a double-appended syllable has no effect on the stemming.) Syllabic contractions should contain two vowels at most.

The vowel (*a, i, u, e, o*) necessarily follows the core letter, or the subscript (*y, r, l, w*, rarely also *yw, rw*) if there is one. Examples are *bam* (*b* is the core letter); *bsgrubs* (*g* is the core letter); *'ga'* (*g* is the core letter); *zhwa* (*zh* is the core letter); *chen* (*ch* is the core letter). If the syllable begins with a vowel, the core letter in our representation is set to be *a* (meaning, we add an extra *a*), which makes *ag* a valid syllable with core letter *a*, vowel *a*, and coda letter *g*. Another example is the syllable *e* that would be represented as having core letter *a* and vowel *e*.

The stem of the syllable consists of the core letter or the stacked letter (which, in turn, consists of the core letter and a superscript, or a subscript, or both), the vowel, and the final letter (if this is found). Syllables can be considered to be stemmically identical if these are consistent, despite additions or omissions of a prescript and/or a postscript.

Under certain circumstances (commonly inflection of verbs), the core letter may be changed. However, the change is not arbitrary, and usually occurs among phonetically "related" letters, such as *k/kh/g; c/ch/j; t/th/d; p/ph/b*.

Possible changes can be found in the vowel, while still retaining the same basic meaning (and the same stem). Most commonly the vowel *o* in verbs changes to *a* and vice-versa, reflecting a change in tense. Since other vowel changes are unfortunately also possible, it seems impossible to identify a pattern. The only viable solution would be to work with a list of verbs and their inflections, or alternatively, to consider a vowel change as substantial (thus failing to recognize the stemmic identity).

1. In the case of an appended syllable, the first vowel is considered to be the main one in the syllable and is placed in the vowel location of the tuple.
2. The second vowel can only be part of one of the 6 possible appended particles. We place the entire particle in the eighth location of the tuple, without breaking it into its component letters, as we are only interested in the complete particle. Any other case is considered a non-Tibetan syllable that is transliterated in Tibetan script, and consequently the syllable is not stemmed.
3. Following the detection of the vowel of the syllable to be stemmed, we place the letters before and after this vowel in the appropriate places in the tuple, according to the constraints for each position. If no legal decomposition is found, the syllable is not stemmed.

The final stage is normalization. As it turns out, there are groups of Tibetan letters that can be replaced one with the other without changing the basic meaning of the syllable. Since we are interested in grouping all syllables that are ultimately stemmically identical into one and the same stem, we normalized all tuples according to the following rules:

A. *a, o* are the same when in the vowel position.
B. *c, ch, j, zh, sh* are the same in the core letter position.
C. *ts, tsh, dz, z* are the same in the core position.
D. *s* in the coda position may be simply omitted.[2]

(We have glossed over a few additional special cases and peculiarities that are dealt with in the stemmer code.)

Once we have the tuple corresponding to the syllable, we extract the components ⟨superscript, ..., final letter⟩ to obtain a quintuple that represents the syllable's stem. For *bsgrub* and *sgrubs*, future tense and imperative of the verb *sgrub*, the stemming process will generate the same stem: *sgrub*.

The stemming tool is available at http://www.cs.tau.ac.il/~nachumd/Tools. html; the code is at https://github.com/lenadank/Tibetan_Stemmer.git.

For the similarity measures and word-segmentation tasks, described in the following sections, each letter is encoded by a number. The particles, as previously mentioned, are encoded as themselves, so overall we have a total of 41 possible values for the various locations in the tuple (29 consonants [excluding *a*], 5 vowels, 6 particles, blank).

4   Learning Similarity

The stemmer, as described above, extracts the information encoded in each Wylie transliterated syllable and makes it explicit. An important task, given two syllables, is to evaluate their stemmic similarity. One can, for example, examine the tuple representation of the syllables and count the number of places in which they differ. While this is a reasonable baseline, it does not take into account the relative importance of each component. This approach also misses the importance of each substitution made. Some substitutions can be considered silent or synonymous; others change the meaning completely; and there is a continuous spectrum in between.

Assessing the relative importance of each substitution by experts is infeasible. We, therefore, use metric learning algorithms for this task. Specifically, we employ Support Vector Machines (SVMs) as described below. The dataset we used for learning close stemmic similarities contains a list of syllables divided into 1521 sets. It was extracted from a list of verb inflections provided by Paul Hacket.

Since this inflection dataset is limited in extent, we model each substitution as two independent changes, and the importance of each substitution, per location in the tuple, is computed as the sum of two learned weights. One weight is associated with the letter in one syllable and the other associated with the parallel letter in the second. This way, instead of a quadratic number of parameters, we have only a linear number.

---

[2] The reason for omitting the coda *s* for the sake of normalization is that in cases where it is added to form the past tense, which results in a syllable that appears to have a stem with coda *s*, we treat this *s* as equivalent to the postscript *s* often added to form the past tense.

4.1   The Learned Model

Given the stemmer's output for two Wylie encoded syllables $x_i, y_i \in \mathbb{R}^5$, we first re-encode it as a more explicit tuple by using an encoding function. Three types of such functions are considered.

In the first type, the encoding is simply the identity function. This is a naïve approach in which the rather arbitrary alphabetic distance affects the computed metric.

The second type encodes each possible letter in each of the five locations (superscript, ..., final letter) as one binary bit. The bit is 1 if the associated location in the stemmed representation has the value of the letter associated with the bit, and 0 otherwise. This representation learns one weight for each letter at each location. If the two syllables x_i and y_i differ in three locations out of the five, the learned model would sum up six weights: each of the three locations would add one weight for each of the two letters involved in the substitution.

The third type of encoding is based on information regarding equivalence groups of letters. In other words, substitutions within each group are considered synonymous. There are five groups with more than one letter:

1. *g, k, kh*
2. *c, ch, j, zh, sh*
3. *d, t, th* close
4. *b, ph, p*
5. *z, dz, tsh, ts*

The rest of the letters form singleton groups. The total number of groups is 21.

Let f be the encoding function. The learned model has a tuple of parameters w, which has the same dimension as $f(x)$, and a bias parameter b. It has the form: $w^\top |f(x_i) - f(y_i)| + b$, that is, a weighted sum of the absolute differences between the encoding functions of the two stemmed syllables.

During training, synonymous and non-synonymous pairs of syllables are provided to the SVM algorithm [3]. Each pair is encoded as a single tuple $|f(x_i) - f(y_i)|$, and an SVM with a linear kernel is used to learn the parameters w and b.

4.2   Evaluation

The dataset contains 1521 sets of verbs and their inflectional forms. The sets are divided into three fixed groups in order to perform a cross validation accuracy estimation. In each cross validation round, two splits are used for training and one for testing. Within each group, all pairs of syllables from within the same set (inflections of the same verb) are used as positive samples. There are 110–140 such pairs in each of the splits. Ten times as many negative samples are sampled.

Table 4 presents the results of the experiments. The area under the ROC curve (AUC) is used to measure classification success. We compare two methods: one does not employ learning and simply observes the Euclidean distance

Table 4. Comparison of the three encoding functions used for metric learning. Results for both the Euclidean (L2) distance and SVM-based metric learning are shown. The reported numbers are mean $\text{AUC} \pm \text{SD}$ over three cross-validation splits.

Method	Naïve	Binary	Equivalence groups
L2 distance	0.7990 ± 0.0149	0.9282 ± 0.0148	0.9534 ± 0.0203
SVM metric learning	0.9129 ± 0.0211	0.9723 ± 0.0212	$\mathbf{0.9808 \pm 0.0154}$

$\|f(x_i) - f(y_i)\|$; the other is based on learning the weights w via SVM. We compare the three functions f described above: (i) Naïve, (ii) Binary, and (iii) Equivalence groups.

As can be seen, the Equivalence group function significantly outperforms the other functions. It is also evident that learning the weights with SVM is preferable to employing a constant weight matrix (which results in a simple Euclidean distance).

5   Word Segmentation

The problem of word segmentation, viz. grouping the syllables into words, is of major importance. Since no spaces or special characters are used to mark word boundaries, the reader has to rely on language models so as to detect the word boundaries.

5.1   Design

The approach we take is based on a flavor of recurrent neural networks (RNNs) called "long short-term memory" (LSTM) [5]. LSTMs have been used in the past for word segmentation of Chinese text [2]. Our work differs from previous work in that we rely on the tuple representation of the syllable, while previous works represent each syllable out of a sizable group of syllables as an atomic unit. Our input tuple is therefore much more compact.

The word-segmentation pipeline consists of the following steps.

First, we represent each syllable using an encoding function that is similar to the Binary function of Sect. 4: each possible letter in each location is assigned a single bit to indicate its existence at this location.

Then, for each syllable, the surrounding syllables are collected to form a context-window of size 5. At the text boundaries, we pad with 0's. This context is represented as a single tuple that results from concatenating five tuples: up to two syllables before, and two syllables after, the current syllable.

Since we consider 5 syllables per time frame, 8 parts, and up to 41 symbols per part, the size of our representation is 1640. As mentioned, when the context-window extends beyond the beginning or end of a sentence, tuples of 0's are used for padding.

The neural network is presented in Fig. 2. It consists of a single LSTM layer with 100 hidden units, followed by a softmax layer. Following the convention of

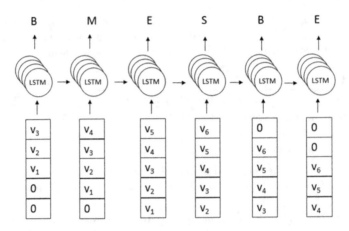

Fig. 2. The LSTM network used for word segmentation.

Table 5. The performance of the word segmentation neural network.

Context size	Classification type	Precision	Recall	F1 score
3 syllables	{B, M, E, S}	0.89	0.88	0.88
	Binary	0.93	0.93	0.93
5 syllables	{B, M, E, S}	0.90	0.90	0.90
	Binary	**0.95**	**0.95**	**0.95**
7 syllables	{B, M, E, S}	0.90	0.90	0.90
	Binary	0.94	0.94	0.94

previous work [2], for each syllable there are 4 possible target labels, indicating whether the syllable is the beginning of a word (B), middle of the word (M), end of the word (E), or constitutes a single-syllable word (S).

An alternative network was also trained, which has only two target labels, 1 when the current syllable is the end of a word and 0 when it's not. This is sufficient for segmenting words – we stop each word after a label of 1 and immediately begin another. Note that in the 4-label network, there is no hard constraint that precludes, for example, a word-middle (M) to appear right after the end of a word (E) or a single syllable word (S). In the binary alternative we propose, consistency is ensured by construction.

5.2   Evaluation

Training data was downloaded from the Tibetan in Digital Communication project (http://larkpie.net/tibetancorpus). The training set consisted of 36,958 sentences and the test set of 9239 sentences. Overall, there are 349,530 words consisting of an average of 1.46 syllables per word.

Training of the network was accomplished using cross-entropy loss and with the Adam learning rate scheduling algorithm [7]. A dropout layer of 50% is added before the LSTM layer.

Table 5 summarizes the performance of the neural network when trained with context-windows sizes of 5 and 7. As can be seen, a context size of 5 syllables works somewhat better, and the proposed binary network outperforms the multilabel network.

The online segmentation tool is available at http://www.cs.tau.ac.il/~nachumd/Tools.html. A screenshot for a sample text is shown in Fig. 3.

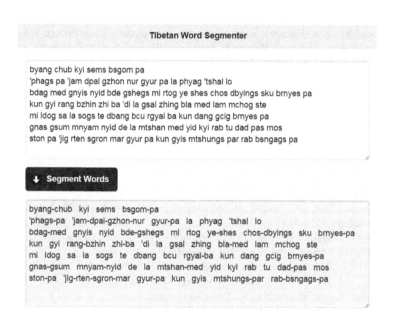

Fig. 3. A screenshot of the online word segmentation application.

6   Conclusion

We have seen the practicality of designing a rule-based stemmer for the syllables of a monosyllabic language like Tibetan. This contributes to an analysis of the morphology of the Tibetan syllable and provides a basis for the development of additional linguistic tools.

We plan on experimenting with the possibility of semi-supervised learning of such a stemmer, and comparing results with this rule-based approach.

The creation of a practical stemming tool for Tibetan made it possible for us to build a reasonable word-segmentation algorithm. We also plan to use it for the development of intelligent search and matching tools for classical Tibetan.

Acknowledgements. We would like to express our deep gratitude to the other participants in the "Hackathon in the Arava" event (held in Kibbutz Lotan, Israel, February 2016; see [1]), who all contributed to the development of new digital tools for analyzing Tibetan texts: Kfir Bar, Marco Büchler, Daniel Hershcovich, Marc W. Küter, Daniel Labenski, Peter Naftaliev, Elad Shaked, Nadav Steiner, Lior Uzan, and Eric Werner. We thank Paul Hacket for crucially providing the necessary data.

This research was supported in part by a Grant (#I-145-101.3-2013) from the GIF, the German-Israeli Foundation for Scientific Research and Development, and by the Khyentse Center for Tibetan Buddhist Textual Scholarship, Universität Hamburg, thanks to a grant by the Khyentse Foundation. N.D.'s and L.W.'s research was supported in part by the Israeli Ministry of Science, Technology and Space (Israel-Taiwan grant #3-10341). N.D.'s research benefitted from a fellowship at the Paris Institute for Advanced Studies (France), with the financial support of the French state, managed by the French National Research Agency's "Investissements d'avenir" program (ANR-11-LABX-0027-01 Labex RFIEA+).

References

1. Almogi, O., Dankin, L., Dershowitz, N., Wolf, L.: A hackathon for classical Tibetan J. Data Min. Digital Humanit. (to appear)
2. Chen, X., Qiu, X., Zhu, C., Liu, P., Huang, X.: Long short-term memory neural networks for Chinese word segmentation. In: Proceedings of the 2015 Conference on Empirical Methods in Natural Language Processing, Lisbon, Portugal, pp. 1197–1206. Association for Computational Linguistics, September 2015. http://aclweb.org/anthology/D15-1141
3. Cortes, C., Vapnik, V.: Support-vector networks. Mach. Learn. **20**(3), 273–297 (1995). https://doi.org/10.1007/BF00994018
4. Hahn, M.: Lehrbuch der klassischen tibetischen Schriftsprache, Indica et Tibetica, 7th edn., vol. 10. Indica et Tibetica Verlag, Marburg (1996)
5. Hochreiter, S., Schmidhuber, J.: Long short-term memory. Neural Comput. **9**(8), 1735–1780 (1997). https://doi.org/10.1162/neco.1997.9.8.1735
6. Huang, H., Da, F.: General structure based collation of Tibetan syllables. J. Comput. Inf. Syst. **6**(5), 1693–1703 (2010)
7. Kingma, D.P., Ba, J.: Adam: a method for stochastic optimization. In: Proceedings of the 3rd International Conference on Learning Representations (ICLR), San Diego, May 2015. http://arxiv.org/pdf/1412.6980v8.pdf
8. Klein, B., Dershowitz, N., Wolf, L., Almogi, O., Wangchuk, D.: Finding inexact quotations within a Tibetan Buddhist corpus. In: Digital Humanities (DH 2014), Lausanne, Switzerland, pp. 486–488, July 2014. http://nachum.org/papers/textalignment.pdf
9. Liu, H., Nuo, M., Wu, J.: Zipf's law and statistical data on modern Tibetan. In: COLING (2014)
10. Wylie, T.V.: A standard system of Tibetan transcription. Harvard J. Asiatic Stud. **22**, 261–267 (1959)
11. Zipf, G.K.: Human Behaviour and the Principle of Least Effort. Hafner Pub. Co., New York (1949)

Part of Speech Tagging for Polish: State of the Art and Future Perspectives

Łukasz Kobyliński(✉) and Witold Kieraś

Institute of Computer Science, Polish Academy of Sciences,
Jana Kazimierza 5, 01-248 Warszawa, Poland
lkobylinski@ipipan.waw.pl, wkieras@uw.edu.pl

Abstract. In this paper we discuss the intricacies of Polish language part of speech tagging, present the current state of the art by comparing available taggers in detail and show the main obstacles that are a limiting factor in achieving an accuracy of Polish POS tagging higher than 91% of correctly tagged word segments. As this result is not only lower than in the case of English taggers, but also below those for other highly inflective languages, such as Czech and Slovene, we try to identify the main weaknesses of the taggers, their underlying algorithms, the training data, or difficulties inherent to the language to explain this difference. For this purpose we analyze the errors made individually by each of the available Polish POS taggers, an ensemble of the taggers and also by a publicly available well-known OpenNLP tagger, adapted to Polish tagset. Finally, we propose further steps that should be taken to narrow down the gap between Polish and English POS tagging performance.

1 Introduction

There is an ongoing discussion whether the problem of part of speech tagging is already solved, at least for English (see e.g. [1]), by reaching the tagging error rates similar or lower than the human inter-annotator agreement, which is ca. 97%. In the case of languages with rich morphology, such as Polish, there is however no doubt that the accuracies of around 91% delivered by taggers leave much to be desired and more work is needed to proclaim this task as solved.

The problem is that while the work on taggers for Polish continues for more than a decade now, the progress is very slow and even reaching the goal of 97% seems very distant. This is in spite of using the latest achievements in machine learning, increasing the size of training data and perfecting the available morphosyntactic dictionaries. The less than perfect quality of automatic POS tagging impacts other NLP tools and the accuracy of other layers of syntactic and semantic annotation generated by these tools.

In this paper, we try to answer the question what is the underlying difficulty in getting closer to the error rates presented by taggers for other languages. For one thing, the task of POS tagging in the case of Polish is much more difficult than in the case of English because of the morphology of the language: the

© Springer International Publishing AG, part of Springer Nature 2018
A. Gelbukh (Ed.): CICLing 2016, LNCS 9623, pp. 307–319, 2018.
https://doi.org/10.1007/978-3-319-75477-2_21

set of all possible tags consists of more than 4 000 choices (ca. 1 500 appear in a manually tagged corpus) versus 30–200 for English. This doesn't answer the question however, because taggers for languages with tagsets of similar size, such as Czech and Slovene, have proved to achieve higher accuracies. We discuss Polish tagset and morphological dictionary in Sect. 2.

If not the language itself, maybe the difficulty lies in the available language resources, or the chosen approach to use them for training the taggers? We briefly present the structure of the National Corpus of Polish, used as the training material for ML methods in Sect. 2.3, discuss the previously proposed taggers in Sect. 3 and show the difficulty in adapting existing methods to Polish in Sect. 4.

We then elaborate on the process of evaluating the quality of taggers in Sect. 5.1, follow with experimental data concerning individual taggers in Sect. 5.2 and combining them into an ensemble of classifiers in Sect. 5.3. In Sect. 6 we discuss the results, the most common mistakes made by the taggers and the problems inherent to the language. Finally, we close with conclusions and perspectives in Sect. 7.

2   Available Resources

2.1   Polish Tagset

There have been several attempts to define a tagset for Polish, usually connected with the development of a reference text corpus, a morphological dictionary, or a tagger. The first formulation of a tagset, which has been used for tagging the IPI PAN corpus,[1] has been proposed in [2]. In this paper we use the more current version of the tagset, proposed for annotating the National Corpus of Polish and described in [3].

2.2   Morfeusz – Morphosyntactic Dictionary

Morfeusz is the most commonly used morphological analyzer for Polish. Although it was recently reimplemented from scratch and significantly enhanced [4], due to technical reasons we were forced to use its previous version [5] as neither the taggers nor the training corpus were adapted to use the newer version of the analyzer.

Morfeusz uses a lexical input obtained from the Grammatical Dictionary of Polish [6], the largest database of Polish inflectional paradigms.[2] The dictionary consists of over 330 000 lexical entries and nearly 7 million wordforms representing over 1100 different inflectional patterns. Its extensive lexical basis goes back to even last decades of 18th century vocabulary which on one hand makes it a desirable resource for morphological analysis, but on the other hand compels to deal with large amounts of archaic, obsolete, dialectal and otherwise stylistically marked lexical entries.

---

[1] IPI PAN corpus was the first large, POS-tagged reference corpus of Polish, now superseded by the National Corpus of Polish.

[2] Now available also on-line at http://sgjp.pl.

2.3   National Corpus of Polish

For training and testing purposes we have used a manually annotated corpus of about 1.2 million words created for National Corpus of Polish project [7]. Contrary to many other resources used for training statistical POS taggers such as Penn Treebank or Prague Dependency Treebank, our training corpus consists not only of newspaper samples but also of fiction and non-fiction, scientific and educational texts, Internet (blogs, fora, Usenet, Wikipedia and other webpages) and oral text samples (media and conversations). Newspapers and magazines constitute only 49% of the corpus. This diversity of data gives us a wider representation of language registers and genres, but may presumably affect both training and evaluation processes (see Sect. 6.4 for a discussion on that topic).

3   Previous Work – POS Taggers for Polish

The first tagger for Polish, proposed by [8], has never been publicly released and is not included in further discussion. TaKIPI tagger, described in [9], assumes a heterogeneous approach to tagging, combining hand-crafted rules with decision trees. TaKIPI is tied to the original, now obsolete IPI PAN corpus tagset and is also excluded from further experiments.

Currently available taggers, using the latest version of the tagset, include: Pantera [10] (an adaptation of the Brill's algorithm to morphologically rich languages), WMBT [11] (a memory based tagger), WCRFT [12] (a tagger based on Conditional Random Fields) and Concraft [13] (another approach to adaptation of CRFs to the problem of POS tagging). Evaluation of performance of a combination of these taggers has been presented in [14].

4   OpenNLP – A Case Study in Adapting a Known Tagger to Polish

One of the questions that may be asked is whether we really need another implementation of a POS tagger, given that so many have already been proposed and in fact several are open sourced and freely available. Such implementations have an unquestionable advantage of being easy to use, supplemented with well-developed user interfaces and (possibly) well tested by multiple users and developers. Unfortunately, such generic tools are not easily adapted to specific languages and associated language resources, or the implemented algorithms do not perform well in case of highly inflective languages, with large tagsets.

An example of such an existing implementation of NLP algorithms, including a POS tagger, is the Apache OpenNLP library.[3] It is a Java-based toolkit, supporting such NLP tasks as tokenization, sentence segmentation, part-of-speech tagging, named entity extraction, chunking, parsing, and coreference resolution.

---

[3] http://opennlp.apache.org/.

Two main machine learning methods are used for solving these tasks: maximum entropy and perceptron. The selection of available ML methods presents another problem when using an existing toolkit. Although maximum entropy has been successfully implemented in many English taggers, several more recent alternatives have been proposed to date (such as CRFs).

We were able to successfully train both a maximum entropy and perceptron-based POS tagger using the OpenNLP toolkit; the accuracy of these models is presented in Table 1 and Fig. 3.[4] As expected, the tagger performs worse than the approaches proposed specifically for Polish. We elaborate on some of our observations below.

Tiered vs One-Pass Classification. Considering the large number of POS tags in Polish tagset, most of the tested taggers divide the task of selecting the correct tag for a word token into several stages, or tiers. For example, the grammatical class is first disambiguated by the WCRFT tagger (1 out of 35 possibilities in the NCP corpus[5]) and after that decision the number of possible combinations of more specific tag parts (e.g. grammatical number, case) is greatly reduced. In the case of OpenNLP, the tags are selected directly, in one pass, which amounts to a problem of selecting 1 out of 1000 possible combinations.

Use of Training Data. The task of tagging is understood as the task of morphosyntactic disambiguation in the case of most of Polish taggers. As such, the models in these taggers are trained to eliminate incorrect possible tags (produced by the morphosyntactic dictionary) for each of the analyzed word tokens. Therefore, during training, not only the correct tag for a particular word is taken into account, but also the set of all possible selections. This helps to make the correct decision in a similar context during tagging. In the case of OpenNLP, only the correct tag is used for training the model.

Use of Morphosyntactic Dictionary. As stated above, previously proposed taggers for Polish rely heavily on morphosyntactic dictionaries and are in fact trained to disambiguate between one of several possibilities generated by the dictionary and not to produce the tags themselves. OpenNLP on the other hand uses the dictionary only for speeding up the beam search algorithm and the model is trained to select one of all possible (previously seen during training) tags. As can be seen from the results in Table 1, this is an advantage in case of tagging unknown words (unknown to the dictionary).

---

[4] As the difference in accuracy between these two approaches turned out not to be statistically significant, we have limited further experiments to maximum entropy models. Trained models available at: http://zil.ipipan.waw.pl/OpenNLP.

[5] In fact, this number is further reduced by the morphosyntactic analyser.

5 Evaluating and Combining the Taggers

5.1 Evaluation Methodology

It is undoubtedly difficult to compare the performance of various approaches to tagging between different languages. That is because of the differences inherent to languages, which were mentioned earlier (e.g. the size of the tagset), or differences in the structure and character of the training and testing material. The problem of evaluating taggers is however much broader and may lead to misconceptions about their real-world performance even when looking at methods proposed for the same language.

There are some obvious conditions that have to remain unchanged to warrant an unbiased comparison: training and testing corpora, additional dictionaries, or the statistical method used to calculate the tagger accuracy. The more subtle decision, often not explicitly stated, is the choice of exact part of the processing pipeline at which the tagger accuracy is measured.

We have decided to continue the line of thought proposed in [15] and evaluate the performance of taggers given plain text as input and measure the accuracy of correct tag assignments to correctly segmented word tokens. This mimics the real-world application of taggers, but hides several stages of processing into one accuracy result (token and sentence segmentation, as well as the morphosyntactic disambiguation itself). Consequently, we also use the accuracy measure proposed in [15], namely the *accuracy lower bound* (Acc_{lower}), which treats all segmentation mistakes as tagger errors. We also distinguish errors made on tokens which are known to morphosyntactic dictionary (Acc_{lower}^{K}) and on tokens for which no morphosyntactic interpretation is provided by the dictionary (Acc_{lower}^{U}).

For the details of data preprocessing we followed the procedure described in [14,15].

5.2 Performance of Individual Taggers

We have firstly re-evaluated all available Polish taggers, using the evaluation methodology described above and language resources described in Sect. 2. Accuracy measures have been calculated by performing ten-fold cross validation of the available training data. The results (presented in Table 1) are on-par with previously published data, but this time we have also included the results for OpenNLP tagger, evaluated using the same methodology.

Table 1. Performance of individual taggers.

n	Tager	Acc_{lower}	Acc_{lower}^{K}	Acc_{lower}^{U}	Training time	Tagging time
1	Pantera	88.95%	91.22%	15.19%	2 624 s	186 s
2	WMBT	90.33%	91.26%	60.25%	548 s	4 338 s
3	WCRFT	90.76%	91.92%	53.18%	27 242 s	420 s
4	Concraft	91.07%	92.06%	58.81%	26 675 s	403 s
5	OpenNLP	87.24%	88.02%	62.05%	11 095 s	362 s

We have also compared tagger efficiency by measuring training and tagging times of each of the methods on the same machine. We used 1.1M tokens both for training and tagging stages and measured the total processing time, including model loading/saving time and other I/O operations (e.g. reading/writing the tokens).

It is worth noting that while the overall accuracy of the OpenNLP tagger is significantly lower than for any other tested approach, it performs better in the case of words unknown to the morphological dictionary. This might suggest that there is room for improvement in the implementations of the best performing methods in the case of such unknown tokens. Usually, a different tagging strategy has to be employed, as the task is different than the usual morphological disambiguation, as in the case of known words.

5.3   Ensemble of Taggers

Next, we tested the hypothesis that the accuracy of an ensemble of taggers increases with each added component tagger, even if its accuracy is lower than the average accuracy of the group. This is because wrong decisions are usually different between taggers and they do not negatively influence the overall accuracy of a voted ensemble. We have indeed observed a slight positive impact of including the OpenNLP tagger into an ensemble of all the tested taggers (see Fig. 1).

In this experiment we have used the setup and strategies described in [14]. The accuracy of an 'oracle' tagger is a hypothetical result of a perfect ensemble voting strategy, in which the correct choice is always made among the tags produced by individual taggers. 'Simple' approach is majority voting, 'weighted' gives advantage to better taggers in case of a draw, while 'per-class' gives advantage to taggers which are known to perform better for a particular grammatical class.

6   Why Are the Taggers Wrong?

6.1   The Most Common Errors

We use the term "part of speech tagging" referring to the process of choosing a proper morphosyntactic interpretation for a given token, which means that it is not restricted to choosing a correct part of speech label, but also a proper lemma and proper values of all grammatical categories of the wordform. In fact, when it comes to literal meaning of the term (i.e. choosing a correct POS label), the problem is rather simple and all tested taggers obtain relatively good results in this task. For nouns (subst), adjectives (adj), numerals (num), past tense verb forms (praet), passive adjectival participle (ppas), active adjectival participle (pact) and pronouns (ppron12 and ppron3) taggers tend to assign a correct part of speech rather than an incorrect one. Significant problems in assigning a correct part of speech label can be observed mostly in the area of grammatically not

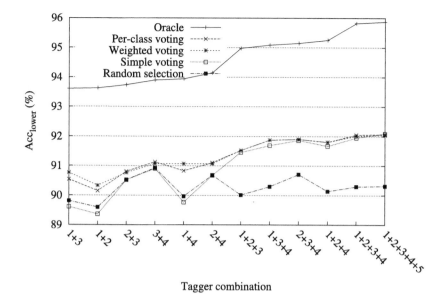

Fig. 1. Accuracy of an ensemble of taggers.

inflected words such as prepositions (prep), particles (qub), adverbs (adv) and conjunctions (conj).[6] However, error rate grows dramatically if one expects the tagger to also choose the correct lemma and all grammatical category values such as: gender, number, case, person, tense, aspect etc. See Fig. 2 for a comparison between the percentage of errors made by selecting an incorrect grammatical class vs. errors in tagging other grammatical categories.

6.2 Homonymy

The need for disambiguation of the morphological analyzer's output arises from homonymy. In general, the more homonymy in a certain language, the more difficult it is to disambiguate. Polish is definitely on the harder side as its average homonymy rate reaches 47%, which means that nearly every second word in a text is morphologically ambiguous.

Problems with homonymous words might be of different nature. One of the homonymy types the most difficult to deal with is syncretism, which is also the most common one. By syncretism we understand homonymy restricted to the inflectional paradigm of a single lexeme.[7] In other words, it means that some tokens can be analyzed as different wordforms of the same lexeme. For example, the noun PIES 'dog' declines by two grammatical numbers (singular and plural)

---

[6] One exception from this general observation are gerunds (ger), which are however systematically homonymous with nouns and thus are extremely difficult to disambiguate not only for taggers, but also for the human annotator.

[7] This phenomenon is typical of fusional languages such as Polish and other Slavonic languages.

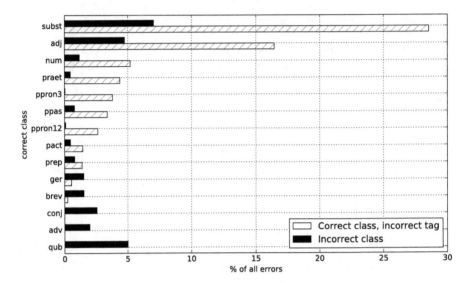

Fig. 2. Tagger errors in choosing the correct grammatical class vs selecting correct tag in a correctly identified class.

and seven grammatical cases, so its full paradigm consists of fourteen wordforms, but only ten orthographically distinct strings, seven wordforms being syncretic. In the case of PIES the syncretic wordforms are: *psa* (genitive or accusative singular), *psie* (locative or vocative singular) and *psy* (nominative, accusative or vocative plural). In fact, every Polish noun and adjective shares this feature to some extent. Most of the syncretisms are systemic (i.e. typical to certain declension classes), but some might be also accidental.

While analyzing the most frequent errors produced by the taggers, one can observe that syncretism is responsible for a significant part of them. Table 2(a) contains a list of most commonly mismatched part of speech tags, which reflects frequent homonyms between grammatical classes like nouns (subst) and adjectives (adj) or conjunctions (conj) and particles (qub). On the other hand, Table 2(b) presents the most commonly mismatched full tags restricted to the noun class, which reflects the most typical syncretisms in noun paradigms characteristic to certain grammatical genders (feminine, neuter and three masculine subgenders named m1, m2, m3). It is easy to observe that these mismatches are usually restricted only to case and sometimes to number, but never to gender (tags in column 1 and 2 are always labeled with the same gender). The first four rows in the table represent two most common syncretisms between singular nominative and singular accusative of masculine inanimate and neuter nouns. Since nominative and accusative are cases typically assigned to subject and object of a sentence, achieving 100% accuracy of morphological disambiguation of those tokens would require at least partial syntactic analysis. This actually shows that the most frequent mistakes in assigning tags to nouns are also those that are most difficult to avoid. All the entries in the table represent syncretisms typical of large classes of Polish nouns.

Table 2. The most common errors made by the taggers.

(a) Errors in grammatical class selection.

tagger	reference	% of errors
adj	subst	1.9422
conj	qub	1.7373
subst	adj	1.5811
adv	qub	1.5518
subst	ger	1.4835
qub	conj	1.4737
subst	brev	1.3664
ger	subst	1.2493
num	adj	1.0541
ppas	adj	1.0541
qub	adv	0.8882
adj	ppas	0.6930

(b) Errors in specific tagging of *subst* class.

tagger	reference	% of errors
sg:nom:m3	sg:acc:m3	1.8641
sg:acc:m3	sg:nom:m3	1.4933
sg:acc:n	sg:nom:n	1.3078
sg:nom:n	sg:acc:n	1.1419
pl:nom:m3	pl:acc:m3	0.8686
pl:acc:f	pl:nom:f	0.7613
pl:nom:f	pl:acc:f	0.6930
pl:acc:m3	pl:nom:m3	0.6637
sg:gen:m1	sg:acc:m1	0.6539
sg:acc:m1	sg:gen:m1	0.5173
pl:nom:n	pl:acc:n	0.4685
sg:gen:f	pl:gen:f	0.4587

Syncretism is not restricted to nouns, but involves also adjectives as well as past forms of verbs. All of these are the sources of common errors of taggers. This proves that syncretism is a serious problem in morphological disambiguation, far more difficult than simple part of speech tagging.

But ambiguous analyses do not necessarily involve only syncretism. Another source of the problem is proper homonymy of inflectional forms of different lexemes – either systemic, serial and motivated by derivational processes or accidental, connecting words derivationally and semantically unrelated. The former was extensively researched by [16] and could be exemplified by masculine nouns differing only in subgender (personal, animate or inanimate) and thus sharing most of the forms in their paradigms, e.g. ADMIRAŁ 'admiral of the fleet' (m1, masculine personal) or 'red admiral butterfly' (m2, masculine animate). The same applies to three homonymous nouns BOKSER 'boxer' which reflect the polysemy of the English word BOXER ('sportsman', 'dog' and 'engine') but is marked on the morphological level by a slightly different inflection. Another example are productive series of pairs such as FIZYK 'physicist' (masculine) and FIZYKA 'physics' (feminine) which systematically share three homonymous forms in their paradigms. Accidental homonymy could be on the other hand illustrated by nouns PALETA (feminine) 'palette/pallet', PALET (masculine) 'payment warrant; archaic legal term', PALETO (neuter) 'coat; archaic loan word'. Each represents a different grammatical gender and thus different inflectional type, but they share some forms, in particular, all three have the same singular locative *palecie*.

An interesting example of this phenomenon is its occasional conjugation with the ambiguity of prepositions' case government. Usually prepositional context helps in disambiguating a noun that is governed by the preposition by ruling out interpretations inconsistent with the case government. However, in some cases the ambiguity of a noun "responds" to the ambiguity of a preposition. Consider a phrase *w krypcie* consisting of two tokens: a preposition W that requires a nominal phrase either in accusative or locative, and a word *krypcie* that could be analyzed either as a locative singular form of the noun KRYPTA 'vault', or an accusative plural of the noun KRYPEĆ 'primitive wooden shoe'. The latter interpretation of the noun is highly unlikely since the word KRYPEĆ is both obsolete and dialectal but it cannot be ruled out on the basis of prepositional government and it calls for other solutions.

Possible linguistic solution to problems illustrated above could be extensive use of non-inflectional information about lexical units extracted from the dictionary, especially about all kinds of stylistic markedness of words (archaisms, dialectalisms, slang etc.) and scope restrictions of their usage (scientific jargon, medical terminology etc.). Also any kind of systematic statistical information about frequencies of word occurrences might improve disambiguation results.

6.3 Lemmatization vs Disambiguation

Another issue that was illustrated by some examples above is the problem of lemmatization. All the tested taggers were aimed at disambiguating morphosyntactic tags, while none of them treated lemmatization as a separate task. In practice the taggers simply ignore the lemmas and take only tags into account. This strategy is reasonable since in most cases it should lead to choosing the correct lemma as well, but sometimes it could result in choosing a completely unlikely lemma before a more probable one. This applies to the example of *w krypcie* shown above – some taggers choose the archaic and rare word KRYPEĆ before stylistically unmarked KRYPTA. The same applies to the nouns OGRÓD 'garden' and OGRODA 'fence' which share the singular locative form *ogrodzie*. Some taggers choose the archaic OGRODA before stylistically unmarked OGRÓD in locative context, which means that in solving such cases they do not take any other significant factor into account and if tags are identical or different but equally justified (as in the case of KRYPEĆ), determining a lemma is more or less a matter of random choice. Avoiding such situations requires an approach in which lemmatization and disambiguation are separate tasks of a tagger as it was suggested in [17]. The basis on which a tagger should resolve a certain lemma is itself a separate issue, but at least two sources of information may turn out useful: text frequency of words and stylistical markedness provided by a dictionary.

6.4 Training Data

A comparison of tagger evaluations between Polish and other languages reveals that there is a significant difference in the structure of National Corpus of Polish, the training material for all presented experiments, and corpora used for training and testing taggers in other languages. For example, English taggers are trained and tested on the part of Penn Treebank which consists exclusively of newspaper articles from the Wall Street Journal. As stated in Sect. 2.3, the NCP consists of a variety of sources, including newspaper articles, but also books, spoken dialogues and data collected from discussion groups on the Internet.

National Corpus of Polish is also smaller than some of the corpora in other languages (1.2M tokens). As such, we wanted to test the hypothesis that (1) the structure of the corpus might influence (negatively) the accuracy of POS tagging and (2) extending the training corpus with more reference data is another possible approach to increasing tagger performance, besides work on the methods themselves.

In order to test the first hypothesis we have evaluated the performance of the OpenNLP tagger on several subcorpora of the NCP, removing data from the sources such as discussion forums, spoken dialogue and books. Each of the subcorpora was chosen to contain roughly the same number of tokens (ca. 54% of the whole corpus), to eliminate the influence of training data size. Based on the experimental data presented in Table 3 we may indeed observe that there is a relationship between the degree of homogeneity of the data and tagging accuracy. This supports the argument that tagging results for Polish are not directly comparable to other languages, for which evaluations are commonly performed on corpora consisting exclusively of newspaper articles.

The influence of the training data size on each of the tested methods has been presented in Fig. 3. In this experiment, we have trained the taggers with

Table 3. Accuracy of tagging vs source of the training and testing data (ten-fold cross-validation).

Train/test data	Acc_{lower}	Acc_{lower}^{K}	Acc_{lower}^{U}	$Avg_{unknown}$
All	85.45%	86.25%	60.01%	3.04%
Without: *internet*	85.54%	86.37%	60.68%	2.90%
w/o: *internet, spoken*	85.71%	86.47%	60.76%	2.96%
w/o: *internet, spoken, books*	86.21%	87.10%	61.83%	3.50%

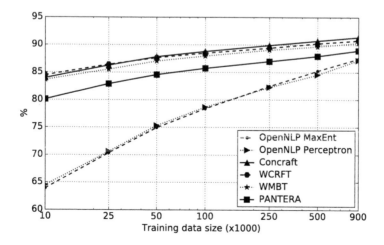

Fig. 3. Learning curve for the tested taggers. Test data size: 100 000 tokens.

randomly drawn subsets of the available training data, increasing data size from 10 000 tokens to 1M tokens, and tested their accuracy on a 100 000 tokens data set. For the best performing taggers, doubling the training data size results in ca. 1 percentage point increase in accuracy.

7   Conclusions and Future Work

In conclusion, we believe that more work concerning Polish language resources is needed to overcome the problem of limited accuracy of POS taggers. Tagging accuracy is directly related to the complexity of this task and Polish is one of the languages with largest tagsets. The accuracy is also directly related to the size of available training data and morphological dictionaries. Some work in this area is already in progress, as the new version of Morfeusz analyzer is under development.

We have also shown that the specific data, which is usually used to evaluate the performance of Polish taggers may negatively impact their results, in comparison with evaluations done for other languages. The linguistic quality and consistency of newspaper articles is usually much higher than that of a text acquired from the Internet, or transcribed dialogues.

In our opinion, future work on POS taggers for Polish should focus on utilizing more of the information available in external language resources (such as stylistic marks of words in the morphological dictionary), tackle the problem of unknown words in more efficient way and also address the problem of lemmatization, which was left out in taggers to date.

Acknowledgment. Work partly financed by the Polish Ministry of Science and Higher Education, a program in support of scientific units involved in the development of a European research infrastructure for the humanities and social sciences in the scope of the CLARIN ERIC consortium and partly financed by Polish National Science Center grant 2014/15/B/HS2/03119.

References

1. Manning, C.D.: Part-of-speech tagging from 97% to 100%: is it time for some linguistics? In: Gelbukh, A.F. (ed.) CICLing 2011. LNCS, vol. 6608, pp. 171–189. Springer, Heidelberg (2011). https://doi.org/10.1007/978-3-642-19400-9_14
2. Przepiórkowski, A., Woliński, M.: The unbearable lightness of tagging: a case study in morphosyntactic tagging of polish. In: Proceedings of the 4th International Workshop on Linguistically Interpreted Corpora (LINC-03), EACL 2003, pp. 109–116 (2003)
3. Przepiórkowski, A.: A comparison of two morphosyntactic tagsets of Polish. In: Koseska-Toszewa, V., Dimitrova, L., Roszko, R. (eds.) Representing Semantics in Digital Lexicography: Proceedings of MONDILEX Fourth Open Workshop, Warsaw, pp. 138–144 (2009)
4. Woliński, M.: Morfeusz reloaded. [18], pp. 1106–1111

5. Woliński, M.: Morfeusz—a practical tool for the morphological analysis of polish. In: Kłopotek, M.A., Wierzchoń, S.T., Trojanowski, K. (eds.) Intelligent Information Processing and Web Mining. AINSC, vol. 35, pp. 503–512. Springer, Heidelberg (2006). https://doi.org/10.1007/3-540-33521-8_55
6. Saloni, Z., Woliński, M., Wołosz, R., Gruszczyński, W., Skowrońska, D.: Słownik gramatyczny języka polskiego, 2. edn. Warszawa (2012)
7. Przepiórkowski, A., Bańko, M., Górski, R., Lewandowska-Tomaszczyk, B. (eds.) Narodowy Korpus Języka Polskiego. Warszawa (2012)
8. Dębowski, Ł.: Trigram morphosyntactic tagger for polish. In: Kłopotek, M.A., Wierzchoń, S.T., Trojanowski, K. (eds.) Intelligent Information Processing and Web Mining. AINSC, vol. 25, pp. 409–413. Springer, Heidelberg (2004). https://doi.org/10.1007/978-3-540-39985-8_43
9. Piasecki, M.: Polish tagger TaKIPI: rule based construction and optimisation. Task Q. **11**, 151–167 (2007)
10. Acedański, S.: A morphosyntactic brill tagger for inflectional languages. In: Loftsson, H., Rögnvaldsson, E., Helgadóttir, S. (eds.) NLP 2010. LNCS (LNAI), vol. 6233, pp. 3–14. Springer, Heidelberg (2010). https://doi.org/10.1007/978-3-642-14770-8_3
11. Radziszewski, A., Śniatowski, T.: A memory-based tagger for polish. In: Proceedings of the LTC 2011 (2011)
12. Radziszewski, A.: A tiered CRF tagger for Polish. In: Bembenik, R., Skonieczny, Ł., Rybiński, H., Kryszkiewicz, M., Niezgódka, M. (eds.) Intelligent Tools for Building a Scientific Information Platform: Advanced Architectures and Solutions. SCI, vol. 467, pp. 215–230. Springer, Heidelberg (2013). https://doi.org/10.1007/978-3-642-35647-6_16
13. Waszczuk, J.: Harnessing the CRF complexity with domain-specific constraints. The case of morphosyntactic tagging of a highly inflected language. In: Proceedings of the 24th International Conference on Computational Linguistics (COLING 2012), Mumbai, India, pp. 2789–2804 (2012)
14. Kobyliński, Ł.: PoliTa: a multitagger for polish. [18], pp. 2949–2954
15. Radziszewski, A., Acedański, S.: Taggers gonna tag: an argument against evaluating disambiguation capacities of morphosyntactic taggers. In: Sojka, P., Horák, A., Kopeček, I., Pala, K. (eds.) TSD 2012. LNCS (LNAI), vol. 7499, pp. 81–87. Springer, Heidelberg (2012). https://doi.org/10.1007/978-3-642-32790-2_9
16. Awramiuk, E.: Systemowość polskiej homonimii międzyparadygmatycznej. Białystok (1999)
17. Radziszewski, A.: Evaluation of lemmatisation accuracy of four polish taggers. In: Proceedings of the LTC 2013 (2013)
18. Calzolari, N., Choukri, K., Declerck, T., Loftsson, H., Maegaard, B., Mariani, J., Moreno, A., Odijk, J., Piperidis, S., eds.: Proceedings of the Ninth International Conference on Language Resources and Evaluation, LREC 2014, Reykjavík, Iceland, ELRA (2014)

Turkish PoS Tagging by Reducing Sparsity with Morpheme Tags in Small Datasets

Burcu Can[1(✉)], Ahmet Üstün[2], and Murathan Kurfalı[2]

[1] Department of Computer Engineering, Hacettepe University,
Beytepe, 06800 Ankara, Turkey
burcucan@cs.hacettepe.edu.tr
[2] Cognitive Science Department, Informatics Institute Middle East Technical
University (ODTÜ), 06800 Ankara, Turkey
{ustun.ahmet,kurfali}@metu.edu.tr

Abstract. Sparsity is one of the major problems in natural language processing. The problem becomes even more severe in agglutinating languages that are highly prone to be inflected. We deal with sparsity in Turkish by adopting morphological features for part-of-speech tagging. We learn inflectional and derivational morpheme tags in Turkish by using conditional random fields (CRF) and we employ the morpheme tags in part-of-speech (PoS) tagging by using hidden Markov models (HMMs) to mitigate sparsity. Results show that using morpheme tags in PoS tagging helps alleviate the sparsity in emission probabilities. Our model outperforms other hidden Markov model based PoS tagging models for small training datasets in Turkish. We obtain an accuracy of 94.1% in morpheme tagging and 89.2% in PoS tagging on a 5K training dataset.

Keywords: Morphology · Syntax · Part-of-speech tagging · Sparsity
Conditional Fandom Fields (CRFs) · Hidden Markov Models (HMMs)

1 Introduction

Turkish is an agglutinating language that builds words by gluing meaning bearing units called morphemes. While gluing morphemes together, vowel harmony and consonant assimilation are intensely applied leading to orthographic transformations in morphemes. For example, the suffix *dir* can be transformed into *dır, dur, dür* depending on the last vowel in the word to which it is being attached. This is called vowel harmony. Moreover, the same morpheme can be transformed into *tir, tır, tur, tür*, this time depending on the last consonant of the word. This is called consonant assimilation. Both vowel harmony and consonant assimilation introduce different realizations of the same morpheme, which are called allomorphs (e.g. *dir, dır, dur, dür, tir, tır, tur, tür* are all allomorphs).

Agglutination already introduces a sparsity problem in natural language processing for especially agglutinating languages. The sparsity problem becomes

© Springer International Publishing AG, part of Springer Nature 2018
A. Gelbukh (Ed.): CICLing 2016, LNCS 9623, pp. 320–331, 2018.
https://doi.org/10.1007/978-3-319-75477-2_22

more crucial when a morpheme has got different realizations. Identifying morphemes that are realizations of each other is the starting point of this work.

Morphological segmentation systems normally provide only the segments of words without any morpheme tags. However, labeled segmentation is required for some natural language processing tasks. For example, in sentiment analysis the Turkish negation suffix *ma* (and its allomorph *me*) needs to be distinguished from the derivational suffix *ma* (and its allomorph *me*) that turns a verb into a noun in order to extract the correct sentiment out. The same also applies for machine translation, question answering, and other natural language processing applications.

Morpheme tagging has become a neglected aspect of morphological segmentation. In this paper, we use conditional random fields (CRF) for morpheme tagging in a weakly-supervised setting. We use the obtained morpheme tags in part-of-speech tagging (PoS tagging) in order to mitigate sparsity in a case when small amount of data is provided. Indeed the sparsity problem is quite severe in PoS tagging for especially agglutinating languages where different methods (e.g. smoothing) have been applied to deal with sparsity. The sparsity is alleviated significantly by using morpheme tags rather than using lexical instances such as words or suffixes.

This paper is organized as follows: Sect. 2 points at the related work in the literature, Sect. 3 describes the CRF model adopted in morpheme tagging and describes the HMMs used in PoS tagging, Sect. 4 presents the experimental results from both tasks and finally Sect. 5 concludes the paper with the remaining future work.

2   Related Work

There has been a substantial amount of work on unsupervised morphological segmentation. Goldsmith [10], Creutz and Lagus [5] build morphological segmentation systems based on minimum description length (MDL). Creutz and Lagus [6] introduce a hidden Markov model (HMM) that employs the probability distributions between different morpheme categories such as prefix, stem, and suffix. Poon et al. [15] introduce a log-linear model for unsupervised morphological segmentation that incorporates MDL-inspired priors.

All of these models provide only morphological segmentations of words and not any morphological tags that identify the morpheme roles within a word. Learning morpheme tags involve distinguishing homophonous morphemes[1] and learning allomorphs. Oflazer [14] introduce derivational boundaries and inflectional groups in Turkish morphological analysis. This is performed by two-level morphology (PC-KIMMO [2,12]) that formulates morphological segmentation via a cascade of finite state transducers by employing morphophonemic alternations. All orthographic and morphophonemic rules are implemented by a set of finite-state automata (FSA) rules. Their model gives a labeled morphological analysis based on these rules.

---

[1] Morphemes with the same surface forms but with different meanings.

Allomorfessor [20] is one of the models that aims to perform morphological segmentation based on allomorphs by modeling mutations between different surface forms of morphemes, namely allomorphs. Can and Manandhar [3] develop an agglomerative hierarchical clustering to find the morpheme classes in an unsupervised setting.

To our knowledge, Cotterell et al. [4] introduce labeled morphological segmentation for the first time in a supervised learning framework without using any rules. They model morphotactics by a semi-Markov model. Different levels of tagsets are introduced that capture different levels of granularity. Our model resembles their model from the aspect of morphological tagging.

Morpheme tags have been used in many natural language processing tasks. El-Kahlout and Oflazer [7] employ morphological tags in order to alleviate the sparsity by matching the Turkish morphemes having the same morphological tag to the same English translation in statistical machine translation task. They address that using morphological tags provides a substantial improvement on the BLUE score.

Morpheme tags have been used in morphological/PoS disambiguation in Turkish language. Ehsani et al. [8] use conditional random fields for disambiguating PoS tags in Turkish by utilizing the morphological tags. They introduce some dependencies between inflectional groups of morphemes in order to simplify the transition probabilities. Sak et al. [16] apply perceptron algorithm for morphological disambiguation. Hakkani-Tur et al. [11] formulate a trigram HMM based on inflectional groups in order to disambiguate morphological parses of a given word. The results show that using the dependencies between inflectional groups of adjacent words improve PoS tagging accuracy. Many of these models select a complete morphological analysis for each word rather than providing a single PoS tag.

Dincer et al. [19] formulate HMMs by emitting suffixes rather than emitting words in order to mitigate the sparsity. However, they do not use any morpheme tags. Our PoS model is mostly similar to their work in this respect. We use morpheme tags in order to cope with the sparsity in emission probabilities rather than using fixed-length endings of words.

3   Model

3.1   Turkish Morphology

Turkish is an agglutinating language that has a productive inflectional and derivational suffixation. This brings the sparsity problem in nlp tasks due to the large vocabulary introduced by the language. The vocabulary size of a corpus having 1 million words becomes 106.547 [11]. In order to deal with the sparsity, a representation that shows inflectional groups and derivation boundaries of the morphological analysis of each word is introduced by Hakkani-Tür et al. [11]. The different morphological analyses of the word *alındı* are given as follows by a two-level morphological analyzer [8]:

al+Verb^DB+Verb+Pass+Pos+Past+A3sg (*it was taken*)

al+Adj^DB+Noun+Zero+A3sg+P2sg+Nom^DB+Verb+Zero+Past+A3sg (*it was your red*)

al+Adj^DB+Noun+Zero+A3sg+Pnon+Gen^DB+Verb+Zero+Past+A3sg (*it was the one of the red*)

alındı+Noun+A3sg+Pnon+Nom (*receipt*)

alın+Verb+Pos+Past+A3sg (*resent*)

alın+Noun+A3sg+Pnon+Nom^DB+Verb+Zero+Past+A3sg (*it was the forehead*)

Here ^DB's denote the derivation boundaries and the rest of the morpheme tags denote the inflectional groups (IGs). Most of the words have more than one morphological analysis in Turkish and the morphological disambiguation aim to find the right morphological analysis of the word given in a specific context.

In this work, we are only using the morpheme tags (both derivational and inflectional) of words in order to find a single PoS tag for each word. We believe that morpheme tags give the best clue for a PoS tag. This is sufficient if we are only interested in syntax but not in the meaning. For example, the analyses of *alındı* that end with *A3sg* can be considered as verbs, whereas the only analysis ending with *Nom* can be considered as a noun. In order to find the morpheme tags we only use the morphotactic features of morphemes within the words, whereas we use contextual features and morphological features in PoS tagging.

3.2   Morphological Tagging by Using CRFs

Conditional Random Fields (CRF) [13] are undirected graphical models that are generally used for segmenting and labeling a given sequence. Unlike HMMs, CRFs are discriminative models that define the conditional distribution $P(Y|M)$ rather than the joint probability distribution $P(M, Y)$, where Y corresponds to the label sequence $Y = \{y_0, y_1, \cdots, y_n\}$ and M corresponds to the input (i.e. observations) sequence $M = \{m_0, m_1, \cdots, m_n\}$. In our case, the label sequence Y refers to the morpheme tags and observation sequence M refers to the morphemes.

The conditional distribution $P(Y|M)$ in our CRF model is given as follows:

$$p(Y|M) = \frac{1}{Z(M)} \prod_{n}^{N} \prod_{i}^{I_n} \lambda F(Y, M) \tag{1}$$

that iterates over the morphemes of each word in the corpus with N words, each having I_n morphemes defined on a feature set F. Here $Z(M)$ is the normalization factor:

$$Z(M) = \sum_{y_i} \prod_{n}^{N} \prod_{i}^{I_n} \lambda F(Y, M) \tag{2}$$

Here λ corresponds to the weight vector for the feature set F. Feature function F consists of two types: state feature functions $s(y, m, i)$ and transition feature functions $s(y', y, m, i)$ where i denotes the input position. State feature function

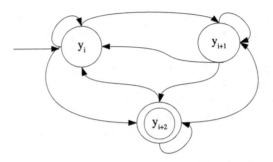

Fig. 1. Our Naive model for the conditional random field. Each state denotes a morpheme tag, where a word w is defined as $w = m_i/y_i + m_{i+1}/y_{i+1} + m_{i+2}/y_{i+2}$.

is non-zero when the label y_i is matched with the label defined in the function, whereas transition functions depend on the label sequence y_{i-1}, y_i.

Our model is given in Fig. 1. We adopt a Naive model where an edge is built between every state pair. Therefore, morpheme tags within the same word are assumed to be dependent on each other, whereas each word is assumed to be independent from the others. Thus, we deal with only morphotactic rules within the same word for morpheme tagging task without using any contextual features.

3.3   Adopting Morphological Tags in PoS Tagging

We use the obtained morpheme tags from the CRF model in order to infer the PoS tags of words. We learn PoS tags according to the following formulation by finding the PoS tag sequence that maximizes the probability for a given sequence of words:

$$argmax_{t_1 \cdots t_n} P(t_1 \cdots t_n | w_1 \cdots w_n) \tag{3}$$

where $t_1 \cdots t_n$ denotes the PoS tags and $w_1...w_n$ denotes the sequence of words. The Bayes' rule is simply applied for the posterior probability as follows:

$$\arg \max_{t_1 \cdots t_n} P(t_1 \cdots t_n | w_1...w_n) = \arg \max_{t_1 \cdots t_n} \frac{P(w_1...w_n | t_1 \cdots t_n) P(t_1 \cdots t_n)}{P(w_1 \cdots w_n)}$$

$$\propto \arg \max_{t_1 \cdots t_n} P(w_1 \cdots w_n | t_1 \cdots t_n) P(t_1 \cdots t_n) \tag{4}$$

where $P(w_1 \cdots w_n)$ is discarded since it is the same for all tag assignments for the given word sequence.

We formulate the posterior probability as a trigram HMM by assuming that each PoS tag depends only on the previous two tags:

$$P(t_1 \cdots t_n) = p(t_1)p(t_2|t_1) \prod_i p(t_i | t_{i-2}, t_{i-1}) \tag{5}$$

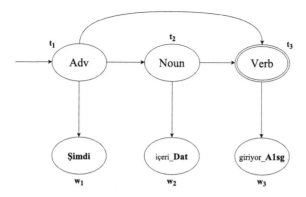

Fig. 2. Our trigram HMM adopted for PoS tagging. The bold units are emitted from the states. The first word *Şimdi* is emitted from t_1, the tag of the last morpheme in the second word *Dat* is emitted from t_2, and the tag of the last morpheme in the third word *A1sg* is emitted from t_3.

We apply interpolation to smooth the transition probabilities in order to rule out zeros in transitions with the equation given below:

$$P_{inter}(t_i|t_{i-n+1}^{i-1}) = \beta_{t_{i-n+1}^{i-1}} P(t_i|t_{i-n+1}^{i-1}) + (1 - \beta_{t_{i-n+1}^{i-1}})P_{inter}(t_i|t_{i-n+2}^{i-1}) \quad (6)$$

which defines an nth-order smoothed model where $P_{inter}(t_i|t_{i-n+1}^{i-1})$ corresponds to the transition probability after interpolation is applied recursively. We estimate the parameters β by tuning our model on a development set.

The sparsity problem also emerges in the emission probabilities. We emit the tag of the last morpheme in the word if the word has more than two segments. Otherwise, the word itself is emitted from the PoS tag as seen in Fig. 2. Therefore, the emission probabilities are estimated as follows:

$$p(w_i|t_i) = \begin{cases} p(y_{n-1}|t_i), & \text{if } w_i = \{m_1/y_1 + \cdots + m_{n-1}/y_{n-1}\} \\ p(w_i|t_i), & \text{otherwise} \end{cases} \quad (7)$$

where y_{n-1} is the morpheme tag of the last suffix in the word. We apply interpolation to smooth $p(w_i|t_i)$ for the words which do not exist in the corpus and cannot be segmented further:

$$P_{inter}(w_i|t_i) = \alpha P(w_i|t_i) + (1 - \alpha)\frac{max(f(w_i), 1)}{N} \quad (8)$$

Here $P_{inter}(w_i|t_i)$ corresponds to the smoothed emission probabilities, $f(w_i)$ is the number of word tokens of type w_i, N is the vocabulary size, and α is the interpolation coefficient.

Viterbi algorithm is applied to find the PoS tag sequence that maximizes the posterior probability given in Eq. 4.

4   Experiments and Results

4.1   Data

We used several different corpora for the experiments. One of them is METU-Sabancı Turkish Treebank [17] that consists of 56k word tokens and 5600 sentences. The dataset includes PoS tags and morphological analyses of the words.

For the additional experiments, in order to compare our CRF model with the Semi-Markov Model by Ryan et al. [4], we used their dataset that consists of 3573 morphologically segmented and tagged word tokens, of which 1987 words belong to the train set and 1586 words belong to the test set.

In order to compare our PoS tagging model with Sak et al. [16], we used their training and test set that are collected from various newspaper archives. This dataset consists of ~800k word tokens and ~47.5k sentences.

For all of the experiments, we used a separate development set that consists of 6K words to tune the interpolation coefficients. We assigned $\alpha = 0.9$ for the emission probabilities, $\beta_1 = 0.6$ (bigram) and $\beta_2 = 0.4$ (unigram) for the bigram transition probabilities, and $\beta_1 = 0.5$ (trigram), $\beta_2 = 0.3$ (bigram), and $\beta_3 = 0.2$ (unigram) for the interpolation used in trigram transitions.

For morpheme tagging experiments, we removed out all punctuation from the datasets and reintroduced the terminal punctuation for PoS tagging task since the word boundaries are crucial for PoS tagging.

4.2   Experiments on Morphological Tagging

In morpheme tagging task, we assume that morphological segmentations of words are provided. We obtained the segmentations and morpheme labels through an open-source morphological analyzer called Zemberek [1] in order to build a train set for the morpheme tagging task. Zemberek defines 84 different morpheme tags on Metu-Sabancı Treebank. We used the open source CRF package [18] for training our own model on our training set. Some of the morphemes belonging to the same morpheme tag obtained from the test set by using the trained CRF model are given in Table 1. The final morpheme tags show that allomorphs can ben learned by our model. For example, la, le, yla, and yle are all allomorphs.

We used Zemberek again in order to create gold sets for the morpheme tagging task. F1 scores for morphological tagging for different sizes of train and test sets obtained from Metu-Sabancı Turkish Treebank are given in Table 2[2]. The F1 score of the model is 80.7% for a training set with 500 words, whereas the F1 score increases up to 94.1% on a 5K training set. Therefore, the F1 score significantly improves on the larger training sets.

We also tested our model on the manually collected newspaper archives, which is much larger than Metu-Sabancı Turkish Treebank. We obtained an F1 score of 93.7% on a 5K train set and 700K test set. This shows that the

---

[2] Precision and recall values are the same because gold sets and results consists of same number of morphemes since we are only doing morpheme tagging and not any segmentation.

Table 1. Some of the morphemes and their tags (with frequencies) obtained from our CRF model.

Morpheme tag	Example morphemes obtained by CRF
Location	da (7439), de (7633), nde (4877), te (1520), nda (8548), ta (1537), çi (2), un (1)
Infinitive	mek (2774), mak (3481)
Inst	la (2218), le (2299), yla (2186), yle (2518)
AfterDoing	yıp (141), yip (104), up (335), üp (106), yup (20), ün (12), ım (24), ümüz (2)
Dative	na (5877), e (7298), ne (4860), ye (2718), ya (2741)
Progressive	ıyor (4738), iyor (6022), uyor (1756), üyor (1343), ıcı (2)
Desire	se (301), sa (659), yacak (8), yecek (3)
Ablative	nden (1959), dan (3053), tan (1061), ndan (3514), ten (874), dik (1)
Narrative	mış (595), müş (276), miş (1640), muş (782), tür (7), tır (30), tikçe (3), ıver (2), tık (1), üver (1), ıcı (2), ün (1), yıver (1), tığ (1), sın (1), çi (1), dür (1), üm (1)

Table 2. Morpheme tagging F1 scores for different training set sizes on Metu-Sabancı Turkish Treebank

Train set size	Test set size	Number of tags	F1 score
500	46440	84	80.71%
1000	45940	84	83.92%
1500	45440	84	88.48%
2000	44940	84	90.39%
3000	43940	84	92.13%
4000	42940	84	93.26%
5000	41940	84	**94.12%**

Table 3. Results of morpheme tagging on manually collected newspaper archives.

Train set size	Test set size	Number of tags	F1 score
5000	720332	88	93.70%

performance of our model does not drop significantly for larger test sets. The results are given in Table 3.

We compared our model with Chipmunk [4] by using their tag set and datasets. The results are given in Table 4. Our CRF model outperforms their model on accuracy, whereas their model outperforms ours on F1 score. However, it should be noted that Chipmunk dataset lacks the derivation morpheme tags, whereas we are also learning derivation morpheme tags in our model.

Table 4. Comparison of Chipmunk [4] and our CRF model for morpheme tagging

	Train set size	Test set size	Accuracy	F1 score
Chipmunk	1987	1586	56.06%	**85.07%**
CRF	1987	1586	**66.62%**	66.62%

Table 5. PoS tagging accuracy scores on Metu-Sabancı Turkish Treebank

Train size	Test size	Tag num.	Word emission acc.	Suffix acc.	Morpheme tag acc.
5025	1017	13	84.85%	86.23%	**88.98%**
18205	1017	13	88.59%	88.69%	**90.95%**
39392	1017	13	89.18%	89.57%	**91.05%**

Table 6. PoS tagging accuracy scores on manually collected newspaper archives

Train size	Test size	Tag num.	Word emission acc.	Suffix acc.	Morpheme tag acc.
5677	1005	13	83.88%	86.66%	**89.25%**
25535	1005	13	90.64%	89.45%	**91.94%**
53829	1005	13	92.43%	91.11%	**92.93%**
106019	1005	13	**94.62%**	91.64%	93.43%
714757	1005	13	**95.44%**	91.34%	93.83%

4.3 Experiments on PoS Tagging

Our PoS tag set consists of 13 major PoS tags [8], that are Adj, Adv, Conj, Det, Interj, Noun, Num, Postp, Pron, Punc, Verb, Ques, Dup.

We tested our model on two different datasets. The first set of experiments were performed on Metu-Sabancı Turkish Treebank. The results are given in Table 5 for different sizes of train/test sets and for different emission types. We provide results for word emissions, last suffix emissions, and the tag of the last morpheme's emissions. For a 5K training set, word emission accuracy is 84.8%, suffix emission accuracy is 86.2%, and morpheme tag emission accuracy is 88.9%. This shows that using morpheme tag emission outperforms both word emissions and last suffix emissions in smaller datasets. The accuracy increases on a ∼40K train set, but still using morpheme tag emissions outperforms using word and last suffix emissions.

The results obtained from the manually collected newspaper archives are given in Table 6. This time using the word emissions outperforms using the last suffix and the last morpheme tag emissions because the sparsity becomes no longer a problem in the larger train sets.

In order to measure the impact of terminal punctuation in PoS tagging, we did two sets of experiments on Metu Sabancı Turkish Treebank. In the first set of experiments, we included the terminal punctuation, whereas in the second set of

Table 7. PoS Tagging accuracy scores for the experiments with/without terminal punctuation on Metu Sabancı Turkish Treebank.

	5K	18K	39K
With terminal punc. - multiple HMM	**88.98%**	**90.95%**	**91.05%**
With terminal punc. - single HMM	88.60%	90.86%	90.96%
Without terminal punc. - multiple HMM	87.51%	89.74%	89.85%
Without terminal punc. - single HMM	86.30%	88.19%	88.53%

Table 8. Comparison of our model with suffix based tagger [19] and the perceptron algorithm [16] on the datasets obtained from Metu Sabancı Turkish Treebank.

	Train set size	Test set size	Accuracy
Suffix-based tagger [19]	5025	1017	84.25%
Perceptron [16]	5025	1017	85.15%
HMM with the last morpheme tag	5025	1017	**88.98%**
Suffix-based tagger [19]	18205	1017	88.90%
Perceptron [16]	18205	1017	86.71%
HMM with the last morpheme tag	18205	1017	**90.95%**

experiments we excluded the terminal punctuation. While including the terminal punctuation, first we built one HMM for each sentence in the training set, second we built only one HMM for the entire corpus where all the words are linked to each other on the same HMM that are separated by terminal punctuation. We obtained an accuracy of 88.9% for multiple HMM approach, whereas we obtained an accuracy of 88.6% for a single HMM approach on 5K train set. In the second set of experiments, we completely excluded the terminal punctuation and repeated the experiments for multiple HMMs and a single HMM. We obtained an accuracy of 87.5% for multiple HMMs, whereas we obtained an accuracy of 86.3% for a single HMM. Results are given in Table 7. It shows that even though terminal punctuation plays an important role in PoS tagging, behaving each sentence as a single HMM by assuming that sentences are independent from each other leads to a slight increase in the accuracy.

We compared our model with Sak et al. [16] and Dincer et al. [19] on Metu Sabancı Turkish Treebank. We used the last 5 letters of each word with a second order HMM to implement the suffix based tagger by Dincer et al. [19], since their model gives the best scores for the last 5 letters. Results are given in Table 8. The results show that our model outperforms the other two models on smaller datasets (i.e. 5K and 18K).

Obtaining data is one major problem in natural language processing tasks. Using small datasets by reducing sparsity is one challenge in natural language processing. Here, we aimed to increase the accuracy of PoS tagging for an agglutinating language on smaller datasets when large datasets are not available. Our results show that it is possible to use a kind of class-based language model by grouping the morphemes according to their syntactic roles within a word by tagging them and then using it for PoS tagging to reduce the sparsity in smaller datasets.

5 Conclusion and Future Work

We introduced a CRF model to tag the morphemes syntactically and a HMM model for PoS tagging that uses these morpheme tags in order to reduce the sparsity in Turkish PoS tagging on smaller datasets. We managed to obtain morpheme tags with F1 score 94.1% on a limited training set by using CRFs. Then, we trained a second-order HMM model with the last morpheme tag of each word emitted from each HMM state in order to perform PoS tagging, contrary to the conventional approach of using words' surface forms emitted from HMM states. The results show that using the last morpheme tags helps dealing with the sparsity especially on small train sets.

We believe that morphological features of the context words will be also informative in morpheme tagging task because Eryiğit et al. [9] shows that using inflectional groups as units in Turkish dependency parsing increases the parsing performance. We leave using the contextual information in morpheme tagging as a future work.

Acknowledgments. This research is supported by the Scientific and Technological Research Council of Turkey (TUBITAK) with the project number EEEAG-115E464 and we are grateful to TUBITAK for their financial support.

References

1. Akın, A.A., Akın, M.D.: Zemberek, an open source NLP framework for Turkic languages. Structure **10**, 1–5 (2007)
2. Antworth, L.E.: PC-KIMMO: A Two-Level Processor for Morphological Analysis. Occasional Publications in Academic Computing, Dallas (1990)
3. Can, B., Manandhar, S.: An agglomerative hierarchical clustering algorithm for morpheme labelling. In: 2013 Proceedings of the Recent Advances in Natural Language Processing, RANLP 2013 (2013)
4. Cotterell, R., Müller, T., Fraser, A., Schütze, H.: Labeled morphological segmentation with semi-markov models. In: Proceedings of the Nineteenth Conference on Computational Natural Language Learning, pp. 164–174. Association for Computational Linguistics, Beijing, July 2015. http://www.aclweb.org/anthology/K15-1017
5. Creutz, M., Lagus, K.: Unsupervised discovery of morphemes. In: Proceedings of the ACL-02 Workshop on Morphological and Phonological Learning, MPL 2002, vol. 6, pp. 21–30. Association for Computational Linguistics, Stroudsburg (2002)
6. Creutz, M., Lagus, K.: Inducing the morphological lexicon of a natural language from unannotated text. In: Proceedings of the International and Interdisciplinary Conference on Adaptive Knowledge Representation and Reasoning, AKRR 2005 (2005)

7. El-Kahlout, D.I., Oflazer, K.: Initial explorations in English to Turkish statistical machine translation. In: Proceedings on the Workshop on Statistical Machine Translation, pp. 7–14. Association for Computational Linguistics, New York City, June 2006. http://www.aclweb.org/anthology/W06-3102

8. Ehsani, R., Alper, M.E., Eryiğit, G., Adalı, E.: Disambiguating main pos tags for Turkish. In: Proceedings of ROCLING - Conference on Computational Linguistics and Speech Processing. Association for Computational Linguistics and Chinese Language Processing (ACLCLP), Taiwan (2012)

9. Eryiğit, G., Nivre, J., Oflazer, K.: Dependency parsing of Turkish. Comput. Linguist. **34**(3), 357–389 (2008)

10. Goldsmith, J.: Unsupervised learning of the morphology of a natural language. Comput. Linguist. **27**(2), 153–198 (2001)

11. Hakkani-Tür, D.Z., Oflazer, K., Tür, G.: Statistical morphological disambiguation for agglutinative languages. Comput. Humanit. **36**(4), 381–410 (2000)

12. Koskenniemi, K.: Two-level morphology: a general computational model for word-form recognition and production. University of Helsinki, Department of General Linguistics (1983)

13. Lafferty, J.D., McCallum, A., Pereira, F.C.N.: Conditional random fields: probabilistic models for segmenting and labeling sequence data. In: Proceedings of the Eighteenth International Conference on Machine Learning, ICML 2001, pp. 282–289. Morgan Kaufmann Publishers Inc., San Francisco (2001)

14. Oflazer, K.: Two-level description of Turkish morphology. In: Proceedings of the Sixth Conference on European Chapter of the Association for Computational Linguistics, EACL 1993, pp. 472–472. Association for Computational Linguistics, Stroudsburg (1993)

15. Poon, H., Cherry, C., Toutanova, K.: Unsupervised morphological segmentation with log-linear models. In: Proceedings of Human Language Technologies: The 2009 Annual Conference of the North American Chapter of the Association for Computational Linguistics, NAACL 2009, pp. 209–217. Association for Computational Linguistics, Stroudsburg (2009)

16. Sak, H., Güngör, T., Saraçlar, M.: Morphological disambiguation of Turkish text with perceptron algorithm. In: Gelbukh, A. (ed.) CICLing 2007. LNCS, vol. 4394, pp. 107–118. Springer, Heidelberg (2007). https://doi.org/10.1007/978-3-540-70939-8_10

17. Say, B., Zeyrek, D., Oflazer, K., Özge, U.: Development of a corpus and a treebank for present-day written Turkish. In: Proceedings of the Eleventh International Conference of Turkish Linguistics, pp. 183–192 (2002)

18. Sha, F., Pereira, F.: Shallow parsing with conditional random fields. In: Proceedings of the 2003 Conference of the North American Chapter of the Association for Computational Linguistics on Human Language Technology, vol. 1, pp. 134–141. Association for Computational Linguistics (2003)

19. Dinçer, T., Karaoğlan, B., Kişla, T.: A suffix based part-of-speech tagger for Turkish. In: Proceedings of the International and Interdisciplinary Conference on Adaptive Knowledge Representation and Reasoning, AKRR 2005 (2005)

20. Virpioja, S., Kohonen, O., Lagus, K.: Unsupervised morpheme analysis with Allomorfessor. In: Peters, C., Di Nunzio, G.M., Kurimo, M., Mandl, T., Mostefa, D., Peñas, A., Roda, G. (eds.) CLEF 2009. LNCS, vol. 6241, pp. 609–616. Springer, Heidelberg (2010). https://doi.org/10.1007/978-3-642-15754-7_73

Part-of-Speech Tagging for Code Mixed English-Telugu Social Media Data

Kovida Nelakuditi$^{(\boxtimes)}$, Divya Sai Jitta, and Radhika Mamidi

Kohli Center on Intelligent Systems (KCIS),
International Institute of Information Technology, Hyderabad (IIIT Hyderabad),
Gachibowli, Hyderabad 500032, Telangana, India
{nelakuditi.kovida,jittadivya.sai}@research.iiit.ac.in,
radhika.mamidi@iiit.ac.in

Abstract. Part-of-Speech Tagging is a primary and an important step for many Natural Language Processing Applications. POS taggers have reported high accuracies on grammatically correct monolingual data. This paper reports work on annotating code mixed English-Telugu data collected from social media site Facebook and creating automatic POS Taggers for this corpus. POS tagging is considered as a classification problem and we use different classifiers like Linear SVMs, CRFs, Multinomial Bayes with different combinations of features which capture both context of the word and its internal structure. We also report our work on experimenting with combining monolingual POS taggers for POS tagging of this code mixed English-Telugu data.

Keywords: Code mixing · Social media data · Part-of-Speech tagging

1 Introduction

Code-mixing refers to the mixing of two or more languages or language varieties in speech [1]. This phenomenon is extended to writing as well. It is the embedding of various linguistic units such as affixes, words, phrases and clauses of one language in the other [1].

India is a land of many languages. People on social media often use more than one language to express themselves. In this paper, we present our work on building a POS tagger for English-Telugu code mixed data collected from social media site Facebook. Telugu is the third most spoken language in the country according to 2001 census[1] and a good percentage of educated Telugu speaking people use English in their daily speech. It could be the use of borrowed words/phrases or code-mixing. We would not make any distinction between borrowing and mixing in this paper. We refer to both the phenomena as code-mixing and handle them alike. Telugu is an agglutinative language belonging to Dravidian language family, with 74 million native speakers. English is the language of

---

[1] http://www.mapsofindia.com/culture/indian-languages.html.

A. Gelbukh (Ed.): CICLing 2016, LNCS 9623, pp. 332–342, 2018.
https://doi.org/10.1007/978-3-319-75477-2_23

instruction in most academic places and is also used for daily communication by both educated and uneducated population of India. Code-mixing happens at a pervasive rate in this setting. In this paper, we discuss the nature of such data, annotation of it and explore different ways of building a POS tagger for it. These methods include use of Machine Learning algorithms and combination of POS taggers of individual languages.

Mixing happens at many levels in English-Telugu code mixed data owing to the rich morphological nature of Telugu [2].

- Full borrowing:
  - Phonological changes: Borrowed words from English or other languages are used in Telugu with a sound change. This is done to achieve nativization.
    Eg: zebra – *jIbrA*[2]
  - Lexical borrowings: Words from English are used in Telugu phrases/sentences. (The inverse phenomenon, though happens, is very rare.)
    Examples:
    1. *nIku* help *cesAnu*.
       English: I helped you.
       In this sentence we can observe that the English word 'help' is used in a Telugu sentence.
    2. *velYlYi* trainu *eVkku*.
       English: go, get on the train.
       In this sentence, we can observe the borrowed word, train is used in the sentence with a suffix, '-u' added at the end. This kind of change happens over a wide range of borrowed English words.
- Partial borrowing:
  - Native suffixation: The words borrowed from English are changed according to the morphological nature of Telugu. Telugu is a postpositional language and agglutinative in nature with morphemes affixing to each other at the end of the root word. The roots can also be borrowed English words, resulting in words with Telugu suffixes and English roots.
    Examples:
    train – *trainu*
    1. trains – *trainlu*
       (*trainu+lu*, 'lu' is the plural marker in Telugu)
    2. 'in train' – *trainlo*
       (*trainu+lo*, 'lo' is the locative marker in Telugu)
    3. 'in trains' – *trainullo* (*trainu+ 'lu'+ 'lo'*)
    4. 'for train' – *trainuki* (*trainu+'ki'*, 'ki' is the beneficiary marker in Telugu)

---

[2] Telugu examples are written in italics and are in wx-format (sanskrit.inria.fr/DATA/wx.html).

- Agglutination in Complex Verbs: A complex verb is a multi-word compound with one noun and a verbal component. Due to agglutinative nature of Telugu, we have complex verbs with borrowed English words acting as nouns that are agglutinated to Telugu verbs.
  Examples:

  1. *luncayiMxi* – lunch+*ayiMxi*
     had lunch
  2. *resultoVcciMxi* – result+*voVcciMxi*
     result is out

- Syntactic Changes: Grammar of one language influences the other.
  1. Word order: English is a SVO language and Telugu is a SOV language. Due to influence of English, unnatural yet comprehensible Telugu constructions are seen in daily usage.
     Example:
     Telugu: *mA sabbulo uMxi nimma Sakwi*
     English: Our soap has the power of lemon
     In the Telugu utterance *'uMxi'* is the verb, *'mA sabbulo'* is the subject of the sentence and *'nimma SAkwi'*, the object. A more natural construction for the above utterance would be:
     Telugu: *mA sabbulo nimma Sakwi uMxi*
     This phenomenon is commonly observed in advertisements and dubbed movies because of the shortcomings in dubbing techniques.
  2. Pro-drop: Telugu is a pro-drop language, whereas English is not. The data is observed to have English sentences with initial subject dropped.
     Example:
     Indian English: will go to the Market after a while.
     Native English: I will go to the Market after a while.

POS tagging is the process of marking up a word in a text as corresponding to a particular part of speech. POS tagging is an important step for many Natural Language Processing Applications like Machine Translation, Dialog systems, parsing etc. Code mixed data from social media is noisy in nature with non-standard spellings and creative style of writing by the users. This nature of social media data poses additional problems. Supervised POS tagging accuracies for English measured on the PennTree bank have converged to around 97.3% [3]. Supervised POS tagging for Telugu[3] has an accuracy of 76.01%. [4] as a part of this work we explore various methods for creating a POS tagger for code-mixed English-Telugu data.

---

[3] ltrc.iiit.ac.in/analyzer/telugu/shallow-parser-tel-3.0.fc8.tgz.

2 Data Annotation

The data is annotated with language labels and POS tags for 6570 words and is shared for NLP Tools Contests: POS Tagging Code-Mixed Indian Social Media Text @ ICON 2015. Further this data is annotated with language labels, normalized form of the word, POS tag of it and chunk level information for 10207 words. We use this data to perform our experiments. This data is annotated by two individual annotators at four levels: which are 1. Language of the word, 2. Correct form of the word i.e. the citation form of the word as available in the dictionary, 3. The part-of-speech tag of the word, 4. Chunk information of the word. We have used the first three levels of the annotated corpus for the experiments described in the paper.

1. Language Identification: Language Identification is the process of identifying the language of a particular word. Every word is tagged with one of the labels: T, E, M, N and R.
   T class has words which belong to Telugu language.
   E class has words which belong to English language.
   N class has named entities.
   M class has words with English roots and Telugu morphological inflections.
   Examples :-
   (a) postlu (post+'lu') posts, here the plural marker, 'lu' from telugu.
   (b) supere (super+'e') super+clitique 'e'
   R class has rest of the words which include words from other languages, URLs, symbols etc.
2. Transliteration/Normalisation: The correct form of the word, i.e. the citation form of the word as available in the dictionary of the corresponding language, is provided along with the word. If the word belongs to T or M then Brahmi script or wx format[4] is used to represent the word, this is the form available in a Telugu dictionary. Brahmi script and wx format have a one to one character mapping. For words belonging to class E, the entries available in a dictionary of English are used.
3. Part-of-speech: Each word is annotated with its part of speech. We use universal POS tagset[5] to tag the data. It acts as a common tagset with 16 tags that can be used for all the languages.
4. Chunking: Chunking is the process of identifying and labeling different types of phrases such as Noun phrase, Verb phrase, Adjectival phrase etc. in a sentence. The annotation is at word level (a word is a segment with a space on both sides) and we annotate each word as beginning (B) or intermediate of a chunk (I), along with type of the phrase. For example, the tag B-NP against a word indicates that word is the starting word of the Noun phrase, and I-NP indicates that it is the intermediate word of Noun phrase. A phrase starts at the beginning tag ('B-') and includes all words between (but not) the word with next beginning tag.

---

[4] wx-format (sanskrit.inria.fr/DATA/wx.html).
[5] http://universaldependencies.org/docs/u/pos/.

Chunking part of the annotation is carried out as we plan to implement chunking in future on code mixed English-Telugu social media data. The experiments described in the paper are on developing a POS tagger and do not concern with the chunking.

The annotation of an example sentence from the corpus, 'plz watch it nd share chusaka nenu cheppanavasarm le meere share chestharuuu' is shown in Table 1 and the statistics of the annotated corpus is described in Table 2.

Table 1. 5-level annotation for an example sentence

plz	E	please	ADV	B-VP
watch	E	watch	VERB	I-VP
it	E	it	PRON	B-NP
nd	E	and	CONJ	B-CP
share	E	share	VERB	B-VP
chusaka	T	cUsAka	VERB	B-VP
neenu	T	nenu	PRON	B-NP
cheppanavasaram	T	ceVppanavarasaraM	NOUN	B-NP
le	T	lexu	VERB	B-VP
meere	T	mere	PRON	B-NP
share	E	share	NOUN	B-NP
cheestaru	T	ceswAru	VERB	B-VP

Table 2. Statistics of the annotated corpus

#words	10207
#sentences	1335
Average length of a sentence	8
#words in class E	4515
#words in class T	4342
#words in class R	379
#words in class M	81
#words in class N	890

3   Related Work

POS taggers on monolingual data give an accuracy of about 97.3% for English [3] and 76.01% for Telugu [4]. They are often seen as sequence labeling problems and have used the context based information in the form of lexical and sub-lexical characteristics of neighboring words. But in code-mixed setting, the context information can be in a different language which makes the understanding difficult.

Work has been done on English-Hindi and English-Bengali data. [5] is a POS tagger developed for English-Hindi code mixed data, where the authors have used monolingual Hindi and English POS taggers and combined the output from the two taggers. [6] describes tagger with combination of two monolingual POS taggers and also four Machine Learning algorithms on English-Hindi code mixed data. Typologically, Hindi is defined as inflectional where as Telugu is defined as agglutinative. [7] as a part of this work we have annotated English-Telugu data of 10,207 words and we have used three different Machine Learning algorithms to build the POS taggers and also experiment with combining POS taggers of individual languages.

Unknown words typically cause problems for POS tagging systems. Words from social media are noisy and fall under the category of OOV words while using existing systems. Normalization is an important and necessary step for proper working of existing POS taggers on this data. [8] developed a POS tagger for twitter data. No such work has been done for Telugu to date. We use CMU's Twitter POS tagger [8] for POS tagging of words identified as belonging to English, this tagger has a normalization module in itself. [9] is a transliteration system built for Indian languages, we use this system for transliteration of Telugu words.

4 Approach

We perform two different kinds of experiments:

1. Machine learning based POS taggers
2. Combining POS taggers of individual languages.

4.1 Machine Learning Based POS Taggers

We use three kinds of Machine Learning algorithms for building the POS tagger viz, Support Vector Machines (SVM), Bayes classification (Bay), Conditional Random Fields (CRF) with different combinations and variations of the following features:

- lexical feature:
  - word (CW).
- sub-lexical features: This set of features include varying length of prefix, suffix, infix character strings derived from the current word and its neighboring words.
  - Prefix, suffix character strings (CPS): Telugu is a postpositional language which is mostly suffixing and English is a prepositional language, due to this reason having prefix, suffix character strings as features helps better understanding of the language and determining the POS tag of it.

- Infix character strings (CI): Telugu is agglutinative, many bound morphemes can fuse to form a word, we include infix character strings to the feature set. This, in addition to suffix character strings can help in finding more number of bound morphemes.
- Presence of postpositions (PSP): We check for the presence of Telugu postpositions at the end of the word and use it as a feature.
- Prefix, suffix character strings of neighboring words (NPS).
- other features:
  - length of the word (WL).
  - neighboring words (N).

4.2   Combining POS Taggers of Individual Languages

Solorio and Liu [10] have proposed to combine POS taggers of individual languages for the code-mixed data. We have used CMU's Twitter POS tagger [8] for English and POS tagger developed at LTRC, which is a part of the shallow parser tool (See footnote 3) for Telugu.

The pipeline of this system is as follows:

1. Language Identification: Language Identification is the process of dividing words into one of the mentioned classes. A CRF model is implemented for this purpose with the following features.
   - lexical feature:
     - word
   - sub-lexical features:
     - Prefix, suffix character strings
     - Infix character strings
     - Presence of Post positions
     - prefix, Suffix character strings of neighboring words
   - other features:
     - length of the word
     - neighboring words
2. Transliteration/normalization: We use CMU pos tagger on English words (M) which reported an accuracy of 89.39% [8], it normalizes English words as a primary step. We use IRTRANS [9] for Telugu words (T) and mixed words (M). This tool is used to convert roman into Telugu script i.e. Brahmi. However, no normalization for words of class M and T is possible at this stage due to lack of resources.
3. POS tagging: We use POS taggers built for monolingual data i.e. we use both English and Telugu POS taggers. We give the words of class T and M to Telugu POS tagger and the words of E and R to English POS tagger. We don't process words belonging to class N as it is directly mapped to proper noun (PROPN). We later map the tagsets of English POS tagger and Telugu POS tagger to universal POS tagger.

5   Experiments and Results

We have used CRF++ tool [11] for implementing CRF, implemented Bayes classification and SVM using Scikit-learn [12].

5.1   Machine Learning Based POS Taggers:

The following Tables 3 and 4 describes the accuracies obtained from three different Machine Learning algorithms namely, CRFs, SVMs and Bayes classification with different combinations of features. Accuracies are reported on three-fold and five-fold cross validation.

Table 3. Results of different feature templates on 3-fold cross validation.

Template	CPS	CI	NW	NPS	PSP	SVM	Bayes	CRF
1	3, 2	3	0	0	Yes	67.78	65.27	73.58
2	3, 2, 1	3	1	3	Yes	65.69	61.46	75.75
3	3, 2, 1	3	2	3	Yes	64.12	59.29	**76.22**
4	3, 2, 1	3	1	3, 2	Yes	60.17	51.96	75.66
5	3, 2, 1	3	1	2	Yes	65.65	63.66	75.90

Table 4. Results of different feature templates on 5-fold cross validation.

Template	CPS	CI	NW	NPS	PSP	SVM	Bayes	CRF
1	3, 2	3	0	0	Yes	70.46	67.34	77.88
2	3, 2, 1	3	1	3	Yes	68.25	63.51	80.98
3	3, 2, 1	3	2	3	Yes	66.47	61.17	**81.40**
4	3, 2, 1	3	1	3, 2	Yes	63.10	53.46	81.05
5	3, 2, 1	3	1	2	Yes	68.48	63.66	80.09

In both the above tables, the following conventions are followed:

- If CPS = k, k length suffix and prefix strings of the current word are taken as a feature.
- If CPS = k1, k2, ... then k1, k2, ... length suffix and prefix strings of the current word are taken as a feature.
- If NPS = k, k length suffix and prefix strings of neighboring words are taken as a feature.
- If NPS = k1, k2, ... then k1, k2, ... length suffix and prefix strings of neighboring words are taken as feature.
- If CI = k, k length infix string form the current word is taken as a feature.

We can observe that template3 gives more accuracy and CRF performs better than SVMs and Bayes classifier. In all the above experiments word length (WL) and current word (CW) are taken as features.

5.2 Combining POS Taggers of Individual Languages:

Language Identification:
Using different combinations and variations of these features, various templates were experimented upon and the accuracies of three-fold cross validation on the dataset are reported in the following Table 5:

Table 5. Results of different feature templates for Language Identification on 3-fold cross validation

Template	CPS	CI	WL	NPS	NW	PSP	CRF
1	3	3	Yes	0	0	No	89.94
2	3, 2, 1	0	No	3	2	No	92.65
3	3, 2, 1	3	Yes	3	2	Yes	93.01
4	3, 2, 1	0	Yes	3	1	Yes	92.76
5	3, 2	3	Yes	3	1	Yes	92.33
6	3	3	No	0	1	No	91.01
7	3, 2, 1	3	No	3	2	No	**93.08**

We can see that template 7 gives the best accuracy and we select this template for further use. After Language Identification, words recognized as belonging to T and M classes are transliterated into Brahmi script using IRTRANS [9]. After transliteration, words belonging to classes M and T are run on POS tagger of Telugu (see Footnote 3) and words belonging to class E and R are run on CMU's Twitter POS tagger for English. We finally map the tagset of Telugu POS tagger [13] and CMU's Twitter POS tagger [8] to universal POS tagset to find POS tags of all the words. It is to be noted that words belonging to class N are mapped to PROPN by this stage. The accuracy of the individual module is 58.66% as tested on the total dataset of 10207 words. The accuracy of the pipeline is 52.37% as tested on training data of 5840 words and testing data of 4367 words.

We observe that including infix strings of length three along with prefix/suffix strings and using the information of presence or absence of postposition in the feature set gives good accuracy as Telugu is a agglutinative language which has multiple morphemes fusing together and we are able to capture the internal structure of the word to some extent using these features. In addition to these, we get the contextual information from neighboring words and we also used features which describe their internal structure. The performance of combination of different language taggers is not so satisfactory compared to the Machine Learning approaches because of three main reasons:

1. Propagation of error - As this is a pipeline system, there is a definite error propagation from Language Identification and Transliteration/Normalisation stages.

2. Non-availability of normaliser for Telugu - Telugu words are complex in constructions. As no normaliser has been built for it to date, the POS tagger of Telugu fails in analysing many words.
3. Loss of context information - With words from different languages in the same sentence, not all the words within a sentence are fed to the POS tagger of a single language as is the case in taggers of monolingual data (in that case POS tagger use the contextual information to perform their best). Hence there is a loss of contextual information in combining POS taggers of individual languages.

6   Conclusion

The paper describes the first step at building shallow parser for English-Telugu code mixed data. Through this paper we present our efforts at attempting various statistical methods for POS tagging of code-mixed social media data. We also create a standard dataset for building supervised models of shallow parsing on this data which we consider as our immediate future work. POS tagging is a necessary and essential step for many NLP applications and the results that we have obtained are encouraging.

References

1. Ayeomoni, M.O.: Code-switching and code-mixing: style of language use in childhood in Yoruba speech community. Nordic J. Afr. Stud. **15**(1), 90–99 (2006)
2. Kosaraju, P., Kesidi, S.R., Ainavolu, V.B.R., Kukkadapu, P.: Experiments on Indian language dependency parsing. In: Proceedings of the ICON10 NLP Tools Contest: Indian Language Dependency Parsing (2010)
3. Manning, C.D.: Part-of-speech tagging from 97% to 100%: is it time for some linguistics? In: Gelbukh, A.F. (ed.) CICLing 2011. LNCS, vol. 6608, pp. 171–189. Springer, Heidelberg (2011). https://doi.org/10.1007/978-3-642-19400-9_14
4. Rao, D., Yarowsky, D.: Part of speech tagging and shallow parsing of Indian languages. In: Shallow Parsing for South Asian Languages, p. 17 (2007)
5. Vyas, Y., Gella, S., Sharma, J., Bali, K., Choudhury, M.: POS tagging of English-Hindi code-mixed social media content. In: EMNLP, vol. 14, pp. 974–979 (2014)
6. Jamatia, A., Gambäck, B., Das, A.: Part-of-speech tagging for code-mixed English-Hindi Twitter and Facebook chat messages. In: Recent Advances in Natural Language Processing, p. 239 (2015)
7. Baskaran, S., Bali, K., Bhattacharya, T., Bhattacharyya, P., Jha, G.N., et al.: A common parts-of-speech tagset framework for Indian languages. In: Proceedings of the LREC 2008. Citeseer (2008)
8. Gimpel, K., Schneider, N., O'Connor, B., Das, D., Mills, D., Eisenstein, J., Heilman, M., Yogatama, D., Flanigan, J., Smith, N.A.: Part-of-speech tagging for Twitter: annotation, features, and experiments. In: Proceedings of the 49th Annual Meeting of the Association for Computational Linguistics: Human Language Technologies: short papers, vol. 2, pp. 42–47. Association for Computational Linguistics (2011)

9. Bhat, I.A., Mujadia, V., Tammewar, A., Bhat, R.A., Shrivastava, M.: IIIT-H system submission for FIRE2014 shared task on transliterated search. In: Proceedings of the Forum for Information Retrieval Evaluation, pp. 48–53. ACM (2014)

10. Solorio, T., Liu, Y.: Part-of-speech tagging for English-Spanish code-switched text. In: Proceedings of the Conference on Empirical Methods in Natural Language Processing, pp. 1051–1060. Association for Computational Linguistics (2008)

11. Kudo, T.: CRF++: yet another CRF toolkit (2005). http://crfpp.sourceforge.net

12. Pedregosa, F., Varoquaux, G., Gramfort, A., Michel, V., Thirion, B., Grisel, O., Blondel, M., Prettenhofer, P., Weiss, R., Dubourg, V., et al.: Scikit-learn: machine learning in Python. J. Mach. Learn. Res. **12**, 2825–2830 (2011)

13. Bharati, A., Sangal, R., Sharma, D.M., Bai, L.: AnnCorra: annotating corpora guidelines for pos and chunk annotation for Indian languages. LTRC-TR31 (2006)

Syntax and Chunking

Analysis of Word Order in Multiple Treebanks

Vladislav Kuboň, Markéta Lopatková[(⊠)], and Jiří Mírovský

Faculty of Mathematics and Physics, Charles University in Prague,
Prague, Czech Republic
{vk,lopatkova,mirovsky}@ufal.mff.cuni.cz

Abstract. This paper gives an overview of results of automatic analysis of word order in 23 dependency treebanks. These treebanks have been collected in the frame of the HamleDT project, whose main goal is to provide universal annotation for dependency corpora; thus it also makes it possible to use identical queries for all the corpora. The analysis concentrates on basic characteristics of word order, the order of three main constituents, a predicate, a subject and an object. A quantitative analysis is performed separately for main clauses and subordinated clauses; further, a presence of an active verb is taken into account – we show that in many languages the subordinated clauses have a slightly different order of words than main clauses; the choice of voice has also an impact on word order.

The development in the field of natural language corpora went through several stages in the course of the last 30 years. The corpora have been growing in size, in the complexity of annotation, in the variety of languages for which corpora exist. On the other hand, due to differences in annotation schemes, language corpora served primarily as valuable monolingual data resources (of course, with an exception of parallel corpora) so far. This unfortunate situation started to change a couple of years ago, when the need for some unified annotation scheme for syntactically annotated corpora (treebanks) became evident.

The idea of unifying dependency treebanks to some common annotation scheme brought several more or less successful attempts for such unified approach. Stanford Dependencies and Stanford Universal Dependencies [2–4],[1] Google Universal Tags [5], Universal Dependencies [6],[2] all these projects constitute individual steps towards the definition of universally accepted annotation schemes for dependency treebanks. In this paper we exploit the annotation developed in the frame of the HamleDT project (Harmonized Multi-Language Dependency Treebank [10]).[3] We are going to show how such common annotation scheme may help in analysis of natural language properties.

We are going to demonstrate it on the analysis of the phenomenon of word order. This phenomenon is very important for theoretical research as well as for

---

[1] http://nlp.stanford.edu/software/stanford-dependencies.shtml.
[2] http://universaldependencies.org/.
[3] https://ufal.mff.cuni.cz/hamledt.

© Springer International Publishing AG, part of Springer Nature 2018
A. Gelbukh (Ed.): CICLing 2016, LNCS 9623, pp. 345–355, 2018.
https://doi.org/10.1007/978-3-319-75477-2_24

practical applications. The freedom of word order to a great extent determines how difficult it is to parse a particular natural language (a language with more fixed word order is typically easier to parse than a language containing, e.g., non-projective constructions).

When concentrating on word order, we study the prevalent order of the verb and its main complements. Natural languages are typically classified into several large groups based on excerptions and careful examination by many linguists [8,9]: For example, Romance languages are typically characterized as SVO (SVO reflecting the order Subject–Verb–Object) languages. English and other languages with a fixed word order typically follow this order of words in declarative sentences; although Czech, Russian and other Slavic languages have a high degree of word order freedom, they still stick to the same order of words in a typical (unmarked) sentence. As for the VSO-type languages, their representatives can be found among semitic (Arabic, classical Hebrew) or Celtic languages, while (some) Amazonian languages belong to the OSV type.

Although all these investigations have been based upon a systematic observation of linguistic material, modern computational linguistics has brought into play much larger resources providing huge volumes of language material (which can be studied by means of automatic tools), and thus it may bring a deeper linguistic insight into the language typology. Thanks to a wide range of linguistic data resources for tens of languages available nowadays, we can easily confirm (or enhance by quantitative clues) the conclusions of traditional linguists. This paper represents a step in this direction.

1 Available Data Resources and Tools

HamleDT (Harmonized Multi-Language Dependency Treebank, [10]) (see Footnote 3) is a compilation of existing dependency treebanks (or dependency conversions of other treebanks), transformed so that they all conform to the same annotation style. These treebanks as well as searching tools are available through a repository for linguistic data and resources LINDAT/CLARIN.[4]

1.1 Corpora

HamleDT integrates corpora for several tens of languages. Wherever it is possible due to license agreements, the corpora are transformed into a common data and annotation format, which enables a user – after a very short period of getting acquainted with each particular treebank – to search and analyze comfortably the data of a particular language.

The HamleDT family of treebanks is based on the dependency framework and technology developed for the Prague Dependency Treebank (PDT),[5] i.e., large syntactically annotated corpus for the Czech language [1]. Here we focus

---

[4] https://lindat.mff.cuni.cz/.

[5] http://ufal.mff.cuni.cz/pdt3.0.

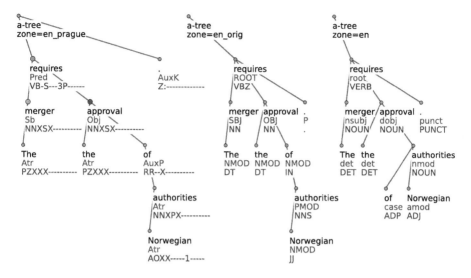

Fig. 1. Three dependency representations of the sentence "The merger requires the approval of Norwegian authorities." in HamleDT 3.0. It is also one of the results of the query from Fig. 2; nodes matching the query are slightly enlarged (in the left tree, nodes "requires", "merger" and "approval").

on the so-called analytical layer, describing surface sentence structure (relevant for studying word order properties). Unfortunately, due to various technical and licensing restrictions, it was not possible to use all treebanks contained in HamleDT. Thus our effort focusses on 23 treebanks with available annotation on this syntactic layer, which still represent a wide variety of languages having various word-order properties.

As an example, Fig. 1 shows three dependency representations for an English sentence in the HamleDT format.[6] Table 1 provides an overview of the languages and the size of the corpora examined in our experiment (first and second columns).

1.2   Querying Tool

The advantage of using a common annotation framework for multiple treebanks also has a very useful consequence – instead of developing tailor-made searching tools we can apply a common tool to all treebanks we are analyzing. In the case of HamleDT, we can use the PML-TQ [7] search tool,[7] originally developed for processing the data from PDT.

---

[6] Data of each treebank in HamleDT are distributed in three annotation schemes – (a) the transformation of the treebank to the praguian style (used in PDT; leftmost in Fig. 1), (b) the original annotation format of the given treebank (or its dependency transformation in case of non-dependency treebanks; in the middle of Fig. 1), and (c) the transformation of the treebank to the Universal Dependencies style (rigthmost in the figure).

[7] https://lindat.mff.cuni.cz/services/pmltq/.

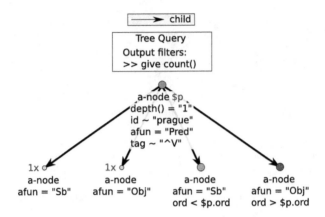

Fig. 2. Visualization of the PML-TQ query

Having the treebanks in the common data format and annotation scheme, the PML-TQ framework makes it possible to analyze the data in a uniform way. A typical user interested in monolingual data can use PML-TQ in an interactive way. Such approach would, of course, not work for our set of 23 treebanks, therefore we have used a command line interface which PML-TQ also provides. This interface makes it possible to create scripts that process a specified set of treebanks automatically.

Let us now give an example of a PML-TQ query used in our analysis. It counts sentences having an SVO word order in the main clause.

```
a-node $p :=
[ depth() = "1", id ~ "prague",
  afun = "Pred", tag ~ "^V",
  1x a-node
    [ afun = "Sb" ],
  1x a-node
    [ afun = "Obj" ],
  a-node
    [ afun = "Sb", ord < $p.ord ],
  a-node
    [ afun = "Obj", ord > $p.ord ] ];
>> give count()
```

The query searches data annotated in the praguian style (id ~ "prague") for sentences containing verbs (tag ~ "^V") with the analytical function of a predicate (afun = "Pred") at the depth of one level below the technical root of the tree (depth() = "1"; i.e., this query focuses on the word order in main clauses, excluding coordinated predicates and disregarding also subordinate clauses). There must be exactly one subject and one object directly depending on the predicate (for the subject: 1x a-node [afun = "Sb"]), the subject must precede the verb (afun = "Sb", ord < $p.ord), and the object must follow it (`afun` = `"Obj"`, `ord` > `$p.ord`). The result of the query is the count of such sentences (`>> give count()`). The visualization of the PML-TQ query can be found in Fig. 2.

2   The Experiment

In order to avoid possible bias caused by a combination of too many language phenomena in complicated sentences, we have decided to exclude all sentences containing coordinated predicates, subjects or objects from our experiment. The phenomenon of coordination is to some extent "orthogonal" to that of word order (especially in dependency-based approaches to a language description); thus the results might have been negatively influenced if coordination of verbs or the coordination of its direct dependents would be allowed.

We have created several queries aiming at a thorough investigation of the phenomenon of the mutual position of a subject, predicate and object. The queries aimed separately at the types of word order in main clauses, in subordinated clauses and in clauses containing active verbs (both main and subordinated ones). In each case we have counted sentences for all six possible combinations (SVO, OVS, VSO, VOS, SOV, OSV – these combinations will be referred to as "full" structures) and for four combinations with either subject or object missing (SV, VS, OV, VO – henceforth "incomplete" structures).[8] Table 1 shows the results counted on main clauses for the six combinations with both subject and object, Table 2 shows the results for sentences without a subject or an object. Similar tables have been collected for main clauses with active verbs, for subordinated clauses and subordinated clauses with active verbs, but for the sake of compactness of this paper, let us mention only some interesting results instead of presenting full tables in the subsequent text.

3   Discussion of Results

The results presented in Tables 1 and 2 may serve as a basis for an estimation of a degree of word order freedom of individual languages. We may design various metrices for such estimation, but it seems quite natural to suppose that the languages with the highest degree of word order freedom will demonstrate the most equal distribution of sentences among all variants of the word order described in our tables. The most equal distribution would also lead to the smallest difference between the two extremes, maximum and minimum. The languages with the lowest difference between the most frequent order and the least frequent one are apparently having the highest degree of word order freedom.

Table 3 shows the rank of individual languages with regard to the word order freedom calculated according to the simple metric proposed in the previous paragraph. If we calculate it on main sentences with "full" structure, i.e. main sentences containing both subject and (exactly one) object, the highest rank belongs

---

[8] In this experiment, we do not distinguish among various sources of incompleteness as e.g. those caused by imperatives, intransitives, pro-drop subjects or objects etc.

Table 1. Relative frequencies for word order variants in the main sentence in 23 studied languages – "full" structures.

Treebank	Number of sentences	Number of occurrences	SVO (%)	OVS (%)	VSO (%)	VOS (%)	SOV (%)	OSV (%)
Catalan	16,786	5,921	65.5	15.3	0.2	0.5	18.2	0.3
German	40,020	18,617	49.8	12.0	35.2	2.8	0.2	0.0
Dutch	13,735	2,646	60.4	15.5	23.5	0.4	0.2	0.1
Portuguese	9,359	2,879	80.7	9.1	1.9	5.1	3.1	0.2
Japanese	17,753	138	0.0	0.0	0.0	0.0	85.5	14.5
Persian	12,455	2,480	15.9	0.2	0.1	0.0	81.1	2.8
Romanian	4,042	1,132	62.1	12.9	0.4	0.8	23.6	0.2
Estonian	1,315	359	85.5	4.7	7.8	1.1	0.8	0.0
Polish	8,227	1,645	71.4	11.5	4.9	5.3	4.1	2.8
Spanish	17,709	5,569	61.3	20.4	1.0	0.5	16.5	0.4
Russian	34,895	6,194	72.2	15.2	1.6	3.5	4.2	3.4
Slovenian	1,936	182	47.8	25.3	4.4	2.2	17.0	3.3
Slovak	63,238	7,794	47.8	22.9	5.5	8.0	12.2	3.6
Arabic	7,664	1,203	22.4	0.2	74.1	3.3	0.0	0.0
English	18,791	8,585	83.1	7.3	0.3	0.0	0.0	9.2
Tamil	600	132	0.0	6.8	0.0	0.0	59.1	34.1
Hindi	13,274	1,490	3.0	0.1	0.0	0.0	93.4	3.5
Latin	3,473	395	25.1	6.8	8.6	4.1	41.3	14.2
Telugu	1,600	254	2.4	3.9	0.0	0.0	69.7	24.0
Czech	87,913	16,862	51.2	21.4	9.6	10.0	5.8	2.0
Bengali	1,279	307	21.5	9.8	0.0	0.0	61.6	7.2
Turkish	5,935	802	3.2	13.3	0.6	0.1	79.2	3.5
Ancient Greek	21,173	1,648	24.6	21.1	5.1	5.2	27.2	16.8

to the two classical languages, Latin and Ancient Greek, closely followed by three Slavic languages (Slovak, Slovenian and Czech) and German. The languages with the most fixed word order are, according to our simple metric, Japanese, Estonian and Hindi.

The rank will change if we apply the same metric to main sentences with "incomplete" structure, i.e., main sentences without a subject or without an object. In addition to the rank of all languages for "incomplete" main sentences, we provide also the difference in a rank between "full" and "incomplete" main structures in Table 3 (the headers "Rank" and "Rank Diff" under the "main clauses") – positive integers indicate more fixed word order, negative integers less fixed word order. Last two columns show the differences in ranking of the word order freedom on subordinate clauses, i.e., the difference in ranking between

Table 2. Relative frequencies for word order variants with missing subject or object in the main sentence in 23 studied languages – "incomplete" structures.

Treebank	Number of sentences	Number of occurrences	SV (%)	VS (%)	OV (%)	VO (%)
Catalan	16,786	6,287	83.5	7.9	4.3	4.3
German	40,020	9,760	45.9	38.8	2.4	12.8
Dutch	13,735	7,222	68.6	27.6	1.2	2.6
Portuguese	9,359	3,372	64.9	11.2	2.8	21.2
Japanese	17,753	2,966	27.0	0.0	73.0	0.0
Persian	12,455	5,435	56.6	0.2	29.2	14.0
Romanian	4,042	2,150	52.3	18.6	10.9	18.2
Estonian	1,315	745	68.9	21.3	1.9	7.9
Polish	8,227	3,881	34.2	21.9	5.9	38.0
Spanish	17,709	6,951	80.5	11.7	1.9	5.9
Russian	34,895	10,825	57.3	30.0	4.6	8.0
Slovenian	1,936	719	42.1	21.4	12.4	24.1
Slovak	63,238	22,115	33.8	24.5	11.1	30.6
Arabic	7,664	3,480	7.5	85.6	0.0	6.9
English	18,791	7,216	92.4	3.6	0.1	3.9
Tamil	600	300	84.3	1.7	14.0	0.0
Hindi	13,274	6,555	51.2	0.1	47.5	1.3
Latin	3,473	885	38.0	16.0	24.3	21.7
Telugu	1,600	912	57.8	2.9	39.1	0.2
Czech	87,913	29,160	43.2	34.8	5.3	16.7
Bengali	1,279	446	65.2	7.0	19.1	8.7
Turkish	5,935	2,391	43.5	3.1	49.9	3.5
Ancient Greek	21,173	3,987	26.1	13.3	29.9	30.8

"full" subordinated structures with respect to "full" main clauses and the rank difference between "incomplete" subordinated sentences and 'incomplete" main ones.

3.1   Rank Differences Between Complete and Incomplete Main Clauses

The rank difference listed in Table 3 helps us to discover interesting cases – namely the languages where the word order freedom in incomplete main clauses substantially changes when compared to the word order freedom of a complete main clause. Although the reasons for this change would require a detailed

Table 3. The rank of languages with regard to the word order freedom – the rank difference (Rank Diff) for "incomplete" main clauses as well as the Rank Diff for "full" subordinated clauses are calculated with respect to the Rank of "full" structures in main clauses, the Rank Diff for "incomplete" subordinated clauses is calculated with respect to the Rank of "incomplete" main clauses.

Language	Main clauses				Subordinated clauses	
	"full"		"incomp."		"full"	"incomp."
	Difference (%)	Rank	Rank	Rank diff	Rank diff	Rank diff
Ancient Greek	22.1	1	1	0	+1	+7
Latin	37.2	2	2	0	+1	+1
Slovak	44.3	3	3	0	+3	+1
Slovenian	45.6	4	4	0	−3	−3
Czech	49.2	5	6	+1	+4	−4
German	49.7	6	8	+2	+7	+12
Tamil	59.1	7	21	+14	+5	−4
Dutch	60.2	8	17	+9	+8	−7
Spanish	60.9	9	19	+10	−1	−7
Bengali	61,6	10	14	+4	+9	+4
Romanian	61.9	11	7	−4	−1	+15
Catalan	65.4	12	20	+8	−7	−11
Polish	68.6	13	5	−8	+1	+9
Telugu	69.7	14	13	−1	+3	+3
Russian	70.6	15	11	−4	−4	−6
Arabic	74.1	16	22	+6	−12	−16
Turkish	79.1	17	9	−8	+4	+10
Portuguese	80.5	18	15	−3	0	0
Persian	81.1	19	12	−7	+1	−1
English	83.1	20	23	+3	+3	−10
Japanese	85.5	21	18	−3	−6	+5
Estonian	85.5	22	16	−6	−15	−9
Hindi	93.4	23	10	−13	−1	+11

knowledge of syntax of all languages concerned, it is nevertheless interesting just to identify the languages which are affected by this phenomenon.

With regard to the rank shift between the word order freedom of complete and incomplete main clauses, the highest increase of word order freedom shows Hindi (the jump from the rank 23 to the rank 10), followed by a group of languages with the rank difference of 7 or 8 (Polish, Turkish and Persian). It is quite interesting that this group is not uniform, it contains languages which are typologically very different. The set of languages demonstrating the rank shift in the opposite direction, from a more free to a more fixed word order in incomplete sentences, also contains one Indian language, Tamil, with a great rank difference (from rank 7 to 21). Three European languages, (Dutch, Spanish and Catalan), exhibit a slightly smaller rank shift of 9, 10 and 8, respectively. It is quite interesting to see that the two closely related languages, Spanish and Catalan, actually have similar properties, in contrast to the properties of Polish mentioned in the previous paragraph which differ a lot from the properties of related Slavic languages as Czech or Slovak, which (unlike Polish) actually have the same word order properties for both complete and incomplete main clauses.

3.2 Rank Differences Between Subordinated and Main Clauses

When comparing the rank of individual languages with regard to the word order freedom between a "full" main and a "full" subordinate clause (the header "Rank Diff" under "full" subordinated clauses), we can also identify two distinctive groups of languages which exhibit a relatively big rank shift. The languages with substantially higher degree of the word order freedom in subordinated clauses are Arabic (from rank 16 for main clauses to rank 4 for subordinated ones) and Estonian (from rank 22 to rank 7).

Among the languages with less free word order in subordinated clauses, there is again one Indian language (Bengali with the rank shift of 9 positions) and, again, Dutch. In case of Dutch we may recall the famous examples of phenomena exceeding the expressive power of context-free languages, namely the subordinated clauses such as *...dat Jan Piet de kinderen zag helpen zwemmen* (... that Jan saw Piet help the children swim) where the Dutch syntax requires a very strict order of words.

Similar interesting observations may be found also in the last two columns of Table 3, but it is obvious that the more detailed is the investigation of rank shifts, the more it depends on the metric chosen for measuring the degree of the word order freedom. Experiments with various metrics would be an interesting research topic for future research.

3.3 The Role of Active Verbs

The same type of observations as in the previous paragraphs may also be performed for various other kinds of sentences. The PML-TQ query language allows for much more subtle specification of sentence filters. Let us just give one more example, not described in Table 3 (the table would be too complex), namely the statistics of how an additional restriction (on the active form of main verbs in both main and subordinated clauses) might change the picture. A relatively big rank shift from a strict word order of main clauses towards more free one in the subordinated ones (both types of clauses containing active verbs) is observed for Portuguese (from rank 17 to the rank 6), Persian (from 19 to 10) and Hindi (from 22 to 14).

The opposite type of shift, towards more fixed word order in subordinated clauses, may be observed in case of Ancient Greek (from rank 3 to rank 11), Catalan (from 7 to 21), Spanish (from 8 to 22) and Japanese (from 14 to 23).

Again, the Catalan and Spanish show very similar properties, while Portuguese has a completely opposite nature in this respect. These simple observations show that the fact whether the clauses are active or passive definitely plays a role for some languages. Further application of additional filters might help to discover other phenomena which affect the freedom of the word order.

If we compare the results for subordinated clauses in general and subordinated clauses containing an active verb, we may notice that for a majority of languages there is actually no difference, the fact whether the clause contains an active verb plays a role only in Romanian, but in a different way than in the previous text. Romanian corpus does not contain any sentences with subordinated clauses containing at the same time an active verb and a complete set of complements, therefore it is impossible to say anything about word order properties of such sentences.

3.4   Final Remarks

Although the results presented above show some very interesting properties, they have certain limits. We should take them more like an indication of possibilities which the whole set of HamleDT treebanks opens than as something final. The detailed interpretation of results obtained by PML-TQ queries is impossible without a certain level of proficiency in languages being observed. For example, in Table 1 we have only 138 Japanese sentences out of the total of 17,753. This extremely low number (compared to the percentages of sentences filtered out for other languages) may indicate some serious problems.[9] It might mean some systematic annotation error or an error in the way how the original annotation was transformed into the HamleDT scheme, or it may have a completely different reason, but without a knowledge of Japanese and/or the detailed knowledge of the original annotation scheme used in the Japanese treebank it is impossible to draw valid conclusions.

4   Conclusions

In the research described in this paper we have made the first step towards a quantitative analysis of 23 treebanks collected and made available through the HamleDT initiative. We have analyzed the word order of available languages represented in HamleDT using the PML-TQ query language. The queries extracted large subsets of all treebanks and provided quantitative data for the analysis of word order freedom and its variations in main and subordinated clauses. We have also suggested a very simple and natural metric of word order freedom and by means of this metric we have identified languages which show different word order properties in different kinds of clauses.

---

[9] The apparent explanation based on a strong pro-drop nature of Japanese seems to be too simplistic as other languages exhibiting the same property (like e.g. Czech) do not show any similar reduction of the number of filtered sentences.

This paper opens new directions for the investigation of additional linguistic phenomena as well, using the same approach based on exploitation of richly syntactically annotated data converted into a common format. By the investigation of a difference between the properties of active and passive sentences we have shown that the PML-TQ queries may be further refined in order to allow deeper analysis of various linguistic phenomena in the future.

Acknowledgments. This work has been using language resources and tools developed, stored, and distributed by the LINDAT/CLARIN project of the Ministry of Education, Youth and Sports of the Czech Republic (project LM2015071).

The work on this project has been partially supported by the LINDAT/CLARIN project of the Ministry of Education, Youth and Sports of the Czech Republic (project LM2015071).

References

1. Bejček, E., Hajičová, E., Hajič, J., Jínová, P., Kettnerová, V., Kolářová, V., Mikulová, M., Mírovský, J., Nedoluzhko, A., Panevová, J., Poláková, L., Ševčíková, M., Štěpánek, J., Zikánová, Š.: Prague Dependency Treebank 3.0. Charles University in Prague, MFF, ÚFAL, Prague (2013). http://ufal.mff.cuni.cz/pdt3.0/
2. de Marneffe, M.C., Dozat, T., Silveira, N., Haverinen, K., Ginter, F., Nivre, J., Manning, C.D.: Universal Stanford Dependencies: a cross-linguistic typology. In: Proceedings of LREC 2014. ELRA, Reykjavik (2014)
3. de Marneffe, M.C., MacCartney, B., Manning, C.D.: Generating typed dependency parses from phrase structure parses. In: Proceedings of LREC 2006. ELRA, Genoa (2006)
4. de Marneffe, M.C., Manning, C.D.: The Stanford typed dependencies representation. In: COLING Workshop on Cross-Framework and Cross-Domain Parser Evaluation. ACL, Manchester (2008)
5. McDonald, R., Nivre, J.: Characterizing the errors of data-driven dependency parsing models. In: Proceedings of EMNLP-CoNLL 2007. ACL, Prague (2007)
6. Nivre, J., de Marneffe, M.C., Ginter, F., Goldberg, Y., Hajič, J., Manning, C.D., McDonald, R., Petrov, S., Pyysalo, S., Silveira, N., Tsarfaty, R., Zeman, D.: Universal dependencies v1: a multilingual treebank collection. In: Proceedings of LREC 2016. ELRA, Portorož (2016)
7. Pajas, P., Štěpánek, J.: System for querying syntactically annotated corpora. In: Proceedings of the ACL-IJCNLP 2009 Software Demonstrations, pp. 33–36. ACL, Suntec, August 2009
8. Sapir, E.: Language: An Introduction to the Study of Speech. Harcourt, Brace and Company, New York (1921). http://www.gutenberg.org/files/12629/12629-h.htm
9. Skalička, V.: Vývoj jazyka. Soubor statí. Státní pedagogické nakladatelství, Praha (1960)
10. Zeman, D., Dušek, O., Mareček, D., Popel, M., Ramasamy, L., Štěpánek, J., Žabokrtský, Z., Hajič, J.: HamleDT: harmonized multi-language dependency treebank. Lang. Res. Eval. **48**(4), 601–637 (2014)

A Framework for Language Resource Construction and Syntactic Analysis: Case of Arabic

Nabil Khoufi[(✉)], Chafik Aloulou, and Lamia Hadrich Belguith

MIRACL Laboratory, ANLP Research Group,
Faculty of Economics and Management, University of Sfax, Sfax, Tunisia
{nabil.khoufi, chafik.aloulou, l.belguith}@fsegs.rnu.tn

Abstract. Language resources such as grammars or dictionaries are very important to any natural language processing application. Unfortunately, the manual construction of these resources is laborious and time-consuming. The use of annotated corpora as a knowledge database might be a solution to a fast construction of a grammar for a given language. In this paper, we present our framework to automatically induce a syntactic grammar from an Arabic annotated corpus (The Penn Arabic TreeBank), a probabilistic context free grammar in our case. The developed system allows the user to build a probabilistic context free grammar from the annotated corpus syntactic trees. It's also offer the possibility to parse Arabic sentences using the generated resource. Finally, we present evaluation results.

Keywords: Language resource · PCFG · Parsing · Arabic language processing
Rule based framework · Viterbi algorithm

1 Introduction

Natural Language Processing (NLP) fields deals with analyzing, understanding and generating the human languages in order to interface with computers in both written and spoken contexts using natural human languages.

Building a generic parsing system (as one of NLP tasks) for Arabic language is a daunting because of morphologic and syntactic richness of Arabic [1]. Parsing systems could be based on statistical models or linguistic resources.

In this paper we focus on the parsing systems using linguistic resources. This approach requires linguistic resources to guide the syntactic analysis. The first method to provide such resources to the parser consists of writing down the language grammar manually. However, manual construction of such linguistic resources is a difficult task to undertake, and is time consuming. Moreover, natural language is far too complex to simply list all the syntactic rules. In addition, it is difficult to exhaustively list lexical properties of words, and lastly, the written grammar has to be validated by Arabic linguists.

A second method to build linguistic resources is the use of treebanks as source of knowledge. Indeed, treebanks, as rich corpora with annotations, provide an easy way to

© Springer International Publishing AG, part of Springer Nature 2018
A. Gelbukh (Ed.): CICLing 2016, LNCS 9623, pp. 356–365, 2018.
https://doi.org/10.1007/978-3-319-75477-2_25

build other linguistic resources, such as extensional and intentional lexicons, syntactic grammars, bilingual dictionaries, etc. This promotes their reuse and makes their implicit information explicit. Another advantage of treebanks is that they are not only developed and validated by linguists, but also submitted to consensus, which promotes their reliability. The possession of such a resource makes it possible to generate new resources based on other formalisms with wide coverage automatically. These resources inherit the original treebank qualities, while improving construction time.

This paper is organized as follows: Sect. 2 is devoted to presenting a comparative study of Arabic linguistic parsers. Section 3 gives an overview of our method for Arabic grammar construction. Section 4 presents the developed system. Experimental results are presented in Sect. 5. Section 6 provides the conclusion.

2 Linguistic Parsers Comparative Study: Case of Arabic

The number of works dealing with parsing Arabic using linguistic resources is very limited when compared to works dealing with other natural languages such as English, Spanish or French. To our knowledge, the majority of works regarding Arabic language parsing use the linguistic approach that yields satisfactory results, but does not attain the English state-of-the-art level yet.

McCord and Cavalli-Sforza [2] developed a Slot Grammar (SG) parser for Arabic (ASG) with new features of SG designed to accommodate Arabic as well as the European languages for which SGs have been built. Slot Grammar is dependency oriented, and has the feature that allows both deep structure (via logical predicate arguments) and surface structure to be shown in parse trees. The authors focused on the integration of BAMA (Buckwalter's Arabic Morphological Analyzer) [3] with ASG, and on a new, expressive SG grammar formalism (SGF) and they illustrated the way SGF is used to advantage in ASG.

The analyzer of Bataineh and Bataineh [4] uses recursive transition networks to build a context free grammar which describes the most common sentences in Arabic. The transition network considers syntax rules as graphs, arcs and labels. These are finite state automata representing rules' transcripts. To represent the maximum of structures, a set of sentences' patterns was also derived from school texts. These patterns are converted to context free rules with the help of Arab linguists. A sentence is accepted by the grammar if it is generated by a complete course (without interruption) of these transitions' networks.

Klein and Manning [5] developed a parser that implements a factored product model, with separate PCFG phrase structure and lexical dependency experts, whose preferences are combined by efficient exact inference, using an A* algorithm. As well as providing an English parser, the parser has been adapted to work with other languages. A Chinese parser based on the Chinese Treebank, a German parser based on the Negra corpus and Arabic parsers based on the Penn Arabic Treebank are also included [6].

Al-Taani et al. [7] constructed a grammar under the CFG formalism (Context Free Grammar) then implemented it in a parser with a top-down analysis strategy. This parser focused on identifying sentence type (Nominal or verbal) and domain words.

The work of [8] presents a simple parser for Arabic sentences. The aim of this parser was to check whether the syntax of an Arabic sentence is grammatically correct by constructing a new, efficient Context-Free Grammar. Alqrainy designed the parser to take advantage of the top-down technique. He used the NLTK (Natural Language ToolKit) tool [9] to build and test the Arabic CFG grammar.

Table 1 summarizes our comparative study of cited Arabic parsers. The comparison is performed using these criteria:

- Grammar formalism,
- Parsing strategy,
- Size of the testing data,
- Results as precision, accuracy or error scores.

Table 1. Comparative study of Arabic parsers

Authors	Grammar formalism	Parsing strategy	Testing data	Results
McCord and Cavalli-Sforza [2]	ASG grammar	Bottom-up	1000 sentences (13–20 words)	72% complete parses (with no guaranty of correctness)
Bataineh and Bataineh [4]	CFG grammar	Top-down	90 sentences	85.4% correct 2.2% incorrect 12.4% rejected
Green and Manning [6]	Human interpretable grammars (PCFG based)	–	ATB 10%	Precision 81.07 Recall 80.27 F-measure 80.67
Al-Taani et al. [7]	CFG grammar	Top-down	70 sentences (2–6 words)	Precision 94%
Alqrainy et al. [8]	CFG grammar	Top-down	105 sentences	Accuracy 95%

3   Our Method for Arabic Grammar Construction

In our work, we have built a Probabilistic Context Free Grammar (PCFG) automatically using an annotated corpus [10]. So we began by presenting the PCFG grammar definitions and then we detail the construction process.

3.1   PCFG Definitions

A probabilistic context-free grammar (PCFG) also called stochastic CFG (SCFG), is an extension of the famous context-free grammar, where a certain probability is assigned to each rule. Probabilistic context-free grammars are defined by a 5-tuplet $< N, T, R, S, P >$ as follows:

- N is a finite set of non-terminal symbols.
- T is a finite set of terminal symbols.
- R is a finite set of rules ri of the form $X \rightarrow Y1Y2 ... Yn$, where $X \in N$, $n \geq 0$, and $Yi \in (N \cup T)$ for $i = 1 ... n$.
- $S \in N$ is a distinguished start symbol.
- P is the set of probabilities pi associated to rules ri where: $\sum P(X \rightarrow Yi) = 1$, $\forall X \in N$ and $Yi \in (N \cup T)$ for $i = 1 ... n$.

Note that some sentences may have more than one underlying derivation in case of the use of a classic CFG and therefore generate several parse trees. Therefore probabilities P in a PCFG are used to produce the most likely parse tree for a given sentence. The probability of a parse tree is obtained by multiplying the probability of each rule used at each node of the tree.

3.2 Description of the Construction Process

Our goal in this section is to automatically build a PCFG grammar from an annotated corpus. This process consists of two steps: The first step is to induce CFG rules from the annotated corpus. The second step is to assign a probability to each induced rule. The application of these two steps allows us to obtain a PCFG. Figure 1 illustrates the workflow of this first process.

Since our construction of the PCFG grammar is based on an annotated corpus, we begin by discussing the corpus we used; then we describe the PCFG induction process in detail.

The PATB Corpus. In our work, we chose to use the well-known corpus, the Penn Arabic Treebank (PATB). This choice was motivated not only by the richness, the reliability and professionalism with which it was developed but also by the syntactic relevance of its source documents (converted to several other Treebank representations). Indeed, its annotations have the advantage of being reliable. This is shown by its efficacy in a large number of research projects in various fields of NLP [11]. The good quality of the text and its annotations is demonstrated by its performance in the creation of other Arabic Treebanks such as the Prague Arabic Dependency Grammar [12] and the Columbia Arabic Treebank [13], which converted the PATB to its syntactic representations in addition to other annotated texts.

Indeed, these annotations were manually elaborated and validated by linguists. Moreover, this treebank is composed of data from linguistic sources written in Modern Standard Arabic. This corpus is also the largest Arabic corpus which integrates syntactic tree files. The use of a large amount of annotated data in a grammar construction process increases the quality of the generated linguistic resource.

The Penn Arabic Treebank (PATB) was developed in the Linguistic Data Consortium (LDC) at the University of Pennsylvania [14]. Texts in the corpus, as with most texts written for adults in Modern Standard Arabic, such as newspaper articles, contain no vowels.

Step 1: CFG Rules Induction. As shown in Fig. 1, the first step of our method is the induction of CFG rules, including duplicates, which will be used in the second step.

A deep study of the PATB allows us to identify the rules that guide the CFG rules induction process. We focused on the morpho-syntactic trees of the PATB and we identified the following induction rules:

- Rule 1: Tree root → Start symbol
- Rule 2: Internal tree node → Non terminal symbol
- Rule 3: Tree word → Terminal symbol
- Rule 4: Tree fragment → CFG rules

We noticed that each parse tree is a sequence of context-free rules and each one has the same symbol "S" at its root. Thus, the root symbol "S" is taken to be the start symbol (S) of the grammar. Non-terminal (N) symbols consist of the set of internal nodes of the whole parse tree. The set of all words seen in the trees (the leaves) compose the set (T) of terminal symbols. Edges between the nodes of the trees are used to induce CFG rules (R).

Step 2: Probabilities Calculation. Once CFG rules had been induced from the PATB with all duplicates, we moved to the second step consisting of the calculation of rule probability (P) to finally obtain the PCFG grammar. Each rule probability is estimated using the following formula:

$$P(X \rightarrow Y) = Count\,(X \rightarrow Y)/Count\,(X)$$

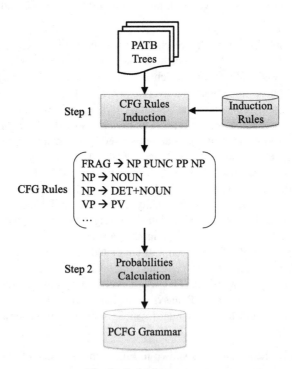

Fig. 1. Induction process

Where:

- $P(X \to Y)$ is the probability associated to the rule $X \to Y$.
- Count $(X \to Y)$ is the number of times the rule $X \to Y$ is seen in the Treebank.
- Count (X) is the count of rules that have the non-terminal X on the left-hand side.

For example, the rule NP $\to$ NP PP is seen 78 times in the PATB and we have counted 1821 rules that have NP on the left-hand side, thus:

$$P \ (NP \to NP \ PP) = 78/1821$$

Generated Resource. After applying our method, we obtained the PCFG grammar. As we mentioned earlier, The PATB is a very rich corpus and it contains many annotations like mood, gender, number, etc. The PATB corpus is annotated using a large set of annotations, which gives a high level of granularity.

For example, the Part of speech annotation tag set contains 498 tags that provide much information like gender, mood, etc. [15]. There are also 22 syntactic category tags and 20 tags that describe semantic relations between tokens. In addition, stop words, which are very numerous in the Arabic language, are also annotated with specific tags.

Incorporation of all this information within the grammar increases its complexity and its size. We chose to reduce the POS tags to the basic tags, which leaves about 70 tags for the Arabic language, to facilitate the use of induced grammars for NLP applications.

We present below some statistics about our PCFG grammar. Table 2 presents the most frequent PCFG syntactic rules generated from the PATB corpus after applying our method and Table 3 presents the overall count of rules (contextual rules and lexical rules).

Table 2. Most frequent rules

Left-Hand symbol (LHS)	NP	VP	S	FRAG	ADJP	UCP	PP
Rule count	1821	1311	1154	360	330	196	150

Table 3. Rule count of the induced PCFG

Contextual rules	Lexical rules	Total
5757	38 901	44 658

4 An Overview of the Developed System

In this section, we present the implementation details and the evaluation results of our method.

4.1   General Architecture

Our system is based on the architecture shown in Fig. 2. More precisely, our tool includes two modules:

- Module 1: The purpose of the first module is to generate the Arabic syntactic grammar (PCFG).
- Module 2: The second one exploits the generated resources to parse Arabic sentences.

To develop our tool, we used java programming language and Viterbi implementation of the NLTK tool.

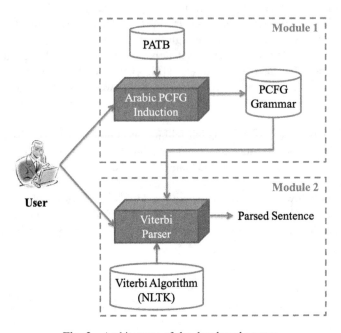

Fig. 2. Architecture of the developed system

4.2   Arabic PCFG Grammar Construction Interface (Module 1)

The objective of this interface is to induce PCFG rules from the PATB. In a first step, we induce CFG rules from the PATB parse trees in a raw format (Buckwalter transliteration). Then we reduce the annotation tag set to keep only useful ones and we convert words from Buckwalter transliteration to Arabic. Finally, we assign probabilities to each rule to obtain the Arabic PCFG grammar. This interface also shows some statistics about the obtained grammar such as the lexical rules count and contextual rules count. Figure 3 presents the PCFG grammar construction interface.

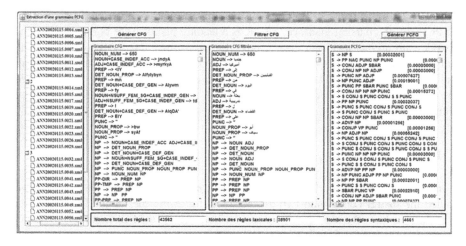

Fig. 3. Arabic PCFG grammar construction interface

4.3 Parsing Interface (Module 2)

Our tool provides to the user the possibility to test the induced grammar. Indeed, the user can analyze syntactically a sentence in modern standard Arabic. The user has to provide a segmented sentence and then to click on the parsing button "Generer l'analyse". Parsing results are then shown in the adequate area in parentheses format and with its probability. The parsing interface is presented in Fig. 4.

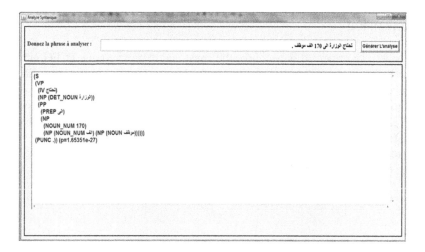

Fig. 4. Parsing interface

Natural Language Toolkit. We used Viterbi implementation of the NLTK tool [9] to test PCFG in the parsing interface. NLTK is a leading platform for building programs to work with human language data. It provides easy-to-use interfaces to over 50 corpora and lexical resources such as WordNet, along with a suite of text processing libraries for classification, tokenization, stemming, tagging, parsing, and semantic reasoning, wrappers for industrial-strength NLP libraries, and an active discussion forum.

5   Evaluation

In order to evaluate the performance of the induced grammar in the parsing task, we divided the PATB corpus into two distinct parts following the "Johns Hopkins 2005 Workshop" splits[1]. We used the major part for induction of the PCFG grammar and the second part for parsing tests.

In order to estimate the performance of our parser, we calculated the Precision, Recall and F-measure for each tested sentence. Then we calculate the macro average of the obtained values. The test set is composed of 1650 sentences extracted from the PATB. The average length of the sentences is about 21 words (the length ranges between 3 and 40 words). Obtained results are exposed in the Table 4.

Table 4. Parsing results

Precision	Recall	F-measure
83.59%	82.98%	83.23%

6   Conclusion

In this paper, we presented our tool for the construction of an Arabic PCFG grammar from the PATB corpus. The implemented method consists of two steps: in the first one we induce CFG rules from PATB parse trees using induction rules. Second we calculated the probability of each rule using the frequency of its occurrence in the corpus. The developed tool allows the user to construct an Arabic PCFG grammar and to test the constructed grammar in parsing task using the Viterbi algorithm.

We tested our system on sentences extracted from the PATB (over 1650 sentences) and we achieved encouraging and satisfactory results (precision: 83.59%, recall: 82.99% and f-measure: 83.24%).

---

[1] http://nlp.stanford.edu/software/parser-arabic-data-splits.shtml.

References

1. Khoufi, N., Boudokhane, M.: Statistical-based system for morphological annotation of Arabic texts. In: Proceedings of the Recent Advances in Natural Language Processing (RANLP 2013), Hissar, Bulgaria, pp. 100–106 (2013)
2. McCord, M.C., Cavalli-Sforza, V.: An Arabic slot grammar parser. In: Proceedings of the 2007 Workshop on Computational Approaches to Semitic Languages: Common Issues and Resources, pp. 81–88. Association for Computational Linguistics (2007)
3. Buckwalter, T.: Buckwalter Arabic morphological analyzer version 2.0 (2004)
4. Bataineh, B.M., Bataineh, E.A.: An efficient recursive transition network parser for Arabic language. In: Proceedings of the World Congress on Engineering, vol. 2, pp. 1–3 (2009)
5. Klein, D., Manning, C.D.: Fast exact inference with a factored model for natural language parsing. In: Advances in Neural Information Processing Systems 15 (NIPS 2002), pp. 3–10. MIT Press, Cambridge (2003)
6. Green, S., Manning, C.D.: Better Arabic parsing: baselines, evaluations, and analysis. In: Proceedings of the 23rd International Conference on Computational Linguistics, pp. 394–402. Association for Computational Linguistics, August 2010
7. Al-Taani, A., Msallam, M., Wedian, S.: A top-down chart parser for analyzing Arabic sentences. Int. Arab J. Inf. Technol. **9**, 109–116 (2012)
8. Alqrainy, S., Muaidi, H., Alkoffash, M.S.: Context-free grammar analysis for Arabic sentences. Int. J. Comput. Appl. **53**(3), 7–11 (2012)
9. Bird, S., Klein, E., Loper, E.: Natural Language Processing with Python. O'Reilly Media Inc., Sebastopol (2009)
10. Khoufi, N., Aloulou, C., Hadrich Belguith, L.: Parsing Arabic using induced probabilistic context free grammar. Int. J. Speech Technol. **19**, 1–11 (2015). https://doi.org/10.1007/s10772-015-9300-x
11. Habash, N.Y.: Introduction to Arabic Natural Language Processing: Synthesis Lectures on Human Language Technologies. Morgan & Claypool Publishers, San Rafael (2010). G. Hirst (Series ed.) 3(1)
12. Hajic, J., Vidová-Hladká, B., Pajas, P.: The Prague dependency treebank: annotation structure and support. In: Proceedings of the IRCS Workshop on Linguistic Databases, pp. 105–114 (2001)
13. Habash, N.Y., Roth, R.M.: CATiB: The Columbia Arabic Treebank. In: Proceedings of the ACL-IJCNLP 2009 Conference Short Papers, pp. 221–224. Association for Computational Linguistics, Stroudsburg, August 2009
14. Maamouri, M., Bies, A., Buckwalter, T., Mekki, W.: The Penn Arabic Treebank: building a large-scale annotated Arabic corpus. In: The NEMLAR Conference on Arabic Language Resources and Tools, pp. 102–109, September 2004
15. Maamouri, M., Bies, A., Kulick, S.: Enhancing the Arabic Treebank: a collaborative effort toward new annotation guidelines. In: Proceedings of the Sixth International Conference on Language Resources and Evaluation (LREC 2008), Marrakech, Morocco, 28–30 May 2008

Enhancing Neural Network Based Dependency Parsing Using Morphological Information for Hindi

Agnivo Saha[(⊠)] and Sudeshna Sarkar[(⊠)]

Department of Computer Science and Engineering, Indian Institute of Technology, Kharagpur, Kharagpur 721302, West Bengal, India
agnivo.saha@gmail.com, sudeshna@cse.iitkgp.ernet.in

Abstract. In this paper, we propose a way of incorporating morphological resources for enhancing the performance of neural network based dependency parsing. We conduct our experiments in Hindi, which is a morphologically rich language. We report our results on two well known Hindi Dependency Parsing datasets. We show an improvement of both Unlabeled Attachment Score (UAS) and Labeled Attachment Score (LAS) compared to previous state-of-the art hindi dependency parsers using only word embeddings, POS tag embeddings and arc-label embeddings as features. Using morphological features, such as number, gender, person and case of words, we achieve an additional improvement of both LAS and UAS. We find that many of the erroneous sentences contain Named Entities. We propose a treatment for Named Entities which further improves both UAS and LAS of our Hindi dependency parser (The parser is available at http://www.cicling.org/2016/data/126/CICLing_126.zip).

1 Introduction

There has been a lot of work in feature based, data driven Dependency Parsing in the last decade (Kubler et al. [1]). Statistical machine learning based methods have been used extensively for dependency parsing. Transition based and graph based methods are two types of statistical methods used for this task. Transition based methods like MALT Parser (Nivre [2]) are data driven and perform quite well on test dataset. Graph based parsers like MST Parser (Mcdonald et al. [3]) construct the maximum spanning tree to find the dependency parse tree. One drawback of such feature based parsers is that they require manually engineered features, which often fail to capture all compositions of features which are important. Often, the number of feature instantiations may become too large.

The advent of neural dependency parsing techniques (Chen and Manning [4]) have improved the accuracy of parsing considerably and address some of the above issues. This is mainly because of the ability of neurons to learn complex compositions of features in the hidden layers. The reduction in the number of features, along with dense representations of word forms have reduced the

© Springer International Publishing AG, part of Springer Nature 2018
A. Gelbukh (Ed.): CICLing 2016, LNCS 9623, pp. 366–377, 2018.
https://doi.org/10.1007/978-3-319-75477-2_26

speed of parsing. The use of features like part-of-speech tags, arc-labels and word-embeddings in a neural network can replace the need of many additional lexicalized resources which are highly sparse. Features like valency, distance and certain cluster based features also improve the accuracy of the parsing (Guo et al. [5]).

Use of word embeddings has found success in various NLP tasks, including constituency parsing (Socher et al. [6]), part-of-speech tagging (Collobert et al. [7]) and dependency parsing (Chen and Manning [4]). Dense representations of word forms allow words which are similar in meaning to have a similar vector representation. There have been quite a lot of work regarding computation of such dense representations of word forms, namely Collobert [8], word2vec [9], GloVe [10].

Hindi is a morphologically rich language. The release of Hindi Dependency Treebank (Bhat et al. [11]) covering 16,649 sentences and 18,847 words has helped in increasing resources which can be used for improving parsers in Hindi. The annotations in this treebank provide information about the part-of-speech tags of the word forms, the syntactic relations between words and the morphological features of the word. Nivre [12] has shown that use of morphological features help in increasing the accuracy of MALT parser to a notable extent. The availability of Hindi Wordnet provides various ontologies which has been used for improving parsing accuracy (Jain et al. [13]). Works done by Nivre [12] and Jain et al. [13] involve manual feature engineering and require a lot of understanding of linguistics. Other works in Hindi Dependency Parsing include use of Combinatory Categorial Grammar (CCG) categories along with MALT Parser (Ambati et al. [14]).

We train a neural network based parser for Hindi by making use of morphological features which are made available in the annotated Universal Hindi Dependency Treebank. The morphological features include the person, number, gender and the case of the word form. We also make use of the Hindi sentences in Wikipedia made available by Faruqui et al. [15] in addition to the Hindi Dependency Treebank and the ICON 2009 tool contest dataset (Palmer et al. [16]) for training word vectors. We find that about 18.7% of the sentences which have at least one error while parsing consist of Named Entities and 6% of the sentences which have at least one error while parsing consist of rare words. Our treatment for Named Entities and rare words improves the accuracy of the Neural Network based Dependency Parser further.

2   Literature Survey

2.1   Dependency Parsing

Dependency Parsing refers to the task of prediction of the Dependency Parse Tree given an input sentence, $s = w_0 w_1 ... w_n$. A dependency tree (Fig. 1) can be denoted by $d = \{(w_h, w_m, l), 0 \leq h \leq n; 0 \leq m \leq n; l \in A_c\}$, where w_h is the head word, w_m is the modifier or the dependent word, A_c is the set of arc labels and l is the dependency relation between the directed edge from w_h to w_m.

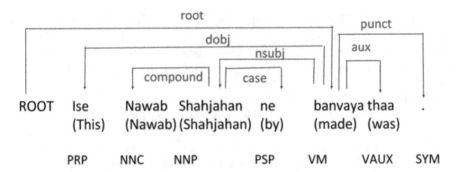

Fig. 1. Labeled Dependency Tree for the sentence "Ise Nawab Shahjahan ne banvaya thaa.". Gloss - "This was made by Nawab Shahjahan".

Two types of statistical dependency parsing have come into existence. Graph based dependency parsing tries to generate the maximum spanning tree in the complete graph between all words in sentences, where the edge weights are computed using features extracted from s. Transition based Dependency Parsing tries to predict the most probable transition sequence taken by the parser to reach some terminal configuration from a starting configuration. Greedy Transition based Dependency Parsers predict the transition to be taken given some current configuration greedily. Zhang et al. [17] showed that Transition based parsers using beam search and global training tend to be more accurate than greedy transition-based Dependency parsers. But Chen and Manning [4] showed that a well-defined neural network based architecture can improve the accuracy obtained by greedy transition-based parsing.

2.2   Transition Based Dependency Parsing

In the arc-standard system (Nivre 2004), a configuration $c = (s, b, A)$ consists of a stack s, a buffer b, and a set of dependency arcs A. The initial configuration has only ROOT in the stack and the buffer contains all the words in the sentence $s = w_0 w_1 ... w_n$ and A is empty. So, the initial configuration is given by $([ROOT], [w_0 w_1 ... w_n], \Phi)$. The terminal configuration is given by $([ROOT], [], A_c)$ where A_c is dependency parse tree of the sentence s. Given, a configuration of the parser, the best possible transition which can be taken by the parser is predicted. Denoting i^{th} top element of stack as s_i and the jth element in the buffer as b_j, the arc-standard system defines the following transitions:-

- Left-arc(l): adds an arc $s_1 \rightarrow s_2$ with label l and removes s_2 from the stack. $|s| \geq 2$.
- Right-arc(l): adds an arc $s_2 \rightarrow s_1$ with label l and removes s_1 from the stack. $|s| \geq 2$.
- Shift(): shifts b_1 from the buffer to the top of the stack. $|b| \geq 1$.

Typical approaches for greedy dependency parsing involves implementing a multi-class classifier (like SVM, Max-Ent, MALT, etc.) for predicting the transition to be taken given the feature vector extracted for the given configuration.

2.3   Neural Network Based Parsers

Constituency Parsing and Dependency Parsing are two types of parsing. Socher et al. [6] built a constituency parser for English using a recursive neural network. They used word embeddings as features to generate the constituency parse tree. They improved their model by including probabilities of a rule $A \rightarrow BC$ extracted by training a PCFG and using it as a part of the scoring function which also includes contribution of the score computed by the recursive neural network, where A, B and C are non-terminal symbols. Chen and Manning [4] built a neural network based model for predicting the transition to be taken given a configuration for the task of dependency parsing. They used Word features, POS features and Label features as input to the neural network.

2.4   Hindi Parsing

Nivre [12] implemented a MALT Parser for dependency parsing in Indian Languages. The Labeled Attachment Score (LAS) improved by almost 10% when he considered morphological features corresponding to the words in the top of the stack and the next top word in the stack. He reported his accuracy on the ICON 2009 dataset (ICON) (Palmer et al. [16]). He achieved a testing Unlabeled Attachment Score (UAS) of 89.8 and LAS of 73.4 considering fine-grained dependency labels. Jain et al. [13] used Hindi Wordnet (Narayan et al. [18]) to extract additional ontology based features for training the MALT parser. They reported their results on the Hindi Dependency Treebank (HDT). They achieved a testing UAS of 92.45 and LAS of 83.88 on the HDT dataset.

3   Objective

We wish to implement a dependency parser which achieves a high accuracy without using deep linguistic resources which are not easily available for different languages. For this purpose, we use the neural network based parsing method as presented by Chen and Manning [4]. We are interested in developing parsers for Indian languages. Since a reasonably good dependency treebank is available for Hindi, we decided to work on developing a Hindi dependency parser. As Hindi Dependency Treebank data is not large, we may make use of additional morphological features that are available to improve the parser.

4   Parsing Model

We consider transition based dependency parsing using the model by Chen and Manning [4]. According to their formulation, given a training sentence s and its dependency parse tree d, the configurations c are generated using d and the transition taken t is learnt given c. In case of unlabeled task, t can be one of the three: Left-arc, Right-arc or Shift. In case of labeled task, t can be Left-arc(l), Right-arc(l) or Shift, where l is the dependency relation or arc label. Thus, in case of labeled task, $2|N_l| + 1$ transitions are possible. The aim is, given a test sentence s_{test}, to predict the dependency tree d_{test}, by predicting transition taken t for intermediate configuration c.

4.1   Feature Templates

In Chen and Manning's [4] work they considered Word Features, POS Features and Label Features. They represented the words, POS tags and the arc-labels as d_1, d_2 and d_3 dimensional vectors, $e_i^w \in \mathbb{R}^{d_1}, e_j^t \in \mathbb{R}^{d_2}, e_k^l \in \mathbb{R}^{d_3}$. The embedding matrices are word embedding matrix, $E^w \in \mathbb{R}^{d_1 X N_w}$, POS embedding matrix, $E^t \in \mathbb{R}^{d_2 X N_t}$ and Arc-Label embedding matrix, $E^l \in \mathbb{R}^{d_3 X N_l}$, where N_w is the size of the vocabulary, N_t is the total number of POS tags, N_l is the total number of arc-labels. $E_p^w, E_p^t, E_p^l, E_p^m$ indicate the word, POS, label and morphological embeddings of the element at position p respectively.

- **Word Features**
    - $E_{s_i}^w, E_{b_i}^w, i = 0, 1, 2$
      s_i and b_i refer to the ith elements respectively in the stack and buffer. $E_{b_i}^w$ refers to the word embedding corresponding to the ith element in the buffer. $E_{s_i}^w$ refers to the word embedding corresponding to the ith element of the stack.
    - $E_{lc1(s_i)}^w, E_{rc1(s_i)}^w, E_{lc2(s_i)}^w, E_{rc2(s_i)}^w, i = 0, 1$
      $lc1$ refers to the leftmost child and $rc1$ refers to the rightmost child. $lc2$ refers to the second child from the left and $rc2$ refers to the second child from the right.
    - $E_{lc1(lc1(s_i))}^w, E_{rc1(rc1(s_i))}^w, i = 0, 1$
      $lc1(lc1(s_i))$ refers to the leftmost child of the leftmost child of the ith element of the stack. $rc1(rc1(s_i))$ refers to the rightmost child of the rightmost child of the ith element of the stack.
- **POS Features**
    - $E_{s_i}^t, E_{b_i}^t, i = 0, 1, 2$
    - $E_{lc1(s_i)}^t, E_{rc1(s_i)}^t, E_{lc2(s_i)}^t, E_{rc2(s_i)}^t, i = 0, 1$
    - $E_{lc1(lc1(s_i))}^t, E_{rc1(rc1(s_i))}^t, i = 0, 1$
- **Label Features**
    - $E_{lc1(s_i)}^l, E_{rc1(s_i)}^l, E_{lc2(s_i)}^l, E_{rc2(s_i)}^l, i = 0, 1$
    - $E_{lc1(lc1(s_i))}^l, E_{rc1(rc1(s_i))}^l, i = 0, 1$

In addition to that, we consider dense representations of morphological features as additional features to the neural network. We have used Chen and Manning's dependency parser as the basis of our model.

- **Morph Features**
  - $E^m_{s_i}, E^m_{b_i} i = 0, 1, 2$
  - $E^m_{lc1(s_i)}, E^m_{rc1(s_i)}, E^m_{lc2(s_i)}, E^m_{rc2(s_i)}, i = 0, 1$
  - $E^m_{lc1(lc1(s_i))}, E^m_{rc1(rc1(s_i))}, i = 0, 1$

4.2   Neural Network Architecture

Chen and Manning [4] mapped the input layer $x = x^w, x^t, x^l$ to a hidden layer with d_h nodes through a cube activation function:

$$h = (W^w_1 x^w + W^t_1 x^t + W^l_1 x^l + b_1)^3 \tag{1}$$

where $W^w_1 \in \mathbb{R}^{d_h X(d_1.n_w)}$, $W^t_1 \in \mathbb{R}^{d_h X(d_2.n_t)}$, $W^l_1 \in \mathbb{R}^{d_h X(d_3.n_l)}$, $n_w = 18, n_t = 18, n_l = 12$.

They proposed the cubic activation function, $g(x) = x^3$ which can easily capture interactions of three input elements, which works good for dependency parsing as shown by Chen and Manning [4].

$$g(w_1 x_1 + ... + w_m x_m + b) = \sum_{i,j,k}(w_i w_j w_k)x_i x_j x_k + \sum_{i,j} b(w_i w_j)x_i x_j + ...$$

A softmax output layer is finally added on the top of the hidden layer for modeling multi-class probabilities $p = softmax(W_2 h)$ of the transitions to be taken given the input configuration, where $W_2 \in \mathbb{R}^{|\tau| X d_h}$.

4.3   Adding Morph Features

We represent the morph features, namely number, person, gender and case as d_4 dimensional vectors, $e^n_i, e^p_j, e^g_k, e^c_r \in \mathbb{R}^{d_4}$. Given a configuration, we extract the Word, POS and label features as extracted by Chen and Manning [4]. Additionally, for each word extracted as feature, we also extract their Morph features which are available in the Treebank. We extract the number, gender, person and the case of the word-form. That accounts for 18 features each for number, gender, person and case. So, we extract 120 features in total. We use the feature NULL for examples which do not have the morphological feature annotated. The new embedding matrix is the morph embedding matrix, $E_m \in \mathbb{R}^{d_4 X(N_n+N_p+N_g+N_c)}$, where N_n is the number of different numbers of the word forms, N_p is the number of different persons of the word forms, N_g is the number of different genders of the word forms and N_c is the number of different cases of the word forms.

We map the input layer $x = x^w, x^t, x^l, x^n, x^g, x^p, x^c$ to a hidden layer h with d_h nodes as:

$$h = (W^w_1 x^w + W^t_1 x^t + W^l_1 x^l + W'^n_1 x^n + W'^g_1 x^g + W'^p_1 x^p + W'^c_1 x^c + b_1)^3 \tag{2}$$

where $W'^n_1 \in \mathbb{R}^{d_h X(d_4.n_n)}$, $W'^g_1 \in \mathbb{R}^{d_h X(d_4.n_g)}$, $W'^p_1 \in \mathbb{R}^{d_h X(d_4.n_p)}$, $W'^c_1 \in \mathbb{R}^{d_h X(d_4.n_c)}$, $n_n = n_g = n_p = n_c = 18$, $d_4 = $ dimension of morph feature embedding used.

The neural network architecture is given in Fig. 2. The input layer corresponds to the word, POS, arc-label and morph features extracted given the current configuration as shown.

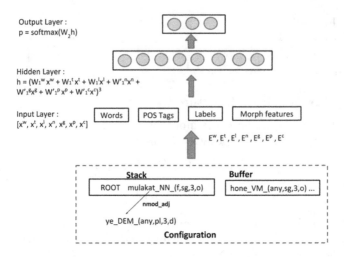

Fig. 2. Neural network model used for dependency parsing.

4.4 Named Entities Treatment

We extract Named Entities such as Person (eg. Riyaz Khan), Location (eg. Vishwapattanam), Organization (eg. Griha Mantralaya), Abbreviation (eg. I.C), Designation (eg. Pradhan Mantri), and Number (eg. 4.5). We observe that these words are present in many of the sentences in HDT which have atleast one error in our prediction. We replace the words belonging to the $NamedEntity_i$ with the tag $NAMED\_ENTITY\_i$ (for example for Person, we replace with the tag $NAMED\_ENTITY\_PERSON$) in the whole corpus. We run the word2vec on the newly generated corpus and for each word present in the $NamedEntity_i$ set, we consider the embedding corresponding to the tag $NAMED\_ENTITY\_i$ as their embedding before training our neural network based dependency parser. We believe that the named entities should have the same embedding, the reason for such a treatment.

4.5 Rare Words Treatment

We consider as rare words the words that occur less than 5 times in the corpus. For each word, we extract all possible (POS tag, arc-label, number, gender, person, case) features corresponding to the word from the annotated corpus (Hindi Dependency Treebank and ICON2009 data). We consider the non-rare words, words with greater than 5 frequency in the corpus with the same set of features as this word. Let the set of non-rare words corresponding to this rare word w_r having same set of (POS, arc-label, number, gender, person, case) features in the corpus be $NR(w_r)$, then the embedding, $v(w_r)$ corresponding to w_r is calculated as,

$$v(w_r) = \frac{\sum_{w_i \in NR(w_r)} v(w_i)}{len(NR(w_r))} \qquad (3)$$

We take the average embedding over all possible non-rare words with same set of features as in the corpus (Table 1).

Table 1. The size of different corpuses used while training word vectors.

Corpus	#Sentences
ICON	3,835
HDT	16,649
Wiki	3,57,669

Table 2. Dataset statistics of Treebank Data.

Dataset	#Train	#Dev	#Test	#Words	#POS	#Labels	Projective (%)
ICON	2,972	543	320	7,074	35	94	81.31
HDT	13,305	1,660	1,684	16,719	34	27	86.28

5 Experiments and Results

5.1 Dataset Statistics

We built a corpus for learning the word vectors to initialize the word embedding matrix for the parser. The corpus consists of 3,78,153 Hindi sentences. For building the corpus, we used the Wikipedia Hindi dump extracted in Faruqui et al. [15]. This accounted for 3,57,669 sentences. In addition, we used the ICON 2009 sentences (Palmer et al. [16]) which accounted for 3,835 sentences and the Hindi sentences present in the Hindi Dependency Treebank [11] which accounted for 16,649 sentences.

We report our results on the Hindi Dependency Treebank (HDT) and ICON 2009 Treebank (ICON). The dataset statistics of HDT and ICON are given in Table 2. The number of morph features in HDT are: #Numbers - 2, #Person - 3, #Gender - 2, #Cases - 7. The percentage of trees which were projective in the training set of HDT was 86.28% and the same for ICON was 81.31%.

5.2 Training Settings

The size of the hidden layer was taken to be 200. We chose Word, POS tag and arc-label features to have 50 dimension embeddings. We varied the dimension of morph features from 10 to 50 in steps of 10. As the total number of labels while training on HDT are 27, so the number of possible transitions are 55. So, the size of the output layer is 55 while training on HDT. While training on ICON data, the size of output layer is 189 as the number of labels present are 94. The POS, Arc-Label and morph embeddings where randomly initialized within

(−0.01, 0.01). For initializing the word embeddings, we used the Continuous Bag of Words Model (CBOW) [9] over the corpus which was implemented as an opensource tool word2vec[1]. The nodes in the hidden layer where dropped with a dropout rate [19] of 0.5. We perform about 20,000 iterations by dividing the training examples into mini batches of size 1,000. We used the precomputation trick as used by Chen and Manning by computing the hidden layer activations for features which occur most number of times.

5.3 Comparison with Existing Parsers

We consider Hindi Wordnet based parser (HWN) [13] as the baseline parser for comparing our results on the HDT data and we consider (Nivre [12]'s) MALT parser along with morph features as our baseline parser for comparing results on ICON data. We also compute the Unlabeled Attachment Score, Labeled Attachment Score and the Label Accuracy for the NNDEP parser [4] (without morph features) and compare with our parser (including morph features). The results of comparison are given in Tables 3 and 4.

- MALT + Morph refers to Nivre [12]'s parser.
- NNDEP refers to Chen and Manning [4]'s parser.
- HWN Parser refers to Jain et al. [13]'s parser.

The LAS values have increased considerably. Using ICON, we achieve a LAS of 79.72 compared to 73.40 achieved by [12] and 78.52 achieved by NNDEP [4] considering fine-grained labels. The improvement in LAS is mainly due to the dense representation of features which help in capturing semantic similarities and information present in the data and provide a better generalization of unseen word-word relationships. It is clear that the additional morph features help in improving the LAS and UAS further. Using HDT, we achieve a LAS of 91.31 compared to 83.88 achieved by HWN parser [13] and 90.96 achieved by NNDEP. The UAS also improved to 94.13 compared to 92.45 achieved by HWN parser and 93.80 achieved by NNDEP.

5.4 Embedding Size of Morph Features

We vary the embedding size of the morph features from 10 to 50 and we report the results obtained. We define NNDEP + MORPH10 to be our parser with size of morph embeddings as 10. Similarly, we define NNDEP + MORPH20 to NNDEP + MORPH50. The comparison of Unlabeled Attachment Score, Labeled Attachment Score and the Label Accuracy on HDT is given in Table 4 and on ICON is given in Table 3. We find that morph embedding size of 30 gives the best results on HDT data, UAS of 94.13, LAS of 91.31 and LA of 94.53. Size 20 gives the best UAS of 88.39 and size 40 gives the best LAS and LA of 79.72 and 82.91 respectively on the ICON data. ICON data has 94 labels compared to HDT data which has only 27 labels. This accounts for the less LAS achieved by the model on ICON data.

---

[1] http://code.google.com/p/word2vec/.

Table 3. Comparison of parsing accuracy with various parsers on ICON 2009 Data. MALT + Morph refers to Nivre [12]'s parser. NNDEP refers to Chen and Manning [4]'s parser.

Parser	UAS (%)	LAS (%)	LA (%)
MALT + Morph	89.8	73.4	75.3
NNDEP	88.21	78.52	81.82
NNDEP + MORPH10	88.33	79.36	82.76
NNDEP + MORPH20	**88.39**	79.45	82.79
NNDEP + MORPH30	88.13	78.75	82.26
NNDEP + MORPH40	88.22	**79.72**	**82.91**
NNDEP + MORPH50	87.81	78.66	81.80

Table 4. Comparison of parsing accuracy with various parsers on Hindi Dependency Treebank Data. HWN Parser refers to Jain et al. [13]'s parser.

Parser	UAS (%)	LAS (%)	LA (%)
HWN Parser	92.45	83.88	86.87
NNDEP	93.80	90.96	94.15
NNDEP + MORPH10	93.89	91.07	94.38
NNDEP + MORPH20	94.01	91.24	94.47
NNDEP + MORPH30	**94.13**	**91.31**	**94.53**
NNDEP + MORPH40	94.08	91.25	94.50
NNDEP + MORPH50	94.06	91.29	94.52
NNDEP + MORPH40 + NE	**94.33**	**91.52**	**94.66**
NNDEP + MORPH40 + Rare	94.20	91.41	94.65
NNDEP + MORPH40 + NE + Rare	94.29	91.41	94.57

5.5 Results

We find that about 18.7% of the erronenous sentences of HDT and about 25.9% of the erroneous sentences of ICON contain words which are Named Entities. We extract the Named Entities using a Named Entity Recognition system developed at Indian Institute of Technology, Kharagpur. Using the Named Entities Treatment as described in earlier section, we find that the UAS improves by 0.25 and LAS improve by about 0.27. The **UAS becomes 94.33** and the **LAS becomes 91.52**, when we use the newly generated word vectors for our model with morph feature dimension as 40. Also, we find that about 6% of the sentences of both HDT and ICON whose parse tree has been predicted in error by our parser contains rare words, words which occur less than 5 times in the large corpus used for training word vectors. Using the Rare Words Treatment on the HDT dataset, we find that the UAS improves by about 0.12 and LAS improves by about 0.16. The **UAS becomes 94.20** and the **LAS becomes 91.41** using word vectors generated after rare words treatment for our model with morph feature dimension as 40.

5.6   Parsing Speed

The Neural Network architecture proposed by Chen and Manning had 48 features accounting for the input's layer size to be 2,400 nodes. The addition of the morph features increase the number of features extracted to 120. So, the speed of parsing a sentence per second reduces as expected. NNDEP's testing speed was found to be 552 sentences/second compared to 425 sentences/second for NNDEP + MORPH10.

6   Conclusions

We find that using morphological features along with word, POS and arc-label features actually improves the accuracy of the dependency parser. In addition, our Named Entity Treatment and treatment of rare words further improves the accuracy of the parser.

Acknowledgement. Professor Sudeshna Sarkar acknowledges DEITY, Government of India for support under the ILMT Project.

References

1. Kübler, S., McDonald, R., Nivre, J.: Dependency Parsing. Synthesis Lectures on Human Language Technologies, vol. 1, pp. 1–127 (2009)
2. Nivre, J.: An efficient algorithm for projective dependency parsing. In: Proceedings of the 8th International Workshop on Parsing Technologies (IWPT). Citeseer (2003)
3. McDonald, R., Pereira, F., Ribarov, K., Hajič, J.: Non-projective dependency parsing using spanning tree algorithms. In: Proceedings of the conference on Human Language Technology and Empirical Methods in Natural Language Processing, pp. 523–530. Association for Computational Linguistics (2005)
4. Chen, D., Manning, C.: A fast and accurate dependency parser using neural networks. In: Proceedings of the 2014 Conference on Empirical Methods in Natural Language Processing (EMNLP), Doha, Qatar, pp. 740–750. Association for Computational Linguistics (2014)
5. Guo, J., Che, W., Yarowsky, D., Wang, H., Liu, T.: Cross-lingual dependency parsing based on distributed representations (2015)
6. Socher, R., Bauer, J., Manning, C.D., Ng, A.Y.: Parsing with compositional vector grammars. In: Proceedings of the ACL Conference. Citeseer (2013)
7. Collobert, R., Weston, J., Bottou, L., Karlen, M., Kavukcuoglu, K., Kuksa, P.: Natural language processing (almost) from scratch. J. Mach. Learn. Res. **12**, 2493–2537 (2011)
8. Bordes, A., Weston, J., Collobert, R., Bengio, Y.: Learning structured embeddings of knowledge bases. In: AAAI (2011)

9. Mikolov, T., Sutskever, I., Chen, K., Corrado, G.S., Dean, J.: Distributed representations of words and phrases and their compositionality. In: Advances in Neural Information Processing Systems, pp. 3111–3119 (2013)

10. Pennington, J., Socher, R., Manning, C.: Glove: global vectors for word representation. In: Proceedings of the 2014 Conference on Empirical Methods in Natural Language Processing (EMNLP), pp. 1532–1543 (2014)

11. Bhat, R.A., et al.: The Hindi/Urdu treebank project. In: Ide, N., Pustejovsky, J. (eds.) Handbook of Linguistic Annotation, pp. 659–697. Springer, Amsterdam (2017). https://doi.org/10.1007/978-94-024-0881-2_24

12. Nivre, J.: Parsing Indian languages with MaltParser. In: Proceedings of the NLP Tools Contest: Indian Language Dependency Parsing, ICON 2009, pp. 12–18 (2009)

13. Jain, S., Jain, N., Tammewar, A., Bhat, R.A., Sharma, D.M.: Exploring semantic information in Hindi wordnet for Hindi dependency parsing (2013)

14. Ambati, B.R., Deoskar, T., Steedman, M.: Using CCG categories to improve Hindi dependency parsing. In: ACL, vol. 2, pp. 604–609 (2013)

15. Faruqui, M., Kumar, S.: Multilingual open relation extraction using cross-lingual projection. arXiv preprint arXiv:1503.06450 (2015)

16. Palmer, M., Bhatt, R., Narasimhan, B., Rambow, O., Sharma, D.M., Xia, F.: Hindi syntax: annotating dependency, lexical predicate-argument structure, and phrase structure. In: The 7th International Conference on Natural Language Processing, pp. 14–17 (2009)

17. Zhang, Y., Clark, S.: Syntactic processing using the generalized perceptron and beam search. Comput. Linguist. **37**, 105–151 (2011)

18. Narayan, D., Chakrabarti, D., Pande, P., Bhattacharyya, P.: An experience in building the indo wordnet-a wordnet for Hindi. In: First International Conference on Global WordNet, Mysore, India (2002)

19. Srivastava, N., Hinton, G., Krizhevsky, A., Sutskever, I., Salakhutdinov, R.: Dropout: a simple way to prevent neural networks from overfitting. J. Mach. Learn. Res. **15**, 1929–1958 (2014)

Construction Grammar Based Annotation Framework for Parsing Tamil

Vigneshwaran Muralidaran[(✉)] and Dipti Misra Sharma

International Institute of Information Technology,
Gachibowli, Hyderabad 500032, Telengana, India
vigneshwaran.m@research.iiit.ac.in, dipti@iiit.ac.in

Abstract. Syntactic parsing in NLP is the task of working out the grammatical structure of sentences. Some of the purely formal approaches to parsing such as phrase structure grammar, dependency grammar have been successfully employed for a variety of languages. While phrase structure based constituent analysis is possible for fixed order languages such as English, dependency analysis between the grammatical units have been suitable for many free word order languages. These approaches rely on identifying the linguistic units based on their formal syntactic properties and establishing the relationships between such units in the form of a tree. Instead, we characterize every morphosyntactic unit as a mapping between form and function on the lines of *Construction Grammar* and parsing as identification of dependency relations between such conceptual units. Our approach to parser annotation shows an average MALT LAS score of 82.21% on Tamil gold annotated corpus of 935 sentences in a five-fold validation experiment.

1 Introduction

A common theoretical assumption in dependency parser implementations is that grammar is a formal system of rules which arranges the morphosyntactic units that make up a sentence. In dependency grammar analysis, both the identified grammatical units and the dependency relations between them are structural or formal in nature. We observed that Dravidian languages, spoken in South India, have interesting morphological properties which can be better characterized from a functional approach to syntax such as *Construction Grammar* [1, 2] or *Cognitive Grammar* [3]. According to the theoretical assumptions of Construction Grammar, every grammatical formative in a language is analyzed as a meaningful mapping between its form and function. We identified that the set of morphological inflections that occur in Dravidian languages can be mapped to a set of meaningful 'construals' i.e. as a form function pairing. We will talk about what these 'construals' are in Sect. 3. Wherever morphemes can be grouped together as one meaningful construal let us call that as one construction unit. Our idea is to identify which morphemes should be grouped together as one construction unit, identify the appropriate construction label for the grouped

© Springer International Publishing AG, part of Springer Nature 2018
A. Gelbukh (Ed.): CICLing 2016, LNCS 9623, pp. 378–396, 2018.
https://doi.org/10.1007/978-3-319-75477-2_27

unit, chunk these construction units and finally identify the dependency relations between these chunks.

We describe the previous works on Parsing in Indian languages and other morphologically rich languages in Sect. 2, the basics of construction based analysis of a text in Sect. 3, the proposed annotation framework in Sect. 4, experiments and results in Sect. 5, error analysis in Sect. 6 and conclusion and future work in Sect. 7.

2   Previous Works

Parsing of natural language texts has been explored in various languages for more than two decades now. Here, we point out to various parsing approaches attempted in Indian languages and other morphologically rich languages so far. Indian languages are morphologically rich, free-word order languages and it is well understood that dependency framework suits better for the analysis of the various grammatical structures of such languages [4–6]. Bharati and Sangal proposed a grammar formalism called as 'Paninian Grammar Framework' [7] that has been successfully used to create treebanks in Indian languages and extensively used for all free word order Indian languages. A simple parser for Indian languages has been proposed by Bharati et al. in a grammar-driven methodology [8]. The proposed methodology was based on Paninian grammatical approach in dependency framework. Prashanth Mannem proposed a bidirectional dependency parser for Hindi, Telugu and Bangla language [9].

Nivre presented the work of optimizing Malt Parser for three Indian languages namely Hindi, Telugu and Bangla in NLP Tool Contest at ICON 2009 [10]. Ambati et al. explored MALT and MST parsers on three Indian languages Hindi, Telugu and Bangla [11]. A statistical syntactic parser for Kannada has been developed based on Penn Treebank structure [12]. Selvam et al. presented a statistical parser for Tamil language using phrase structure hybrid language model [13]. In 2011, Ramasamy and Žabokrtský discussed their results on rule-based and corpus based approaches in Tamil dependency parsing. They report an accuracy of more than 74% for the unlabeled task and more than 65% for labeled tasks on a small corpus size of 204 sentences with 2961 words [14]. Straka et al. report an LAS score of 69.7% and UAS score of 78.3% on the Universal Dependencies Treebank [15] for Tamil parsing using 600 sentences.

Kumari in 2012 presented a hybrid approach by combining the output of both the MALT and MST in an intuitive way and showed that this can perform better than both the parsers [16]. In 2011, a constraint based Hybrid dependency parser for Telugu was reported [17], with an LAS score of 68.06% with 1119 training data size. For other morphologically rich languages across the world, average LAS scores vary from 91.83% (German language with the largest training set) to around 83% (for Hebrew and Swedish with smallest training data) amongst the languages in SPMRL shared task 2013 [18]. Some of the state-of-the-art

Table 1. Label accuracy scores (LAS) reported for various Indian languages

Language	LAS	Training size
Hindi	90.99%	12041
Telugu	68.06%	1119
Bangla	79.81%	1000
Tamil	69.7%	600

parsing results reported for a few Indian languages in data-driven approach[1] to the best of our understanding are shown in Table 1.

3   Our Approach

We take inspiration from the theoretical perspectives about syntax in *Cognitive Grammar* [3]. On the outset, we propose that there are two dimensions for understanding any syntactic unit functionally as a construction schema. (i) Composition - what concepts make up a construction schema (ii) Interaction - what kind of dependencies that it has with other such construction schemas. The four basic kind of interactions are shown below in Fig. 1. To make a fully functional characterization of syntax, the following construction schemas are defined.

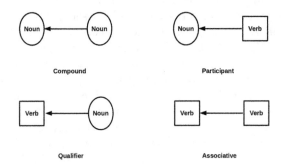

Fig. 1. Basic types of interactions in discourse world

1. Any morphosyntactic unit that is conceived as outlining a *thing* irrespective of its internal complexity is a 'noun' schema. Anything which is conceived as outlining a *relation* is a 'verb' schema.
2. A relation can be conceived as outlined through time or immediately perceived as a configuration. We refer to the former interpretation as a 'process' and the latter as a 'status'.

---

[1] Hindi, Telugu and Bangla accuracies reported are experimented with Computational Paninian Grammar Framework [7]. The Tamil accuracy reported is based on the Universal Treebank results reported by Straka et al. [15].

3. If a relation attributes its relational properties to a noun, it is called as 'qualifier' schema.
4. If a relation attributes its relational properties to another relation, it is called as 'associative' schema.
5. If a thing is conceived to be an integral part of another thing, it is said to be in 'combinative' schema.
6. If a thing is conceived to be an integral part of a relation, it said to be in 'participant' schema.

The above mentioned conceptualizations suggest that every syntactic unit is reducible to two fundamental concepts namely 'thing' and 'relation'. (Refer to [3] to understand these in detail). The four possible ways they are conceived to interact in the speaker's conceptual world are the basis for all kinds of complex discourse interactions. These four kinds of interactions are shown in Fig. 1. The direction of the arrow points to the dependent unit in the interaction. These conceptualizations dynamically made by the speaker are called 'construals' and the syntactic patterns that act as symbolic mappings to these construals are called construction schemas. It is these construction schemas that we intend to learn in our approach and it is between these construction schemas that the dependency relations are established. Let us understand these ideas with some examples.

1. I saw a **long** beach. *Status Qualifier schema*
2. The beach was **long**. *Status Pronominal schema*
3. The boy **who lives** near our home, is playing in the beach. *Process Qualifier schema*
4. I met **my** friend in the conference. *Status Qualifier schema*
5. **Three** people entered the classroom. *Status Qualifier schema*
6. **The fact that I spoke** to him on his birthday made my friend happy. *Process Pronominal schema*
7. The book is **mine**. *Status Pronominal schema*

In the above examples, the entities which invoke some construction schemas are highlighted in boldface. For example various syntactic units such as *long (adjective), who lives (relative clause), my (personal pronoun), three (numeral quantifier)* in examples 1, 3, 4 and 5 exhibit a common pattern of describing or attributing some relational property on a noun. Such a discourse 'construal' of attributing some property on a *thing* is the construction pattern that we call as *Qualifier Schema*. Whereas the same adjective *long* in example 2, the predicate form of genitive case pronoun *mine* in 7, construction such as *the fact that I spoke* in example 6 exhibit the discourse 'construal' of creating referring expressions. For instance one can say 'Contrary to my expectation, the beach **was long**. That surprised me'. As you can see, the predicate of *being long* is an idea that you can anaphorically refer to as 'that'. Such a discourse construal of 'referring expressions' are what we call as *Pronominal schema*.

We call it 'pronominal' because a *qualifier* attributed on a generic *thing* creates a referring expression. The whole expression now will stand 'in place of' a generic *thing* in a discourse world and thus 'pronominal'. If the referring expression is from a *process* relation it is *Process pronominal* or if it is from *status* relation then it is *Status pronominal*. Thus we see that, it is not the formal structural properties that we want to capture in our grammatical analysis but the functional concepts. Neither are they entirely semantic from an information point of view. Rather these construals are dynamic viewpoints that the speaker adopts while describing a given semantic information.

The observation we made was that these kinds of discourse construals are directly encoded through regular morphological inflections in Dravidian languages. For example in Tamil, whenever a *Qualifier Schema* is construed in discourse, morphologically it is expressed as a relative participial inflection (RP) of a verb. Tamil lacks lexical adjectives such as *broad, beautiful, long* and expresses these ideas as verbal participial inflections such as *agalam*-**An-a** *(breadth-become-RP)*, *azhag*-**An-a** *(beauty-become-RP)*, **nIND-a** *(elongate-RP)* with RP inflection in each case. But the same expressions, while functioning as predicates, take an additional pronominal suffix and become *agalam*-**An-a-du** *(breadth-become-RP-PronSuffix)*, *azhag*-**An-a-du** *(beauty-become-RP-PronSuffix)*, **nIND-a-du** *(elongate-RP-PronSuffix)* - thus creating referring expressions. These RP inflections and pronominal suffixes analogously occur in other parts of speech such as genitive markers, relative clauses, qualifiers, quantifiers and so on. We want to exploit these morphological regularities by mapping them to their construction schema patterns. The Table 2 lists out the various formal grammatical units and their mapping to the construction patterns that we identified. As the mappings mentioned in Table 2 are directly available as morphological regularities in Tamil, they can be easily identified for annotation. The following example demonstrates this idea.

Table 2. Mapping the formal grammatical units to construction schemas

Grammatical units	Construction schemas
Adjectives, genitive case markers, relative clauses, appositions, quantifiers as noun modifiers, subordinate clauses as noun modifiers	Qualifier Schema (Both process and status)
Predicate adjectives, genitives as predicates, rel.clause referring expressions, predicate quantifiers, subordinate clauses signaling entity-event relations	Pronoun Schema (Both process and status)
Case markers, prepositions, subordinate and coordinate conjunctions, complementizers, adverbs	Status Associative
Discourse event-event relations, grammatical aspects, modality, complex predicates	Process associative; Subtypes are: conjunctive, concursive, conditional, infinitive
Case assigned nouns	Participant
Nouns which are part of multiword expressions	Combinative

(1)   veLiy-uRavu   **kuRittu**   inRu   vivAdam   naDaipe-tR-adu.
      External-affairs **regarding**   today   debate   happen-PST-NonHum.SG

      'Regarding (our) external relations, a debate happened (in parliament) today'

(2)   veLiy-uRavu   **kuRitt-a**   vivAdam
      External-affairs **regarding-RP**   debate

      'The debate (which is) about external affiars'

(3)   vivAdam   veLiy-uRavu   **kuRitt-a-du**
      Debate   External-affairs **regarding-RP-PronSuffix**

      'The debate is about external affiars'

As you can see in the above examples, even function words such as postpositions are expressed through verb morphology in Tamil. **kuRittu (about)** inflects as a relative participial form (RP) **kuRitta** when it attributes its property on a noun but takes a pronominal inflection **kuRittadu** when it acts as a predicate. These and other such morphological regularities that consistently map to construal functions form the basis of our annotation labels.

4   Annotation Scheme

A gold annotation of dependency parsing in this Construction Grammar (CG) Framework involves two stages: (a) Processing the raw text into morphosyntactic patterns for which Construction labels have to be annotated (b) Grouping these construction units into chunks and then annotating the dependency relations between these chunks. The list of Construction schemas and the dependency relations that can exist between them are mentioned in Tables 3 and 4. A simple example is given below to illustrate the annotation scheme:

(4)   timuka   talaivar-um   mun-nAL   mudalamaiccar-um-**An-a**   karuNAnidhi
      DMK   leader-also   pre-day   CM-also-**become-RP**   Karunanidhi
      nEtRu   seidiyALargaL-ai   sandit-tu   pEs-in-Ar.
      yesterday   reporters-acc   met-conjunctive   speak-PST-honorific

      'Mr. Karunanidhi, the DMK leader and the ex-Chief Minister, met the reporters yesterday and spoke (with them)'

In the above example, there is a proper noun *Karunanidhi* which is described by the apposition phrase *the DMK leader and the ex-Chief Minister* in English. Functionally this invokes a Qualifier schema because the noun is being described by the apposition phrase. In fact, in the above Tamil sentence this qualifier schema is encoded morphologically by the status verb 'Agu (become)' with RP inflection *Ana* that is highlighted above. The verb is not a process verb that describes an event, but only a status verb that describes a configuration. Therefore, in our annotation scheme the above raw text is annotated as follows:

(5)   timuka   talaivarum   munnAL   mudalamaiccarum   Ana   karuNAnidhi
      NN   NN   NN_COMB   NN   ST_QUAL   NNP
      nEtRu   seitiyALargaLai   sandittu   pEsinAr.
      NST   NN   PR_CONJ   PR_FIN

Table 3. Tagset for construction schemas in Tamil

Tag Name	Construction Schema
CC	Coordinating Conjunction
ECHO	Echo Word
NN	Noun
NN_COMB	Nouns in Combinative Schema
NST	Spatio Temporal Noun
OPER	Operator
PRON	Pronoun
PSP	Postposition
PSP_PRON	Postposition in Qualifier schema
PSP_QUAL	Postposition in Qualifier schema
QT_CONC	Quotative concursive
QT_COND	Quotative conditional
QT_CONJ	Quotative conjunctive
QT_FIN	Quotative complete
QT_PRON	Quotative Pronoun

Tag Name	Construction Schema
QT_QUAL	Quotative Qualifier
RDP	Reduplication
ST_CONC	Status Concursive
ST_COND	Status Conditional
ST_CONJ	Status Conjunctive
ST_FIN	Status Complete
ST_PRON	Status Pronoun
ST_QUAL	Status Qualifier
SYM	Symbol
UNK	Unknown token
PR_CONC	Process Concursive
PR_COND	Process Conditional
PR_CONJ	Process Conjunctive
PR_FIN	Process Complete
PR_PRON	Process Pronoun
PR_QUAL	Process Qualifier

Table 4. Dependency relations between construction schemas

Tag Name	Construction Schema
ccof	Coordination conjunction
conc:man	Concursive- Manner
conc:seq	Concursive- Event concurrence
cond	Conditional
concess	Concession
conj:gram	Conjunctive- Grammatical
conj:man	Conjunctive- Manner
conj:seq	Conjunctive- Event continuance
inf:gram	Infinitive- Grammatical
inf:man	Infinitive- Manner
inf:seq	Infinitive- Purpose
k1	karta karaka
k1s	karta samanadhikarana
k2	karma karaka
k2s	karma samanadhikarana
k3	karana karaka

Tag Name	Construction Schema
k4	sampradana karaka
k4a	anubhava karta
k5	apadana karaka
k7	vishayadhikarana
k7a	adhikarana extended
k7p	deshadhikarana
k7t	kaladhikarana
nmod:stat	Status Qualifier
pof	Complex Predicate
r6	Genitive case
ras	Associative relation
rh	Reason relation
main	Attachment to ROOT node
rt	Purpose relation
sent_adv	Process Conditional
status	Status functions in discourse
vmod:stat	Status Associative

As a first stage, the given raw text is split into construction patterns like above (note that we have split the unit *Ana*, which is usually formally analysed as an adjectivializer, from the raw text since it invokes a construction pattern of Qualifier). Most of the tokens in Tamil are already directly mappable to construction schemas in the above example. The labels are shown below each token. In this way all the sentences are processed such that the tokens resulting after this preprocessing will be construction units that are ready to be labelled. The comprehensive list of construction labels and their meanings are shown in Table 3. After this labelling is done, the labeled construction units are chunked based on whether the consecutive construction units are *usage based syntactic freezes* or not. For example *multi-word expression (MWEs), numeric quantifiers modifying a noun, demonstrative adjectives modifying a noun* are the three instances where based on usage in Tamil, the consecutive construction labels can be safely grouped in a single chunk. As an example the sample labelled sentence shown in 5 is chunked in the following manner:

(6) (timuka   talaivarum)   (munnAL   mudalamaiccarum)   (Ana)
     (NN   NN)   (NN_COMB   NN)   (ST_QUAL)
     (karuNAnidhi) (nEtRu) (seitiyALargaLai) (sandittu)   (pEsinAr).
     (NNP)   (NST)   (NN)   (PR_CONJ)   (PR_FIN)

In the above sentence, two units *timuka* and *talaivarum* are grouped together as one chunk because it forms a noun-compound MWE which is a syntactic freeze as per its usage. i.e. You can chunk them together and only the head of the chunk is going to participate in the dependency relationship. After chunking, the final dependency analysis is shown in Fig. 2. It can be seen that the above dependency analysis uses the *karaka* relations such as k1, k2, k7 as discussed in Computation Paninian Grammar (CPG) that has been successfully applied for Indian languages [7]. This is because the *karaka relations* such as *karta, karma, karana* etc. are already syntactico-semantic concepts that describe the meaningful role that a noun is conceived to play in an action denoted by a verb. These relations are well understood from traditional Sanskrit grammar and are extremely suitable for analysing relatively free-word ordered languages

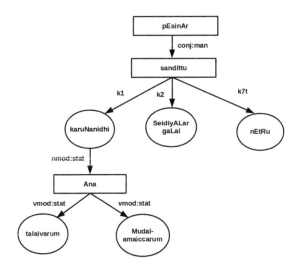

Fig. 2. Parsed output

like Indian languages. Roughly we can say that the *karaka* roles are analogous to theta roles with the difference that these are not purely semantic from the information point of view but 'syntactico-semantic'. In that sense *karaka* roles are different from theta/ semantic roles. Refer to [6, 7] to understand these more. In other words *karaka* theory provides mapping between form (vibhakti markers) and construed meaning (karaka relations) which exactly is what Construction schemas are about. Hence we are not going to reinvent the wheel but rather use these traditional notions in the general Construction Grammar formulation. As a part of the larger picture, these karaka roles exemplify the 'participant' interaction that is shown in Fig. 1. In what way is the proposed Construction Grammar annotation different from the Computational Paninian Grammar annotation? The answer can be summarized as following:

- We use all the *karaka* relations as used in CPG as these are 'construals' of how a noun can interact meaningfully with a verb.
- Additionally, we treat the function words that can occur in a language as *status* verbs i.e. postpositions, adjectives, adverbs, complementizers etc. Instead of analyzing them as formal parts of speech, we treat them as relations which describe atemporal configurations and to that extent they are all verbs in *status* interpretation. (Refer to Sect. 3 and [3] to understand this better). In fact, Tamil uses verbal morphology to derive these function words.
- We also handle various other meaningful discourse construals such as *Qualifier, Pronominal, Conjunctive, Concurrent, Condition and Infinitive* schemas which are necessary to understand the peculiarities in Dravidian syntax. Refer to [19–24] to understand some of the syntactic issues in Dravidian languages. How our discourse construals can explain these issues in a meaningful way is a theoretical enterprise that we are working on which is beyond the scope of the current work. Basically these discourse construals are included in our annotation framework as we hypothesize that they are inevitable to fully characterize Dravidian syntax meaningfully. Furthermore, we also hypothesize that these discourse construals are directly machine-learnable by way of morphological regularities in these languages.
- We also handle 'combinative' interaction in Fig. 1 by means of chunking multi-word expressions, noun compounds as one chunk.
- In short, we are treating the *karaka* relations as one of the four types of discourse interactions that can happen between *things* and *relations* in the speaker's discourse world. Other three types of interactions are also captured though construction labels and chunking in our scheme.

In the third point above, we have mentioned a few discourse construals and pointed out that they are inevitable to understand certain syntactic peculiarities in Dravidian languages. Out of these, we have discussed about *Qualifier* and *Pronominal* schemas with illustrations in Sect. 3. We will just briefly describe the other four schemas namely *Conjunctive, Concurrent, Condition, Infinitive*. These four schemas signify the *associative* interactions discussed in Fig. 1 i.e. how one relation interacts with another relation. For instance, if two events take place in discourse, they are perceived in the following ways.

1. Event 1 is perceived as having continuance with event 2 - *Conjunctive schema*
2. Event 1 is perceived as being concurrent/parallel with event 2 - *Concurrent schema*
3. Event 1 is perceived as a condition/situation for describing event 2 - *Conditional schema*
4. Event 1 is perceived as yet to happen from the point of view of event 2 - *Infinitive schema*

All these four 'construals' are directly encoded in the verb morphology of Dravidian languages. Every Dravidian verb has four basic non-finite inflections that directly correspond to these construals that we mentioned above. It is these construals that are skilfully exploited by the languages to create subordinate clauses, grammatical aspects, modalities etc. by way of creating a series of non-finite verbs ending with one finite-verb. The bottom line is: formally different syntactic phenomena such as subordinate clause formation, grammatical aspects, modalities are constructed using the same above four construals. Again learning these construals instead of their formal properties is straightforward and suitable for these languages, because these discourse level construals are directly marked in morphology. For example in Tamil, *conjunctive* inflection of a verb occurs in clauses showing discourse sequence as well as in grammatical aspects. The *infinitive* inflection occurs in clauses showing goal/purpose as well as to describe moods and modalities. The *concurrent* inflection occurs in clauses conceiving concurrence as well as in describing manner adverbs. The *conditional* inflection occurs in conditional clauses, situated descriptions, in complement clauses which are based on some condition etc. The following examples serve to illustrate the point.

(7) rAman   vITTuk-ku   **pO-y**   paDam   pArt-t-An
     Raman   *home-dat*   **go-conjunctive**   *movie* *see-pst-agr*
     'Ram went home and watched a movie'

(8) rAman   munb-E   paDatt-ai   **pArt-tu**   iruk-kiR-An
     Raman *before-only* *movie-acc*   **see-conjunctive**   *exist-pres-agr*
     'Ram has already seen the movie'

Note that the same conjunctive form of a verb is morphologically used by the language to indicate both clausal sequences and grammatical aspects. Usually, in formal analyses this is explained through grammaticalization theory. But in construction grammar, it is explained as the cognitive construal that underlies the conjuctive participial inflections. (Refer to [25] for cognitive operations underlying conjunctive inflection and how it relates to grammatical aspect). Discussing in detail why the same conjunctive, infinitive form are used for event - event relations as well as for aspects and modalities, with relevant grammatical evidences is a task we are not undertaking in this work. What is relevant for our parsing task at hand is, we treat every one of these non-finite verbal inflections as separate construction schemas and the dependency parser has to learn whether the relation that holds between them is event relation or merely grammatical and so on.

Thus with this understanding of the concepts that we have discussed, we formulated the construction schema labels and the dependency labels. The complete set of Construction labels in our annotation scheme is listed in Table 3. The complete set of dependency relations that can exist between these construction schemas are listed in Table 4. To understand the *karaka* related concepts such as *karta samanadhikarana, deshadhikarana* etc. please refer to the dependency annotation guidelines [26]. Other concepts such as conjunctive, concurrent, conditional, infinitive and their subfunctions namely grammatical/manner/event relations in discourse are already discussed. Extended adhikarana (k7a) is a *karaka* construal that is specific to Tamil. Volitionality conceived upon the giver or receiver in a transaction event is given the label k7a.

5   Experiments and Results

To verify if the interactions between the proposed conceptual schemas are true and suitable for learning dependencies, we collected a set of 935 sentences for our annotation experiments. Out of these, 354 sentences are taken from newspaper data that we crawled from various online newspapers in Tamil. The crawled data had a total of approximately 5 lakh sentences (5,17,421 sentences). This crawled corpus is raw, uncleaned and unprocessed. The remaining 581 sentences are taken from a portion of the gold standard annotated training data for Tamil full parser, which is available as part of ILMT consortium[2]. The gold annotation[3] of the 581 sentences was available on CPG framework. These sentences are from various domains such as tourism, health, religion etc. While the former 354 sentences are general newspaper sentences that could be on any topic, the latter 581 sentences are collected from domain specific data and therefore the type of sentences, length of sentences could differ.

We annotated all the 935 sentences with the proposed tags and dependencies. We used MALT parser[4] for the purpose of our experiments. The parser settings are of Ambati et al. [27]. The features given to the MALT parser are the stem and morphological inflections. Experiments were conducted for the following test scenarios to verify the performance of the parser.

1. Varying the sources of data (AUKBC or Newspaper data)
2. Varying the size of the training data for the given source
3. Varying the gold annotation scheme (existing CPG annotation vs our proposed annotation)
4. Combining both the data and verifying the performance with the proposed annotation

Since we took only 581 sentences from AUKBC data (CPG annotation) and annotated those sentences in our proposed CG framework, the parser evaluation

---

[2] Indian Language Machine Translation Project funded by DIT, Government of India.
[3] The gold annotation was carried out by AU-KBC Research Centre, Chennai.
[4] http://www.maltparser.org/download.html.

Table 5. Parser accuracy test cases

Ann. type	Data size	Data type	Avg. No. of chunks in a sentence	Avg. LAS
CPG	176	AUKBC	5.1	56.02%
CG	176	AUKBC	5.8	72.98%
CPG	581	AUKBC	5.2	60.57%
CG	581	AUKBC	5.9	82.24%
CG	354	Crawled data	10.6	72.93%
CG	354+176	(Crawled+AUKBC) data	9.0	74.96%
CG	354+581	(Crawled+AUKBC) data	7.7	82.21%

that can be compared between the two annotation schemes for only this amount of data. Therefore the fourth testing scenario on 935 sentences reports only the parsing accuracy on our proposed annotation scheme. In every one of these test cases, a five fold validation was done. The results are reported in Table 5. The annotation type *CG* refers to the proposed Construction Grammar framework and *CPG* refers to the Computation Paninian Grammar framework. The average number of chunks in a sentence mentioned in the table is the length of a sentence in terms of number of chunks in it. As it can be seen, the number of chunks are not the same between the two annotation schemes. In fact since we split a token wherever status interpretations of verbs could be made, our annotation scheme almost always has more number of chunks in a sentence than the CPG annotation. For instance, note that for the data size of 176 sentences, the average number of chunks in CPG is 5.1 whereas in the proposed CG scheme it is 5.8. The average length of sentences is more in the crawled newspaper data than the AUKBC data. The last three rows show that as the average number of chunks in a sentence in the corpus decreases, the accuracy increases which is only expected because there will be lesser long distance dependencies. Every row that is shown in the result is the average of five fold validation performed on the data. For every given data size we chose 80% for training and 20% for testing. The average LAS accuracy of the parser on a total of 935 sentences is 82.21%.

5.1 Partial Evaluation Results of Dependency Labels and Attachment

Technically the LAS scores are themselves not directly comparable because the annotation labels and linguistic scheme are different. Therefore, we will show the partial evaluation results of dependency labels which reveal as to how within a given linguistic framework various labels are learnt consistently.

The precision and recall of dependency labels and attachment taken together for a training size of 176 sentences annotated using the proposed method is reported in Table 6. Table 7 shows the precision and recall of dependency relation + attachment for the training data size of 176 using the CPG annotation. Similarly, Tables 8 and 9 report the precision and recall of dependency labels + attachment for the training size of 581 sentences, annotated using CG framework

Table 6. Precision and recall of DepRel and attachment; 176 sentences; CG annotation

Deprel	Gold	Correct	System	Recall %	Precision %	Deprel	Gold	Correct	System	Recall %	Precision %
adv	8	6	7	75.00	85.71	k7	13	10	14	76.92	71.43
ccof	3	0	0	0.00	NaN	k7a	1	0	0	0.00	NaN
conj:gram	8	7	7	87.50	100.00	k7p	2	0	0	0.00	NaN
conj:man	1	0	0	0.00	NaN	k7t	1	0	3	0.00	0.00
inf:gram	9	8	8	88.89	100.00	main	36	36	36	100.00	100.00
k1	33	28	51	84.85	54.90	nmod:stat	5	5	5	100.00	100.00
k1s	2	2	2	100.00	100.00	pof	5	1	1	20.00	100.00
k2	12	5	6	41.67	83.33	r6	13	10	11	76.92	90.91
k2s	1	1	1	100.00	100.00	rsp	1	0	0	0.00	NaN
k3	2	0	0	0.00	NaN	sent adv	3	1	1	33.33	100.00
k4	6	1	1	16.67	100.00	status	2	2	2	100.00	100.00
k5	2	0	0	0.00	NaN	vmod:stat	21	20	36	95.24	55.56

Table 7. Precision and recall of DepRel and attachment; 176 sentences; CPG annotation

Deprel	Gold	Correct	System	Recall %	Precision %
adv	12	7	14	58.33	50.00
ccof	13	9	12	69.23	75.00
jjmod	1	0	0	0.00	NaN
k1	29	21	47	72.41	44.68
k1s	2	0	1	0.00	0.00
k2	22	10	16	45.45	62.50
k3	1	0	0	0.00	NaN
k4	5	2	2	40.00	100.00
k7	5	0	3	0.00	0.00
k7a	1	0	0	0.00	NaN

Deprel	Gold	Correct	System	Recall %	Precision %
k7p	5	2	9	40.00	22.22
k7t	5	0	2	0.00	0.00
lwg_psp	2	2	3	100.00	66.67
main	35	34	35	97.14	97.14
nmod	5	0	0	0.00	NaN
r6	11	7	10	63.64	70.00
rh	1	0	0	0.00	NaN
rsp	1	0	0	0.00	NaN
sent- adv	1	0	0	0.00	NaN
undef	3	0	0	0.00	NaN
vmod	3	2	9	66.67	22.22

Table 8. Precision and recall of DepRel and attachment; 581 sentences; CG annotation

Deprel	Gold	Correct	System	Recall %	Precision %	Deprel	Gold	Correct	System	Recall %	Precision %
adv	23	18	19	78.26	94.74	k5	2	2	3	100.00	66.67
ccof	13	12	13	92.31	92.31	k7	49	39	47	79.59	82.98
conj:gram	28	26	29	92.86	89.66	k7a	2	0	0	0.00	NaN
conj:man	7	5	9	71.43	55.56	k7p	9	4	5	44.44	80.00
conj:seq	3	0	0	0.00	NaN	k7t	9	2	7	22.22	28.57
inf:gram	37	36	37	97.30	97.30	main	115	113	116	98.26	97.41
inf:man	1	0	1	0.00	0.00	nmod:stat	16	14	14	87.50	100.00
inf:seq	3	0	0	0.00	NaN	pof	18	13	18	72.22	72.22
k1	110	98	140	89.09	70.00	r6	29	25	25	86.21	100.00
k1s	16	14	15	87.50	93.33	ras	0	0	4	NaN	0.00
k2	32	20	24	62.50	83.33	rsp	1	0	0	0.00	NaN
k2s	1	0	0	0.00	NaN	rt	1	0	0	0.00	NaN
k3	9	9	9	100.00	100.00	sent adv	11	4	4	36.36	100.00
k4	17	13	14	76.47	92.86	status	11	10	11	90.91	90.91
k4a	0	0	2	NaN	0.00	vmod:stat	90	83	99	92.22	83.84

Table 9. Precision and recall of DepRel and attachment; 581 sentences; CPG annotation

Deprel	Gold	Correct	System	Recall %	Precision %	Deprel	Gold	Correct	System	Recall %	Precision %
adv	40	33	46	82.50	71.74	k7t	17	5	9	29.41	55.56
ccof	43	32	48	74.42	66.67	lwg_psp	10	4	4	40.00	100.00
jjmod	2	0	0	0.00	NaN	main	116	114	116	98.28	98.28
k1	104	77	154	74.04	50.00	nmod	13	1	5	7.69	20.00
k1s	15	2	3	13.33	66.67	nmod_relc	3	0	3	0.00	0.00
k2	63	22	59	34.92	37.29	r6	32	24	28	75.00	85.71
k2g	4	2	2	50.00	100.00	ras- k1	2	0	0	0.00	NaN
k2p	1	0	0	0.00	NaN	ras- k2	2	0	0	0.00	NaN
k2s	1	0	0	0.00	NaN	ras- neg	2	0	0	0.00	NaN
k3	6	2	2	33.33	100.00	rd	2	0	0	0.00	NaN
k4	18	12	24	66.67	50.00	rh	4	1	2	25.00	50.00
k5	1	1	1	100.00	100.00	rsp	4	0	0	0.00	NaN
k7	20	6	8	30.00	75.00	sent- adv	4	2	5	50.00	40.00
k7a	1	0	0	0.00	NaN	undef	8	0	0	0.00	NaN
k7p	27	22	39	81.48	56.41	vmod	18	15	25	83.33	60.00

and CPG framework respectively. The dependency label counts are not readily comparable even for the same label because the chunks and the relationships that exist between them are different in the two frameworks. For instance, in CG output shown in Table 8 the gold count of vmod:stat is 90 when the training size was 581 sentences. But in CPG output shown in Table 9, the gold count of vmod is just 18. This discrepancy can be understood if we notice that the ccof count in the CPG annotation is 43 but in CG it is just 13. i.e. what are analyzed as conjunction of two entities in CPG are analyzed as list of entities associating with a status verb in CG. Conjunctions are actually done by particles such as 'um' whose syntactic behaviour is different from a proper conjunction like 'and' in English. It semantics is similar to MO particle of Japanese well discussed in literature [28]. That explains why there is more vmod:status in CG parser output. Overall, it can be seen that the dependency label + attachment precision and recall are comparatively better in our proposed CG annotation consistently.

5.2   Reasons for Better Learning

There are two reasons for better learning of dprel relations and overall LAS accuracy.

- Our dependency relations are directly learnable from morphological inflections
- The form-function pairing analysis is able to generalize the peculiar constructions that occur in Tamil syntax better.

The dependency labels that we are using in our annotation scheme are not coarser but fine grained. For instance, what is just a 'vmod' in CPG annotation is annotated with fine-grained labels such as 'conj:gram', 'conj:man', 'conj:seq', 'conc:man', 'conc:seq' etc. Yet these labels are learnt better because these labels are directly inferred from the morphological inflections given as a feature to

MALT parser. The fact that construction analysis performs better across 5 fold experiments shows that most of the functional properties of the language are directly encoded in morphology.

By mapping the morphological features to meaningful construals, we are able to explain the peculiar morpho-syntactic constructions in Tamil, which are otherwise difficult or impossible to characterize. Take for example a construction like the one shown below:

(9) nIND-a-d-um   kaLaippu   mikk-a-d-um-An-a
     elongate-RP-PronSuffix-also   *tiredness*   *exceed-RP-PronSuffix-also-became-RP*
     payaNam   toDar-nd-adu
     journey   continue-pst-agr

     'The long and arduous journey continued'

As it can be seen there are no pure adjectives in Tamil and therefore an expression like *long and arduous* is rendered by a complex expression *nINDadum kaLaippu mikkadumAna* whose literal morphological glosses are given above. Conventionally, the parts of speech and chunking in CPG framework is as follows:

1. ((nINDadum NN)) - Noun chunk
2. ((kaLaippu NN)) - Noun chunk
3. ((mikkadumAna JJ) (payaNam NN)) - Noun chunk
4. ((thodarndhadhu VB)) - Verb chunk

Note that the word 'nINDadum' is treated as a noun because morphologically it exhibits the property of a noun with pronominal suffix. But in terms of dependency analysis, 'nINDadum' and 'kaLaippu mikkadumAna' should be analyzed as adjectives that together modify the noun 'payaNam'. However, since chunk is structurally defined in CPG as minimal non-recursive unit of analysis [29] the second adjective 'mikkadumAna' and the noun 'payaNam' together are by definition grouped together as one Noun chunk. Now once it is chunked like this, there is no way to learn that there are two adjectives in the expression that are modifying a noun. The most likely dependency tree that the parser will end up learning with this configuration would be as shown in Fig. 3. This is a type of error that cannot be handled no matter how much training data is provided because the problem here has a theoretical basis where there is a mismatch between what the morphology of the language says and what formal syntactic category the word stands for.

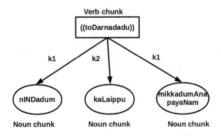

Fig. 3. Wrong dependency that is likely to be learnt

In construction grammar approach, this problem is easily handled because we map the word 'nINDadum' as *Status Pronominal Schema*, 'kaLaippu' as *Noun Schema*, 'mikkadum' as *Status Pronominal Schema*, 'Ana' as *Status Qualifier Schema*. Thus the three words in the raw text are treated as four construction units that map to their meaningful 'construals' and now the dependency relations are established between them as shown in Fig. 4. It might look that after all the final dependency relations are the same as in CPG, so what is the CG contribution here? Recognizing 'Ana' not as a formal adjectivializer suffix but a *Status Qualifier* schema that can take 'configurations' as its arguments allows us to chunk the tokens differently and build the dependency tree shown in Fig. 4.

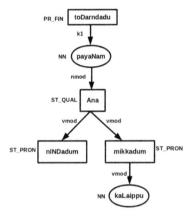

Fig. 4. Dependency analysis according to CG framework

This and many other expressions where the language uses its morphological properties to create 'verbiness' are very well captured by following a construction based approach. In the next section, we will discuss about what kinds of errors commonly occur in parser output, analyse those error scenarios and what improvements can be introduced to handle those errors.

6   Error Analysis

We performed a five fold validation of parser output for varying training sizes ranging from 354 to 935 sentences and checked what are the most frequent errors that are encountered by the parser trained with the proposed CG framework. The Table 10 shows the top five frequent errors across different folds of iteration when the training data size was 935 sentences.

Table 10. Five frequent drel errors in different folds of iteration for 935 sentences training data

Fold 1			Fold 2			Fold 3			Fold 4			Fold 5		
Gold	System	Frequency	Gold	System	Frequency	Gold	System	Frequency	Gold	System	Frequency	Gold	System	Frequency
k1	k2	11	k1	k2	16	k1	k2	11	k1	k2	20	k1	k7t	13
k1	pof	11	k1	pof	11	pof	k1	8	k7	k7p	12	k7	k7p	9
k7	k7p	6	pof	k2	6	k7	k7t	6	k1	nmod	11	k1	pof	6
conj:seq	conj:man	6	k4	rt	5	k1	pof	6	pof	k1	10	conj:man	conj:seq	6
k7	k7t	5	k7	k7p	4	vmod:stat	k4	5	k1	vmod	9	k7	k7t	5

The most frequent errors that are observed in the parser output are as follows:

1. k1 is misidentified as k2 - karta (*agent* roughly speaking) is misidentified as karma (*patient* roughly speaking)
2. pof (Complex predicate) is misidentified as k1 (doer)
3. k7p (spatial location) is misidentified as k7 (general locative)
4. k4 (goal) instead of rt (purpose)
5. conj:gram (grammatical) instead of conj:man (manner)

The most frequent error is: the gold label is k1 while the system identifies it as k2 and vice versa. This usually happens because it is quite common in Tamil that the accusative case is not morphologically marked on a noun and the language is relatively free word ordered. The following sentence is taken from one of the test file in which the parser misidentified k2 as k1. It illustrates the point.

(10)   vinAyaga   cadurttik-ku   **paDam**   veLiyiDuv-a-du   kamal-in   tiTTam
        *Vinayaka   chaturthi-dat   **movie**   release-RP-pron   Kamal's   plan*

'It is Kamal's plan to release **the movie** during Vinayaka Chaturthi (a religious festival)'

In the above example, the highlighted word 'paDam' does not morphologically show the accusative marker. Because the parser is trained on morphological features and 'k1' and 'k2' are potential candidates in the similar context, these two tags are the most misidentified. The confusion between 'pof' and 'k1' is only expected because the noun which is a part of a complex predicate can be easily confused as a participant of the verb. The other two frequent errors are due to the granularity of the dependency relation that should be identified. *k7p* is more granular than *k7* and therefore difficult to learn. Same is true for *conj:gram* and *conj:man*. The other interesting candidate is 'rt (purpose)' being misidentified with 'k4 (goal)'. Because conceptually purpose and goal are metaphorically directed towards some object, the same fourth case marker is used in Tamil to denote these two functions. Hence it is difficult to tell apart one from another. Since these error scenarios typically involve making distinctions between granular relations, with more training data these can be learnt better.

7   Conclusions and Future Work

In this work, we have discussed that by exploiting the morphological regularities in Tamil and by mapping these morphological forms to meaning on the theoretical basis of Construction Grammar, we are better able to learn the morpho-syntactic peculiarities in Tamil data. Since these 'construals' are actually learned through morphological features consistently across 5 folds for varying training data size, it shows that there is a merit in applying form-function pairing as a means of syntactic analysis. Though we have hinted at these morphological properties with a few illustrations, we are working on a full theory in which the syntactic problems that are discussed in linguistic literature [19–24] can be explained based on the functional 'construals' underlying these peculiar morphological inflections. In future, we are looking forward to apply this idea of form-function pairing for other morphologically rich languages as well, since most meaningful information about the syntax comes from morphology in these languages.

References

1. Goldberg, A.E.: Construction Grammar. Wiley Online Library (2002)
2. Fried, M., Östman, J.O.: Construction grammar. In: Construction Grammar in a Cross-Language Perspective (2011)
3. Langacker, R.W.: Cognitive Grammar: A Basic Introduction. Oxford University Press, Oxford (2008)
4. Shieber, S.M.: Evidence against the context-freeness of natural language. In: Savitch, W.J., Bach, E., Marsh, W., Safran-Naveh, G. (eds.) The Formal Complexity of Natural Language. Studies in Linguistics and Philosophy, vol. 33, pp. 320–334. Springer, Heidelberg (1985). https://doi.org/10.1007/978-94-009-3401-6_12
5. Melčuk, I.A.: Dependency Syntax: Theory and Practice. SUNY Press, Albany (1988)
6. Bharati, A., Chaitanya, V., Sangal, R., Ramakrishnamacharyulu, K.: Natural Language Processing: A Paninian Perspective. Prentice-Hall of India, New Delhi (1995)
7. Bharati, A., Sangal, R.: Parsing free word order languages in the Paninian framework. In: Proceedings of the 31st Annual Meeting on Association for Computational Linguistics, pp. 105–111. Association for Computational Linguistics (1993)
8. Bharati, A., Gupta, M., Yadav, V., Gali, K., Sharma, D.M.: Simple parser for Indian languages in a dependency framework. In: Proceedings of the Third Linguistic Annotation Workshop, pp. 162–165. Association for Computational Linguistics (2009)
9. Mannem, P.: Bidirectional dependency parser for Hindi, Telugu and Bangla. In: Proceedings of NLP Tools Contest: Indian Language Dependency Parsing, ICON 2009, India (2009)
10. Nivre, J.: Parsing Indian languages with MaltParser. In: Proceedings of the NLP Tools Contest: Indian Language Dependency Parsing, ICON 2009, pp. 12–18 (2009)
11. Ambati, B.R., Gadde, P., Jindal, K.: Experiments in Indian language dependency parsing. In: Proceedings of the NLP Tools Contest: Indian Language Dependency Parsing, ICON 2009, pp. 32–37 (2009)
12. Antony, P., Warrier, N.J., Soman, K.: Penn treebank-based syntactic parsers for South Dravidian languages using a machine learning approach. Int. J. Comput. Appl. **7**, 14–21 (2010)

13. Selvam, M., Natarajan, A., Thangarajan, R.: Structural parsing of natural language text in Tamil using phrase structure hybrid language model. Int. J. Comput. Inf. Syst. Sci. Eng. **2008**, 2–4 (2008)

14. Ramasamy, L., Žabokrtský, Z.: Tamil dependency parsing: results using rule based and corpus based approaches. In: Gelbukh, A.F. (ed.) CICLing 2011. LNCS, vol. 6608, pp. 82–95. Springer, Heidelberg (2011). https://doi.org/10.1007/978-3-642-19400-9_7

15. Straka, M., Hajic, J., Straková, J., Hajic Jr., J.: Parsing universal dependency treebanks using neural networks and search-based oracle. In: International Workshop on Treebanks and Linguistic Theories (TLT 2014), p. 208 (2014)

16. Kumari, B.V.S., Rao, R.R.: Hindi dependency parsing using a combined model of Malt and MST. In: 24th International Conference on Computational Linguistics, p. 171. Citeseer (2012)

17. Kesidi, S.R., Kosaraju, P., Vijay, M., Husain, S.: A constraint based hybrid dependency parser for Telugu. Int. J. Comput. Linguist. Appl. **2**, 53 (2011)

18. Seddah, D., Tsarfaty, R., Kübler, S., Candito, M., Choi, J., Farkas, R., Foster, J., Goenaga, I., Gojenola, K., Goldberg, Y., et al.: Overview of the SPMRL 2013 shared task: cross-framework evaluation of parsing morphologically rich languages. Association for Computational Linguistics (2013)

19. Amritavalli, R., Jayaseelan, K.: Finiteness and negation in Dravidian. In: The Oxford Handbook of Comparative Syntax, pp. 178–220 (2005)

20. Amritavalli, R.: Separating tense and finiteness: anchoring in Dravidian. Nat. Lang. Linguist. Theory **32**, 283–306 (2014)

21. McFadden, T., Sundaresan, S.: Finiteness in south Asian languages: an introduction. Nat. Lang. Linguist. Theory **32**, 1–27 (2014)

22. Jayaseelan, K.A.: The serial verb construction in Malayalam. In: Dayal, V., Mahajan, A. (eds.) Clause Structure in South Asian Languages. Studies in Natural Language and Linguistic Theory, vol. 61, pp. 67–91. Springer, Heidelberg (2004). https://doi.org/10.1007/978-1-4020-2719-2_3

23. Jayaseelan, K.: Coordination, relativization and finiteness in Dravidian. Nat. Lang. Linguist. Theory **32**, 191–211 (2014)

24. Herring, S.C.: Aspect as a discourse category in Tamil. In: Annual Meeting of the Berkeley Linguistics Society, vol. 14 (2011)

25. Karmakar, S., Kasturirangan, R.: Cognitive processes underlying the meaning of complex predicates and serial verbs from the perspective of individuating and ordering situations in bānlā. In: Proceedings of the First International Conference on Intelligent Interactive Technologies and Multimedia, pp. 81–87. ACM (2010)

26. Bharati, A., Husain, D.S.S., Bai, L., Begam, R., Sangal, R.: Anncorra: Treebanks for Indian languages, guidelines for annotating Hindi treebank (version-2.0) (2009)

27. Ambati, B.R., Husain, S., Nivre, J., Sangal, R.: On the role of morphosyntactic features in Hindi dependency parsing. In: Proceedings of the NAACL HLT 2010 First Workshop on Statistical Parsing of Morphologically-Rich Languages, pp. 94–102. Association for Computational Linguistics (2010)

28. Szabolcsi, A.: What do quantifier particles do? Linguist. Philos. **38**, 159–204 (2015)

29. Bharati, A., Sangal, R., Sharma, D.M., Bai, L.: Anncorra: Annotating corpora guidelines for POS and chunk annotation for Indian languages. LTRC-TR31 (2006)

Comparative Error Analysis of Parser Outputs on Telugu Dependency Treebank

Silpa Kanneganti[(⊠)], Himani Chaudhry, and Dipti Misra Sharma

Kohli Center on Intelligent Systems (KCIS),
International Institute of Information Technology,
Hyderabad (IIIT Hyderabad) Gachibowli, Hyderabad 500032, Telangana, India
{silpa.kanneganti,himani}@research.iiit.ac.in, dipti@iiit.ac.in

Abstract. We present a comparative error analysis of two parsers - MALT and MST on Telugu Dependency Treebank data. MALT and MST are currently two of the most dominant data-driven dependency parsers. We discuss the performances of both the parsers in relation to Telugu language. We also talk in detail about both the algorithmic issues of the parsers as well as the language specific constraints of Telugu. The purpose is, to better understand how to help the parsers deal with complex structures, make sense of implicit language specific cues and build a more informed Treebank.

1 Introduction

In this paper we present a comparative error analysis of two dependency parsers - MALT [1] and MST [2], on Telugu Treebank data. We discuss the performances of both the parsers in relation to Telugu language, to better understand language specific workings of each of the models as well as the nuances of the language much better. The purpose is, to figure out better ways to help the parsers deal with complex structures, make sense of implicit language specific cues and build a more informed Treebank. McDonald and Nivre [3] talk in detail about the errors of data driven parsing models, Husain [4] reports error analysis of MALT on Hindi language data. MALT and MST are currently two of the most dominant data-driven dependency parsers. MALT adopts a local, greedy, transition-based approach, and MST takes to global, exhaustive, graph based modeling. While they are neck and neck when it comes to parsing accuracy, their approaches could not be more different from each other. Further, Vempaty et al. [5] discuss their observations with regard to Telugu language, in their attempt to build a Telugu treebank.

2 The Parsers

MALT [1] is a classifier based Shift/Reduce parser, that employs transition-based inference algorithms [6] for parsing. History based feature models are used for predicting the next parser action [7] and Support vector machines to map histories

© Springer International Publishing AG, part of Springer Nature 2018
A. Gelbukh (Ed.): CICLing 2016, LNCS 9623, pp. 397–408, 2018.
https://doi.org/10.1007/978-3-319-75477-2_28

to parser actions [8]. MST uses Maximum Spanning Tree algorithm [9],[10] for non-projective parsing and Eisner's algorithm for projective parsing [11]. It uses large-margin structured learning [12] as the learning algorithm.

3   The Data

3.1   Telugu Language

Telugu, a morphologically, syntactically complex language, is highly inflectional and agglutinative. It is a nominative-accusative language, with SOV as its default word order. The verbs exhibit a rich inflectional morphology. Hence it encodes various grammatical categories like tense, case, gender, number, person, negatives, imperatives etc. morphologically [13].

3.2   The Treebank Data

The data set we use is from the ICON10 parsing contest Husain et al. [14]. The data, comprising of 1451 sentences was annotated using the Computational Pāṇinian Grammar (CPG) dependency annotation scheme [15],?. We use inter-chunk dependency trees. A description of the dependency relations used in the CPG framework has been provided by Chaudhry et al. [16]. Some of the issues that arise while trying to parse a free word order language like Telugu are [4]

- difficulty in extracting relevant linguistic features
- lack of explicit cues
- long distance dependencies
- complex linguistic phenomena

3.3   Error Classification

The error analysis is carried out on MALT and MST parsers trained on Telugu treebank based on 4 properties: Sentence length, Arc length, Arc type, Arc depth [3,4]. While analyzing errors we tried to neglect the ones caused by algorithmic limitations, learning and data sparsity issues. Our focus is on the errors that could be occurring due to lack of robust features and complex structures too difficult to learn.

3.4   Experiment Setup

For experiment purposes, we are using MALT[1][1] with parser settings from Ambati et al. [17] and best settings from Ambati et al. [18] for MST[2]. We are using maxent tool [3] for labeling MST parsed trees. The train and the test set

---

[1] MALT version 1.8.1.
[2] MST version 0.5.0.
[3] http://homepages.inf.ed.ac.uk/lzhang10/maxent.html.

Table 1. Accuracies

	LAS		UAS		LS	
	MALT	MST	MALT	MST	MALT	MST
5fold	67.36%	67.24%	90.5%	89.1%	69.3%	68.6%
Best	76.14%	75.84%	95.63%	92.18%	77.36%	78.62%

contain 1161, 290 sentences respectively. Error classification is based on the test data. Table 1 depicts the accuracies of 5fold cross validation and the train and test set being used for the analysis. The evaluation is based on LAS[4], although LAS, UAS[5] and LS[6] are being reported.

4 Error Analysis

Error Analysis in this paper is discussed in two parts. In Sect. 5.1 we talk about linguistic errors that are common and different between both the parsers as well as Telugu specific parsing issues, while in Sect. 5.2 we talk about algorithm related parser statistics and errors of both the parsers.

4.1 Linguistic Analysis of Errors

In this section we discuss the linguistic aspect of the errors of both the parsers. Below, we discuss the types and causes of these errors.

4.1.1 Errors Common to Both the Parsers
Complements and Adjuncts:
Ambiguity between a subject and its complement due to lack of identification of post position.

(1)   neVlalaku piMdam velaMwa   uMtuMdi
      In-months embryo   size-of-a-finger is

      'In months the embryo is the size of a finger.'

    The post-position 'aMwa' could be a cue to identify 'velaMwa' as the subject compliment 'k1s' here. Instead, it is being marked as another subject 'k1' along with 'piMdam', which is the actual 'k1' in the sentence, as seen in example 1. The post-position, in turn is perhaps not being identified due to the language's tendency to manifest post-positions agglutinated with the word.

---

[4] LAS – Labeled Attachment Score.
[5] UAS – Unlabeled Attachment Score.
[6] LS - Labeled Score.

(2)   neVlalaku piMdam velaMwa   uMtuMdi
      In-months embryo size-of-finger is

      'In months the embryo is the size of a finger.'

Being a null subject language, Telugu tends to drop the pronominal subject of a finite clause. Also, agglutination of the subject marker with the verb further adds to the complexity of the sentence structures. We observe that in cases where the inherent subject is absent, the object is being misannotated as the subject.

(3)   oVkati   wulasiki iccAdu
      one(thing) to-Tulasi he-gave

      'He gave one to Tulasi.'

Due to the absence of inherent subject and agglutination of the subject indicator 'du' (he) with the verb, in example 3 the NP 'oVkati' is being misidentified as subject.

Further, not just Telugu verbs agglutinate with the subject markers such as 'ru' (they), 'du' (he), 'di' (she/it), but both, Telugu nouns and verbs can also agglutinate conditionals such as 'we' (if - example 6) and emphatic markers such as 'lone' (itself - example 5)

(4)   illu   KAlYI ceSAdu ani ceVppAru
      house vacate he-did that they-said

      'They said that he vacated the house.'

(5)   yavvanaMlone kAlYlu cewulu AdawAyi
      In-youth-itself legs   hands move

      'youth is when legs and hands move.'

In example 5, 'yavvanaMlone' that is,'yavvanaM' (youth) + 'lone' (in-itself) is a case of agglutination of the emphatic marker 'lone' with the noun 'yavvanaM'.

In example 6, instead of identifying the relations of 'OVwwidi' (stress) as subject (k1) and of 'padi' (put/falls) as the adjunct reason (rh), it is mislabeling them as the 'noun of a complex predicate' (pof), and 'verb modifier' (vmod), respectively

(6)   manasumIxa oVwwidi padiwe   vaswuMxi
      on-the-heart stress   if-put/falls comes

      'It comes when the heart is stressed out.'

Adjuncts:

Quantifiers (QF) and Quotatives (UT) are being mislabeled as 'karma' instead of verb modifiers.

(7) waggipowuMxi ani   veVlYlipoyAdu
     will-go-away   saying he-left

'He left saying it will go away.' In example 7 'ani' (quotative that) is mislabeled with the object label (k2), instead of being identified as a verb modifier and being labeled 'vmod'.

Noun Modifiers:

(8) maMxula RApuku poyAdu
     medical to-shop he-went

'he went to the medical shop.'

In example 8 'maMxula' (Medical) is a modifier of the noun 'RApu-ku' (to-shop) and should be annotated as the noun-modier (nmod). Instead, it is being marked object (k2) of 'poyAdu' (he went).

Genitives:
In Telugu, the genitive marker is often covert (dropped), and the relation must be inferred. In such cases, error in identifying a genitive relation between a subject (noun) and its complement (noun) is observed.

(9) bidda walli   kadupulo peVruguwuMxi
     baby mother stomach grows

'Baby grows in its mom's stomach.'

In example 9, the noun 'walli' does not have an overt genitive marker. Thus the parsers are unable to infer the relation between 'walli' (mother) and 'kadupulo' (stomach) as a genitive relation, and marks it as a child of the verb 'peVruguwuMxi' (grows).

(10) rAmulu kUwuru ramA caxuvukuMxi
      Ramulu daughter Rama studied

'Ramulu's daughter Rama studied.'

In the phrase 'rAmulu kUwuru' in 10, 'rAmulu' (Ram) has a genitive relation with 'kUwuru' (daughter) but the parsers are unable to infer it in the absence of an explicit case marker.

However, if the sentence had an overt genitive marker, say 'yoVkka' ('s), as in example 11 it would perhaps be easier for the parsers to identify the relation correctly.

(11) rAmulu yoVkka kUwuru ramA caxuvukuMxi
      rAmulu 's   daughter ramA studied

'Ramulu's daughter Rama studied.'

Complex Predicates:
We note that the parsers are erroneously annotating noun components of complex predicates with other relations, such as location, subject, object, etc.

(12)   maMxulu vAdiwe rakwapotu axupulo uMtuMxi ani weVlusukunnAru
       medicines if-used blood-pressure in-control will-be that-they have-learnt

'They have learnt that if medicines are used blood pressure can be kept in control.'

In example 12, 'axupulo' is annotated as a location, with the label (k7p), instead of being marked as a part of (pof) the complex predicate. Thus, more context and semantic information is needed to handle Complex predicates.

Conjunctions:
Broadly, the word 'ani' in Telugu, either occurs as a subordinating conjunct or as a complementizer (that). It is a Quotative (UT) quite loaded semantically. Instances of 'ani' are being misidentified as the head of the sentences, as seen in example 13

(13)   nIku   nagalu   ceyiswAnu   ani javAbu uMxi
       for-you jewellery i-will-get-made ani answer is-there

'"I will get jewellery made for you" the answer is there.'

Erroneous Errors:
Parser tree error are seen to be occurring in the output broadly as occurrence of multiple children with same dependency label.

Error in differentiating between a subject and its complement, due to lack of identification of the post-position, as seen (and discussed) in example 2 also falls under the purview of the parse tree error type.

4.1.2   Errors Different to both the Parsers
As discussed elsewhere in the paper, each of the parsers performs better over the other, for some categories. Below, we report some such errors:

Complements:

(14)   nIvu beVMga peVttukovaxxu
       you worry   don't-keep
       glt 'you don't worry.'

In example 14, the NP beVMga 'worry', instead of being annotated, the object of the verb is being marked with the relation 'part-of' (pof) to indicate that it is the Nominal component of a Complex predicate by MST. Whereas MALT is able to annotate such cases correctly, by and large.

Noun Modifiers:

(15)   sAkRi   saMwakaM uMxi
       witness signature   is-there
       'The signature of the witness is there.'

In example 15, 'sAkRi' (witness), a modifier of the nominal node 'saMwakaM' (signature), is being annotated as the subject of the verb 'uMxi' (is there) by MALT. MST is annotating such cases correctly. Thus, in such cases, not only does the verb have two children annotated with the relation 'kartā' (k1), (making it a case for 'Erroneous errors'), noun modifiers are also being incorrectly annotated. (MALT is seen to prefer longer arcs over more convenient shorter arcs in case of nmods and genitives).

Genitives:

(16) gexe   pAlu wIsi   pattukupoyevAdu
      buffalo milk after-taking he-used-to-take-with-him

'After taking the buffalo milk, he used to carry it with him.'

We see in example 16, that though 'gexe' (buffalo) has a genitive relation with the Noun 'pAlu' (milk), it is being incorrectly annotated as a place/location (k7p) by MALT. MST, on the other hand, is seen to annotate such cases correctly.

In Sect. 4.1 we discussed some examples of the parser outputs comparing the performance of the two parsers. Though there are other categories in which one of the parsers performs better than the other, it is beyond the scope of the paper to discuss all of the examples.

4.2   Parser Specific Errors

In this section we discuss the algorithm specific statistics and errors of both the parsers. The purpose of this is to analyze the type of sentences each parser is comfortable with. The parameters for this are Sentence Length, Arc Length, Arc Depth and Arc Type, four major factors known to contribute to errors in data driven dependency parsing [3]. Due to lack of Non-projective sentences in Telugu treebank data [19], we are not discussing Non-projectivity in this paper.

4.2.1   Sentence Length

Table 2 shows the percentage of errors for both the parsers relative to sentence length.

Observations:
As can be noted from the above Table 2, MALT performs better than MST on shorter sentences. The reason for this could be that the greedy inference algorithm that Malt parser employs has to make less parsing decisions with shorter sentences as compared to longer sentences. For the longer sentences, the near exhaustive search algorithm that MST uses is more effective. Also, the rich feature representation of MALT helps with it having less error propagation issues than MST parser [3].

Table 2. Parser errors for different sentence lengths

S.L	MALT	MST
1	0	0
2	38	34
3	56	58
4	68	70
5	65	75
6	100	100
7	100	100
8	0	100

4.2.2   Arc Length

Arc length corresponds to the linear distance between a head and its child. Table 3 shows the number of labeled arcs each parser is getting wrong over all arc lengths against the type of the arcs.

- Most of the errors for both MALT and MST are at Arc lengths 1 and 2.
- For MALT, 4% of arcs with length 3 and 4, 34% with length 1 and 23% with length 2 are incorrect. For MST, 7% arcs with length 3, 5 % with length 4, 35% with length 1 and 25% with length 2 are incorrect.
- More than 95% of verbal Complement and Adjunct errors are being shared by lengths 1 and 2 for both the parsers

Table 3. Parser errors for different Arc lengths

	MALT					MST				
Arc.length	1	2	3	4	5–8	1	2	3	4	5–8
Main(290)	1	0	0	2	0	1	0	0	1	0
Complements(379)	68	27	5	0	0	62	34	8	1	0
Adjuncts(183)	22	23	1	1	0	34	25	4	1	0
Noun Mods(10)	10	0	0	0	0	4	0	0	0	0
Adj Mods(0)	0	0	0	0	0	0	0	0	0	0
Apposns(0)	0	0	0	0	0	0	0	0	0	0
Genitive(16)	5	0	0	0	0	2	0	0	0	0
Conj(25)	0	0	0	0	0	4	0	0	0	0
Comp Preds(32)	27	0	0	0	0	25	0	0	0	0
Rel Clause(4)	3	0	0	0	0	4	0	0	0	0
Coord(8)	0	2	0	0	0	0	0	0	0	0
Subord(0)	0	0	0	0	0	0	0	0	0	0

- Both report Noun modification, Genitive, Complex Predicate and Relative clausal errors at length 1.
- Intra-clausal coordination errors do not show in MALT, but MST is getting 16% of them wrong at length 1.
- While MALT reports Inter-clausal coordination errors at length 2 MST does not have any.

Observations:
From Table 3, MALT performs better than MST for short arcs. This can be explained by the fact that shorter arcs are created before the longer arcs in MALT's greedy parsing algorithm and hence are less prone to error propagation [3]. Given that the average sentence length in the treebank is relatively small, nothing conclusive can be said about longer arcs.

4.2.3   Arc Depth

The depth of an edge is the level at which it is situated in the tree. Table 4 shows the number of labeled arcs each parser is getting wrong over all arc depths against the type of the arcs.

- Most of the errors occur at Arc depth 1 and 2 for both the parsers.
- For MALT, 25% and 33 % of arcs with depth 3 and 4 respectively, 21% with depth 1 and 29% with depth 2 are incorrect. For MST 22% arcs with depth 3, 55% with depth 4, 23% with depth 1 and 27% with depth 2 are incorrect.
- For both the parsers around 95% of verbal complement errors are shared between Arc Depths 1 and 2.
- 85% of Adjunct errors are at depth 1, for MALT, MST stands at 81%.
- Both parsers report all the nmod errors at depth 2.

Table 4. Parser errors for different Arc depths

	MALT					MST				
Arc.Depth	1	2	3	4	5–8	1	2	3	4	5–8
Complements	79	15	6	0	0	85	15	4	1	0
Adjuncts	40	5	0	2	0	52	8	1	3	0
Noun Mods	0	10	0	0	0	0	4	0	0	0
Adj Mods	0	0	0	0	0	0	0	0	0	0
Apposns	0	0	0	0	0	0	0	0	0	0
Genitive	0	4	1	0	0	0	1	1	0	0
Conj	0	0	0	0	0	0	4	0	0	0
Comp Preds	24	2	1	0	0	22	2	1	0	0
Rel Clause	0	2	0	1	0	3	0	1	0	0
Coord	2	0	0	0	0	0	0	0	0	0
Subord	0	0	0	0	0	0	0	0	0	0

- All the Genitive errors for both occur at arc depths 2 and 3 with MALT showing more errors at depth 2 than MST.
- Intra-clausal coordination errors do not crop up in MALT parser, MST is reporting all of its errors at depth 2.
- Both the parsers are reporting most of their Complex Predicate errors at depths 1.
- All Relative clausal errors at for MALT are at depths 2 and 4, while MST is showing them at 1 and 3.
- MALT reports Inter-clausal coordination errors, at depth 1 while MST does not have any.

Observations:
Table 4 shows that for arcs close to the root, MST is more precise than MALT and vice-versa for arcs further away from the root. This could be because the dependency arcs further away from the root are being constructed early in the parsing algorithm of MALT [3]. MALT's reduced likelihood of error propagation and rich feature representation help the parser in this case.

4.2.4   Arc Type
Arc type is defined by its dependency label. Appendix 1 [4] lists the different types of labels that are used to annotate the treebank. Given their number, they are classified into coarser classes Husain [4] (Table 5).

Table 5. Percentages of Arc type errors

Arc.Type	Telugu	
	MALT	MST
Main	1	.6
Complements	25	27
Adjuncts	25	35
Noun modifiers	100	40
Adj modifiers	0	0
Apposition	0	0
Genitives	31	12
Conjunctions	0	16
Complex predicates	84	78
Relative clauses	75	100
Coordination	25	0
Subordination	0	0

4.2.5   Observations

It can be inferred from Table 5 that MALT performs better on Main, complements, Adjuncts, Intra-clausal coordination and Rel clause type of arcs while MST is better on Noun Modifiers, Genitives, Complex Predicates, Inter-clausal coordination arcs. It is surprising that MST is doing better with classes that require shorter arcs like Noun modifiers and Genitives given that MALT is meant to perform better for shorter arcs. Due to lack of implicit cues in Telugu, the words that are failing to attach as modifiers are being connected to roots. MALT over-predicting root modifiers could be blamed for this. Also MST doing better with Inter-clausal errors, could be due to the their arc lengths being usually big and close proximity to the root [3].

5   Conclusion and Future Work

We presented the differences in errors caused by both MALT and MST parsers for Telugu Language. Through the insights we gained from these experiments, we hope to build a better parsing system as well as a more informed treebank for Telugu. In the future we plan to extend these experiments to other Indian languages - Hindi, Bengali and Urdu etc.

Acknowledgment. We thank Riyaz Ahmad Bhat, Vigneshwaran Muralidharan and Irshad Ahmad Bhat for their assistance and comments that greatly improved the manuscript.

References

1. Nivre, J., Hall, J., Nilsson, J., Chanev, A., Eryigit, G., Kübler, S., Marinov, S., Marsi, E.: Maltparser: a language-independent system for data-driven dependency parsing. Nat. Lang. Eng. **13**, 95–135 (2007)
2. McDonald, R., Pereira, F., Ribarov, K., Hajič, J.: Non-projective dependency parsing using spanning tree algorithms. In: Proceedings of the Conference on Human Language Technology and Empirical Methods in Natural Language Processing. HLT 2005, Stroudsburg, PA, USA, Association for Computational Linguistics,pp. 523–530 (2005)
3. McDonald, R.T., Nivre, J.: Characterizing the errors of data-driven dependency parsing models. In: Proceedings of the 2007 Joint Conference on Empirical Methods in Natural Language Processing and Computational Natural Language Learning EMNLP-CoNLL, pp. 122–131 (2007)
4. Husain, S., Agrawal, B.: Analyzing parser errors to improve parsing accuracy and to inform tree banking decisions. Linguistic Issues in Language Technology, 7 (2012)
5. Vempaty, C., Naidu, V., Husain, S., Kiran, R., Bai, L., Sharma, D.M., Sangal, R.: Issues in analyzing Telugu sentences towards building a Telugu treebank. In: Gelbukh, A. (ed.) CICLing 2010. LNCS, vol. 6008, pp. 50–59. Springer, Heidelberg (2010). https://doi.org/10.1007/978-3-642-12116-6_5
6. Nivre, J.: Inductive dependency parsing. Text, Speech and Language Technology, vol. 34. Springer, Netherlands (2006)

7. Black, E., Jelinek, F., Lafferty, J., Magerman, D.M., Mercer, R., Roukos, S.: Towards history-based grammars: using richer models for probabilistic parsing. In: Proceedings of the Workshop on Speech and Natural Language. HLT 1991, Stroudsburg, PA, USA, Association for Computational Linguistics, pp. 134–139 (1992)

8. Kudo, T., Matsumoto, Y.: Japanese dependency analysis using cascaded chunking. In: Proceedings of the 6th Conference on Natural Language Learning, vol. 20. COLING 2002, Stroudsburg, PA, USA. Association for Computational Linguistics, pp. 1–7 (2002)

9. Chu, Y.J., Liu, T.H.: On shortest arborescence of a directed graph. Sci. Sinica **14**, 1396 (1965)

10. Edmonds, J.: Optimum branchings. J. Res. Natil Bur. Stan. B **71**, 233–240 (1967)

11. Eisner, J.M.: Three new probabilistic models for dependency parsing: an exploration. In: Proceedings of the 16th Conference on Computational Linguistics, vol. 1. Association for Computational Linguistics, pp. 340–345 (1996)

12. McDonald, R., Crammer, K., Pereira, F.: Online large-margin training of dependency parsers. In: Proceedings of the 43rd Annual Meeting on Association for Computational Linguistics. Association for Computational Linguistics, pp. 91–98 (2005)

13. Garapati, U.R., Koppaka, R., Addanki, S.: Dative case in Telugu: a parsing perspective. In: Proceedings of the Workshop on Machine Translation and Parsing in Indian Languages (MTPIL 2012), COLING 2012, pp. 123-132, Mumbai (2012)

14. Husain, S., Mannem, P., Ambati, B.R., Gadde, P.: The ICON-2010 tools contest on Indian language dependency parsing. In: Proceedings of ICON-2010 Tools Contest on Indian Language Dependency Parsing, ICON, vol. 10, pp. 1-8. Citeseer (2010)

15. Bharati, A., Chaitanya, V., Sangal, R., Ramakrishnamacharyulu, K.: Natural Language Processing: A Paninian Perspective. Prentice-Hall of India, New Delhi (1995)

16. Chaudhry, H., Sharma, H., Sharma, D.M.: Divergences in English-Hindi parallel dependency treebanks. DepLing **2013**, 33 (2013)

17. Ambati, B.R., Husain, S., Nivre, J., Sangal, R.: On the role of morphosyntactic features in Hindi dependency parsing. In: Proceedings of the NAACL HLT 2010 First Workshop on Statistical Parsing of Morphologically-Rich Languages. Association for Computational Linguistics, pp. 94–102 (2010)

18. Ambati, B.R., Gadde, P., Jindal, K.: Experiments in Indian language dependency parsing. In: Proceedings of the ICON-2009 NLP Tools Contest: Indian Language Dependency Parsing, pp. 32–37 (2009)

19. Bhat, R.A., Sharma, D.M.: Non-projective structures in Indian language treebanks. In: The 11th International Workshop on Treebanks and Linguistic Theories, Edições Colibri, pp. 25–30 (2012)

Gut, Besser, Chunker – Selecting the Best Models for Text Chunking with Voting

Balázs Indig[1,2(✉)] and István Endrédy[1,2]

[1] MTA-PPKE Hungarian Language Technology Research Group,
50/a Práter Street, Budapest 1083, Hungary
{indig.balazs,endredy.istvan}@itk.ppke.hu
[2] Faculty of Information Technology and Bionics,
Pázmány Péter Catholic University, 50/a Práter Street, Budapest 1083, Hungary

Abstract. The CoNLL-2000 dataset is the de-facto standard dataset for measuring chunkers on the task of chunking base *noun phrases (NP)* and arbitrary phrases. The state-of-the-art tagging method is utilising TnT, an HMM-based *Part-of-Speech tagger (POS)*, with simple majority voting on different representations and fine-grained classes created by lexicalising tags. In this paper the state-of-the-art English phrase chunking method was deeply investigated, re-implemented and evaluated with several modifications. We also investigated a less studied side of phrase chunking, i.e. the voting between different currently available taggers, the checking of invalid sequences and the way how the state-of-the-art method can be adapted to morphologically rich, agglutinative languages.

We propose a new, mild level of lexicalisation and a better combination of representations and taggers for English. The final architecture outperformed the state-of-the-art for arbitrary phrase identification and NP chunking achieving the F-score of 95.06% for arbitrary and 96.49% for noun phrase chunking.

Keywords: Phrase chunking · Voting · IOB labels
Multiple IOB representations · Sequential tagging · HMM · MEMM
CRF

1 Introduction

For chunking in English, the current state-of-the-art tagging method is 'SS05' [12]. It was tested on the *CoNLL-2000 dataset* [14] and achieves an F-score of 95.23% for chunking base noun phrases (NP) and 94.01% for arbitrary phrases. Its concept is based on the lexicalisation of Molina and Pla [6] and voting between multiple data representations (see details in Sect. 2.1). However, the paper leaves multiple questions unanswered. How well does a basic tagger (originally developed for POS tagging), like TnT [2], maintain well-formedness of chunks? How converters handle ill-formed input? And so on. In this paper, we also investigate the independent impact of voting without lexicalisation to F-scores. We also evaluate multiple taggers (with different underlying methods), with respect to the number of invalid sequences created during tagging and converting between representations.

© Springer International Publishing AG, part of Springer Nature 2018
A. Gelbukh (Ed.): CICLing 2016, LNCS 9623, pp. 409–423, 2018.
https://doi.org/10.1007/978-3-319-75477-2_29

2   Introduction to Chunking

The task of applying tags to each token in a sentence consecutively is called *sequential tagging*. In general, the tagger tries to assign labels to (neighbouring) tokens correctly. The well-known special cases of this task include Part-of-Speech tagging, Named-Entity Recognition (NER) and chunking. In the latter two *IOB tags* are used to determine a *well-formed one level bracketing* on the text. In each sentence each token has an assigned label, which indicates the *beginning (B)*, *inside (I)*, *end (E)* of a chunk. One may distinguish the *outside (O)* of a sought sequence and one-token-long, *single (S or 1)* chunks. In addition, each marked sequence (except the outside labels) may have a type that corresponds to the task (*typed* or *untyped* case).

Table 1. Multiple IOB representations: An example sentence from the training set represented with five different IOB label sets

Word	IOB1	IOB2	IOE1	IOE2	IOBES
These	I	B	I	E	S
include	O	O	O	O	O
,	O	O	O	O	O
among	O	O	O	O	O
other	I	B	I	I	B
parts	I	I	I	E	E
,	O	O	O	O	O
each	I	B	I	I	B
jetliner	I	I	E	E	E
's	B	B	I	I	B
two	I	I	I	I	O
major	I	I	I	I	O
bulkheads	I	I	I	E	E
,	O	O	O	O	O

2.1   Representation Variants

Numerous representations exist, which try to catch information differently (see Table 1 and [15] for details). There are variants which are basically the same. We try to follow the most convenient form of these. For example, there exists the *BILOU format*[1] (B=B, I=I, L=E, O=O, U=S) or the *bracket variant* ([=B, I=I,]=E, O=O, []=S) which are equivalent to the *SBIEO*[2] format. In this paper, we prefer the name *IOBES*. *IOB2* format is commonly referred to as *IOB* or *BIO* or *CoNLL format*. There also exists the *Open-Close* notation (O+C or OC for short), which is roughly the same as bracket variant ([=B, I=O,]=E, O=O, []=S). In the untyped case, however, where only one type of chunks is sought

---

[1] Begin, Inside, Last, Outside, Unique.
[2] Single, Begin, Inside, End, Outside.

one can not make a difference between the outside and the inside chunks, so the whole representation relies on the right positioning of opening and closing tags, which makes this representation very fragile. Besides, there are two inferior representations the *Inside-Outside* notation (IO) and the *prefixless* notation, where the tag consists only of the chunk type. These variants can not distinguish between subsequent chunks of the same type, but is easy to use (for searching nonconsecutive chunks) due to their simplicity.

2.2   Conversion of (Possibly Invalid) IOB Sequences

POS taggers and IOB taggers might have many characteristics in common, but they are significantly different. IOB tags have an important substantial property: *well-formedness*, which relate tags at the intra-token level and can easily be inspected, while POS tags have no such property. In the literature, there is no mention of the proper handling of invalid sequences, especially during conversion between formats. However, lexicalising increases the number of invalid tag sequences assigned by simple taggers, because as the number of labels grows the sparse data problem arises.

At the time of writing this paper, we only found one tool that is publicly available and able to convert between multiple IOB representations: *IOBUtils*[3], which is the part of the *Stanford CoreNLP tools* written by Christopher Manning [5]. This converter is written in *JAVA 1.8* and has no external interface to use by its own. It can handle labelsets IOB1, IOB2, IOE1, IOE2, IO, SBIEO/IOBES and BILOU with or without type (e.g. B-VP) and it seemed robust and fault-tolerant. IOBUtils also has some similarities with the official evaluation script of CONLL-2000 task. First, it converts the chunks from the IOB labels into an intermediate representation and then transforms them into the required IOB labelset.

The original source of the SS05 approach (see Sect. 2.3 for details) served as another, independent implementation written in *Perl*. We still decided to create our own tool for conversion (inspired by the Perl version) in *Python 3* (to enable us to verify thoroughly the speed and robustness of the method). We also examined IOBUtils, a converter based on a rather different concept, jointly with the original SS05 and our approach. The latter two converters use the fact that two neighboring tags of each label and the current tag can unambiguously determine the result of the conversion. This method seemed to be very clumsy and fragile compared to IOBUtils because a big number of corner cases needed to be handled properly.

In our measurements, we compared the three converters on a realistic data which could be malformed (the intermediate stages of the SS05 algorithm). To help later reproduction and application of our results, we made all three converters available freely along with our whole pipeline.

---

[3] https://github.com/stanfordnlp/CoreNLP/blob/master/src/edu/stanford/nlp/sequences/IOBUtils.java.

2.3   State-of-the-Art Chunker for English

For chunking English, the CoNLL-2000 shared task [14] is the de-facto standard for measuring and comparing taggers. The current state-of-the-art method, *SS05* [12], uses this dataset with every 10th sentence separated from the training set for development. We used this method as our baseline.

The concept of SS05 has three basic steps. First, the *lexicalisation*: every IOB label is augmented with POS tags and with words which are more frequent than threshold of 50[4] (see Table 2 for details). Second, *the conversion and tagging*: the lexicalised tags converted to the five IOB formats, then tagging is performed with each format separately with TnT tagger [2]. In the third step, the five outputs are converted to a common format and *voted* by simple majority voting and the resulting tag becomes the final output. This method yields an F-score of 95.23% for NPs and 94.01% for arbitrary phrases. We mainly followed this concept, if it is not indicated otherwise.

Table 2. Lexicalisation: every IOB label is augmented with the POS tag, and (above a given frequency threshold that was set to 50 in SS05) also with the word as well (Full), and our lighter, mild version with less labels (just words) where just the labels of the frequent words is modified. We use '+' for separator because it is easier to parse than '-' used originally in SS05.

	Unlexicalised		Lexicalised			
	Original format		Full		Mild (just words)	
Word	**POS**	**IOB Label**	**POS**	**IOB Label**	**POS**	**IOB Label**
Rockwell	NNP	B-NP	NNP	NNP+B-NP	NNP	B-NP
said	VBD	O	VBD	O	VBD	O
the	DT	B-NP	the+DT	the+DT+B-NP	the+DT	the+DT+B-NP
agreement	NN	I-NP	NN	NN+I-NP	NN	I-NP

We were given the original Perl source code of SS05[5], which helped to understand better the workflow and the undisclosed details. Our first step was to reproduce their results, but the package did not contain the specific version of TnT that they were using, so we applied a potentially different version of the original Brants implementation that we had access to. The review of their code resulted several bug fixes (in the Perl code), and a somewhat parallel (highly extended), refactored Python code that finally could reproduce their results, but with smaller F-scores.

---

[4] The authors did not disclose why they chosen 50 as threshold and whether they observed invalid sequences or not.

[5] The full, original Perl source code is available at our github page https://github.com/ppke-nlpg/SS05 with the permission of the authors.

3   Adaptation of Lexicalisation (to Agglutinative Languages)

For English SS05 is a fast (but not freely available) method and has the clear advantage over other competing taggers. The question comes naturally: can it be used for agglutinative languages as well? The lexicalisation part of the method generates finer label classes by adding lexical information to the POS tags and labels of frequent words and POS tags to the labels of non-frequent words (see Table 2). Due to this procedure, there is an 18 times increase in the number of labels (from 23 to 422). Unfortunately, this high number of classes makes it impossible to use or reasonably slows down most of the taggers available for English, because of the exponential growth of learning time that can be observed for example in *Maximum Entropy-based (ME) learners*, which have been shown effective for Hungarian maximal NP chunking [3]. In ME training a multi-label learning problem is broken down into a number of independent binary learning problems (one per label) which makes the training process exponentially slower.

This problem with a high number of classes holds particularly true for agglutinative languages like Hungarian. If we want to apply the SS05 algorithm to agglutinative languages (to enjoy its benefits on speed and performance compared to other taggers), we can not use lexicalisation or we rule out most of the (state-of-the-art) taggers because of the *slow-down effect* due to the highly increased number of classes. For example the number of tags describing a token morphosyntactically in Hungarian in the Humor tagset [7] exceeds 200, which is one order of magnitude greater than the 36 Penn POS tags used for English.

Lexicalisation, combined with the large number of frequent words and the forms of IOB tags, makes the number of labels so high that it can not be handled easily with any tagger, e.g. Brant's TnT tagger has the upper limit of 2048 tags and ME taggers are also unsuitable because of the exponential slow-down mentioned above. This fact makes the method unsuitable for complex languages such as Hungarian.

In order to be able to tackle the challenge arousing from the high number of labels in agglutinative languages, we made experiments with both zero and mild lexicalisation. In the latter case just the frequent words (referenced as *just words*) and their labels are lexicalised[6], the other labels are left untouched (see Table 2 for example) instead of the original *full lexicalisation* where the labels of the non-frequent words were augmented with their POS tag. (See Table 2 for details.)

4   Multiple Taggers Voted

Since the results of SS05 were published, many good taggers have been made available. The original authors wanted to show the impact of their method on an

---

[6] The label of those word and POS combinations, that were above some threshold were augmented with the word and POS tag. (See Table 2 for details.) The threshold was selected to be 50 by following the original SS05 paper.

ordinary tagger, but left the evaluation of better taggers to the reader. As TnT is being a surprisingly fast POS tagger, we have no doubt that their method was extremely fast and efficient.

Our primary goal was to create a freely available, fast, and adaptable solution, that performs in pair or better than SS05 for English and can be evaluated on other (agglutinative) languages as well to measure the impact of voting on different representations with a wide spectrum of taggers. Therefore, we gathered freely available taggers with different inner-workings and compared them to TnT at the aforementioned levels of lexicalisation.

The following sections contain the brief introductions of the used taggers. We also wanted to evaluate the performance of some taggers, that represent the state-of-the-art on an agglutinative language to examine if they yield some additional improvement to our tagging task.

4.1   TnT

TnT is originally a POS tagger created by Thorsten Brants in 2000 [2]. The program is not freely available, closed source and written in C/C++. The underlying method uses second-order Hidden Markov Machine (HMM) with an extra check for boundary symbols at the end of sentences to prevent *'loose end'*. Additionally, the program uses many trickery solution disclosed partly in the original paper. The program also uses a *guesser*, that tries to model suffixes of rare unseen words, but this feature has no role in this experiment. We used a different version of this program from the one used originally for SS05, but with default settings replicating the original experiment.

4.2   NLTK-TNT

In search of a free substitute of TnT we found *Natural Language Toolkit for Python (NLTK)* [1] which implements a basic variant of TnT. This framework is written in Python and the TnT implementation lacks the handling of Unknown words (but has the API for a drop-in replacement, that was not necessary for this experiment) and using a simple HMM augmented with the checking of boundary symbols at the end of a sentence. The whole program is implemented in Python bearing in mind the simplicity instead of speed, therefore the tagging phase is very slow and unknown tags are substituted with the *Unk* symbol.

4.3   PurePOS

PurePOS [10] is a freely available, fast substitute of TnT, including most of the *advanced features* found in TnT. The program is implemented in Java and its speed is comparable with TnT. PurePOS is the state-of-the-art POS tagger for Hungarian [10], but to our knowledge it was never used for a task other than POS tagging. On its input the tagger needs word, stem and label tuples. As stem we used the word itself to make it impossible for the suffixguesser to have any impact on the experiments. We also experimented with the words instead of POS-tags and the combination of words and POS tags without success.

4.4   CRFsuite

We also tested *CRFsuite* [9], a first-order *Conditional Random Field* tagger. Conditional Random Fields (CRFs) is a popular method for general sequential tagging and CRFsuite is a fast, freely available implementation of first-order CRFs in C++. As input, one can define features for each token. The author has his own featureset for CoNLL-2000 task, which was evaluated as well (referred to as 'official CRFSuite'). There is also a slow-down with the increasing number of labels, but the running times are still feasible.

4.5   HunTag3

To evaluate different methods in our experiments we used *HunTag3* [3], a maximum entropy markov model (MEMM) tagger. The program consists of a simple Maximum Entropy (ME) unigram model combined with a first- or second-order Viterbi decoder. The user can define advanced features for tagging and the program can handle more than 2 million features. Its authors observed zero invalid tag sequence with it for Hungarian and English [3]. The program is written in *Python 3* and uses standard tools (SciPy [4], NumPy [16], Scikit-learn [11]) internally, therefore it is considerably fast without lexicalisation, but due to the maximum entropy approach many classes make the program exponentially slow. Therefore we did not use this tagger on lexicalised data.

The authors of *HunTag3* made it possible to use CRFSuite as an external tagger after the advanced featurization of *HunTag3*. We also included this method to our experiments. *HunTag3* (including CRFSuite as an external tagger) is successfully surveyed for *maximal Noun Phrase (NP) chunking*[7] making *HunTag3* the current state-of-the-art chunker for Hungarian [3]. In the same paper, the program is also used for Hungarian named-entity recognition (NER) and for English chunking on the CoNLL-2000 dataset too[8] [3].

5   The Test Bench

The original CoNLL-2000 data had to be prepared for processing: the development set of every tenth sentence was stripped and the set of frequent words was generated from the development set with the original threshold frequency of 50. We implemented an unified wrapper for each tagger and converter, which made it possible to use each tagger in conjunction with each converter for each representation on each lexicalisation level. The tagged data then were delexicalised and converted to the voting format, voted and then finally evaluated. We kept the evaluated data at each intermediate step, so we could get an insight of what is really happening and why. Furthermore, we also surveyed whether voting

---

[7] Where the top level NP in the parse tree is the sought.
[8] *HunTag3* with CRFsuite in this combination outperformed *HunTag3*, but did not overcome the state-of-the-art.

more independent taggers boosts F-scores or not. Voting was defined on different dimensions: for instance, voting between more taggers in one IOB format or between all the IOB formats with one tagger[9].

During the implementation we found that a good converter can be a crucial part of the system, and we found that not every bug could be squashed by testing every possible way of conversion on the original gold standard data.

6   Results

First we tried to reproduce the results of SS05 with the code we got. Since TnT was not the part of the package, we could not reproduce the exact same numbers of SS05 even with the original code. The numbers we got were significantly lower, because of the multiple bugs we found and fixed in the converting routines. The reproduced numbers, with the fixed programs (as the part of the full experiment) are in [bracketed typewriter style]. See Tables 3, 5, 6 and 8.

6.1   Lexicalisation Alone

To measure the gain on voting, we ran each tagger on each representation and lexicalisation level solely. We found that *CRFSuite with the official features* unanimously took the lead, and it turned out, that the best lexicalisation level was the mild lexicalisation (*just words*), where only the frequent words were lexicalised. See Tables 3, 4 and 5.

Table 3. We tested each tagger on their own with **no lexicalisation**. The reproduced results of SS05 are in [bracketed typewriter style], the best F-scores are in *italics*. In all cases *official CRFSuite* performed best.

	TnT	NLTK TnT	HunTag3 bigram	HunTag3 trigram	HunTag3 CRFSuite	Official CRFSuite	PurePOS
IOB1	[82.23]	82.10	91.73	92.00	92.41	*92.84*	84.28
IOB2	[84.27]	84.84	92.40	92.75	92.84	*93.40*	84.92
IOE1	[78.83]	78.75	91.85	92.05	92.10	*92.92*	84.39
IOE2	[81.45]	81.81	92.37	92.81	92.77	*93.25*	86.75
IOBES	[86.95]	87.58	93.26	93.47	93.41	*93.79*	87.85

---

[9] The full source code of the test bench is available freely at https://github.com/ppke-nlpg/gut-besser-chunker.

Table 4. We tested each tagger on their own with **mild lexicalisation (just words)**. The best F-scores are in *italics* and all F-scores above 94% are **bold**. In all cases *official CRFSuite*, performed best.

	TnT	NLTK TnT	HunTag3 bigram	HunTag3 trigram	HunTag3 CRFSuite	Official CRFSuite	PurePOS
IOB1	87.39	87.33			93.20	*94.13*	88.33
IOB2	88.67	88.69			93.85	*94.70*	88.82
IOE1	87.06	87.00			93.35	*94.09*	88.50
IOE2	88.95	89.14			**94.13**	*94.61*	90.27
IOBES	90.23	90.66			**94.28**	*94.94*	91.04

Table 5. We tested each tagger on their own with **full lexicalisation**. The reproduced results of SS05 are in [bracketed typewriter style], the best F-scores are in *italics* and all F-scores above 94% are **bold**. In all cases *official CRFSuite*, performed best.

	TnT	NLTK TnT	HunTag3 bigram	HunTag3 trigram	HunTag3 CRFSuite	Official CRFSuite	PurePOS
IOB1	[91.12]	91.00			92.64	*93.65*	91.42
IOB2	[91.33]	91.32			93.21	*94.03*	91.34
IOE1	[91.17]	91.04			92.94	*93.65*	91.35
IOE2	[91.36]	91.40			93.44	*94.12*	91.58
IOBES	[91.43]	91.61			93.35	*94.16*	91.65

6.2 Different Representations Voted Against Each Other

Each training and test data were converted to all representations with all lexicalisation levels, tagged with each tagger and we selected each five representation separately and the remaining representations were converted by each converter and voted by simple majority voting (as in SS05). We also examined the performance of the converters, but apart from some bugs, there were no significant

Table 6. We tested each tagger with simple majority voting with **no lexicalisation**. The reproduced results of SS05 are in [bracketed typewriter style], the best F-scores are in *italics*. In all cases *official CRFSuite* performed best.

	TnT	NLTK TnT	HunTag3 bigram	HunTag3 trigram	HunTag3 CRFSuite	Official CRFSuite	PurePOS
IOB1	[84.40]	84.64	92.60	92.83	93.11	*93.42*	85.47
IOB2	[84.47]	84.70	92.69	92.84	93.11	*93.45*	85.52
IOE1	[84.46]	84.70	92.62	92.84	93.09	*93.39*	85.50
IOE2	[84.44]	84.74	92.66	92.81	93.12	*93.42*	85.52
IOBES	[85.50]	85.64	93.03	93.17	93.32	*93.67*	86.11

performance differences. We found that the mild (just words) lexicalisation performed best along with the official CRFSuite tagger and voting did not improved overall results compared to sole taggers. See Tables 6, 7, 8 and 9.

In many cases during voting there was a draw between the taggers (2×3 or 2×2 identical votes and one tagger with a third opinion) and the number of votes were equal. The original paper and code of SS05 method has not clarified how these situations were handled. We used lexical ordering on tags by default because the code suggested us this was happened. Although we think

Table 7. We tested each tagger with simple majority voting with **mild lexicalisation** (**just words**). The best F-scores are in *italics* and all F-scores above 94% are **bold**. In all cases *official CRFSuite*, performed best.

	TnT	NLTK TnT	HunTag3 bigram	HunTag3 trigram	HunTag3 CRFSuite	Official CRFSuite	PurePOS
IOB1	88.58	88.65			**94.15**	*94.68*	89.19
IOB2	88.65	88.72			**94.17**	*94.70*	89.23
IOE1	88.63	88.72			**94.14**	*94.68*	89.23
IOE2	88.59	88.68			**94.18**	*94.70*	89.26
IOBES	89.27	89.36			**94.51**	*95.06*	89.77

Table 8. We tested each tagger with simple majority voting with **full lexicalisation**. The reproduced results of SS05 are in [bracketed typewriter style], the best F-scores are in *italics* and all F-scores above 94% are **bold**. In all cases *official CRFSuite*, performed best.

	TnT	NLTK TnT	HunTag3 bigram	HunTag3 trigram	HunTag3 CRFSuite	Official CRFSuite	PurePOS
IOB1	[91.73]	91.63			93.76	*94.33*	91.77
IOB2	[91.74]	91.64			93.75	*94.32*	91.77
IOE1	[91.73]	91.66			93.74	*94.31*	91.77
IOE2	[91.74]	91.67			93.75	*94.33*	91.78
IOBES	[92.18]	92.08			93.96	*94.65*	92.20

Table 9. Average gain on voting compared to sole tagging on each representation for each tagger. The table shows no real difference (<1%) with lexicalisation and for taggers other than TnT.

	TnT	NLTK TnT	HunTag3 bigram	HunTag3 trigram	HunTag3 CRFSuite	CRFSuite Official	PurePOS
No lex.	1.908	1.868	0.398	0.282	0.444	0.23	−0.014
Just words	0.284	0.262			0.468	0.27	−0.056
Full lex.	0.542	0.462			0.676	0.466	0.39

not handling draws in voting by not setting any *tie breaking rule* is a generally bad idea as it makes the method unpredictable as this important detail is left to the implementer. Therefore we set the following tie breaking rule: in case of tie the first voter is right. In Sect. 6.2 IOB2 was always chosen to be first as the least converted representation achieving no significant change in results. But in Sect. 6.2 *CRFSuite tagger with its official features* was chosen to be first causing the results to unanimously improve with about 1% for each representation (which we consider to be an artefact).

6.3 Different Taggers Voted Against Each Other

We voted different taggers against each other in each representation on each lexicalisation level to examine if using more taggers can add information to voting or not. None of the scores performed as good as *CRFSuite official* on its own. See Table 10 for details.

Table 10. Different taggers were voted against each other in each representation on each lexicalisation level. The best F-scores are in *italics* and all F-scores above 94% are **bold**. In all cases *mild lexicalisation (just words)*, performed best.

	No lexicalisation	Just words lexicalised	Full lexicalisation
IOB1	91.89	*93.55*	93.45
IOB2	92.61	***94.11***	93.81
IOE1	92.00	*93.56*	93.51
IOE2	92.75	***94.36***	**94.04**
IOBES	93.32	***94.60***	**94.03**

6.4 Converters

At the time of writing this paper there is only one converter available (CoreNLP IOBUtils). We implemented our own version and used the original SS05 code too. After comparing the three converters we found that IOBUtils is a robust and error correcting converter. We fixed a lot of bugs in the other two implementations, reducing the differences between them. Our converter was able to count invalid tag sequences[10]. This made possible it to show how lexicalisation affect the number of invalid sequences on different levels of lexicalisation. See Table 11 for details.

---

[10] By counting we mean that we counted the tags with invalid neighbours.

Table 11. Number of invalid sequences (no lex./just words/full lex.): we can observe, that by tagger, representation and lexicalisation the numbers differ in wide ranges, but it is clear that lexicalising makes the taggers harder to produce valid sequences.

	TnT	NLTK TnT	HunTag3 CRFSuite	CRFSuite official	PurePOS
IOB1	168/234/319	148/230/313	286/260/266	306/294/304	197/274/317
IOB2	423/662/634	490/658/633	0/19/168	0/14/111	0/4/49
IOE1	0/1/1	0/2/2	4/13/11	0/2/2	0/0/0
IOE2	174/107/205	187/84/215	0/44/254	0/12/158	3/16/76
IOBES	862/805/985	647/702/898	2/95/865	2/51/521	2/22/210

6.5   Summary

Albeit voting does not add much to the F-scores, official CRFSuite tagger and mild lexicalisation (just words) voted on IOBES achived the best results both in arbitrary phrase chunking and NP chunking. See Table 12 for details. However, if we consider the training times we can not ignore the fact that on one hand, the lexicalisation is the best performing factor of the SS05 method and can not be omitted, but on the other hand, it is also the factor that slows down the whole algorithm.

The cornerstone of the lexicalisation is the right frequency threshold and lexicalisation rules. In our experiments we used the ones given by the SS05 algorithm as we only wanted to select the best tagger and also to eliminate the unneccessary part of the lexicalisation to be able to adapt easier for agglutinative languages. The current threshold is not a well-founded, language agnostic number. We think that, the deep investigation of the lexicalisation threshold could improve performance and also would optimise the training times. Unfortunately, searching for the right lexicalisation constant in conjunction with the best tagger would made the training times infeasible.

Given the above fact, the overhead caused by the voting is insignificant. Especially if one just want to train the tagger only once, because the tagging and voting not demand considerably more time.

Table 12. Summary of final F-scores, which outperformed the previous state-of-the-art results (+1% improvement)

Method	Arbitrary phrase chunking	NP chunking
SS05 [12]	94.01	95.23
SS05 (as we could reproduce)	91.74	93.99
Official CRFSuite + mild lex. (just words)	**95.06**	**96.49**

7   Conclusion

We reimplemented the state-of-the-art SS05 chunking method and asked for the authors for their source code to determine the parts of the algorithm that were undisclosed before in the original paper. On the one hand, we were able to understood their workflow. On the other hand, we could fix numerous bugs in their program and rerun all their tests. We tested a mild level of lexicalisation (just words) in conjunction with more taggers with different inner-workings. We tested multiple IOB label converters and fixed their bugs.

Our mild lexicalisation in conjunction with CRFSuite performed best across the taggers and lexicalisations. With voting these scores were further improved, however we found that simple majority voting in general does not add much to a combination of a good tagger with the appropriate level of lexicalisation[11].

We found that voting between all the examined taggers with simple majority voting can neither outperform the results of the best tagger alone in any representation nor on any level of lexicalisation. We examined the adaptation of the system to agglutinative languages with far more possible labels at lexicalisation and checked the number of invalid sequences created by taggers which has an influence on converters and therefore on the final result. We made our pipeline including the highly extended version of the SS05 algorithm fully available, to ease later reproduction and fine-tuning of our method.

8   Future Work

In English one must distinguish between *arbitrary phrase identification* where most of the tokens belongs to one of the many chunk types and the *base Noun Phrase (NP) chunking task* where the only sought type is the lowest level NP-s. In Hungarian the most sought chunk types are the *maximal Noun Phrases (NP)* where the top level NP in the parse tree is needed. In this paper we only could measure our method on the first two, but we think that in the future all three of the task could be examined adding the named-entity recognition (NER) task for both English and Hungarian as well. All of the aforementioned task can be investigated in the same way by the same programs evaluated here in terms of voting[12], but this task spans beyond this paper.

We think there is much room for improvement on the fine-tuning of the lexicalisation such as using word classes generated by word embedding, a nowadays popular method, for lexical categorisation [13] and use the same technique to improve the corpus quality [8].

We also think that the possibility of using decision trees or other meta-learners to the voting task should be considered as all the needed tools are available including the problem of the large number of draws during the voting. Additionally, there is also room for experimenting with other available taggers as with our mild lexicalisation less tagger is ruled out by the large number of labels.

---

[11] Which makes the whole method more adaptable to agglutinative languages.

[12] As the full pipeline is publicly available.

Acknowledgments. We would like to thank professor Anoop Sarkar [12] for his cooperation and providing the original Perl code and letting us to use and distribute it with our own. This helped a lot to understand their algorithm better.

References

1. Bird, S., Klein, E., Loper, E.: Natural Language Processing with Python. O'Reilly Media Inc., Sebastopol (2009)
2. Brants, T.: TnT: a statistical part-of-speech tagger. In: Proceedings of the Sixth Conference on Applied Natural Language Processing, pp. 224–231. Association for Computational Linguistics (2000)
3. Endrédy, I., Indig, B.: HunTag3: a general-purpose, modular sequential tagger - chunking phrases in English and maximal NPs and NER for Hungarian. In: 7th Language & Technology Conference, Human Language Technologies as a Challenge for Computer Science and Linguistics, pp. 213–218. Poznań: Uniwersytet im. Adama Mickiewicza w Poznaniu (2015)
4. Jones, E., Oliphant, T., Peterson, P., et al.: SciPy: Open source scientific tools for Python (2001). http://www.scipy.org/
5. Manning, C.D., Surdeanu, M., Bauer, J., Finkel, J., Bethard, S.J., McClosky, D.: The Stanford CoreNLP natural language processing toolkit. In: Proceedings of 52nd Annual Meeting of the Association for Computational Linguistics: System Demonstrations, pp. 55–60 (2014). http://www.aclweb.org/anthology/P/P14/P14-5010
6. Molina, A., Pla, F.: Shallow parsing using specialized HMMs. J. Mach. Learn. Res. **2**, 595–613 (2002)
7. Novák, A.: A new form of humor - mapping constraint-based computational morphologies to a finite-state representation. In: Proceedings of the Ninth International Conference on Language Resources and Evaluation (LREC 2014). ELRA, Reykjavik, Iceland (2014)
8. Novák, A.: Improving corpus annotation quality using word embedding models. Polibits **53**, 49–53 (2016). https://doi.org/10.17562/PB-53-5
9. Okazaki, N.: CRFsuite: a fast implementation of Conditional Random Fields (CRFs) (2007). http://www.chokkan.org/software/crfsuite/
10. Orosz, G., Novák, A.: Purepos 2.0: a hybrid tool for morphological disambiguation. In: RANLP, pp. 539–545 (2013)
11. Pedregosa, F., Varoquaux, G., Gramfort, A., Michel, V., Thirion, B., Grisel, O., Blondel, M., Prettenhofer, P., Weiss, R., et al.: Scikit-learn: machine learning in python. J. Mach. Learn. Res. **12**, 2825–2830 (2011)
12. Shen, H., Sarkar, A.: Voting between multiple data representations for text chunking. In: Kégl, B., Lapalme, G. (eds.) AI 2005. LNCS (LNAI), vol. 3501, pp. 389–400. Springer, Heidelberg (2005). https://doi.org/10.1007/11424918_40
13. Siklósi, B.: Using embedding models for lexical categorization in morphologically rich languages. In: Gelbukh, A. (ed.) Computational Linguistics and Intelligent Text Processing: 17th International Conference, CICLing 2016, Konya, Turkey, April 3-9, 2016. Springer International Publishing, Cham (2016)
14. Tjong Kim Sang, E.F., Buchholz, S.: Introduction to the CoNLL-2000 shared task: Chunking. In: Proceedings of the 2nd Workshop on Learning Language in Logic and the 4th Conference on Computational Natural Language Learning, ConLL 2000, vol. 7, pp. 127–132. Association for Computational Linguistics, Stroudsburg, PA, USA (2000). http://dx.doi.org/10.3115/1117601.1117631

15. Tjong Kim Sang, E.F., Veenstra, J.: Representing text chunks. In: Proceedings of the Ninth Conference on European Chapter of the Association for Computational Linguistics, pp. 173–179. Association for Computational Linguistics (1999)
16. Van Der Walt, S., Colbert, S., Varoquaux, G.: The NumPy array: a structure for efficient numerical computation. Comp. Sci. Eng. **13**(2), 22–30 (2011)

Named Entity Recognition

A Deep Learning Solution to Named Entity Recognition

V. Rudra Murthy$^{(\boxtimes)}$ and Pushpak Bhattacharyya

Indian Institute of Technology Bombay, Mumbai, India
{rudra,pb}@cse.iitb.ac.in

Abstract. Identifying named entities is vital for many Natural Language Processing (NLP) applications. Much of the earlier work for identifying named entities focused on using handcrafted features and knowledge resources (feature engineering). This is a barrier for resource-scarce languages as many resources are not readily available. Recently, Deep Learning techniques have been proposed for various NLP tasks requiring little/no hand-crafted features and knowledge resources, instead the features are learned from the data. Many proposed deep learning solutions for Named Entity Recognition (NER) still rely on feature engineering as opposed to feature learning. However, it is not clear whether the deep learning system or the engineered features are responsible for the positive results reported. This is in contrast with the goal of deep learning systems *i.e.,* to learn the features from the data itself. In this study, we show that a feature learned deep learning system is a viable solution to NER task. We test our deep learning systems on CoNLL English and Spanish NER datasets. Our system is able to give comparable results with the existing state-of-the-art feature engineered systems for English. We report the best performance of 89.27 F-Score for English when comparing with systems which do not use any handcrafted features or knowledge resources. Evaluation of our trained system on out-of-domain data indicate that the results are promising with the reported results. Our system when tested on Spanish NER achieves the best reported F-Score of 82.59 indicating its applicability to other languages.

Keywords: Deep learning · Named Entity Recognition
Recurrent Neural Network · Convolutional Neural Network

1 Introduction

Named Entity Recognition has been an important task in NLP. Various problems like Information Extraction, Question Answering, Machine Translation requires identifying named entities. Existing Supervised NER systems typically require large amounts of training data. Success of many of these systems depend on handcrafted features and knowledge resources in the form of gazetteers, part-of-speech taggers etc. This poses a major constraint for resource poor languages.

© Springer International Publishing AG, part of Springer Nature 2018
A. Gelbukh (Ed.): CICLing 2016, LNCS 9623, pp. 427–438, 2018.
https://doi.org/10.1007/978-3-319-75477-2_30

A diverse set of machine learning algorithms have been applied for tackling NER as part of CoNLL 2003 NER Shared Task Challenge [21]. The shared task saw use of many statistical models like Hidden Markov Models, Maximum Entropy Models, Conditional Random Fields (CRF), Voted Perceptrons, Recurrent Neural Networks. Many of the participants reported results on NER using system combination. Much of the focus was on using handcrafted features in the form of gazetteers, part-of-speech tags, affixes, capitalization features, chunk tags. The inclusion of these features and also system combination achieved best results on the shared task challenge.

Deep learning systems, which use multiple layers of neural network have shown promising results in various applications. The advantage offered by the deep learning technique is the requirement for little/no handcrafted features for these tasks. On the contrary, traditional machine learning algorithms typically rely on knowledge resources and lots of handcrafted features. The deep learning systems typically learn useful representations (feature learning) in an unsupervised way on a large unlabeled data which are then used as features in the supervised task.

Deep Learning techniques have also been successfully applied in various NLP tasks [3,4,16,17]. Word embeddings [10,11] are continuous low-dimensional dense vector representations of words learned in an unsupervised way. These word embeddings have been shown to capture various syntactic and semantic information about the word. Such representations when used in mainstream NLP applications have shown to perform better or achieve comparable performance compared to existing systems.

In this paper, we explore the use of a complete deep learning based approach for NER. We use Long Short Term Memory (LSTM) variant of Recurrent Neural Network along with Convolutional Neural Network (CNN) to train a NER system. The system uses only pre-trained word embeddings as input and and the decoder is much simpler compared to the decoder used in Senna [4]. Our experiments indicate that a feature learned deep learning approach is able to achieve closer to state-of-the-art results.

2   Related Work

System based on the combination of various machine learning algorithms was the best performing system [8] in CoNLL Shared task for English. This system used various features like part-of-speech tags, affixes, orthographic information, gazetteers, chunk information etc. Later a system based on multi-task approach [1] further improved the performance. The approach used a handful of features like part-of-speech tags, affixes, tokens in a syntactic window chunk etc.

A complete deep learning system, Senna [4] was proposed to solve various NLP problems. The model proposed to use a time-delay neural network which ran over words and a decoder layer calculating sentence-level likelihood using transition matrix at the top to find the best tag sequence at the sentence level. The model used pre-trained word embeddings and during training of the NER system the word embeddings were updated. The only features used by the system were pre-trained word embeddings, uppercase information and gazetteer list.

A new way to train phrase embeddings was proposed in the context of NER [12]. These phrase embeddings are then used with other features to train an NER system. The system used two CRFs, where the output from first CRF is given as input to the second CRF. This system achieves the best reported results on CoNLL Shared task data.

Bidirectional LSTMs [9] were also tested for NER task. They observe that vanilla Bidirectional LSTMs have the disadvantage of not being able to capture tag dependencies. They use a decoder layer similar to [4] at the top to capture the tag dependencies.

Another deep learning solution, CharWNN [14], to automatically retrieve relevant features from the character sequence forming a word in the context of part-of-speech tagging was proposed. They send the character sequence through a CNN and obtain character level features. These character level features are then augmented to the word embeddings. The model was later applied to Portuguese and Spanish NER [5].

A common theme emerging from most of the systems is feature engineering. Existing deep learning systems proposed also rely on feature engineering. Deep Learning approach with no feature engineering for NER as a viable solution needs to be explored. This is important because for many resource poor languages, there is unavailability of resources required and the need for creating handcrafted features. In this paper we study the feasibility of such a complete deep learning approach on English and Spanish NER. We also compare our approach with CharWNN [5] for Spanish NER as theirs is a complete deep learning approach.

3 Deep Learning NER

Given a sequence of word-entity label pairs i.e., $D = (X, Y)$ where $X = (x_1, \ldots, x_n)$ is the sequence of words in a sentence and $Y = (y_1, \ldots, y_n)$ is the corresponding tags. The task is to find the best possible named entity tag sequence t^* for a given sequence of words as in Eq. 1.

$$\underset{t^*}{\operatorname{argmax}} \; P(t|X) \tag{1}$$

This involves estimating the parameters of the conditional probability $P(Y|X)$. The conditional probability can be decomposed as in Eq. 2.

$$P(y_1, \ldots, y_n | x_1, \ldots, x_n) = \prod_{i=1}^{N} P(y_i | x_1, \ldots, x_n, y_{i-1}) \tag{2}$$

LSTMs have traditionally been favored for problems with sequential nature. For NER task the modeling using LSTM is given in Eq. 3,

$$P(y_1, \ldots, y_n | x_1, \ldots, x_n) = \prod_{i=1}^{N} P(y_i | g(x_1, \ldots, x_i)) \tag{3}$$

where g is a LSTM which extracts relevant features by looking at current word and all the previous words.

Since the above modeling does not take into account the information from the right context Bidirectional LSTM [15] is preferred as it is able to capture information from both the directions.

$$P(y_1, \ldots, y_n | x_1, \ldots, x_n) = \prod_{i=1}^{N} P(y_i | g(x_1, \ldots, x_i), h(x_i, \ldots, x_n)) \qquad (4)$$

Here both g and h are LSTMs and we use the same LSTM for both forward direction as well as backward direction. We call this architecture Bi-LSTM.

This is the simplest LSTM model usually used for Sequence Labeling tasks. The major disadvantage with the above approach is the independence assumption between successive tags i.e., the model does not account for $P(y_i | y_{i-1})$. For example in NER, modeling that *I-PER* tag always follows *B-PER* tag is important. Modeling of this dependence is crucial for successful application of LSTMs for sequence labeling task.

Later BI-LSTM-CRF [9] was proposed which used a CRF like layer which was added on top of LSTM layer to obtain the best tag sequence. They show that Bidirectional LSTM performs relatively poorly and the accuracy increases with the addition of CRF like top layer.

In our work, we instead use a *teacher training* model to capture the tag dependencies. We use Feedforward Neural network as the decoder. This decoder takes in representation from the Bidirectional LSTMs as well as the correct previous tag as input. This kind of architecture is similar to the decoder in Neural Machine Translation system [19]. During testing, Viterbi decoding is used to find the best possible tag sequence.

$$P(y_1, \ldots, y_n | x_1, \ldots, x_n) = \prod_{i=1}^{N} P(y_i | g(x_1, \ldots, x_i), h(x_i, \ldots, x_n), y_{i-1}) \qquad (5)$$

The architecture of the model is as shown in Fig. 1. The input to the system is pre-trained word embeddings. Forward LSTM reads the entire source sequence one word at a time left-to-right. Similarly backward LSTM reads the words from right-to-left. The hidden state of both the forward and backward LSTMs for a particular word is concatenated. Additionally true previous tag is concatenated and given to the decoder for predicting the NER tag.

3.1   Character nGram Features

The bidirectional LSTM model described above relies heavily on pre-trained word embeddings. It is clear that various character-level features like suffixes, uppercase information, presence of non-alphanumeric characters help in NER task. Unlike existing systems which augments these handcrafted features into the word embedding we follow the path of learning these features from the data [5, 14]. The major difference from our approach and theirs is the use of multiple

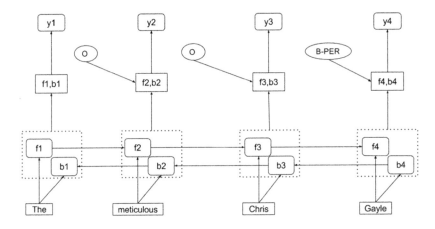

Fig. 1. Architecture of bidirectional LSTM for NER

region sizes. Unlike CharWNN [5, 14] where region size is kept to 5 characters, we extract features from convolutional layers for each region size from 1 to 4. The convolutional layer is followed by a max-pooling layer. These character-level features bring in additional information along with the word embeddings. The intuition for having parallel convolutional layers of varying region size is to extract relevant nGram character features. For example, convolutional layer looking at unigram characters, learns presence of uppercase characters and presence of non-alphanumeric characters. Convolutional layer looking at trigram characters tries to extract relevant trigram character features and need not worry about presence/absence of uppercase characters.

The architecture of CNN layer which runs over a unigram and extract multiple features is as shown in Fig. 2. Unlike CharWNN [5, 14] we do not have a common character lookup table before the convolutional layer.

The convolutional layer which runs over unigrams extracts character embedding but they serve a different purpose. By not having a common character lookup table we directly look for relevant ngram character sequences. The features extracted are augmented into the pre-trained word embeddings. We call this system *Feature Learned NER*.

3.2 Sparse Word Embeddings

The current neural embeddings are dense and uninterpretable. These dense embeddings capture many modalities of the word. Extracting relevant features from this dense representation for a particular task may require a complex model. To tackle this we would like to have a relatively sparse representation for the word but capturing much of the information from the dense representation.

There has been some work to obtain interpretable sparse representations from the word embeddings [7]. The approach uses sparse coding to obtain sparse interpretable word representations. Unlike the previous work which looks for

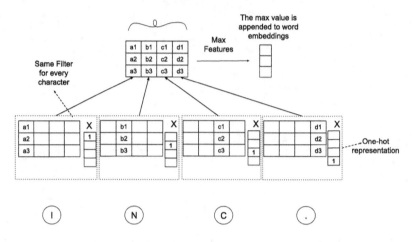

Fig. 2. Character level CNN for NER

interpretable word representations we use a relatively sparse word representations on NER task to evaluate its usefulness.

We learn a sparse word representation by sending the pre-trained word embedding through an autoencoder layer. We use Rectified Linear Units (ReLU) at the hidden layer followed by dropout layer [18]. By using ReLUs we hope to achieve sparsity in the hidden layer representation and dropout layer brings in regularization into the model. The sparse hidden layer representation obtained replaces the word embeddings in our model.

4 Experimental Setup

We begin the section by describing the datasets used in our experiments and the description of various hyper-parameters used in our experiments. The hyper-parameters are chosen by doing a random grid search and evaluation on the development set. In all our experiments the pre-trained word embeddings are not updated.

4.1 Word Embeddings

We have used the pre-trained Glove word embeddings [13]. For our experiments we choose the 50 dimensional word embeddings trained on Wikipedia and Gigaword. The word embeddings are available for around 400 K words. We also test using word embeddings from Senna [4] in our experiements. Since, senna word embeddings are fine-tuned for the NER task, we believe that the results would be better than glove word embeddings. Interesting observation would be performance of senna word embeddings when used on out-of-domain tasks. For Spanish NER, we use the publicly available Spanish word embeddings[1] [2]. The word embeddings are of 300 dimensional and trained using word2vec [10,11] tool.

---

[1] http://crscardellino.me/SBWCE/.

4.2   Corpus

We evaluate our system on the standard CoNLL 2003 English NER Shared Task data [21]. The data is distributed in the IOB format with B-XXX tag used when two mentions of the same entity appear next to each other. The data identifies 4 types of tags, Person, Location, Organization and Miscellaneous. This data is converted to standard IOB format where every entity begins with the tag B-XXX followed by I-XXX if it is a multiword named entity.

We test our trained model on out-of-domain datasets. MUC-7 dataset is the common choice for reporting out-of-domain performance. It is annotated for named entities like Person, Location and temporal entities like Date, Time and Number expressions like monetary units. We choose MUC7 Formal run for comparison with existing systems. Since the dataset has no Miscellaneous tags as in CoNLL 2003 data, we follow the same procedure as [22] where we keep only *Person, location, Organization* tags and all other tags as non-named entities (*O* tag). Similarly after prediction is done, we replace all predicted *Misc* tags by *O*.

We also test our Deep Learning model on Wiki50 Dataset [23]. The Wiki50 Dataset contains annotations for both Named Entities and Multiwords. The dataset was created from fifty wikipedia articles from different domains. We convert all non-named entity tags to *Others*. The Named Entity tags are converted into IOB Format. We consider the entire Wiki50 dataset for testing.

We compare the performance of our model with various existing systems. We compare the results of our model on out-of-domain datasets and the reported performance by other systems. For Wiki50 Dataset and MUC7 formal run, we run Senna [4] without tokenization and report the results.

For Spanish NER, we use the CoNLL 2002 Spanish corpus which was used for CoNLL 2002 Shared Task [20]. The data is similar to the English CoNLL 2003 data. The official splits are used as training, development and test files.

4.3   Parameter Setup

The number of hidden layer neurons were set to 200 in all our models. For Convolution layer at the character layer we considered ngrams of 1,2,3 and 4. Each convolutional layer extracted 15 sets of features from the character layer. The features extracted from convolution layers which run over ngram characters of 1,2,3 and 4 are concatenated and augmented to word embeddings. If any of the word had less than 4 characters we pad it with a special symbol. Pre-trained word embeddings are not updated in our model. For word with no word embeddings we kept it's corresponding word embedding as all zeroes. We used Adagrad [6] in all our experiments for optimization. After every epoch the cost on development set was monitored and when the cost increased, the learning rate was halved. We used Backpropagation Through Time algorithm without truncation for training. We observed that batch size of 1 gave the best results. We used dropout layer after the bidirectional LSTM layer in all our models.

For getting sparse word representation, the Glove word embeddings are sent through an autoencoder layer. We keep the number of hidden layer neurons to 500. The hyper-parameter space was between 100 to 600 dimensions.

Our goal was not to learn an interpretable sparse representation so we did not try using large number of hidden neurons. The training was stopped for 4 epochs to prevent the model from learning an identity function. This learned representation gave the best performance on the CoNLL English Shared task development set. The obtained representation was also evaluated on syntactic and semantic analogy tasks [11] and the results were lower than reported by [13].

For Spanish NER, we set the number of hidden layer neurons to 150. There are 4 parallel convolutional layers running over character ngrams of 1, 2, 3 and 4 and extracting 20 features each.

5 Results

In this section we analyze the obtained results. The Table 1 discusses the results obtained from Knowledge Lean Deep Learning models along with baseline systems and Deep Learning based state-of-the-art systems.

Table 1. CoNLL English NER shared task test results

System	F1 (%)
Senna [4] (no gazetteers)	88.67
Huang [9] (no gazetteers)	88.83
Passos [12]	**90.90**
Bi-LSTM	81.10
Bi-LSTM + Transition	83.30
Bi-LSTM + Character features	85.87
Feature learned NER (CNN Single Width)	88.19
Feature learned NER	88.90
Feature learned NER (Senna)	89.20
Feature learned NER + Sparse Embedding	89.27

The results obtained from the Vanilla Bi-LSTM with only pre-trained word embeddings are much lower compared to existing systems. The major source for these errors were the confusion with non-named entities and named entities. Another major source was missing word embeddings for many words forcing the network to predict correct tag only looking at the context. Adding transition features (previous tag information) into the system improves the results but the improvements in the results were not satisfactory.

Adding a Convolutional layer over the character sequence and using features obtained after max pooling along with word embeddings boosted the results for the Vanilla Bi-LSTM. The convolutional layer looked at ngrams of 3 characters.

Combining previous tag information with character level features improved the results further and brought into the regime of existing systems. Now instead of running a single convolutional layer over 3 characters having parallel convolutional layers each running over n characters (with $n = 1, 2, 3, 4$) gave a F-Score of 88.9. Using Senna embeddings as input we see an increase in F-Score to 89.2. The results indicate that our system is able to give comparable results to feature engineered systems.

Learning a sparse representation for words over using pre-trained word embeddings gave the best performance of our system. The results obtained are comparable with state-of-the-art systems. When we consider systems which does not use any handcrafted features or knowledge resources, our reported results are the best with a F-Score of 89.27. The results indicate that a complete deep learning approach is a viable solution for resource scarce languages.

The following table reports the results for Spanish NER. We report the results only for Feature Learned NER system which uses multiple character features along with transition features from *teacher training*. The results indicate that our feature learned deep learning model is able to achieve state-of-the-art performance for Spanish NER (Table 2).

Table 2. Results on Spanish NER: CoNLL 2002 test data

System	F1 (%)
CharWNN [5]	82.21
Feature learned NER	**82.59**

Out-of-domain Results. Here we report the results of our model on two out-of-domain datasets. The results are provided in Tables 3 and 4. Here we compare only those systems which were trained on CoNLL Shared Task dataset and the trained model was tested on MUC7 datasets. For Wiki50 dataset, we report the results only for Senna.

Table 3. Wiki50: Out-of-domain results

System	F1 (%)
Senna (with gazetteers)	54.26
Feature Learned NER	54.53
Feature learned NER + Sparse Embedding	52.34
Feature Learned NER (Senna)	**55.51**

On MUC7 and Wiki50 datasets, the results are lower compared to existing systems. On analysis on Wiki50 data, we find that a significant number of named entities do not have any corresponding word vectors. The statistics of unknown words present in different datasets for *Feature Learned NER* model is given in table 5. This makes our system to rely heavily on learned character features. As our model tries to learn the character level features in a supervised way it might not generalize well.

Table 4. MUC7 formal run: out-of-domain results

System	F1 (%)
CRF + Glove [13]	82.2
Senna (with gazetteers)	79.66
CRF [22]	**82.71**
Feature learned NER	80.53
Feature learned NER + Sparse Embedding	79.60
Feature learned NER (Senna)	79.49

Table 5. Statistics on known and unknown words in test set

Dataset	CoNLL	MUC7	Wiki50
Present + Tagged correct	37909	50145	70096
Present + Incorrect tagged	5430	4909	9345
Absent + Correct	3051	4227	16089
Absent + Incorrect	276	145	4838

Sparse word representations lost some of the information present in the original word embedding. This was also evident when these representations were tested for syntactic and similarity analogy tasks. This could be the reason for the poor performance on out-of-domain task. Our model outperforms Senna system on both MUC7 dataset and Wiki50 dataset.

6   Conclusion

In this paper, we have demonstrated that a feature learned deep learning system is a viable approach for NER. Our experiments show that a feature learned deep learning system gives comparable results with the existing state-of-the-art systems for English NER. When systems which do not perform feature engineering are considered, we achieve the best F-Score on CoNLL English NER task. The performance of our system on out-of-domain task is also encouraging. Best F-Score is observed for Spanish NER using our feature learned deep learning approach. This is an encouraging result for the applicability of a feature learned deep learning based NER system for resource scarce languages. We believe that learning both character-level features and gazetteer features in an unsupervised way is the way to go for improving the performance of this NER system on both in-domain and out-of-domain tasks. We would like to study the performance of our NER system on resource scarce and morphologically rich languages which presents a different challenge.

References

1. Ando, R.K., Zhang, T.: A framework for learning predictive structures from multiple tasks and unlabeled data. J. Mach. Learn. Res. **6**, 1817–1853 (2005)
2. Cardellino, C.: Spanish Billion Words Corpus and Embeddings, March 2016
3. Cho, K., van Merrienboer, B., Gülçehre, C., Bahdanau, D., Bougares, F., Schwenk, H., Bengio, Y.: Learning phrase representations using RNN encoder-decoder for statistical machine translation. In: Proceedings of the 2014 Conference on Empirical Methods in Natural Language Processing, EMNLP 2014, pp. 1724–1734 (2014)
4. Collobert, R., Weston, J., Bottou, L., Karlen, M., Kavukcuoglu, K., Kuksa, P.: Natural language processing (almost) from scratch. J. Mach. Learn. Res. **12**, 2493–2537 (2011)
5. dos Santos, C., Guimaraes, V., Niterói, R.J., de Janeiro, R.: Boosting named entity recognition with neural character embeddings. In: Proceedings of NEWS 2015 the Fifth Named Entities Workshop, p. 9 (2015)
6. Duchi, J., Hazan, E., Singer, Y.: Adaptive subgradient methods for online learning and stochastic optimization. J. Mach. Learn. Res. **12**, 2121–2159 (2011)
7. Faruqui, M., Tsvetkov, Y., Yogatama, D., Dyer, C., Smith, N.: Sparse overcomplete word vector representations. In: ACL 2015 (2015)
8. Florian, R., Ittycheriah, A., Jing, H., Zhang, T.: Named entity recognition through classifier combination. In: Proceedings of the Seventh Conference on Natural Language Learning (CONLL 2003) at HLT-NAACL 2003, vol. 4, , pp. 168–171. Association for Computational Linguistics (2003)
9. Huang, Z., Xu, W., Yu, K.: Bidirectional LSTM-CRF models for sequence tagging. CoRR, abs/1508.01991 (2015)
10. Mikolov, T., Sutskever, I., Chen, K., Corrado, G.S., Dean, J.: Distributed representations of words and phrases and their compositionality. In Advances in neural information processing systems, pp. 3111–3119 (2013)
11. Mikolov, T, Yih, W.-T., Zweig, G.: Linguistic regularities in continuous space word representations. In: HLT-NAACL, pp. 746–751 (2013)
12. Passos, A., Kumar, V., McCallum, A.: Lexicon infused phrase embeddings for named entity resolution. In: CoNLL-2014, p. 78 (2014)
13. Pennington, J., Socher, R., Manning, C.D.: Glove: global vectors for word representation. In: Proceedings of the 2014 Conference on Empirical Methods in Natural Language Processing (EMNLP 2014), pp. 1532–1543 (2014)
14. Santos, C.D., Zadrozny, B.: Learning character-level representations for part-of-speech tagging. In: Jebara, T., Xing, E.P. (eds.) Proceedings of the 31st International Conference on Machine Learning (ICML 2014), and JMLR Workshop and Conference Proceedings, pp. 1818–1826 (2014)
15. Schuster, M., Paliwal, K.K.: Bidirectional recurrent neural networks. IEEE Trans. Signal Process. **45**(11), 2673–2681 (1997)
16. Socher, R., Huang, E.H., Pennin, J., Manning, C.D., Ng, A.Y.: Dynamic pooling and unfolding recursive autoencoders for paraphrase detection. In: Advances in Neural Information Processing Systems, pp. 801–809 (2011)
17. Socher, R., Brody, H., Manning, C.D., Ng, A.Y.: Semantic compositionality through recursive matrix-vector spaces. In: Proceedings of the 2012 Joint Conference on Empirical Methods in Natural Language Processing and Computational Natural Language Learning, pp. 1201–1211. Association for Computational Linguistics (2012)

18. Srivastava, N., Hinton, G., Krizhevsky, A., Sutskever, I., Salakhutdinov, R.: Dropout: a simple way to prevent neural networks from overfitting. J. Mach. Learn. Res. **15**(1), 1929–1958 (2014)
19. Sutskever, I., Vinyals, O., Le, Q.V: Sequence to sequence learning with neural networks. In: Advances in Neural Information Processing Systems, pp. 3104–3112 (2014)
20. Tjong Kim Sang, E.F.: Introduction to the conll-2002 shared task: language-independent named entity recognition. In: Proceedings of the 6th Conference on Natural Language Learning, COLING 2002, Vol. 20, pp. 1–4. Association for Computational Linguistics (2002)
21. Tjong Kim Sang, E.F., De Meulder, F.: Introduction to the conll-2003 shared task: Language-independent named entity recognition. In: Proceedings of the Seventh Conference on Natural Language Learning (CONLL 2003) at HLT-NAACL 2003, pp. 142–147. Association for Computational Linguistics (2003)
22. Turian, J., Ratinov, L., Bengio, Y.: Word representations: a simple and general method for semi-supervised learning. In: Proceedings of the 48th Annual Meeting of the Association for Computational Linguistics, ACL 2010, pp. 384–394. Association for Computational Linguistics (2010)
23. Vincze, V., Nagy I., Berend, G.: Multiword expressions and named entities in the wiki50 corpus (2011)

Deep Learning Approach for Arabic Named Entity Recognition

Mourad Gridach$^{(\boxtimes)}$

Department of Computer Science, High Institute of Technology,
Ibn Zohr University, Agadir, Morocco
m.gridach@uiz.ac.ma

Abstract. Inspired by recent work in Deep Learning that have achieved excellent performance on difficult problems such as computer vision and speech recognition, we introduce a simple and fast model for Arabic named entity recognition based on Deep Neural Networks (DNNs). Named Entity Recognition (NER) is the task of classifying or labelling atomic elements in the text into categories such as Person, Location or Organization. The unique characteristics and the complexity of the Arabic language make the extraction of named entities a challenging task. Most state-of-the-art systems use a combination of various Machine Learning algorithms or rely on handcrafted engineering features and the output of other NLP tasks such as part-of-speech (POS) tagging, text chunking, prefixes and suffixes as well as a large gazetteer. In this paper, we present an Arabic NER system based on DNNs that automatically learns features from data. The experimental results show that our approach outperforms the model based on Conditional Random Fields by 12.36 points in F-measure. Moreover, our model outperforms the state-of-the-art by 5.18 points in Precision and gets very close results in F-measure. Most importantly, our system can be easily extended to recognize other named entities without any additional rules or handcrafted engineering features.

1 Introduction

Named Entity Recognition (NER) can be described as the task of labelling or identifying atomic elements in the text into categories such as Person, Location, Organization, etc. from large corpora. Named entities can be classified into three top-level categories according to DARPA's Message Understanding Conference: entity names, temporal expressions and number expressions (Chinchor et al. 1999).

- **Named Entities**: to determine proper names including Person, Organization and Location Names.
- **Temporal Expressions**: to determine absolute temporal expressions including Date and Time.
- **Number Expressions**: to determine two type of numeric expressions including Money and Percentage.

© Springer International Publishing AG, part of Springer Nature 2018
A. Gelbukh (Ed.): CICLing 2016, LNCS 9623, pp. 439–451, 2018.
https://doi.org/10.1007/978-3-319-75477-2_31

As mentioned earlier, in this paper, we focus on Arabic NER considered as the most useful for other natural language processing applications and largely discussed by researchers in this field. In addition, we will use the standard definition formulated in the shared task of the Conferences on Computational Natural Language Learning (CoNLL). In their sixth and seventh editions, Named Entity Recognition was defined to be the task of labelling proper names in a text and classify them into five categories: Person, Location, Organization, Miscellaneous and Other.

Named entity recognition plays a vital role in many other NLP applications. In Machine Translation (MT), Babych and Hartley (2003) have shown that using Named Entities help the MT system to improve the translation task. In Text Clustering, Toda and Kataoka (2005) have argued that integrating a NER system in their clustering system improved the performance and allowed them to outperform the existing state-of-the-art system. In Question Answering (QA), Greenwood and Gaizauskas (2007) have shown that using a NER system in their QA model improves its performance. Moreover, Information Retrieval (IR), document and news searching, semantic parsers, part of speech taggers and thematic meaning representations could all use NER to improve their performance.

Recently, Arabic NER task has been the subject of discussion between many researchers in the NLP field. The most Arabic NER systems fall into three categories: rule-based approach, Machine Learning based approach and hybrid approach. Each approach has its advantages and shortcomings. Rule-based NER approach relies on a set of handcrafted rules extracted and verified by experts in linguistics (Shaalan and Raza 2009; Elsebai et al. 2009). One shortcoming of these systems is the dependency to the experts, which means that the linguist must check any rule added to the system that is considered as time-consuming. The second category of NER systems relies on Machine Learning based approach (Benajiba et al. 2007b; Benajiba and Rosso 2008a; Benajiba et al. 2008b). These systems have shown significant improvement in terms of coverage and robustness, but they require a set of manually handcrafted engineering features. To develop these systems, handcrafted rules, huge dictionaries (gazetteers) and external parsers (POS taggers and chunkers) are needed. In addition, the task to come up with good features requires additional knowledge in the domain and can be time-consuming. Finally, the hybrid approach which takes the advantages of rule-based and Machine Learning-based approaches (Abdallah et al. 2012; Shaalan and Oudah 2014). It achieves better results than the previous approaches, but it is too complex because it relies on a set of handcrafted features and grammatical rules.

Arabic is very complex language compared to European languages. Therefore, building Arabic NLP applications and in particular NER is very challenging. The first challenging characteristic is the agglutinative nature of Arabic words: each word consists of combination of prefixes, stem and suffixes that results in very complex morphology (AbdelRahman et al. 2010). Moreover, Arabic language has rich morphology because its words are highly inflectional and derivational which results in different level of ambiguity. The third challenging task is the lack of capitalization. For Arabic language, capitalization is not a feature unlike the most European languages, which add more complexity to recognize named entities. Finally, most of the linguistic resources are not available for research, unlike other languages such as English, French or German. As we know, the first available Arabic NER dataset developed for research purposes was ANERcorp (Benajiba and Rosso 2007a).

In recent years, experiments show that Deep Neural Networks (DNNs) are extremely powerful machine learning models that have achieved great success in many various applications and difficult problems such as speech recognition (Dahl et al. 2012; Hinton et al. 2012) and object recognition (Ciresan et al. 2012; Krizhevsky et al. 2012; Le et al. 2012; LeCun et al. 1998). The power of these models comes from the ability to perform arbitrary parallel computation for a number of steps. Furthermore, many recent works showed that DNNs can achieve state-of-the-art in various Natural Language Processing applications. These applications include language modelling (Bengio et al. 2003), sentiment analysis (Socher et al. 2013; dos Santos and Gatti 2014), paraphrase detection (Socher et al. 2011) and more recently in the field of statistical machine translation (SMT) (Cho et al. 2014; Sutskever et al. 2014). In addition, in the presence of large labelled training set, DNNs can use supervised backpropagation algorithm for training (Rumelhart et al. 1986). Consequently, supervised backpropagation algorithm will optimize the network parameters and find the solution whenever a good parameter setting of the network existed.

While spending most of the time in designing suitable features to build an Arabic NER system is labor-intensive and requires the presence of an expert in linguistics, it also makes the system too complex to be adapted to new domains since the linguistic resources may not be available to achieve this task. Motivated by the success of deep neural networks in various NLP applications, we introduce an Arabic NER system based on DNNs. We train the model to automatically learn good features without any hand-features design. In our knowledge, we are the first to use DNNs to build an Arabic NER system.

We use word embeddings as introduced by (Mikolov et al. 2013) to pre-train our word vectors. They are word representations that are learned in unsupervised fashion. Using word embeddings shows a significant improvement in performance of the system. Represented in high dimensional space, these representations are able to capture many properties in different levels such as syntactic, morpho-syntactic and semantic levels. Therefore, words with the same characteristics or properties are close to each other in space representation.

The main contributions of this paper are the following:

– Use Deep Neural Networks to learn automatically good features;
– The model doesn't use any handcrafted engineering features which is time-consuming;
– The model doesn't use any kind of linguistic rules;
– The system doesn't use any kind of large gazetteers.
– Show that using word representations improve the system performance;
– Outperform or get close results to the previous Arabic NER systems based on different approaches.
– Can be easily extended to recognize other types of named entities.

The experimental results confirm that using deep learning approach is effective. We use ANERcorp dataset for training and testing our model. Experiments show that our model outperforms the model based on Conditional Random Fields (Benajiba et al. 2008a) by 12.36 points in F-measure. Moreover, our model outperforms the state-of-the-art system (Abdallah et al. 2012) by 5.18 points in Precision and gets very close results in F-measure.

The rest of this paper is structured as follows. In Sect. 2, we present our novel approach based on DNNs and in Sect. 3 we describe the main experimental results. Finally, we give a brief overview on some previous work on Arabic Named Entity Recognition in Sect. 4 and we conclude and discuss future works in Sect. 5.

2   Our Approach

In deep learning community, the year 2006 was the real breakthrough in this field, and it has emerged as a new research area of machine learning (Hinton et al. 2006; Bengio 2009). In the past several years, the algorithms and models developed by deep learning community had huge impact on many fields such as Natural Language Processing (Collobert et al. 2011; Socher et al. 2011; Mikolov et al. 2011), Computer Vision (Ciresan et al. 2012), Speech Recognition (Hinton et al. 2012; Deng et al. 2013) and more recently Statistical Machine Translation (called Neural Machine Translation) (Cho et al. 2014; Sutskever et al. 2014). Therefore, deep learning systems have won several contests in pattern recognition and machine learning.

Motivated by recent success of deep neural networks in many fields including NLP, we decided to develop a new model based on deep architectures to recognize Arabic named entities. As we will show in the experiments section, the model gave astonishing results because when we evaluate it on ANERcorp dataset, it outperforms the baseline and gets close results to the state-of-the-art system. In the next section, we present a method called word embeddings that we used to pre-train our word vectors. Next, we describe our approach to recognize Arabic named entities using deep neural networks. Finally, we describe two optimization methods used to increase the performance of our system.

2.1   Word Embeddings

Word embeddings (or word representations) are representation of words with real values in high dimensional space typically induced via neural language models (Bengio et al. 2003; Collobert and Weston 2008). These models can be learned from unlabelled text data. Many experiments have been shown that using word embedding improve significantly various NLP tasks such as part-of-speech (POS) tagging (Huang et al. 2012), chunking (Turian et al. 2010), dependency parsing (Pennington et al. 2014), syntactic parsing (Finkel et al. 2008; Täckström et al. 2012), and more recently, in different way, for neural machine translation (Cho et al. 2014).

It should be noted that word2vec developed by (Mikolov et al. 2013) is the most widely used approach. Instead of capturing co-occurrence counts directly, word2vec model tries to predict surrounding words of each word in a window of fixed length.

This model became very popular because it is faster to train and can be easily incorporate a new sentence, document or add new word to the vocabulary. These representations are very powerful at encoding dimensions of similarity. To train our word vectors, we used the skip-gram model, which is a various approach of word2vec model. The aim of this model is to find word representations to predict the surrounding words in a sentence or a document (Mikolov et al. 2013).

Therefore, the task of testing similarity (syntactic or semantic) between words can be solved quite well just by doing vector subtraction in the embedding space. The following example shows a syntactic similarity between words:

- vector("word") – vector("words") ≈ vector("model") – vector("models") ≈ vector ("family") – vector("families").

For semantic similarity, we have:

- vector("shirt") – vector("clothing") ≈ vector("chair") – vector("furniture")
- vector("king") – vector("man") ≈ vector("queen") – vector("woman").

The last example can be read as: "king" to "man" is like "queen" to "woman".

2.2   System Description

Our Arabic NER system is inspired by the successful deep neural network architecture presented by (Collobert et al. 2011). In this section, we describe the architecture of our Arabic NER system. The model is based on deep learning that can outperform previous best models in the literature without any use of handcrafted engineering features or knowledge from experts in linguistics. As far as we know, we are the first to use deep learning architectures to recognize named entities in Arabic texts.

In order to classify Arabic words into five classes (Person, Organization, Location, Miscellaneous and a null-class (O) for words that do not represent a named entity (most words fall into this class)), we begin by pre-trained our word vectors using word2vec model (Mikolov et al. 2013). The result is a matrix $M \in R^{d \times |V|}$ called the embedding matrix where d is the dimension of word vectors and V is the size of the vocabulary.

Our deep neural network has 3 layers: the input layer, hidden layer and an output layer. In the input layer, we explicitly represent context as a window consisting of center word concatenated with its immediate surrounding neighbours. Let x_t be the center word to be classified into named entity. Thus, the window around this word will be represented with the vector $v^{(t)}$ as follows:

$$v^{(t)} = Mx_{t-C}, \ldots, Mx_{t-1}, Mx_t, Mx_{t+1}, \ldots, Mx_{t+C} \tag{1}$$

The inputs $x_{t-C}, \ldots, x_{t-1}, x_t, x_{t+1}, \ldots, x_{t+C}$ are one-hot vectors (just indices) into the embedding matrix $M \in R^{d \times |V|}$. In this case, the number of words surrounding the center word is "2C" words, where $C \in \mathbb{N}$. When we multiply the embedding matrix with a particular one-hot vector x_t, we get its vector representation, which is a d-dimensional vector. Thus, each column of the embedding matrix M is the vector representation for a particular word in the vocabulary.

In the next step, the vector $v^{(t)} \in R^{(2C+1)d}$ is passed to the hidden layer hl, where the following linear transformation is applied: $hl(v^{(t)}) = Wv^{(t)} + b_1$, where the matrix of weights W and the bias b1 are the model parameters to be learned by the network. Learning feature combinations from the word representations of the window will be the goal of the hidden layer (hl). Then, the output of hl is passed through a non-linearity also called the activation function to enable the network to learn non-linear discriminative functions. We used "tanh" function as our non-linearity. Therefore, we get the following equation:

$$h = \tanh\left(Wv^{(t)} + b_1\right) \tag{2}$$

Finally, h is passed to the output layer (ol) where the following linear transformation is applied: $ol(h) = Uh + b_2$. The matrix of weights U and the bias b2 are model parameters to be learned by the network. Then, the softmax function is applied to get probabilities.

$$\hat{y} = softmax(Uh + b_2) \tag{3}$$

We will use the cross-entropy as the loss function for this model. It should be noted that theoretically, a squared error function could be used, but in that case, the optimization tricks developed for softmax function will not be applied. So, the loss function will be:

$$J(\theta) = -\sum_{k=1}^{5} y_k \log \hat{y}_k \tag{4}$$

Where $y \in R^5$ (5 classes) is a one-hot label vector and θ is the set of all model parameters to be learned by the neural network.

2.3　Advanced Optimization and Tricks

Weight Decay. Training deep neural networks based on the maximum likelihood (ML) estimation or the cross-entropy error may be overtrained. One of the ways to control the capacity of DNNs is using the weight-decay (also called the Gaussian prior) (Hinton 2012). This method penalizes large weights using penalties or constraints on their squared values (L2 regularization) or absolute values (L1 regularization).

The experiments show that if we augment our cost function with weight-decay regularization, then it will avoid the parameters from blowing up or begin to be highly correlated. Thus, it leads to classifiers with better generalization capacity because parameter weights will be pushed close to zero (see the next section for more details about the experimental results). In this paper, we use L2 regularization. Then, the regularization term will be:

$$J_{reg}(\theta) = \frac{\lambda}{2}\left[\sum_{ij} W_{ij}^2 + \sum_{i'j'} U_{i'j'}^2\right] \tag{5}$$

Where W and U are model parameters, λ is the regularization parameter and θ refers to all the model parameters. After adding the regularization term, the combined loss function to optimize will be:

$$J(\theta) = -\sum_{k=1}^{5} y_k \log \hat{y}_k + \frac{\lambda}{2}\left[\sum_{i,j} W_{ij}^2 + \sum_{i'j'} U_{i'j'}^2\right] \qquad (6)$$

Initialize Parameters from a Uniform Distribution. (LeCun et al. 1998) showed that the initial values of the parameters can have an important effect in training deep neural networks. The best method that works well in practice is to choose the initial parameters randomly from a centered uniform distribution. It will avoid neurons to become too correlated and ending up in poor local minima. Empirical results show that choosing the initial values of the parameters randomly from the uniform distribution $[-\varepsilon, \varepsilon]$ for a certain value of ε works well in practice (see the next section for more details about the results). Let M be a matrix of dimension (m, n), then we choose the value of ε such that:

$$\varepsilon = \frac{\sqrt{6}}{\sqrt{m+n}} \qquad (7)$$

3   Experiments

We applied our model to recognize named entities in Arabic texts. In this section, we begin by presenting the dataset used for training and testing our model. Next, we discuss some training details. Finally, we report the accuracy of our model and discuss some experimental results.

3.1   Dataset Details

The lack of available resources for training and testing Arabic NER systems faces many researchers in this field. To train and test our model, we use ANERcorp which is a dataset developed by (Benajiba and Rosso 2007a) for Arabic Named Entity Recognition task. This dataset contains more than 150 000 words annotated for this task (11% of the tokens are NEs). ANERcorp is suitable for supervised learning algorithms including deep neural networks. The tokens have been chosen from both news wire and other web resources. The authors manually annotated the dataset. More details about ANERcorp dataset can be found in (Benajiba and Rosso 2007a; Benajiba et al. 2007b).

3.2   Training Details

Before training our model, we performed some pre-processing steps: we begin by padding each sentence with begin and end tokens (<s> and </s>). These two tokens have their own word vector embeddings. We replace unknown words with a special token UUNNKK. We note that we will have five model hyperparameters to be tuned during training (d: dimension of word vectors, h: number of hidden layers, C: the context window, λ: regularization parameter and α: the learning rate). Finally, we convert each digit into special representation (e.g., 3.142 will be represented as DG. DGDGDG), which also has its own word vector embeddings.

To learn the parameters θ of the network, we use stochastic gradient descent (SGD) and backpropagation algorithm to compute the gradients. We use Python programming language with some libraries for highly scientific computing (NumPy, etc.) to develop the code of our model[1]. After several experiments of tuning hyperparameters, we got the best results by choosing the dimension of hidden units h = 100, a learning rate α = 0.05, the dimension of the word embedding d = 50, the regularization parameter λ = 0.001, the width of the context window C = 5. Using C = 5 instead of C = 3 has the advantage to capture multiword expressions like famous Arabic proper names "عبد الرحمان" (abderahman), "عبد القادر" (abdelkader), etc.

3.3 Experimental Results

In this section, on the one hand, we show how using word2vec model improves the system performance. On the other hand, we show that augmenting the cost function with a weight-decay will improve the system performance. Finally, using a uniform distribution to randomly initialize the model parameters has an impact on the system performance. In the second part of this section, we compare our model with a baseline and state-of-the-art system.

We carried out five experiments to evaluate our system using ANERcorp dataset. The first experiment was done using the deep neural networks without use of any trick (weight-decay, initialize parameters from a uniform distribution and word2vec). This model is referred to here as DLANER. The second experiment was done using the first model (DLANER) and randomly initializes parameters using the Uniform distribution (DLANER + UD). The third experiment was done using the first model (DLANER) with weight-decay regularization (DLANER + WD). The fourth experiment combines the second and the third model (DLANER + UD + WD). The final experiment uses the fourth model with a pre-trained word vectors using word2vec (DLANER + UD + WD + word2vec).

Table 1 shows the results of the five experiments carried out using ANERcorp dataset. From this experiments, it can be seen that combining initialization trick using the uniform distribution with a certain value of ε, weight-decay regularization and word vectors pre-trained with word2vec in the last experiment (DLANER + UD + WD + word2vec) achieves a score of 88.64 in F-measure and outperforms all the previous models.

Table 2 shows the comparative results between our deep learning approach and the baseline (Benajiba et al. 2008a) and state-of-the-art system (Abdallah et al. 2012). From this comparison, we can see that our system outperforms the baseline by 12.36 points in

---

[1] The code will be released right after the paper will be accepted.

Table 1. Evaluation results of the proposed model

Models	Precision	Recall	F-measure
DLANER	94.36	81.63	87.24
DLANER + UD	95.03	81.95	87.57
DLANER + WD	96.62	80.65	87.49
DLANER + UD + WD	95.76	81.95	88.31
DLANER + UD + WD + word2vec	95.76	82.52	88.64

Table 2. Comparative results performance between our model, baseline and state-of-the-art system.

Models	Precision	Recall	F-measure
CRF approach	85.89	69.34	76.28
NERA system	90.58	**87.05**	**88.77**
Our approach	**95.76**	82.52	88.64

F-measure. Moreover, our model outperforms the-state-of-the-art system by 5.18 points in Precision and gets very close results in F-measure.

In order to understand where the system performance comes from, we will use a famous method used in computer vision where the individual neurons learn to detect edges, shapes, etc. in images. In natural language processing, we are going to look at the word, which precedes the center word in order to get sense of how our deep neural network learns to recognize named entities. Table 3 shows the results.

Table 3. Words preceding the center word

Location	Organization	Person
شمال	منظمة	الدكتور
بمدينة	معهد	الشيخ
حي	حلف	ولد
المملكة	باتحاد	التشيكي
بمغادرته	لشركة	عبد
بالدار	لجامعة	ماريك
شبه	مجلة	مي
كوريا	الحنين	ليرنر
والولايات	مماثل	وعبد
تقويم	سي	سالم

It appears that the neural network learns to detect the preceding word of the most named entities. As an example, the neural network will learn, with high probability, that the word following the word "منظمة" (organization) will be an organization. The same learning process will be applied to the word following "الدكتور" (Doctor) which will be a person.

4 Related Work

Although there has been a substantial amount of work in the area of Arabic Named Entity Recognition, nearly all of these papers have used two main methods: rule-based method and machine learning-based method. More recently, a third method is taking place called hybrid method, which takes advantages of both rule-based and machine learning-based methods.

Rule-based NER approaches are considered among the earliest methods used for Arabic NER. They use a set of handcrafted linguistic rules to classify named entities. In addition, they use dictionaries or gazetteers as additional resources to the previous rules. As shown in (Shaalan and Raza 2009), the authors developed a rule-based NER system for modern standard Arabic (MSA) by implementing a set of rules as regular expressions and a filtering mechanism that mainly focuses on rejecting incorrect NEs based on a blacklist. This system achieves a performance of 87.7% F-measure for the Person, 85.9% for Location, and 83.15% for Organization. Other systems use features extracted from other NLP systems such as morphological analysers with pattern matching to recognize Arabic named entities (Elsebai et al. 2009). When evaluated on a dataset of 700 news articles extracted from Aljazeera television website, this system achieves 89% in F-measure.

The second approach used for Arabic NER relies on machine learning algorithms. The most used algorithms are Maximum Entropy (ME), Conditional Random Fields (CRF), Support Vector Machines (SVM), Hidden Markov Models (HMM) and Decision Trees (DT). To build these systems, they need to come up with a set of handcrafted engineering features to classify input text depending on annotated corpora. For example, (Benajiba et al. 2007b) uses n-grams and maximum entropy to develop an Arabic NER system. Later, (Benajiba and Rosso 2008a) uses Conditional Random Fields to develop ANERsys with the following set of features: part of speech (POS) tags, Base Phrase Chunks (BPC), gazetteers, and nationality information. (Benajiba et al. 2008b) developed another Arabic NER system based on SVM approach. The authors used a set of features including POS tagging, gazetteers, nationality and corresponding English capitalization.

The last approach called the hybrid approach combines rule-based and machine learning-based approaches. The state-of-the-art results were produced by (Shaalan and Oudah 2014) where the system contains a rule-based NER component producing NE labels and machine learning-based processor using features from the last component. In contrast, we were not able to reproduce their results because we could not get access to the rules used by the system. (Abdallah et al. 2012) proposes a hybrid system (called integrated approach) that integrates machine learning with rule-based systems. The system yields an F-score of 88.77% when tested on ANERcorp dataset.

5 Conclusion and Future Works

In this paper, we propose a new approach for designing an Arabic NER system. Our approach is based on deep neural networks (DNN) where we added some optimization tricks to increase the performance of the model. We also observe the impact of word embedding on the performance of our system. Our system achieves an F-measure of 88.64% on the test set. We show that our proposed model improves performance and outperforms the baseline model and gets very close results to the state-of-the-art system on recognizing Arabic named entities. In addition, it does not require handcrafted engineering features, grammatical rules or additional data processing. Finally, our model is flexible because it is easy to add other type of named entities to be recognized such as temporal named entities, sports NEs, politics NEs, etc.

In the future works, we would like to test our model in another dataset with more named entities. Moreover, we would like to use the same model to recognize Arabic Named Entities in social media (twitter as an example).

References

Abdallah, S., Shaalan, K., Shoaib, M.: Integrating rule-based system with classification for arabic named entity recognition. In: Gelbukh, A. (ed.) CICLing 2012. LNCS, vol. 7181, pp. 311–322. Springer, Heidelberg (2012). https://doi.org/10.1007/978-3-642-28604-9_26

Rahman, S.A., Elarnaoty, M., Magdy, M., Fahmy, A.: Integrated machine learning techniques for Arabic named entity recognition. Int. J. Comput. Sci. Issues (IJCSI) 7, 27–36 (2010)

Babych, B., Hartley, A.: Improving machine translation quality with automatic named entity recognition. In: Proceedings of EACL-EAMT, Budapest (2003)

Benajiba, Y., Rosso, P.: Anersys 2.0: conquering the NER task for the Arabic language by combining the maximum entropy with POS-tag information. In: IICAI, pp. 1814–1823 (2007a)

Benajiba, Y., Rosso, P., BenedíRuiz, J.M.: ANERsys: an arabic named entity recognition system based on maximum entropy. In: Gelbukh, A. (ed.) CICLing 2007. LNCS, vol. 4394, pp. 143–153. Springer, Heidelberg (2007b). https://doi.org/10.1007/978-3-540-70939-8_13

Benajiba, Y., Rosso, P.: Arabic named entity recognition using conditional random fields. In: Workshop on HLT & NLP within the Arabic World. Arabic Language and Local Languages Processing: Status Updates and Prospects (2008a)

Benajiba, Y., Diab, M., Rosso, P.: Arabic named entity recognition: an SVM-based approach. In: Proceedings of Arab International Conference on Information Technology (ACIT 2008), pp. 16–18 (2008b)

Bengio, Y., Ducharme, R., Vincent, P., Janvin, C.: A neural probabilistic language model. J. Mach. Learn. Res. 3, 1137–1155 (2003)

Bengio, Y.: Learning deep architectures for AI. Found. Trends Mach. Learn. 2(1), 1–127 (2009)

Chinchor, N., Brown, E., Ferro, L., Robinson, P.: Named entity recognition task definition. In: MITRE and SAIC (1999)

Cho, K., van Merrienboer, B., Gulcehre, C., Bougares, F., Schwenk, H., Bengio, Y.: Learning phrase representations using RNN encoder-decoder for statistical machine translation. In: EMNLP, October 2014

Ciresan, D., Meier, U., Schmidhuber, J.: Multi-column deep neural networks for image classification. In: CVPR (2012)

Collobert, R., Weston, J.: A unified architecture for natural language processing: deep neural networks with multitask learning. In: Proceedings of the 25th International Conference on Machine Learning, pp. 160–167. ACM (2008)

Collobert, R., Weston, J., Bottou, L., Karlen, M., Kavukcuoglu, K., Kuksa, P.: Natural language processing (almost) from scratch. J. Mach. Learn. Res. 12, 2493–2537 (2011)

Dahl, G.E., Yu, D., Deng, L., Acero, A.: Context-dependent pre-trained deep neural networks for large vocabulary speech recognition. IEEE Trans. Audio Speech Lang. Process. **20**(1), 30–42 (2012). Special Issue on Deep Learning for Speech and Language Processing

Deng, L., Li, J., Huang, J.-T., Yao, K., Yu, D., Seide, F., Seltzer, M., Zweig, G., He, X., Williams, X., Gong, Y., Acero, A.: Recent advances in deep learning for speech research at Microsoft. In: Proceedings of International Conference on Acoustics Speech and Signal Processing (ICASSP) (2013)

dos Santos, C., Gatti, M.: Deep convolutional neural networks for sentiment analysis of short texts. In: Proceedings of COLING 2014, the 25th International Conference on Computational Linguistics, Technical Papers, pp. 69–78. Dublin City University and Association for Computational Linguistics, Dublin, Ireland, August 2014

Elsebai, A., Meziane, F., Belkredim, F.Z.: A rule based persons names Arabic extraction system. Commun. IBIMA **11**(6), 53–59 (2009)

Finkel, J.R., Kleeman, A., Manning, C.D.: Efficient, feature-based, conditional random field parsing. In: Proceedings of ACL (2008)

Greenwood, M., Gaizauskas, R.: Using a named entity tagger to generalise surface matching text patterns for question answering. In: Proceedings of the Workshop on Natural Language Processing for Question Answering (EACL03) (2007)

Hinton, G., Osindero, S., Teh, Y.-W.: A fast learning algorithm for deep belief nets. Neural Comput. **18**, 1527–1554 (2006)

Hinton, G., Deng, G., Yu, D., Dahl, G., Mohamed, A., Jaitly, N., Senior, A., Vanhoucke, V., Nguyen, P., Sainath, T., Kingsbury, B.: Deep neural networks for acoustic modeling in speech recognition. IEEE Signal Process. Mag. **29**(6), 82–97 (2012)

Hinton, G.: Neural networks for machine learning. Coursera, video lectures (2012)

Huang, E.H., Socher, R., Manning, C.D., Ng, A.Y.: Improving word representations via global context and multiple word prototypes. In: Proceedings of ACL (2012)

Krizhevsky, A., Sutskever, I., Hinton, G.E.: ImageNet classification with deep convolutional neural networks. In: NIPS (2012)

Le, Q.V., Ranzato, M.A., Monga, R., Devin, M., Chen, K., Corrado, G.S., Dean, J., Ng, A.Y.: Building high-level features using large scale unsupervised learning. In: ICML (2012)

LeCun, Y., Bottou, L., Bengio, Y., Haffner, P.: Gradient based learning applied to document recognition. Proc. IEEE **86**(11), 2278–2324 (1998)

Mikolov, T., Kombrink, S., Burget, L., Černocký, J., Khudanpur, S.: Extensions of recurrent neural network language model. In: Proceedings of ICASSP (2011)

Mikolov, T., Sutskever, I., Chen, K., Corrado, G., Dean, J.: Distributed representations of words and phrases and their compositionality. In: NIPS, pp. 3111–3119 (2013)

Pennington, J., Socher, R., Manning, C.: GloVe: global vectors for word representation. In: Proceedings of the 2014 Conference on Empirical Methods in Natural Language Processing (EMNLP), October 2014, pp. 1532–1543 (2014)

Rumelhart, D.E., Hinton, G.E., Williams, R.J.: Learning internal representations by error propagation. In: Symposium on Parallel and Distributed Processing (1986)

Shaalan, K., Raza, H.: NERA: named entity recognition for Arabic. J. Am. Soc. Inform. Sci. Technol. **60**(8), 1652–1663 (2009)

Shaalan, K., Oudah, M.: A hybrid approach to Arabic named entity recognition. J. Inf. Sci. **40**(1), 67–87 (2014)

Socher, R., Huang, E.H., Pennington, J., Ng, A.Y., Manning, C.D.: Dynamic pooling and unfolding recursive autoencoders for paraphrase detection. In: Advances in Neural Information Processing Systems 24 (2011)

Socher, R., Perelygin, A., Wu, J.Y., Chuang, J., Manning, C.D., Ng, A.Y., Potts, C.: Recursive deep models for semantic compositionality over a sentiment treebank. In: Conference on Empirical Methods in Natural Language Processing (2013)

Sutskever, I., Vinyals, O., Le, Q.V.: Sequence to sequence learning with neural networks. In: NIPS (2014)

Täckström, O., McDonald, R., Uszkoreit, J.: Cross-lingual word clusters for direct transfer of linguistic structure. In: Proceedings of NAACL (2012)

Toda, H., Kataoka, R.: A search result clustering method using informatively named entities. In: Proceedings of the 7th Annual ACM International Workshop on Web Information and Data Management (WIDM), pp. 81–86. ACM Press (2005)

Turian, J., Ratinov, L.-A., Bengio, Y.: Word representations: a simple and general method for semi-supervised learning. In: Proceedings of ACL (2010)

Hybrid Feature Selection Approach
for Arabic Named Entity Recognition

Miran Shahine[1(✉)] and Mohamed Sakre[2]

[1] Arabic Academy for Science,
Technology and Maritime Transport, Cairo, Egypt
miran.shahine@gmail.com
[2] Shorouk Academy, Cairo, Egypt
m.sakre@sha.edu.eg

Abstract. Named Entity Recognition (NER) task has drawn a great attention in the research field in the last decade, as it played an important role in the Natural Language Processing (NLP) applications; In this paper, we investigate the effectiveness of a hybrid feature subset selection approach for Arabic Named Entity Recognition (NER) which is presented using filtering approach and optimized Genetic algorithm. Genetic algorithm is utilized through parallelization of the fitness computation in order to reduce the computation time to search out the most appropriate and informative combination of features for classification. Support Vector Machine (SVM) is used as the machine learning based classifier to evaluate the accuracy of the Arabic NER through the proposed approach. ANER is the dataset used in our experiments which is presented by both language independent and language specific features in Arabic NER; Experimental results show the effectiveness of the feature subsets obtained by the proposed hybrid approach which are smaller and effective than the original feature set that leads to a considerable increase in the classification accuracy.

Keywords: Named entity recognition · Natural Language Processing
Feature subset selection · Genetic algorithm · Machine learning
Support vector machines

1 Introduction

Named entity recognition (NER) is the task of detecting names in a text and classify them into a predefined categories such as Person, Location and Organization names [1], it acts as an important module in many Natural Language Processing (NLP) applications such as Information Retrieval [2], Question Answering [3], Machine Translation [4] and recently its importance appears in Opinion mining, which employ it as a preprocessing step to boost their performance. The performance of machine learning NER systems depends on features being used in the training and testing phases. Feature selection is crucial and critical step to improve the classifier's performance as it acts as an optimization problem. By extracting as much information as possible from a given data set while using relatively small number of features, then we can save significant time needed for classification and build models that generalize better for unseen data

© Springer International Publishing AG, part of Springer Nature 2018
A. Gelbukh (Ed.): CICLing 2016, LNCS 9623, pp. 452–464, 2018.
https://doi.org/10.1007/978-3-319-75477-2_32

points. So the choice of certain features can affect accuracy of the learned classification algorithm in a brute manner, as if the language is not expressive or informative enough, it will fail to be determined. Named Entity (NE) identification in Arabic language is a more challenging task compared to other languages due to a number of facts such as: (i) Arabic has no Capital letters, which will be an obstacle in identifying NEs properly which represent an important feature in other languages as proper nouns, (ii) Non Standard Written Text may not include diacritics, which can lead to ambiguous different meaning [5, 6], (iii) The complex morphology of the language due to the agglutinative nature of the language [7, 8]. The word consists of three parts which are: prefix, lemma and suffix. A one Arabic word can be translated into a phrase in English. For example the word "وسيخرجونها" (wasayukhrijwnaha, And-they-will-get-it-out), this can cause a sparse data that will lead to overtraining and poor performance, (iv) Inflected verbs which shared functions as both verbs and Personal NEs, (v) Word different meanings depending on the context it lies in, consider the word الشرق الأوسط which means' the middle east' as it can differ in its meaning depending on which context the word it appears in, it can be a Location or Organization name [9] and (vi) The limitation of the number of free resources is another issue to take into consideration as most of the corpora found to be free for research are not annotated as well as the Arabic gazetteers which tends to be rare, there are just one publicly free reliable corpus which is ANER which developed by [10], Therefore the complexity of Arabic language causes a complexity in defining the accurate NEs and their types, which put the burden on the features used in order to get more information about each word in an effective manner to solve such problem.

Feature selection is a crucial task due to the complexity related to the interaction of features. A Relevant feature can become redundant when working with other features and will not add additional information related to the target concept [12]. Therefore, an optimal feature subset should be obtained to overcome this issue. The feature selection task is challenging also because of the large search spaces. The size of the search space increases exponentially with respect to the number of available features in the data set [13]. Therefore implying exhaustive search is impractical and costly, so In order to tackle feature selection problems, an efficient global search technique is needed. Another issue is to trigger, If all feature combinations are selected therefore the feature space becomes high dimensional and thus the learning algorithm that is used will be slowed down unnecessarily due to the large number of dimensions of feature space, while also experiencing lower classification accuracies due to learning irrelevant information. Another problem that can be tackled by feature selection is the computational time. The computational time has a directly proportional relation with the size of the feature subset, because using a large number of features can increase the search space and hence the time needed for learning. So it will increase the complexity without increasing the system's accuracy. Furthermore, the irrelevant features can raise the over-fitting problem which can be observed due to an irrelevant feature taken place in the learning process. Large and redundant available feature sets are useless and can even worsen the performance, therefore selecting the right and relevant features that can contribute to the accuracy of the learning algorithm by reducing the potential of over-fitting and clearly discriminate between different named entity classes is a challenging task that can change the whole performance path. Feature selection is the procedure that can play this role and select small and most effective feature subset from the original large sets.

Despite the recent progress in Arabic NER, the effort has been dispersed in several directions related to rule based and hybrid approaches and the usage of all available varieties of features which recommend a prior knowledge of the language, without considering the faced performance problems that can be accompanied with such number and type of these features and its impact on the whole NER performance. Most NER systems use additional features, such as POS tags, shallow parsing information and gazetteers [14]. The capability of finding better subsets of informative features in terms of reduced size and increased classification performance can provide robust and reliable framework.

In this paper, a Hybrid approach for automated feature selection is proposed for solving the Named Entity Recognition (NER) task for Arabic. Moreover, we will also consider the cardinality of the obtained feature subset as performance indicator. The approach consists of both filter and Genetic wrapper feature selection steps, This hybrid selection process is repeated with different feature set sizes and dataset characteristics, The results of the study enabled us to discover which features or feature combinations are better identifiers for the Arabic NER task.

2 Background

2.1 Feature Selection

Feature selection (FS) is the process of selecting an optimal subset of relevant features and removes irrelevant, redundant and noisy features from the original set. Feature selection therefore is the process of choosing small but effective of available features. It is one of the most important factors for the success of Machine Learning (ML) tasks. A poor feature set may degrade the performance of the system, so when these irrelevant and noisy features are not involved in the learning process, the ML algorithms can focus on the most important aspects of data and build more simpler and accurate data models. Also the best fitting features will contribute to the accuracy of the model with least dimensions, thereby tackling the curse of dimensionality problem. However reducing the size of the feature subset will accelerate the learning algorithm but the size of feature subsets is not always proportional with the accuracy of the learning algorithm that works on it. Thus, the current Paradigm in feature selection is based on balancing between minimizing the number of feature and maximizing the accuracy. Two approaches often used are forward selection [15] and backward selection [16]. The forward selection is an iterative selection method which starts with no features in the model, It tests the addition of each candidate feature and the ones that mostly minimizes error and improves the model the most on the validation sets are added. This process repeats until there is no significant decrease in the error rate no improvement achieved. The backward selection starts with all available candidate variables. And then iteratively remove (delete) the ones which increases the error rate, this process repeats until no further improvement is observed. The disadvantage of these strategies is that if a feature is already removed, it cannot be reconsidered anymore. This treatment is not efficient in case some features are correlated in recognizing an entity. Two main methods for feature selection are considered; Filter methods that rank each feature according to some univariate metric and certain statistical criteria to evaluate the features' relevance with regard to the class labels of samples, only the features with the highest ranking are selected and retained while the remaining low ranking features are discarded. Filters rely only on the training data characteristics to select the features without involving any learning algorithm; they are simple and fast and can be scaled to larger databases with larger number of features. Unlike filters, wrapper methods select a feature subset using a learning algorithm as part of the evaluation function, a heuristic search technique is taken place to generate different feature subsets, and then an induction algorithm is used to evaluate the features and to estimate the accuracy. Wrappers are more efficient when it comes to the predictive accuracy than filters but they are more computationally intensive and may lead to an over-fitting problem.

Some Arabic NER researchers manually designed features and used all of them without selecting optimal subsets like [10] while others incorporate additional features in order to enhance their system's performance [17, 18] investigate many sets of features and explore them all individually and in combination in an attempt to get an optimal subset, and explore individual features and rank them according to their impact, while [8] explore the features in set of groups with rule-based decision included as one of the feature groups, they examined the contribution of ML-based features and the exclusion of the morphological features, But in practice, these approaches are computationally not feasible.

3 Arabic NER Feature Space

Since NER systems are significantly affected by the language and for some languages the difference in performance is more than 20 percent [19] That's why the NER task for Arabic is different from English and other European languages this is due to the linguistic differences and the rich morphological characteristics of the Arabic language, The feature set is represented by the combination of n number of features describing a particular token that can be represented as n-dimensional vector during training, In this section we discuss the feature sets that have been used in our experiments along with the necessary details:

- Contextual window feature: Since extracting the features for only the focus token cannot yield the best performance, due to the exclusion of the valuable informative features of the context that will be missed. The context is defined as a window of ± n tokens from NE of interest [6]. If the 2 previous tokens are considered In order to capture titles and descriptive nouns such as, "السيد"(Alsayd, Mr), there will be a problem appeared with the exclusion of the 2 succeeding, as we found by experiment that there are some indicators which are not considered as a part of a particular NE and appears to help in NE identification, for instance, the company indicator "ذات مسئولية محدودة" (Dhat Mas}wulyap Mahdwudap, LLC) which appears frequently after NE of type Organization, so we added the 2 succeeding words to be (2 1 2).

- Word Length: it's a binary valued feature checks whether the length of the token is less than a predetermined threshold (set to 3) value in our experiments. As we observed that words with less than 3 letters are most considered to be non NE.
- N-gram: By following [20], the authors extracted 12 character n-gram features which represent the leading and tailing characters of the word (unigrams, bigrams and trigrams).
- Gazetteers: Boolean feature to indicate whether the word in a gazetteer or not, ANER gazetteers are used as they are available freely online for public use, According to [10] they are built manually from different sources:
  - Person gazetteer: 2,309 tokens.
  - Location gazetteer: 1,950 names.
  - Organization gazetteer: 262 names.
- Stop word: This feature indicates whether the current word is a stop word or not, they are turned off because they considered noisy and captured by other features.
- Morphological Features: Arabic is a language with rich and complex morphological structure, to extract the morphological features MADA toolkit was used [21]:
  - Aspect: Arabic has three aspects for verbs: perfective, imperfective and imperative, since none of the NEs is verbal, it will be a binary feature, which indicates if a token is aspect or not, and thus a non NE is determined.
  - Case: includes three cases Arabic nominals: nominative, accusative and genitive.
  - Gender: Arabic nominals can include gender information, According to MADA, it's a binary feature and the possible values for this feature are masculine (MASC) and feminine (FEM).
  - Number: represents the indication of grammatical number of tokens i.e. singular, dual or plural.
  - Mood: includes three Arabic moods for imperfective verbs: indicative, subjunctive or jussive.
  - State: three states are available: definite, indefinite, or construct.
  - Voice: an indication for passive and active voice.
  - Clitics: the clitics which are attached to the stem.
  - Gloss word's Translation: it represents the English translation (gloss) provided by MADA as if the word starts with a capital letter.
- Output feature: it refers to the output class which represents the word NE type.
- Part of Speech (POS) information: Stanford POS tagger used to indicate the POS tags in our experiments for its demonstrated good performance.
- Projected English Capitalization: it is developed by [11]. it indicates whether the word corresponds to a capitalized English word in MADA's Arabic-English lexicon. But they construct a mapping between English and Arabic Wikipedia titles connected by cross-lingual links in article metadata in an attempt to increase the recall, Six features to indicate whether the current word, the previous word, or the combination of the two map to a capitalized English term [11].

4 Hybrid Feature Selection Approach

The aim of the hybrid approach is to combine the strengths of both methods, which is less computationally expensive than the wrapper and possibly more accurate than the filter.

4.1 Filtering Based Techniques

In the first stage, a filter is implemented as a preprocessing step to act as an additional heuristic variable without affecting the evaluation done by the wrapper approach, the features are selected using CHI2 and information gain based filter methods, each of them is able to point out a different list of ranked features, which are ranked according to the strength of their correlation to a given class, A feature with a high ranking value indicates a higher discrimination of this feature compared to other classes which means that the feature contains more potentially useful information.

CHI square. The chi square test is based on the difference between the observed and the expected values for two events. In feature selection we use it to test whether the occurrence of a particular term (feature) and the occurrence of a target class are independent. Then the features are ranked according to their relevance and this will lead to discarding the insignificant features automatically. The chi square statistic is defined as:

$$X^2(t, c) = \sum_{t \in \{0,1\}} \sum_{c \in \{0,1\}} \frac{\left(N_{t,c} - E_{t,c}\right)^2}{E_{t,c}} \tag{1}$$

Where N is the observed frequency and E is the expected frequency for each term t and class c.

Information Gain. Measures how much information the presence or absence of aterm contributes to making the correct classification decision for any class. Information gain is a symmetrical statistical measure that is; the amount of information gained about the feature after class observation is equal to the amount of information gained about class after the feature observation. This is a desirable to determine the features inter-correlation. Its formula is as follows:

$$IG(t, c) = -\sum_{i=1}^{M} P(c) \log P(c) + P(t) \sum_{i=1}^{M} P(c|t) \log P(c|t)$$
$$+ P(\bar{t}) \sum_{i=1}^{M} P(c|\bar{t}) \log P(c|\bar{t}) \tag{2}$$

Where t is the term with respect to the class c, and M is the number of classes, $P(c)$ is the class c probability, $p(t)$ and $p(\bar{t})$ represent the probability of the presence and the absence of term t, respectively, $p(c|t)$ and $p(c|\bar{t})$ are the probability of the class c given the presence and absence of term t, respectively.

In the proposed system, only the shared features of both filter's feature list are considered and construct the feature pool which go on through the following steps:

1. The filters are carried out on the original dataset. The results will be two ordered lists of ranked features in descending order of relevance.
2. According to a fixed threshold T, we consider only the T top-ranked features from each list.
3. Feature pool is constructed in two ways, the first, is to construct the feature pool by the shared features of both lists, constructing a feature sub-space, the second, is to use all the features selected by both filters combined together in an attempt to test and analyze the effectiveness such combination.

At the next stage, the feature subspace is explored by a wrapper that uses a genetic algorithm as a search strategy.

4.2 Genetic Based Wrapper

A genetic algorithm [22] is a probabilistic methodology for solving optimization problems based on the biological evolution process which uses the concept of natural selection. In recent years GA has been adopted as a wrapper method due to the high performance it provides and the relative insensitivity to noise, another advantage of a GA is that it can avoid local optima and provide multi criteria optimization functions. Moreover, GAs are more effective in exploring feature spaces of high dimensionality [00]. The proposed approach follows the standard genetic algorithm procedures with the use of its operators.

Population Initialization. The initial population P (0) is created randomly. A population size of 30 is used in our experiment. As from empirical studies, a population size of size 30 and 100 is usually recommended.

Coding Scheme and Individual Representation. According to the genetic analogy the individuals of the population is represented in chromosomes which is composed of several variables (which called genes). The GA maintains a population of N chromosomes which represents the possible solutions that need to compete for the best fitness score. A method is needed to encode potential solutions to that problem in a certain form, so a simple and effective binary coding scheme is used, as the length of the individual represents the candidate features in the sub space.

- Suppose that we identify the total number of features as F.
- Therefore the length of the chromosome (individual) which will represent them is F. in which the total number of available features are 33 (F = 33).
- The chromosome is encoded by N-bit binary vector.
- The variables of each chromosome are initialized randomly to either 1 or 0.
- If the ith position of a chromosome is 1 so that feature represented or participated in constructing the classifier, otherwise it will be 0 which means that the feature isn't participating in the classifier constructed.

Fitness Computation. The total features to be considered in our NER task is 33. The size of feature's subsets is not proportional to the accuracy of the learning algorithm. To evaluate the fitness of an individual in the population (a feature subset), the following fitness function is used in our GA:

$$Fitness(ind_i) = accuracy(ind_i) \tag{3}$$

We tried to prevent repeating the fitness tests by storing each individual's fitness in the memory; so that if those individuals appear in future generations, the system would not re-evaluate them. Therefore the fitness of each individual in the population is only computed when it is not in the memory. This strategy is followed in an attempt to decrease the computation time of the fitness function. The only problem that would appear is the memory overflow so a fixed size is used to store these individuals. And the lowest fitness individual will be removed from memory whenever the memory is full.

Genetic Operators. Genetic algorithms use the main principals of evolution such as reproduction, selection, crossover, and mutation (known as genetic operators):

- *Selection:* After deciding on the encoding scheme, the next decision to make is how the selection will be performed, The main idea behind selection is that better individuals get higher chances to survive and selected as parents, proportional to their fitness, so the fitter the chromosome, the more it's likely to be selected to reproduce, parent selection method used is Roulette wheel selection to implement fitness proportionate selection strategy. It is used to probabilistically select individuals from a population for later breeding.
- *Crossover:* is the recombination phase, Crossover is a critical genetic operator that allows new solution regions in the search space to be explored, produces two new off-springs by copying selected bits from each parent, The cross-over point is chosen randomly, so that the first i bits of the two offspring are contributed by one parent and the remaining bits by the other parent. Here, we use the normal single point crossover.
- *Mutation:* Flip Mutation with a probability of 0.02 is used, which means that every bit in the chromosome has a 2 percent chance of being flipped from 1 to 0 or vice versa. In which features will randomly be added and removed from the subsets.
- *Termination Condition:* the evolution will stop after 30 generations.

4.3   Induction Algorithm

The induction algorithm is used to create the classification model and evaluation of feature subsets. The choice of induction algorithm is independent of the genetic algorithm. Therefore, different induction algorithms can be flexibly incorporated into the proposed method. SVM is the used algorithm in all experiments of the proposed approach due to its superior classification performance, the basic flowchart of the hybrid feature selection system is shown in Fig. 1 below.

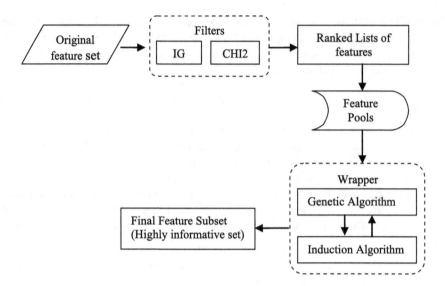

Fig. 1. Proposed hybrid approach

5   Experimental Analysis

5.1   Experimental Setup

A series of experiments were conducted to evaluate the performance of the proposed system in order to extract the various types of NEs. ANERcorp dataset has been used for evaluation as well as comparison with existing systems, that's developed by [10] with 150,286 tokens.

We split it to 80/20 split for training and testing, respectively. As [20] reported the highest value results on the dataset with that split.

In the first stage, we set the threshold T = 22 and choose the following filtering methods: Information Gain and Chi-squared to obtain the relevancy scores of the terms in each class, the resulted lists included the first ranked features, namely the subset T22.

Second Stage: The second stage considers the GA mechanism of feature selection and the classifier accuracy of the mechanism. We proposed support vector machines (SVM) for the accuracy evaluation; parameter tuning is needed for GA in order to optimize its configuration. Different values tested for (i) Population number (10 up to 30) (ii) generation number (10 up to 50) (iii) crossover rate (0.6, 0.9, and 1) (iv) mutation rate (0.005, 0.01, and 0.02).

We experimented different window sizes in our baseline feature set that ranges from (−1/+1 to −4/+4). The performance of the proposed hybrid approach is evaluated and compared with the state of the art baseline results in literature of [18, 20].

5.2   Results and Discussion

First stage results in the following feature pools: FP1 = 18 features which represents the shared features of the two filters and FP2 = 22 features which represents the combined set of features in both ranked lists, results are shown in Table 1.

Table 1. Feature pools

	Feature pools
Shared features	10 N-gram,Gazetteers, Projected English Capitalization, window size
Combined features	12 N-gram, Gazetteers, Projected English Capitalization, POS, window size, word length

Both feature pools fed to the GA-SVM to select the optimal subset of features, First, we tested different values of the Population size starting from 10 up to 50, and then we observed that the best results were obtained when the value assumed by this parameter is 30. Values less than 30 make the algorithm to converge to a local optimum, while values greater than that made the computational load increases considerably. When it comes to the number of generations, we considered values starting from 10 up to 30, because more than that can exhaust the machine without more improvement to the performance. Whilst crossover operator had the best results when the value was 0.9 and did not achieve significant variations when the values change. Finally, the probability of mutation with the value 0.02 gave the best results, in addition, exceeding 0.02 value results in intensive computational load.

The pool FP1 that represents the shared features seems to define the most effective search space, as its feature subset that is obtained by the GA achieved better F-measure than that with the combined features FP2, both subsets resulted from the GA includes the shared subsets of features. We found that a context size of one previous token and one subsequent token (i.e. window size is 3) achieves the best performance in this task.

The first subset SUB1 obtained by GA appears to have improved performance compared to the second subset SUB2. Our approach Slightly superior, leading to 1 point improvement in overall F-measure compared to baseline in literature and outperforms Benajiba's CRF sys by 6 points. Therefore, the effectiveness of GA-SVM is confirmed by finding small subsets of informative features, and irrespective to the size

Morphological features have been excluded from both GA obtained subsets, which emphasize the work of [20] that using n-gram is enough to reach a better performance, but the results show that working with gazetteers can improve the recall along with the capitalization valuable feature extracted from the cross lingual titles.

Still the recall needs more improvement for both subsets which emphasize the construction of other features like larger gazetteers and cross lingual resources such as DBpedia in order to overcome the problems related to low recall.

Finally, Table 6 summarizes the recognition accuracy, in terms of precision, recall, and F-measure, achieved by FS hybrid approach and the other two NER. All systems have been evaluated against Benajiba's ANERcorp (Tables 2, 3, 4 and 5).

Table 2. Feature subsets obtained by GA-SVM

GA-SVM		
Feature pool	FP$_1$	FP$_2$
Subset size	14	17
F-measure	**82**	**77**

Table 3. Baseline results on ANER dataset by(Abdul-Hamid and Darwish, 2010)

	Precision	Recall	F-measure
LOC	93	83	88
ORG	84	64	73
PERS	90	75	82
Overall	89	74	81

Table 4. Performance over SUB1

	Precision	Recall	F-measure
LOC	95	84	89
ORG	87	67	76
PERS	88	75	81
Overall	**90**	**75**	**82**

Table 5. Performance over SUB2

	Precision	Recall	F-measure
LOC	93	86	89
ORG	83	60	70
PERS	84	65	73
Overall	**87**	**71**	**77**

Table 6. Compared results on ANERcorp

	Precision	Recall	F-measure
Benajiba et al.,2008	86	69	76
Abdul- Hamid and Darwish,2010	89	74	81
Our Hybrid FS	**90**	**75**	**82**

6   Conclusion and Future Work

In this Paper, we have presented a hybrid feature selection approach for the Arabic NER task; our aim was to discover the feature combinations that act as better identifiers for the NE class and whether there is a correlation among the useful features. It is composed of a filter and wrapper selection stages, as filtering eliminates irrelevant features and helps defining the most effective search space for the GA, to evaluate the proposed framework we conducted experiments on ANER dataset. Experimental results show that using filter prior to the GA can generate smaller feature subset than the original one, but with a higher F-score for the NER system. The resulted feature set yielded improved results over those in the literature related to ANER dataset with as much as 1 point F-measure improvement for overall F-measure. So it obtained comparable or better results than other selection methods proposed in Arabic NER literature and showed that the hybrid approach is robust and effective in finding small and effective subsets that can enhance the NER system performance. We plan to develop our system using different strategies such as Particle Swarm Optimization and Ant Colony for feature selection in attempt to get optimal feature subset with more reduced computational time.

References

1. Nadeau, D., Sekine, S.: A survey of named entity recognition and classification. Lingvisticae Investigationes. **30**(1), 3–26 (2007)
2. Benajiba, Y., Diab, M., Rosso, P.: Arabic named entity recognition: a feature-driven study. IEEE Trans. Audio Speech Lang. Process. **17**(5), 926–934 (2009)
3. Abouenour, L., Bouzoubaa, K., Rosso, P.: IDRAAQ: new Arabic question answering system based on query expansion and passage retrieval. CLEF (Online Working Notes/Labs/ Workshop) (2012)
4. Babych, B., Hartley, A.: Improving machine translation quality with automatic named entity recognition. In: Proceedings of the 7th International EAMT workshop on MT and other Language Technology Tools, Improving MT through other Language Technology Tools: Resources and Tools for Building MT, pp. 1–8. Association for Computational Linguistics (2003)
5. Zaghouani, W.: RENAR: A rule-based Arabic named entity recognition system. ACM Trans. Asian Lang. Inf. Process. (TALIP) **11**(1), 2 (2012)
6. Shaalan, K.: Rule-based approach in Arabic natural language processing. Int. J. Inf. Commun. Technol. (IJICT) **3**(3), 11–19 (2010)
7. Shaalan, K., Raza, H.: NERA: named entity recognition for arabic. J. Am. Soc. Inf. Sci. Technol. **60**(8), 1652–1663 (2009)
8. Oudah, M., Shaalan, K.: A pipeline Arabic named entity recognition using a hybrid approach. In: Proceedings of the 24th International Conference on Computational Linguistics, COLING 2012, pp. 2159–2176. India (2012)
9. Meselhi, M.A., Bakr, H.M.A., Ziedan, I., Shaalan, K.: A novel hybrid approach to arabic named entity recognition. In: Shi, X., Chen, Y. (eds.) CWMT 2014. CCIS, vol. 493, pp. 93–103. Springer, Heidelberg (2014). https://doi.org/10.1007/978-3-662-45701-6_9
10. Benajiba, Y., Rosso, P., BenedíRuiz, J.M.: ANERsys: an arabic named entity recognition system based on maximum entropy. In: Gelbukh, A. (ed.) CICLing 2007. LNCS, vol. 4394, pp. 143–153. Springer, Heidelberg (2007). https://doi.org/10.1007/978-3-540-70939-8_13
11. Mohit, B., Schneider, N., Bhowmick, R., Oflazer, K., Smith, N.A.: Recall-oriented learning of named entities in Arabic Wikipedia. In: Proceedings of the 13th Conference of the European Chapter of the Association for Computational Linguistics, pp. 162–173. Association for Computational Linguistics (2012)

12. Dash, M., Liu, H.: Feature selection for classification. Intell. Data Anal. **1**(3), 131–156 (1997)
13. Guyon, I., Elisseeff, A.: An introduction to variable and feature selection. J. Mach. Learn. Res. **3**, 1157–1182 (2003)
14. Ratinov, L., Roth, D.: Design challenges and misconceptions in named entity recognition. In: Proceedings of the Thirteenth Conference on Computational Natural Language Learning, pp. 147–155. Association for Computational Linguistics (2009)
15. Pahikkala, T., Airola, A., Salakoski, T.: Speeding up greedy forward selection for regularized least-squares. In: Proceedings of the Ninth International Conference on Machine Learning and Applications, ICMLA 2010, pp. 325–330 (2010)
16. Kabir, M.M., Shahjahan, M., Murase, K..: A backward feature selection by creating compact neural network using coherence learning and pruning. In: Proceedings of JACIII 11, pp. 570–581 (2007)
17. Benajiba, Y., Rosso, P.: Anersys 2.0: conquering the ner task for the Arabic language by combining the maximum entropy with pos-tag information. In: IICAI, pp. 1814–1823 (2007)
18. Benajiba, Y., Rosso, P.: Arabic named entity recognition using conditional random fields. In: Workshop on HLT & NLP within the Arabic World. Arabic Language and Local Languages Processing: Status Updates and Prospects (2008)
19. Benajiba, Y., Diab, M.T., Rosso, P.: Using language independent and language specific features to enhance Arabic named entity recognition. Int. Arab J. Inf. Technol. **6**(5), 463–471 (2009)
20. Abdul-Hamid, A., Darwish, K.: Simplified feature set for Arabic named entity recognition. In: Proceedings of the 2010 Named Entities Workshop (NEWS 2010), pp. 110–115, Stroudsburg, PA, USA. Association for Computational Linguistics (2010)

Named-Entity-Recognition (NER) for Tamil Language Using Margin-Infused Relaxed Algorithm (MIRA)

Pranavan Theivendiram, Megala Uthayakumar[(✉)], Nilusija Nadarasamoorthy,
Mokanarangan Thayaparan, Sanath Jayasena, Gihan Dias, and Surangika Ranathunga

Department of Computer Science Engineering, University of Moratuwa, Moratuwa, Sri Lanka
{pranavan.11,megala.11,nilu.11,mokaranagan.11,
sanath,gihan,surangika}@cse.mrt.ac.lk

Abstract. Named-Entity-Recognition (NER) is widely used as a foundation for Natural Language Processing (NLP) applications. There have been few previous attempts on building generic NER systems for Tamil language. These attempts were based on machine-learning approaches such as Hidden Markov Models (HMM), Maximum Entropy Markov Models (MEMM), Support Vector Machine (SVM) and Conditional Random Fields (CRF). Among them, CRF has been proven to be the best with respect to the accuracy of NER in Tamil. This paper presents a novel approach to build a Tamil NER system using the Margin-Infused Relaxed Algorithm (MIRA). We also present a comparison of performance between MIRA and CRF algorithms for Tamil NER. When the gazetteer, POS tags and orthographic features are used with the MIRA algorithm, it attains an F1-measure of 81.38% on the Tamil BBC news data whereas the CRF algorithm shows only an F1-measure of 79.13% for the same set of features. Our NER system outperforms all the previous NER systems for Tamil language.

Keywords: NER · NLP · Tamil · NE · CRF · Margin-Infused relaxed algorithm
MIRA

1 Introduction

Named-Entity-Recognition (NER) is a task of identifying named entities in a given text. In other words, NER is used to locate and classify elements in a text into predefined categories such as the names of persons, organizations, locations, date/time, quantities, currency values and percentages. It is a subtask in information extraction. NER is an important precursor in many Natural Language Processing (NLP) tasks such as document summarization, intelligent document access, speech related tasks, machine translation and question answering systems [1].

NER systems built before were based on machine-learning techniques such as HMM (Hidden Markov Model), MEMM (Maximum-Entropy Markov Model), SVM (Support Vector Machine) and CRF (Conditional Random Fields). Among all those, CRF is the most widely used technique. Stanford NER system also uses the CRF algorithm [2]. Most researchers have preferred the CRF algorithm in the past over other multi-class machine learning algorithms such as HMM and MEMM since HMM

© Springer International Publishing AG, part of Springer Nature 2018
A. Gelbukh (Ed.): CICLing 2016, LNCS 9623, pp. 465–476, 2018.
https://doi.org/10.1007/978-3-319-75477-2_33

suffers from the dependency problem and data sparsity problem and MEMM suffers from label bias problem [3]. CRF gave the best prediction accuracy among the machine learning algorithms.

In addition to the supervised learning techniques, some attempts on building NER systems using semi-supervised learning techniques have been made [4, 5]. Those attempts are based on the active learning approach, which reduces the need of annotated corpus by 80% while maintaining the performance. The success of the system purely depends on informativeness, representativeness and diversity of the selected corpus. Selection of parameters such as batch size and lambda of function for the action learning is quite hard task.

There are 3 existing NER systems for Tamil language. Those systems are implemented using algorithms such Expectation-Maximization [7], SVM [3] and CRF [3, 6]. Previous Tamil NER systems used a limited number of features and these systems are not available for public usage. This has slowed down the Tamil NLP related development. Our research aims to overcome these limitations and to accelerate Tamil NLP related developments by facilitating the identification of 5 different named entities: individual, place, organization, count and time expressions. These tags are selected because person, location, organization, numerical expression and time expression are considered as main Named Entities and increasing the tag set will influence the accuracy of the NER system negatively.

For the named entity classification task, we employ the Margin-Infused Relaxed Algorithm (MIRA). MIRA is a new multi-class algorithm, which has been shown as promising and having potential to be better than most of the other multi-class classification algorithms [8, 9]. This is the first time MIRA is used for Tamil NER. Results show that our MIRA-based NER system gives better performance than all previous Tamil NER approaches. Our system makes use of a basic Part-Of-Speech (POS) tagger and a morphological analyzer (both implemented by us) as an integral part of the system.

This paper consists of five sections. The next section discusses previous work on building NER systems for Tamil. Section 3 describes our approach. Section 4 gives evaluation results of our system, and the final section gives the details about future work and concludes the paper.

2   Related Work

In the past, three different types of techniques have been used for building NER systems. Those are machine-learning techniques [1, 10, 11] grammar based techniques [13] and hybrid based approaches [12]. However, Tamil NER systems have been built only using machine learning techniques [3, 6] and hybrid based approaches [7].

In the context of Tamil NER systems, Vijaykrishna and Shobha [6] built a tourism domain specific NER system for Tamil Language using the CRF algorithm. However they have only used the noun phrases for training and testing. Geetha and Pandian [7] developed a generic NER system for Tamil using the Expectation Maximization (EM) algorithm. Malarkodi et al. [3] created a generic NER system for Tamil language using the CRF algorithm and SVM algorithm separately to compare the performance of the two algorithms. That research revealed CRF outperforms SVM in the context of Tamil NER. Among those generic NER systems for Tamil language, the best F1-measure obtained so far is 71.68% [3]. None of these Tamil NER systems is available for public usage and some of those systems use Romanization to handle the complexity introduced by some Tamil letters that are made of two or more Unicode code points.

Previous Tamil NER systems used word, POS [3, 6, 7], chunk [3] and patterns (for date and time, in particular) [6] as features.

3   Our Approach

This section describes our approach and the important components of the Tamil NER system. Figure 1 gives the high-level view of the system. Prefix and gazetteer lists are newly used features in our system with respect to the past Tamil NER systems. Our system has three main components: morphological analyzer, part of Speech (POS) tagger and the Named Entity tagger. Morphological analyzer is used to get the stem of the word and a noun/verb tag for the word based on contextual rules.

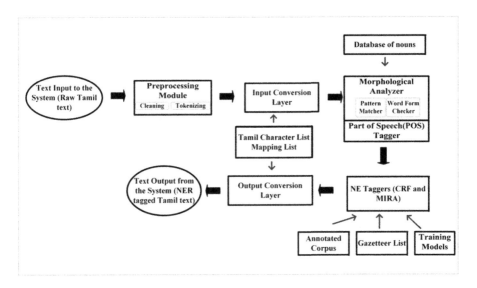

Fig. 1. Overview of the system

3.1   Corpus and Word List Databases

Corpus and the list of words play a vital role in the accuracy and performance of the system. System accuracy increases with the increase of the corpus size.

An annotated corpus of 125,000 words is used for training and testing of the Tamil NER system. This corpus is created from the Tamil BBC newspaper articles.

3.2 Pre-processing

The system uses text files, online resources and websites as an input to the system. It extracts the content and sends them to the preprocessing module. At the preprocessing stage, the content is cleaned and noisy data are removed. Then, the content is tokenized using the Stanford tokenizer [14]. Stanford tokenizer differentiates the sentences by using the "." as an indicator. However, "." is used in some other instances other than the end of the sentence. Following are the some usages of "." other than the end of the sentence:

- Titles - திருமதி (Mrs.), திரு (Mr.)
- Initials - எம்.என்.பெர்னாண்டோ (M. N. Fernando)
- Abbreviation of organization names, political party names, etc. - த.தே.கூ (T. N. A)

In order to overcome above-mentioned misinterpretations in detecting sentences, some rules were incorporated in to the system during the preprocessing phase and the mistakes are corrected.

In Tamil, following are some obvious rules:

1. Sentences do not end with a single letter word
2. Initials and titles are finite and well defined set in Tamil, so we created a list of titles and initials to check those in a given text

Therefore, we incorporated those rules for correcting the problems caused by tokenization.

3.3 Input-Output Conversion Layer

Most of the letters of Tamil Language consist of two Unicode points. Handling one Tamil letter as two Unicode characters gave problems in morphological analysis. Input-Output Layer makes it possible to deal with one Tamil letter as one object. In addition, it gives the opportunity to find the vowel and consonant of compound characters.

For example, க் +ஆ = கா ; if we have the character கா, we can infer consonant க் and vowel ஆ.

As shown in Fig. 1 the raw text is fed to the input conversion layer. After all the NE related processing is done on the objects, the objects are converted back to Unicode format using output conversion layer.

3.4 Morphological Analyzer

Morphological analyzer is a program to analyze the internal structure of the word. It uses different features, language properties and pattern matching. As Tamil is a morphologically rich and agglutinative language (one word affixing with two or more morphemes is known as agglutinative nature), finding the stem of the word is difficult. Most of the words are pinned with morphemes.

A partial morphological analyzer is enough for the NER system. In the context of the Tamil NER system, the morphological analyzer is used for 2 purposes.

1. To identify the stem of the word, in order to reduce the complexity of handling inflected forms of words
2. To derive noun/verb tags for a given word to be used in contextual rules. In addition to that these tags are also used as features for NER algorithms

Our morphological analyzer uses a rule-based approach, which returns only the stem word when the input word is given. It also identifies whether the word is a noun or a verb. For the purpose of rule-based morphological analysis, 100,000 noun stems and verb stems were extracted from the Tamil WordNet using a web crawler. A noun and verb database is created from them. Having all the patterns of a word in the corpus is impossible due to the agglutinative nature of Tamil language. Therefore, the corpus only keeps the stem of the words. Therefore, before checking with the corpus, we find the stem of the word using some rule-based and pattern-matching approaches. Linguistic preprocessing is used to extract the stem from the word.

3.5   Gazetteer and Rule-Based Module

A Gazetteer is created with 100,000 names of people from Sri Lanka. It is a collection of names of Sinhala, Tamil and individuals from other ethnic groups in Sri Lanka. That is, some names are original Tamil names (written originally in Tamil) and others are transliterated from Sinhala (the main language in Sri Lanka). Some of these transliterated names lead to inaccuracies in the Tamil context. For example, the following Sinhala name is conflicting with a Tamil word (which is actually not a name in Tamil). The highlighted portion of the following name caused the contextual problems.

– **இந்து** தரங்கணீசமரரத்ன கொடிகார
(**Indhu** Tharangani Samararathna Kodikara)
"Indhu" means "Hinduism" in Tamil

Such conflicting names are very rare; therefore, probability of such names occurring in a text of interest is very low. Therefore, such names were removed from the Gazetteer.

The clue words, which are usually useful to find named entities at the first step, are identified and are used to create a rule-based module, which can generate a feature for the Machine Learning model. In the rule-based module, we used Gazetteer of names and clue words for Person, Organizations, Places, and Count and compared them with the text to be tagged. If the clue words are found in the given text, we indicate that as a feature to the machine learning based tagger.

Organizations have about 100 clue words, which are usually used at the end of the organization names.

Eg: கூட்டுத்தாபனம் (Corporation), **திணைக்களம்** (Department)

For persons, we have used title and salutation words. The following are some example of title words.

– **திரு** (Mr), **திருமதி** (Mrs), **செல்வி** (Miss)

There will be a high chance that a name of a person following a title word. For places, we have added the main cities and names of all the countries of the world to the clue words list of places. For count (numbers), literals used to represent numbers and decimals are added to the clue words list.

3.6   Part-of-Speech (PoS) Tagger

Part Of Speech tagging is marking up a word in a text with a corresponding predefined part of speech tag. PoS is a useful feature in NER because most of the named entities, especially names are proper nouns. Hence, there is a high chance that a noun can be a Named Entity.

Stanford PoS tagger [15] is used to build a Tamil PoS tagger. The standard PoS definition has 32 tags [16]. Classification into higher number of classes ended up in higher error rates. Hence, tags set is defined in manner where 21 tags are used. 80,000 words are tagged manually and used as training data for the tagger. These tags are NN, NNC, RB, VM, SYM, PRP, JJ, NNP, PSP, QC, VAUX, DEM, UT, QF, NEG, QO, CC, WQ, INTF, NNPC and RBP[1].

3.7   Named-Entity Tagger

This part is implemented using the MIRA classifier. We also implemented it using CRF as well, in order to compare with the performance of MIRA. Both the classifiers are trained using CRF++ tool [17]. Following are the tags used for tagging the named entities in the given text.

– INDIVIDUAL
– PLACE
– ORGANIZATION
– TIME
– COUNT

The following are the features considered for building the classifiers.

– Contextual length (window size)
– Part-of-Speech (POS) tag
– Noun and verb phrases derived from the morphological analyzer
– Gazetteers
– Surefire rules
  Example: We have used the following as surefire rules in order to facilitate the identification of the NEs in a text.

---

[1] NN - Noun, NNC - Compound Noun, RB - Adverb, VM - Verb Main, SYM - Symbol, PRP - Personal Pronoun, JJ - Adjective, NNP - Pronoun, PSP - Prepositions, QC - Quantity Count, VAUX - Verb Auxiliary, DEM - Determiners, QF - Quantifiers, NEG - Negatives, QO - Quantity Order, WQ - Word Question, INTF - Intensifier, NNPC - Compound Pro Noun.

Surefire rules:

- For individuals: there is a high chance that a name of an individual follows after titles such as Mr., Mrs., Prof., Dr.
- For organizations: there is a set of starting and ending words such as station, department, organization, university and so on.

- Prefixes and suffixes

  E.g. Prefix for "University": U, Un, Uni,..
  Suffix for "University": y, ty, ity,….

Some NEs start with certain prefix. For example, **சிவகாமி** (Sivagamy), **சிவநாதன்** (Sivanathan), **சிவநேசன்** (Sivanesan) are some names which starts with the same prefix "**சிவ** (Siva)". Similar logic applies for suffix as well.

- Orthographic features

  Orthographic features are like a pattern, which match a tokenized word. In the following examples, "X" is used as a placeholder to represent Tamil characters whereas digits are denoted with the letter "N". Other special characters are used as it is.

E.g.: மாலா   → XX

  போனாள்   → XXX

  45   → NN

In particular, orthographic feature is very helpful in identifying date and time. For instance, NN-NN-NNNN pattern shows it is a date.

- Length of word

Gazetteer, surefire rules, prefix, suffix, orthographic features and length of words are novel features used by us (no other Tamil NER systems used them). These features are selected based on previous successful researches on other languages [1, 8]. Part of our research involves identification of best feature combination for Tamil NER system. However, we were not able to use some of the features used in other languages due to language and resource constraints. Consider following two examples.

1. Most of the English NER systems use capitalized first letter of a word as a feature for their NE systems as English has a capitalization concept for proper nouns. However, it is not applicable for Tamil language, which does not have the capitalization concept.
2. Stanford NER [2] uses distributed similarity feature, which is based on similarity between words. In order to use this feature, an annotated clustered corpus is required. However, that type of resource is not available for Tamil language.

3.8 MIRA Based Tagger

MIRA has been proven to be better than CRF in the context of some other languages [9], and for some other languages CRF and MIR performs equally [18]. Bengali Named Entity Recognition system had a better performance when using MIRA algorithm rather than CRF when tested using South and South East Asian Languages (NERSSEAL) shared task data. MIRA based English Named Entity Recognition system had similar performance as CRF algorithm when tested with CoNLL-2003 data set. Tamil belongs to Dravidian language family while Bengali belongs to Indo-Aryan family and English belongs to Indo-European family, but Dravidian languages share strong areal feature with Indo-Aryan languages [19] and they do not show any significant connection with Indo-European languages. Therefore, there is a high chance that Tamil NER may show similar performance as Bengali NER. Therefore, we decided to test the performance of MIRA algorithm in the context of Tamil NER.

MIRA is an online algorithm, which is based on error minimization. It makes use of a matrix to build a model. In each iteration, different matrices are considered by making a small change to the parameters of the earlier matrix and the matrix that makes the lowest error is selected as the final matrix. Likewise, iterations are continued throughout the training data and final matrix is discovered. Pseudo code of MIRA algorithm [8] is given below,

$Initialize: Set\ M \neq 0\ M \in R^{kxn}$

$Loop: For\ t = 1, 2, \ldots, T$

- $Get\ a\ new\ instance\ \overline{x^t}$

- $Predict\ \hat{y^t} = argmax_r \left\{ \overline{M_r}.\overline{x^t} \right\}$
- $Get\ a\ new\ label\ y^t$
- $Find\ \overline{\tau^t}\ that\ solves\ the\ following\ optimizatio\ problem:$

$$min_\tau = \frac{1}{2} \sum ||\overline{M_r} + \tau_r \overline{x^t}||_2^2$$
$$Subject\ to: (1)\ \tau_r \leq \delta_{r,y^2}\ for\ r = 1, \ldots, k$$

$$(2) \sum_{r=1}^{k} \tau_r = 0$$

- $Update: \overline{M_r} \leftarrow \overline{M_r} + \tau_r^t \overline{x^t}\ for\ r = 1, 2 \ldots, k$

$$Output = H(\bar{x}) = argmax_r \left\{ \overline{M_r}.\bar{x} \right\}$$

3.9 Conditional Random Field (CRF) Based Tagger

Another classifier used by this system is based on the CRF algorithm. We have selected CRF since it has shown good performance over other techniques in Tamil NER as stated before.

Lafferty et al. [20] defines the CRF algorithm as follows:

Let G = (V, E) be a graph such that, $Y = \left(Y_v \right)_{v \in V}$, so that Y is indexed by the vertices of G. Then (X, Y) is a conditional random field in case, when conditioned on X, the random variables Yv obey the Markov property with respect to the graph:

$p(Y_v | X, Y_w, w \neq v) = p(X, Y_w, w \sim v)$, where $w \sim v$ means that w and v are neighbors in G.

The joint distribution over the label sequence Y given X has the form,

$$p_\theta \propto \exp \sum_{e \in E,k} \lambda_k f_k(e, y|_e, X) + \sum_{e \in V,k} \mu_k g_k(v, y|_v, X)$$

Where x is a data sequence, y a label sequence, and $y|S$ is the set of components of y associated with the vertices in sub graph S. CRF predicts the label for the given data based on the probability of a label given the features.

4 Evaluation

The POS Tagger was trained using the 80,000 words from the Tamil BBC News. Precision measures for POS tagger is given in Table 1.

Table 1. Result of the POS tagger

Number of words	Correct predictions	Precision %
747	707	94.65
1,169	1,104	94.44
1,567	1,473	94.00

125,000 words collected from Tamil BBC News are used as the data set for Tamil NER system. Evaluations were performed using 10-fold cross validation. For each instance of testing, 112,500 words are used for training the model and the testing is done using the remaining 12,500 words. Table 2 gives the comparison of MIRA algorithm and CRF algorithm with respect to various features.

Table 2. Comparison of MIRA and CRF algorithms

Features used	F1 - measure	
	MIRA	CRF
With only word features	61.0%	46.6%
Word features + POS	72.20%	68.34%
Word features + noun and verb tags derived from morph	72.25%	57.37%
Word features + POS + noun and verb tags derived from morph	70.93%	67.37%
Word features + POS + noun and verb tags derived from morph + gazetteer + surefire rules	78.45%	77.53%
Word features + POS + noun and verb tags derived from morph + gazetteer + surefire rules + suffix	76.73%	76.15%
Word features + POS + noun and verb tags derived from morph + gazetteer + surefire rules + prefix	80.73%	79.13%
Word features + POS + noun and verb tags derived from morph + gazetteer + surefire rules + orthographic features	81.38%	76.76%

It is clear that the MIRA algorithm outperforms the CRF algorithm in most of the instances, based on the overall F1-measure of the system. With the above testing, we found that the optimal features for MIRA model are window size 3, PoS, noun and verb tags derived from morph, gazetteers, surefire rules, prefix of length 4 and orthographic features. Optimal features for CRF model are window size 3, POS tag, noun and verb tags derived from morph, gazetteer, surefire rules and prefix length 2. Our tests revealed that the suffix feature is not suitable to be used in conjunction with the gazetteer feature as it negatively affects the F1-measure. Table 3 gives the evaluations results of individual entities for CRF and MIRA models with the best feature combinations of respective algorithms.

Table 3. Comparison of MIRA and CRF algorithm for different entities

Named entity	Precision (in %)		Recall (in %)		F1-measure (in %)	
	MIRA	CRF	MIRA	CRF	MIRA	CRF
INDIVIDUAL	85.12	**86.00**	**84.68**	77.36	**84.90**	81.45
ORGANIZATION	85.92	**93.33**	**67.78**	62.22	**75.78**	74.67
COUNT	**90.45**	82.84	78.92	**82..84**	**84.29**	82.84
PLACE	**95.30**	93.94	**69.25**	67.39	**80.22**	78.48
TIME	95.08	**100**	**71.60**	64.20	**81.69**	78.20
OVERALL	90.37	**91.22**	**74.45**	70.80	**81.38**	79.13

With the above table, it is very clear that MIRA outperforms CRF in most of instances and the test results reveal that when making predictions, MIRA focuses more on recall and the CRF focuses more on precision.

Table 4 gives the comparison of our approach with the previous attempts. Our system shows an increase of 9% in F1-measure with respect to previous generic NER systems and our system outperforms the domain focused NER system with a small margin.

Table 4. Comparison of our approach against previous Tamil NER systems

Approaches	Precision	Recall	F1-measure
Shobha and Vijakrishna`s tourism domain NER (2008)	88.52%	73.71%	80.44%
Geetha and Pandian`s generic NER (2008)	83.01%	64.70%	72.70%
Malarkodi et al. generic NER (2012)	71.28%	70.51%	70.68%
Our generic NER (2016)	**90.37%**	**74.45%**	**81.38%**

5   Conclusion

This paper presented a Named Entity Recognition (NER) system for Tamil language using MIRA. By trying out different features, we have found the optimum combination of features to get a F1- measure of 81.38%. Optimum set of features are word interval, PoS tag, noun and verb tags derived from morph, gazetteer list, sure-fire rules, prefixes and orthographic features. When compared with CRF, the algorithm traditionally used for Tamil NER, MIRA classifier out-performs it in most of the instances for Tamil NER.

As future work, we expect to support more Named Entity types such as currency and percentage. In addition to that, we expect to expand the corpus and to identify further features in order to increase the accuracy of the system.

Acknowledgement. We would like to thank AU-KBC research centre of Chennai, Forum for Information Retrieval Evaluation (FIRE) and Department of Registrations of Persons Sri Lanka for providing us necessary language resources and tools to carry out this research.

References

1. Nadeau, D., Sekine, S.: A survey of named entity recognition and classification. Lingvisticae Investig. **30**(1), 3–26 (2007)
2. Finkel, J.R., Grenager, T., Manning, C.: Incorporating non-local information into information extraction systems by gibbs sampling. In: Proceedings of the 43rd Annual Meeting on Association for Computational Linguistics, pp. 363–370 (2005)
3. Malarkodi, C.S., Pattabhi, R.K., Sobha, L.D.: Tamil NER–coping with real time challenges. In: 24th International Conference on Computational Linguistics, pp. 23–38 (2012)
4. Laws, F., Schätze, H.: Stopping criteria for active learning of named entity recognition. In: Proceedings of the 22nd International Conference on Computational Linguistics-Volume 1, pp. 465–472 (2008)
5. Shen, D., Zhang, J., Su, J., Zhou, G., Tan, C.-L.: Multi-criteria-based active learning for named entity recognition. In: Proceedings of the 42nd Annual Meeting on Association for Computational Linguistics, p. 589 (2004)
6. Vijayakrishna, R., Sobha, L.: Domain focused named entity recognizer for tamil using conditional random fields. In: Proceedings of the IJCNLP-08 Workshop on NER for South and South East Asian Languages, pp. 59–66 (2008)
7. Pandian, S., Pavithra, K.A., Geetha, T.: Hybrid three-stage named entity recognizer for tamil. In: The Sixth Annual Conference on Informatics and Systems (INFOS), pp. 45–52 (2008)
8. Crammer, K., Singer, Y.: Ultraconservative online algorithms for multiclass problems. J. Mach. Learn. Res. **3**, 951–991 (2003)
9. Banerjee, S., Naskar, S.K., Bandyopadhyay, S.: Bengali named entity recognition using margin infused relaxed algorithm. In: Sojka, P., Horák, A., Kopeček, I., Pala, K. (eds.) TSD 2014. LNCS (LNAI), vol. 8655, pp. 125–132. Springer, Cham (2014). https://doi.org/10.1007/978-3-319-10816-2_16
10. Cunningham, H., Tablan, V., Roberts, A., Bontcheva, K.: Getting more out of biomedical documents with GATE's full lifecycle open source text analytics. PLoS Comput. Biol. **9**(2), e1002854 (2013)
11. Bird, S.: NLTK: the natural language toolkit. In: Proceedings of the COLING/ACL on Interactive Presentation Sessions, pp. 69–72 (2006)
12. Ekbal, A., Haque, R., Das, A., Poka, V., Bandyopadhyay, S.: Language independent named entity recognition in indian languages. In: IJCNLP, pp. 33–40 (2008)
13. Mikheev, A., Moens, M., Grover, C.: Named entity recognition without gazetteers. In: Proceedings of the Ninth Conference on European Chapter of the Association for Computational Linguistics, pp. 1–8 (1999)
14. Manning, C., Surdeanu, M., Bauer, J., Finkel, J., Bethard, S., McClosky, D.: The stanford CoreNLP natural language processing toolkit. In: Proceedings of 52nd Annual Meeting of the Association for Computational Linguistics: System Demonstrations, pp. 55–60 (2014)

15. Toutanova, K., Klein, D., Manning, C.D., Singer, Y.: Feature-rich part-of-speech tagging with a cyclic dependency network. In: Proceedings of the 2003 Conference of the North American Chapter of the Association for Computational Linguistics on Human Language Technology-Volume 1, pp. 173–180 (2003)
16. Dhanalakshmi, V., Shivapratap, G., Soman Kp, R.S.: Tamil POS tagging using linear programming. Int. J. Recent Trends Eng. **1**(2), 166–169 (2009)
17. Kudo, T.: CRF++: Yet another CRF toolkit, CRF++: Yet Another CRF toolkit (2005). https://taku910.github.io/crfpp/. Accessed 24 Jan 2016
18. Crammer, K., McDonald, R., Pereira, F.: Scalable large-margin online learning for structured classification. In: NIPS Workshop on Learning With Structured Outputs (2005)
19. Krishnamurti, B.: The Dravidian Languages. Cambridge University Press, Cambridge (2003)
20. Lafferty, J., McCallum, A., Pereira, F.C.: Conditional random fields: Probabilistic models for segmenting and labeling sequence data. In: ICML 2001 Proceedings of the Eighteenth International Conference on Machine Learning, pp. 282–289 (2001)

Word Sense Disambiguation and Anaphora Resolution

Word Sense Disambiguation Using Swarm Intelligence: A Bee Colony Optimization Approach

Saket Kumar and Omar El Ariss[(✉)]

Department of Computer and Mathematical Sciences,
Pennsylvania State University, Harrisburg, PA, USA
{sxk516,oelariss}@psu.edu

Abstract. Word Sense Disambiguation (WSD) is a problem of figuring out the correct sense of a word in a given context. We introduce an unsupervised knowledge-source approach for word sense disambiguation using a bee colony optimization algorithm that is constructive in nature. Our algorithm, using WordNet, optimizes the search space by globally disambiguating a document by constructively determining the sense of a word using the previously disambiguated words. Heuristic methods for unsupervised word sense disambiguation mostly give less importance to the context words while determining the sense of the target word. In this paper, we put more emphasis on the context and the part of speech of a word while determining its correct sense. We make use of a modified simplified Lesk algorithm as a relatedness measure. Our approach is then compared with recent unsupervised heuristics such as ant colony optimization, genetic algorithms, and simulated annealing, and shows promising results. We finally introduce a voting strategy to our algorithm that ends up further improving our results.

Keywords: Word sense disambiguation · Bee colony optimization
Semantic relatedness · Lesk algorithm · Unsupervised WSD · WordNet

1 Introduction

Humans can infer meaning from various conflicting dentitions through the use of not only the definition of a word but also based on their experiences and the text's context and domain. A word has one or more different senses attached to it. Each and every sense of a word is represented by a definition, a list of synonym, and an example of sense usage. Word Sense Disambiguation (WSD) is the process of assigning the most appropriate sense of a word in a particular context. The functional importance of WSD lies in processing the sequence of words by pinpointing their meaning without the need for human intervention. It is crucial for many applications such as translation, summarization, and information retrieval.

Algorithms for WSD fall into two main groups, supervised and unsupervised. Supervised approaches for WSD make use of machine learning algorithms such as classification and feature recognition. The resultant classification models are trained on an annotated corpus, which is a collection of data that has been tagged correctly by a

© Springer International Publishing AG, part of Springer Nature 2018
A. Gelbukh (Ed.): CICLing 2016, LNCS 9623, pp. 479–495, 2018.
https://doi.org/10.1007/978-3-319-75477-2_34

linguist. The process of manual sense annotation of words to create a large corpus of examples is laborious and requires the presence of a linguist. On the other hand unsupervised approaches use information from an external knowledge source to determine the exact sense of these words. The knowledge source can be a machine readable dictionary, or an organized network of words or semantics [1]. In addition, the knowledge source should be periodically updated as there are domain changes, new senses are introduced, and new words are added. These changes impact the semantic networks and require them to be updated.

WSD falls in the class of AI-complete problems, which means it is at least as hard as most of the difficult problems in AI, and by parallel comparison, in complexity theory it is NP-complete [2]. There are many distinct reasons why WSD is a difficult problem [2]:

- It is difficult to define the senses of words and the level of detail represented by a particular sense with respect to a sense usage.
- It is difficult to determine if the word should be disambiguated for a more generic sense or for a finer sense in a given context.
- It is difficult to determine how much context to use to achieve the most accurate disambiguation.

In this paper, we introduce an unsupervised knowledge-source approach for WSD. The algorithm is based on Bee Colony Optimization (BCO) and we will refer to it in this paper as BCA-WSD. The population-based approach exploits the relationship between the word and the context in which it occurs. This is done by strengthening the relationship between the senses and the surrounding words through the use of the senses' parts of speech, definitions, entailments, synonyms, and attributes. The algorithm's knowledge source is a database of semantic networks called WordNet 2.1 [3].

BCA-WSD endeavors to solve the WSD as an optimization problem by maximizing the relatedness measure between all the words in a document. We first generate artificial bees based on the word chosen to represent the hive. The bees then travel to other words in the document and based on their path, assign senses to other words. Furthermore and in order to improve the results, we introduce a voting strategy to our BCA-WSD.

Our algorithm uses for evaluation a dataset from SemEval 2007 [4], an international workshop on semantic evaluations. The test data is composed of five different documents ranging from general topics to more specific topics. The results are compared against an ant colony optimization (ACO) algorithm, a genetic algorithm (GA), a simulated annealing (SA), and the Most Frequent Sense (MFS) algorithm. MFS acts as a baseline for all the approaches.

There are two main contributions for our work. The first one lies in the way a heuristic solves the problem. The main difference between our approach and other heuristic approaches for WSD is that ours is constructive in nature. Previous heuristic approaches initially start with randomly generated solution and improve on it, while our algorithm starts with one disambiguated word and then works on solving the problem by disambiguating one additional word at a time. This process is similar to a human disambiguating a sentence or a document.

The second contribution lies in the relatedness measure that is used. Most heuristic approaches use original Lesk algorithm [5], which gives less importance to the context words while determining the sense of the target word. In this paper, we put more emphasis on the context and the part of speech of the words while determining their correct senses. We make use of the simplified Lesk algorithm [6]. This relatedness measure gives more importance to context of the target words than any other relatedness measure. We also extend the gloss, while using the simplified Lesk algorithm, by adding additional relation words from WordNet. The extension of glosses gives higher probability of context words when used in simplified Lesk algorithm.

The rest of the paper is structured as follows. In Sect. 2, we describe the previous work done on WSD. In Sect. 3 we discuss the general Bee Colony Algorithm approach. Section 4 introduces the basic Bee Colony Algorithm with respect to WSD. Section 5 evaluates the proposed approach, and then compares our results with other approaches such as ACO, GA, SA and MFS. Finally we conclude the paper while focusing on some future directions.

2 Related Work

There are two main approaches to solve WSD: supervised and unsupervised [7]. In the supervised approach, the available data is divided, in an appropriate ratio, into training data and test data. The WSD task then becomes a classification problem. This approach requires a large corpus of data that is tagged with correct senses. Classifiers such as decision trees, Naive Bayes, neural networks, and support vector machines are popular for this approach. It has been observed that supervised WSD methods provide better results than unsupervised approaches for WSD [2, 4]. The main disadvantage of the supervised approach is the need for sense-annotated data.

Supervised WSD systems that are built on a particular domain and then applied on other domains usually summer from a drop of accuracy. Chan and Ng [8] proposed an adaptation of WSD system to new domains by introducing the concept of active learning, count merging and expectation-maximization. Their approach helps to leverage the accuracy of the WSD system while reducing the annotation effort. This is done by assigning weights to the examples in the training set and determining their predominant senses. On the other hand, Zhu and Hovy [9] addressed the problem of over-sampling and under-sampling of active learning methods when applied to WSD systems. They proposed a bootstrap-based approach for over-sampling, which yields above par accuracy than the normal over-sampling approaches. They also proposed to determine the stopping condition of an active learning for WSD systems based on the max-confidence and min-error techniques.

The unsupervised approach has the advantage of not depending on hand-annotated data [2]. The general idea of this approach is based on the fact that the correct sense of a target word will have similar words when used in the same domain. For example, the target word "code" in the domain of computer science may have the words "computer" and "program" in the surrounding text. The WSD problem is then reduced into measuring the similarity of the context words with the sense in question. Instead of learning from a hand-annotated data, it makes use of a knowledge source such as a dictionary or a semantic network. In this paper, we are going to focus on the unsupervised approach for determining the correct senses of the target words.

Michael Lesk introduced a dictionary-based algorithm [5] that counts the number of overlapping words between the definition of the target word sense and all the senses of the surrounding words in context. An overlapping word is defined as the common word that occurs in two sets of words. This process is repeated for all the senses of the target word, where stop words are not considered as overlapping words. In the original Lesk algorithm, when a target word is disambiguated with its surrounding words, the definition and examples of each sense of the target word are compared to the definition and examples of each sense of the surrounding words. The sense for the target word with the maximum overlapping of words is considered to be the correct sense.

Cowie [10] proposed a simulated annealing (SA) approach that uses Longman's Dictionary of Contemporary English (LDOCE) as the external knowledge source. This approach works on each sentence individually in a text. The relatedness measure in this approach depends upon the existence of the surrounding words in a particular gloss of a sense. Initially, a configuration of senses is prepared for a particular sequence of words. Then the configuration is changed by selecting a random word and its random sense. The probability of keeping the configuration of senses fixed depends upon the total score of all the senses relative to the previous configuration score.

Gelbukh [11] proposed a genetic algorithm (GA) approach that uses the original Lesk algorithm as the similarity measure. It enriches the glosses by using the synonym relation between the words. It also takes into account the linear distance between two words in the context window period when disambiguating each other. This approach stresses the fact that a sense of a particular word is determined by the surrounding words in its context and aims to maximize the notion of coherence.

Schwab [12] proposed the use of Ant Colony Optimization to solve WSD, which we reference in this thesis as ACA-2011. This approach uses a variant of the original Lesk algorithm where glosses are expanded using the relations in WordNet. Each word is represented as a component in a vector. The algorithm starts with a graph. The graph is built based on the text structure where each word is linked to its predecessor sentence, and each sentence is linked to its predecessor paragraph. ACA-2011 performed better than the original Lesk algorithm (brute force approach) using a limited amount of time.

Schwab [13] proposed another use of Ant Colony Optimization, which we refer to in this paper as ACA-2012. ACA-2012 was an adaptation of ACA-2011 with the inclusion of majority voting approach that performed better than ACA-2011 in terms of efficiency and correctness.

Nguyen [1] proposed another Ant Colony Optimization algorithm for WSD, which we refer to as TSP-ACO. This approach endeavors to solve the WSD problem by relating it to the Travelling Salesman Problem (TSP). It uses the original Lesk algorithm in a vector model space where each gloss is extended by the relations in WordNet and the glosses from the extended WordNet and Wikipedia. This approach also differs from other unsupervised approach by using a combination of knowledge sources instead of a single knowledge source.

3   Bee Colony Algorithm

The subfield of artificial intelligence that is based on the research of the behavior of common individuals in different decentralized systems is known as swarm intelligence [14]. Insects in nature have a very subtle way of adjusting to the environment. They make use of their large population to build a decentralized system. They fetch, share, and make collective decisions for flourishing their colonies. There are two vital properties of swarm intelligence: self-organization and division of work. These properties help in solving complex distributed problems. Self-organization can be characterized in terms of positive information, negative information, fluctuation, and multiple interactions. Delegation and the division of work are based on the skills of the insect at the local level. The adaptation of swarm intelligence to the search space based on the low-level communications helps in building a global solution [15].

Bee Colony Optimization (BCO) is a meta-heuristic method introduced by Teodorovic [14]. BCO is derived from the notion of joint intelligence between the bees in nature. It is a bottom-up iterative approach where multiple bees, called scout bees, are created based on the problem on hand and the swarm intelligence of the bees is used to solve the problem. Each bee in a single iteration tries to constructively solve the problem. In other words, the number of scout bees depicts the number of solutions in the search space that the algorithm is simultaneously building. The bees then communicate with each other to collect and share information about their partial solutions. The scout bees with the best partial solutions are called recruiter bees. They get the privilege to advertise their solution to the rest of the scout bees. The non-recruiter bees will then decide whether to continue working on their current partial solution or follow one of the recruiter bees' solutions. When the algorithm terminates, the final result will be the bee with the best solution.

In the first phase, the hive and the scout bees are initialized. In the forward phase each scout bee makes one or more local steps from a fixed amount of possible steps, adds them to the bee's current path (the partial solution), and determines the quality (fitness value) of its path (Scout bee's solution). The next step is to sort the scout bees based on their quality. The backward phase consists of picking the recruiter bees using a selection function. Selection of the recruiter bees can be one of the many approaches such as roulette wheel, tournament selection or elitism. Next, for every scout bee initially in the hive, the probability of the bee being loyal to its path is calculated. If the scout bee is loyal it continues with its own path and if it is not loyal it picks up the path of one of the recruiter bees. The above-mentioned process is iterated by the bees in two steps: a forward and a backward phase alternating one after another. The above process is repeated until there is no improvement in the solution or a predefined maximum value is attained. Figure 1 shows the alternating forward and backward phase.

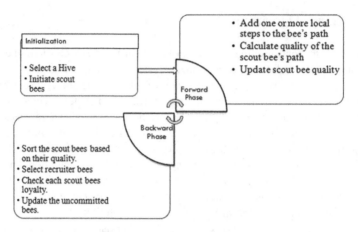

Fig. 1. Forward/Backward phase

4   Word Sense Disambiguation Using Bee Colony Optimization Algorithm

In BCA-WSD algorithm, one word among all the words to be disambiguated is randomly selected as the hive. The number of senses of this selected word, which is now the hive, determines the number of scout bees. The path of a bee represents the words that were disambiguated. Initially this path contains the hive word as its first word. When a bee visits a new target word to disambiguate, it determines the extended sense of the new target word with the help of the words in the path it traversed. This target word, with its disambiguated sense, is added to the bee's path. The bee returns back to the hive and the forward/backward phase repeats again. While the bees move to and fro between the hive and the target words, they iteratively build an appropriate sequence of context words as their path. This help in disambiguating the next target word visited by the bee with a wider context.

The BCA-WSD algorithm has three main phases: initialization, forward, and backward phase. During initialization, we determine the frequency of the target words in the gloss, example, hypernyms, entailments, and attributes of the other target words. In the forward phase, the bee travels to different target words picking the most appropriate extended sense of the target word based upon the path it traversed. In the backward phase, the path quality of all the bees is calculated and then used to decide the effective path to be explored. The forward and backward phases are alternated until all target words are assigned an extended sense. The main activities of BCA-WSD and their interactions are shown in Fig. 2 and are discussed in more details below.

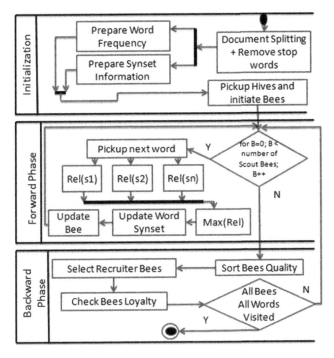

Fig. 2. BCA-WSD overview

4.1   Initialization

The first step in the initialization phase is the document splitting process; for each document that we need to disambiguate, we split it into groups of n sentences. For each group of n sentences we split the target words into:

- All the target words that are nouns or adjectives
- All the target words that are verbs or adverbs.

The document splitting process is shown in Fig. 3. The second step is to run the BCA-WSD algorithm on each group of sentences separately. Senses with scores less than a certain threshold are not considered. We choose to iterate over a single instance more than once. The next iteration runs on the result of the previous one. We first remove the stop words. The input now only contains words that need to be disambiguated. These target words can have multiple senses for the same parts of speech and can vary for different parts of speech. If the sense of a particular target word for the tagged parts of speech does not exist then we retrieve the cluster sense of that particular target word. For each target word we assume we have its lemma and POS. We prepare a global bag of words using all the senses of the target words. This process is repeated for each and every target word. The next step is to count the frequency of all the target word lemmas in the global bag. The global bag is filled with words from the gloss, hypernyms, entailments, and attributes of the particular senses.

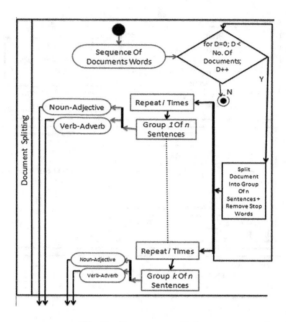

Fig. 3. Document splitting

We can imagine that the frequency denotes the closeness of the target word to the hive. The higher the frequency, the closer the target word is to the hive, and the more probable it is to be chosen during the forward phase. The next step is to prepare the extended information for all senses of the target words. All the stop words are removed from glosses, hypernyms, entailments, and attributes of the particular sense while preparing the extended sense. The extension of a sense is prepared initially because fetching information from the WordNet database is a costly process and we can avoid repeated transactions for the same sense. The last step in the initialization phase is to randomly select a target word from the input set of target words to act as the hive. We retrieve the senses of the hive based on its POS. Scout bees corresponding to each sense of the hive are generated. Each bee has its own properties such as the bee's path and its path quality.

4.2   Forward Phase

The simplified Lesk algorithm [6] is a variation of the original Lesk algorithm, where it disambiguates the target word by using the surrounding context. Context words are checked for their presence in the definition and examples of each sense of the target word. The simplified Lesk algorithm does not use the definitions and examples of context words. For example, suppose the words "computer," "code," and "instruction" occur together in a sentence. Suppose further that the word "instruction" is the target word for which the sense needs to be labelled and it is been used as a noun; the words "code" and "computer" are the surrounding words. For simplicity, we are going to consider the definitions without the example sentences. There are five distinct noun senses for the word "instruction" [3]:

- *instruction1* – a message describing how something is to be done
  (Hypernym) – what a communication that is about something is about
  (Pertainyms) – of or relating to or used in instruction
- *instruction2* – the activities of educating or instructing; activities that impart knowledge or skill
  (Hypernym) – any specific behavior
  (Pertainyms) – providing knowledge
- *instruction3* – the profession of a teacher
  (Hypernym) – the profession of teaching (especially at a school or college or university)
  (Pertainyms) – of or relating to or used in instruction
- *instruction4* – (computer science) a line of code written as part of a computer program
  (Hypernym) – (computer science) the symbolic arrangement of data or instructions in a computer program or the set of such instructions
- *instruction5* – a manual usually accompanying a technical device and explaining how to install or operate it

There are three distinct noun and two distinct verb senses for the word "code" [3]:

- *code1* – a set of rules or principles or laws (especially written ones)
- *code2* – a coding system used for transmitting messages requiring brevity or secrecy
- *code3* – (computer science) the symbolic arrangement of data or instructions in a computer program or the set of such instructions
- *code4* – attach a code to
- *code5* – convert ordinary language into code

There are two distinct noun senses for the word "computer" [3]:

- *computer1* – a machine for performing calculations automatically
- *computer2* – an expert at calculation (or at operating calculating machines)

The simplified Lesk algorithm checks for the existence of the surrounding words "code" and "computer" in the definition of instruction1. A similar step is done for instruction2, instruction3, instruction4, and instruction5. Instruction1, instruction2, instruction3, and instruction5 have score of zero. Instruction4 has a score of two since both the surrounding words "code" and "computer" appear in its definition. The sense instruction4 has the largest number of overlapping words and is therefore chosen as the correct sense for the target word "instruction."

We can see the process of maximizing the relatedness between context words and the senses of the target word as an optimization problem [16]. Equation 1 shows the optimization function. Given a context C that is composed of a sequence of words $\{w_1, w_2, w_3 \ldots w_n\}$ of context length n, we assume a target word to be w_t, where $1 \leq t \leq n$. Let us assume that w_t has m possible senses s_{ti}, where $1 \leq i \leq m$. The goal is to establish a sense s_{ti} that has the maximum relatedness measure between the sense s_{ti} and all the other words (i.e., context words of the text). We extend the sense s_{ti}

to an extended sense es_{ti} by adding to it other relevant information that we describe below. We use the simplified Lesk algorithm [6] as the relatedness measure (rel) between a context word and an extended sense es_{ti}.

$$WSD_{OPT} = \arg max_{i=1}^{m} \left(\sum_{j=1, j\neq i}^{n} rel(w_j, es_{ti}) \right) \tag{1}$$

The output of the optimization function is the score of the extended sense index of the target word w_t that is best related to the context words. The words in the definition and examples of a sense are called a gloss. The sense with the most appropriate meaning tends to have the maximum number of context words in its gloss. Here, the glosses of the senses are further extended with additional information based on the part of speech (POS) of the target word. This process is similar to the work done in [17]. The following information are added to es_{ti}:

- When the POS of the target word is a noun – We include the glosses and examples for: (1) the adjectives that are pertainyms of that particular noun sense, and (2) all the directly related hypernyms of the noun sense; these terms represent the more general meaning of the noun sense.
- When the POS of the target word is a verb – We include the glosses and examples for: (1) the verb entailments that are implied by a particular verb sense, and (2) all the directly related hypernyms of the verb sense which represent the more general meaning of the verb sense.
- When the POS of the target word is an adjective - We include the glosses and examples for the nouns for which the pertainym is that particular adjective sense.

In our extension of senses, the simplified Lesk algorithm checks for the existence of the surrounding words "code" and "computer" in the definition, examples, hypernyms and pertainyms of *instruction1*. A similar step is done for *instruction2*, *instruction3*, *instruction4*, and *instruction5*. *Instruction1*, *instruction2*, *instruction3*, and *instruction5* have score of zero. Instruction4 has a score of two since both the surrounding words "code" and "computer" appear in its definition and "computer" occurs once in the hypernym. We do not consider the repetition of the words while scoring. The sense *instruction4* has the largest number of overlapping words and is therefore chosen as the correct sense for the target word "instruction."

The first step for a bee in the forward phase is to choose a new word to add to its path. Each bee prepares a roulette wheel based on the frequency of the target words calculated in the initialization phase for all the unvisited target words. If the target word has a frequency of 0, then we add one to its frequency so that this word has a chance, even if it is a slight one, to be picked. Each bee picks a target word from the roulette wheel to visit randomly. After picking the word, the bee uses the optimization function of Eq. 1 to assign an appropriate extended sense to the picked target word. The bee adds the new disambiguated word to its path. Using a roulette wheel gives higher chances to the target words with high-frequency value. The words in the bee's path act as the context words and are used in the disambiguated process. The scout bee checks the words in its path for their existence in the extended senses of the target word.

The extended sense with the maximum overlap of words with the path words is assigned as the correct sense to the target word. If the score of both extended senses is equal or zero then we pick the most frequent extended sense in that case. After the extended sense for the target word is determined, the next step is to add this target word to the bee's path. The quality of the path is also updated with the chosen extended sense score to reflect the quality of the bee's path. Each bee which is initiated from the hive goes in different directions by picking up different words in the vicinity, where the words with higher frequencies are closer to the hive. Each bee, upon reaching a target word, finds the correct sense of that target word based on the path it has pursued.

The overview of the forward phase can be seen in Fig. 4. In this example the word "model" represents the hive, while each bee represents one of the three senses of "model." There are eight words to disambiguate (excluding the hive word). Each of the three bees selects the next word to disambiguate using the roulette wheel. In this case, *Bee1*, *Bee2*, and *Bee3* select "computer," "task," and "need," respectively, and disambiguate them. The circular ring denotes the single-forward phase. The words on the ring denote the target words to be disambiguated in that particular phase. Each bee eventually visits all the target words in the search space.

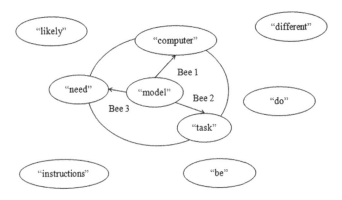

Fig. 4. Forward phase overview

4.3   Backward Phase

In this phase we sort the bees based on their path qualities (their fitness value), where the best bee is placed at the beginning. Then we determine whether the bees will:

(a) become recruiters
(b) abandon their path and follow one of the recruiter's path
(c) be loyal to their path

We define the Recruitment Size (*Rec*) based on the number of bees in the hive. For example, for a hive of seven bees we define *Rec* as three as we want to explore combination of target words which gives maximum WSD_{OPT}. We pick the bees with the highest path quality as recruiters. Each bee then decides whether to stay loyal to its path or abandon its path. Equation 2 is used to calculate the loyalty probability of the bth bee:

$$ployalty_b^{u+1} = e^{-\left(\frac{o_{max}-o_b}{x}\right)} \ where \ \ o_b = \frac{qual_b - qual_{min}}{qual_{max} - qual_{min}}, x = \sqrt[4]{u} \tag{2}$$

Where u denotes the number of iterations of the forward phase, x defines the fourth root of u u., $qual_{max}$ defines the maximum overall value of the partial solution among all bees, where a partial solution here refers to all the target words that are disambiguated by a bee in previous phases, $qual_{min}$ defines the minimum overall value of the partial solution, $qual_b$ defines the path quality of the bth bee with the partial solution, o_b is the normalized value of the partial solution generated by the bth bee, and o_{max} is the maximum overall normalized value of the partial solution generated among all bees. As the number of forward phases increases, the probability for a bee to change its path decreases.

We have modified the equation for the loyalty probability that was proposed by the authors of the BCO algorithm [18, 19] to fit the problem of word sense disambiguation. We introduced x, which is the Fourth root of u, while calculating the loyalty probability since we are disambiguating a single target word in one forward phase and the number of target words can grow as much as five hundred. Higher values of u.u resulted in very few bees changing their path. Therefore, we made the value of u smaller to give the bees a higher chance to change their paths in the early stages of the algorithm than in later stages.

The next step is for uncommitted bees to pick which of the recruiters bees to follow. Equation 3 is used for calculating the recruitment probability of the bth uncommitted bee:

$$precruitment_b = \frac{o_b}{\sum_{k=1}^{Rec} o_k} \tag{3}$$

where o_b the normalized value of the partial solution is generated by the bth scout bee, and o_k denotes the normalized value of the kth recruiter bee partial solution. Using a roulette wheel, the uncommitted bees can pick any one of the recruiter bees to follow. The recruitment probability can be compared to the waggle dance that the bees perform in nature.

The overview of the backward phase can be seen in Fig. 5. *Bee1*, based on its fitness value, decides to stay loyal to its path and moves to the next target word "likely." *Bee2* decides to stay loyal to its path and moves to the next target word "be." *Bee3* decides to abandon its path *Bee3PreviousPath* and changes its path to that of *Bee1*. It then moves to the next target word "different."

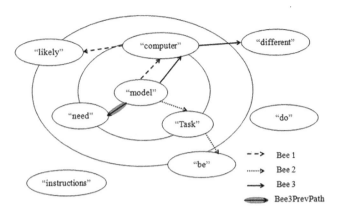

Fig. 5. Backward phase overview

5 Experimentations and Results

In this section we report the results of our BCA-WSD algorithm on WordNet 2.1 and compare all the results with some of the previous WSD systems. We also propose a new variation to our algorithm by including a voting strategy. This modified version, which we call VBCA-WSD, yields better results.

5.1 Dataset

We use SemEval-2007 Task 07: Coarse-Grained English All-Words Task as our test data. This data set is composed of five documents related to fields like journalism, book review, travel, computer science and biography. The first three documents were obtained from the Wall Street Journal corpus [20]. The computer science document is obtained from Wikipedia and the biography document is an extract from Amy Steedman's Knights of the Art, a biography of Italian painters. Documents three and four consist of 51.87% of the total target words [4]. In this Semantic Evaluation (SemEval) task, linguists annotated the sense of each of the target words by picking the most appropriate sense from the coarse-grained sense inventory of WordNet. Each of the five documents is an annotated set of target words. The target word is tagged with its respective part of speech (noun, verb, adverb or adjective). The lemma of the target word is also tagged along with original content. Table 1 gives an idea of the number of target words in each document.

5.2 Results

In order to evaluate the effectiveness of our approach to compare it with other unsupervised WSD methods we use precision, recall and F-measure of the results. We used the sense inventory and scorer provided by the SemEval-2007 Task 07 organizers [4]. We kept the F-measure value, although it does not add any additional information, in order to compare our work with other related work.

Table 1. Target words in each document.

Article	Domain	Target words
D001	Journalism	368
D002	Book review	379
D003	Travel	500
D004	Computer science	677
D005	Biography	345
Total		2269

The results of BCA-WSD algorithm are shown in Table 2. As we are splitting each document into a group of 10 sentences, and we apply our algorithm to each group individually, we expect the selected extended sense to have a higher score than the threshold score (TS) of five. We iterate five times over the initial data set to improve the performance. We consider ten sentences as the context. One important observation to conclude is the importance of disambiguation of target words based on their parts of speech separately. This approach shows very good results that are comparable to the baseline (MFS). BCA-WSD performed well for documents d001 and d002 that are related to more general topic. BCA-WSD performed on par compared to the baseline for documents d003. BCA-WSD did not performed so well for documents d004 and d005 that are related to specific topics.

Table 2. BCA-WSD.

Article	Attempted	Precision	Recall	F-measure
D001	100	86.141	86.141	86.141
D002	100	83.641	83.641	83.641
D003	100	77.800	77.800	77.800
D004	100	74.446	74.446	74.446
D005	100	74.783	74.783	74.783
Total	100	78.669	78.669	78.669

5.3   VBCA-WSD

BCA-WSD gave different results every time we ran it. This is normal due to its probabilistic nature and due to the randomness of choosing one target word that becomes the hive. To use this variation in results to our advantage, we run BCA-WSD algorithm for 200 times on separate instances of the input. Then we perform voting on the results to determine the senses with the highest agreement. We call the voting algorithm VBCA-WSD as shown in Fig. 6.

The results of VBCA-WSD algorithm are shown in Table 3. After running the algorithm two hundred times, we then we perform voting on the results to determine the extended senses with the highest agreement. Documents d001 and d004 show improvement in performance compared to that of BCA-WSD. Documents d002 and d005 show lower performance compared to that of BCA-WSD. The final result, an

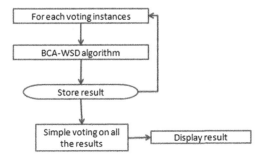

Fig. 6. VBCA-WSD

Table 3. VBCA-WSD.

Article	Attempted	Precision	Recall	F-measure
D001	100	86.685	86.685	86.685
D002	100	82.322	82.322	82.322
D003	100	77.800	77.800	77.800
D004	100	77.253	77.253	77.253
D005	100	73.913	73.913	73.913
Total	100	79.242	79.242	79.242

F-measure of 79.242, shows a good improvement over the results of our initial algorithm without voting.

5.4   Comparison to Related Work

We finally compare BCA-WSD and VBCA-WSD with the unsupervised WSD systems such as: Ant Colony Optimization (ACA) [1, 12, 13], Genetic Algorithm (GA) [11], Simulated Annealing (SA) [10], and the Most Frequent Sense (MFS), which is the baseline. We restricted our comparison to the related work by focusing on unserpvised metaheuristic techniques. For the sake of comparing our results to the related work, we did not use the Senseval datasets since all the previous work only used Semeval data. The results of our algorithm and the related work are shown in Table 4, and are all based on WordNet 2.1. The WSD systems are arranged in order of high to low result.

VBCA-WSD approach outperformed other WSD systems. It shows the benefits of exploiting WSD problem using segregation of parts of speech and a certain degree of threshold in a global context. None of the previous approaches have segregated words on parts of speech. None of the previous approaches have segregated words on parts of speech. The previous WSD systems that are based on heuristics algorithm did not manage to outperform the baseline (MFS). BCA-WSD puts lot of emphasis on the context on which a word sense is disambiguated.

Table 4. Comparison of WSD systems.

System	Attempted	Precision	Recall	F-measure
VBCA-WSD	100	79.24	79.24	79.24
MFS	100	78.89	78.89	78.89
ACA-2012 voting	100	78.76	78.76	78.76
BCA-WSD	100	78.67	78.67	78.67
TSP-ACO	99.80	78.50	78.10	78.30
ACA-2012	100	77.64	77.64	77.64
ACA-2011	100	74.35	74.35	74.35
SA	100	74.23	74.23	74.23
GA	100	73.98	73.98	73.98

6 Conclusion

In this paper we proposed a bee colony optimization algorithm to solve the word disambiguation problem. Virtual bees were successful in disambiguating the words based on the paths the bees traveled. VBCA-WSD used a modified simplified Lesk algorithm in order to give more importance to the context during the disambiguation process. Using words in the bee's path helps in relating the words to their context. Our results were compared with six other approaches and one baseline, which is the most frequent sense. Our algorithm yielded better results using a single a knowledge source compared to that of ACA-2012 which used a combination of knowledge sources. Disambiguating the words based on noun-adjective group, verb-adverb group showed greater efficiency. In addition, the introduction of the majority strategy boosted our results further.

For future work, we intend to use a combination of knowledge sources. This will allow us to exploit the relationship between verbs and adverbs based on their occurrence with each other and will help us improve the relatedness measure. In addition, we plan to investigate the use of parts of speech, different relatedness measures and their combinations in the optimization function.

References

1. Nguyen, K.-H., Ock, C.-Y.: Word sense disambiguation as a traveling salesman problem. Artif. Intell. Rev. **40**, 405–427 (2011)
2. Navigli, R.: Word sense disambiguation: a survey. ACM Comput. Surv. **41**, 10:1–10:69 (2009)
3. PrincetonUniversity: WordNet. http://wordnet.princeton.edu/
4. Navigli, R., Litkowski, K.C., Hargraves, O.: SemEval-2007 Task 07: coarse-grained english all-words task. In: Proceedings of 4th International Workshop in Semantic Evaluations, pp. 30–35 (2007)
5. Lesk, M.: Automatic sense disambiguation using machine readable dictionaries. In: Proceedings of the 5th Annual International Conference on Systems Documentation - SIGDOC 1986, pp. 24–26 (1986)

6. Mihalcea, R.: Knowledge-based methods for WSD. In: Word Sense Disambiguation Algorithms and Applications, pp. 107–131 (2007)
7. Jurafsky, D., Martin, J.H.: Speech and language processing: an introduction to natural language processing, computational linguistics, and speech recognition. Speech Language Process. An Introduction to Natural Language Processing Computational Linguistic Speech Recognition, 21, pp, 0–934 (2009)
8. Chan, Y.S., Ng, H.T.: Domain Adaptation with Active Learning for Word Sense Disambiguation, pp. 49–56, Computational Linguistics, Prague (2007) (in Press)
9. Zhu, J., Hovy, E., Rey, M.: Active Learning for Word Sense Disambiguation with Methods for Addressing the Class Imbalance Problem. Computational Linguistics, pp. 783–790 (2007)
10. Cowie, J., Guthrie, J., Guthrie, L.: Lexical disambiguation using simulated annealing. In: Proceedings of the 14th Conference Computational Linguistics COLING 1992, vol. 1, pp. 359–365 (1992)
11. Gelbukh, A., Sidorov, G., Han, S.-Y.: Evolutionary Approach to Natural Language Word Sense Disambiguation through Global Coherence Optimization. WSEAS Trans. Comput. 2(1), 257–265 (2003)
12. Schwab, D., Guillaume, N.: A global ant colony algorithm for word sense disambiguation based on semantic relatedness. Highlights in Practical Applications of Agents and Multiagent Systems, pp. 257–264. Springer, Heidelberg (2011)
13. Schwab, D., Goulian, J., Tchechmedjiev, A., Blanchon, H.: Ant colony algorithm for the unsupervised word sense disambiguation of texts: comparison and evaluation. In: Proceedings of COLING 2012, pp. 2389–2404 (2012)
14. Teodorović, D.: Bee Colony. Innov. Swarm Intell. 248, 39–60 (2009)
15. Bonabeau, E., Dorigo, M., Theraulaz, G.: Swarm Intelligence: From Natural to Artificial Systems. Oxford University Press, New York (1999)
16. Patwardhan, S., Pedersen, T.: Using WordNet-based context vectors to estimate the semantic relatedness of concepts. In: 11th Conference on European Chapter Association Computational Linguistics, Vol. 1501, pp. 1–8 (2006)
17. Schwab, D., Tchechmedjiev, A., Goulian, J., Nasiruddin, M., Sérasset, G., Blanchon, H.: GETALP system: propagation of a Lesk measure through an ant colony algorithm. In: Second Joint Conference on Lexical and Computational Semantics (*SEM), Proceedings of the Seventh International Workshop on Semantic Evaluation (SemEval 2013), Vol. 2, pp. 232–240 (2013)
18. Teodorović, D., Lucic, P., Markovic, G., Dell'Orco, M.: Bee colony optimization: principles and applications. In: 8th Seminar on Neural Network Applications in Electrical Engineering, pp. 151–156 (2006)
19. Markovi, G.Z., Teodorović_ca, D.B., Aćimović-Raspopović, V.S.: Routing and wavelength assignment in all-optical networks based on the bee colony optimization. AI Commun. - Netw. Anal. Nat. Sci. Eng. 20(4), 273–285 (2007)
20. Charniak, E., Blaheta, D., Ge, N., Hall, K., Hale, J., Johnson, M.: Bllip 1987–89 WSJ Corpus Release 1 (2000)

Verb Sense Annotation for Turkish PropBank via Crowdsourcing

Gözde Gül Şahin[✉] [iD]

Department of Computer Engineering, Istanbul Technical University,
34469 Istanbul, Turkey
isguderg@itu.edu.tr

Abstract. In order to extract meaning representations from sentences, a corpus annotated with semantic roles is obligatory. Unfortunately building such a corpus requires tremendous amount of manual work for creating semantic frames and annotation of corpus. Thereby, we have divided the annotation task into two microtasks as verb sense annotation and argument annotation tasks and employed crowd intelligence to perform these microtasks. In this paper, we present our approach and the challenges on crowdsourcing verb sense disambiguation task and introduce the resource with 5855 annotated verb senses with 83.15% annotator agreement.

Keywords: Turkish PropBank · Verb sense disambiguation
Crowdsourcing · Semantic annotation

1 Introduction

Recently increasing interest on extracting semantic information from sentences has triggered new literature on shallow and deep semantic parsing. Extracting shallow meaning is usually achieved in two steps, first by identifying the predicates then detecting the complements of predicates and their relation to its predicate. That relation is referred to as a semantic role and this task is named as semantic role labeling (SRL), a.k.a shallow semantic parsing. SRL has many potential applications such as machine translation, information extraction and question answering.

PropBank [16] is one of the most commonly used semantic resource for SRL, among other resources such as FrameNet [2] and VerbNet [19]. The semantically annotated corpora associated with PropBank, help researchers to specify SRL as a task, furthermore is used as training and test data for supervised machine learning methods. An examplary sentence with predicate *go*, annotated with PropBank annotation scheme is given below:

[What flights]$_{Arg1}$ **go** [from Seattle]$_{Arg3}$ [to Boston]$_{Arg4}$?

© Springer International Publishing AG, part of Springer Nature 2018
A. Gelbukh (Ed.): CICLing 2016, LNCS 9623, pp. 496–506, 2018.
https://doi.org/10.1007/978-3-319-75477-2_35

The predicate "go" in the sentence above is framed with "go.01 (motion)" sense, which has the arguments, a.k.a semantic roles *Arg1 (entity in motion/goer)*, *Arg2 (extent)*, *Arg3 (start point)* and *Arg4 (end point, end state of Arg1)*. In **Prop-Bank (PB)**, the core arguments that are specific to the predicate sense are numbered from Arg0 to Arg5, where temporary, non-core arguments such as manner, temporal, extent are labeled with temporary tags ArgM-MNR, ArgM-TMP and ArgM-EXT accordingly. PropBank simplifies semantic roles, but defines neither relations between verbs nor all possible syntaxes for each verb.[1] Increasing popularity of semantic parsing led to development of PropBanks for other languages such as Arabic [23], Chinese [22], Hindi/Urdu [4], Brazilian [7], Portuguese [5] and Finnish [11] to the best of our knowledge.

Crowdsourcing is a thriving approach that can be used for building various linguistic resources. Many NLP tasks such as corpora creation [14], named entity recognition [24], multilingual semantic textual similarity [1] and grammatical error detection [13] have employed crowdsourcing. In study by [20] it has been shown that crowdsourced data is high quality and enhances NLP systems. Such studies have exploited crowdsourcing field and have inspired tutorials [6] and best practice guidelines [17] to be written on the subject. In spite of high employment rates of crowdsourcing platforms for NLP tasks, it can not be easily utilized by complex semantic tasks such as semantic role annotation. Fortunately, recent increasing interest on semantics provoked the community to exploit crowdsourcing on semantic annotation tasks. Studies on annotation of frame elements (FE) [9,10] and building a large semantically corpus with help from crowd intelligence [3] are successful examples of human aid in semantic annotation.

Manually building a corpus annotated with Turkish PropBank verb frames [12,18] is a tedious work. Unfortunately, there are not enough resources to perform corpus annotation with sophisticated annotation tools and numerous trained annotators. Due to low resources for building Turkish PropBank, we have decided to divide annotation task into two microtasks that can be crowdsourced: Verb Sense Annotation and Argument Labeling. In verb sense annotation task, people are asked to disambiguate the meaning of the verbs in the sentences from a morphologically and syntactically analyzed corpus, where in argument labeling task, they are asked to label the arguments of the previously annotated verb senses. In this paper, we focus on crowdsourcing process of verb sense disambiguation. Most common platforms of human intelligence tasks are Amazon Mechanical Turk[2] and CrowdFlower[3]. CrowdFlower's support for low-resourceful languages such as Turkish, ease of use due to its mark up language with conditional logic ability, large number of natives from wide variety

---

[1] In PropBank only Arg0 and Arg1 are associated with a specific semantic content. Arg0 is used for actor, agent, experiencer or cause of the event; Arg1 represents the patient, if the argument is affected by the action, and theme, if the argument is not structurally changed.

[2] https://www.mturk.com.

[3] https://crowdflower.com.

of cultures and improved quality control system has made it the right platform for us to perform Turkish verb sense annotation task.

2   Data Preparation

We have used the revised METU-Sabancı Treebank (MST) [15][4], that has been developed with morphological analysis techniques explained in [21] and has been renamed as ITU-METU-Sabancı Treebank (IMST). The representation model used by MST is shown in Fig. 1.

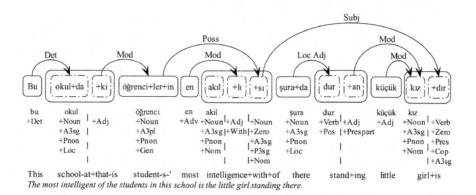

Fig. 1. Dependency links in a Turkish sentence (taken from [8]), where + sign shows morpheme boundaries, rectangle indicates word boundaries and dashed rectangle shows IG boundaries within words. Morphological analyzer results are entered below the IGs.

Turkish is among languages that has rich derivational morphology as can be seen from the examplary sentence given in Fig. 1. Here, each word is splitted into inflectional groups (IG), separated by derivational boundaries (DB). Each IG is treated as if it was a separate word and annotated with its own part of speech and other necessary tags. Consider the word *duran* from Fig. 1 which has two IGs: The first IG, dur (to stand), is a Verb, and the second IG is an adjective derived from the first IG with the morpheme *an* as (*dur+an* = standing). In our study, we aim to annotate all verbs in any form such as the predicate *duran* and this representation simplifies detecting verbs in different (adjective, noun or adverb) forms. Recent study on framing of verbs for Turkish PropBank has reported 759 annotated verb roots and 1262 annotated senses [12,18]. In order to reduce costs of the verb sense disambiguation task following preprocessing steps have been performed before the corpus has been outsourced to crowd:

- The predicates with only one sense have been eliminated because they are automatically annotated,

---

[4] METU-Sabancı Treebank is a morphologically and syntactically analyzed balanced corpus with 5635 sentences.

– The predicate *ol (to be/become/happen)* has been eliminated since it is mostly used as light verb or copula, hence can be semi-automatically annotated with the help of dependency type to its head.[5]

3   Job Design

The design of input rows is given in Table 1. Here, only the first two senses of the predicate is shown however verb sense annotation task includes all senses that exist in Turkish PropBank. Each sense is included in a separate column as SENSEWEXi, where $i \leq 15$. In addition to the columns given in Table 1, additional attributes such as unit id, state (finalized, judgable, golden), number of judgements done for this unit and agreement value between 0 and 1 are appended automatically by CrowdFlower.

Table 1. Input row for crowdsourcing task

Field	Definition	Example
SID	Sentence Id	3
SENT	Sentence	Ona herşeyimi **verdim** (En) I **gave** him everything
PRED	Predicate	ver (En) to give
PRNO	The order of the predicate	2
SENSEWEX1	First sense of the predicate and a usage example	Transfer etmek, iletmek (Yemlerini ben **veririm**) (En) To transfer, transmit (I **give** them the bait)
SENSEWEX2	Second sense of the predicate and a usage example	Tespit etmek (İsim **vermek**, randevu **vermek**) (En) To fix, establish (To **give** a name, to **fix** a date)

Tasks with varying number of options such as verb sense annotation require dynamic rendering of questions. We have designed each sense in Turkish Prop-Bank to be a new radio button. Since not all predicates have the same number of senses, we have dynamically rendered a new radio button when the column SENSEWEXi is not equal to "*N.A*". The required code snippet written in CML (CrowdFlower Markup Language) is shown below.

---

[5] In Light Verb Constructions (LVC) with the verb *ol*, nominal dependent is linked with *MWE* dependency type to the predicate *ol* and in English PropBank copula is not annotated.

```
{% if SENSEWEX1 != 'N.A' %}
<cml:radio label={{SENSEWEX1}}''></cml:radio>
{% endif %}
```

A screenshot from the designed task is given in Fig. 2. Here, the predicate that we are interested is shown with **Eylem (Predicate)** in its root form. In cases when the same predicate occurs more than once in the same sentence, the occurrence order is written near the predicate. Below the predicate, the sentence from the treebank where the predicate takes place is shown with **Cümle (Sentence)** tag. Then the taskers are asked to choose the sense with the closest meaning to the predicate's meaning among the senses in Turkish PropBank. In addition to the senses, the option **Hiçbiri (None)** is offered to the annotators. English translations for each field in the design is written with (En) tag.

Eylem: bit
(En)Predicate: to finish
Cümle: Masal da burada bitmiş .
(En)Sentence: So the fairy tale ended here.
Lütfen en yakın anlamı seçiniz: (En) Please choose the closest meaning:

  ○ Tükenmek, son bulmak (Dün akşam param bitmişti.) (En) To end.
  ○ Çok sevmek (Ben böyle sese biterim.) (En) To love so much
  ○ Hiçbiri (En) None

Fig. 2. One of the questions shown to the taskers

4   Quality of Annotation

Crowdsourcing is an attractive option for researchers working on languages with low resources only if the quality of annotation can be assured. In this work, it is accomplished in three steps:

- Quiz contributors: They must pass the quiz to start working on the task,
- Train contributors: Explain the correct answer of each test question and supply clear instructions,
- Remove underperformers: Eliminate contributors who repeatedly fail test questions.

In Quiz mode, before the contributors can start working on actual task, we want them to answer five test questions. If they give wrong answers to more than one question, they are automatically removed from the task since their confidence level drops below 70%, which is the minimum confidence level accepted for our task. Due to our high confidence level requirement, 40% of the contributors were eliminated in Quiz Mode.

One of the most important parts of the task is training of annotators. According to best practice guidelines [17] and CrowdFlower documents, instructions should be written in a very **direct**, **clear** format and should include various examples for different scenarios. Moreover, giving an overview of the reasons why such task is offered motivates the taskers. We have briefly presented the steps of the task as follows:

- **Read** the predicate.
- **Read** the complete sentence.
- **Detect** the meaning of the predicate.
- **Read** the meaning definitions and example sentences given in options.
- **Choose** the correct sense.
- **Choose None** if the meaning does not exist.

Furthermore, to prepare the taskers for challenging cases we have included annotation examples for light verbs and metaphorically used verbs.

Second part of training annotators is to supply correct answers and the reasons if they fail a test question. At the bottom of the test question modification/monitoring window shown in Fig. 3, *Reason* text field can be seen. We have paid attention to fill those fields, so that the contributors can be informed whenever they miss a question.

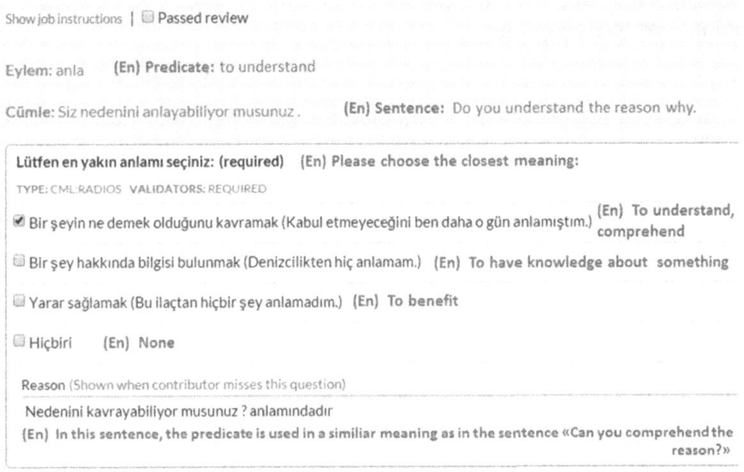

Fig. 3. Test question preparation view with "Reason" and "Passed Review" fields. Checkbox design allows marking of multiple answers for test question.

Finally, the contributors who passed the Quiz Mode are constantly monitored, by randomly putting test questions a.k.a gold units inside the actual work. We have configured our task to ask 5 questions per page, where 1 of the questions is a test question. Confidence levels of all annotators are updated as they answer random test questions and similarly if their performance drops below 70%, they are removed from the task. 27% of the annotators that pass the quiz mode, have been removed during work Mode. Recommended amount of golden units is at least 10% of total rows in the task, thus we have prepared 585 number of test questions from unambiguous sentences. Taskers are allowed to contest to gold units if they think the gold answer is wrong. When the number of contests on a particular test question exceeds the threshold, CrowdFlower platform warns the task owner about that test question and asks it to be reviewed. As shown in Fig. 3, *Passed Review* button is used to indicate that the contested question has been reviewed by the expert.

ID	JUDGMENTS	% MISSED	% CONTESTED ●	ENABLED	ACTIONS
760702306	6			ON	Show Details
760665504 (passed)	3			OFF	Show Details
760702328	8			ON	Show Details
760685396 (passed)	4			OFF	Show Details

Fig. 4. Monitoring view of test questions: "Missed" and "Contested" indicates the ratio of missing/contesting the test question. "On/Off" Button can be used for including/excluding the test questions.

Some active contributors may memorize test questions during annotation process. Therefore, the task should be constantly monitored and test questions should be alternately enabled and disabled during annotation. In Fig. 4, test question monitoring and enable/disable buttons are shown. Another solution to prevent active contributors from memorizing test questions is to set maximum number of judgements per contributor roughly to be less than numberOfTestQuestions $\times 4^6$. Maximum number of judgement per contributor is set to 10% of total rows by default.

We have configured our task to have three judgements per question, accept only native Turkish speakers and level 2 and 3 contributors as shown in Table 2. In addition to taskers from external channels, we have allowed internal team members to contribute to the task.

One of the workers' rights issues that is raised by the use of mechanised labour is low wages (below \$2 per hour). The average trusted judgement time is calculated as 17 s, and payment per page is 5 cents, thus average earning per hour can be calculated as \$2,11. Moreover, we have rewarded the contributors with a confidence level above 80% and a sufficient number of judgement count, with a bonus which equals to the average earning per hour.

At the end of data collection, a method to measure the agreement on an answer and calculate the safety of accepting crowd's answer is necessary. The consolidation of one or more contributor responses into a summarized result is referred to as **aggregation**. We have used the default aggregation method that

---

[6] Number of questions per page - 1.

Table 2. Configuration of verb sense annotation task

Setting	Value
Judgment per row	3
Rows per page	5
Payment per page	5 cents
Contributers' level (1–3)	2
Channels	Internal and External
Geography/Language	Only Turkish
Minimum confidence of contributors	70%

returns a single top result and its confidence value between 0 and 1. This value is calculated by weighting the agreement by contributor trust.

5   Results

As a result, 5855 rows have been annotated and 18123 judgements have been made. 265.9 rows have been annotated per hour and all annotation process took 68 h. More than 100 taskers contributed from 39 different cities of Turkey. The overall annotator agreement is calculated as 83.15% and the total cost of the job was 277$. The maximum amount of judgment made by one tasker is less than 800, which is only 4.44% of the job, as shown in Fig. 5. In Fig. 6, distribution of judgements with respect to the contributors' confidence is given. This figure shows that quality control mechanism of CrowdFlower eliminated the contributors with a confidence level lower than 70% which led to small amount of low-confident judgements. After completion of this task, the rows with confidence lower than 0.7 (~2000 rows) and ones that were aggreed as *None* (~700 rows) have been revised.

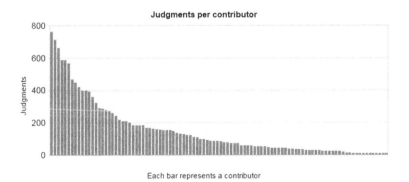

Fig. 5. Judgement per Contributor

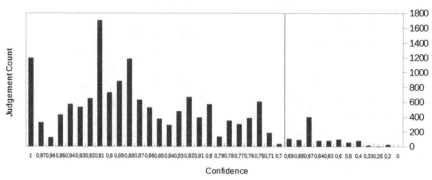

Fig. 6. Judgement vs. Confidence. The vertical red line marks the confidence level 0.7 (Color figure online)

6   Conclusion

In this work, we have presented an approach to perform semantic annotation with crowdsourcing techniques, based on previous work on Turkish PropBank frames and sophisticated quality controlling mechanisms supplied by Crowd-Flower platform. As the crowdsourcing mechanism seems promising even for low-resourceful languages such as Turkish, we additionally plan to crowdsource the annotation of arguments of verb senses in near future.

By completion of the argument labeling task, we would have been creating necessary resources for Turkish to be included in the task of semantic role label-ing and driving the NLP community for building semantic role labelers that will have wider coverage of language families. Apart from semantic role labeling task, the linguistic resource explained in this paper, can be used separately for word sense disambiguation tasks.

References

1. Agirre, E., Banea, C., Cardie, C., Cer, D., Diab, M., Gonzalez-Agirre, A., Guo, W., Mihalcea, R., Rigau, G., Wiebe, J.: Semeval-2014 task 10: multilingual semantic textual similarity. In: Proceedings of the 8th International Workshop on Semantic Evaluation (SemEval 2014), pp. 81–91 (2014)
2. Baker, C.F., Fillmore, C.J., Lowe, J.B.: The Berkeley FrameNet project. In: Pro-ceedings of the 36th Annual Meeting of the Association for Computational Linguis-tics and 17th International Conference on Computational Linguistics, ACL 1998, vol. 1, pp. 86–90. Association for Computational Linguistics, Stroudsburg (1998). https://doi.org/10.3115/980845.980860
3. Basile, V., Bos, J., Evang, K., Venhuizen, N.: Developing a large semantically annotated corpus. In: LREC, vol. 12, pp. 3196–3200 (2012)

4. Bhatt, R., Narasimhan, B., Palmer, M., Rambow, O., Sharma, D.M., Xia, F.: A multi-representational and multi-layered treebank for Hindi/Urdu. In: Proceedings of the Third Linguistic Annotation Workshop, ACL-IJCNLP 2009, pp. 186–189. Association for Computational Linguistics, Stroudsburg (2009). http://dl.acm.org/citation.cfm?id=1698381.1698417
5. Branco, A., Carvalheiro, C., Pereira, S., Silveira, S., Silva, J., Castro, S., Graça, J.: A PropBank for Portuguese: the CINTIL-PropBank. In: LREC, pp. 1516–1521 (2012)
6. Callison-Burch, C., Ungar, L., Pavlick, E.: Crowdsourcing for NLP. In: Proceedings of NAACL 2015. North America Association for Computational Linguistics (2015)
7. Duran, M.S., Aluísio, S.M.: Propbank-Br: a Brazilian treebank annotated with semantic role labels. In: LREC, pp. 1862–1867 (2012)
8. Eryiğit, G., Nivre, J., Oflazer, K.: Dependency parsing of Turkish. Comput. Linguist. **34**(3), 357–389 (2008)
9. Fossati, M., Giuliano, C., Tonelli, S.: Outsourcing FrameNet to the crowd. In: ACL, vol. 2, pp. 742–747 (2013)
10. Fossati, M., Tonelli, S., Giuliano, C.: Frame semantics annotation made easy with DBpedia. In: Crowdsourcing the Semantic Web (2013)
11. Haverinen, K., Kanerva, J., Kohonen, S., Missilä, A., Ojala, S., Viljanen, T., Laippala, V., Ginter, F.: The Finnish proposition bank. Lang. Resour. Eval. **49**(4), 907–926 (2015)
12. İşgüder, G.G., Adalı, E.: Using morphosemantic information in construction of a pilot lexical semantic resource for Turkish. In: Proceedings of Workshop on Lexical and Grammatical Resources for Language Processing, pp. 46–54. Association for Computational Linguistics and Dublin City University, Dublin, August 2014. http://www.aclweb.org/anthology/W14-5807
13. Madnani, N., Tetreault, J., Chodorow, M., Rozovskaya, A.: They can help: using crowdsourcing to improve the evaluation of grammatical error detection systems. In: Proceedings of the 49th Annual Meeting of the Association for Computational Linguistics: Human Language Technologies: Short Papers, vol. 2, pp. 508–513. Association for Computational Linguistics (2011)
14. Negri, M., Mehdad, Y.: Creating a bi-lingual entailment corpus through translations with mechanical turk: $100 for a 10-day rush. In: Proceedings of the NAACL HLT 2010 Workshop on Creating Speech and Language Data with Amazon's Mechanical Turk, pp. 212–216. Association for Computational Linguistics (2010)
15. Oflazer, K., Say, B., Hakkani-Tür, D.Z., Tür, G.: Building a Turkish treebank. In: Abeillè, A. (ed.) Treebanks: Building and Using Parsed Corpora. Text, Speech and Language Technology, vol. 20, pp. 261–277. Springer, Dordrecht (2003). https://doi.org/10.1007/978-94-010-0201-1_15
16. Palmer, M., Gildea, D., Kingsbury, P.: The proposition bank: an annotated corpus of semantic roles. Comput. Linguist. **31**(1), 71–106 (2005)
17. Sabou, M., Bontcheva, K., Derczynski, L., Scharl, A.: Corpus annotation through crowdsourcing: towards best practice guidelines. In: Proceedings of LREC (2014)
18. Sahin, I.G.G.: Framing of verbs for Turkish PropBank. In: Proceedings of Turkic Computational Linguistics, TurCLing 2016, 17th International Conference on Intelligent Text Processing and Computational Linguistics, CICLING 2016 (2016)
19. Schuler, K.K.: VerbNet: a broad-coverage, comprehensive verb lexicon. Doctoral dissertation, University of Pennsylvania (2005)

20. Snow, R., O'Connor, B., Jurafsky, D., Ng, A.Y.: Cheap and fast–but is it good?: evaluating non-expert annotations for natural language tasks. In: Proceedings of the Conference on Empirical Methods in Natural Language Processing, pp. 254–263. Association for Computational Linguistics (2008)
21. Sulubacak, U., Eryiğit, G.: A redefined Turkish dependency grammar and its implementations: a new Turkish web treebank & the revised Turkish treebank. In: Proceedings of Turkic Computational Linguistics, TurCLing 2016, 17th International Conference on Intelligent Text Processing and Computational Linguistics, CICLING 2016 (2016)
22. Xue, N., Palmer, M.: Adding semantic roles to the Chinese treebank. Nat. Lang. Eng. **15**(1), 143–172 (2009)
23. Zaghouani, W., Diab, M., Mansouri, A., Pradhan, S., Palmer, M.: The revised Arabic PropBank. In: Proceedings of the Fourth Linguistic Annotation Workshop, LAW IV 2010, pp. 222–226. Association for Computational Linguistics, Stroudsburg (2010). http://dl.acm.org/citation.cfm?id=1868720.1868756
24. Zhai, H., Lingren, T., Deleger, L., Li, Q., Kaiser, M., Stoutenborough, L., Solti, I.: Web 2.0-based crowdsourcing for high-quality gold standard development in clinical natural language processing. J. Med. Internet Res. **15**(4), e73 (2013)

Coreference Resolution for French Oral Data: Machine Learning Experiments with ANCOR

Adèle Désoyer[1,2,3,4,5], Frédéric Landragin[1,2,3,4(✉)] [iD], Isabelle Tellier[1,2,3,4],
Anaïs Lefeuvre[6,7], Jean-Yves Antoine[6], and Marco Dinarelli[1,2,3,4]

[1] Lattice, CNRS, ENS, Paris, Orléans, France
{frederic.landragin,marco.dinarelli}@ens.fr
[2] Université de Paris 3, Paris, Orléans, France
[3] Université Sorbonne Paris Cité, Paris, Orléans, France
isabelle.tellier@sorbonne-nouvelle.fr
[4] PSL Research University, Paris, Orléans, France
[5] Modyco, CNRS, Université Paris Ouest – Nanterre La Défense,
Nanterre, Orléans, France
adele.desoyer@gmail.com
[6] LIFAT, CNRS, Université François Rabelais de Tours, Tours, Orléans, France
jean-yves.antoine@univ-tours.fr
[7] LIFO, Université d'Orléans, Orléans, France
anais.halftermeyer@univ-orleans.fr

Abstract. We present CROC (Coreference Resolution for Oral Corpus), the first machine learning system for coreference resolution in French. One specific aspect of the system is that it has been trained on data that come exclusively from transcribed speech, namely ANCOR (ANaphora and Coreference in ORal corpus), the first large-scale French corpus with anaphorical relation annotations. In its current state, the CROC system requires pre-annotated mentions. We detail the features used for the learning algorithms, and we present a set of experiments with these features. The scores we obtain are close to those of state-of-the-art systems for written English.

Keywords: Mention-pair model · Dialogue corpus
Coreference resolution · Machine learning

1 Introduction

Coreference Resolution has now become a classical task in NLP. This task consists in identifying coreference chains of *mentions* in texts. Supervised machine learning approaches are now largely dominant in this domain, but they require annotated corpora. No such corpus was available for French so far. In this paper, we first describe ANCOR, the first large-scale French corpus annotated with coreferent mentions. It is made of transcribed oral data, which is a specificity

© Springer International Publishing AG, part of Springer Nature 2018
A. Gelbukh (Ed.): CICLing 2016, LNCS 9623, pp. 507–519, 2018.
https://doi.org/10.1007/978-3-319-75477-2_36

relatively to corpora available for other languages. We then present CROC, a baseline system which has been learned with ANCOR, and a variant. Their performances are very close to those observed on coreference resolution challenges for English.

2   The ANCOR Corpus

We present ANCOR, a French corpus annotated with coreference relations which is freely available and large enough to serve the needs of data-driven approaches in NLP. With a total of 488,000 lexical units, ANCOR is among the largest coreference annotated corpora available at present and the only one of comparable size in French.

The main originality of this resource lies in the focus on spoken language. Nowaday systems using NLP for Information retrieval or extraction, for text summarization or even for machine translation have mainly been designed for written language. Oral language presents some interesting specificities, such as the absence of sentence units, the lack of punctuation, the presence of speech disfluencies, and, obviously, the grammatical variability of utterances. See [26] for a more detailed list of oral specific features which make French oral processing a big challenge.

2.1   Presentation of the Corpus

The ANCOR corpus is made of spoken French and it aims at representing a certain variety of spoken types. It integrates three different corpora that were already transcribed during previous research projects (Table 1). The first and larger one has been extracted from the ESLO corpus, which collects sociolinguistic interviews [16]. This corpus can be divided in two sub-corpora (ESLO-ANCOR and ESLO-CO2), corresponding to two distinct periods of recordings. It is characterized by a low level of interactivity. On the opposite, OTG and Accueil_UBS concern highly interactive Human-Human dialogues [18]. These two corpora differ by the media of interaction: direct conversation for the first one, phone call for the other one. Conversational speech (OTG and Accueil) only represents 7% of the total because of the scarcity of such free resources in French. All corpora are freely distributed under a Creative Commons license.

Table 1. ANCOR source corpus types and characterization

Corpus	Speech type	Interactivity	Size	Duration
ESLO	Interview	Low	452,000 words	27,5 h
ESLO_ANCOR			*417,000 words*	*25 h*
ESLO_CO2			*35,000 words*	*2.5 h*
OTG	Task oriented conversational speech	High	26,000 words	2 h
Accueil_UBS	Phone conversational speech	High	10,000 words	1 h

2.2   Annotation Scheme

The corpus has been fully annotated by hand on the Glozz platform [15]. Glozz produces a stand-off XML file structured according to a DTD that was specifically designed for ANCOR (it has also been translated in the MMAX2 format for portability purposes). This stand-off annotation allows a multi-layer work and enrichments through time.

The scope of annotation takes into account all noun phrases (NP from now) including pronouns but strictly restricts to them. As a result, the annotation scheme discards coreferences involving verbal or propositional mentions which have been annotated in the OntoNotes corpus. This restriction was mainly intended to favor data reliability by focusing on easily identifiable mentions [12].

Another specificity of the scheme is the annotation of isolated mentions. NPs are annotated even if they are not involved in any anaphoric relation, and this is a real added value for coreference resolution since the detection of singletons is known to be a difficult task [20].

We followed a detailed annotation scheme in order to provide useful data for deep linguistic studies and machine learning. Every nominal group is thus associated with the following features:

- Gender,
- Number,
- Part of Speech (only mentions have been annotated with these features, the corpus does not provide any morpho-syntactic annotation level on its own for other tokens),
- Definition (indefinite, definite, demonstrative or expletive form),
- PP: inclusion or not in a prepositional phrase,
- NE: Named Entity Type, as defined in the Ester2 coding scheme [8],
- NEW: discourse new vs. subsequent mention.

Coders were asked to link subsequent mentions with the first mention of the corresponding entity (discourse new) and to classify the relation among five different types of coreference or anaphora:

- **Direct coreference:** coreferent mentions are NP with the same lexical head.
- **Indirect coreference:** NP coreferent mentions with distinct lexical head (schooner ... vessel).
- **Pronominal anaphora:** the subsequent coreferent mention is a pronoun.
- **Bridging anaphora:** non coreference, but the subsequent mention depends on its antecedent for its referential interpretation (meronomy for instance: the schooner ... its bowsprit).
- **Bridging pronominal anaphora:** the subsequent mention is a pronoun. Its interpretation depends on its antecedent but the two mentions are not coreferent (for instance: the hostel ... they are welcoming).

This annotation scheme is quite similar to previous works on written language [5,28]. Since ANCOR represents the first large coreference corpus available for

French, it is important that the resource should concern researchers that are working on written documents too. Unlike [9], we did not distinguish between several sub-categories of bridging anaphora. We consider such a refined taxonomy to exceed the present needs of NLP while introducing a higher subjectivity in the annotation process. For the same reasons, we did not consider the relation of near-identity proposed in [20]. Recent experiments have shown that near-identity leads to a rather low inter-coders agreement [3].

2.3   Distributional Data

This section gives a general outline of the annotated data, to roughly show what should be found in the resource.

Table 2 details how the mentions and relations are distributed among the sub-corpora. With more than 50,000 relations and 100,000 mentions, ANCOR should fulfill the needs of representativity for linguistic studies and machine learning experiments. Table 3 shows that the repartition of nominal and pronominal entities is noticeable stable among the four corpora and leads to a very balanced overall distribution (51.2% vs. 48.8%).

Table 2. Content of the different sub-corpora

Corpus	Number of mentions	Number of relations
ESLO	106,737	48,110
ESLO_ANCOR	97,939	44,597
ESLO_CO2	8,798	3,513
OTG	7,462	2,572
Accueil_UBS	1,872	655
Total	116,071	51,337

Table 3. Mentions: distributional information

Entities	Nominal	Pronouns	% of NE
ESLO_ANCOR	51.8	48.4	66.3
ESLO_CO2	49.4	50.6	52,4
OTG	47.5	52.5	48.6
Accueil_UBS	48.5	51.5	43.3
Total	51.2	48.8	59.8

This observation certainly results from a general behavior of French speakers: pronominal anaphora are indeed an easy way for them to avoid systematic repetitions in a coreference chain.

In addition, ANCOR contains around 45,000 annotated Named Entities (NE). Therefore, it should stand for a valuable resource for NE recognition applications. 26,722 NE have been annotated as persons, 3,815 as locations, 1,746 as organizations, 1,496 as amounts, 1,390 for time mentions and 1,185 as products (Table 4).

Table 4. Most frequent named entities in ANCOR

Person	Location	Organization	Amount	Time	Product
26,722	3,815	1,746	1,496	1,390	1,185

Table 5. Relations: distributional percentages

Direct	Indirect	Pronominal	Bridging	Bridging pronominal
38,2	6,7	41,1	9,8	1,0

Table 6. Coreference/anaphora: distributional percentages

Corpus	ESLO_Ancor	ESLO_CO2	OTG	Accueil_UBS	Total
Direct	41,1	35,2	39,7	40,5	38,2
Indirect	7,3	11,2	6,1	7,5	6,7
Pronoun anaphora	43,9	38,2	46,4	46,0	41,1
Bridging anaphora	10,4	14,4	13,5	11,0	9,8
Pronoun bridging	0,9	1,0	3,3	0,6	1,0

Finally, Table 5 presents the distribution of coreference/anaphora relations. Once again, strong regularities between the sub-corpora are observed. In particular, direct coreference and pronominal anaphora are always prevalent. ANCOR contains around 20,000 occurrences of direct coreference and pronominal anaphora which are always prevalent through the corpus.

2.4 Annotation Reliability Estimation

The estimation of data reliability is still an open issue on coreference annotation. Indeed, the potential discrepancies between coders frequently lead to alignment mismatches that prevent the direct application of standard reliability measures [1,15,19]. We propose to overcome this problem by assessing separately the reliability of (1) the delimitation of the relations and (2) the annotation of their types. More precisely, three experiments have been conducted:

1. Firstly, we have asked 10 experts to delimitate the relations on an extract of ANCOR. These coders were previously trained on the annotation guide. We computed, on the basis of every potential pair of mentions, standard agreement measures: κ [4], α [11] and π [22]. This experiment aims above all at evaluating the degree of subjectivity of the task rather than the reliability of the annotated data, since the experts were not the coders of the corpus.

2. On the contrary, the second experiment concerned the annotators and the supervisor of the corpus. We asked them to re-annotate an extract of the corpus. Then we computed intra-coders agreement through a comparison to what they really performed on the actual corpus. This experiment aims at providing an estimation of the coherence of data (Table 6).

3. Finally, we asked our 10 first experts to attribute one type to a selection of relations that were previously delimited in the ANCOR corpus. We then computed agreement measures on the resulting type annotation.

Table 7. Agreement measures for the ANCOR corpus

Agreement	κ	π	α
Delimitation: inter-coder agreement	0.45	0.45	0.45
Delimitation: intra-coder agreement	0.91	0.91	0.91
Type categorization: inter-coder agreement	0.80	0.80	0.80

We observe on Table 7 very close results with the three considered reliability metrics (no difference before the 4th decimal). This is not surprising since we consider a binary distance between classes. The inter-coder agreement on delimitation is rather low (0.45). One should however note that this measure should be biased by our discourse-new coding scheme. Indeed, if a disagreement only concerns the first mention of a coreference chain, all the subsequent relations will unjustifiably penalize the reliability estimation. Further measures to come with the chain coding scheme will soon give an estimation of this potential bias. Anyway, this rather low agreement suggests that the delimitation task is highly prone to subjectivity, even when coders are trained. In particular, a detailed analysis of confusion matrices shows that most discrepancies occur between the delimitation of a bridging anaphora and the decision to not annotate a relation. Besides, this kind of disagreement appears to be related to personal idiosyncrasies. On the contrary, the results become very satisfactory when you consider intra-coders agreement (0.91). This means that our coders followed a very coherent strategy of annotation, under the control of the supervisor. This coherence is, in our opinion, an essential guarantee of reliability. Lastly we observed very good agreements on the categorization task (0.80), which reinforce our decision not to consider near-identity or detailed bridging types.

3   Machine Learning for Coreference Resolution

Coreference Resolution has become a classical task for NLP challenges, e.g. those organized by MUC (http://www.itl.nist.gov/iaui/894.02/related_projects/muc/proceedings/muc_7_toc.html), ACE (http://www.itl.nist.gov/iad/mig//tests/ace/), SemEval (http://semeval2.fbk.eu/semeval2.php?location=tasks) or CoNLL (http://conll.cemantix.org/2011/ & then http://conll.cemantix.org/2012/). But none of these challenges included French corpora. For French, as no labelled data were available before ANCOR, only hand-crafted systems have been proposed so far [13,27]. We rely instead on machine learning approaches. In this section we present our system, named CROC for "Coreference Resolution for Oral Corpus". It only treats the co-reference task. We thus suppose that every mention has already been recognized and associated with its specific features (see Sect. 2.2). The system was trained on the ANCOR_Centre corpus, using the WEKA machine learning platform [29].

3.1   Brief State of the Art

Several approaches have been proposed to reformulate coreference resolution as a machine leaning problem. The first and simpler one is the *pairwise* approach which proposes to classify every possible pair of referring mentions as *co-referential* or *not*. This approach assumes that referring mentions are provided (as we do in this paper) and requires a post-processing to build global chains from a set of local pairs. In order to do so, [17,23] apply a *Closest-First* strategy, which attaches a mention to its closest (on the left) co-referring other mention, whereas [2,24] propose a *Best-First* strategy, taking into account "co-referential probabilities".

Twin-candidate models [31] are variants of the pairwise approach in which the classification is applied to *triples* instead of pairs: an anaphoric mention and two candidates for its antecedent (the result being either *first* or *second* depending on which of the two candidates is the selected antecedent). criteria between candidates. Other more sophisticated models such as the *Twin-candidate* [31], *mention-ranking* [6] or *entity-mention* [30] have also been proposed. Our coreference resolution system is a baseline, it will thus use the *pairwise* and *Closest-First* strategies.

3.2   Representation of the Data in CROC

We have developed CROC as a baseline system which follows the pairwise and closest-first strategies. Pairwise systems rely on a good representation of pairs of mentions. In state of the art models, this representation is usually based on the classical set of *features* proposed in [23], augmented by those of [17]. For our experiments, we used all of these features when they are available in the corpus, plus some new ones we designed. The added features concern speakers and speech turns: they are specific to oral data (in particular to dialogues). One of our purposes is to evaluate the impact of these oral-specific features on the results. For each candidate pair of mentions (i,j), our set of features includes (cf. also Table 10):

Table 8. Results of *end-to-end* systems

System	Language	Corpus	MUC	B3	CEAF	BLANC
[23]	English	MUC-7	60.4	—	—	—
[17]	English	MUC-7	63.4	—	—	—
[25]	English	ACE-2003	67.9	65.9	—	—
[24]	English	MUC-7	62.8	79.4	—	—
[10]	English	ACE-2004	67.0	77.0	—	—
[12]	English	CoNNL-2012	68.8	54.56	50.20	—
[13]	French	Heterogeneous	36	69.7	55	59.5

Table 9. Results of systems starting with pre-annotated mentions

System	Language	Corpus	MUC	B3	CEAF	BLANC
[31]	English	MUC-7	60.2	—	—	—
[14]	English	ACE-2	80.7	77.0	73.2	77.2
[7]	English	ACE-2	71.6	72.7	67.0	—
[2]	English	ACE-2004	75.1	80.8	75.0	75.6
CROC	**French**	**ANCOR**	**63.45**	**83.76**	**79.14**	**67.43**
One-Class SVM	**French**	**ANCOR**	**61.73**	**84.58**	**80.41**	**69.66**

1. features characterizing each mention i and j:
   - at the morphological level: is it a pronoun? is it a definitive SN? is it a demonstrative SN?
   - at the enunciative level: is it a new mention?
   - at the semantic level: is it a named entity? of which type? Note that no freely available reliable semantic network is available for French, so no other semantic feature was used.
2. *relational* features, characterizing the pair:
   - at the lexical level: are the mentions strictly equal? partly equal?
   - at the morphosyntactic level: do they agree in gender? in number? Note that, in French, even if personal pronouns like "il" (he), "elle" (she)... agree in gender and number with their antecedent, possessive pronouns like "son", "sa"... (his, her...) agree with the noun they introduce and not with the referred antecedent.

- at the spatial level: how many characters/tokens/mentions/speech turns separate them?
- at the syntactic level: is one of the mentions included in the other one?
- at the contextual level: are their preceding/next tokens the same?
- at the enunciative level: are they produced by the same speaker?

3.3   Baseline Results

From the initial corpus, we kept 60% of data for learning, 20% for development, and 20% for test. In order to estimate the influence of the learning corpus size, we distinguished three sets: a small one (71,881 instances), a medium one (101,919 instances) and a big one (142,498 instances). In these sets, 20% of instances are coreferent pairs that are directly extracted from the corpus, and 80% are not-coreferent pairs (negative examples). We also tested different sets of features. In particular, we distinguished three sets: a first one that includes all features, a second one with only relational features, and a third one with all the features that are not linked to oral specificities. A last source of variation concerned the machine learning algorithm used: we tried decision trees, SVM (SMO with default parameters), and Naive Bayes using the WEKA platform.

Experiments involving development data showed that the best-performing model is the one calculated by SVM on small training set of data described by all features. Test data are submitted to this model, and the results are filtered by the *Closest-First* method, retaining only the closest antecedent if several pairs involving a mention were found coreferent. We present in Tables 8 and 9 the results of some state-of-the-art coreference resolution systems, for the four metrics dedicated to coreference resolution. Oral-specific features do not significantly improve the results.

3.4   One-class SVM

One of the main problems when using the *pairwise* approach is that, in order to train the binary classification model, artificial negative instances must be generated. Since there is no information to decide whether a pair of not-coreferent mentions is plausible or not, all possible pairs must be generated. The number of such pairs is polynomial in the length of a given mention set in a text, and this in turn means that negative instances are by far more numerous than positive instances. Since this may create a problem of unbalanced representation of positive and negative classes in SVM, heuristics have been proposed to filter out part of negative instances [17,23]. Despite such heuristics, negative instances are still much more than positive ones.

In order to overcome this problem we investigated the use of models which do not need negative instances. One such model still belongs to the SVM family, namely *One-class SVM* [21]. One-class SVM only needs positive instances, and instead of separating positive and negative instances from each other, separates positive instances from the origin. In order to make a comparison with our

Table 10. CROC complete feature set

	Features	Definitions	Possible values
1	m_1_TYPE	syntactic category of m_1	{N, PR, UNK, NULL}
2	m_2_TYPE	syntactic category of m_2	{N, PR, UNK, NULL}
3	m_1_DEF	definition of m_1	{UNK, INDEF, EXPL, DEF_SPLE, DEF_DEM}
4	m_2_DEF	definition of m_2	
5	m_1_GENDER	gender of m_1	{M, F, UNK, NULL}
6	m_2_GENDER	gender of m_2	{M, F, UNK, NULL}
7	m_1_NUMBER	number of m_1	{SG, PL, UNK, NULL}
8	m_2_NUMBER	number of m_2	{SG, PL, UNK, NULL}
9	m_1_NEW	is a new entity introduced by m_1?	{YES, NO, UNK, NULL}
10	m_2_NEW	is a new entity introduced by m_2?	{YES, NO, UNK, NULL}
11	m_1_EN	entity type of m_1	{PERS, FONC, LOC, ORG, PROD, TIME, NO, AMOUNT, UNK, NULL, EVENT}
12	m_2_EN	entity type of m_2	
13	ID_FORM	are m_1 and m_2 forms identical?	{YES, NO, NA}
14	ID_SUBFORM	are there identical sub-forms?	{YES, NO, NA}
15	INCL_RATE	tokens covering ratio	REAL
16	COM_RATE	common tokens ratio	REAL
17	ID_DEF	is there definition equality?	{YES, NO, NA}
18	ID_TYPE	is there type equality?	{YES, NO, NA}
19	ID_EN	is there named entity type equality?	{YES, NO, NA}
20	ID_GENDER	is there gender equality?	{YES, NO, NA}
21	ID_NUMBER	is there number equality?	{YES, NO, NA}
22	DISTANCE_MENTION	distance (number of mentions)	REAL
23	DISTANCE_TURN	distance (number of speech turns)	REAL
24	DISTANCE_WORD	distance (number of words)	REAL
25	DISTANCE_CHAR	distance (number of characters)	REAL
26	EMBEDDED	embedding of m_2 in m_1?	{YES, NO, NA}
27	ID_PREVIOUS	are previous tokens identical?	{YES, NO, NA}
28	ID_NEXT	are next tokens identical?	{YES, NO, NA}
29	ID_SPK	are speakers identical?	{YES, NO, NA}
30	ID_NEW	are discursive status identical?	{YES, NO, NA}

baseline, we trained such a model with exactly the same data and features. The results are shown in Table 9, line *One-Class SVM*. Our research in this direction is in progress, but we can see that currently results obtained with this approach are roughly equivalent to baseline results.

4 Conclusion and Perspective

Most current researches on coreference resolution concern written language. In this paper, we presented experiments that were conducted on ANCOR, a large French corpus based on speech transcripts, annotated with rich information and coreference chains. This corpus represents the first significant effort to provide sufficient coreference training data in French for machine learning approaches. We described CROC, a baseline approach for automatic coreference resolution on French, as well as another machine learning approach based on one-class SVM. Our first results are roughly equivalent to state-of-the-art performances, which suggests that standard ML approaches for coreference resolution should apply satisfactory on spoken language.

For further investigation, we plan to more carefully study the impact of the various corpus origins on the final results. Does the speech type and/or the level of interactivity influence the way co-reference chains are built in dialogues? To better compare our results with the state of the art, other more complex learning models also need to be tested on these data. And finally, to provide a real *end-to-end* system, we have to automatically identity the mentions and their specific features, as a pre-processing step.

Acknowledgments. This work was supported by grant ANR-15-CE38-0008 ("DEMOCRAT" project) from the French National Research Agency (ANR), and by APR Centre-Val-de-Loire region ("ANCOR" project).

References

1. Artstein, R., Poesio, M.: Inter-coder agreement for computational linguistics. Comput. Linguist. **34**(4), 555–596 (2008)
2. Bengtson, E., Roth, D.: Understanding the value of features for coreference resolution. In: Proceedings of EMNLP 2008, pp. 236–243 (2008)
3. Broda, B., Niton, B., Gruszczynski, W., Ogrodniczuk, M.: Measuring readability of polish texts: baseline experiments. In: Proceedings of the Ninth International Conference on Language Resources and Evaluation, Reykjavik, Iceland (2014)
4. Cohen, J.: A coefficient of agreement for nominal scales. Educ. Psychol. Meas. **20**, 37–46 (1960)
5. van Deemter, K., Kibble, R.: On coreferring: coreference in MUC and related annotation schemes. Comput. Linguist. **26**(4), 629–637 (2000)
6. Denis, P.: New learning models for robust reference resolution. Ph.D. thesis, University of Texas at Austin (2007)
7. Denis, P., Baldridge, J.: Specialized models and ranking for coreference resolution. In: Proceedings of the Conference on Empirical Methods in Natural Language Processing, EMNLP 2008, pp. 660–669 (2008)
8. Galliano, S., Gravier, G., Chaubard, L.: The ester 2 evaluation campaign for the rich transcription of French radio broadcasts. In: Proceedings of Interspeech (2009)
9. Gardent, C., Manuélian, H.: Création d'un corpus annoté pour le traitement des descriptions définies. TAL **46**(1), 115–139 (2005)

10. Haghighi, A., Klein, D.: Coreference resolution in a modular, entity-centered model. In: Human Language Technologies: The 2010 Annual Conference of the North American Chapter of the Association for Computational Linguistics, pp. 385–393 (2010)

11. Krippendorff, K.: Content Analysis: An Introduction to Its Methodology. SAGE Publications Inc., Thousand Oaks (2004)

12. Lassalle, E.: Structured learning with latent trees: a joint approach to coreference resolution. Ph.D. thesis, Université Paris Diderot (2015)

13. Longo, L.: Vers des moteurs de recherche intelligents: un outil de détection automatique de thèmes. Ph.D. thesis, Université de Strasbourg (2013)

14. Luo, X., Ittycheriah, A., Jing, H., Kambhatla, N., Roukos, S.: A mention-synchronous coreference resolution algorithm based on the bell tree. In: Proceedings of the 42nd Annual Meeting on Association for Computational Linguistics (2004)

15. Mathet, Y., Widlöcher, A.: Une approche holiste et unifiée de l'alignement et de la mesure d'accord inter-annotateurs. In: Actes de TALN, pp. 1–12. ATALA (2011)

16. Muzerelle, J., Lefeuvre, A., Schang, E., Antoine, J.Y., Pelletier, A., Maurel, D., Eshkol, I., Villaneau, J.: Ancor_centre, a large free spoken French coreference corpus: description of the resource and reliability measures. In: Proceedings of the Ninth International Conference on Language Resources and Evaluation, Reykjavik, Iceland (2014)

17. Ng, V., Cardie, C.: Improving machine learning approcahes to corefrence resolution. In: Proceedings of ACL 2002, pp. 104–111 (2002)

18. Nicolas, P., Letellier-Zarshenas, S., Schadle, I., Antoine, J.Y., Caelen, J.: Towards a large corpus of spoken dialogue in French that will be freely available: the parole publique project and its first realisations. In: Proceedings of LREC (2002)

19. Passonneau, R.J.: Computing reliability for coreference annotation. In: Proceedings of LREC, pp. 1503–1506 (2004)

20. Recasens, M.: Coreference: theory, resolution, annotation and evaluation. Ph.D. thesis, University of Barcelona (2010)

21. Schölkopf, B., Platt, J.C., Shawe-Taylor, J.C., Smola, A.J., Williamson, R.C.: Estimating the support of a high-dimensional distribution. Neural Comput. **13**(7), 1443–1471 (2001)

22. Scott, W.: Reliability of content analysis: the case of nominal scale coding. Public Opin. Q. **19**, 321–325 (1955)

23. Soon, W.M., Ng, H.T., Lim, D.C.Y.: A machine learning approach to coreference resolution of noun phrases. Comput. Linguist. **27**(4), 521–544 (2001)

24. Stoyanov, V., Cardie, C., Gilbert, N., Riloff, E., Buttler, D., Hysom, D.: Reconcile: a coreference resolution research platform. Technical report, Cornell University (2010)

25. Stoyanov, V., Gilbert, N., Cardie, C., Riloff, E.: Conundrums in noun phrase coreference resolution: making sense of the state-of-the-art. In: Proceedings of the Joint Conference of the 47th Annual Meeting of the ACL and the 4th International Joint Conference on Natural Language Processing, pp. 656–664 (2009)

26. Tellier, I., Eshkol, I., Taalab, S., Prost, J.P.: POS-tagging for oral texts with CRF and category decomposition. Res. Comput. Sci. **46**, 79–90 (2010)

27. Trouilleux, F.: Identification des reprises et interprétation automatique des expressions pronominales dans des textes en français. Ph.D. thesis, Université Blaise Pascal (2001)

28. Vieira, R., Salmon-Alt, S., Schang, E.: Multilingual corpora annotation for processing definite descriptions. In: Proceedings of PorTAL (2002)

29. Witten, I.H., Frank, E., Trigg, L., Hall, M., Holmes, G., Cunningham, S.J.: Weka: practical machine learning tools and techniques with java implementations (1999)
30. Yang, X., Su, J., Lang, J., Tan, C.L., Liu, T., Li, S.: An entity-mention model for coreference resolution with inductive logic programming. In: Proceedings of ACL 2008, pp. 843–851 (2008)
31. Yang, X., Zhou, G., Su, J., Tan, C.L.: Coreference resolution using competition learning approach. In: Proceedings of ACL 2003, pp. 176–183 (2003)

Arabic Anaphora Resolution
Using Markov Decision Process

Fériel Ben Fraj Trabelsi[1(✉)], Chiraz Ben Othmane Zribi[1(✉)],
and Saoussen Mathlouthi[2]

[1] National School of Computer Sciences,
Manouba University, Manouba, Tunisia
{Feriel.benfraj,Chiraz.BenOthmane}@riadi.rnu.tn
[2] Faculty of Science of Bizerte, Carthage University, Tunis, Tunisia
Saw.mathlouthi@gmail.com

Abstract. The anaphora resolution belongs to the attractive problems of the NLP field. In this paper, we treat the problem of resolving pronominal anaphora which are very abundant in Arabic texts. Our approach includes a set of steps; namely: the identification of anaphoric pronouns, removing the non-referential ones, identification of the lists of candidates from the context surrounding the identified anaphora and choosing the best candidate for each anaphoric pronoun. The last two steps could be seen as a dynamic and probabilistic process that consists of a sequence of decisions and could be modeled using a Markov Decision Process (MDP). In addition, we have opted for a reinforcement learning approach because it is an effective method for learning in an uncertain and stochastic environment like ours. Also, it could resolve the MDPs. In order to evaluate the proposed approach, we have developed a system that gives us encouraging results. The resolution accuracy reaches up to 80%.

Keywords: Anaphora resolution · Antecedent · Arabic language
Markov Decision Process (MDP)

1 Introduction

The aim of the anaphora resolution is to identify the most plausible antecedent among a list of possible candidates. Such antecedent should be interpreted as referent of an anaphora that appears before (sometimes after) it in the discourse. The anaphora resolution is a very important task and it is one of the most active researches for the NLP filed. In fact, it is required in several NLP applications, as for instance, the word sense disambiguation, parsing and machine translation. It is also necessary for a lot of other research fields such as the information retrieval.

Several studies have been made in this area for several languages using different approaches that we classify into three main categories:

- Symbolic approaches based on syntactic, semantic and even pragmatic knowledge, which are general. The independency of corpora allows the application of these knowledge to different types of texts and, also, their adaptation to different languages. The approaches of this category are numerous [1–8].

© Springer International Publishing AG, part of Springer Nature 2018
A. Gelbukh (Ed.): CICLing 2016, LNCS 9623, pp. 520–532, 2018.
https://doi.org/10.1007/978-3-319-75477-2_37

- Corpus-based approaches are more recent. The used knowledges are extracted from relevant data volumes called corpus. Using these corpora, the resolution approach was considered in different ways, such as the statistics extraction of the collocations' models [9] and the training of machine learning algorithms, as for instance, the research based on a binary classification [10], the second based on decision trees [11] and the third that uses memory-based learning [12].
- Hybrid approaches incorporate both linguistic and corpus-based knowledge such as for the researches of [13, 14]. These approaches combine different types of knowledge that provide higher coverage for both classes of the previous approaches. But, the disadvantage, in this case, is the relatively high time resolution.

Furthermore, the researches on resolving Arabic anaphora are few. We mention the research study of Mitkov [5] who uses his robust algorithm for resolving English anaphora and adapts it to the Arabic language. It is basically a rule-based approach that uses preference criteria to track down the antecedent among a list of potential candidates. We also quote the research of Elghamry et al. [15] who make use of a statistical and dynamic algorithm to resolve Arabic anaphora using a web corpus. It involves only three criteria; namely the collocation evidence, the Recency and bands. The Recency considers that the candidates that belong to the same sentence as the anaphor are the most important to refer such anaphor. For the ultimate criterion, the idea consists on dividing iteratively the research space into bands (–20 initially window) in order to reduce the number of candidates. The third research that we mention is of Abolohom and Omar [16] who propose an approach that consists of two steps. A rule-based step is applied when identifying anaphora and extracting the candidates list. A machine learning classifier (K-Nearest-Neighbor) is used for the resolution step where, a set of varied criteria is used.

However, the anaphora phenomenon is very common in Arabic texts. Anaphora has different forms and types, we note:

– Pronominal anaphora consists on taking back a sentence's component using a pronoun. According to the pronoun category, there are different sub-categories; namely enclitics, personal pronouns, demonstrative pronouns and relative ones.
– Lexical anaphora allows the recovery of a term through a noun or a nominal phrase (NP).
– Verbal anaphora uses the verb فعل (*to do*) to indicate action.
– Comparative anaphora presents a relationship of similarity, complementarity or comparison between the anaphora and its antecedent.

This variety of anaphora types leads us to conclude that it is impossible to treat all of them in the same way. In fact, the most abundant anaphora, in the Arabic texts, belong to the pronominal class. This observation is justified by the statistics of [17] made over texts of various fields (newspapers, technical manuals, educational books and news). The authors noticed that the percentage of pronominal anaphora among all the anaphora types exceed 89% for all the categories of texts. Consequently, in this paper, we have chosen to treat this type of anaphora.

In addition, the problem of anaphora resolution gets worse as the Arabic language presents a varied set of specificities that makes it a complex task. We could mention, as examples: clitics agglutination, grammatical ambiguity, extensive lengths of sentences and the flexibility of the components order within sentences.

In this paper, we present a novel approach for resolving Arabic anaphoric pronouns using Markov Decision Process. Indeed, the environment of our problem contains the list of candidates and the preference criteria that will be used for predicting the most plausible antecedent. The anaphora resolution approach that we propose seeks to optimize a sequence of decisions that aim to consolidate the choice of a candidate or to weaken it in order to find the best one. This environment is uncertain because choosing the best antecedent is closely attached to a number of information derived from criteria that are probable. In addition, the environment is dynamic since the importance of the criteria varies when changing the texts styles.

Since the proposed resolution system is probabilistic and dynamic, it could be seen as a Markov Decision Process (MDP) that consists of states, actions, transitions and reward functions. The reinforcement learning has a very interesting method for solving MDP as the interaction with the environment reinforces the actions, and so could give the best solution.

The remainder of this paper is organized as follows: Sect. 2 is devoted to the description of the preliminary resolution steps. Section 3 presents the criteria that we choose to desmabiguate among the candidates. Section 4 describes the details of the proposed approach. In Sect. 5, we present the experimental evaluation of the implemented system with the results discussion. In Sect. 6, we conclude the paper with some remarks and future prospects.

2 Pre-resolution Steps

Our approach starts with some preliminary steps; namely the identification of pronouns, the elimination of the non-anaphoric ones. For each anaphora, we extract a list of antecedent candidates. Subsequently, this list is filtered to leave only those that are the most suitable with the target anaphora.

The input text is morpho-syntactically analyzed [18] and grammatically tagged [19]. Thus, the identification of anaphora is carried out using the grammatical parts-of-speech.

After the identification of the anaphora, the elimination of pleonastic pronouns consists on removing the pronouns that correspond to the specific model:

adjective+ال+من إنه (It is + adjective) as for the example إنه من الصعب (It is difficult to) or إنه من المناسب (It is suitable to).

The deictic pronouns are also eliminated. These correspond to the 1st and 2nd personal pronouns that specify the communication partners and their sense returns to their own uses. We mention, for example the sentence أنا أعد الطعام وأنت ترتب البيت (I prepare the meal and you organize the house). Deictic pronouns include also spatio-temporal pronouns الآن (now) or هنا (here).

The identification of candidates differs with the variety of pronouns' types. Thus, a relative pronoun is always anaphoric and its antecedent is the NP that is often located immediately before it. Personal and demonstrative pronouns may have several possible candidates. Therefore, the system identifies any NP that exists around the anaphora in a window of [–n, n], where n is the number of sentences to consider before and after the current sentence. The value of n may be fixed during experiments.

In a subsequent step, the candidates are filtered using the gender and number agreements. This criterion prefers candidates which agree in gender and number with the anaphora. However, there are some limited exceptional cases. Thus, we notice the plural of non-human and inanimate entities (male or female) which admit a singular female anaphora as for the example: الكلاب أيقظتنا بنباحها (*The dogs woke us because of their barking*). The enclitic ها (*her*) which is singular female refers the plural not human word الكلاب (*the dogs*). This step is very important since it reduces significantly the size of the list of candidates to retain only the most likely ones. Therefore, we ensure that unlikely candidates will not be present throughout the research process of the correct antecedent. Consequently, we not lose in terms accuracy but we will win the execution time.

Once the anaphora are identified and their candidate lists are generated and filtered, we have to seek the best antecedent of each anaphora from the list of possible candidates. In order to solve this problem, we propose a learning model that uses the Markov Decision Process (MDP) for making the criteria selection. The MDP receives, as input, anaphora with its context, the list of possible candidates and a set of criteria.

3   Criteria Description

The evaluation of candidates is based on several criteria which aim is to distinguish between the candidates. There are two types of criteria: linguistic criteria and corpus based ones. Nevertheless, it should be noted that the importance of these criteria in the anaphora resolution is largely dependent on the context. In fact, for a couple (anaphora, candidate), some of the mentioned criteria can be checked and, thus contribute to the resolution. Consequently, they will have more or less importance in this task. Other criteria do not have the same impact if, for example, they are not checked in the target context. Thus, the criteria do not necessarily participate together in the resolution of a particular anaphora.

3.1   Linguistic Criteria

The linguistic criteria were chosen based on:

- Linguistic ascertainments regarding anaphora referents and made on a set of literary texts,
- Arabic language Characteristics,
- Inspiration of previous researches [4–6].

These criteria are numerous. The first chosen criterion is the *definition* that considers that the defined NPs are preferred over those undefined. In fact, for some types of anaphoric pronouns, the antecedent is necessary defined as for the demonstrative and relative pronouns. The second criterion is the *topic* which favors the subjects of the sentences (current and/or previous) over other candidates, since according to our linguistic ascertainments anaphora often refers the active agents in the sentences (subject, or vice-subject for the passive voice).

Furthermore, it is semantically logical that the closest candidate is the most salient. Therefore, the higher salience must be assigned to the candidates who are in the same sentence. The value of salience decreases whenever the antecedent is increasingly away from the target anaphora. Thus, the third criterion is the *distance* between the anaphora and the antecedent. However, the head of the section admits always a high salience, as its probability to be the antecedent of an anaphora is also high. Indeed, the agent subject that starts the speech often remains the center of interest throughout the section. So, the *head of section* is also one criterion.

Besides, the *proper noun* acts usually as the topic of a sentence. Therefore, it must be a suitable antecedent for different anaphora in the same speech. This criterion considers that proper nouns are privileged compared to other candidates.

For a particular style of texts; especially the technical manuals, some *patterns of sentences* give the privilege to candidates which respect the following patterns:

- V – NP/anaphora as for the sentence اضغط على الزر لرفع الصوت، اضغط عليه أكثر
  (*Hold down the button to increase the volume, hold down it more*)
- V_1 NP … conjunction V_2 anaphora (conjunction V_3 anaphora) as for the following sentence اقطع الزبدة، أذبـها و أضفـها للعجين (*Cut the butter, let it melt and add it to the dough*).
- V_1 NP … conjunction anaphora V_2 … as for instance: بقيت خديجة إلى منتصف الليل وهي تصنع الحلويات (*Khadija remained until midnight to prepare desserts*).

Finally, the *sequential models* constitute a criterion that favors the first nominal groups in the following construction:

Connector V1 NP1, V2 NP2. (Sentence) Connector V3 **Anaphora**, V4 NP4; where the connector indicates a goal or an explanation (…. حتى ،كي ،لـ) (*for*).

Unlike the patterns of sentences, this criterion should be valid for any text style.

3.2   Corpus-Based Criteria

The corpus-based criteria are, rather, inspired from previous work, especially that of [20], without forgetting to check their linguistic interest in the resolution of Arabic anaphora. We consider two criteria that belong to this category.

Firstly, the *patterns of co-occurrence* are based on statistics which are generated from textual corpus. Indeed, collocations may be the solution when choosing between several candidates. The process consists on replacing the anaphora by its possible candidates. Then, it verifies which is the most frequently present in the anaphora context. For calculating the probability of co-occurrence, we use the following formula of mutual information:

$$IM = \log_2 \frac{a}{(a+b)(a+c)}$$

where a is the number of occurrences of the couple of words (m_1, m_2), b is the number of couples that involve m_1 as the first element of the couple and m_2 is not its second element. The parameter c is the number of couples where m_2 is the second element of the couple and m_1 is not the first.

Secondly, the *repetition* criterion considers that the noun (or NP) which is repeated two or more times in the same section is a preferred candidate. Indeed, the words and more precisely their lemmas tend to be repeated in the same text [21]. The probability of repeating each lemma is defined by the following formula:

$$P(lemma_i) = \frac{\text{number of occurrences of the } i^{th}\text{lemma}}{\text{total number of lemmas}}$$

4 Anaphora Resolution Approach

According to the uncertainty of the preference criteria for the disambiguation among antecedents, the MDP is applied twice. Indeed, we use MDP for choosing the antecedents. Then, for each antecedent we applied other MDP for choosing sets of criteria. These MDP are resolved using reinforcement learning.

4.1 MDP for Choosing Antecedents and Criteria

Anaphora processing is performed in a sequential manner. The resolution of each anaphora is modeled by a MDP for choosing the candidates. Such process is defined as follows:

- The initial sate S_0 that includes the anaphora and its context.
- The set of states $S = \{S_1, S_2, ..., S_n\}$, where n is the number of likely candidates. These states present the anaphora and its context after choosing a candidate. All these states are final.
- The set of actions $A = \{A_1, A_2, ..., A_n\}$. Every action refers to the selection of a candidate. Therefore, the choice of the candidate i advances the system from the state S_0 to the state S_i.
- The transition function T enables the progression of the system from one state to another at each executed action. Initially, the transitions from the state S_0 to the states S_i are all equiprobable as well as the candidates.

The reward function r_i is the reward value received at the state S_i and affected to the action A_i. This award corresponds to the score of the candidate evaluation according to the selected combination of criteria which should maximize the evaluation gain of the selected candidate.

The MDPs used for the criteria selection are nested into those of the selected candidates. To select the criteria, we collect them into subsets; so that the number of elements in each subset varies between 1 and m, where m is the maximum number of criteria. For example, with three criteria c_1, c_2 and c_3, the generated subsets are: $\{c_1\}$, $\{c_2\}$, $\{c_3\}$, $\{c_1, c_2\}$, $\{c_1, c_3\}$, $\{c_2, c_3\}$ and $\{c_1, c_2, c_3\}$. The total number k of the sub-sets of criteria is estimated to be the sum of all possible combinations of 1 among m criteria. It is defined using the following formula:

$$k = \sum_{i=1}^{m} C_m^i = \sum_{i=1}^{m} \frac{m!}{i!(m-1)!}$$

Therefore, the sub-MDP applied in the criteria selection for each state of the candidate i, where $i \in [1, n]$, is defined as follows:

- The initial state S_i that contains the anaphora, its context and the ith candidate.
- The set of states $S = \{S_{i1}, S_{i2}, ..., S_{ik}\}$; where k is the number of all possible combinations of m criteria. Each state S_{ij} contains the anaphora, the ith candidate and the jth combination of criteria, where $j \in [1, k]$.
- The set of actions $A = \{A_{i1}, A_{i2}, .. A_{ik}\}$ where each action A_{ij} denotes the choice of the jth combination of criteria to evaluate the ith candidate.
- The transition function T enables the progression of the resolution to every action performed. The probability of transition from the state S_i to the state S_{ij} depends only on the action and the current state.
- The reward function r_{ij} is the reward value received at the state S_{ij} and affected to each action A_{ij}.

The sub-MDP starts at the initial state S_i and progresses to the state S_{ij} after executing a set of actions based on the various possible combinations of criteria. In addition, each criterion has a fixed score r_i that illustrates its confidence and a variable probability p_i that denotes its relevance when identifying the correct candidate. This process is described by the Fig. 1.

Anaphora resolution consists on solving the MDP of the candidates' choice. To solve this MDP, we must first solve the sub-MDPs of all possible candidates. Therefore, we propose to use reinforcement learning approach in order to learn an optimal strategy for the MDP and its sub-MDPs. This approach offers a process of trials and errors to find optimal strategies.

4.2   Reinforcement for Candidates Evaluation

The reinforcement learning is an attractive method for solving PDMs since the learning process needs only a scalar reinforcement signal. Inspired by the reinforcement learning that aims to maximize a value function, we consider the following assumptions:

- The reward is fixed, immediate and depends on the criterion type.
- The value function associates to each final state a reinforcement measure that represents a reward. This measure depends on the criteria and the probabilities of their selection.

Our aim is to maximize this value function by a set of tests using different set of probabilities that could be assigned to the criteria. At the sub-MDP which models the criteria selection, we try to find for each candidate the combination of criteria that

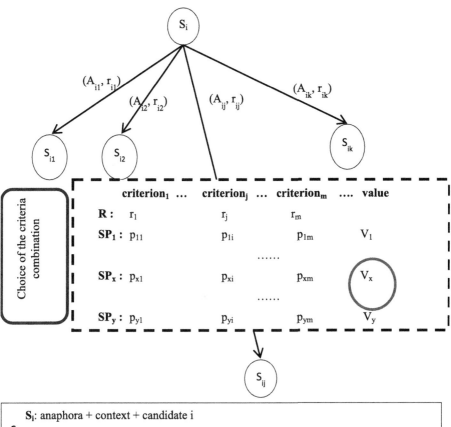

Fig. 1. Sub-MDP for choosing the criteria combinations of the candidate i

maximizes the gain. We are convinced that the relevance of the criteria is uncertain. Therefore, we propose to associate random sets of probabilities to the criteria and then define the best set that maximizes the value function. For each set of probabilities, we seek the combination of criteria that maximizes the value function V. This function value depends on the probabilities p_i and the awards r_i. It is defined by the following formula:

$$V = \sum_{i=1}^{l} ri * pi$$

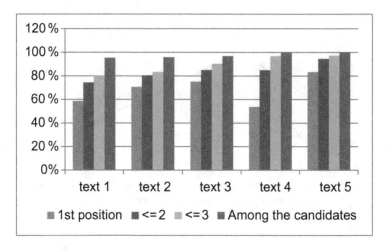

Fig. 2. Position of the correct antecedent in the list of candidates

5   Experimental Evaluation

At the beginning of this section, we describe the corpus test. Then, we present the evaluation of our anaphora resolution approach. Thus, we consider the standard evaluation metrics; namely the accuracy, recall and f-measure, and we add two others:

- The exactness is the percentage of anaphora for which the correct antecedent belongs to the list of candidates.
- The position is the average position of the correct antecedent among the list of candidates after using our system.

5.1   Test Corpus

Our test corpus consists of literary texts which we extracted from the textbook of the 8th year of the Tunisian basic education. This corpus contains 2610 words, 1649 phrases, 354 sentences (having an average size of 7 words) that compose 38 subsections (having an average size of 9 sentences). It covers 420 anaphoric pronouns. For our resolution approach, we considered the personal, demonstrative and relative pronouns. The majority of pronouns are personal (80%). Demonstrative pronouns are much less used (12%). However, relative pronouns are rarely used (4%).

Besides, we have decomposed this corpus into five texts according the size of the sub-sections (in terms of sentences) and the numbers of words in the sentences.

5.2   Evaluation Results

In order to measure the difficulty of the resolution task, we count the minimum, maximum and average numbers of candidates per anaphora. The lists have at least one candidate. However, their sizes could reach 17 candidates per anaphora. The average is equal to 6 candidates. Thus, the percentage of ambiguous anaphora, i.e. those having at least 2 candidates, is relatively high (92.26%). These statistics are obtained after the filtering process made based on gender and number agreements. The filtering process eliminates the undesirable candidates. The table below presents the results of the resolution (Table 1).

Table 1. Evaluation results

Corpus	Accuracy	Recall	F-measure	Exactness	Position
Text 1	58.62	57.30	57.95	95.40	1.91
Text 2	70.83	67.11	68.91	95.83	1.52
Text 3	75.27	71.43	73.29	96.77	1.45
Text 4	53.76	51.55	52.63	100	1.64
Text 5	83.33	78.95	81.08	100	1.22

We note that the resolution accuracy varies according to the text type. This variation explains that the difficulty of the task depends on the nature of the text i.e. the morpho-syntactic, semantic and pragmatic structures of the speech. In addition, we conclude that the 5th text has the best performance values with an accuracy of 83.33% and a recall equal to 78.95%. Thus, the F-measure of that text reaches 81.08%. The sub-sections in this text have the smallest sizes. The exactness of this text that reaches 100% shows that for each anaphora, the correct antecedent is still in the list of possible candidates. Furthermore, the average position of the correct candidate is very close to the order 1 while noting that the average position does not exceed the value 2.

To highlight a possible improvement in the results, we present the anaphora resolution rates for different orders of the correct candidate. The figure below provides the details of the resolution for each text. It illustrates the accuracy of our resolution system for different possible cases, namely: the correct antecedent is in the 1st position of the list of candidates, it is among the top two candidates (≤ 2), it is among the top three candidates (≤ 3) or it belongs to the list of candidates.

Statistics show that our approach has solved up to 83.33% of anaphora in one of the corpus texts. Moreover, this value could be improved as for 94.44% of the anaphora; the correct antecedent is among the top two candidates. Moreover, in many cases, the value function of the correct antecedent is very similar to that of the first element of the list. Statistics show also that most of the correct antecedents are chosen among the top three candidates of the lists.

When evaluating the different stages of resolution, we found that there are lots of anaphora which are not correctly resolved. These failures are mainly due to the following reasons:

- The best antecedent could be far from the anaphora and the window of possible candidates
- The presence of successive anaphora, especially in the case of relative pronouns. The following example shows two anaphora whose the second refers the first. هو الذي وضع بيده هذه الشوكة (*It is the one who puts by himself this fork*). In the case of this sentence, the relative pronoun الذي (*who*) refers another pronoun هو (*he*) and these two pronouns refer together another further antecedent in the previous discourse.

- The candidate whose identification requires information from the pragmatic level, i.e. derived from the understanding of the general context of the text as for the example below.

ولم يجد من يتركه عند ابنه غير كلب داجن يملكه فتركه عند الصبي وأغلق عليهما البيت (*He did not find who to leave with his son except a domestic dog. He leaved it with the boy and closed them home*)

In this example, the anaphora هما (*them*) refers the antecedent that is composed of the two NP (*a domestic dog*) كلب داجن and الصبي (*the boy*) that does not appear as a single entity in the text but it is deduced from two distant phrases.

Finally, we conclude that the literary nature of our corpus texts affects the resolution performance. In fact, the length of the sentences and the diversity of their morphologic, syntactic and semantic structures make the resolution more difficult.

5.3 Comparison with Other Works

To point out the relevance of our work among other research studies which process the same problem, we decide to compare our results to those of Mitkov [5], Elghamry et al. [15] and Abolohom and Omar [16]. These works processed anaphora in Arabic (Table 2).

Table 2. Our system vs other Arabic anaphora resolution systems

Research studies	Corpus type	Accuracy	Recall	F-measure
[5]	Technical reports	95.2%	–	–
[15]	Gold standard	78%	100%	87.6%
[16]	Quranic corpus	71.7%	78.2%	74.8%
Our research	Literary texts	83.33%	78.95%	81.08%

Our system gives satisfactory results compared to other research studies on Arabic anaphora resolution. The most efficient research study is the oldest one [5]. Its experiments were made on a restricted domain (technical texts) and it uses a little corpus test. The two others have very similar results to those of our approach. Nevertheless, the comparison between the different research studies using these measures is not objective. It depends on the type of the used approach (symbolic, corpus-based or hybrid), the test corpus and its size.

In addition, all these works involve a set of criteria and use them together for the disambiguation between the candidates. However the originality of our approach is its dynamism since it involves a set of various criteria and it aims to search through the sub-set of criteria that help to predict the correct candidate depending on the context of a given anaphora. Thus, for a given context, our system could use only a little sub-set of our criteria; *i.e.* the most relevant according to this context.

6   Conclusion

Our developed system leads us to note that success rates vary from one text to another according to their types. For one of the tested texts, our approach provides a resolution accuracy equal to 83% and a recall equal to 79% and an F-measure of the order of 81%. The exactness of the same text (100%) for each anaphora shows that the correct candidate is still in the list of possible candidates. However, the failures of the resolution are mainly due to the morpho-syntactic and semantic diversities of the literary texts such as the sizes of the sentences. Therefore, we conclude that the characteristics of the Arabic language, especially, literary texts make the anaphora resolution task more and more arduous. However, the obtained results are satisfactory. But, we prospect to evaluate our system by using other types of texts such as technical manuals or journalistic texts and adding other preference criteria.

References

1. Wada, H.: Discourse processing in MT: problems in pronominal translation, Helsinki, Finland (1990)
2. Lappin, S., Leass, H.J.: An Algorithm for Pronominal Anaphora Resolution, London (1994)
3. Nakaiwa, H., Shirai, S., Ikehara, S., Kawaoka, T.: Extrasentential resolution of Japanese zero pronouns using semantic and pragmatic constraints (1995)
4. Baldwin, B.: CogNIAC: high precision coreference with limited knowledge and linguistic resources, Madrid, Spain (1997)
5. Mitkov, R.: Robust pronoun resolution with limited knowledge. School of Languages and European Studies, Wolverhampton WV1 1SB, United Kingdom (1998)
6. Bittar, A.: Un algorithme pour la résolution d'anaphores événementielles. UFR de Linguistique, Université Paris 7 Denis Diderot (2006)
7. Nouioua, F.: Heuristique pour la résolution d'anaphores dans les textes d'accidents de la route. Institut Galilée, Université Paris 13, Villetaneuse (2007)
8. Gelain, B., Sedogbo, C.: La résolution d'anaphore à partir d'un lexique-grammaire des verbes anaphoriques, France (2010)
9. Dagan, I., Itai, A.: Automatic processing of large corpora for the resolution of anaphora references, Helsinki, Finland (1990)
10. Connoly, D., Burger, J.D., Day, D.S.: A machine learning approach to anaphoric reference, Manchester, UK (1994)
11. Aone, C., Bennet, S.W.: Applying machine learning to anaphora resolution, Berlin (1996)
12. Preiss, J.: Machine Learning for Anaphora Resolution (2001)
13. Mitkov, R.: Anaphora resolution: a combination of linguistic and statistical approaches, Lancaster, UK (1996)
14. Weissenbacher, D., Nazarenko, A.: Identifier les pronoms anaphoriques et trouver leurs antécédents: l'intérêt de la classification bayésienne. Laboratoire d'Informatique de Paris-Nord, Université Paris-Nord, Villetaneuse, France (2007)
15. Elghamry, K., Al-Sabbagh, R., El-Zeiny, N.: Arabic anaphora resolution using web as corpus. In: Proceedings of the Seventh Conference on Language Engineering, Cairo, Egypt (2007)
16. Abolohom, A., Omar, N.: A hybrid approach to pronominal anaphora resolution in Arabic. J. Comput. Sci. 11(5), 764–771 (2015). https://doi.org/10.3844/jcssp.2015.764.771

17. Hammami, S., Belguith, L., Hamouda, A.B.: Arabic Anaphora Resolution: Corpora Annotation with Coreferential Links, Sfax, Tunisie (2009)
18. Ben-Othmane, C.: De la synthèse lexicographique à la détection et à la correction des graphies fautives arabes. Université de Paris XI, Orsay (1998)
19. Ben-Othman-Zribi, C., Torjmen, A., Ben-Ahmed, M.: A multi-agent system for POS-tagging vocalized Arabic texts. Int. Arab J. Inf. Technol. **4**, 322–329 (2005)
20. Nasukawa, T.: Robust method of pronoun resolution using full-text information, Kyoto, Japan (1994)
21. Ben Othmane, C., Ben Ahmed, M.: Le contexte au service des graphies fautives arabes. In: TALN 2003, Batz-sur-Mer (2003)

Arabic Pronominal Anaphora Resolution Based on New Set of Features

Souha Mezghani Hammami[✉] and Lamia Hadrich Belguith

ANLP Research Group-MIRACL Laboratory, Faculty of Economic Sciences
and Management, B.P. 1088, 3018 Sfax, Tunisia
{souha.mezghani,l.belguith}@fsegs.rnu.tn

Abstract. In this paper, we present a machine learning approach for Arabic pronominal anaphora resolution. This approach resolves anaphoric pronouns without using linguistic or domain knowledge, nor deep parsing. It relies on some features which are widely used in the literary for other languages such as English. In addition, we propose new features specific for Arabic language. We provide a practical implementation of this approach which has been evaluated on three data sets (a technical manual, newspaper articles and educational texts). The results of evaluation shows that our approach provide good performance for resolving the Arabic pronominal anaphora. The measures of F-measure are respectively 86.2% for the genre of technical manuals, 84.5% for newspaper articles and 72.1% for the literary texts.

Keywords: Anaphora resolution · Pronominal anaphora
Novel features · Arabic language · Machine learning

1 Introduction

Anaphora resolution is extremely important for many natural language applications (i.e., machine translation, automatic summarization, information retrieval, question answering, etc.), since they are conditioned by the correct identification of co-referent elements in the text [9]. The automatic anaphora resolution began since the sixties mainly with the works of [2,20] which handled the pronominal anaphora because, on one hand, it is the most spread in texts and on the other hand, because it is the anaphora type which is relatively the easiest to detect and to solve [9]. Generally, the pronominal anaphora resolution (where the anaphor is a pronoun) consists first of all in identifying the anaphoric pronouns (elimination of the non-referential pronouns) and then in establishing a list of possible candidates (where the candidate is a noun phrase) to which different information can be attributed. Finally, it is a question of identifying the best possible antecedent of the pronoun among a list of candidates. In this context, research work in automatic anaphora resolution has explored many approaches for indo-European languages and has consolidated practical systems. However, few works have been done in developing anaphoric resolution system involving

© Springer International Publishing AG, part of Springer Nature 2018
A. Gelbukh (Ed.): CICLing 2016, LNCS 9623, pp. 533–544, 2018.
https://doi.org/10.1007/978-3-319-75477-2_38

Arabic. In fact, to our knowledge, there is only three research works on Arabic anaphora resolution done by [4,10,21]. The works related to anaphora resolution in Arabic language are still at the developing stage due to the following facts: Arabic language resources are very limited, i.e. annotated corpus, morphological analyzers, Part-of-Speech (PoS) taggers, named entity (NE) taggers, parsers etc. are not readily available. In addition, Arabic language presents specific complexities which make the work on Arabic anaphora resolution more difficult than other languages such as complex morphology, complex sentence structure and omission of diacritics (short vowels).

This work propose a machine learning approach for Arabic anaphora resolution (third personal and relative pronouns) operating on real data and exploiting the minimum of linguistic knowledge. In addition, this approach handle the pronominal anaphora in varied nature of corpus and is not limited to a particular domain. We present the following main contributions in this paper:

- A proposal of novel features that are appropriate for resolving anaphoric pronouns in Arabic.
- A machine learning approach for pronominal anaphora resolution in Arabic texts that are only based on a part-of-speech tagger output.
- An annotation of an Arabic corpora with anaphoric links for training and test.
- An evaluation of the whole system.

In the following section, we present different types of pronominal anaphora that will be resolved. In Sect. 3, the state-of-the-art of anaphora resolution will be summarized. Section 4 describes the proposed machine learning approach. Our approach falls under the category of the "knowledge poor" approaches which do not rely extensively on linguistic and domain knowledge. In Sect. 5, the evaluation results are presented. These results are achieved based on three data sets: a technical manual, newspaper articles and educational texts. Finally, in Sect. 6, we present our conclusions and some future works.

2   Arabic Pronouns

Anaphora is a linguistic relation between two textual entities which is defined when a textual entity (the anaphor) refers to another entity of the text which usually occurs before (the antecedent). When the anaphor refers to an antecedent and both have the same referent in the real world, they are called coreferential. Although, coreference and anaphora are two different concepts, in reality, they most often co-occur except in some cases (e.g., verb anaphora). In this work, we have essentially focused on the pronominal anaphora which is the most frequent in texts [5,9]. Hereafter, we present different types of anaphoric pronouns that we will be resolved (personal and relative pronouns). Personal pronouns can be divided into three subclasses, namely:

- Nominative disjoint pronoun: هو /huwa/ هي /hiya/ هما /humaA/ هم /humo/ هنّ /hunã/

- Accusative disjoint pronoun: إِيَّاه /iyahu/ إِيَّاها /iyahaa/ إِيَّاهما /iyahumaa/ إِيَّاهم /iyahum/ إِيَّاهنَ /iyahonna/
- Genitive and accusative joint pronoun: ـه /hu/ ـها /haA/ ـهما /humaA/ ـهمْ /humo/ ـهنَ /hunã/

Although, the pronominal anaphora definition is the same in all natural languages, the characteristics of an anaphoric pronoun can be different from one language to an other. It is generally due to the nature of the studied language. We did a comparative study between anaphoric pronouns in Arabic and those in English. As a result, we noticed some differences between Arabic and English pronominal systems such as the morphology, grammatical cases, number, etc.

- Third personal pronouns can be disjoint (e.g., هي /hiya/ *she or it*, هو /huwa/ *he or it*) or joint pronouns (e.g., ـه /hu/ *he or it*, ـها /hA/ *she or it*). Joint pronouns should be attached to a noun (زوجته /zawojatihu/ *his wife*), a verb (تره /tarahu/ *saw him*) or a preposition (لها /lahaA/ *to her*).
- We do not have different forms for the third personal pronouns depending in the grammatical case. Indeed, the pronoun (ـه /hu/ *he, his or him*) can be used as subject pronoun (إنه /Inhu/ *he*), object pronoun (تره /tarahu/ *saw him*) or possessive pronoun (زوجته /zawojatihi/ *his wife*).
- The third personal pronouns can be used as demonstrative pronouns (e.g., هو الرئيس قادم /huwa Alr a}iysu qaAdimN/ *He, the president, is coming*).
- There is a dual form for the pronouns (i.e., هما /humaA/).
- The singular feminine pronoun (e.g., هي /hiya/ *she*) can refers to a plural non-human unit.

These specificities of Arabic pronouns prove how the Arabic language differs from other languages such as English because of its complex structure and the fact that it is a morphologically rich language. Consequently, Arabic anaphora resolution is more complex as we have to consider specificities of the Arabic language for each step of the resolution. Indeed, the identification of arabic pronouns is not always straightforward and the use of morphological analysis is very essential to solve the ambiguities. For example, in the word (وجه /wajohu/ *face*) the letter (ـه /hu/) is a part of this word while in the word (كتابه /kitaAbahu/ *his book*) it represents a pronoun. In other cases, the morphological analysis is insufficient to solve these ambiguities, so it's necessary to use a morph-syntactic disambiguation. For example, in the sentence (فهم الولد الدَرس /fahima Alwaladu Ald arosa/ *The boy understood the lesson*), the word (فهم /fahima/ *understood*) is a verb, however in the sentence (فهم يلعبون /fahumo yaloEabuwna/ *and they play*) it is an anaphoric pronoun (هم /hum/ *they*) attached to the coordination conjunction (ف /fa/ *and or then*).

3 Related Work in Arabic Anaphora Resolution

Many research have been performed in the field of anaphora and coreference resolution and especially in the field of pronominal resolution. These research works are based on two main approaches (symbolic approach and knowledge-poor approach) according to the way antecedents are computed and tracked down. The symbolic approach exploits domain and linguistic knowledge (i.e., where syntactic, semantic information, world knowledge and centering theory, were used) [3,6,13,17]. These linguistic information are difficult both to represent and to process, and require considerable human input. On the contrary, the knowledge-poor approach is based on statistical [8,14] or machine learning techniques [12,15,16,19]. While research work in automatic anaphora resolution has explored many approaches for indo-European languages and has consolidated practical systems Arabic anaphora resolution engines are demonstrated only in few works. To our knowledge, there are only three research works on Arabic anaphora resolution done by Elghamry et al. [4], Mitkov et al. [10] and Zitouni et al. [21]. We will give a brief description of these works.

The Mitkov's approach [10] is the first work which was interested to the Arabic pronominal anaphora resolution. It requires only an identification of the noun phrase candidates without using a complete syntactic analysis and it relies on a set of indicators such as definiteness, heading, collocation, referential distance, term preference, etc. The indicators assign a positive or negative score (2, 1, 0 or −1) to the antecedents and the candidate with the highest composite score is proposed as an antecedent. The approach initially developed and tested for English, but it has been subjected to some modifications to work with Arabic. The evaluation for Arabic showed a very high "critical success rate" as well. The robust approach used without any modification scored a "critical success rate" of 78.6%, whereas the improved Arabic version scored 89.3%.

Zitouni et al. [21] presented a statistical-based mention detection system. The system is statistical and built around maximum entropy, and works under the ACE 2004 framework. The coreference resolution system is similar to the one described in Luo et al. [7]. In fact, the approach is modified to accommodate the special characteristics of the Arabic language. The maximum entropy algorithm is used to compute the probability of an active mention "m" linking with an in-focus partial entity "e". The authors introduced two types of features: entity-level and mention-pair. The entity-level features capture some characteristics such as gender and number agreement, while the mention-pair features encode the lexical features. The results of the evaluation showed that the used syntactical information allowed the improvement of the coreference system performances (from 73.2 % to 74.6 %).

In order to overcome the problem of the absence of arabic NLP resources, Elghamry et al. [4] propose an anaphora resolution algorithm based on a statistical approach for Arabic unrestricted texts, which uses the minimum of linguistic information. Indeed, no syntactic information is necessary, just a space of search with words is used. The algorithm thus makes a search in a space of 20 words. It uses the word bands notion to reduce the space of search and consequently, reduce the number of candidates by dividing the words into two word bands. The band with the highest score is chosen and is divided again into two bands of less size. The band score corresponds to the sum of the conditional probability of the bi-grams; each bi-gram consists of the pronoun and its candidate. The algorithm is based on the Web as corpus to determine the frequency of bi-grams and then to measure the conditional probability of every bi-gram. The algorithm reported a precision rate of 78%, F-measured performance of 87.6%, and recall rate of 100%, all measured according to a standard set of 5000 pronouns.

4   A Machine Learning Approach to Arabic Pronoun Resolution

In this work, we adopt a corpus based and machine learning approach to perform anaphora resolution for Arabic, a resource poor language. We identify and implement several features for this task. We design and evaluate our system using three different arabic data sets. One of the main advantages of this approach is that it is not relying on domain knowledge. In addition, this approach operates without a parser, it is only based on PoS tagger output. In this section, we will detail our approach. Firstly, we explain proposed features for pronoun resolution. Then, we describe the Arabic annotated corpus with anaphoric links and finally, we present the implementation of the system.

4.1   Our Features

In a system which uses vector based machine learning techniques, we need to define a set of features to fill the feature vectors. Each anaphora resolution system uses its own set of features which must be compatible with the technique used, the studied language and the purpose of the system. Choosing the right set of features is necessary for improving the performance of the system. In addition, these features must be generic enough to be used across different domains. In our system, each pair of pronoun/candidate is described with a set of 25 features (see Table 1). They are features commonly used for anaphora resolution plus features specific to the Arabic language. We distinguish two classes of features: classical features and novel features which we describe in the following sections.

4.1.1   Classical Features

Classical features are widely used in the literature for different languages such as French and English which specify e.g. the morphological characteristics, type, grammatical case and number of repetition. For these features, some instances contain one value for the antecedent and one for the anaphor. Other instances describe the relation between antecedent and anaphor in terms of distance (in words and sentences).

- cd_POS specify the candidate category which can be a noun, a proper noun or a pronoun.
- cd_TYPE specify the candidate type which can be an adjective (e.g. الصغيرة /AlSgyrp/ *the little*), a number (e.g. الإثنان /AlIvnAn/ *two*), an adjective

Table 1. Our features

Feature	Definition	Value
Classical features		
cd_POS	Candidate category	Noun, proper noun, pronoun
cd_TYPE	Candidate type	Adjective, number, adjective number, comparison, human or latin
cd_GENDER	Candidate gender	Masculine, feminine
cd_NUMBER	Candidate number	Singular, dual, plural
cd_CASE	Candidate case	Nominative, accusative, genitive, other
cd_is_DEF	Candidate definiteness	True, false
cd_is_NP	Candidate is a noun phrase	True, false
cd_COUNT_Text	Candidate count in text	#Times Cd appears in the text
cd_COUNT_PARAG	Candidate count in paragraph	#Times Cd appears in the paragraph which contains pronoun
cd_SUBJECT	Candidate is subject	True, false
cd_in_PP	Candidate in prepositional phrase	True, false
pr_TYPE	Pronoun type	Joint, Not-joint, relative
pr_GENDER	Pronoun gender	Masculine, feminine, masculine/feminine
pr_NUMBER	Pronoun number	Singular, dual, plural
JOINT_pr_TYPE	Joint pronoun type	Verb, noun, preposition
SENT_DIST_pr_cd	Sentence distance	#Sentences between pronoun and candidate
WORD_DIST_pr_cd	Word distance	#Words between pronoun and candidate
Novel features		
PREVIOUS_pr_TYPE	Previous pronoun type	Joint, Not-joint, relative, ?
PREVIOUS_pr_GENDER	Previous pronoun gender	Masculine, feminine, masculine/feminine, ?
PREVIOUS_Pr_NUMBER	Previous pronoun number	Singular, dual, plural, ?
JOINT_PREVIOUS_pr_TYPE	Previous joint pronoun type	Verb, noun, preposition, ?
DIST_pr	Word distance	#Words between current pronoun and previous pronoun
AGGREE_NOUN_TYPE	Noun type agreement	True, false
AGGREE_VERB_PERSON	Verb person agreement	True, false
NOT_cd_SENTENCE	No candidates in sentence where occur the pronoun to be resolved	True, false

number (e.g. الأول /AlOwl/ *the first*), a comparison (e.g. الأصغر /AlOSgr/ *the smallest*), human (e.g. محمد /mHmd/ *Mohamed*) or Latin. This last type is added because our corpus of technical manual contains English words which can represent a candidate for a pronoun.

- cd_GENDER and cd_NUMBER specify the gender and the number of the candidate NP.
- cd_CASE specify the case of the candidate NP. Its possible values are nominative, accusative, genitive or other if is not identified.
- cd_is_NP it takes the value true if the candidate is composed of more than one word.
- cd_COUNT_Text and cd_COUNT_PARAG determine the times that candidate appears in the text and in the paragraph which contains pronoun. This feature include repeated noun phrases which may often be preceded by definite articles or demonstratives. Also, a sequence of noun phrases with the same head counts as lexical reiteration (e.g. البنت الجميلة /Albnt Aljmylp/ *the beautiful girl*, هذه البنت /h*h Albnt/ *this girl*).
- cd_is_DEF its possible values are true or false. We regard a noun phrase as definite if the head noun is modified by a definite article الـ /Al/ *the*, or by demonstrative pronouns (e.g., هذا /h*A/ *this*) or by possessive pronouns (e.g., كتابه /ktAbh/ *his book*) or if a noun phrase is formed by an annexation (e.g., باب البيت /bAb Albyt/ *the door of the house*).
- cd_SUBJECT it takes the value true if the candidate is the first noun phrase in a sentence.
- cd_in_PP indicates if the candidate belongs or not to a prepositional phrase (PP). It takes the value true or false. This value is determined by verifying if the candidate is preceded by a preposition.
- pr_GENDER and pr_NUMBER specify the gender and the number of the pronoun.
- pr_TYPE specify the pronoun type which can be a joint pronoun, a disjoint pronoun or a relative pronoun.
- JOINT_pr_TYPE specify the joint pronoun type which can be joint to a verb, a noun, or a preposition.
- SENT_DIST_pr_cd and WORD_DIST_pr_cd specify the number of sentences and words between a pronoun and its antecedent.

4.1.2 Novel Features

We propose novel features which are specific for Arabic language but could be used and tested as well for different domains of texts. Most of these features describe the pronoun which precedes the pronoun to be resolved in terms of gender, number, type, type of joint pronoun and distance between pronoun (pr) and the previous pronoun (previous_pr). In Arabic, two successive pronouns which have the same gender and number but not the same type refer in most of cases to the same antecedent, as shown in (1), where the pronouns (ها /hA/ *it*) and (التي /Alty/ *which*) refer to the same antecedent (اللعبة /AllEbp/ *the toy*).

(1) كسرتُ اللعبة التي اشتريتها /ksrtu AllEbp Alty A$trythA/ *I have broken the toy which you bought*

- PREVIOUS_pr_GENDER and PREVIOUS_pr_NUMBER specify the gender and the number of the previous pronoun.
- PREVIOUS_pr_TYPE specify the previous pronoun type which can be a joint pronoun, a disjoint pronoun, a relative pronoun or ? if does not exist pronoun which precedes the current pronoun.
- JOINT_PREVIOUS_pr_TYPE specify the previous joint pronoun type which can be joint to a verb, a noun, a preposition or ? if does not exist pronoun which precedes the current pronoun.
- AGGREE_NOUN_TYPE this feature is only applicable if the pronoun and the previous pronoun are joint pronouns to a noun. It takes the value true if it is the same noun type for the noun in which the pronoun is jointed to it and the one in which the previous pronoun is jointed to it.
- AGGREE_VERB_PERSON this feature is only applicable if the pronoun and the previous pronoun are joint pronouns to a verb. It takes the value true if it is the same person verb for the verb in which the pronoun is jointed to it and the one in which the previous pronoun is jointed to it.
- DIST_pr specify the number of words between a pronoun and a previous pronoun.
- NOT_cd_SENTENCE It takes the value true if there is no candidate in the sentence which contains the pronoun.

4.2   Collected and Annotated Corpora

One of the major Arabic anaphora resolution problems is the scarcity of annotated corpora with anaphoric links. For this reason, we have developed an Arabic annotated corpus with anaphoric links which can be used for the training and for the test of our pronoun resolution approach. Our corpus is composed of texts in different fields drawn from Arabic newspapers, technical manual and educational books. The corpus size is 104155 words and it is constituted of a three data sets belonging to different varieties and not having the same structure nor the same objective. Indeed, the role of newspaper articles is essentially informative, while, schoolbooks have an educational purpose. On the other hand, the technical manual proposes instructions about the use of the mobile phone.

- Newspaper articles (economy, sport, tourism, etc.) which have an informative role,
- Literary texts from different genres and by different authors which are mainly obtained from Arabic schoolbooks.
- Technical manual for a mobile phone which presents instructions to the user.

By referring to the results in Table 2, we notice that the density of pronouns in the literary texts is widely higher than of the technical manual or the newspaper articles. On the other hand, we remark that pronouns are more frequent in

Arabic texts than in texts for other language such as French. In fact, Tutin [18] indicates that, in French language, the most higher pronoun density is attenuated in literary texts and it is just equal (4.32%).

Table 2. Pronoun density in our corpus

Corpus	Words	Pronouns	Density (in%)
Newspaper articles	29185	2016	6.91%
Literary texts	33595	3355	9.99%
Technical manual	41375	1119	2.7%

Table 3. Distribution of third personal and relative pronouns reported in our corpus

Corpus	Pronouns	Disjoint pr	(in%)	Joint pr	(in%)	Relative pr	(in%)
Newspaper articles	2016	215	10.67%	1411	70%	390	19.34%
Literary texts	3355	197	5.87%	2921	87%	237	7.1%
Technical manual	1119	46	4.11%	767	68.54%	305	27.25%

Table 3 presents the relative distribution of pronouns by type, observed in our corpus. The distribution of different types of pronouns depends on the corpus (and on the domain of the texts). Indeed, relative pronouns are more frequent in the technical manual (27.25%) while the using rate of these pronouns does not exceed the value (7.1%) in literary texts. But, we noticed that the relative importance of joint personal pronouns is very high for most kinds of texts, because they constitute the largest share of pronouns in texts.

4.3 Generation of Feature Vectors

In order to apply our approach to unrestricted arabic texts, it has been necessary to use some language processing (NLP) tools. The goal of these NLP tools is to provide the necessary information about each pronoun and candidate for subsequent generation of features in the training examples. They consist of a tokenizer, a part-of-speech tagger and a noun phrase identification tool.

In this work, we used the STAr tokenizer for Arabic texts [1] and the MADA part-of-speech Tagger [11] to compute the attribute values for a given pronoun occurrence. STAr segments the input text into paragraphs and sentences based on a contextual analysis of the punctuation marks and some particles, such as the coordination conjunctions. MADA (Morphological Analysis and Disambiguation for Arabic) is an utility that, given raw Arabic text, adds as much lexical and morphological information as possible by disambiguating in one operation part-of-speech tags, lexemes, diacritizations and full morphological analyses [11]. We also implemented a module for noun phrase identification based on some rules which are manually extracted.

After the processing modules, instances of training corpus are created as follows: we generate a positive training instance from each pair Pronoun/Antecedent from the annotated text. Negative instances are created by pairing the pronoun with any candidate occurring in the same paragraph which contains this pronoun. These instances are created as feature vectors and are used to train a classifier. Finally, the classifier has to decide, in terms of given features, whether the candidate is antecedent for the pronoun or not.

5 Experimental Evaluation

In order to learn anaphora resolution decisions, we experimented with WEKA tool[1] four learning algorithm (J48, JRip, NB and SMO)[2]. Then, we introduce three experiments with different set of features to illustrate how the set of features affect the performance of our system. On one hand, we perform experiments with only the above mentioned classical feature combinations. On the other hand, we have added the novel features which are specific for Arabic language (presented in Sect. 4.1). Finally, we use only the 8 novel features. The results were tested using 10-fold cross-validation on our three data sets (technical manual, newspaper articles and literary texts).

Table 4. Classification results

	Classifier	Technical manual			Newspaper articles			Literary texts		
		Pr	R	F	Pr	R	F	Pr	R	F
with 17 features	J48	84.2	84.4	84.3	83.2	82	82.6	75	69.5	72.1
	JRip	82.1	85.1	83.5	80.4	83.4	81.8	70	55.8	62.1
	Naive Bayes	75.3	88.5	81.4	66.5	93.3	77.7	40.7	89.2	55.9
	SMO	76.4	88.9	82.2	77.7	83.3	80.4	69.2	41.7	52
with 25 features	J48	85.4	87	86.2	85.8	83.1	84.5	74.3	67.7	70.9
	JRip	85.8	85.4	85.6	81.9	84.3	83.1	70.9	58.6	64.1
	Naive Bayes	78.3	87.8	82.8	69.7	90.6	78.8	42.5	86.2	57
	SMO	80.9	89.4	84.9	81.7	79	80.3	70.3	44.9	54.8
with 8 features	J48	85.6	79.5	82.4	97.7	72.7	83.3	84.9	43.1	57.2
	JRip	82.5	72.7	77.3	99.3	50.9	67.3	71.5	23.1	34.9
	Naive Bayes	66.2	66.2	66.2	55.4	20.9	30.3	37.9	20.9	27
	SMO	71.5	69.5	70.5	67.8	13.3	22.3	63.5	4.6	8.5

The obtained results are summarized in Table 4 where the performance is reported in terms of precision, recall and F-measure. These results show, on one

---

[1] Weka: http://weka.wikispaces.com/.

[2] The parameters for these classifiers are those of Weka's default settings.

hand, that the complexity of the anaphora resolution varies according to the type of the handled corpus. Indeed, the value F-measure is lower in the literary texts (72.1%) compared with those obtained for the corpora technical manual (86.2%) and newspaper articles (84.5%). It is not difficult to understand that this complexity is due to (i) the high density of the pronouns in the literary texts and to (ii) the unimportant number of the relative pronouns which are more easier to solve than the third personal pronoun. In addition, these results show that our system has achieved noticeable improvement when we add novel features for technical manual and newspaper articles but is not often the case of literary texts. We can explain the not relevance of the novel features in the literary texts by the fact that these features are more effective if the pronoun and the one who precedes it are not of the same type. This counterpart, the literary texts does not present a big variation in the use of the pronouns. Indeed, the use of the joint pronouns is dominant compared with other types. Besides, we notice that the use of only 8 features provides a good precision value compared with other sets of features but it obtains a low recall. This involves that using novel features alone is insufficient but they have a beneficial effect on performance when adding them to other features.

6   Conclusion

In this paper, we have proposed a machine learning-based method for Arabic anaphora resolution especially when the anaphor is a pronoun and the antecedent is a noun phrase. In order to evaluate our anaphora resolution system in unrestricted texts, only lexical and morphological information were used. Basing on these information, we proposed 8 new specific features for Arabic language in addition to the 17 classical features used in other languages. Then, we evaluated the contribution of novel features on three data sets (a technical manual, newspaper articles and educational texts). Therefore, the experimental results show that identifying features for a pronoun which occur just before the pronoun to be solved is effective in both technical manual and newspaper articles. Indeed, The value of F-measure obtained by, for example, JRip algorithm in a technical manual is improved by almost 2% (from 83.5% to 85.6%).

In future, we aim to propose an hybrid approach for anaphoric pronoun resolution to improve the performance of our system.

References

1. Belguith, H., Baccour, L., Mourad, G.: Segmentation de textes arabes base sur l'analyse contextuelle des signes de ponctuations et de certaines particules. In: 12me confrence sur le Traitement Automatique des Langues Naturelles, Dourdan France, pp. 451–456 (2005)
2. Bobrow, D.: A question-answering system for high school algebra word problems. In: Proceedings of AFIPS Conference (1964)
3. Carbonell, J., Brown, D.: Anaphora resolution: a multi-strategy approach. In: Proceedings of the 12th International Conference (1988)

4. Elghamry, K., Al-Sabbagh, R., El-Zeiny, N.: Arabic anaphora resolution using web as corpus. In: Proceedings of the Seventh Conference on Language Engineering. Cairo, Egypt (2007)

5. Hammami, M., Belguith, H., Ben Hamadou, A.: Anaphora in Arabic language: developing a corpora annotating tool for anaphoric links. In: 9th International Arab Conference on Information Technology ACIT2008, Hammamet, Tunisia (2008)

6. Hobbs, J.: Resolving pronoun references. Lingua **44**, 339–352 (1978)

7. Luo, X., Zitouni, I.: Multi-lingual coreference resolution with syntactic features. In: Proceedings of the seventh conference on Language Engineering, pp. 660–667, Cairo, Egypt (2005)

8. Mitkov, R.: Robust pronoun resolution with limited knowledge. In: Proceedings of the 18th International Conference on Computational Linguistics (COLING98)/ACL98 Conference, Montreal, Canada, pp. 869–875 (1998)

9. Mitkov, R.: Anaphora Resolution. Longman, New York (2002)

10. Mitkov, R., Belguith, H., Malgorzata, S.: Multilingual robust anaphora resolution. In: Proceedings of the 3rd Conference on Empirical Methods in Natural Language Processing, Granada, Spain, pp. 7–16 (1998)

11. Rambow, O., Habash, N., Roth, R.: Mada+tokan: a toolkit for Arabic tokenization, diacritization, morphological disambiguation, pos tagging, stemming and lemmatization. In: Proceedings of the 2nd International Conference on Arabic Language Resources and Tools. Cairo, Egypt, April 2009

12. Recasens, M., Hovy, E.: A deeper look into features for coreference resolution. In: Lalitha Devi, S., Branco, A., Mitkov, R. (eds.) DAARC 2009. LNCS (LNAI), vol. 5847, pp. 29–42. Springer, Heidelberg (2009). https://doi.org/10.1007/978-3-642-04975-0_3

13. Rich, E., LuperFoy, S.: An architecture for anaphora resolution. In: Proceedings of the Second Conference on Applied Natural Language Processing, Austin, Texas (1988)

14. Sobha, L., Patnaik, B.: Vasisth: an anaphora resolution system for Indian languages. In: Proceedings of International Conference on Artificial and Computational Intelligence for Decision, Control and Automation in Engineering and Industrial Applications, Monastir, Tunisia (2000)

15. Sobha, L.D., Vijay, S.R., Pattabhi, R.R.: Resolution of pronominal anaphors using linear and tree CRFs. In: 8th DAARC, Faro, Portugal (2011)

16. Soon, W., Ng, H., Lim, D.: A machine learning approach to coreference resolution of noun phrases. Comput. Ling. **27**(4), 521–544 (2001)

17. Strube, M., Hahn, U.: Functional centering: grounding referential coherence in information structure. Comput. Ling. **27**(4), 309–344 (1999)

18. Tutin, A.: A corpus-based study of pronominal anaphoric expressions in French. In: Proceedings of DAARC 2002 (2002)

19. Wick, M., Singh, S., McCallum, A.: A discriminative hierarchical model for fast coreference at large scale. In: Proceedings of ACL 2012 (2012)

20. Winograd, T.: Understanding Natural Language. Academic Press, New York (1972)

21. Zitouni, I., Luo, X., Florian, R.: A statistical model for Arabic mention detection and chaining, pp. 199–236. CSLI Publications, Center for the Study of Language and Information, Stanford (2010)

Semantics, Discourse, and Dialog

GpSense: A GPU-Friendly Method for Commonsense Subgraph Matching in Massively Parallel Architectures

Ha-Nguyen Tran and Erik Cambria[⊠] [iD]

School of Computer Science and Engineering,
Nanyang Technological University, Singapore, Singapore
{hntran,cambria}@ntu.edu.sg

Abstract. In the context of commonsense reasoning, spreading activation is used to select relevant concepts in a graph of commonsense knowledge. When such a graph starts growing, however, the number of relevant concepts selected during spreading activation tends to diminish. In the literature, such an issue has been addressed in different ways but two other important issues have been rather under-researched, namely: performance and scalability. Both issues are caused by the fact that many new nodes, i.e., natural language concepts, are continuously integrated into the graph. Both issues can be solved by means of GPU accelerated computing, which offers unprecedented performance by offloading compute-intensive portions of the application to the GPU, while the remainder of the code still runs on the CPU. To this end, we propose a GPU-friendly method, termed *GpSense*, which is designed for massively parallel architectures to accelerate the tasks of commonsense querying and reasoning via subgraph matching. We show that GpSense outperforms the state-of-the-art algorithms and efficiently answers subgraph queries on a large commonsense graph.

1 Introduction

When communicating with each other, people provide just the useful information and take the rest for granted. This 'taken for granted' information is what is termed 'commonsense' – obvious things people normally know and usually leave unstated. Commonsense is not the kind of knowledge we can find in Wikipedia, but it consists in all the basic relationships among words, concepts, phrases, and thoughts that allow people to communicate with each other and face everyday life problems. Commonsense is an immense society of hard-earned practical ideas, of multitudes of life-learned rules and exceptions, dispositions and tendencies, balances and checks.

Commonsense computing [3] has been applied in many different branches of artificial intelligence, e.g., personality detection [19], handwritten text recognition [26], multimodality [18], and social data analysis [6]. In the context of sentic computing [2], in particular, commonsense is represented as a semantic network of natural language concepts interconnected by semantic relations.

© Springer International Publishing AG, part of Springer Nature 2018
A. Gelbukh (Ed.): CICLing 2016, LNCS 9623, pp. 547–559, 2018.
https://doi.org/10.1007/978-3-319-75477-2_39

This kind of representation presents two major implementation issues: performance and scalability, both due to the fact that many new concepts learnt through crowd-sourcing [5] are continuously integrated into the graph. These issues are also the crucial problems of querying and reasoning over large-scale commonsense knowledge bases (KBs). The core function of commonsense querying and reasoning is subgraph matching which is defined as finding all matches of a query graph in a database graph. Subgraph matching is usually a bottleneck for the overall performance because it involves subgraph isomorphism which is known as an NP-complete problem [7].

Many proposed methods for subgraph matching problem are backtracking algorithms [8,9,12,24], with novel techniques for filtering candidates sets and re-arranging visit order. However, all of them use only small database graphs, and thus, there still remains the question of scalability on large graphs. Some recent methods utilize indexing techniques to deal with large graphs, through building the index may take long time and large memory space [22,27,28]. Another approach is based on distributed computing [1,23]. By finding results on many machines simultaneously, those algorithms are able to deal with large-scale graphs. However, an issue with these methods is that the communication between a large number of machines is costly.

Graphics Processing Units (GPUs) have recently become popular computing devices because of their massive parallel. Such basic graph operations as breadth-first search [10,13,16], shortest path [10,15], and minimum spanning tree [25] on large graphs can be implemented on GPUs efficiently. The previous backtracking methods for subgraph matching, however, cannot be straightforwardly applied to GPUs due to their inefficient uses of GPU memories and SIMD-optimized GPU multi-processors [14].

In this paper, we propose *GpSense*, an efficient and scalable method for solving the subgraph matching problem on large commonsense KBs. GpSense is based on a *filtering-and-joining* strategy which is designed for massively parallel architecture of GPUs. In order to optimize the performance, we utilize a series of optimization techniques which contribute to increase the GPU occupancy, reduce workload imbalance and especially enhance commonsense reasoning tasks. The rest of the paper is structured as follows: Sect. 2 introduces the background of the subgraph matching problem and filtering-and-joining approach; Sect. 3 discusses how to transform commonsense KBs to directed graphs; Sect. 4 describes the GPU implementation in details; experiment results are shown in Sect. 5; finally, Sect. 6 concludes the paper.

2   Subgraph Matching Problem

2.1   Problem Definition

A graph G is defined as a 4-tuple (V, E, L, l), where V is the set of nodes, E is the set of edges, L is the set of labels and l is a labeling function that maps each node or edge to a label in L. We define the size of a graph G is the number of edges, $size(G) = |E|$.

Definition 1 (Subgraph Isomorphism). *A graph* $G = (V, E, L, l)$ *is subgraph isomorphic to another graph* $G' = (V', E', L', l')$, *denoted as* $G \subseteq G'$, *if there is an injective function (or a match)* $f: V \to V'$, *such that* $\forall \ (u, v) \in E$, $(f(u), f(v)) \in E'$, $l(u) = l'(f(u)), l(v) = l'(f(v))$, *and* $l(u, v) = l'(f(u), f(v))$.

A graph G is called a subgraph of another graph G (or G is a supergraph of G), denoted as $G \subseteq G'$ (or $G' \supseteq G$), if there exists a subgraph isomorphism from G to $G'a$.

Definition 2 (Subgraph Matching). *Given a small query graph* Q *and a large data graph* G, *subgraph matching problem is to find all subgraph isomorphisms of* Q *in* G.

2.2 GPU Approach for Subgraph Matching

In this subsection, we introduce an approach to solve the subgraph matching problem on General-Purpose Graphics Processing Units (GPGPUs). The approach is based on a *filtering-and-joining* strategy which is specially designed for massively parallel computing architecture of modern GPUs. The main routine of the GPU-based method is depicted in Algorithm 1.

Algorithm 1: GPUSubgraphMatching (q(V, E, L), g(V', E', L'))

Input: query graph q, data graph g
Output: all matches of q in g

1 P := generate_query_plan(*q, g*);
2 **forall the** *node* $u \in P$ **do**
3   **if** *u is not filtered* **then**
4   c_set(u) := identify_node_candidates(*u, g*);
5   c_array(u) := collect_edge_candidates(*c_set(u)*);
6   c_set := filter_neighbor_candidates(*c_array(u), q, g*);
7 refine_node_candidates(*c_set, q, g*);
8 **forall the** *edge e (u,v)* $\in E$ **do**
9   EC(e) := collect_edge_candidates(*e, c_set, q, g*);
10 M := combine_edge_candidates(*EC, q, g*);
11 **return** M

The inputs of the algorithm are a query graph q and a data graph g. The output is a set of subgraph isomorphisms (or matches) of q in g. In the method, we present a match as a list of pairs of a query node and its mapped data node. Our solution is the collection M of such lists. Based on the input graphs, we first generate a query plan for the subgraph matching task (Line 1). The query plan contains the order of query nodes which will be processed in the next steps. The query plan generation is the only step that runs on the CPU. After that, the main procedure will be executed in two phases: filtering phase (Line 2–7) and joining phase (Line 8–10). In the filtering phase, we filter out node candidates which cannot be matched to any query nodes (Line 2–6).

After this task there still exists a large set of irrelevant node candidates which cannot contribute to subgraph matching solutions. The second task continues pruning this collection by calling the refining function *refine_node_candidates*. In such a function, candidate sets of query nodes are recursively refined until no more candidates can be pruned. The joining phase then finds the candidates of all data edges (Line 8–9) and merges them to produce the final subgraph matching results (Line 10).

Query Plan Generation: *generate_query_plan* procedure is to create a good node order for the main searching task. It first picks a query node which potentially contributes to minimize the sizes of candidate sets of query nodes and edges. Since we do not know the number of candidates in the beginning, we estimate it by using a node ranking function $f(u) = \frac{deg(u)}{freq(u.label)}$ [9,23], where $deg(u)$ is the degree of a query node u and $freq(u.label)$ is the number of data nodes having the same label as u. The score function prefers lower frequencies and higher degrees. After choosing the first node, *generate_query_plan* follows its neighborhood to find the next nodes which has not been selected and is connected to at least one node in the node order. The process terminates when all query nodes are chosen.

The Filtering Phase: The purpose of this phase is to reduce the number of node candidates and thus decrease the amount of edge candidates as well as the running time of the joining phase. The filtering phase consists of two tasks: initializing node candidates and refining node candidates. In order to maintain the candidate sets of query nodes, for each query node u we use a *boolean* array, $c\_set[u]$, which has the length of $|V'|$. If $v \in V'$ is a candidate of u, *identify_node_candidates* sets the value of $c\_set[u][v]$ to *true*. The *filter_neighbor_candidates* function, however, will suffers the low occupancy problem since only threads associated with *true* elements of $c\_set[u]$ works while the other threads are idle. To deal with the problem, *collect_node_candidates* collects *true* elements of $c\_set[u]$ into an array $c\_array[u]$. As a result, each running thread can easily be mapped to a candidate of u. After that the *filter_neighbor_candidates* function will filter the candidates of nodes adjacent to u based on $c\_array[u]$. This device function follows a warp-based execution approach. The details of these parallel functions will be discussed in Sect. 4.

The Joining Phase: Based on the candidate sets of query nodes, *collect_edge_candidates* function collects the edge candidates individually. The routine of the function is similar to *filter_neighbor_candidates*, but it inserts an additional part of writing obtained edge candidates to candidate edge arrays. In order to output the candidates to an array, we employ the *two-step output scheme* [12] to find the offsets of the outputs in the array and then write them to the corresponding positions. *combine_edge_candidates* merges candidate edges using a warp-centric fashion to produce the final subgraph matching solutions.

3   Commonsense Knowledge as a Graph

In this section, we discuss how a commonsense KB can be naturally represented as a graph and how such a KB can be directly transformed to a graph representation.

3.1   Commonsense Knowledge Graph

Instead of formalizing commonsense reasoning using mathematical logic [17], some recent commonsense KBs, e.g., SenticNet [4], represent data in the form of a semantic network and make it available to be used in natural language processing. In particular, the collected pieces of knowledge are integrated in the semantic network as triples of the format $< concept\text{-}relation\text{-}concept >$. By considering triples as directed labeled edges, the KB naturally becomes a directed graph. Figure 1 shows a semantic graph representation of a part of commonsense knowledge graph for the concept *cake*.

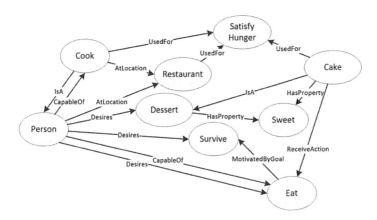

Fig. 1. Commonsense knowledge graph

3.2   Commonsense Graph Transformation

This subsection describes how to directly transform a commonsense KB to a directed graph. The simplest way for transformation is to convert the KB to a flat graph using direct transformation. This method maps concepts to node IDs and maps relations to labels of edges. Note that the obtained graph contains no node labels because each node is mapped to a unique ID. Tables 1 and 2 show the mapping from concepts and relations of the commonsense KB in Fig. 1 to node IDs and edge labels. The transformed graph from the KD is depicted in Fig. 2.

In the general subgraph matching problem, all nodes of a query graph q are variables. In order to produce the subgraph isomorphisms of q in a large

Table 1. Node mapping table

Concept	Node ID
Person	v_0
Cook	v_1
Restaurant	v_2
Dessert	v_3
Survive	v_4
Eat	v_5
Satisfy Hunger	v_6
Sweet	v_7
Cake	v_8

Table 2. Edge label mapping table

Relation	Edge label
IsA	r_0
CapableOf	r_1
AtLocation	r_2
Desires	r_3
UsedFor	r_4
HasProperty	r_5
MotivatedBy	r_6
ReceiveAction	r_7

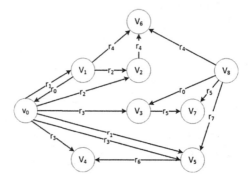

Fig. 2. Direct transform of Commonsense KB

data graph g, we must find the matches of all query nodes. Unlike the general problem, the query graphs in commonsense querying and reasoning tasks contain two types of nodes: concept nodes and variable nodes.

A concept node can only be mapped to one node ID in the data graphs while a variable node may have many node candidates in the data graph. Similarly, query edges are also categorized into variable edges and labeled edges. Figure 3 illustrates the conversion of a commonsense query to a directed query graph.

In the sample query transformation, the query concepts *Person* and *Satisfy Hunger* correspond to two data nodes with IDs of v_0 and v_6. Two query relations *IsA* and *UsedFor* are mapped to edge labels r_0 and r_4. The query graph also contains 2 variable edges ?x, ?y and 3 variable nodes ?a, ?b, ?c. The direct transformation is a common and simple approach to naturally convert a semantic network to a directed graph.

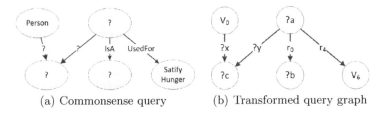

(a) Commonsense query   (b) Transformed query graph

Fig. 3. Direct transformation of Commonsense query

4   Commonsense Subgraph Matching

In this section, we introduce the complete implementation of our commonsense subgraph matching method, namely *GpSense*, on large-scale commonsense KBs using GPUs. In order to support commonsense querying and reasoning, optimization techniques are also applied to GpSense.

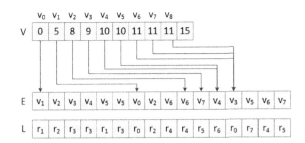

Fig. 4. Graph representation of the data graph in Fig. 2

4.1   Graph Representation

In order to maintain and efficiently process the data graph $G(V, E)$ on GPUs, we use three array structures: an array whose size is identical to the size of V plus one, termed *nodes array*, and another array consisting of adjacency lists of all nodes in V, termed *edges array*. The nodes array has pointers to the adjacency lists of the nodes in the edges array. The additional, the last element of the nodes array indicates the length of the edges array. The last array with the size of $|E|$ stores the labels of all edges in the data graph. Figure 4 shows the representation of the graph illustrated in Fig. 2 in the GPU memory.

The advantage of the data structure is that nodes in the adjacency list of a node are stored next to each other in the GPU memory. During GPU execution, consecutive threads can access consecutive elements in the memory. Therefore, we can avoid the random access problem and decrease the accessing time for GPU-based methods consequently.

4.2   GPU Implementation

In this subsection, we describe the implementation of parallel functions such as *collect_node_candidates*, and *filter_neighbor_candidates* in detail. These functions are based on two optimization techniques: occupancy maximization to hide memory access latency and warp-based execution to take advantage of the coalesced access and to deal with workload imbalance between threads within a warp.

collect_node_candidates: The purpose of this device function is to collect the candidates of a query node u in the *boolean* array $c\_set[u]$ to a candidate array $c\_array[u]$. The output of this function will maximize the GPU occupancy (i.e., maximize the number of running threads) for the next procedures. GpSense executes the task by adopting a stream compaction algorithm [11] to gather elements with the *true* values in $c\_set[u]$ to the output array $c\_array[u]$. The algorithm employs prefix scan to calculate the output addresses and to support writing the results in parallel. The example of collecting candidate nodes of $?c$ is depicted in Fig. 5. By taking advantage of c_array, candidate nodes v_1, v_2, v_3, v_4, v_5 can easily be mapped to consecutive active threads. As a result, our method achieves a high occupancy.

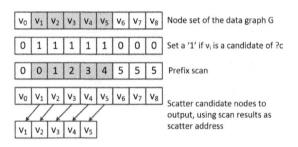

Fig. 5. Collect candidate nodes of $?c$

filter_neighbor_candidates: GpSense follows the adjacent edges of u to filter the candidates of query nodes connected to u. The step might suffer from warp divergence because of the diverse sizes of adjacency lists of $u's$ candidates. To overcome the problem, we employ a *coarse-grained* and *warp-based* method inspired by Hong et al. [13]. In this approach, an entire warp is responsible for the adjacency list of a candidate. Figure 6 shows an example of filtering candidate nodes of $?a$ based on the candidate set of $?c$, $C(?c) = \{v_1, v_2, v_3, v_3, v_5\}$. Each candidate of $?c$ is mapped to a warp to filter the candidates of its adjacency node $?a$.

4.3   Optimization Techniques

In this subsection, we introduce optimizations that we apply to enhance the efficiency of the subgraph matching problem in commonsense querying and reasoning.

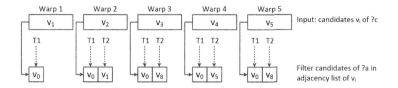

Fig. 6. Filter candidates of ?a based on candidate set of ?c

Modify the query plan based on the properties of commonsense queries. First, unlike query graphs in general subgraph matching problems, commonsense query graphs contain concept nodes and variable nodes. We only need to find the matches of nodes in a subset of variable nodes, termed *projection*. Second, nodes of a commonsense knowledge graph are not labeled. They are mapped to node IDs. Therefore, the frequency of a concept node in a query is 1 and that of a variable node is equal to the number of data nodes. As a result, the ranking function used for choosing the node visiting order cannot work for commonsense subgraph matching.

Using the above observations, we make a modification for generating the node order as follows: we prefer picking a concept node u with the maximum degrees as the first node in the order. By choosing u, we can minimize the candidates of variable nodes connected to u. The next query node v will be selected if v is connected to u and the adjacency list of v consists of maximum number of nodes which is not in the order among the remain nodes. We continue the process until edges connected to nodes in the node order can cover the query graph.

Use both incoming and outgoing graph representations: An incoming graph is built based on the incoming edges to the nodes while an outgoing graph is based on the outgoing edges from the nodes. The representation of Commonsense graph in Fig. 4 is an example of outgoing graph representation. Given a query graph in Fig. 3, assume that we only use an outgoing graph as the data graph. Based on the above query plan generator, node v_0 is the first node in the order. After that we filter the candidates of ?c based on v_0. Since ?c does not have any outgoing edges, we have to pick ?a as the next node and find its candidates by scanning all the data graphs. There are some issues in this approach: (1) We need to spend time to scan all the data graph nodes. (2) The number of candidates can be very large because the filtering condition is weak. To overcome the problem, we use an incoming graph along with the given outgoing graph. By using the additional graph, candidates of ?a can be easily filtered based on the candidate set of ?c. The number of candidates of ?a, therefore, is much smaller than that in the previous approach. Consequently, GpSense can reduce a large amount of intermediate results during execution which is one of the most crucial issues for GPU applications.

5   Experiment Results

We evaluate the performance of GpSense in comparison with state-of-the-art subgraph matching algorithms, including VF2 [8], QuickSI (QSI) [22], GraphQL (GQL) [12] and TurboISO [9]. The experiments are conducted on SenticNet and its extensions [20,21]. The query graphs are extracted from the data graph by picking a node in SenticNet and following BFS fashion to achieve other nodes. We choose nodes in the dense area of SenticNet to ensure that the obtained queries are not just trees. The runtime of the CPU-based algorithms is measured using an Intel Core i7-870 2.93 GHz CPU with 8 GB of memory. Our GPU algorithms are tested using CUDA Toolkit 6.0 running on the NVIDIA Tesla C2050 GPU with 3 GB global memory and 48 KB shared memory per Stream Multiprocessor. For each of those tests, we execute 100 different queries and record the average elapsed time. In all experiments, algorithms terminate only when all subgraph matching solutions are found.

The first set of experiments is to evaluate the performance of GpSense on SenticNet and compare it with state-of-the-art algorithms. SenticNet is a commonsense knowledge graph of about 100,000 vertices which is primarily used for sentiment analysis. In this experiment, we extract subsets of SenticNet with the size varying from 10,000 nodes to 100,000 nodes. All the data graphs can fit into GPU memory. The query graphs contain 6 nodes.

Figure 7 shows that GpSense clearly outperforms VF2, QuickSI and GraphQL. Compared to TurboISO, our GPU-based algorithm obtains the similar performance when the size of the data graphs is relatively small (i.e., 10,000 nodes). However, when the size of data graphs increases, GpSense is more efficient than TurboISO.

Figure 8a shows the performance results of GpSense and TurboISO on the query graphs whose numbers of nodes vary from 6 to 14. Figure 8b shows their performance results when the node degree increases from 8 to 24, where the num-

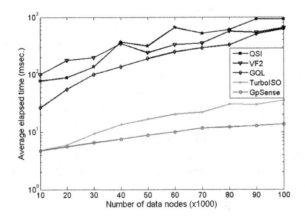

Fig. 7. Comparison with state-of-the-art methods

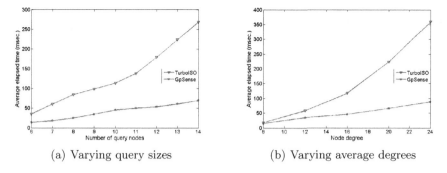

(a) Varying query sizes   (b) Varying average degrees

Fig. 8. Comparison with TurboISO

ber of query nodes is fixed to 10. As shown in the two figures, the performance of TurboISO drops significantly while that of GpSense does not.

This may be due to the fact that the number of recursive calls of TurboISO grows exponentially with respect to the size of query graphs and the degree of the data graph. In contrast, GpSense with the large number of parallel threads can handle multiple candidate nodes and edges at the same time, thus the performance of GpSense remains stable.

6 Conclusion

In this paper, we introduced an efficient GPU-friendly method for answering subgraph matching queries over large-scale commonsense KBs. Our method, GpSense, is based on a *filtering-and-joining* approach which is suitable to be executed on massively parallel architecture of GPUs. Along with efficient GPU techniques of coalescence, warp-based and shared memory utilization, GpSense provides a series of optimization techniques which contribute to enhance the performance of subgraph matching-based commonsense reasoning tasks. Experiment results show that our method outperforms previous backtracking-based algorithms on CPUs and can efficiently answer subgraph matching queries on large-scale commonsense KBs.

References

1. Brocheler, M., Pugliese, A., Subrahmanian, V.S.: COSI: cloud oriented subgraph identification in massive social networks. In: International Conference on Advances in Social Networks Analysis and Mining (ASONAM), pp. 248–255. IEEE (2010)
2. Cambria, E., Hussain, A.: Sentic Computing: A Common-Sense-Based Framework for Concept-Level Sentiment Analysis. Socio-Affective Computing, vol. 1. Springer, Switzerland (2015)

3. Cambria, E., Hussain, A., Havasi, C., Eckl, C.: Common sense computing: from the society of mind to digital intuition and beyond. In: Fierrez, J., Ortega-Garcia, J., Esposito, A., Drygajlo, A., Faundez-Zanuy, M. (eds.) BioID_MultiComm2009. LNCS, vol. 5707, pp. 252–259. Springer, Heidelberg (2009). https://doi.org/10.1007/978-3-642-04391-8_33

4. Cambria, E., Poria, S., Bajpai, R., Schuller, B.: SenticNet 4: A semantic resource for sentiment analysis based on conceptual primitives. In: COLING, pp. 2666–2677 (2016)

5. Cambria, E., Rajagopal, D., Kwok, K., Sepulveda, J.: GECKA: game engine for commonsense knowledge acquisition. In: The Twenty-Eighth International Flairs Conference, pp. 282–287 (2015)

6. Cambria, E., Mazzocco, T., Hussain, A., Eckl, C.: Sentic medoids: Organizing affective common sense knowledge in a multi-dimensional vector space. In: Liu, D., Zhang, H., Polycarpou, M., Alippi, C., He, H. (eds.) ISNN 2011. LNCS, vol. 6677, pp. 601–610. Springer, Heidelberg (2011). https://doi.org/10.1007/978-3-642-21111-9_68

7. Cook, S.A.: The complexity of theorem-proving procedures. In: Proceedings of the Third Annual ACM Symposium on Theory of Computing, pp. 151–158. ACM (1971)

8. Cordella, L.P., Foggia, P., Sansone, C., Vento, M.: A (sub) graph isomorphism algorithm for matching large graphs. IEEE Trans. Pattern Anal. Mach. Intell. **26**(10), 1367–1372 (2004)

9. Han, W.-S., Lee, J., Lee, J.-H.: Turbo ISO: towards ultrafast and robust subgraph isomorphism search in large graph databases. In: Proceedings of the 2013 ACM SIGMOD International Conference on Management of Data, pp. 337–348. ACM (2013)

10. Harish, P., Narayanan, P.J.: Accelerating large graph algorithms on the GPU using CUDA. In: Aluru, S., Parashar, M., Badrinath, R., Prasanna, V.K. (eds.) HiPC 2007. LNCS, vol. 4873, pp. 197–208. Springer, Heidelberg (2007). https://doi.org/10.1007/978-3-540-77220-0_21

11. Harris, M., Sengupta, S., Owens, J.D.: GPU Gems 3 - Parallel prefix sum (scan) with CUDA, Chap. 39. NVIDIA Corporation (2007)

12. He, H., Singh, A.K.: Graphs-at-a-time: query language and access methods for graph databases. In: Proceedings of the 2008 ACM SIGMOD International Conference on Management of Data, pp. 405–418. ACM (2008)

13. Hong, S., Kim, S.K., Oguntebi, T., Olukotun, K.: Accelerating CUDA graph algorithms at maximum warp. ACM SIGPLAN Not. **46**, 267–276 (2011)

14. Jenkins, J., Arkatkar, I., Owens, J.D., Choudhary, A., Samatova, N.F.: Lessons learned from exploring the backtracking paradigm on the GPU. In: Jeannot, E., Namyst, R., Roman, J. (eds.) Euro-Par 2011, Part II. LNCS, vol. 6853, pp. 425–437. Springer, Heidelberg (2011). https://doi.org/10.1007/978-3-642-23397-5_42

15. Katz, G.J., Kider Jr., J.T.: All-pairs shortest-paths for large graphs on the GPU. In: Proceedings of the 23rd ACM SIGGRAPH/EUROGRAPHICS Symposium on Graphics Hardware, pp. 47–55. Eurographics Association (2008)

16. Merrill, D., Garland, M., Grimshaw, A.: Scalable GPU graph traversal. ACM SIGPLAN Not. **47**, 117–128 (2012)

17. Mueller, E.T.: Commonsense Reasoning: An Event Calculus Based Approach. Morgan Kaufmann, San Francisco (2014)

18. Poria, S., Cambria, E., Howard, N., Huang, G.-B., Hussain, A.: Fusing audio, visual and textual clues for sentiment analysis from multimodal content. Neurocomputing **174**, 50–59 (2016)

19. Poria, S., Gelbukh, A., Agarwal, B., Cambria, E., Howard, N.: Common sense knowledge based personality recognition from text. In: Castro, F., Gelbukh, A., González, M. (eds.) MICAI 2013, Part II. LNCS (LNAI), vol. 8266, pp. 484–496. Springer, Heidelberg (2013). https://doi.org/10.1007/978-3-642-45111-9_42
20. Poria, S., Gelbukh, A., Cambria, E., Das, D., Bandyopadhyay, S.: Enriching SenticNet polarity scores through semi-supervised fuzzy clustering. In: IEEE International Conference on Data Mining Workshops (ICDMW), pp. 709–716 (2012)
21. Poria, S., Gelbukh, A., Cambria, E., Yang, P., Hussain, A., Durrani, T.S.: Merging SenticNet and WordNet-affect emotion lists for sentiment analysis. In: IEEE International Conference on Signal Processing (ICSP), vol. 2, pp. 1251–1255. IEEE (2012)
22. Shang, H., Zhang, Y., Lin, X., Yu, J.X.: Taming verification hardness: an efficient algorithm for testing subgraph isomorphism. Proc. VLDB Endow. $1(1)$, 364–375 (2008)
23. Sun, Z., Wang, H., Wang, H., Shao, B., Li, J.: Efficient subgraph matching on billion node graphs. Proc. VLDB Endow. $5(9)$, 788–799 (2012)
24. Ullmann, J.R.: An algorithm for subgraph isomorphism. J. ACM (JACM) $23(1)$, 31–42 (1976)
25. Vineet, V., Harish, P., Patidar, S., Narayanan, P.: Fast minimum spanning tree for large graphs on the GPU. In: Proceedings of the Conference on High Performance Graphics 2009, pp. 167–171. ACM (2009)
26. Wang, Q.-F., Cambria, E., Liu, C.-L., Hussain, A.: Common sense knowledge for handwritten Chinese text recognition. Cogn. Comput. $5(2)$, 234–242 (2013)
27. Zhang, S., Li, S., Yang, J.: Gaddi: distance index based subgraph matching in biological networks. In: Proceedings of the 12th International Conference on Extending Database Technology: Advances in Database Technology, pp. 192–203. ACM (2009)
28. Zhao, P., Han, J.: On graph query optimization in large networks. Proc. VLDB Endow. $3(1$–$2)$, 340–351 (2010)

Parameters Driving Effectiveness of LSA on Topic Segmentation

Marwa Naili, Anja Chaibi Habacha$^{(\boxtimes)}$, and Henda Hajjami Ben Ghezala

RIADI laboratory, National School of computer Science (ENSI),
University of Manouba, 2010 Manouba, Tunisia
maroua.naili@riadi.rnu.tn, {anja.habacha,Henda.Benghezala}@ensi.rnu.tn

Abstract. Latent Semantic Analysis (LSA) is an efficient statistical technique for extracting semantic knowledge from large corpora. One of the major problems of this technique is the identification of the most efficient parameters of LSA and the best combination between them. Therefore, in this paper, we propose a new topic segmenter to study in depth the different parameters of LSA for the topic segmentation. Thus, the aim of this study is to analyze the effect of these different parameters on the quality of topic segmentation and to identify the most efficient parameters. Based on extensive experiments, we showed that the choice of LSA parameters is very sensitive and it has an impact on the quality of topic segmentation. More important, according to this study, we are able to propose appropriate recommendation for the selection of parameters in the field of topic segmentation.

Keywords: LSA · Topic segmentation · LSA parameters

1 Introduction

Topic segmentation is an important technique which aims to identify topic boundaries in textual document by dividing it into topically homogeneous blocks. This technique is used in several Natural Language Processing (NLP) tasks such as information retrieval and text summarizing. Therefore many topic segmenters have been presented such as TextTiling [11] and C99 [4]. These segmenters are based on internal resources like lexical repetition. So to improve them, some researchers have used complementary semantic knowledge by using external resources like Larousse thesaurus [13], a co-occurrence network [8], the Latent Dirichlet Allocation method [16] and the Latent Semantic Analysis (LSA) method [2,3,5]. This latter has shown promising results in topic segmentation. But while many researchers have studied in depth the LSA method in information retrieval [6,7], text categorization [15,17,18] and automatic essay scoring and evaluation [9,23], no proper study had been conducted on the effect of LSA parameters on topic segmentation. Therefore, two main areas will be studied in this paper. The first one is about an extensive literature study on LSA and its parameters in different fields. The second one deals with the impact of these

© Springer International Publishing AG, part of Springer Nature 2018
A. Gelbukh (Ed.): CICLing 2016, LNCS 9623, pp. 560–572, 2018.
https://doi.org/10.1007/978-3-319-75477-2_40

different parameters on topic segmentation by proposing a new topic segmenter named ToSe-LSA. Therefore, we present a deep investigation of these different parameters in order to identify the most efficient ones and to recommend a relevant tuning of parameters in the field of topic segmentation. Moreover, we carry out a systematic analysis for LSA parameters selection by comparing them to related works in different fields.

This paper is organized as follows: the next section presents an overview on LSA. The Sect. 3 deals with the proposed topic segmenter ToSe-LSA. Experimental results and discussion are reported in Sect. 4. The Sect. 5 presents an comparison between our results and related works in different fields. The conclusion and future work are presented in Sect. 6.

2 Latent Semantic Analysis

2.1 LSA Process

After being patented in 1988, LSA was firstly used by Deerwester et al. [6] and Dumais [7] as a new method for automatic indexing and retrieval. Since then, LSA was used in different fields such as text mining [1,15,18], educational technology [9,22,23], big data analytic [12], lexical game [14] and topic segmentation [2,3,5]. LSA is a powerful automatic statistical technique for extracting semantic knowledge. It is based on the idea that there is a latent link between words and their contexts (sentences, paragraphs or documents) and it consists of two main steps. The first step is the construction of a term-document matrix M by using a collection of documents. M is a rectangular matrix with a size of $m*n$. The rows of M correspond to m terms, the columns correspond to n documents and $M[i,j]$ is equal to the frequency of term i in sentence j. The second step of LSA is the Singular Value Decomposition (SVD). In fact, the matrix M will be decomposed into three matrices ($M = U * S * V^T$): U and V^T are two orthogonal matrices and S is a diagonal matrix. Then, an approximation of the matrix M will be done by using only the k largest singular values from S and their corresponding singular vectors from U and V^T ($M_k = U_k * S_k * V_k^t$). So the matrix M_k will be considered as the reduced semantic space. Therefore, to measure the similarity between two terms (or documents), we just need to measure the angle between the two vectors that correspond to these terms (or documents). We note that terms (respectively documents) are represented by the row (respectively column) vectors of M_k.

2.2 Parameters of LSA

The LSA method is based on three groups of parameters:

– **Pre-processing:** is a current step in several applications such as information retrieval and topic segmentation. It includes different operations such as stop words elimination, stemming, local frequency setting (frequency of terms in the document) and global frequency setting (frequency of terms in the corpus). Several researchers have insisted on the importance of the pre-processing step such as Nakov et al. [18] and Wild et al. [23]. Nakov et al. [18] conducted four tests on the Bulgarian language: without pre-processing, with stop words elimination, with stemming and with a combination between the two last ones. As result, they proved that the elimination of stop words improves the results of LSA and that the stemming operation can be over-looked. On the other hand, Wild et al. [23] conducted the same tests but for the German language. As result they proved also that the elimination of stop words improves the results of LSA but the stemming operation has a negative effect.

- **Local and global weighting:** After the construction of the term-document matrix, a set of transformation must be performed to normalize terms fre-quencies. This transformation can be expressed as the product of local and global weighting. The local weighting takes into account the frequency of terms in a particular document while the global weighting is related to terms frequencies in the whole corpus. As shown in Table 1, the most used func-tions for local weighting are TF, $BinaryTF$ and $LogTF$. Whereas for the global weighting, the most used functions are IDF, GF-IDF, $Entropy$ and $Normalization$. Several tests have been made to determine the best com-bination of local and global weighting. Dumais [7] tested the combinations between TF and $LogTF$ for local weighting and IDF, GF-IDF, $Entropy$ and $Normalization$ for global weighting. As result, he proved that $LogTF$ and $Entropy$ are the best choice for the English language. Furthermore, Nakov et al. [17,18] performed the same tests on the two languages English and Bulgarian and he proved the results of Dumais [7]. But, they also proved that the IDF improves the performance of LSA. Lintean et al. [15] also presented a deep study of this parameter by evaluating $Binary$, TF and $LogTF$ for local weighting and $Bianry$, IDF and $Entropy$ for global weighting. The experiments revealed that best results are obtained by $LogTF$, $binary$ and IDF. Wild et al. [23] performed the same tests as Nakov et al. [17,18]. They found that the effect of the local weighting is insignificant while for the global weighting it is the opposite. Guillermo et al. [9] tested only IDF and $Entropy$ for global weighting and proved that it improves the LSA performance.
- **The dimension of the semantic space:** The choice of dimensionality is considered as the most influent parameter on LSA. After applying the SVD on the original term-document matrix, a reduced matrix is reconstructed by using only k singular values. These values are the largest and most important in the matrix S. The objective of this reduction is to obtain an approxima-tion of the space to use the most important information and to reduce the noise problem. Therefore, the real issue in LSA is the choice of k. Dumains [7] proved that the reduction of the semantic space improves the LSA method and they showed that the right choice of k is equal to 300. However, Wiemer-Hastings et al. [21] questioned the effectiveness of the reduction of semantic space and they proved that there is no difference between using the full seman-tic space or the reduced version. Like Dumais [7], Wild et al. [23] proved the importance of the choice of dimensionality and they reported a new method

to unify it regardless of the size of the used corpus. In fact, they used a percentage of the cumulated singular values and they showed that the best choice is between 40% and 50%. Guillermo et al. [9] used the same approach as Wild et al. [23] and they proved that the best results are obtained with the two percentages 40% and 60%.

Table 1. Weighting functions.

Local weighting function: LW		
Name	Equation	Description
TF	$LW_{ij} = TF_{ij}$	TF_{ij} is the frequency of term t_i in document d_j
Binary TF	$LW_{ij} = 1$ if $TF_{ij} > 0$ or 0 if not	$LW_{ij} = 1$ if the frequency of t_i in d_j is superior or equal to 1
Log TF	$LW_{ij} = \log TF_{ij} + 1$	TF_{ij} is the frequency of t_i in d_j
Global Weighting function: GW		
IDF	$GW_{ij} = 1 + \log \frac{N}{d(i)}$	$d(i)$ is the number of document which contains t_i
GF-IDF	$GW_{ij} = \frac{g(i)}{d(i)}$	g(i) is the frequency of t_i in corpus
Entropy	$GW_{ij} = 1 + \sum_j \frac{p_{ij} \log p_{ij}}{\log N}$	$p_{ij} = \frac{TF_{ij}}{\sum_j TF_{ij}}$
Normalization	$GW_{ij} = \frac{1}{\sqrt{\sum_j TF_{ij}}}$	TF_{ij} is the frequency of t_i in d_j

2.3   Study of LSA Parameters in the Topic Segmentation

LSA was used twice on topic segmentation by Choi et al. [5] and Bestgen [2]. Choi et al. [5] studied only the impact of the choice of dimensionality by testing five values of k (100, 200, 300, 400 and 500). As result, they proved that the best value of k is 500. For the rest of parameters, they didn't argue their choices. For example, for the pre-processing step, they eliminated the stop words and terms with local frequency equal to 1. For the frequency matrix transformation, they used the TF for local weighting and IDF for global weighting. Bestgen [2] studied only some operations of the pre-processing step such as uncommon words (global frequency inferior or equal to 2, 10 and 20) elimination, stemming and the specificity of the corpus. A corpus is specific if the semantic space is constructed from the same test corpus otherwise it is not specific. As results, he proved that the quality of topic segmentation increases if we use a specific corpus and if we remove words with global frequency inferior or equal to 2. He also proved that stemming has a negligible effect on topic segmentation. For the rest of parameters, Bestgen [2] used $LogTF$ for the local weighting and $Entropy$ for global weighting and k equal to 300. According to these related works, it is clear that there is a lack of research on LSA for the topic segmentation and that there is no general rule to initialize the parameters of LSA. Therefore in this work, we will use LSA to add semantic knowledge to topic segmentation. More important, we will study the impact of LSA parameters on the quality of results.

3   Proposed Topic Segmenter: ToSe-LSA

To use the LSA on topic segmentation, we propose a new topic segmenter named ToSe-LSA. As shown in Fig. 1, ToSe-LSA goes throw five steps. The first step is the pre-processing which includes three operations: word extraction, stop word elimination and stemming by using Porter Stemmer. The second step is the construction of the frequency dictionary. Each sentence is presented by a vector. And each vector is composed of terms (term i), their stem (t_i), their frequency (Ft_i) and their corresponding vector from the semantic space (Vt_i) which is constructed by LSA method. We note that only the terms that belong to the semantic space will be presented by their corresponding vector. The third step is the construction of the similarity matrix in which we will compare all pairs of all terms by measuring the angle between their corresponding vectors. For the rest of terms the cosine value will be equal to 0. Therefore the similarity between two sentences will be calculated by using Eq. 1. The two last steps are based on C99 [4] which is one of the most efficient endogenous segmenters [2,8,13]. The forth step is the construction of the rank matrix based on a 11×11 rank mask. The rank corresponds to the number of neighboring elements which belongs to the 11×11 matrix and have a lower similarity value. Finally, the fifth step is the topic boundaries identification by using Reynar's maximization algorithm [20].

$$Sim(S_1, S_2) = \frac{\sum_{t_i \in S_1 \cap SS} \sum_{t_j \in S_2 \cap SS}(Ft_i * Ft_j * cos(Vt_i, Vt_j))}{\sum_{t_i \in S_1}(Ft_i) * \sum_{t_j \in S_2}(Ft_j)} \tag{1}$$

With S_1 and S_2 correspond to sentences 1 and 2; SS corresponds to the Semantic Space; Ft_i and Ft_j correspond to the frequency of terms t_i and t_j; Vt_i and Vt_j correspond to the vectors of t_i and t_j in SS.

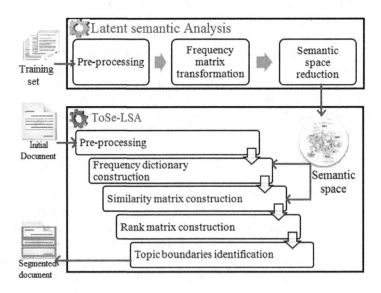

Fig. 1. TSS-LSA process based on the different parameters of LSA

One of the originalities of this approach is that the LSA method is employed in two levels: the construction of the frequency dictionary and of the similarity matrix which is not the case in [2,3,5]. They only integrate LSA for the similarity calculation. Moreover, for ToSe-LSA we used a parametric approach to easily varying the LSA parameters. Therefore we have different semantic spaces which depend on the choice of parameters. Also, ToSe-LSA is independent of the used corpus which means that we can test any corpus regardless of the one used to construct the semantic space.

4 Empirical Results

4.1 Test Corpora

To evaluate the effect of LSA on topic segmentation, we used two English corpora based on ACM digital library[1]. The first one is a training corpus used to construct the latent semantic space. It contains 16832 words (11020 stems). This corpus deals with several topics of computer science such as artificial intelligence, information systems, networks and security. The second one is a test corpus which contains 7313 words (4512 stems). This corpus is employed to test ToSe-LSA segmenter based on artificial documents. These laters correspond to a serial concatenation of two documents.

4.2 Experiment Results

As mentioned at the beginning of this paper, one of our main goals is to study the effect of the different parameters of LSA on topic segmentation. Therefore, we conducted a series of experiments by varying the parameters of LSA. For the pre-processing step, we eliminated the stop words and we used the stem of each terms. More important, we are focused on local and global frequencies setting. Thus, we studied three different values for the local frequency ($LF= 1$, 2 and 3) and the global frequency ($GF= 1$, 2 and 3). For the frequency matrix transformation, we used different functions for the local weighting ($LW = TF$ and $LogTF$) and global weighting ($GW = IDF$ and $Entropy$). To reduce the semantic space, we used the same method as Wild et al. [23]. In fact, we used different percent of the cumulated singular values (40%, 50%, 60%, 70% and 80%). To report the results of these experiments, we used the WindowDiff metric which measures the error rate by using a sliding window.

– **Local and global frequencies:**

The Fig. 2 contains 5 histograms which are plotted as a function of the couple: local and global frequencies (LF, GF). Each histogram is associated with a specific dimension of the semantic space and shows the variation of WindowDiff for different combinations of local and global weighting functions ($TF * IDF$,

---

[1] ACM home. http://dl.acm.org/.

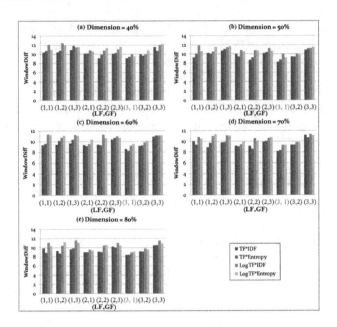

Fig. 2. WindowDiff according to the local and global frequency.

$TF * Entropy$, $LogTF * IDF$ and $LogTF * Entropy$). As shown in Fig. 2, the lowest values of WindowDiff are achieved by the couple of local and global frequencies (3,1) regardless of the dimension of the semantic space. Moreover, the biggest error rates are almost obtained by the couples: (1,1), (1,2) and (3,3).

- **Local and global weighting:**

In Fig. 3, 9 subfigures are presented according to the couple local and global frequencies (LF, GF). For each one, the variation of WindowDiff values are plotted according to the dimension of the reduced semantic spaces. And each one of these curves corresponds to a specific combination of the local and global weighting functions. As shown in Fig. 3, we can note that the two curves of $TF * IDF$ and $TF * Entropy$ are always below the other curves, except for some points. And regardless of the values of the local and global frequencies (LF, GF), the smallest error rates are obtained by $TF * IDF$ or $TF * Entropy$. Furthermore, the highest error rates are obtained by $LogTF * IDF$ and $LogTF * Entropy$. Also, we note that the curves of $TF * IDF$ and $TF * Entropy$ are close which is the same for the two curves of $LogTF * IDF$ and $LogTF * Entropy$. So we can remark that regardless of the local weighting, the error rate decreases by using IDF or $Entropy$. And by fixing the global weighting, the WindowDiff values decrease by using TF.

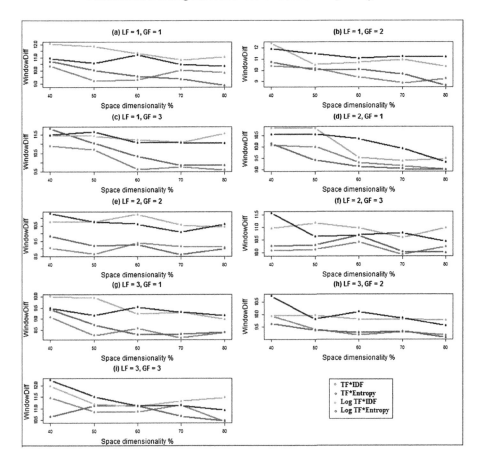

Fig. 3. WindowDiff according to the local and global weighting.

- **Dimension of the semantic space:**

The choice of dimension, which correspond to the k largest singular values, is very important. Therefore we describe, in Fig. 4, the different curves of the singular values according to local and global frequencies and to local and global weighting functions. Each curve shows the top 100 singular values sorted in descending order. As first remark, it is clear that the curves of singular values vary according to the local and global weighting. For example, the most important values are obtained by $TF*IDF$ and $TF*Entropy$. For the others combination ($LogTF*ID$ and $LogTF*Entropy$), the singular values are less important and very close. Also we can note that, the curves of $TF*Entropy$, $LogTF*IDF$ and $LogTF*Entropy$ became stable at some point between 70 and 80. Therefore, to reduce the semantic space, we have to use a specific percentage of the semantic space that contains these largest singular values. As shown in Fig. 3, best results are obtained by percentages between 70% and 80%. Yet the worst results are obtained by percentages equal to 40% and 50%. Thus, if we eliminate less than 70% of the semantic space we can loose important information and if we keep more than 80% we will have a noise problem.

4.3   Discussion of Results

In this paper, we have studied the different parameters of LSA on topic segmentation by conducting 180 tests for each document of the test corpus. As first result, we proved that the different parameters of LSA have significant impact on topic segmentation. In fact, by increasing the global frequency, the quality of topic segmentation decreases. However, the biggest values of local frequency improve the results of topic segmentation. Therefore, the best result is obtained by the couple ($LF = 1, GF = 3$). For the term-document matrix transformation, we proved that the two functions IDF and $Entropy$ increase the quality of topic segmentation regardless of the used local weighting. Also, by using the local weighting function $LogTF$, the quality of topic segmentation decreases compared to the use of TF. This result is the same for the two global weighting IDF and $Entropy$. So we can conclude that the local and global weighting are independent. And the best results are obtained by combining TF with IDF or $Entropy$. The choice of dimensionality is delicate and has an important impact on topic segmentation. In this study, we showed that, by using the highest percentage of the cumulated singular values (70%, 80%), the quality of topic segmentation increased. As conclusion, the best combination of LSA parameters is: local frequency = 3, global frequency = 1, local weighting = TF, global weighting = IDF and dimension of the semantic space = 70%.

5   Comparison with Related Works

According to this study, we were able to define the best combination of LSA parameters for Topic Segmentation. As shown in Table 2, if we compare our results with related works in different fields (Information Retrieval:IR, Essay Grading:EG and Text Categorization:TC), we can find some similarities and differences. For example, concerning the global weighting, we found the same results as Nakov et al. [17], Guillermo et al. [9] and Lintean et al. [15]. Yet, for the local weighting, Dumais [7], Nakov et al. [17] and Lintean et al. [15] proved that the best choice is the $LogTF$. For Wild et al. [23], the impact of the local weighting can be overlooked which is the contrary of our results. Moreover, for the choice of dimensionality we didn't find the same result as Wild et al. [23] and Guillermo al al. [9]. But still, the biggest values of dimension give always better results. These differences can be explained by the fact that the field of study has an impact on the choice of LSA parameters. Furthermore, if we compare our results with related works on Topic Segmentation (TS), we can say that some choices can be doubted. We mention the choice of Bestgen [2] for the local weighting which we have showed that it decreases the quality of topic segmentation. Also he used only 300 of the singular values like Dumais [7]. But

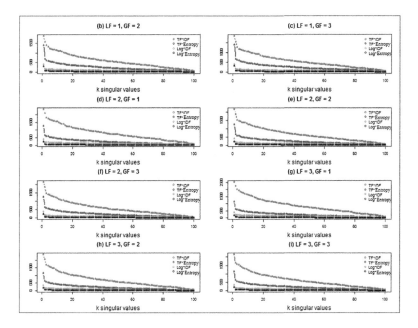

Fig. 4. Singular values.

it can be a wrong choice because as we showed the field of application has an effect on the LSA parameters. Therefore using choices made in other field can't be always the right solution and it might have some unpredictable impact. On the other hand, like Bestgen [2], we proved the advantage of using the smallest global frequency. Moreover, if we compare our results with the work of Choi et al. [5], for the dimensionality, they showed that using k equal to 500 improves the quality of segmentation. However, this choice is not relative to the corpus size and not based on a general rule. Therefore, we generalized this choice by using a percentage of the cumulated singular values. Also we note that they used the same corpus for the construction of the semantic space and to test the quality of topic segmentation which could confound the obtained results. That is why, we used two different corpora. Finally, we note that we have studied all LSA parameters which, up to our best knowledge, is not the case for other studies. According to this study, we can identify some general rules. The first one is about the use of global frequency which is better to be small. The second one is about the choice of local and global weighting: it is better to use TF for local weighting and IDF or Entropy for global weighting. The third one is about the reduction of space by choosing the value of k. In general, it is recommended to use the biggest value between 60% and 80%.

Table 2. Comparison with LSA's parameters related works.

Works	Field	Pre-processing		Weighting		Dimension of space
		LF	GF	LW	GW	
Dumais [7]	IR	-	-	TF, LogTF	Entopy, IDF GF-IDF, Normalization	k = 300
Wiemer-Hastings et al.[21]	IR	-	-	-	-	Without effect
Nakov et al. [17,18]	TC	-	-	TF, LogTF	Entropy, IDF GF-IDF, Normalization	k in [50..1500]
Lintean et al. [15]	TC	-	-	TF, Binary, LogTF	Binary, IDF Entropy	-
Wild et al.[23]	EG	-	-	TF, LogTF BinaryTF	Entropy, IDF Normalization	30%, 40%, 50%
Guillermo et al.[9]	EG	-	-	-	Entropy, IDF Normalization	20%, 40%, 60%, 100%
Choi et al.[5]	TS	-	-	TF	IDF	k = 100, 200, 300, 400, 500
Bestgen [2]	TS	-	2, 10, 20	LogTF	Entropy	k = 300
Our work	TS	1, 2, 3	1, 2 3	TF, LogTF	IDF, Entropy	40%, 50%, 60,% 70%, 80%

6   Conclusion and Future Work

In this work, we conducted a deep study to investigate the efficiency of LSA parameters and their impact in the domains of topic segmentation. Therefore, we presented a deep description of LSA and its parameters. Then we proposed a segmenter named ToSe-LSA which uses the semantic similarity (LSA) as well the lexicon and syntax (lexical cohesion) ones for the topic segmentation. Based on extensive experiments, we proved that LSA improves the quality of topic segmentation especially with the right choice of parameters which are: local frequency = 3, global frequency = 1, local weighting = TF, global weighting = IDF and Entropy and dimension of the semantic space = 70% and 80%. From these results and according to related works, we can draw some generalized rules to get better results with LSA. For example, it is better not to vary the value of global frequency. For the global weighting, the best functions are IDF and Entropy. Also, better results are made by reducing the semantic space between 60% and 80%. To go further, we should study the LSA method on Arabic topic segmenters [10]. Moreover, we will add another level to the topic segmentation which is the conceptual level by using a ConceptNet-based semantic parser [19].

Acknowledgments. We would like to show our special gratitude to professor emeritus **Mouhamed Ben Ahmed**, whose has invested his effort in guiding this research and contributed with his precious suggestions, support and encouragement.

References

1. Bechet, N., Chauche, J., Roche, M.: EXPLSA an approach based on syntactic knowledge in order to improve LSA for conceptual classification task. In: CICLing 2008, vol. 33, pp. 213–224 (2008)
2. Bestgen, Y.: Improving text segmentation using latent semantic analysis: a reanalysis of choi, wiemer-hastings and moore. Computat. Linguist. **32**, 5–12 (2006)
3. Bestgen, Y., Pierard, S.: Comment evaluer les algorithmes de segmentation thematique? essai de construction d'un mmateriel de reference. In: TALAN 2006, pp. 407–414 (2006)
4. Choi, F.Y.Y.: Advances in domain independent linear text segmentation. In: Proceedings of NAACL, pp. 26–33 (2000)
5. Choi, F.Y.Y., Wiemer-Hastings, P., Moore, J.: Latent semantic analysis for text segmentation. In: Proceedings of EMNLP, pp. 109–117 (2001)
6. Deerwester, S., Dumais, S.T., Furnas, G.W., Landauer, T.K., Harshman, R.: Indexing by latent semantic analysis. J. Am. Soc. Inf. Sci. **41**, 391–407 (1990)
7. Dumais, S.: Enhancing performance in latent semantic indexing (LSI) retrieval. Technical Report TM-ARH017527, Bellcore, Morristown, NJ (1992)
8. Ferret, O.: Improving text segmentation by combining endogenous and exogenous methods. In: International Conference RANLP, Borovets, Bulgaria, pp. 88–93 (2009)
9. Guillermo, J.B., Jose, A.L., Ricardo, O., Inmaculada, E.: Latent semantic analysis parameters for essay evaluation using small-scale corpora. J. Quant. Linguist. **17**(1), 1–29 (2010)
10. Habacha, A.C., Naili, M., Sammoud, S.: Topic segmentation for textual document written in arabic language. Procedia Comput. Sci. **35**, 437–446 (2014). KES-2014 Gdynia, Poland, September 2014
11. Hearst, M.A.: Texttiling: segmenting text into multi-paragraph subtopic passages. Comput. Linguist. **23**(1), 33–64 (1997)
12. Kundu, A., Jain, V., Kumar, S., Chandra, C.: A journey from normative to behavioral operations in supply chain management: a review using latent semantic analysis. Expert Syst. Appl. **42**(2), 796–809 (2015)
13. Labadié, A., Prince, V.: Lexical and semantic methods in inner text topic segmentation: a comparison between C99 and transeg. In: Kapetanios, E., Sugumaran, V., Spiliopoulou, M. (eds.) NLDB 2008. LNCS, vol. 5039, pp. 347–349. Springer, Heidelberg (2008). https://doi.org/10.1007/978-3-540-69858-6_40
14. Lafourcade, M., Zampa, V.: PTICLIC: a game for vocabulary assessment combining JEUXDEMOTS and LSA. In: CICLing 2009, pp. 1–7 (2009)
15. Lintean, M., Moldovan, C., Rus, V., McNamara, D.: The role of local and global weighting in assessing the semantic similarity of texts using latent semantic analysis. In: FLAIRS Conference, pp. 235–240 (2010)

16. Misra, H., Yvon, F., Jose, J.M., Cappe, O.: Text segmentation via topic modeling: an analytical study. In: Proceedings of the 18th ACM Conference on Information and Knowledge Management, pp. 1553–1556 (2009)
17. Nakov, P., Popova, A., Mateev, P.: Weight functions impact on LSA performance. In: Recent Advances in Natural Language Processing - RANLP 2001, Tzigov Chark, Bulgaria (2001)
18. Nakov, P., Valchanova, E., Angelova, G.: Towards deeper understanding of the LSA performance. In: Recent Advances in Natural Language Processing - RANLP 2003 (2003)
19. Poria, S., Agarwal, B., Gelbukh, A., Hussain, A., Howard, N.: Dependency-based semantic parsing for concept-level text analysis. In: Gelbukh, A. (ed.) CICLing 2014. LNCS, vol. 8403, pp. 113–127. Springer, Heidelberg (2014). https://doi.org/10.1007/978-3-642-54906-9_10
20. Reynar, J.C.: Topic Segmentation : Algorithms and Applications. Ph.D. thesis, University of Pennsylvania (1998)
21. Wiemer-Hastings, P., Wiemer-Hastings, K., Graesser, A.: How latent is latent semantic analysis? In: Proceedings of the Sixteenth International Joint Congress on Artificial Intelligence, pp. 932–937. Morgan Kaufmann, San Francisco (1999)
22. Wild, F., Haley, D., Bulow, K.: Using latent-semantic analysis and network analysis for monitoring conceptual development. JLCL **26**, 9–21 (2011)
23. Wild, F., Stahl, C., Stermsek, G., Yoseba, K.P., Neumann, G.: Factors influencing effectiveness in automated essay scoring with LSA. In: AIED 2005, pp. 947–949. The Netherlands, Amsterdam (2005)

A New Russian Paraphrase Corpus.
Paraphrase Identification and Classification
Based on Different Prediction Models

Ekaterina Pronoza(✉) and Elena Yagunova(✉)

Saint Petersburg State University, 7/9 Universitetskaya nab., St. Petersburg 199034, Russia
katpronoza@gmail.com, iagounova.elena@gmail.com

Abstract. Our main objectives are constructing a paraphrase corpus for Russian and developing of the paraphrase identification and classification models based on this corpus. The corpus consists of pairs of news headlines from different media agencies which are extracted and analyzed in real time. Paraphrase candidates are extracted using an unsupervised matrix similarity metric: if the metric value satisfies a certain threshold, the corresponding pair of sentences is included in the corpus. These pairs of sentences are further annotated via crowdsourcing. We provide a user-friendly online interface for crowdsourced annotation which is available at http://paraphraser.ru. There are 7480 annotated sentence pairs in the corpus at the moment, and there are still more to come. The types and the features of these sentence pairs are not introduced to the annotators. We adopt a 3-classes classification of paraphrases and distinguish precise paraphrases (conveying the same meaning), loose paraphrases (conveying similar meaning) and non-paraphrases (conveying different meaning).

Keywords: Paraphrase identification · Low-level features · Lexical features
Semantic features · Matrix similarity metric

1 Introduction

In this paper we introduce a new publicly available Russian paraphrase corpus and our results on the development of paraphrase classification and identification models based on this corpus.

This research is part of our work on the ongoing project ParaPhraser.ru [38] which is dedicated to paraphrase extraction, identification and generation.

The corpus consists of pairs of news headlines from different media agencies which are extracted and analyzed in real time. Paraphrase candidates are extracted using an unsupervised matrix similarity metric: if the metric value satisfies a certain threshold, the corresponding pair of sentences is included in the corpus. These pairs of sentences are further annotated via crowdsourcing. We provide a user-friendly online interface for crowdsourced annotation which is available at http://paraphraser.ru. There are 7480 annotated sentence pairs in the corpus at the moment, and there are still more to come. We do not introduce the types and the features of these sentence pairs to the annotators.

© Springer International Publishing AG, part of Springer Nature 2018
A. Gelbukh (Ed.): CICLing 2016, LNCS 9623, pp. 573–587, 2018.
https://doi.org/10.1007/978-3-319-75477-2_41

The corpus is mainly intended to be used for Information Extraction and Text Summarization. We believe that the former task requires semantically equivalent, or precise, paraphrases while the latter one demands roughly similar ones (so-called loose paraphrases). That is why we adopt a 3-classes classification of paraphrases and distinguish precise paraphrases (conveying the same meaning), loose paraphrases (conveying similar meaning) and non-paraphrases (conveying different meaning).

In paraphrase community, paraphrase identification and classification methods usually employ machine learning techniques, and in our research we adopt the same approach [11, 12]. As part of the development of paraphrase identification and classification models, we experiment with different types of prediction models. These models are based on various similarity measures which come from both semantic and paraphrase communities. At the moment we work with shallow (surface) measures based on the overlap between the sentences as well as with semantic measures: dictionary-based and distributional ones. The prediction models are actually the feature sets produced by the described types of sentence similarity measures and used in the classification task (paraphrase identification task can also can be considered a classification task, with 3 classes of paraphrases reduced to 2). Thus, the similarity measures are not compared directly (for example, via their correlation with paraphrase classes). We compare and analyze their utility as features in the paraphrase classification task. Each type of similarity measures produces a feature set, and the performance of these feature sets is evaluated and compared.

2 Related Work

A detailed overview of all the paraphrase corpora existing is beyond the scope of this paper. A thorough and insightful review of different sentential paraphrase datasets can be found in [43] where the authors present recommendations on paraphrase corpora construction and raise a number of important problems to the community [28].

Due to the aim of our research and space limitations we focus on the well-known Microsoft Research Paraphrase Corpus (MSRP) and on The Paraphrase Database, a publicly available resource of Russian paraphrases. MSRP is definitely the one which greatly inspired research in paraphrase community. It was constructed as a broad-domain corpus of sentential paraphrases which would be amenable to statistical machine translation (SMT) techniques (5801 pairs of English sentences collected from news clusters and annotated by 2 experts) [18]. An initial set of paraphrases is extracted using Levenshtein distance. The authors consider first 3 sentences of the articles and apply several criteria to their length and lexical distance between the sentences, the resulting dataset is extracted using SVM with morphological, lexical, string similarity and composite features. Although MSRP is widely used as the gold standard in the paraphrase extraction methods, it is often criticized by researchers for its loose definition of paraphrase, for its 2-way annotation, high lexical overlap, etc.

While constructing a Russian corpus, we try to solve the problem of paraphrase ambiguity by distinguishing 2 types of paraphrases: precise and loose ones. We have 3-way annotation: precise paraphrases, loose paraphrases and non-paraphrases. As for the lexical overlap problem, we consider this overlap acceptable and even helpful in our case. Russian is a language with free word order, and pairs of sentences which consist of the same words put in different order could be used for learning syntactic patterns for paraphrase generation.

As mentioned above, there already exists a large publicly available Russian paraphrase resource, namely, the dataset published by Ganitkevich et al. as part of The Paraphrase Database project (PPDB) [22]. The authors collected an impressively large database of paraphrases on word-, phrase- and syntactic levels. Syntactic level paraphrases are annotated with nonterminal symbols (constituents, in terms of phrase structure grammar) and contain placeholders which can be substituted with any paraphrase matching its syntactic type. In addition, all types of paraphrases are annotated with count and probability-based features.

The training data for Russian is substantial in PPDB (over 2 million sentence pairs), and the resulting dataset is large as well. It is collected from the corpora typically used in SMT: CommonCrawl, Yandex 1M corpus and News Commentary. The authors use a language independent method to extract paraphrases from parallel bilingual texts: paraphrases are found in a single language by "pivoting" over a shared translation in another language. Such approach was introduced by [5] and since then it has been successfully applied by many researchers. Such grouping leads to the rapid growth of the dataset, and, with a number of available morphological parsers today, it seems unnecessary. We believe that we should use language-specific methods (in contrast with language-independent ones) when dealing with a morphologically rich language.

Unlike other paraphrase resources, our corpus is not intended to be a general-purpose one. According to our tasks (IE and TS) we collect it from the news texts. The corpus consists of sentential paraphrases, and lower level paraphrase pairs can be extracted from it using any of the existing methods (e.g., SMT methods).

Based on the previous research on semantic similarity measures and features for paraphrase identification, we can classify existing sentence similarity measures into the following groups:

1. Shallow (based on string or lexical overlap).
2. Semantic (based on the semantic structure of the sentences, using external semantic resources).
3. Syntactic (based on the syntactic structure of the sentences).
4. Distributional (based on vector space models, or distributional semantic models).

Shallow similarity measures are the earliest similarity measures used in paraphrase and semantic communities. They are mainly based on the overlap of words, phrases or characters [17–19, 29]. We also classify metrics originating from machine translation, like BLEU [25, 40, 55], as shallow, if they are based on the surface forms of words and do not employ any semantic resources. Other shallow features include edit distance between the sentences [10, 40, 55], sentence length difference [25, 52], the length of the longest common subsequence [13, 29, 40, 55], the number of matching proper names and cardinal numbers [13, 29], etc.

Most dictionary-based semantic similarity measures use WordNet [18] or WordNet-like resources and exploit synonymy or hypernymy relations [33]. Roughly speaking, the similarity between the words can be calculated as the length of the shortest path between them in the WordNet graph. A comprehensive study of different WordNet-based measures can be found in [6]. Such measures are more sophisticated than the shallow ones, but they have evident limitations because they are strongly dependent on the quality and coverage of the corresponding semantic resources.

Another approach (syntactic measures) is applied in [16, 42, 49, 52] etc. and implies the use of dependency pairs in the features for paraphrase identification (e.g., dependency relations overlap calculated as precision and recall, edit distances between syntactic parse trees, etc.). [30, 48] propose to calculate the similarity between texts as the similarity between their respective syntactic n-grams using tree edit distance. Syntactic measures allow us to capture even more linguistic phenomena than shallow and dictionary-based semantic ones. However, syntactic measures use the output of a syntactic parser, and they unavoidably propagate errors of the parser.

Distributional semantic measures can serve as an alternative to both semantic and syntactic measures as they can predict semantic similarity without analyzing the deep structure of the sentences. For example, in [1] it is shown that on high-complexity datasets like Microsoft Research Paraphrase Corpus (MSRP) [18] and the third recognizing textual entailment challenge (RTE3) dataset [8] overlap-based and distributional measures perform better than syntactic and dictionary-based semantic ones. The distributional approach is based on the supposition that semantically close words occur in similar contexts. Distributional models can be classified into count-based (e.g., based on Latent Semantic Analysis (LSA)) and predictive ones (e.g., skip-grams and bag-of-words models in the recursive neural network implementation in Word2vec [31, 32]). Some authors argue that predictive models outperform count-based ones on a wide range of semantic tasks (see [6] for example) while others disagree. In our experiments we employ skip-gram model implemented in Word2vec as they are very convenient to use and can be trained iteratively.

In paraphrase identification community, the described similarity measures are commonly used as features in the classification task. State-of-the-art results on MSRP are usually achieved by using a combination of different types of similarity measures as features for the classifier and/or by tuning distributional features, e.g., by using discriminative TF-IDF weighting (TF-KLD) [25] together with fine-grained linguistic features or discriminative TF-IDF weighting for both words and phrases with smoothing based on the K nearest neighbours (KNN) algorithm (TF-KLD-KNN) [54]. In this paper we focus on the analysis of the performance of different types of similarity measures as features in paraphrase classification task, and tuning them is beyond our task at the moment. Instead, our aim is to provide the comparison of similarity measures with respect to their effectiveness of the corresponding features on different types of sentence pairs.

A study of different similarity measures for paraphrase identification was already conducted, for example, in [2]. The authors considered three types of measures: overlap-based measures (i.e., shallow), linguistic measures and the measures based on vector space model with TF-IDF weighting and cosine distance. These three measures were tested against three different datasets for English with two paraphrase classes: paraphrases and non-paraphrases.

There also exist more recent studies discussing paraphrasing (many of them come from plagiarism detection). For example, in [3] paraphrase identification problem is considered as part of the plagiarism detection task, and the authors analyze the types of paraphrases which underlie plagiarism acts. It is shown that there is a correlation between linguistic complexity of the paraphrases and the performance of plagiarism detection systems. In [22] content- and citation-based approaches to plagiarism detection in scientific publications are compared, and the conclusion is made that a combination of the methods can be beneficial. In another work [23] a comparative evaluation of different models for cross-language plagiarism detection is conducted. The models are based on different types of text similarity measures, and a simple model based on character n-grams is reported to be the best to choice to compare texts across languages if they are syntactically related.

Unlike other studies, we focus on the paraphrase classification task itself, and unlike [2] where the approach is most similar to ours, we consider a substantially extended set of shallow features, and over 40 different vector distances other than the cosine distance. In vector space model we adopt a bag-of-words approach but do not apply any weighting. We believe that the preliminary analysis should be conducted to select the appropriate weighting scheme. We also distinguish semantic and syntactic linguistic similarity measures (and focus on the semantic ones in this paper). As we work with Russian, which is a morphologically rich language, our semantic features are also extended: they combine the information about synonymy relations and word formation families.

3   Data

Our paraphrase corpus consists of Russian sentence pairs collected from news headlines. Several Russian media sources are parsed every day in real time, their headlines are compared using an unsupervised similarity metric described in [38] and candidate pairs are included in the corpus. They are further evaluated via crowdsourcing and labeled as precise, loose or non-paraphrases. At the moment there are 7480 sentence pairs (1759 precise, 3043 loose and 2678 non-paraphrases). Most negative instances are represented by the unsuccessful paraphrase candidates, but some, although not being paraphrase candidates according to the unsupervised similarity metric, are still added to the corpus to make it more balanced. Both types of negative instances are considered non-paraphrases when rejected by the annotators (i.e., labeled as non-paraphrases).

The corpus is expected to have high degree of word overlap in the paraphrases. It is not surprising because, firstly, the language of news reports does not vary much. News headlines, from which the corpus is collected, are even more laconic and their style is similar in different media sources. Secondly, pairs of headlines are included into the corpus based on the values of the similarity metric which incorporates both semantic similarity and string-level similarity between the sentences. This metric extends the one proposed in [20] and can also be called a variant of soft cosine similarity measure [46, 47]. For a general-purpose paraphrase corpus, high word overlap could be a serious drawback, but in our case the corpus is created for the use in information extraction and text summarization where the data is often represented by the news texts.

4   Methodology and Tools

We analyze sentence similarity measures by utilizing them as feature sets in the 3-classes paraphrase classification problem. Each similarity measure produces several features (which are described in [39] in detail) and forms a paraphrase prediction model. We experiment with 3 prediction models: shallow model, dictionary-based semantic model and distributional model. Every instance in the dataset is represented by a pair of sentences from the paraphrase corpus, which is to be labeled as "1" for precise para-phrases, "0" for loose paraphrases or "−1" for non-paraphrases. To evaluate the perform-ance of the feature sets, we select a classifier, split our corpus into training (80%) and test (20%) sets, run the classifier and calculate average weighted precision, recall and F1-score. Due to space limits we cannot provide a detailed description of the training process. More detailed information can be found in [39].

At this stage of the research we do not focus on the choice of the best possible classifier – we only intend to use the one which is good enough for our task, and that is why we selected Gradient Tree Boosting (GTB) [21]. Ensemble classifiers (like GTB and Random Forests [9] usually consisting of Decision Trees), are often more powerful than standard classifiers (and they were, in our preliminary experiments) and, unlike standard classifiers, do not impose restrictions like linear separability on the data.

All our experiments, including classifiers training, feature selection and parameters tuning, are conducted using scikit-learn [37] and pandas [34]. To calculate feature values which involve lemmatization or POS-tagging, we use TreeTagger [45]. For distribu-tional semantic features Word2vec skip-gram model [31] is trained on the news corpora from 4 different sources. These corpora consist of about 4.3 million sentences, 65.8 million tokens and contain news reports from 2012 and 2013 (which are not included in our paraphrase corpus). The context window size is 7 words (i.e., (−7, +7)), frequency threshold equals 4, and the dimensionality of word vectors is set to 300.

5   Results. Discussion

To evaluate our prediction models, we split the dataset into training (80%) and test (20%) sets, run the same instance of GTB for every feature set, and calculate average weighted precision, recall and F1-score (we mostly focus on the latter scores), as described in Sect. 4. The results for different feature sets (i.e., similarity measures) are presented in Table 1.

According to the F1 scores shown in Table 1, shallow features perform better than dictionary-based semantic features, which, in their turn, are better than distributional semantic features. The best scores are achieved on the combination of shallow, dictionary-based and distributional semantic and extended cosine features (they are given in bold in Table 1).

As it was shown in [39], the 3 considered prediction models often fail to predict paraphrase class when specific linguistic phenomena take place in the sentence pairs (e.g., presupposition, metaphor, etc.). As our annotators are naïve Russian speakers and are not given any specific instructions concerning the annotation process, they often mark sentences as precise or loose paraphrases based on their general knowledge of the current affairs in the world.

Table 1. Evaluation of Different Feature Sets

Feature set	Precision, %	Recall, %	F1-score, %
shallow	62.42	61.04	60.67
semantic	62.41	59.28	58.78
distrib	61.22	58.25	57.16
distrib + cosine	60.63	57.69	56.54
distrib + cosine_ext	61.02	58.17	57.22
shallow + semantic	63.75	62.15	62.02
shallow + distrib	63.49	61.83	61.61
shallow + distrib + cosine	63.42	61.67	61.42
shallow + distrib + cosine_ext	64.04	62.39	62.19
shallow + semantic + distrib	63.72	62.23	62.04
shallow + semantic + distrib + cosine	64.05	62.87	62.68
shallow + semantic + distrib + cosine_ext	**65.73**	**63.90**	**63.66**
shallow + semantic + cosine	63.82	62.55	62.31
shallow + semantic + cosine + ext	64.42	62.95	62.72

At this stage of the research we look deeper into the annotation of the corpus. We analyze the degree of the agreement of the annotators and try to understand what problems the annotators face while making their decisions.

Firstly, we calculate Cohen's Kappa for each pair of users (among those who annotated at least 20 sentence pairs). Secondly, we look into the pairs of sentences on which our annotators mostly disagree to see which linguistic phenomena these sentence pairs contain. According to the results of Kappa calculation, the level of agreement between different pairs of annotators is quite low: it does not exceed 60%. We believe that it is caused by the nature of the data (news texts).

In Fig. 1 a graph representing the level of agreement between the annotators is shown: the thickness of each edge corresponds to the Kappa value for the pair of users (i.e., nodes of the graph). Size and colour saturation of each node corresponds to the degree of the node. The graph is constructed using yEd graph editor.[1]

---

[1] https://www.yworks.com/products/yed .

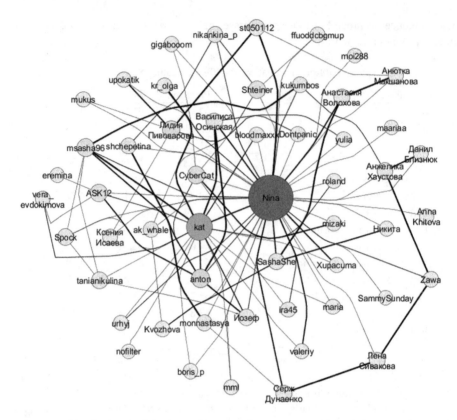

Fig. 1. Level of Agreement between the Annotators: Cohen's Kappa

To extract the sentence pairs on which most annotators disagree, we calculate entropy of the annotators' scores for each sentence pairs and then select those with maximum entropy values. The percentage of various linguistic phenomena occurring in such sentence pairs is presented in Table 2 (for the sample of 100 sentence pairs with maximum entropy values). We believe that these linguistic phenomena are the most

Table 2. Percentage of Different Linguistic Phenomena among the Most Disagreed-On Sentence Pairs (Top-100)

Linguistic Phenomenon	%	Linguistic Phenomenon	%
Different content	68	Synonymy	11
Presupposition	43	Context knowledge	11
Syntactic synonymy	28	Different time	8
Quotation	22	Metonymy	6
Phrasal synonymy	20	Transliteration	3
Reordering	19	Metaphor	2
Numeral	12		

important ones for the configuration of the human text comprehension variability and methods of influencing the addressee (which is one of the goals of the media).

It is shown in Table 2 that different content in the sentences is the most widely spread linguistic phenomenon among those sentence pairs on which our annotators disagree. It occurs in 68% of the sentence pairs in our sample and is probably caused by the annotators' inattentiveness while making their decision. There are different content focuses for different annotators, so it is not only annotators' mistakes. And presupposition, which is a frequent phenomena in the sample, is also not the result of the annotators' mistakes. It is rather a problem of the crowdsourced annotation in general.

Presupposition and referential focus is a very complex phenomenon [4, 7, 44]. In this paper we distinguish two main types of presupposition: semantic (or logical) and pragmatic. The first one is more formal [36][2].

Presupposition has always been a widely discussed topic in both philosophical and linguistic traditions. Thus, the 'Fregean' view of presupposition, and presuppositional failure, according to which there is a logical relation between sentences (or statements) such that if p presupposes q, p has no truth value (or the truth value 'zero') whenever q is not true, has enjoyed? a widespread popularity among linguists [14]. It is one of the simplest variants of definition (ex., a more orthodox two-valued logics). Thus, it was subject to attacks from several scholars, for instance, [3, 27, 53] (see [14, 15]). [50, 51] paved the way from the truth of presuppositions to **projection under negation**, and then linguists themselves found many other projection diagnostics for presuppositions: conditionals, question, modals, etc ([35, 36]). And this definition is the most convenient for our research.

The great part of the semantic presupposition issues can be automatically described with the help of the ontology and thesaurus resources.

The second type of presupposition is a pragmatic one; in ([23, 24, 26]) a pragmatic definition of presupposition was put forward, based on the so called Common Ground condition. This approach is duly criticized in [41].

There are many attempts to define the pragmatic presupposition (or presuppositions). We discuss three subtypes here. One of these definitions focus us on the base of the knowledge (for annotators, for instance); the more complete one focus us on the base of the knowledge and opinion mining; the last one depends on the different communicative situations. The identification of the first subtype may be based on the ontology and thesaurus resources; the next one needs the most complicated and specialized semantic resources (and it is possible only for the partial solution). The third subtype is not so common because it must include very different information (often without formalization) (see Table 3).

---

[2] The notion of presupposition is the most important notion that came into linguistics from logic ([36]).

Table 3. Misclassification Examples

#Sentences	True class	Shal.	Sem.	Distrib + cosine + cosine_ext
1 Осужденному за взрыв на Манежной площади отменили приговор. //The sentence against the convict for the bombing at Moscow's Manezh Square has been cancelled.// Верховный суд отменил приговор за взрыв на Манежной площади. / The Supreme Court has cancelled the sentence for the bombing at Moscow's Manezh Square.	1	0	0	0
2 Порошенко: в Донбассе введут военное положение при атаке на силовиков. / Poroshenko: in Donbass martial law will be imposed to respond to the attack on the security forces.// Порошенко пообещал ответить на наступление ополченцев военным положением. / Poroshenko promised to respond to the attack from the militia with martial law.	1	−1	0	−1
3 В Альпах разбился снегоход с российскими туристами. / In the Alps, a snowmobile with Russian tourists crashed.//Группа туристов разбилась на снегоходе в Альпах. /A group of tourists crashed on a snowmobile in the Alps.	1	0	0	−1
4 В Москве задержан очередной фигурант "болотного дела"./ In Moscow, another person involved in "swamp case" is arrested.//«Росузник» сообщил о задержании нового фигуранта «болотного дела». /"Rosuznik" informed about the detention of a new person involved in "swamp case".	0	0	−1	0
5 В аэропорту Казани пассажирский Airbus задел хвостом маяк./ At the airport of Kazan a passenger Airbus hurt a localizer with its tail.// Пассажирский самолет в аэропорту Казани задел хвостом курсовой маяк. / Passenger aircraft at the airport of Kazan hurt a localizer with its tail.	0	1	0	0
6 В Приднестровье создали Стабфонд для нужд президента и КГБ. / In Transnistria, the Stabilization Fund has been created for the needs of the President and the KGB.// Приднестровье создало Стабфонд за счет российского газа. / Transnistria has created the Stabilization Fund at the cost of the Russian gas.	0	−1	−1	−1
7 Боевики ИГ взяли на себя ответственность за теракты в Афганистане. / IG militants claimed responsibility for the attacks in Afghanistan.// "ИГ" взяло на себя ответственность за двойной теракт в Афганистане. /"IG" claimed responsibility for the double bombing in Afghanistan.	0	1	1	1
8 В Приднестровье создали Стабфонд для нужд президента и КГБ. / In Transnistria, the Stabilization Fund has been created for the needs of the President and the KGB. // Приднестровье создало Стабфонд за счет российского газа. / Transnistria has created the Stabilization Fund at the cost of the Russian gas.	0	−1	−1	−1
9 В Турции за коррупцию арестованы сыновья трех министров. / In Turkey three sons of ministers have been arrested for corruption.// Турецкая полиция задержала сыновей трех министров. / Turkish police have detained three sons of ministers.	0	−1	0	−1

Class 1 (Precise Paraphrases). Sentence pairs 1–3 show that all the three feature types fail to detect or to understand pragmatic presupposition (see "convict" and "Supreme Court" in #1: only a convicted person can be a subject to the cancelled sentence, and the court is supposed to cancel the sentence; in #2 "Donbass" as the reference to the place of action in the first sentence is also obvious, especially for the Russian speaker, if the action concerns the martial law imposed to respond to the attack by the militia) as well as syntactic synonymy combined with word and phrase-level synonymy and reordering. We believe that in #3 precise paraphrase class is disputable. For a naïve Russian speaker "tourists" might be identical to "Russian tourists" in a news report due to the presupposition phenomenon, especially if it is a Russian news report; sure it is not a general truth, but the prediction based on general expectations of our group annotators.

Class 0 (Loose paraphrases (conveying similar meaning)). Sentence pairs 4–7 show that all the three feature types fail to detect or to understand presupposition or to understand context or to understand communication phrase structure.

In #6 there is a "difficult" sentence pair: understood metaphorically, the sentences might be considered somewhat similar, however, such understanding requires large amounts of general knowledge and it is extremely hard to teach a machine to distinguish such subtle meanings. The main part of the difficulty concerned on the different variants of the communication phrase structure choice: ex. <u>In Transnistria, the Stabilization Fund has been created</u>/ *for the needs of the President and the KGB* ‖ <u>Transnistria has created the Stabilization Fund</u>/ *at the cost of the Russian gas*. The main part of the information structure from point of view annotators - marked by bold, the unimportant part – marked by italic. This is reason of the difference of the definition of the "true class" (annotators result) and the tree prediction classes. It was similar with #8: "<u>Transnistria has created the Stabilization Fund</u>" and "<u>In Transnistria, the Stabilization Fund has been created</u>" are the most important parts in the phrase structure. The similar situation is about #9, but structure of the phrase is more simple. What is more important: *Turkish police have detained three sons of ministers* or the reason (that *they have been arrested* <u>for corruption</u>)?

Sentence pairs 4 demonstrates the low interest (and knowledge) of our annotators in the problems of the Snow Revolution and Bolotnaya's situation.

Actually #4 is as disputable as #3 as the decision was evidently made by the annotators based on their pragmatic presupposition (i.e., prior knowledge about the world). In #7 the sentences are even more overlapping than in #5, and in this case all the 3 types of features are mistaken, failing to detect the minor difference between the sentences. In #9 the sentences are smaller and less overlapping. They also contain synonyms and words from the same families. Perhaps, that is why the class 0 is only correctly predicted by dictionary-based model while others misclassified the pair in #9 as non-paraphrase. Pair #4 is similar to #9, but it also contains different named entities which negatively affects the results of the dictionary-based semantic features (other types of features are correct here). These named entities express the place of action and the source of information (which is not always important for the annotators or readers of a news report) respectively, while the described action is the same in two sentences. The most interesting cases are sentence pairs #4 and #7.

To eliminate the problem of inattentive annotation and partly of presupposition we are going to select the users which marked sentence pairs with different presuppositions or different content as precise paraphrases and discard the scores of such "unreliable" users while calculating paraphrase class of the sentence pairs (at the moment paraphrase class of a sentence pair is calculated as the median of the scores of all users who marked the sentence pair).

6 Conclusion and Future Work

We have presented the results of our work on the creation of a Russian paraphrase corpus and the experiments with different prediction models for paraphrase classification. The experiments were conducted as part of the crowdsourcing project ParaPhraser.ru. In this project we automatically collect Russian sentential paraphrases from the news headlines and work on the development of a paraphrase identification model. The corpus is freely available and quite representational: it can already be used in various studies related to paraphrases and semantics in general (and we use it ourselves to develop a paraphrase identification model). The work on the project is going on, and the corpus is constantly increasing in size.

We have compared three prediction models based on different classes of similarity measures: shallow measures (based on string/word/phrases overlap), dictionary-based semantic measures (employing external semantic resources) and distributional semantic measures (based on vector space model). The measures were used as features in paraphrase classification task. As part of our previous work, we revealed factors which complicate not only the paraphrase classification task for our models but also the annotation task for the users. Such factors are complex linguistic phenomena like presupposition and metaphor. In this paper we focused on the level of agreement between the users and analyzed the types of sentences on which the annotators mostly disagree. According to the results of the experiments, disagreement between the users is mostly caused by both the quality of their annotation and linguistic phenomena like presupposition. In the nearest future we plan to improve the annotation by discarding the scores of the "unreliable" users.

Acknowledgments. We would like to thank Lilia Volkova for her invaluable help. The authors also acknowledge Saint-Petersburg State University for the research grant 30.38.305.2014.

References

1. Abusch, D.: Presupposition triggering from alternatives. J. Semant. **27**(1), 37–80 (2010)
2. Achananuparp, P., Hu, X., Shen, X.: The evaluation of sentence similarity measures. In: Song, I.-Y., Eder, J., Nguyen, T.M. (eds.) DaWaK 2008. LNCS, vol. 5182, pp. 305–316. Springer, Heidelberg (2008). https://doi.org/10.1007/978-3-540-85836-2_29
3. Allwood, J.: Negation and the strength of presuppositions or there is more to speaking than words. In: Logical Grammar Reports 2 (1972). Revised version in Dahl, Ö. (eds.) Logic, Pragmatics, and Grammar, 1975, Department of Linguistics, University of Göteborg

4. Apresjan, V.: Semanticheskaya struktura slova I ego vzaimodeystviye s otritsaniem (In Russian). In: Computational Linguistics and Intellectual Technologies, Papers from the Annual International Conference "Dialogue", Bekasovo, 26–30 May 2010, Issue 9(16), pp. 13–19. RGGU, Moscow (2010)
5. Bannard, C., Callison-Burch, C.: Paraphrasing with bilingual parallel corpora. In: Proceedings of the 43rd Annual Meeting of the ACL, pp. 597–604. ACL (2005)
6. Baroni, M., Dinu, G., Kruszewski, G.: Don't count, Predict! a systematic comparison of context-counting vs. context-predicting semantic vectors. In: Proceedings of the 52nd Annual Meeting of the Association for Computational Linguistics, pp. 238–247. ACL (2014)
7. Boguslavskij, I.: Sfera deystviya leksicheskih edinits. Yazyki Russkoy Kultury, Moscow (1996). (In Russian)
8. Braslavski, P., Ustalov, D., Mukhin, M.: A spinning wheel for YARN: user interface for a crowdsourced thesaurus. In: Proceedings of the Demonstrations at the 14th Conference of the European Chapter of the Association for Computational Linguistics, pp. 101–104. Gothenburg, Sweden (2014)
9. Breiman, L.: Random forests. Mach. Learn. **45**(1), 5–32 (2001)
10. Brockett, C., Dolan, B.: Support vector machines for paraphrase identification and corpus construction. In: Proceedings of the 3rd International Workshop on Paraphrasing, pp. 1–8 (2005)
11. Burrows, S., Potthast, M., Stein, B.: Paraphrase acquisition via crowdsourcing and machine learning. ACM Trans. Intell. Syst. Technol. **4**(3), 43 (2013)
12. Callison-Burch, C.: Paraphrasing and Translation. Institute for Communicating and Collaborative Systems, School of Informatics, University of Edinburgh (2007)
13. Chitra, A., Kumar, S.: Paraphrase identification using machine learning techniques. In: Proceedings of the 12th International Conference on Networking, VLSI and Signal Processing, World Scientific and Engineering Academy and Society (WSEAS), Stevens Point, pp. 245–249 (2010)
14. Dahl, Ö.: In defense of a strawsonian approach to presupposition. In: Klein, W., Levelt, W. (eds.) Synthese Language Library. Crossing the Boundaries in Linguistics. Studies Presented to Manfred Bierwisch, vol. 13, pp. 191–200. D. Reidel Pub. Co. (1981)
15. Dahl, Ö.: Topic-comment structure revisited. In: Dahl, Ö. (ed.) Topic and Comment, Contextual Boundness and Focus, pp. 1–24. Helmut Buske, Hamburg (1974)
16. Das, D., Smith, N.A.: Paraphrase identification as probabilistic quasi-synchronous recognition. In: Proceedings of the Joint Conference of the 47th Annual Meeting of the Association for Computational Linguistics and of the 4th International Joint Conference on Natural Language Processing of the Asian Federation of Natural Language Processing, pp. 468–476. ACL (2009)
17. Dice, L.R.: Measures of the amount of ecologic association between species. Ecology **26**(3), 297–302 (1945)
18. Dolan, W.B., Quirk, C., Brockett, C.: Unsupervised construction of large paraphrase corpora: exploiting massively parallel news sources. In: Proceedings of the 20th International Conference on Computational Linguistics, Geneva, Switzerland (2004)
19. Eyecioglu, A., Keller, B.: ASOBEK: Twitter paraphrase identification with simple overlap features and SVMs. In: Proceedings of the 9th International Workshop on Semantic Evaluation (SemEval 2015), pp. 64–69 (2015)
20. Fernando, S., Stevenson, M.: A semantic similarity approach to paraphrase detection. In: Computational Linguistics UK (CLUK 2008) 11th Annual Research Colloqium, pp. 45–52 (2008)

21. Friedman, J.: Greedy function approximation: a gradient boosting machine. Ann. Stat. **29**(5), 1189–1232 (2001)

22. Ganitkevitch, J., Callison-Burch, C.: The multilingual paraphrase database. In: Proceedings of the Ninth International Conference on Language Resources and Evaluation (LREC 2014), European Language Resources Association (ELRA), pp. 4276–4283. Reykjavik, Iceland (2014)

23. Heim, I.: Presupposition projection and the semantics of attitude verbs. J. Semant. **9**(3), 183–221 (1992)

24. Heim, I.: Presupposition projection. In: van der Sandt, R. (ed.) Reader for the Nijmegen Workshop on Discourse Processes, University of Nijmegen (1990). http://semanticsarchive.net/Archive/GFiMGNjN/Presupp%20projection%2090.pdf

25. Ji, Y., Eisenstein, J.: Discriminative improvements to distributional sentence similarity. In: Proceedings of the Conference on Empirical Methods in Natural Language Processing (EMNLP), pp. 891–896. Seattle (2013)

26. Karttunen, L.: Presuppositions and linguistic context. In: Theoretical Linguistics, vol. 1, pp. 181–194. Mouton De Gruyter, Berlin (1974)

27. Kempson, R.: Presupposition and the Delimitation of Semantics. University Press, Cambridge (1975)

28. Knight, K., Marcu, D.: Summarization beyond sentence extraction: a probabilistic approach to sentence compression. Artif. Intell. **139**(1), 91–107 (2002)

29. Kozareva, Z., Montoyo, A.: Paraphrase identification on the basis of supervised machine learning techniques. In: Salakoski, T., Ginter, F., Pyysalo, S., Pahikkala, T. (eds.) FinTAL 2006. LNCS (LNAI), vol. 4139, pp. 524–533. Springer, Heidelberg (2006). https://doi.org/10.1007/11816508_52

30. McClendon, J.L., Mack, N.A., Hodges, L.F.: The use of paraphrase identification in the retrieval of appropriate responses for script based conversational agents. In: Proceedings of the Twenty-Seventh International Florida Artificial Intelligence Research Society Conference, pp. 196–201. The AAAI Press, Menlo Park (2014)

31. Mikolov, T., Chen, K., Corrado, G., Dean, J.: Efficient Estimation of Word Representations in Vector Space (2013). http://arxiv.org/abs/1301.3781/

32. Mikolov, T., Sutskever, I., Chen, K., Corrado, G.S., Dean, J.: Distributed representations of words and phrases and their compositionality. In: Advances in Neural Information Processing Systems, pp. 3111–3119. Curran Associates Inc. (2013)

33. Miller, G., Fellbaum, C.: Wordnet: An Electronic Lexical Database. MIT Press, Cambridge (1998)

34. Python Data Analysis Library (pandas). http://pandas.pydata.org

35. Paducheva, E.V.: Egotsentricheskie valentnosti I dekonstruktsiya govoryashego. In: Voprosy yazykoznaniya 2011, no. 3, pp. 3–18 (2011). http://lexicograph.ruslang.ru/TextPdf1/egocentricals.pdf. (In Russian)

36. Paducheva, E.: Presuppositions and Semantic Typology of Projective Meanings. http://lexicograph.ruslang.ru/TextPdf1/PROJECTION-POSTER.pdf

37. Pedregosa, F., Varoquaux, G., Gramfort, A., Michel, V., Thirion, B., Grisel, O., Blondel, M., Prettenhofer, P., Weiss, R., Dubourg, V., Vanderplas, J., Passos, A., Cournapeau, D., Brucher, M., Perrot, M., Duchesnay, E.: Scikit-learn: machine learning in Python. J. Mach. Learn. Res. **12**, 2825–2830 (2011). http://scikit-learn.org

38. Pronoza, E., Yagunova, E., Pronoza, A.: Construction of a Russian paraphrase corpus: unsupervised paraphrase extraction. In: Proceedings of the 9th Summer School in Information Retrieval and Young Scientist Conference (2015, in press)

39. Pronoza, E., Yagunova, E.: Comparison of sentence similarity measures for russian paraphrase identification. In: Artificial Intelligence and Natural Language and Information Extraction, Social Media and Web Search FRUCT Conference (AINL-ISMW FRUCT), pp. 74–82. IEEE, Piscataway (2015)

40. Rajkumar, A., Chitra, A.: Paraphrase recognition using neural network classification. Int. J. Comput. Appl. **1**(29), 42–47 (2010)

41. Roberts, C., Simons, M., Beaver, D., Tonhauser, J.: Presupposition, conventional implicature, and beyond: a unified account of projection. In: Proceedings of the Workshop on Presupposition, ESSLLI 2009, Bordeaux, Universite de Bordeaux, Bordeaux (2009)

42. Rus, V., McCarthy, P.M., Lintean, M.C.: Paraphrase identification with lexico-syntactic graph subsumption. In: Proceedings of the Twenty-First International FLAIRS Conference, pp. 201–206. The AAAI Press, Menlo Park (2008)

43. Rus, V., Banjade, R., Lintean, M.: On paraphrase identification corpora. In: Proceedings of the Ninth International Conference on Language Resources and Evaluation (LREC-2014), pp. 2422–2429. European Language Resources Association (ELRA), Reykjavik (2014)

44. Russell, B.: Mr. Strawson on referring. Mind **66**, 385–389 (1957)

45. Schmid, H.: Improvements in part-of-speech tagging with an application to german. In: Proceedings of the ACL SIGDAT-Workshop, vol. 21, pp. 1–9. ACL (1995)

46. Sidorov, G.: Non-linear Construction of N-grams in Computational Linguistics: Syntactic, Filtered, and Generalized N-grams. Sociedad Mexicana de Inteligencia Artificial, Mexico (2013). (In Spanish)

47. Sidorov, G., Gelbukh, A., Gómez-Adorno, H., Pinto, D.: Soft similarity and soft cosine measure: similarity of features in vector space model. Computación y Sistemas **18**(3), 491–504 (2014)

48. Sidorov, G., Gómez-Adorno, H., Markov, I., Pinto, D., Loya, N.: Computing Text Similarity using Tree Edit Distance, NAFIPS 2015 (2015, accepted)

49. Socher, R., Huang, E.H., Pennington, J., Ng, A.Y., Manning, C.D.: Dynamic pooling and unfolding recursive autoencoders for paraphrase detection. In: Proceedings of the Conference on Neural Information Processing Systems, vol. 24, pp. 801–809. MIT Press, Cambridge (2011)

50. Stalnaker, R.: Presuppositions. J. Philos. Logic **2**, 447–457 (1973)

51. Stalnaker, R.: Common ground. Linguist. Philos. **25**(5–6), 701–721 (2002)

52. Wan, S., Dras, M., Dale, R. Paris, C.: Using dependency-based features to take the "Parafarce" out of paraphrase. In: Proceedings of the Australasian Language Technology Workshop, pp. 131–138 (2006)

53. Wilson, D.: Presupposition and Non-Truth-Conditional Semantics. Academic Press, London (1975)

54. Yin, W., Schutze, H.: Discriminative phrase embedding for paraphrase identification. In: Proceedings of Human Language Technologies: The 2015 Annual Conference of the North American Chapter of the ACL, Denver, Colorado, 31 May–5 June, pp. 1368–1373. ACL (2015)

55. Zhang, Y., Patrick, J.: Paraphrase identification by text canonicalization. In: Proceedings of the Australasian Language Technology Workshop, pp. 160–166 (2005)

Constructing a Turkish Corpus for Paraphrase Identification and Semantic Similarity

Asli Eyecioglu[1(✉)] and Bill Keller[2(✉)]

[1] Department of Computer Enginnering, Bartin University,
74100 Bartin, Turkey
aozmutlu@bartin.edu.tr
[2] Department of Informatics, University of Sussex, Brighton BN19QJ, UK
billk@sussex.ac.uk

Abstract. The Paraphrase identification (PI) task has practical importance for work in Natural Language Processing (NLP) because of the problem of linguistic variation. Accurate methods should help improve performance of key NLP applications. Paraphrase corpora are important resources in developing and evaluating PI methods. This paper describes the construction of a paraphrase corpus for Turkish. The corpus comprises pairs of sentences with semantic similarity scores based on human judgments, permitting experimentation with both PI and semantic similarity. We believe this is the first such corpus for Turkish. The data collection and scoring methodology is described and initial PI experiments with the corpus are reported. Our approach to PI is novel in using 'knowledge lean' methods (i.e. no use of manually constructed knowledge bases or processing tools that rely on these). We have previously achieved excellent results using such techniques on the Microsoft Research Paraphrase Corpus, and close to state-of-the-art performance on the Twitter Paraphrase Corpus.

Keywords: Paraphrase identification · Turkish · Corpora construction
Knowledge-lean · Paraphrasing · Sentential semantic similarity

1 Introduction

Paraphrase identification (PI) may be defined as "the task of deciding whether two given text fragments have the same meaning" [22]. The PI task has practical importance for Natural Language Processing (NLP) because of the need to deal with the pervasive problem of linguistic variation. Accurate methods should help improve the performance of NLP applications, including machine translation, information retrieval and question answering, amongst others. Acquired paraphrases have been shown to improve Statistical Machine Translation (SMT) systems [5, 24, 28], for example.

To support the development and evaluation of PI methods, the creation of paraphrase corpora plays an important part. This paper describes the construction and characteristics of a new paraphrase corpus, for Turkish. The Turkish Paraphrase Corpus[1] (TuPC) is comprised of pairs of sentences, together with associated semantic

---

[1] TuPC can be downloaded from: http://aslieyecioglu.com/data/.

© Springer International Publishing AG, part of Springer Nature 2018
A. Gelbukh (Ed.): CICLing 2016, LNCS 9623, pp. 588–599, 2018.
https://doi.org/10.1007/978-3-319-75477-2_42

similarity scores based on human judgements. The corpus thus supports study of PI methods that make simple, binary judgements (i.e. paraphrase v. non-paraphrase) or that are capable of assigning finer-grained judgements of semantic similarity.

To our knowledge, this is the first such corpus made available for Turkish. The corpus will be of value as a novel resource for other researchers working on PI and semantic similarity in Turkish. It should also be of wider interest, to researchers investigating paraphrase and cross-linguistic techniques. We report the result of initial PI experiments which may serve as a baseline for other researchers. Our approach is novel in using 'knowledge lean' techniques. By 'knowledge lean' we mean that we do not make use of manually constructed knowledge bases or processing tools that rely on these. As far as possible, we work with just tokenised text and statistical tools.

The work described here continues the authors' earlier investigations into the extent to which PI techniques requiring few external resources may achieve results comparable to methods employing more sophisticated NLP processing tools and knowledge bases. We have already demonstrated that excellent results can be obtained using combinations of features based on lexical and character n-gram overlap, together with Support Vector Machine (SVM) classifiers. The approach has been shown to perform well on the Microsoft Research Paraphrase Corpus (MSRPC) and recently attained near state-of-the-art performance on the Twitter Paraphrase Corpus (TPC).

In the rest of the paper we briefly introduce current work on paraphrase corpora and then describe the methods that were used to construct the TuPC. Our methodology should be of interest to researchers considering building paraphrase corpora for other languages. This may be particularly the case for languages with few existing language processing resources. We then describe our approach to PI using knowledge-lean techniques and report on initial experiments with the new corpus.

2 Paraphrase Corpora

The creation of paraphrase corpora in recent years has been important in promoting research into methods for PI. Paraphrase corpora provide a basis for training and evaluation of models of paraphrase. The MSRPC [11, 30] was constructed by initially extracting candidate sentence pairs from a collection of topically clustered news data. A classifier was used to identify a total of 5801 paraphrase pairs and "near-miss" non-paraphrases. These were inspected by human annotators and labeled as either paraphrase (3900) or non-paraphrase (1901). The Plagiarism Detection Corpus (PAN) was constructed by aligning sentences from 41,233 plagiarised documents. It is made available by Madnani et al. [23] for the use of PI tasks, publishing the initial results of their experiment. The Semeval-2015 Task1, "Paraphrase and Semantic Similarity in Twitter" [35] involves predicting whether two tweets have the same meaning. Training and test data are provided in the form of the TPC [34], which is constructed semi-randomly and annotated via Amazon Mechanical Turk by 5 annotators.

Paraphrase corpora have been constructed for languages other than English. The TICC Headline Paraphrase Corpus [33] is a collection of English and Dutch, while the Hebrew Paraphrase Corpus [32] implements a scoring scheme with sentence pairs labelled as paraphrase, non-paraphrase and partial paraphrase. WiCoPaCo, [26] is a corpus for French, which includes text normalization and corrections. However, this data is not constructed solely for paraphrasing tasks, and no scoring scheme is supplied. Previously, a Turkish Paraphrase Corpus [10] has been reported. However, the corpus is not widely available and does not provide any negative instances or scoring scheme. Consequently, it is less useful for research purposes.

The Microsoft Research Video Description Corpus [7] consists of multilingual sentences generated from short videos that are described with one sentence by Mechanical Turk workers. A multi-lingual paraphrase databases PPDB [18] has been constructed using pivot language techniques [2] and recently expanded to more than 20 languages. It does not currently provide sentential paraphrases, but a substantial quantity of phrasal paraphrases is available, in many languages.

3 A Paraphrase Corpus for Turkish

Turkish belongs to the Altaic languages family of western and central Asia that are named under the Turkic languages. Turkish is a highly inflected and agglutinative language. The productive use of affixes is very typical, either to change the meaning or the stress of a word. Turkish and English use the same Latin alphabet, except for a few letters. Words are space-separated and follow subject-object-verb order.

Our strategy for creating TuPC combines the methodologies from previously constructed corpora MSRPC [30] and TPC [34]. We automatically extracted and paired sentences from daily news websites. Candidate pairs were then hand-annotated by inspection of their context. In contrast to MSRPC, but like TPC, candidate pairs were scored according to semantic similarity.

3.1 Data Collection

We implemented a simple web crawler to extract plain text and cluster it according to a pre-selected list of sub-topics. A list of URLs was gathered from a website that links to most daily Turkish newspapers. A further list of URLs for each newspaper was then extracted. Widely used heading tags were identified by looking at categories on each website. For instance, the latest news can be found under the heading "last minute" on one site, and "latest" on another. A subset of popular headings was established to extract topic related news. Some examples are:

[haber, gundem, guncel, sondakika, haberler, turkiye, haberhatti,...,]

This step serves as an initial filter that limits the topics to "news", "latest", "sports" and so on, rather than including articles related to "travel", "fashion" and other miscellaneous texts. For each category, we gathered all news from the different news sites. We collected in two phases, between 4th and 14th May 2015 and then between 17th and 23rd June 2015. This was because the news sites tend not to update all of their content each day, and that can lead to duplicate data. Approximately 10k lines of text on a daily basis were clustered according to topic.

The resulting data are texts with html mark-up. These were made in-line by removing the html formatting. Duplicate lines were then removed and a sentence segmentation tool [3] was trained on a small collection of Turkish text to be used on our dataset. The tool was very successful in splitting paragraphs into sentences, but manual correction was also applied in some cases. We believe that the sentence seg-menter tool can be trained on a larger collection of Turkish text for better results. The resulting text is not adjusted for case and numbers and named entities are not replaced with generic names (unlike in MSRPC).

3.2 Generating Candidate Pairs

To generate candidate pairs we first removed any sentence considered too short (less than 5 words) or overly long (greater than 40 words). This criterion is adapted from the methodology adopted for the construction of the MSRPC. Next, all pairs of sentences were considered as candidate paraphrase pairs and initially filtered according to the following criteria, which are based on previous methods used for paraphrase corpora construction. A candidate pair is removed if:

1. Lexical overlap is less than five words between a pair; or
2. Absolute difference of length is more than seven between a pair

Our text has relatively high lexical overlap due to the presence of named entities, relatively long sentences and the prevalence of stop words. Consequently, the lexical overlap criterion adopted for the MSRPC (fewer than four words) is too stringent in our case. Our second criterion is similar to MSRPC's word-based length difference of sentences, which is defined as 66.6%. For the TPC, tweets of length less than three are filtered out and the remaining candidate pairs are pruned with Jaccard similarity score, if it is less than 0.5. The filtering criteria that we applied are generally comparable to MSRPC due to the similarity of source data. Finally, to make the selection process of candidate paraphrases easier, we further filtered out any pairs with at least three overlapping words, after removing stop words.

The filtering process is enough to exclude pairs that are unfavourable to be selected as candidates. In addition, once two sentences are paired, both are excluded from further pairs. In this way, each sentence is used only once. Despite this, we obtained a great number of candidate pairs. Approximately 5K were obtained on a daily basis and so we handpicked a final set of 1002 candidate pairs.

3.3 Similarity Scoring

Our annotation method follows that used in creating the TPC. Rather than simply label sentence pairs as paraphrase or non-paraphrase, a finer-grained, semantic similarity score is assigned. This annotation scheme is richer and more general, since semantic similarity scores can be converted to paraphrase judgments (but not vice-versa). We followed the guidelines provided for the semantic similarity task [1] and annotators scored sentence pairs directly, according to the criteria shown in Table 1. A similar scoring scheme has been adopted for the TPC. However, in that case, scores were generated based on counts of binary judgments made by a number of annotators.

Eight native speakers of Turkish were recruited as annotators. The data were split into two halves and different groups of four annotators judged each half. We prepared a

Table 1. Sentential Semantic Similarity scores for candidate paraphrase pairs

5 - Identical	Completely equivalent; mean the same
4 - Close	Mostly equivalent; some small details differ
3 - Related	Roughly equivalent; some important information differs/missing
2 - Context	Not equivalent, but share some details
1 - Somewhat related	Not equivalent, but are on the same topic
0 - Unrelated	On different topics

set of annotation guidelines to explain the scoring process. The guidelines introduced the similarity scores, along with an example pair for each score and a short explanation. The examples were also chosen from multiple samples by asking three different native speakers who gave exactly the same score for those pairs.

In order to further clarify the task, a small preliminary experiment was completed prior to annotation. Two short videos were prepared. These were adapted from Microsoft Research (MSR) Video Description Corpus [7], where the objective was to generate a paraphrase evaluation corpus by asking multilingual Mechanical Turk reviewers to describe the videos. Providing our annotators with the description of other reviewers for the same videos is a heuristic approach designed to familiarize them with the task. They summarized the videos with one sentence and, afterwards, we gathered all the annotators' descriptions. Then we showed them how the interpretation or wording for the same video can be different.

The annotators noted that this small experiment gave a better understanding of the scoring task. Video description sentences were collected and also included in the set of guidelines. After completing this preliminary experiment, sentence pairs were sent to each annotator. We did not impose a time restriction for the full task because annotators noted that some pairs were confusing and required more time.

For the purpose of experimenting with simple PI, the assigned scores were also converted to binary labels. First, the scores of each annotator were converted to binary labels by taking sentence pairs marked as identical (5), close (4) and related (3) as positive (i.e. paraphrase) and those marked context (2), somewhat related (1) and unrelated (0) are as negative (i.e. non-paraphrase). The number of positive and negative decisions for each instance may be summarised as a pair. For example, (1, 3) shows that only one annotator tagged the pair as a paraphrase, while the remaining three labeled it a non-paraphrase. In Table 2, we show the criteria of the binary judgment that is based on the number of annotators' answers. Note that where there are equal numbers of positive and negative decisions, we consider a pair 'debatable'. This approach is similar to TPC labeling method, which is also based on the agreement between annotators.

For semantic similarity, we also provide mean scores. These range between 1.75 and 3.00 for debatable pairs, whereas positive pairs are higher than 3.00 and negative pairs are scored less than 1.75. Additionally, the criteria defined in Table 2 can be interpreted in a range between 0 and 1 as follows: (4, 0): 1.0; (3, 1): 0.75; (2, 2): 0.50; (1, 3): 0.25 and (0, 4): 0.0.

Table 3 presents three sample pairs from the data. Each pair is shown with the scores of 4 different annotators and the average similarity scores. The debatable pair has been scored (4, 2, 3, 0) by four annotators, and the average similarity is 2.25.

Table 2. The criteria of binary judgement based on the number of annotator's answers

Number of answers	Judgement
(4, 0); (3, 1)	Positive (1)
(0, 4); (1, 3)	Negative (0)
(2, 2)	Debatable

Table 3. Sample pairs from TuPC

Value	Scores	Pair of Sentences
Debatable	(4, 2, 3, 0) Average (2.25)	İşadamı Ethem Sancak, Aydın Doğan ve Ertuğrul Özkök ile ilgili "Bazı şeyleri açıklarsam Türkiye'de barınamazlar" dedi 24 TV'de konuşan İşadamı Ethem Sancak "Doğan Grubu, Aydın Doğan ve Ertuğrul Özkök'le ilgili çarpıcı açıklamalar yaptı
Positive	(3, 4, 5, 5) Average (4.25)	Çekilişte 10 rakamı isabet eden 15 kupondan 13'ünün Muğla'nın Yatağan ilçesindeki bayilere yatırıldığı ortaya çıktı 10 rakamı isabet eden 15 kupondan 13'ü Muğla'nın Yatağan ilçesindeki bayi ya da bayilerden yatırıldı
Negative	(1, 3, 0, 0) Average (1.00)	Toplam konut satışları içerisinde ipotekli satış payının en yüksek olduğu il ise yüzde 52,9 ile Kars oldu Toplam konut satışları içinde ilk satışın payı yüzde 45,4 oldu

3.4   Inter-annotator Agreement

Assigning similarity scores was a challenging task for the annotators. They noted that the difficulty lay not in deciding whether or not two sentences were semantically similar, but in determining the precise degree of similarity. To understand how consistent the annotators were with one another we first applied Cohen's Kappa [8] as a measure of inter-annotator agreement between pairs of annotators. In general, a Kappa coefficient value of 1 indicates strong agreement between annotators, whereas a result of 0 shows an agreement by chance. A negative value (<0) indicates no agreement. Interpretation of intermediate values is subject to debate [19] and we report Landis and Koch's [21] agreement interpretation scores. Kappa scores in our case ranged from 0.23 ("fair agreement") to 0.56 ("moderate agreement").

Cohen's Kappa can only be used for pairs of annotators and so cannot reflect the inter-rater reliability of the full dataset. To compute agreement across all annotators, we used Fleiss's Kappa [17]. We aggregated the data by concatenating the two halves. Although different annotators scored each part of the data, Fleiss's Kappa relies only on the number of scores given to each instance. We calculated the statistic on the full dataset for both semantic similarity and binary paraphrase judgments. The results are reported in Table 4.

The results show "moderate agreement" at the level of binary paraphrase judgements. We note that this is entirely consistent with the inter-rater reliability amongst annotators of about 0.40 reported for the TPC [34]. The results show "slight agreement" with regard to semantic equivalency, on the other hand. The lower score is to be

Table 4. Fleiss Kappa score is computed based on the two different judgment criteria

Scoring	Fleiss Kappa (%)
Degree of semantic equivalency (0–5)	0.17
Binary judgment (1, 0, Debatable)	0.42

expected, as the binary judgment is a more coarse-grained measure of semantic equivalence. Lower agreement for the finer-grained scoring confirms the annotators' sense that it is harder to be precise about degree of similarity. Despite this, inter-rater agreement for binary judgment demonstrates that there is a broad consensus between annotators' scores.

3.5   Corpus Statistics

The TuPC comprises 1002 sentence pairs labelled for both sentential semantic similarity and paraphrase. After converting scores to binary labels, we obtained 563 positive, 285 negative and 154 debatable pairs. Excluding the 154 debatable pairs, TuPC has 848 sentence pairs that can be used for the PI task. A breakdown of the TuPC according to agreement between 4 annotators is shown in Table 5.

Table 5. TuPC data statistics

	Agreement	Number of pairs	Value
Positive	**(4, 0)**	376	563
	(3, 1)	187	
Debatable	**(2, 2)**	154	154
Negative	**(1, 3)**	151	285
	(0, 4)	134	

There are various ways to split a dataset into train and test sets. TPC has a relatively small test set as compared its training set (838 and 11530 sentence pairs respectively, after removing debatable pairs). The percentage of the train and test sets for MSRPC is 70.3% and 29.7%, respectively, while for PAN it is approximately 77% and 23%. For TuPC we selected 60% (500 pairs) for training and 40% (348 pairs) for testing. TuPC contains 339 positive and 161 negative sentence pairs in the train set; and 224 positive and 124 negative pairs in the test set. Note that TuPC was shuffled randomly before splitting the data into train and test set. A naive baseline obtained by labelling every sentence pair as positive (i.e. paraphrase) yields an accuracy of 0.6639 and an F-Score of 0.798.

4   Knowledge-Lean Paraphrase Identification

Much PI research makes use of existing NLP tools and other resources. Duclaye et al. [12] exploits the NLP tools of a question-answering system. Finch et al. [16], Mihalcea et al. [27], Fernando and Stevenson [15], Malakasiotis [25], and Das and Smith [9] employ lexical semantic similarity information based on resources such as WordNet [14]. A number of researchers have investigated whether state-of-the-art results can be obtained without use of such tools and resources. Socher et al. [31] trains recursive neural network models that learn vector space representations of single words and multi-word phrases. Blacoe and Lapata [4] use distributional methods to find compositional meaning of phrases and sentences. Lintean and Rus [22] consider overlap of word unigrams and bigrams. Finch et al. [16] combines several MT metrics and uses them as features. Madnani et al. [23] combines different machine translation quality metrics and outperforms most existing methods.

The work reported in this paper is part of on-going research that aims to investigate the extent to which knowledge-lean techniques may help to identify paraphrases. By knowledge-lean we mean that little or no use is made of manually constructed knowledge bases or processing tools that rely on these. As far as possible, we work with just text. We previously presented a knowledge-lean approach to identifying twitter paraphrase pairs using character and word n-gram features by employing SVM classifiers. Our system was ranked first out of 18 submitted systems as part of SemEval-2015 Task 1: *Paraphrase and semantic similarity in Twitter* [35]. We demonstrated that better results can be obtained using fewer but more informative features. Our solution already outperforms the most sophisticated methods applied on the TPC [13], and competitive results are obtained on the MSRPC and PAN. In the following we report initial experiments applying these techniques to the Turkish Paraphrase Corpus.

4.1 Overlap Features

Text pre-processing is essential to many NLP applications. It may involve tokenising, removal of punctuation, part-of-speech tagging, morphological analysis, lemmatisation, and so on. For identifying paraphrases, this may not always be appropriate. Removing punctuation and stop words, or performing lemmatisation results in a loss of information that may be critical in terms of PI. We therefore keep text pre-processing to a minimum. The TuPC includes punctuation and spelling errors but these have not been hand-corrected. We tokenise by splitting at white space. Lowercasing was also performed, as for other corpora with which we have experimented.

We consider different representations of a text as a set of tokens, where a token may be either a word or a character n-gram. For the work described here we restrict attention to word and character unigrams and bigrams. Use of a variety of machine translation techniques [23] that utilise word n-grams motivated their use in representing texts for this task. In particular, word bigrams may provide potentially useful syntactic information about a text. Character bigrams, on the other hand, allow us to capture similarity between related word forms. Possible overlap features are constructed using basic set operations:

Size of union (U): the size of the union of the tokens in the two texts of a candidate paraphrase pair.

Size of intersection (N): the number of tokens common to the texts of a candidate paraphrase pair.

In addition we consider **Text size (S)**. For a given pair of sentences, feature S_1 is the size of the set of tokens representing the first sentence and S_2 is the size of the second sentence. Knowing about the union, intersection or size of a text in isolation may not be very informative. However, for a given token type, these four features in combination provide potentially useful information about similarity of texts. The four features (U, N, S_1 and S_2) are computed for word and character unigrams and bigrams. This yields a total of eight possible overlap features for a pair of texts, plus four ways of measuring text size. Each data instance is then a vector of features representing a pair of tweets.

4.2   Method

We trained SVM classifiers, using implementations from scikit-learn [29]. We report results obtained using Support Vector Classifier (SVC), which was adapted from *lib-svm* [6] by embedding different kernels. We have experimented with both linear and Radial Basis Function (RBF) kernels. Linear kernels are known to work well with large datasets and RBF kernels are the first choice if a small number of features is applied [20]. Both cases apply to our experimental datasets. Both linear and RBF classifiers are used with their default parameters.

In keeping with earlier experiments, we have applied a simple scaling method to features. This is a form of standardisation also called "z-score" in statistics, in which the transformed data variable has a mean of 0 and a variance of 1. Subtracting the mean, μ_x, from the feature vector, x, and dividing each of those features by its standard deviation, σ, scales features, and a new feature vector, $\hat{X}$ is obtained (Eq. 1). Apart from this *Simple Scaler*, features are kept as they are.

$$\hat{X} = \frac{x - \mu_x}{\sigma} \tag{1}$$

In all experiments, 10-fold cross validation was applied. We combined the whole dataset (train/test) and calculated the feature values. These features were split into 10 different sets after applying simple scaling. Each of 10 sets is used as test set against the remaining 9 sets as training data. Both the linear and RBF kernels of SVM are experimented with for character and word unigrams and bigrams.

4.3   Results

In Table 6, C1 and C2 each denote four features (U, N, S_1 and S_2) produced by character unigrams and bigrams, respectively. Similarly, W1 and W2 each denote four features generated by word unigrams and bigrams. Combinations such as C1W2 represent eight features (those for C1 plus those for W2) and the notation C12 abbreviates the combination of both C1 and C2, etc.

The presence of the C2 features is observed to lead to the best results. Indeed, C2 alone produces the best result overall, with an accuracy of 77.5 and an F-Score of 83.7, when combined with the linear SVM classifier. This comfortably beats the 'naïve' baseline. The C2 features also provide the best results for the RBF kernel. The results

Table 6. Character and word n-grams results on TuPC

SVM (Linear kernel)				SVM (RBF kernel)				
Features	Acc.	Pre.	Rec.	F-sc.	Features	Acc.	Pre.	Rec.	F-sc.
C1	66.4	66.4	100.0	79.8	C1	68.8	68.7	97.3	80.5
C2	**77.5**	**80.8**	**86.9**	**83.7**	C2	76.4	78.8	88.3	83.3
W1	73.6	75.5	89.3	81.8	W1	72.2	75.4	86.5	80.5
W2	71.2	72.7	90.9	80.8	W2	71.3	73.4	89.2	80.5
C1C2	76.8	80.5	86.0	83.1	C1C2	75.5	77.6	88.8	82.8
W1W2	73.7	76.1	88.5	81.7	W1W2	73.7	76.7	86.9	81.4
C1W1	72.6	75.1	88.1	81.0	C1W1	72.4	74.7	88.6	81.0
C2W2	76.7	80.3	86.1	83.1	C2W2	76.4	78.7	88.5	83.3
C1W2	71.1	73.2	89.4	80.4	C1W2	71.6	72.9	91.3	81.0
C2W1	76.4	79.6	86.9	83.0	C2W1	76.3	78.7	88.3	83.2
C12W12	77.5	81.0	86.5	83.6	C12W12	74.5	77.0	88.1	82.1

are consistent with our earlier work with MSRPC, PAN and TPC, where the inclusion of character bigram overlap features also helps achieve the best results. We hypothesize that measuring overlap of character bigrams provides a way of detecting similarity of related word-forms and thus performs a similar function to stemming or lemmatization. For MSRPC, PAN and TPC, comparable 10-fold cross-validation experiments have shown that the feature combination C2W1 is robust in yielding optimal results. The combination C2W1 gives best results for the PI task when used with either the linear (MSRPC) or RBF (PAN and TPC) kernels. For both PAN and TPC performance is at state-of-the art level for PI.

5 Conclusion

We have described the creation of a paraphrase corpus for Turkish. As far as we are aware, the TuPC is unique in providing both positive and negative instances and is currently the only paraphrase corpus for PI and semantic textual similarity available in Turkish. It may be noted that the TuPC is relatively small compared to other paraphrase corpora. However, the methods used to create it provide for a diverse set of paraphrase examples.

We have detailed the methods used to gather raw data from news websites and to score candidate pairs. These methods will be of value to others interested in creating paraphrase corpora and could be applied to generate additional paraphrase data for the TuPC. The main obstacle to producing data is the relatively time-consuming process of scoring candidate pairs. In future we may investigate the possibility of developing methods for crowdsourcing, for example through games for linguistic annotation.

A novel knowledge-lean approach to PI using character and word n-gram features and SVM classifiers has been presented. Our approach has already been shown to outperform more sophisticated methods applied to the TPC, and competitive results have also been obtained for the MSRPC and PAN. The results obtained for the TuPC cannot be compared directly to those for the other paraphrase corpora, but it is notable that the same features (in particular, overlap of character bigrams) yield the best results. The performance of our approach on the TuPC provides other PI researchers with a more realistic and challenging baseline than the naïve comparator of labeling all instances as paraphrases. We are now investigating methods for determining semantic similarity and intend to report on our results in another paper.

References

1. Agirre, E., et al.: Semeval-2012 task 6: a pilot on semantic textual similarity. In: Proceedings of the 6th International Workshop on Semantic Evaluation, in conjunction with the First Joint Conference on Lexical and Computational Semantics, pp. 385–393 (2012)
2. Bannard, C., Callison-Burch, C.: Paraphrasing with bilingual parallel corpora. In: Proceedings of the 43th Annual Meeting on Association for Computational Linguistics, pp. 597–604 (2005)
3. Bird, S., et al.: Natural Language Processing with Python. O'Reilly Media Inc., Newton (2009)
4. Blacoe, W., Lapata, M.: A comparison of vector-based representations for semantic composition. In: Proceedings of the 2012 Joint Conference on Empirical Methods in Natural Language Processing and Computational Natural Language Learning (EMNLP-CoNLL 2012), pp. 546–556 (2012)
5. Callison-Burch, C. et al.: Improved statistical machine translation using paraphrases. In: Proceedings of the main conference on Human Language Technology Conference of the North American Chapter of the Association of Computational Linguistics (HLT-NAACL 2006), pp. 17–24 (2006)
6. Chang, C., Lin, C.: LIBSVM: a library for support vector machines. ACM Trans. Intell. Syst. Technol. **2**(3), 1–27 (2011)
7. Chen, D.L., Dolan, W.B.: Collecting highly parallel data for paraphrase evaluation. In: Proceedings of the 49th Annual Meeting of the Association for Computational Linguistics: Human Language Technologies (HLT 2011), pp. 190–200 (2011)
8. Cohen, J.: A coefficient of agreement for nominal scales. Educ. Psychol. Meas. **20**(1), 37–46 (1960)
9. Das, D., Smith, N.A.: Paraphrase identification as probabilistic quasi-synchronous recognition. In: Proceedings of the Joint Conference of the 47th Annual Meeting of the ACL and the 4th International Joint Conference on Natural Language Processing of the AFNLP: ACL-IJCNLP 2009, pp. 468–476 (2009)
10. Demir, S., et al.: Turkish Paraphrase Corpus. In: Proceedings of the Eight International Conference on Language Resources and Evaluation (LREC 2012), pp. 4087–4091 (2012)
11. Dolan, B., et al.: Unsupervised construction of large paraphrase corpora. In: Proceedings of the 20th international conference on Computational Linguistics - COLING 2004, NJ, USA, p. 350–es (2004)
12. Duclaye, F., et al.: Using the web as a linguistic resource for learning reformulations automatically. In: Proceedings of the Third International Conference on Language Resources and Evaluation (LREC 2002), Las Palmas, Canary Islands, Spain, pp. 390–396 (2002)
13. Eyecioglu, A., Keller, B.: ASOBEK : Twitter Paraphrase Identification with Simple Overlap Features and SVMs. In: Proceedings of the 9th International Workshop on Semantic Evaluation (SemEval 2015), Denver, Colorado, pp. 64–69 (2015)
14. Fellbaum, C.: WordNet: An Electronic Lexical Database. MIT Press, Cambridge (1998)

15. Fernando, S., Stevenson, M.: A semantic similarity approach to paraphrase detection. In: Proceedings of the 11th Annual Research Colloquium of the UK Special Interest Group for Computational Linguistics, pp. 45–52 (2008)
16. Finch, A. et al.: Using Machine Translation Evaluation Techniques to Determine Sentence-level Semantic Equivalence. In: Proceedings of the Third International Workshop on Paraphrasing (IWP 2005), pp. 17–24 (2005)
17. Fleiss, J.L.: Measuring nominal scale agreement among many raters. Psychol. Bull. **76**, 378–382 (1971)
18. Ganitkevitch, J., et al.: PPDB : the paraphrase database. In: Proceedings of NAACL-HLT, Atlanta, Gerogia, pp. 758–764 (2013)
19. Gwet, K.L.: Handbook of Inter-rater Reliability. Advanced Analytics, Gaithersburg (2012)
20. Hsu, C.-W., et al.: A Practical Guide to Support Vector Classification. BJU Int. **101**(1), 1396–1400 (2008)
21. Landis, J.R., Koch, G.G.: The measurement of observer agreement for categorical data. Biometrics **33**(1), 159–174 (1977)
22. Lintean, M., Rus, V.: Dissimilarity kernels for paraphrase identification. In: Proceedings of the 24th International Florida Artificial Intelligence Research Society Conference, Palm Beach, FL, pp. 263–268 (2011)
23. Madnani, N., et al.: Re-examining machine translation metrics for paraphrase identification. In: Proceedings of the 2012 Conference of the North American Chapter of the Association for Computational Linguistics: Human Language Technologies (NAACL-HLT 2012), PA, USA, pp. 182–190 (2012)
24. Madnani, N., et al.: Using paraphrases for parameter tuning in statistical machine translation. In: Proceedings of the Second Workshop on Statistical Machine Translation (WMT 2007), Prague, Czech Republic (2007)
25. Malakasiotis, P.: Paraphrase recognition using machine learning to combine similarity measures. In: Proceedings of the ACL-IJCNLP 2009 Student Research Workshop, Suntec, Singapore, pp. 27–35 (2009)
26. Max, A., Wisniewski, G.: Mining naturally-occurring corrections and paraphrases from wikipedia's revision history. In: Proceeding of LREC, pp. 3143–3148 (2010)
27. Mihalcea, R., et al.: Corpus-based and knowledge-based measures of text semantic similarity. In: Proceedings of the 21st national conference on Artificial intelligence- Volume 1, pp. 775–780. AAAI Press (2006)
28. Owczarzak, K., et al.: Contextual bitext-derived paraphrases in automatic MT evaluation. In: StatMT 2006, Stroudsburg, PA, USA, pp. 86–93 (2006)
29. Pedregosa, F., Varoquaux, G., Gramfort, A., Michel, V., et al.: Scikit-learn: Machine Learning in Python. http://scikit-learn.org/stable/
30. Quirk, C., et al.: Monolingual machine translation for paraphrase generation. In: EMNLP-2014, pp. 142–149 (2004)
31. Socher, R., et al.: Dynamic pooling and unfolding recursive autoencoders for paraphrase detection. In: Advances in Neural Information Processing Systems, pp. 801–809 (2011)
32. Stanovsky, G.: A Study in Hebrew Paraphrase Identification. Ben-Gurion University of Negev (2012)
33. Wubben, S. et al.: Creating and using large monolingual parallel corpora for sentential paraphrase generation, pp. 4292–4299 (2010)
34. Xu, W.: Data-driven approaches for paraphrasing across language variations. New York University (2014)
35. Xu, W., et al.: SemEval-2015 Task 1: paraphrase and semantic similarity in Twitter (PIT). In: Proceedings of the 9th International Workshop on Semantic Evaluation (SemEval) (2015)

Evaluation of Semantic Relatedness Measures for Turkish Language

Ugur Sopaoglu[1(✉)] and Gonenc Ercan[2]

[1] Department of Computer Engineering, Çankaya University,
Eskisehir Yolu 29. Km Etimesgut, Ankara, Turkey
sopaoglu@cankaya.edu.tr
[2] Institute of Informatics, Hacettepe University, Beytepe, Ankara, Turkey
gonenc@cs.hacettepe.edu.tr

Abstract. The problem of quantifying semantic relatedness level of two words is a fundamental sub-task for many natural language processing systems. While there is a large body of research on measuring semantic relatedness in the English language, the literature lacks detailed analysis for these methods in agglutinative languages. In this research, two new evaluation resources for the Turkish language are constructed. An extensive set of experiments involving multiple tasks: word association, semantic categorization, and automatic WordNet relationship discovery are performed to evaluate different semantic relatedness measures in the Turkish language. As Turkish is an agglutinative language, the morphological processing component is important for distributional similarity algorithms. For languages with rich morphological variations and productivity, methods ranging from simple stemming strategies to morphological disambiguation exists. In our experiments, different morphological processing methods for the Turkish language are investigated.

Keywords: Semantic relatedness · Distributional similarity
Lexical semantics

1 Introduction

When asked, a human will surely argue that the word pair "car" and "wheel" is closer to each other semantically than the pair "car" and "chicken". A semantic relatedness function tries to quantify this sense of closeness. It is an interesting problem for artificial intelligence researchers as different natural language processing tasks can utilize such measures.

Measuring semantic relatedness between words, requires a form of information source. One approach is to rely on structured general knowledge-bases manually built by human experts, such as Roget's Thesaurus [1] and Wordnet [2]. However building and maintaining such knowledge-bases requires arduous work. As an alternative information source large volumes of unstructured text can be used, i.e. by using co-occurrence statistics of words. These methods use the distributional semantic hypothesis [3]. Research on measuring semantic relatedness

© Springer International Publishing AG, part of Springer Nature 2018
A. Gelbukh (Ed.): CICLing 2016, LNCS 9623, pp. 600–611, 2018.
https://doi.org/10.1007/978-3-319-75477-2_43

for the English language dates back to 1950s [4]. While some work investigates semantic relatedness in different languages (e.g. German) [5] to the best of our knowledge, no work exists for an agglutinative language. We first construct evaluation resources for Turkish. Furthermore, we compare the performance of semantic relatedness algorithms that uses Wordnet or Vector Space Models (VSM).

The dimensions of semantic space in VSMs is the set of words occurring in the corpus. Thus, how words are processed and tokenized is an important parameter of the algorithm. In English, it is shown that there is no significant difference between using an error-prone stemmer or an accurate lemmatizer [6]. In recent studies, it is shown that with good morphological processing strategies the results of semantic relatedness in German and Greek languages [7] can be improved. In this work, we further investigate the performance of different semantic relatedness measures in an agglutinative language like Turkish.

2   Semantic Relatedness Measures

An information source is required to measure the relatedness of terms. We use two different sources in this research. First, a manually built knowledge-base, Turkish Wordnet is used. Second, co-occurrence statistics is used, i.e. a raw text corpus is used as the information source.

2.1   Wordnet-Based Measures

Budanitsky and Hirst [8] compares different Wordnet-based measures for the English language. In this work, three Wordnet-based measures are evaluated for the Turkish language. Wu and Palmer [9] uses the maximum depth common ancestor of the two-word senses in the Wordnet taxonomy to measure the semantic relatedness. A similar measure is defined by Leacock and Chodorow [10], which directly uses the shortest path between the two senses. Finally, information theoretic measure defined by Resnik [11] is used. These three measures are re-implemented for the Turkish Wordnet.

2.2   Co-occurrence Based Measures

Corpora based semantic relatedness measures depend on the distributional semantic hypothesis, which states that words occurring in similar contexts are semantically related. A context can be realized as a document, paragraph or sentence in a computational model. A computational model estimated from infinitely many discourse observations can represent semantic relatedness of words accurately. For this reason, a method able to accommodate large datasets should be preferred.

Vector Space Models (VSM) [12] is one such method that can be used to model semantic relatedness by observing word distributions in raw text corpora. VSM considers different contexts as a vector in a $|V|$ dimensional space, where V is the vocabulary and dimensions are the words in V. For information retrieval in search engines context is defined as documents and the corpora is represented as a sparse document-term matrix.

For semantic relatedness using documents as context may result in a high noise-to-signal ratio, as a document can be formed of different topics. However, considering a smaller context becomes computationally demanding as number of sentences or paragraphs in a large corpus can be in the order of billions. Lund and Burgess [13] uses term-by-term matrices to build a semantic hyperspace of terms. In this hyperspace each dimension of ith term w_i is the co-occurrence frequency of w_i with another word w_j. Using this method, the scope of the observation context can be degraded as small as few words while keeping the computational requirements in manageable levels.

Our implementation of the distributional hypothesis semantic relatedness measure is based on Rapp [14]. First we create a raw co-occurrence matrix C of $|V| \times |V|$ dimensions, by passing a sliding window of size W through the corpora, where C_{ij} is the number of times words w_i and w_j occur in the same window of size W. Given the raw co-occurrence matrix it is possible to quantify semantic relatedness between words using vector similarity functions. However, raw co-occurrence counts are misleading since some words are more common than others and such words can occur with almost any other word. Thus, the co-occurrence counts are transformed using a weighting function. The C matrix is transformed to weighted A matrix by using the following function:

$$A_{ij} = log(1 + C_{ij}) - \sum_{k=0}^{(V)} p(w_i w_k) log(p(w_i, w_k)) \tag{1}$$

where, $p(w_i, w_k)$ is the probability of observing both w_i and w_k. Note that this is the same weighting function used in Rapp [14], which is a modified version of the function used in Landauer et al. [15]. Matrix A can directly be used with full dimensions as a semantic relatedness function, however both efficiency and effectiveness significantly improves by reducing the number of dimensions in A with Singular Value Decomposition (SVD). SVD decomposes the original matrix to three components as follows.

$$A_{m \times n} = U_{m \times r} \Sigma_{r \times r} V_{r \times n}^T \tag{2}$$

U and V matrices store the left and right singular vectors respectively. Σ matrix stores the singular values in its diagonal in descending order. The eigenvalues of the AA^T matrix are square of the singular values of A. Only retaining the top k singular values and vectors, and truncating the matrices to k yields a projection from $|V|$ to k dimensional space. This dimension reduction is referred to as SVD truncation. SVD truncation produces a projection which minimizes the Frobenius norm distance between the original matrix A and its projection to k-dimensional space A_k'. In other words the dimension reduction is performed in a way to keep the most prominent relationships in the full-dimensional matrix. This can be considered as a noise removal strategy. In the context of semantic relatedness this noise can be common, as words that are related to each other only in a specific local context will tend to co-occur in a small subset of the corpora. Following noise removal the matrix A'_k stores a k-dimensional semantic space of words. The cosine similarity between the vectors of A'_k is returned as the final semantic relatedness scores.

Morphological Variations. In agglutinative languages, for instance Turkish, the number of lemmas in the dictionary is usually low. However through the use of affixes it is possible to derive new words. Agglutinative languages are very productive in terms of creating new words from existing ones. Sak et al. [16] reports that even in a very large corpus of about 490M words adding a 1M-word text adds a new 5,539 word forms to the dictionary. They also show that as the size of the corpora increases the number of suffix combinations exceeds the number of stems found in the corpora. Given these properties how to form the dictionary for tasks related to lexical semantics becomes an important research question.

Table 1. Morphological parses of the word "ekmek"

(a) Ekmek (Bread)+Noun+A3Sg+Pnon+Nom
(b) Ek (to plant)+Verb+Pos^DB+Noun+Infl+A3Sg+Pnon+Nom

A morpheme is a unit in morphology used to create words in a language. Morphological structure of a language defines how morphemes can be combined and how some morphemes are modified in this process. Table 1 shows two morphological parses of the word "ekmek", where the first one shows the parse of the noun which means 'bread' and the second shows the verb 'to plant'. The morphology tags are added to the root of the word with + sign, each tag shows the form of the word, e.g. A3Sg signifies that the word is in singular first person form. When a derivational suffix is appended the "^DB" is used to indicate the derived words boundary and part of speech.

Table 2. Ambiguity in morphology level

(1) Tarlaya tohum **ekmek** (**plant** seeds to field)
(2) Fırından **ekmek** almak (buy **bread** from oven)

Even though morphology structure is defined inside the word boundaries, obtaining the intended morphological parse depends on the context. For example in Table 2 the word "ekmek" in the first sentence is used in the form (b) of Table 1, while it is used in form (a) in the second sentence. This ambiguity can only be resolved using the sentence level features.

In order to build the dictionary for the VSM, three different alternatives exist. Stemming algorithms are the simplest morphological processing techniques that can be used in semantic relatedness measures. A stemming algorithm simply strips common suffixes from words without using a lexicon, regardless of whether this procedure produces a valid word or not. On the other hand morphology parsers use a lexicon and a more detailed analysis to determine the morphemes of words. Finally the most expensive alternative is a lemmatizer, which usually depends on a morphology parser with an additional component for disambiguating the correct morphological parse of words depending on the sentence.

The Turkish stemmer used in this research is a multi-stage stemmer implemented as a deterministic finite automata (DFA) [17]. While the suffixes are included in the automata, system does not store a lexicon. Thus, it is not possible to determine if the output is a valid word in Turkish or not. However this method can be implemented efficiently using a small finite state machine. We will refer to this morphological processing system as STEMMER in the following discussions.

A morphological analyser requires a lexicon containing the words in the language and their corresponding grammatical classes. A word can be a noun, verb, adjective and adverb. Some affixes are applied to a certain class, for instance "la" derivational suffix is applied to nouns converting the word to verb e.g. "tuz-la". In morphologically complex languages such as Turkish, Finnish or Hungarian two level finite state transducers (FST) are employed for morphological processing [18]. Turkish morphology is modelled with two level FST by Oflazer [19]. Sak et al. [20] rebuilds a two-level FST with an updated rule-set and lexicon. We will refer to this system as TRFST in the following discussions. The open source morphological processing library Zemberek [21] uses a trie to store the lexicon and tries to match the words in the lexicon after applying different affixation rules. We will refer to this system as ZEMBEREK in the following discussion.

When parsing a raw text file, it is common to encounter words with multiple valid parses. This is known as morphological ambiguity. The morphological analysers ZEMBEREK and TRFST suffer from this ambiguity problem. In ZEMBEREK, the most frequent lemma of the word is selected as the correct parse [21]. Although the details are not provided, i.e. it is not clear from which corpora these statistics are collected from, the frequency of each lemma is estimated using a tagged corpus for Turkish. In a similar way TRFST uses the probability of generating the word parse. This probability is estimated using a morpheme level language model estimated from a corpora [22].

Finally we have experimented with morphological disambiguation using the averaged perceptron algorithm [23]. The perceptron algorithm uses features depending on the root with class of the word, morpheme tags in words and surface level form of words. They define 17 different features spanning a 3 word window adjacency. We will refer to this system as DISAMB in further discussions.

3   Created Turkish Language Evaluation Resources

As this work is a pioneering work on semantic relatedness for Turkish language, the evaluation resources are created. In addition, a method using the Turkish Wordnet is proposed to investigate semantic relatedness functions in more detail.

3.1   Word Association

The most straightforward task for semantic relatedness is the word association task, which compares the score produced by the system to human judgements. The word association dataset produced by Rubenstein and Goodenough [4] is a gold standard word association dataset for English. The dataset consists of 65-word pairs with different levels of relatedness. 51 human subjects are asked to assign a score for each word pair, where 5 represents highly related, and 1 represents unrelated. The effectiveness of a system is the correlation between the system-generated score and average of human judgements. Correlation is calculated using the Pearson's correlation values.

Table 3. Example word pairs for the word association task

TR word pair	EN word pair	TR mean score	EN mean score
araba-yolculuk	car-journey	2.95	1.55
bilge-büyücü	sage-wizard	0.89	2.46
birader-delikanlı	brother-lad	2.20	2.41
öğlen-akşam	midday-noon	3.00	3.68
kıyı-ağaçlık	shore-woodland	1.11	0.90
takı-mücevher	gem-jewel	3.79	3.94
sırıtma-delikanlı	grin-lad	0.30	0.88
yemek-meyve	food-fruit	2.42	2.69

A list of 101 Turkish word pairs is formed, by first translating the 65 English word pairs and extending the set with new 36 word pairs. Human judgement scores are measured from 76 volunteer annotators. The average inter-annotator correlation is 0.762, which is close but slightly lower than 0.80 reported for the English language. The correlation between the human judgement scores of Turkish and English word pairs is 0.82 indicating that although the language is changed relatedness is preserved. Table 3 lists a subset of this dataset, containing the average of human scores for both English and Turkish.

3.2   Clustering Purity

The cognitive abilities of humans in categorizing words is a task investigated in psychology [24], where the aim is to detect norm categories that are commonly described with similar example words. A recent work trying to predict the categories of words by monitoring the brain activity uses a dataset of 12 categories with 5 words in each category [25]. A dataset for Turkish is built by translating these categories and words. For evaluation, words are clustered using the semantic relatedness scores. Best scoring function should be able to produce clusters that are closer to the original categories. The accuracy of the clustering is measured using the clustering purity measure. Let the original categories be $\psi = \{\psi_1, \psi_2, ..., \psi_k\}$ and the output of clustering algorithm be the set of clusters $C = \{c_1, c_2, ..., c_k\}$.

$$purity(\psi, C) = \frac{1}{N} \sum_i max_j |\psi_j \cap C_i| \tag{3}$$

where N is the number of words. If the clustering purity is equal to 1 this means that the two sets ψ and C are equal to each other. Bullinaria and Levy [6] use this dataset to evaluate semantic relatedness scoring functions for English. In order to use this dataset in Turkish semantic relatedness evaluation, words are translated to Turkish.

3.3   Wordnet Recall

Wordnet can be used as ground truth for evaluation, as it is formed by linguists manually. Following this motivation, the word pairs with high semantic relatedness scores are compared to Wordnet relationships. The recall of Wordnet synonyms, hypernyms/hyponyms, meronyms and siblings are calculated for the k-nearest neighbours of words in the semantic space. This not only evaluates the accuracy of the semantic relatedness function but also provides an ability to investigate the types of relationships in the semantic space.

4   Results

We will present the results in three subsections, in order to answer three major research questions. In the first section, Wordnet and corpus statistics based methods are compared. In the second, different Turkish morphology processing methods are evaluated in the semantic relatedness task. Finally we assess the kind of lexical relationships that neighbouring words in the semantic space may have.

4.1   Comparison of Knowledge-Based and Knowledge-Lean Methods

The WordNet semantic relatedness functions are designed assuming that the sense graph is connected. However in Turkish, this is not valid as there are components in the sense graph. If two senses are in different components the semantic relatedness is not defined. To resolve this problem, when the word pairs reside in different components its semantic relatedness score is assumed to be 0.

Table 4. Comparison of WordNet and co-occurence statistics

Semantic rel. method	Word assoc.	Clustering purity
Wu & Palmer	0.65	0.7894
Leacock & Chodorow	0.55	0.7719
Resnik	0.59	0.6140
Full Sem. Space	0.4812	0.6491
SVD Sem. Space	0.7627	0.7368

Table 4 shows the results for both word association and clustering purity tasks. Full dimensional semantic space (using the matrix A) achieves the worst results. This can be attributed to the relatively small-size of the Turkish Wikipedia corpus. The results of this method may improve when a substantially larger corpus is used. When SVD dimension reduction is applied results are improved significantly (using paired t-test).

When the WordNet based measures are compared to SVD Semantic Space there are conflicting results. In the word association task SVD semantic space method is significantly better. However in the clustering purity task the WordNet based methods achieve better results. This contradiction is partly due to the fact that in clustering purity categories, the words sharing a category are usually similar terms rather than related words. As demonstrated by Le et al. [26], taxonomy based methods achieve better results in similarity tasks. We report a similar finding in our experiments, with the clustering purity task.

4.2   Morphological Processing Variations

In order to investigate which morphology tools to use in semantic space creation, experiments with four morphology tools are explained in Sect. 2.2 are conducted. Starting from the least computationally demanding method $STEMMER$ to most $DISAMB$ semantic spaces are built. Experiments with different window sizes W are performed, however we will only report the results of $W = 5$ as it achieves the best results in most of the configurations. Table 5 shows the results of different morphological processing algorithms in clustering purity and word association tasks.

Table 5. Comparison of morphology algorithms

Morphology algorithm	Clustering purity	Word association
STEMMER	0.6140	0.6548
ZEMBEREK	0.7193	0.7645
TRFST	0.6842	0.7643
DISAMB	0.6491	0.7390

Table 6. T-test results for **Word Association/Clustering Purity**

	STEMMER	ZEMBEREK	TRFST	DISAMB
STEMMER	−/−	+/+	+/−	+/+
ZEMBEREK	+/+		−/+	+/+
TRFST	+/−	−/+		+/+
DISAMB	+/+	+/+	+/+	

Table 6 shows whether the results of Word Association and Clustering Purity experiments are statistically significant or not. The '+' symbol indicates a significant difference between the specified methods in Table 5. If the difference is not significant using 95% confidence factor, it is marked with the '−' symbol. Each cell of the table compares algorithms, the first symbol is used for word association and the other is for clustering purity. The lightweight stemming approach achieves lower results in both tasks. The performance of more sophisticated morphological analysis, such as $ZEMBEREK$ or $TRFST$, using the most common parse heuristic is significantly better than $STEMMER$ and $DISAMB$ with a confidence factor of 0.95. While the difference between the stemmer and morphological analysers is not surprising, using a morphological disambiguation procedure degrades the performance. Thus, the best results are obtained with a parser using a simple most common lemma heuristic as the disambiguation procedure. This can be attributed to having more dense co-occurrence vectors, instead of having too sparse vectors for uncommon lemmas.

4.3 WordNet Recall

Figure 1 shows the average recall values of each Wordnet relationship for different k values, where k is the number of neighbours considered in the semantic space. This plot resembles the shape of the logarithm function, in which as k becomes larger, the increase in recall value diminishes. The recall value becomes saturated, indicating that words related to each other are usually the nearest neighbours in the semantic space. As expected, stronger relationship type synonymy is more focused on most similar neighbours when compared to weaker relationships like Hyponymy, Meronymy and Siblings. The siblings are more scattered in the semantic space as the growth of the plot is closer to linear. Rei and Briscoe [27] uses a similar technique to detect hyponyms in vector space and report a precision of 20% when $k = 1$, our result is similar to this with an 18%. Furthermore, 20% of nearest neighbours are siblings (sharing a hypernym). When $k = 1$, i.e. the nearest neighbour of each word w_i is selected, 47% of the time it is also defined as directly related to w_i in WordNet. When the performance of semantic space nearest neighbours are compared with random selection, when $k = 500$ only up to six percent of the synonyms can be covered by random selection, on the other hand about 60% of synonyms are covered by the semantic space nearest neighbours.

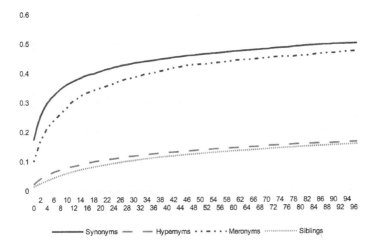

Fig. 1. Semantic space nearest neighbours and WordNet relationships

WordNet stores the words and their associated meanings in reverse sorted order of commonality. The most common sense used in discourse for a word is stored as the first sense of the word. Using this information it is possible to investigate if the semantic space is biased towards the most common sense. As expected since the most common sense is observed more in the data, the semantic space is biased towards these senses. For the nearest neighbours, about 86.29% of the relationships found in WordNet are related to the most common sense of the word.

5 Conclusion

Lexical semantics is a hot research area well studied for the English language. This work investigated the strategies for using semantic relatedness in an agglutinative language like Turkish. Our results show that even with the difficulties in Turkish related to morphological richness, similar results obtained for English are possible. Our comparisons show that even with the Turkish Wordnet, which only contains about 15 thousand nouns, it is possible to perform semantic categorization task with high accuracy. As it is the case for English, the Wordnet based measures achieve lower scores than semantic spaces in the word association task.

As Turkish is an agglutinative language, the level of morphological processing is potentially an important decision determining both efficiency and effectiveness of a method. While the performance of stemming algorithms are significantly lower than morphology analysis tools with lexicons, using a morphological disambiguation tool does not seem to improve the effectiveness. Performing more involved morphological disambiguation decisions decreases the semantic relatedness scores. This can be viewed from a statistical point-of-view as choosing the probably correct lemmas in the sentences not only is accurate most of the times, but also accumulates frequencies in fewer vectors rather than loosing frequency information by creating too sparse word vectors. Considering both effectiveness and efficiency using ZEMBEREK or TRFST algorithms seems to be a good choice for building semantic spaces and word embeddings.

Our comparison of Wordnet relationships with semantic space nearest neighbours reveals that an important portion of (about half) semantic space neighbours are also directly related in Wordnet. It is important to note inherited hyponymy/hypernymy or meronymy/holonymy are not included in this count. So, even larger overlap between the two resources can be expected. An interesting finding is that nearest neighbour of a word in semantic space is more probably a hyponym/hypernym sibling rather than hyponyms/hypernyms or even synonym of the word.

Acknowledgments. This work is partially supported by The Scientific and Technological Research Council (TUBITAK) under contract number EEAAG-114E776.

References

1. Jarmasz, M., Szpakowicz, S.: Roget's thesaurus and semantic similarity. In: Recent Advances in Natural Language Processing III: Selected Papers from RANLP 2003, pp. 212–219 (2004)
2. Fellbaum, C.: WordNet: An Electronic Lexical Database (Language, Speech, and Communication). illustrated edition edn. (1998)
3. Turney, P.D., Pantel, P., et al.: From frequency to meaning: vector space models of semantics. J. Artif. Intell. Res. **37**(1), 141–188 (2010)
4. Rubenstein, H., Goodenough, J.B.: Contextual correlates of synonymy. Commun. ACM **8**(10), 627–633 (1965)
5. Zesch, T., Gurevych, I., Mühlhäuser, M.: Comparing wikipedia and German wordnet by evaluating semantic relatedness on multiple datasets. In: Human Language Technologies: The Conference of the North American Chapter of the Association for Computational Linguistics; Companion Volume, Short Papers, pp. 205–208. Association for Computational Linguistics (2007)
6. Bullinaria, J.A., Levy, J.P.: Extracting semantic representations from word co-occurrence statistics: stop-lists, stemming, and SVD. Behav. Res. Methods **44**(3), 890–907 (2012)
7. Zervanou, K., Iosif, E., Potamianos, A.: Word semantic similarity for morphologically rich languages. In: Proceedings of the LREC (2014)
8. Budanitsky, A., Hirst, G.: Evaluating wordnet-based measures of lexical semantic relatedness. Comput. Linguist. **32**(1), 13–47 (2006)
9. Wu, Z., Palmer, M.: Verbs semantics and lexical selection. In: Proceedings of the 32nd Annual Meeting on Association for Computational Linguistics, pp. 133–138. Association for Computational Linguistics (1994)
10. Leacock, C., Chodorow, M.: Combining local context and WordNet similarity for word sense identification. In: WordNet: An Electronic Lexical Database, vol. 49, no. 2, pp. 265–283 (1998)
11. Resnik, P.: Using information content to evaluate semantic similarity in a taxonomy. arXiv preprint arXiv:cmp-lg/9511007 (1995)

12. Pedersen, T., Patwardhan, S., Michelizzi, J.: WordNet:: similarity - measuring the relatedness of concepts. In: Demonstration Papers at HLT-NAACL, pp. 38–41. Association for Computational Linguistics (2004)
13. Lund, K., Burgess, C.: Producing high-dimensional semantic spaces from lexical co-occurrence. Behav. Res. Methods Instrum. Comput. **28**(2), 203–208 (1996)
14. Rapp, R.: Discovering the senses of an ambiguous word by clustering its local contexts. In: Classification - The Ubiquitous Challenge, pp. 521–528. Springer (2005)
15. Landauer, T.K., Dumais, S.T.: A solution to Plato's problem: the latent semantic analysis theory of acquisition, induction, and representation of knowledge. Psychol. Rev. **104**(2), 211 (1997)
16. Sak, H., Güngör, T., Saraçlar, M.: Resources for Turkish morphological processing. Lang. Resour. Eval. **45**(2), 249–261 (2011)
17. Eryigit, G., Adali, E.: An affix stripping morphological analyzer for Turkish. In: Proceedings of the IASTED International Conference on Artificial Intelligence and Applications, Innsbruck, Austria, pp. 299–304 (2004)
18. Koskenniemi, K.: Two-level model for morphological analysis. In: IJCAI, vol. 83, pp. 683–685 (1983)
19. Oflazer, K.: Two-level description of Turkish morphology. Lit. Linguist. Comput. **9**(2), 137–148 (1994)
20. Sak, H., Güngör, T., Saraçlar, M.: A stochastic finite-state morphological parser for Turkish. In: Proceedings of the ACL-IJCNLP 2009 Conference Short Papers, pp. 273–276. Association for Computational Linguistics (2009)
21. Akın, A.A., Akın, M.D.: Zemberek, an open source nlp framework for turkic languages. Structure **10**, 1–5 (2007)
22. Sak, H., Saraclar, M., Gungor, T.: Morphology-based and sub-word language modeling for Turkish speech recognition. In: 2010 IEEE International Conference on Acoustics Speech and Signal Processing (ICASSP), pp. 5402–5405. IEEE (2010)
23. Sak, H., Güngör, T., Saraçlar, M.: Morphological disambiguation of Turkish text with perceptron algorithm. In: Gelbukh, A. (ed.) CICLing 2007. LNCS, vol. 4394, pp. 107–118. Springer, Heidelberg (2007). https://doi.org/10.1007/978-3-540-70939-8_10
24. Battig, W.F., Montague, W.E.: Category norms of verbal items in 56 categories a replication and extension of the Connecticut category norms. J. Exp. Psychol. **80**(3p2), 1 (1969)
25. Mitchell, T.M., Shinkareva, S.V., Carlson, A., Chang, K.-M., Malave, V.L., Mason, R.A., Just, M.A.: Predicting human brain activity associated with the meanings of nouns. Science **320**(5880), 1191–1195 (2008)
26. Le, M.N., Fokkens, A.: Taxonomy beats corpus in similarity identification, but does it matter? In: RProceedings of Recent Advances in Natural Language Processing, p. 346
27. Rei, M., Briscoe, T.: Looking for Hyponyms in Vector Space. In: CoNLL, pp. 68–77 (2014)

Using Sentence Semantic Similarity to Improve LMF Standardized Arabic Dictionary Quality

Wafa Wali[✉], Bilel Gargouri[✉], and Abdelmajid Ben Hamadou[✉]

MIR@CL Laboratory, Faculty of Economics and Managment of Sfax,
University of Sfax, Sfax, Tunisia
{wafa.wali,bilel.gargouri}@fsegs.rnu.tn,
abdelmajid.benhamadou@isimsf.rnu.tn

Abstract. This paper presents a novel algorithm to measure semantic similarity between sentences. It will introduce a method that takes into account of not only semantic knowledge but also syntactico-semantic knowledge notably semantic predicate, semantic class and thematic role. Firstly, semantic similarity between sentences is derived from words synonymy. Secondly, syntactico-semantic similarity is computed from the common semantic class and thematic role of words in the sentence. Indeed, this information is related to semantic predicate. Finally, semantic similarity is computed as a combination of lexical similarity, semantic similarity and syntactico-semantic similarity using a supervised learning. The proposed algorithm is applied to detect the information redundancy in LMF Arabic dictionary especially the definitions and the examples of lexical entries. Experimental results show that the proposed algorithm reduces the redundant information to improve the content quality of LMF Arabic dictionary.

1 Introduction

Some applications of Natural Language processing have a requirement for an efficient method to calculate semantic similarity between sentences. Many different approaches have been applied on computing semantic similarity between sentences. Some of them consider semantic information based on semantic similarity between words using the WordNet database, the statistical corpus and the Support Vector Model (SVM) [4, 8, 13, 18]; Others incorporates the syntactic information to compute sentences similarity like syntactic structure [14, 19, 24, 26], the word order [5, 16] or the common grammatical categories [15, 25] which can be with by pattern-matching. However, these methods are evaluated on English databases where the accuracy can decrease if they are practical to others languages. Besides, some semantic knowledge notably semantic class and thematic role and syntactic-semantic knowledge especially semantic predicate are not considered in calculating sentences similarity. Indeed, the semantic predicate provides an interaction mechanism between the syntactic behavior, the

© Springer International Publishing AG, part of Springer Nature 2018
A. Gelbukh (Ed.): CICLing 2016, LNCS 9623, pp. 612–622, 2018.
https://doi.org/10.1007/978-3-319-75477-2_44

discourse model and the real word knowledge. Also, the semantic predicate is constituted of semantic arguments which are characterized by semantic class and thematic role. The proposed algorithm addresses the limitations of these existing approaches by using semantic predicate, semantic class and thematic role. It measures the semantic similarity via the synonymy relations between words of sentences. And the syntactico-semantic similarity via the semantic predicate between sentences. Both semantic and syntactic information play a role in conveying the meanings of sentences. Thus, the overall sentence is defined as a combination of lexical similarity (common words), semantic similarity (synonymy words) and syntactico-semantic similarity (common semantic arguments) using a supervised learning. The rest of this paper is organized as follows. Section 2 introduces related works. Section 3 outlines the proposed algorithm. Section 4 carries out experiments to detect redundancies in LMF standardized Arabic dictionaries, and the final section gives the conclusion and some perspectives.

2   Related Work

In this section, the related works that achieve the best results so far in sentence similarity problem are presented as follows. Jiang and Conrath [7] introduced an approach for measuring semantic similarity distance between words and concepts wherein a lexical taxonomy structure is combined with corpus statistical information. Lin [17] presented the idea of measuring the similarity between two objects based on an information theoretical approach. Turney [27] described Latent Relational Analysis (LRA), a method for computing semantic similarity based on the semantic relations between two pairs of words. Li et al. [16] also proposed a similarity measure that translates each sentence in a semantic vector by using a lexical database and a word order vector. Then, it creates a ratio to weight the significance between semantic and syntactic information. Mihalcea et al. [18] presented a method for measuring the semantic similarity of texts, using corpus based and knowledge-based measures of similarity. Islam and Inkpen [5] reported on a Semantic Text Similarity (STS) measure using a corpus-based measure for semantic word similarity and a modified version of the Longest Common Subsequence (LCS) string matching algorithm. Ramage et al. [22] proposed an algorithm that aggregates relatedness information via a random walk over a graph constructed from WordNet. Gad and Kamel [3] proposed a semantic similarity based model (SSBM) that computes semantic similarities by exploiting WordNet. Furthermore, Pedersen [20] presented a through comparison between similarity measures for concept pairs based on Information Content (IC). Oliva et al. [19] reported on a method, called SyMSS, for computing short-text and sentence semantic similarity. The method considers that the meaning of a sentence is made up of the meanings of its separate words and the structural way the words are combined. Bollegala et al. [1] suggested an empirical method to estimate semantic similarity using page counts and text snippets retrieved from a web search engine for two words. Furthermore, Lintean and Rus [23]exploited word-to-word semantic similarity metrics for estimating the semantic similarity at sentence level. Šaric et al. [24] presented a system consisting of two major components for determining the semantic similarity of short texts using a support vector regression model with multiple features measuring word overlap similarity and syntax similarity. However, the methods presented below are evaluated on the English databases where the accuracy can decreased if they are applied to other languages. Besides in these methods, some semantic knowledge such as semantic class, thematic role and semantic predicate are not take into account in measuring sentence similarity. In the following, we will detail the proposed sentence similarity method.

3   The Proposed Sentence Similarity Method

The proposed method derives the sentence similarity from lexical, semantic and syntactico-semantic information that are contained in the compared sentences. The overall sentence similarity is computed from lexical, semantic and syntactico-semantic similarity [29] using a supervised learning [28]. In fact, a sentence is considered to be a sequence of words each of which carries useful information. Sentences considered in this paper are assumed to 10 words. The task is to establish a computational method that is able to compute the similarity between sentences. As shown in the Fig. 1, the proposed algorithm is divided into two phases: the similarities scores and the similarities ponderation.

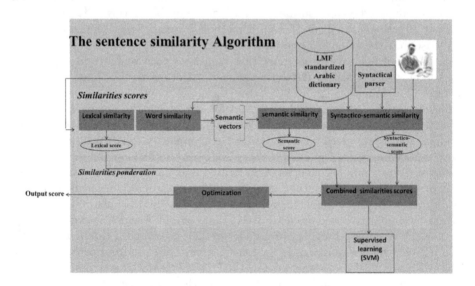

Fig. 1. Sentence similarity computation diagram

3.1   Similarities Score

3.1.1   Lexical Similarity Between Sentences

In this module, we aim to calculate the lexical similarity that is based on the number of stems that occur in both sentences after removal the stop words, such as the punctuation signs. The lexical similarity relies on the assumption that similar sentences have the same stem, which is efficient only if both sentences are of sufficient lengths. The lexical similarity is computed as follows:

$$LS(S1, S2) = \frac{SC}{WS1 + WS2 - SC} \tag{1}$$

Where SC is the number of common stems between S1 and S2, WS1 is the stems number of sentence S1 and WS2 is the stems number of sentence S2.

3.1.2   Word Similarity

Given two words: W1 and W2, we need to find the semantic similarity of Sw(W1, W2) for these two words. In LMF standardized Arabic dictionary [11], each sense of word is presented in the class Sense and has the relations pointers to others senses of lexical entries through the class sense Relation, such as synonymy, antonymy, etc. We can find the semantic similarity between W1 and W2 via the common synonymy list that shares the two synonymies lists for the compared words and that are extracted from the LMF standardized Arabic dictionary. A formula for word similarity is defined as follows:

$$WS(W1, W2) = \frac{SC}{SW1 + SWS2 - SC} \tag{2}$$

Where SC is the number of common synonyms between the synonymous of W1 and the synonymous of W2, SW1 is the number of synonymous of the word w1 and SW2 is the number of synonymous of the word w2.

3.1.3   Semantic Similarity Between Sentences

This module focuses on the computation of the semantic similarity between sentences. The computation process takes the word similarity into consideration. The process starts by forming a joint word set that containing all distinct words between S1 and S2. Since inflectional morphology may make a word appear in a sentence with different form that conveys specific meaning for the specific context, we use a lemmatized form. Then, each sentence is readily represented by the use of a joint word set as follows. The vector derived from the joint word set is called the semantic vector, denoted by T. Thus, the joint word set for two sentences S1:الوقت الممتد من الفجر الى غروب الشمس in English "The time from dawn to sunset" and S2:زمن مقداره من طلوع الشمس الى غروبها in English "The time from sunrise to sunset" is T=طلوع ,مقدار,زمن,شمس ,غروب ,الى ,فجر ,من ,ممتد,وقت. Each element in this vector corresponds to a word in the joint word set, so its dimension equals the number of words in the joint word set. The value of an element of the semantic vector, Ti (i = 1, 2, ..., m), is determined by the semantic similarity of the corresponding word to a word in the sentence. Take S1 as example:

Case 1: If wi appears in S1, then Ti is set to 1.

Case 2: If wi is not contained in S1, a semantic similarity score is computed between wi and each word in the sentence S1, using the Eq. (2). Thus, the most similar word in S1 to wi is that with the highest similarity score. Having applied the procedure for S1 and S2, the semantic vectors for are T1, T2 respectively. For the example sentence pair, we have:

$T1 = \{0, 0, 2/3, 1, 1, 1, 1, 1, 1, 1, 1\}$,

$T2 = \{1, 1, 1, 1, 1, 1, 0, 1, 0, 2/3\}$

After that, the semantic similarity between sentences is calculated based on derived semantic vectors using a cosine coefficient as described in the following formula.

$$SS(S1, S2) = \frac{T1.T2}{\|T1\|.\|T2\|} \tag{3}$$

Where T1 is the semantic vector of the sentence S1 and T2 is the semantic vector of the sentence S2.

3.1.4   Syntactico-Semantic Similarity Between Sentences

In this module, syntactic structure for each sentence is derived using a syntactic parser. Moreover, semantic predicate is determined by an expert for each sentence. LMF standardized Arabic dictionary [11] is then used to check the matching between the syntactic structure and the semantic predicate in order to extract the semantic argument properties notably the semantic class and the thematic role. Indeed, it would be interesting to consider the idea of semantic arguments properties because it was not treated in previous works. Thus, these properties provide real word knowledge in local discourse. Syntactico-semantic similarity between sentences is computed using the Jaccard coefficient [6]. This measure is based on the number of common semantic arguments. In fact, semantic arguments are considered similar if they have same properties in terms of semantic class and thematic role. The syntactico-semantic similarity function between sentences S1 and S2 is defined as follows:

$$SMS(S1, S2) = \frac{AC}{AS1 + AS2 - AC} \tag{4}$$

Where AC is the number of common semantic arguments, AS1 is the semantic arguments number of sentence S1 and AS2 is the semantic arguments number of sentence S2.

3.2   Similarities Ponderation

3.2.1   Combined Similarity Scores

Both of the semantic and syntactico-semantic pieces of information play a role in comprehending the meaning of sentences. Thus, the sentence similarity is based on the aggregation of the scores defined above, namely lexical, semantic and syntactico-semantic similarity. The aggregation combines between them in linear function using α, β and γ that decide how much lexical, how much semantic and how much syntactico-semantic knowledge contribute to the overall similarity computation.

$$Sim(S1, S2) = \alpha * SL(S1, S2) + \beta * SS(S1, S2) + \delta * SSM(S1, S2) \qquad (5)$$

The aim of the turning parameters α, β and γ is to turn the contribution of each score into the final score. In this work, these parameters are processed using the supervised learning. In fact, the supervised learning is based on the Support Vector Machine (SVM) algorithm especially the Sequential Minimal Optimization (SMO) function [21]. This function solves the optimization problems and it is faster, easier to implement and requires a reduced memory space [9].

3.2.2 Optimization

This phase is essential in any learning process and it aims to refine the classification generated by equation of the SMO function in order to have a final score. All parameters are empirically set in this paper, 0.2 for lexical similarity score, 0.45 for semantic similarity score and 0.35 for syntactico-semantic similarity score. Apart, from the turning parameters α, β and γ; there is another parameter that will be indicated by an expert after the sentence similarity calculation. It is a threshold for sentence similarity decision as 0.8 for similar sentence threshold.

4   Applying Sentence Similarity to Detect the Redundancy in LMF Arabic Dictionary

The original motivation for the development of our sentence similarity measure came from the redundancy detection in LMF standardized Arabic dictionary [11]. Indeed, the redundancy affects the definitions and the examples related to a sense of lexical entry. Thus, the elimination process of redundant definitions and examples in LMF Arabic dictionary is a task that ensures the optimization of dictionary linguistic quality. Indeed, the presence of repetitive sentences reduces the required knowledge quality because there are redundancies of the same information. To detect the redundancy in LMF standardized Arabic dictionary, we proposed the following algorithm that will treat all senses of lexical entries having more definitions and examples and we applied the similarity sentences function proposed below between them. The proposed algorithm is presented below:

```
FOR ALL lexical Entries DO
nb=0; /*number of definitions related to a sense*/
nb1=0; /*number of examples related to a sense*/
 FOR ALL SENSE DO
   IF SENSE Is Not Empty THEN
/* Definitions in LMF Arabic dictionary situate in Definition class*/
/* examples in LMF Arabic dictionary situate in Context class*/
```

```
   nb:=Count(Definition);
   nb1:=count(Context);
  END IF
/*Detection redundancy of definitions*/
  IF (nb>1) THEN
  i:=0;
  T[]:=NULL;
  FOR ALL Definition DO
   T[i]:= Definition;
   i:=i+1;
  END FOR
  i:=0
  FOR ALL T DO
     j:=i+1;
        FOR ALL T DO
        IF (T[j]=NULL) BREAK;
        END IF
        Score:=Sim(T[j], T[i]);
        IF(score>0.8)THEN Write (Definitions similar)
        END IF
        j:=j+1;
        END FOR
        i:=i+1;
    END FOR
  END IF
/*Detection redundancy of examples*/
IF (nb1>1) THEN
   i:=0;
   T[]:=NULL;
   FOR ALL Context DO
    T[i]:= Context;
    i:=i+1;
   END FOR
   i:=0
   FOR ALL T DO
      j:=i+1;
         FOR ALL T DO
         IF (T[j]=NULL) BREAK;
         END IF
         Score:=Sim(T[j], T[i]);
         IF(score>0.8)THEN Write (Examples similar)
         END IF
         j:=j+1;
         END FOR
         i:=i+1;
     END FOR
    END IF
   END FOR
END FOR
```

5   Experiments and Results

The sentence similarity measure is evaluated through its integration in a specific application, such as redundancy detection in LMF standardized Arabic dictionary [11]. LMF standardized Arabic dictionary [11] is a lexical resource conforms to the LMF standard ISO-24613 [2]. Most the content of this dictionary was obtained by an automatic conversion of the Arabic dictionary "Al-Ghany" available in HTML format on the Internet (http://www.alfaseeh.com accessed on 01/12/2006) [10]. The model of this dictionary covers all lexical levels, such as morphological, flexion paradigms [12], syntactic and semantic. This LMF Arabic dictionary contains about 38000 lexical entries among them 10800 verbs and 3800 roots. Furthermore, this dictionary contains in total 34000 lexical entries including 62157 senses, 62310 definitions and 71376 examples. Besides, in the LMF Arabic dictionary, we have 449 senses having more than a definition and 9171 senses having more than an example. Thus, the experimentation of our sentence similarity measure is reported on 22020 sentences whose 21000 are the examples and 1020 are the definitions. In order to evaluate our similarity measure, we collected human ratings for the similarity of pairs of sentences following existing designs for sentence similarity measures. The participants consisted of 5 native speakers of Arabic educated to graduate level or above. Indeed, the participants were asked to complete a questionnaire, rating the similarity of meaning of the sentence pairs on the scale from 0.0 (minimum similarity) to 4.0 (maximum similarity). Table 1 shows a portion of the testing result, where all human similarity score are delivered as the mean score of each pair and have been scaled into the range [0..1], for comparison with our sentence measure similarity. To evaluate the performance of our sentence similarity measure, we used the following metrics:

- Recall = It is the number of the of the determined relevant sentences divided by the existing number of sentences:

$$Recall = \frac{Number\text{-}of\text{-}pairs\text{-}detected\text{-}correctly\text{-}as\text{-}similar\text{-}by\text{-}our\text{-}measure}{Number\text{-}of\text{-}sentences\text{-}in\text{-}dictionary} \tag{6}$$

- Precision = It is the number of the determined relevant sentences divided by the number of returned sentences:

$$Precision = \frac{Number\text{-}of\text{-}pairs\text{-}detected\text{-}correctly\text{-}as\text{-}similar\text{-}by\text{-}our\text{-}measure}{Number\text{-}of\text{-}pairs\text{-}detected\text{-}by\text{-}our\text{-}measure} \tag{7}$$

- F-measure = It is the geometric mean of precision and recall and expresses a trade off between those two measures:

$$F\text{-}measure = 2 * \frac{Precision.Recall}{Precision + Recall} \tag{8}$$

All experiments are performed using the $\alpha = 0.2$, $\beta = 0.45$ and $\delta = 0.35$ parameters, which are empirically determined with respect to Eq. 5. Table 1 shows the recall, decision and F-measure obtained by the proposed method on 11010 pairs of sentences. Thus, our method yields into a better result (F-measure = 86.38%).

These results provide strong support for the utility of the syntactico-semantic knowledge notably the semantic predicate, semantic class and thematic role and the semantic knowledge especially the synonymy relations between lexical entry senses in the process of computing sentence similarity. Considering the promising performance of this proposed measure, further studies, some of which are underway needed, to investigate its application to other languages such as French and English.

Table 1. System results

	Recall	Precision	F-measure
Our system	81.56%	91.82%	86.38%

6   Conclusion

This paper presented a method for determining similarity measure between two sentences from semantic and syntactico-semantic information that they contain. The aggregate function Sim(S1, S2), defined in Eq. 5, combines the lexical, semantic and syntactico-semantic similarity in a linear way. Our proposal considers the impact of semantic similarity in sentences. The semantic similarity measure takes into account the common number of synonymy relations between lexical entries senses that are extracted from LMF Arabic dictionary. Indeed, the LMF Arabic dictionary describes morphologic, syntactic and semantic knowledge about lexical entries; and ensures the relation between semantic and syntactic level via Predicative representation class. The proposed method also takes the shared semantic arguments especially semantic class and thematic role into consideration between sentences extracted from LMF Arabic dictionary. These pieces of knowledge are extracted from Semantic Argument class after the determining process of the syntactic behavior (derived from syntactic parser) and the semantic predicate (derived from an expert) link. To investigate the value of the proposed method in real applications, it was applied to detect the information redundancies in LMF standardized Arabic dictionary notably the definitions and the examples related to same sense. An algorithm for incorporating a sentence similarity measure was proposed in order to detect the information redundancy and to improve the LMF standardized Arabic dictionary quality. Our algorithm realizes an excellent F-measure (86.38%) for 11010 pairs of sentences. In the future work, we aim to extend the detection of information redundancy on the definitions and the examples belonging to different senses. Also, we aim to integrate others semantic relationships, such as hyponymy in the computing process sentence similarity.

References

1. Bollegala, D., Matsuo, Y., Ishizuka, M.: A web search engine-based approach to measure semantic similarity between words. IEEE Trans. Knowl. Data Eng. **23**(7), 977–990 (2011)
2. Francopoulo, G., George, M., Calzolari, N., Monachini, M., Bel, N., Pet, M., Soria, C., et al.: Lexical markup framework (LMF). In: Proceedings of LREC, vol. 6 (2006)
3. Gad, W.K., Kamel, M.S.: New semantic similarity based model for text clustering using extended gloss overlaps. In: Perner, P. (ed.) MLDM 2009. LNCS (LNAI), vol. 5632, pp. 663–677. Springer, Heidelberg (2009). https://doi.org/10.1007/978-3-642-03070-3_50
4. Gunasinghe, U.L.D.N., De Silva, W.A.M., de Silva, N.H.N.D., Perera, A.S., Sashika, W.A.D., Premasiri, W.D.T.P.: Sentence similarity measuring by vector space model. In: 2014 International Conference on Advances in ICT for Emerging Regions (ICTer), pp. 185–189, December 2014
5. Islam, A., Inkpen, D.: Semantic text similarity using corpus-based word similarity and string similarity. ACM Trans. Knowl. Discov. Data (TKDD) **2**(2), 10 (2008)
6. Jaccard, P.: Etude comparative de la distribution florale dans une portion des Alpes et du Jura. Impr. Corbaz (1901)
7. Jiang, J.J., Conrath, D.W.: Semantic similarity based on corpus statistics and lexical taxonomy. arXiv preprint cmp-lg/9709008 (1997)
8. Jingling, Z., Huiyun, Z., Baojiang, C.: Sentence similarity based on semantic vector model. In: 2014 Ninth International Conference on P2P, Parallel, Grid, Cloud and Internet Computing (3PGCIC), pp. 499–503, November 2014
9. Sathiya Keerthi, S., Shevade, S.K., Bhattacharyya, C., Murthy, K.R.K.: Improvements to Platt's SMO algorithm for SVM classifier design. Neural Comput. **13**(3), 637–649 (2001)
10. Khemakhem, A., Elleuch, I., Gargouri, B., Ben Hamadou, A.: Towards an automatic conversion approach of editorial Arabic dictionaries into LMF-ISO 24613 standardized model. In: Proceedings of the Second International Conference on Arabic Language Resources and Tools, Cairo, Egypt (2009)
11. Khemakhem, A., Gargouri, B.: ISO standard modeling of a large Arabic dictionary. Nat. Lang. Eng. (2015). https://doi.org/10.1017/S1351324915000224
12. Khemakhem, A., Gargouri, B., Abdelwahed, A., Francopoulo, G.: Modélisation des paradigmes de flexion des verbes arabes selon la norme lmf-iso 24613
13. Lee, C.M., Chang, J.W., Hsieh, T.C., Chen, H.H., Chen, C.H.: Similarity measure based on semantic patterns. Adv. Inf. Sci. Serv. Sci. (AISS) **4**(18), 10 (2012)
14. Lee, M.C., Chang, J.W., Hsieh, T.C.: A grammar-based semantic similarity algorithm for natural language sentences. Sci. World J. (2014)
15. Li, L., Hu, X., Hu, X., Wang, J., Zhou, Y.-M.: Measuring sentence similarity from different aspects. In: 2009 International Conference on Machine Learning and Cybernetics, vol. 4, pp. 2244–2249. IEEE (2009)
16. Li, Y., McLean, D., Bandar, Z.A., O'shea, J.D., Crockett, K.: Sentence similarity based on semantic nets and corpus statistics. IEEE Trans. Knowl. Data Eng. **18**(8), 1138–1150 (2006)
17. Lin, D.: An information-theoretic definition of similarity. In: ICML 1998, pp. 296–304 (1998)
18. Mihalcea, R., Corley, C., Strapparava, C.: Corpus-based and knowledge-based measures of text semantic similarity. In: AAAI 2006, pp. 775–780 (2006)

19. Oliva, J., Serrano, J.I., del Castillo, M.D., Iglesias, A.: SyMSS: a syntax-based measure for short-text semantic similarity. Data Knowl. Eng. **70**(4), 390–405 (2011)
20. Pedersen, T.: Information content measures of semantic similarity perform better without sense-tagged text. In: Human Language Technologies: The 2010 Annual Conference of the North American Chapter of the Association for Computational Linguistics, pp. 329–332. Association for Computational Linguistics (2010)
21. Platt, J.C.: 12 fast training of support vector machines using sequential minimal optimization. In: Advances in Kernel Methods, pp. 185–208 (1999)
22. Ramage, S., Rafferty, A.N., Manning, C.D.: Random walks for text semantic similarity. In: Proceedings of the 2009 Workshop on Graph-Based Methods for Natural Language Processing, pp. 23–31. Association for Computational Linguistics (2009)
23. Rus, V., Lintean, M.: A comparison of greedy and optimal assessment of natural language student input using word-to-word similarity metrics. In: Proceedings of the Seventh Workshop on Building Educational Applications Using NLP, pp. 157–162. Association for Computational Linguistics (2012)
24. Šarić, F., Glavaš, G., Karan, M., Šnajder, J., Bašić, B.D.: TakeLab: systems for measuring semantic text similarity. In: Proceedings of the First Joint Conference on Lexical and Computational Semantics, vol. 1, Proceedings of the Main Conference and the Shared Task, vol. 2, Proceedings of the Sixth International Workshop on Semantic Evaluation, pp. 441–448. Association for Computational Linguistics (2012)
25. Shan, J., Liu, Z., Zhou, W.: Sentence similarity measure based on events and content words. In: Fuzzy Systems and Knowledge Discovery, pp. 623–627 (2009)
26. Ştefănescu, D., Banjade, R., Rus, V.: A sentence similarity method based on chunking and information content. In: Gelbukh, A. (ed.) CICLing 2014. LNCS, vol. 8403, pp. 442–453. Springer, Heidelberg (2014). https://doi.org/10.1007/978-3-642-54906-9_36
27. Turney, P.: Measuring semantic similarity by latent relational analysis (2005)
28. Wali, W., Gargouri, B., Ben Hamadou, A.: Supervised learning to measure the semantic similarity between Arabic sentences. In: Núñez, M., Nguyen, N.T., Camacho, D., Trawiński, B. (eds.) ICCCI 2015. LNCS (LNAI), vol. 9329, pp. 158–167. Springer, Cham (2015). https://doi.org/10.1007/978-3-319-24069-5_15
29. Wali, W., Gargouri, B., Ben Hamadou, A.: Using standardized lexical semantic knowledge to measure similarity. In: Buchmann, R., Kifor, C.V., Yu, J. (eds.) KSEM 2014. LNCS (LNAI), vol. 8793, pp. 93–104. Springer, Cham (2014). https://doi.org/10.1007/978-3-319-12096-6_9

Multiword Expressions (MWE) for Mizo Language: Literature Survey

Goutam Majumder[1], Partha Pakray[1], Zoramdinthara Khiangte[2],
and Alexander Gelbukh[3(✉)]

[1] National Institute of Technology, Aizawl, Mizoram, India
goutam.nita@gmail.com, parthapakray@gmail.com
[2] Pachhunga University College, Aizawl, Mizoram, India
zoramdintharakhiangte@gmail.com
[3] CIC, Instituto Politécnico Nacional, Mexico City, Mexico
gelbukh@gelbukh.com
http://www.parthapakray.com
http://www.gelbukh.com

Abstract. We examine the formation of multi-word expressions (MWE) and reduplicated words in the Mizo language, basing on a news corpus (reduplication is a repetition of a linguistic unit, such as morpheme, affix, word, or clause). To study the structure of reduplication, we follow lexical and morphological approaches, which have been used for the study of other Indian languages, such as Manipuri, Bengali, Odia, Marathi etc. We also show the effect of these phenomena on natural language processing tasks for the Mizo language. To develop an algorithm for identification of reduplicated words in the Mizo language, we manually identified MWEs and reduplicated words and then studied their structural and semantic properties. The results were verified by linguists, experts in the Mizo language.

Keywords: Multi-word expressions · Reduplication · Mizo language
Onomatopoetic sounds · Natural language processing

1 Introduction

The Mizo language (*Mizo ţawng*) is spoken mainly in the north-eastern state of Mizoram in India, where it is an official language, as well as in Chin State in Burma and the Chittagong Hill Tracts of Bangladesh. Mizoram is one of the 29 states of India, with capital city Aizawl. The name "Mizoram" is derived from "mi" (people), "zo" (hill) and "ram" (land), meaning "land of the hill people". A Mizo historian K. Zawla writes: "The term is a combination of two words Mi and Zo. Mi means people and Zo means hill. They were known as the Mizo because, before they migrated to the present land, they settled in hills of Burma and its surroundings areas for more than two hundred years or more. Hence the term Mizo emerged and became popular during this period" [23]. Located in the northeastern India, Mizoram is the southernmost landlocked state sharing borders with three of the seven sister states Tripura, Assam, Manipur, as well as with the neighboring countries Bangladesh and Myanmar.

© Springer International Publishing AG, part of Springer Nature 2018
A. Gelbukh (Ed.): CICLing 2016, LNCS 9623, pp. 623–635, 2018.
https://doi.org/10.1007/978-3-319-75477-2_45

According to 2011 Census of India [6], Mizoram's population is one million (the second least populated state of India) [7].

The Mizo people are an ethnic group native to north-eastern India; groups of people in western Myanmar and eastern Bangladesh also speak the Mizo language. In a wider sense, the name Mizo includes all the ethnic groups of Kuki-Chin language family under the Tibeto-Burman group of languages, who share common traditions, customs, and beliefs [23]. The Mizo language belongs to the Kukish branch of the Sino-Tibetan language family, under Kamarupan group; see Fig. 1.

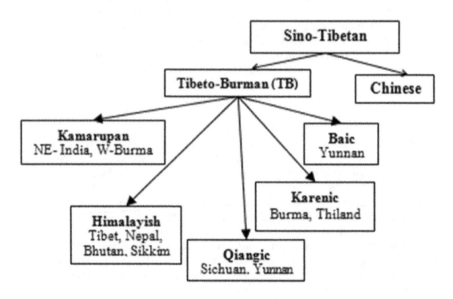

Fig. 1. Sino-Tibetan language tree

There are eleven tribes of Mizo people, the largest one being Lushei (Lushai, Lusai); Mizo is the modern name of Lusai language [17]. The Mizo language is inherited from Pawi, Paite, and Hmar [1]. P. Sarmah and C. Wiltshire investigated the different tones and morpho-tonology characteristics of the Mizo language and showed that it has four tones: high, low, rising, and falling [2–5]. Because of the tonic nature of this language, multiword expressions (MWEs) play an important role in almost all NLP tasks for the Mizo language, such as machine translation, information retrieval, question answering etc. They are particularly helpful in parsing, where the sequence of words forming the MWE is treated as a single word with a single part-of-speech (POS) tag. While in NLP-based systems, words of a sentence are considered as a lexical unit, this does not hold for MWEs [16].

In this work, we analyze different multiword expressions along with reduplication phenomena and their semantic structure, along with their impact on the automatic analysis of the Mizo language. Related work on MWEs is presented in Sect. 2. In Sect. 3, different characteristics of the Mizo language are discussed. In Sect. 4, semantic structure of various types of reduplication in Mizo is studied. In Sect. 5, conclusions are drawn and future work is outlined.

2   Related Work

Sag et al. [8] indicated that MWEs are one of the key problems for natural language processing (NLP) and machine translation (MT). MWEs are idiomatic in nature, which creates problem for MT; it is difficult to find the corresponding MWEs in the target language. Two promising approaches have been found for treatment of MWEs in MT: example-based MT, where one has to find the MWEs in the target language, and role-based approach. It is difficult to define translation rules for MWEs, but their proper treatment is very important in automatic processing any language.

Chakraborty and Bandyopadhyay [9] introduced the study of MWEs in Bengali language for the analysis of the writings of Rabindranath Tagore. They used a rule-based approach. The task was divided in two phases. In the first phase, identification of MWEs for the Bengali language was done at expression level; in the second phase, semantic classification was done. They also addressed the task of identification of the positive and negative sense for Bengali language.

Nongmeikapam et al. [10] identified the MWEs and reduplicated multiword expressions for another Indian language, Manipuri. Their motivation was improvement of POS tagging for the Manipuri language. They identified MWEs of different types, such as complete reduplication words, partial reduplication words, echo reduplication words, mimic reduplication words, double reduplicated words and semantically reduplicated words.

Parimalagantham [11] studied structural reduplication in the Tamil and Telugu languages and identified lexical and morphological reduplication for these two languages. In this work, the lexical category task was further sub-divided into three categories: echo formation, compounding and word reduplication. The third category, word reduplication, was further subdivided into three categories: complete, partial, and discontinuous word reduplication. The classification also included the information on whether the reduplication class of the word is changes. For the morphological reduplication, four categories were grouped into expression category: sound symbolism, mimic words, iconicity, and onomatopoeia.

Balabantaray and Lenka [12] reported reduplication in the Odia language. In Odia language, MWEs were found in different forms [9–11]. Apart from that, some other categories were also reported basing on positive and negative sense. All such categories have been grouped into morphological group where a free morpheme or a bound morpheme is copied for reduplication.

Venkatapathy et al. [15] identifies the 'Noun+Verb' expression in Hindi and also rate those multi-word expressions in between 1 to 4. In this study it has been reported that Hindi is a collection of large number of MWEs among them few are idiomatic and others are literal expressions. Maximum Entropy Model (MaxEnt) has been used to measure their relative compositionality automatically. These MaxEnt integers represent various features in multi-word expression in Hindi. Finally these features were used to mapping the 'Noun+Verb' expressions into Verb-Noun in English. Accuracy of identification for MWEs in different languages has been listed in Table 1.

Table 1. Accuracy of identification of mwes in different Indian language.

Authors	Language	Result (percentage)	
Chakraborty and Bandyopadhyay [9]	Bengali	Precision	92.82
		Recall	91.50
		F-Score	92.15
Nongmeikapam et al. [10]	Manipuri	Precision	73.15
		Recall	78.22
		F-Measure	75.60
Balabantaray and Lenka [12]	Odia	80.00 (checked by a linguist)	
Asad [13]	Marathi	N/A	
Parimalagantham [11]	Tamil and Telugu	N/A	
Venkatapathy et al. [15] identification of Hindi+Verd expressions	Hindi	Maximum Entropy Model: 76.03 by annotators	

3   Characteristics of Mizo Language

3.1   Periods of Mizo Language

The following periods of the Mizo language can be identified.

Pre-Christianization period (1860–1894) [18]: Although the Mizo alphabet proper was created in 1st April 1894, written Mizo literature can be said to start from the publication of Progressive Colloquial Exercises in the Lushai Dialect by Thangliana (which is the Mizo name of Thomas Herbert Lewin) in 1874. In this book he wrote down three Mizo folktales Chemtatrawta, Lalruanga and Kungawrhi with their English translations, and included some Mizo words with their English meaning [19].

Early Period (1894–1920): During this period, most of the literary work was produced by the missionaries. After the evolution of Mizo alphabet schools were established. On 22 October 1896 the first Mizo language book was published under the title "Mizo Zir Tir Bu" (lit. Mizo primer). This was a book on Christian religion and morality based on Christianity.

Middle Period (1920–1970): In this period, the rise of prominent writers such as Liangkhaia, who published hundreds of articles in the monthly Kristian Tlangau and authored Mizo chanchin (in two volumes) which contained, besides a coherent treatment of Mizo history, a number of ancient chants and festive songs which he collected from various sources.

Modern Period (1970–present): The Mizo Academy of Letters started awarding its Book of the Year in 1989. The academy also awards lifetime achievement in Mizo literature. Some of the most prominent writers during this period are James Dokhuma (1932–2008), Khawlkungi (1927–2015), B. Lalthangliana (1945–present), Siamkima Khawlhring (1938–1992), Rev Dr Zairema (1917–2008), R.L. Thanmawia (1954–

present), C. Laizawna (1959–present), Laltluangliana Khiangte (1961–present), Zikpuii Pa (1929–1994), Vanneihtluanga (1959–present), etc. In literature many such records were found [20, 21].

3.2 Impact of English Language in Mizoram

After the evolving of Mizoram state as monolingual, the people of this state tend towards to use of local language than Hindi or English. But people having a good skill on English language get higher valued among the citizens of this state [22].

Present scenario of learning of English language: At present, various problems arise in the teaching and learning of English as a second language in Mizoram and the communicative ability or the proficiency level is still very poor. The various problems in the teaching and learning of English in Mizoram may be listed as under:

- *Lack of exposure:* People get less chance to speck in English, because Mizo's have a homogeneous unilingual society. But there is no opportunity for the learners to get exposed to the communicative forms of English. For most the English classroom is the only opportunity for exposure to the language. So, only limited opportunities are available for the use of English and to hear people speak in English. This results in the poor communication ability of the learners.
- *Difference with English language:* Syntactically Mizo language is different form English language.
- *Mizo is a tonal language:* Another problem faced by Mizo's while learning English is that Mizo language is a tonal language where variations in tone and intonation pattern can change the meaning of words and utterances. Four different levels of tones have been used by Mizo people, which make the difference meaning of same word. One equivalent of the English word "earth" is "lei" in Mizo with the use of a particular tone. With different tones "lei" can portray different meanings like "buy", "tongue", "lop-sided" and "bridge".
- *Learning of English pronunciation:* Another problem of Mizo learners begins with English phonology. The real sound or power of a letter of the English alphabet is not always the same as the name of the letter would indicate, and the real sound can only be learnt from a teacher. And sometime the teachers themselves are not trained in phonetics. Many words are mispronounced because they are unaware of the phonetic symbols in words such as bouquet, buffet, bury, alumni, arrears, debut, debris, indict, etc.
- *Facing problem in English spelling system:* Mizo learners face another problem in the English spelling system. The many peculiarities found in the English spelling system create confusion in the Mizo learners, because the spelling system is quite unlike the one followed in Mizo in which words are pronounced in a different way. Silent letters are also a source of confusion to learners of English as a second language like in the words such as psychology, receipt, debt, coup, psalm, receipt, etc.
- *Lack of innovation:* Another problem faced by the students in the learning of English in Mizoram is that the teachers lack innovation. They teach English like any

other subject. They do not give extra effort for the students to enhance their communicative ability, to be able to use in real life situations.

Importance of English language in Mizoram: The objective of English teaching in Mizoram is not simply to make the learners learn the four language skills: listening, speaking, reading and writing, but to enable them to play their communicative roles effectively.

4   Classification of Reduplication and MWEs in Mizo Language

4.1   Meaning of MWEs and Reduplication

In literature definition of MEWs are expressed in different ways. Sag et al. [8] define that it is an idiosyncratic expression has no boundary and the idiosyncratic expression can be any one of them lexical, statistical, syntactic, semantic or pragmatic. On the other side Baldwin et al. [14] stated that a sequence of word which is act as a single unit also called MWEs in linguistic.

On the other hand, reduplication can be stated that a new word is formulated by a part or a whole of the root word by some morphological process [12].

Identification of reduplication words. To detection of re-duplication words two consecutive words W_1 and W_2 is considered. The detection of reduplication words can be divided into following steps:

- Step 1: In this step raw text files have been prepared, which contains Mizo news information of different category. Each line contains a single sentence in the text files.
- Step 2: After the preparation of text files tokenization has been done and each token i.e. word has been stored into the text file.
- Step 3: After tokenization two consecutive words have been taken and passed over a system, which contains different rules for detecting MWEs.
- Step 4: In the final step, based on rules has been divided into different groups. The system flow for detection of MWEs is shown in Fig. 2.

For **complete re-duplication**, two consecutive words with same characters have been considered.

For **partial re-duplication**, two affix conditions such as prefix and suffixes have been considered. During this identification stage third category of affix i.e. infix also considered but no words with this category has been found and it is also justified by a linguistic expert.

For **discontinuous re-duplication**, again two consecutive words are considered. But for this category algorithm only considered words of length 3 to 4 and only look for changing a vowel among the words.

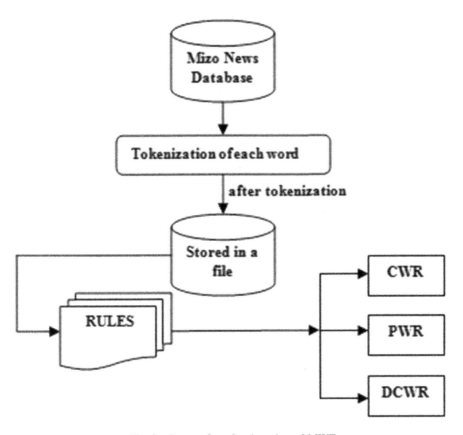

Fig. 2. System flow for detection of MWEs.

Algorithms for detection of CWR, PWR and DCWR are shown below.

Algorithm for detection of CWR

Step 1: Initialization
        w1 = first word
        w2 = second word
        both w1 and w2 are stored as a character array

Step 2: *if* **length** of w1 *and* w2 **equal then**
    Step 2a: *if* each character sequence is **equal** *then*
            continue
    Step 2b: *else break*

Step 3: *if* till the end of *Step 2a* is **true** in w1 and w2 **then**
    Step 3a: detected word is CWR

Step 4: Save the word.

Algorithm for detection of PWR

Step 1: Initialization
w1 = first word
w2 = second word
both w1 and w2 are stored as a character array

Step 2: *if* **length** of w2 **greater than** w1 *then*

Step 3: *for* **PWR as prefix category**
Step 3a: *if each character of* w1
is present in w2 from **starting then**
    Step 3ab: Save the word.
Step 3b: *else break.*

Step 4: *for* **PWR as suffix category**
Step 4a: *if each character of* w1
is present in w2 from **end then**
Step 4ab: Save the word.
Step 4b: *else break.*

Algorithm for detection of CWR

Step 1: Initialization
w1 = first word
w2 = second word
*both w1 and w2 stored as a character array and w1, w2 are
in same size*

Step 2: *if length* of (w1 **and** w2) equal to (3 **OR** 4) *then*

Step 3: **Save the location** of a character **vowel type.**
        I as location in w1
        J as location in w2

Step 4: if I **equal to** J *then*
    Step 4a: if character **in** w1[I] **and** w2[J] **not equal** *then*

Step 5: Save the word.

4.2 Division of MWES

Based on the lexical expressions MWEs are categories as follows:

Echo Formation: In this form an MWEs is represented by a pair of words where the second word does not have any meaning and the second word is simply an echo word.

Compound: Compound MWEs are also found, in which two words of opposite meaning forming a new word.

Word Reduplication: Word reduplication can be further divided into the following sub-categories:

- *Complete Word Reduplication (CWR):* In this form of reduplication words are formed by simply repetition of the base or root form without of phonological or morphological variations. The reduplication word can occur as noun, adjective, adverb, wh-question type, verbs, command and request. But in Mizo language all CWR belongs to either adverb or adjective category. It is also found that class changing is not available in Mizo i.e. all CWR remains in same class. Examples of CWRs are given in Table 2.

Table 2. List of complete word reduplication in Mizo language

Examples	Meaning	In Mizo sentence	Translation into English
ngei ngei	Without fail and certainly	Lo kal **ngei ngei** rawh	*Come without fail*
chu chu	That	**Chu chu** a tha ber	*That's the best one*
mai mai	Simply	Khati khan ti **mai mai** suh	*Don't just simply do like that*
kher kher	Particular	Chu thil **kher kher** ah chuan ka tel ve lo	*I am not involved in such a particular thing*
reng reng	Often, frequently, constantly	Hetiang thil te chu a thleng **reng reng** alawm	*This thing often/constantly happened*
sup sup	Group	An kal **sup sup**	*They walk in a group*
hlawk hlawk	Hurriedly	A kal **hlawk hlawk**	*He walks hurriedly*
zawt zawt	Rapidly	A tawng **zawt zawt** mai	*He talks rapidly*
her her	Giggling	A nui **her her**	*She is giggling*
dem dem	Steadily	A luang **dem dem**	*It flows steadily*

- *Partial Word Reduplication (PWR):* In this form of reduplication a part of base form is copied and the base form may be a syllable. After the syllable part any alphabets may be added for partial reduplication. The repetition can be takes place at any position of the word as initial, middle or final. In the partial reduplication along with the root word a suffix or a prefix or an infix can be added to form a new word. Partial reduplication words based on suffix and prefixes are listed in Table 3. Word as infix is not available in this language.
- *Discontinuous (DCWR):* In this form of reduplication, the vowel has a role in forming a new word with the present syllable by replacing itself with another vowel. Examples based on changes of a vowel are listed in Table 4.

Morphological Reduplication: Morphological reduplication words are also expressive in nature and words imitate a sound and the unit also refers a sound. These expressive words also categories into three groups as follows:

Table 3. List of affixes used for partial word reduplication in Mizo language

Type	Words in suffix/prefix	Meaning	After PWR word	Meaning
As prefix	*ni*	Day	*ni* – tin	Everyday
	ni	Sun	*ni* – eng	Sunlight
	kawr	Shirt	*kawr* – lum	Warm clothes
	mit	Eyes	*mit* – tui	Tears
	ziak	Write	*ziak* – tu	Writer
	chhiar	Read	*chhiar* – tu	Reader
	ho	Team/Group	*ho* – tu	Leader
	sual	Bad	*sual* – na	Badness
	thatchhia	Lazy	*thatchhiat* – na	Laziness
	fin	Wise	*fin* – na	Wisdom
As suffix	*zu*	Wine	khawi – *zu*	Honey
	thla	Moon	hlim – *thla*	Shadow
	kawr	Shirt	ke – *kawr*	Trouser
	mei	Tail	kang – *mei*	Fire
	hal	Burn	tui – *hal*	Thirsty
	thali	Vegetables	thli – *thali*	Inspect
	sang	Tall	ngai – *sang*	To Admire
	rit	Heavy	phur – *rit*	Burdens
	mit	Eye	khei – *mit*	Ankle
	mu	Sleep	silai – *mu*	Bullet

Table 4. List of discontinuous word reduplication (DCWR) in Mizo language

Main word	Word after discontinuity	Meaning	Used in Mizo	English Translation
rem	*rum*	Very messy/dirty	I pindan chu a va hnawk **rem rum** ve	Your room is very messy
chek	*chuk*	Is blur/not clear	Vawiin khua chu chum a va zingnasa **chek chuk** ve	Today's weather is very foggy
dik	dak	To go to some place and give attention to something	Ngati nge iv a rawn **dik dak** lo ve?	Why don't you come and pay attention?
dek	duk	To describe how deep something is	He tuichhunchhauh chu a va thuk **dek duk** ve	The well is very deep

- *Sound Symbolism:* an onomatopoetic sounds which reflects the properties of an external word. All words start with a common sound symbolism like 'gl' in English (glow and gleam) and also suggest that combination of 'gl' – conveys the idea of sheen and smoothness. Not available in Mizo language.

- *Mimic Words:* sound of any object by repeating the whole root word but the morphemes are onomatopoetic. These sounds express emotions and natural. Such words are listed in Table 5.

Table 5. List of mimic words in Mizo language

Mimic words	Meaning (sound of something)	A sentence in Mizo
vuk vuk	Sound of wind blowing	Thil chu a thaw ri **vuk vuk** mai
kerh kerh (as sound)	Used to describe a sound made by brook of a river	Lui chu a luang ri **her her** mai. (**as used while written**)
dur dur	Used to describe a sound of a thunder	Khawpui chu a ri **dur dur** mai

- *Iconicity:* a sound which represents some idea such as occurs through onomatopoeia. For this task it is not found.

Semantic MWEs: The both word have semantic meaning. Semantic words like English 'traffic-signal' and others are also used in Mizo as same as English.

Details of experiments. For detection of reduplication words one year i.e. 2014 Mizo news data has been considered for this task. This dataset contains news texts of different categories like sports, national and international political news, cultural, etc. Data was collected from a daily published local newspaper agency '*Vanglaini*' [24]. Before detection data was pre-processed and clean by removing all html tags and kept in a text file. The whole dataset is composed of 1 billion words and also contains duplicate entry. Identification and detection process for MWEs have been discussed in Sect. 4. Automatic detections and frequency of MWEs in one year Mizo news corpus has been shown in Table 6.

Table 6. Frequency and automatic detection of MWEs in news corpus data

Type	Word count
Main document words	1, 19, 20, 594
CWRs	1004
PWRs	631

5   Conclusions and Future Work

In this paper, we presented an overview of multi-word expressions in the Mizo language and algorithms for their detection.

In our future work, an automatic annotation tool will be designed for part-of-speech tagging in Mizo language. For this purpose detection of MWEs is one of the first and most important steps to improve the performance of POS tagging. In future a stemming algorithm will be developed and for stemming purpose suffixes and prefixes of PWRs will be helpful. Standard IR matrices like Precision, Recall and F-Score will be used for evaluation purpose.

Acknowledgments. This work presented here under the research project Grant No. YSS/2015/ 000988 and supported by the Department of Science & Technology (DST) and Science and Engineering Research Board (SERB), Govt. of India at NIT Mizoram. We would like to acknowledge National Institute of Technology Mizoram for providing the research environment and sponsorship to carry out this work. Also, we are thankful to Mr. Jereemi Bentham and Mr. Sunday Lalbiknia, students of CSE dept. of NIT Mizoram for their help.

References

1. Lalthangliana, B.: 'Mizo tihin ṭawng a nei lo' tih kha
2. Sarmah, P., Wiltshire, C.: An acoustic study of Mizo tones and morpho-tonology, unpublished
3. Chhangte, L.: A preliminary grammar of the Mizo language, Master's thesis. University of Texas, Arlington (1986)
4. Fanai, L.: Some aspects of the auto segmental phonology of English and Mizo, M.Litt. Dissertation, CIEFL, Hyderabad (1989)
5. Fanai, L.: Some aspects of the lexical phonology of Mizo and English: an auto segmental approach, Ph.D. dissertation, CIEFL, Hyderabad (1992)
6. http://en.wikipedia.org/wiki/Hunterian_transliterationdia_does_not_exist
7. Pakray, P., Pal, A., Majumder, G., Gelbukh, A.: Resource building and parts-of-speech (POS) tagging for the Mizo language. In: 2015 Fourteenth Mexican International Conference on Artificial Intelligence (MICAI), pp. 3–7. IEEE (2015)
8. Sag, I.A., Baldwin, T., Bond, F., Copestake, A., Flickinger, D.: Multiword expressions: a pain in the neck for NLP. In: Gelbukh, A. (ed.) CICLing 2002. LNCS, vol. 2276, pp. 1–15. Springer, Heidelberg (2002). https://doi.org/10.1007/3-540-45715-1_1
9. Chakraborty, T., Bandyopadhyay, S.: Identification of reduplication in Bengali corpus and their semantic analysis: a rule-based approach. In: Proceedings of 23rd International Conference on Computational Linguistics, pp. 73–76 (2010)
10. Nongmeikapam, K., Nonglenjaoba, L., Nirmal, Y., Bandyopadhyay, S.: Reduplicated MWE (RMWE) helps in improving the CRF based Manipuri POS tagger. Int. J. Inf. Technol. Comput Sci. **2**(1), 45–59 (2012)
11. Parimalagantham, A.: A Study of Structural Reduplication in Tamil and Telugu. Doctoral dissertation. http://www.languageinindia.com/aug2009/parimalathesis.html
12. Balabantaray, R.C., Lenka, S.K.: Computational model for reduplication in Odia. Int. J. Comput. Linguist. Nat. Lang. Process. **2**(2), 266–273 (2013)
13. Asad, M.: Reduplication in modern Maithili. J. Lang. India **14**(4), 28–58 (2015)
14. Baldwin, T., Bannard, C., Tanaka, T., Widdow, D.: An empirical model of multiword expressions decomposability. In: Proceedings of the ACL Workshop on Multiword Expressions: Analysis, Acquisition and Treatment, vol. 18, pp. 89–96, July 2003
15. Venkatapathy, S., Agrawal, P., Josh, A.K.: Relative compositionality of Noun+Verb multi-word expressions in Hindi. In: Proceedings of ICON Conference on Natural Language Processing, Kanpur (2005)
16. Becker, J.D.: The phrasal lexicon. In: Proceedings of Theoretical Issues of NLP, Workshop in CL, Linguistics, Psychology and AI, Cambridge, pp. 60–63 (1975)

17. The Sino-Tibetan Language Family_STEDT.htm. Accessed 5 Aug 2015
18. Chawngthu, T.: Mizo thuhlaril hmasawn dan part – I, September 2011. Misual.com
19. Khiangte, L.: Thuhlaril, 2nd Edn. (1997)
20. Lalthangliana, B.: Mizo Literature, 2nd Edn. (2004)
21. Chawngthu, T.: Mizo thuhlaril hmasawn dan part – II. Misual.com
22. Lalsangpuii, M.A.: The problems of English teaching and learning in Mizoram. J. Lang. India **15**(5) (2015)
23. Zoramdinthara: Mizo Fiction: Emergence and Development. Ruby Press & Co., New Delhi, pp. 1–2 (2013)
24. http://www.vanglaini.org/

Classification of Textual Genres Using Discourse Information

Elnaz Davoodi, Leila Kosseim[(✉)], Félix-Hervé Bachand, Majid Laali,
and Emmanuel Argollo

Department of Computer Science & Software Engineering,
Concordia University, Montreal, Canada
{e_davoo,m_laali}@encs.concordia.ca, leila.kosseim@concordia.ca,
felixherve@gmail.com, emmanuel.argollo@gmail.com

Abstract. This papers aims to measure the influence of textual genre on the usage of discourse relations and discourse markers. Specifically, we wish to evaluate to what extend the use of certain discourse relations and discourse markers are correlated to textual genre and consequently can be used to predict textual genre. To do so, we have used the British National Corpus and compared a variety of discourse-level features on the task of genre classification.

The results show that individually, discourse relations and discourse markers do not outperform the standard bag-of-words approach even with an identical number of features. However, discourse features do provide a significant increase in performance when they are used to augment the bag-of-words approach. Using discourse relations and discourse markers allowed us to increase the F-measure of the bag-of-words approach from 0.796 to 0.878.

1 Introduction

Well-written texts are composed of textual units that are connected to each other via discourse relations. Such relations (e.g. CAUSE, CONDITION) communicate an inference intended by the writer and allow the creation of coherent connections between textual units. Discourse relations can be made explicit through discourse markers such as *but, since, because*, etc. or can be left implicit, when no explicit cue phrase is used to indicate the relation.

Previous work such as [1,4,18,26] has shown a correlation between the use of discourse relations and certain textual dimensions, such as genre, level of formality and level of readability. For example, Webber [26] has shown that the distribution of discourse relations in the PDTB corpus [19] is influenced by the textual genre; that is, texts from different genres tend to contain more of discourse relations than others.

The goal of this paper is to provide more insight on these preliminary investigations and measure the influence of textual genre on the usage of discourse relations and discourse markers on a larger scale. Specifically, we wish to evaluate

© Springer International Publishing AG, part of Springer Nature 2018
A. Gelbukh (Ed.): CICLing 2016, LNCS 9623, pp. 636–647, 2018.
https://doi.org/10.1007/978-3-319-75477-2_46

to what extend the use of discourse relations and discourse markers are corre-lated to textual genre and consequently can be used to predict textual genre. To do so, we have used the British National Corpus [2] which contains naturally occurring texts organized into various textual genres and compared a variety of discourse-level features on the task of genre classification.

2 Previous Work

In the literature, the term *genre* is often used to refer to slightly different concepts and is used in variety of domains such as linguistics, music, web documents, etc. [6]. In the context of written texts, Lee [16] and Swales [23] define textual genre using external criteria such as the type of intended audience, the communica-tive purpose, the activity type, etc. However, because these external criteria are difficult to detect automatically, efforts have been made to define textual genre using internal linguistic and structural properties which are easier to detect and measure [5,10,12].

Automatic genre classification is typically based on machine learning tech-niques that use structural and linguistic properties of texts. Previous work on automatic genre classification have generally followed two main approaches: lin-guistic analysis and frequency-based techniques. Karlgren and Cutting [11] used word and character statistics, part-of-speech (POS) frequencies as well as func-tion word counts as features sets. They used the Brown corpus [7] and performed the classification in three genres. These features achieved a 73% accuracy on 4 genres. Kessler et al. [12] investigated the influence of four sets of structural (e.g. POS frequencies), lexical (e.g. word frequencies), character level (e.g. punctua-tion and delimiter frequencies) and derivative features (i.e. ratio and variation of lexical and character-level features) on automatic genre classification. They pointed out that their feature set achieves a higher performance than Karlgren et al. features [11]; however there is no evidence that this improvement is statis-tically significant. In addition to focusing on a wide range of linguistic features, Freund et al. [8] used a bag-of-word document representation; however the data for this experiment was collected from the Internet which made the data set biased and evaluation was hard to reproduce. Instead of considering all of the words in the documents, Stamatatos et al. [22] considered only the most com-mon words in English as well as punctuation marks as a feature set and classified the Wall Street Journal articles into its four genres. The most common words in English were extracted from the BNC corpus. Four genres from the Wall Street Journal formed the corpus, but only 13 to 20 samples per genre were used to get the error rate of less than 7%.

To our knowledge, very little research has explored the influence of discourse-level properties on automatic genre classification. Webber [26] has investigated the influence of textual genre on the usage of discourse relations within the PDTB corpus [19]. Although the corpus was rather small and skewed (1902 documents in the *news* genre, 104 documents in *essays*, 55 in *summaries*, and 49 in *letters*), she showed that the distribution of discourse relations is influenced by the textual genre. Moreover, it was pointed out that the genre appears to be a predictive feature for labelling discourse relations, especially when there is no lexical cue to signal the relation. More recently, Bachand et al. [1] studied the usage of discourse relations across textual genres, and also across sub-topics of each genre. In their investigation, a wide range of corpora across various textual genres were used (e.g. [3,19,20,24,25]). According to their corpus study, certain discourse relations are more likely to occur in certain textual genres, and further down, in certain sub-topics of these specific genres.

As a follow-up to these recent works, we wanted to investigate if the differences in discourse-level usage noted by Webber [26] and Bachand et al. [1] were sufficient to be used as features for a textual genre classification task.

3   Methodology

3.1   The Corpus

In order to perform textual genre classification, we used a subset of the British National Corpus [2]. The British National Corpus (BNC) is a collection of English documents (100 million words) from various sources, both written and spoken. This corpus was selected because it is significant in size and is already divided into 8 textual genres. In this work, we only considered 4 of these genres, as we only deal with written documents (as opposed to spoken) and the definition of some classes is rather broad. The subset of the BNC that we used is composed of 2,179 documents (about 60,000,000 words) divided in 4 different textual genres:

1. Academic Prose (ACPROSE) is composed of documents containing specialized explanations in a specific field of study, such as research papers, academic theses, and studies. These types of documents are typically segmented into distinct sections, such as abstract, methodology, discussion, and results, each making use of various discourse structures.
2. Fiction (FICTION) contains documents that follow the general structure of a fictional story. It should be noted, that the structure of narrative fiction is not very strict.
3. News (NEWS) contains documents which retell series of events. These are typically news articles, recaps of sporting events, or political editorials.
4. Non-academic Prose and Biographies (NONAC) have a similar communicative purpose to academic prose. However, whereas academic prose targets audiences at the university-level, non-academic prose targets audiences with general knowledge. The NONAC genre is also mostly divided into sections, with the exception of biographies, with each section having its own discourse structure.

Table 1 summarizes the subset of the BNC used in our experiments. As Table 1 shows, the corpus is somewhat balanced both with respect to the number of documents (with NONAC being more frequent), and with respect to the number of words (with NEWS being generally shorter).

Table 1. Statistics on the BNC sub-corpus used

	Textual genre				Total
	ACPROSE	FICTION	NEWS	NONAC	
Number of documents	497	452	486	744	2, 179
Number of words	15, 715, 469	15, 806, 443	4, 300, 672	24, 064, 370	59, 886, 954
Average no of words/document	31, 621	34, 970	19, 158	32, 345	27, 484

3.2   Discourse Features

In order to extract discourse-level information, the documents from the BNC subset were parsed using the PDTB End-to-End Discourse Parser [17]. Several publicly available discourse parsers could have been used (eg. [14,17]). We chose the End-to-End parser as it is the most commonly used parser and provides local discourse-level information such as the type of discourse relations (i.e. *implicit* or *explicit*), the name of the relation (known as its *sense*) and the discourse marker when applicable. The End-to-End parser uses the PDTB [19] set of discourse relations organised into 3 levels of granularity: 4 relations at level 1, sub-divided into 12 relations at level 2 themselves sub-divided into 23 relations at level 3. For the purpose of this work, we considered the 12 relations[1] at level 2 for which the End-to-End parser achieves the best performance. In addition, to tag discourse markers, the End-to-End parser uses the set 100 discourse markers from the PDTB.

Several features were used in order to evaluate the influence of discourse-level information on textual genre:

Discourse Relations (DR) at level-2 of the PDTB were used as the first set of features. As indicated above, these were extracted from the documents using the End-to-End Discourse parser [17]. For this feature, we used all the relations identified by the End-to-End parser regardless of how these were realized in the documents: *explicitly* through a discourse marker or *implicitly*[2]. This gave rise to 12 features.
For the value of each DR feature, we used the Log Likelihood ratio as defined by [21]. This measure was used as it indicates how many times each DR is more likely to occur in a specific genre as opposed to another.

Discourse Markers (DM) can be used to signal several discourse relations. For example, the marker *since* can signal a TEMPORAL or a CAUSAL relation. According to [15], markers can signal on average 3.05 different relations in the PDTB. The DR feature above takes into account both explicitly stated relations (those signalled via a discourse marker), as well as implicit relations

---

[1] Which include ASYNCHRONOUS, SYNCHRONOUS, CAUSE, CONDITION, CONTRAST, CONCESSION, CONJUNCTION, INSTANTIATION, RESTATEMENT, ALTERNATIVE, EXCEPTION and LIST.

[2] We experimented with using only explicit relations and only implicit relations, but the results were not conclusive.

(those that are not signaled by a marker). In order to focus only on explicit relations and to minimize the effect of mislabelled discourse relations, we also used discourse markers (as tagged by the End-to-End parser) as a feature set. Previous work have used different sets of discourse markers (e.g. [13]); however, to ensure consistency with the previous feature, we used the list of 100 discourse markers used in the PDTB corpus [19].

Here also, we used the Log Likelihood ratio to calculate how many times more likely each DM is in each genre. Each document is represented as a vector of 100 features (one for each DM), and the value of each feature is its Log Likelihood ratio.

It is important to note that the discourse markers that we used, actually mark a relation. Indeed, some cue phrases (such as *and*) may be used to signal a discourse relation (e.g. CONJUNCTION) or may be used in a non-discourse marking role. Section 4.3 analyses the difference between the use of discourse markers (that do signal a discourse relation) as opposed to cue phrases (that may not signal a relation).

Bag-of-words (BOW) To compare the effectiveness of the above features, we used a standard bag-of-words approach as a baseline. Words were extracted after case-folding, stemming, and digit and punctuation removal. In addition, for comparative purposes, we also used the Log Likelihood ratio of words across the four genres and only considered the words that had a Log Likelihood ratio up to 100 times less than the highest Log Likelihood ratio. This gave rise to 2,233 features.

Table 2 summarizes statistics on the discourse-level features extracted from our BNC sub-corpus. As Table 2 shows, the NEWS genre seems to contain significantly less DRs, but recall from Table 1 that it also contains less words. On average, this genre contains more DRs (1 DR every 7.30 words) compared to the other genres (1 DR every 13 or 18 words for the other genres). Another interesting remark is that the NEWS genre seems to use more discourse markers (on average, 1 word out of 22 is a marker) whereas the other genres have a marker every 44 to 50 words.

Table 2. Discourse-level features in the BNC sub-corpus used

	Textual genre				Total
	ACPROSE	FICTION	NEWS	NONAC	
Number of Discourse Relations (DR)	836,861	1,236,677	591,463	1,335,213	4,000,214
Number of Discourse Markers (DM)	343,170	352,772	197,395	498,281	1,391,618
Number of Cue Phrases (CP)	1,190,664	1,148,681	285,221	1,804,345	4,428,911
Ratio no of words/DR	18.78	12.82	7.30	18.18	15.15
Ratio no of words/DM	47.61	45.45	22.22	50.00	43.48

4   Results and Analysis

To perform the classification task, we used 3 classifiers provided by WEKA [9]: Multinominal Naïve Bayes, Decision Tree, and Random Forest. The first two classifiers were used as a baseline; while the Random Forest was investigated for its properties of reducing overfitting.

4.1   Initial Results

Table 3 shows the results obtained in terms of precision, recall, and F-measure using 10-fold cross validation. The last column of Table 3 indicates if there is a statistically significant decrease ($\Downarrow$) or no difference ($=$) in F-measure compared to the bag-of-words (BOW) model.

As Table 3 shows, the best results are consistently obtained using the bag-of-words features, regardless of the classifier used. This set of features is however, much larger than the others (2,233 vs 100 and 12). The performance of discourse markers (DM), with only 100 features, is very close to that of the BOW. In the case of the Random Forest classifier, it even achieves the same performance as BOW[3]. Finally, discourse markers (DM) achieve a better F-measure with all three classifiers, than discourse relations (DR).

Table 3. Initial results of the classification task

Classifier	Features	# Features	P	R	F	Stat. sign.
Naïve Bayes	BOWtop2233	2233	0.761	0.745	0.733	
	DM	100	0.682	0.662	0.653	$\Downarrow$
	DR	12	0.550	0.517	0.511	$\Downarrow$
Decision Tree	BOWtop2233	2233	0.757	0.746	0.748	
	DM	100	0.695	0.695	0.695	$\Downarrow$
	DR	12	0.629	0.629	0.629	$\Downarrow$
Random Forest	BOWtop2233	2233	0.816	0.798	0.796	
	DM	100	0.797	0.800	0.797	$=$
	DR	12	0.717	0.717	0.715	$\Downarrow$

4.2   Influence of the Feature Size

As shown in Table 3, the BOW features achieve the best results; however, it has two major drawbacks: First, it constitutes a much larger feature set than the other two approaches, and second, the actual words used as features need to be identified for each corpus and hence these features are tailored for the corpus at hand. On the other hand, discourse relations and discourse markers both constitute a small feature set and the features are fixed for all corpora.

---

[3] The difference between the two F-measures is not statistically significant.

To investigate this further, we performed additional experiments, but this time, we reduced the number of features so as to be at par across all experiments. Specifically, we took:

1. BOWtop12: the top 12 most discriminating words.
2. BOWrandom12: 12 random words, to be used as a baseline.
3. BOWtop100: the top 100 most discriminating words.
4. BOWrandom100: 100 random words, to be used as a baseline.
5. DMtop12: the top 12 most discriminating discourse markers.

The results are shown in Tables 4 and 5. Not surprisingly, random words always achieve the lowest performance (equivalent to picking the most frequent genre). Most importantly, the tables show that even when the cardinality of the features sets are identical, the BOW approach still outperforms discourse relations and discourse markers.

Table 4. Results of classification tasks using 12 features

Classifier	Features	# Features	P	R	F	Stat. sign.
Naïve Bayes	BOWtop12	12	0.679	0.617	0.610	
	DR	12	0.550	0.517	0.511	⇓
	DMtop12	12	0.598	0.607	0.573	⇓
	BOWrandom12	12	0.347	0.250	0.227	⇓
Decision Tree	BOWtop12	12	0.689	0.682	0.682	
	DR	12	0.629	0.629	0.629	⇓
	DMtop12	12	0.622	0.622	0.619	⇓
	BOWrandom12	12	0.322	0.349	0.266	⇓
Random Forest	BOWtop12	12	0.729	0.720	0.717	
	DR	12	0.717	0.717	0.715	⇓
	DMtop12	12	0.675	0.669	0.661	⇓
	BOWrandom12	12	0.457	0.351	0.266	⇓

4.3 Discourse Markers Versus Cue Phrases

Recall from Sect. 3.2, that to be considered a discourse marker, cue phrases need to mark a discourse relation. For example, the cue phrase *since* was considered as a discourse marker only if it marked a discourse relation (i.e. CAUSE). If the *since* did not mark a relation, as in:

(1) Equitable of Iowa Cos., Des Moines, had been seeking a buyer for the 36-store Younkers chain *since* June, when it announced its intention to free up capital to expand its insurance business[4].

---

[4] The example is taken from the PDTB [19].

Table 5. Results of classification tasks using 100 features

Classifier	Features	# Features	P	R	F	Stat. sign.
Naïve Bayes	BOWtop100	100	0.723	0.705	0.681	
	DMtop100	100	0.682	0.662	0.653	⇓
	BOWrandom100	100	0.425	0.411	0.382	⇓
Decision Tree	BOWtop100	100	0.746	0.737	0.739	
	DMtop100	100	0.695	0.695	0.695	⇓
	BOWrandom100	100	0.532	0.511	0.504	⇓
Random Forest	BOWtop100	100	0.818	0.805	0.805	
	DMtop100	100	0.797	0.800	0.797	⇓
	BOWrandom100	100	0.544	0.525	0.519	⇓

then it was not counted as a feature. The intuition behind this was to avoid adding noisy features as these cue phrases are often grammatical words. The disadvantage, however, is that a discourse parser is required to parse the documents in advance to identify discourse relations and markers. To evaluate if the use of such a discourse parser was really necessary, we compared the use of both cue phrase (CP) and discourse markers (DM). Hence we performed the same genre classification task again but replaced discourse markers (DM) with their corresponding cue phrases (CP) without verifying if these were used to mark a discourse relation of not. As shown in Table 6, using all cue phrases does not provide as much information as using only the cue phrases that signal a discourse relation. This seems to show that it is not the cue phrase per se that is a discriminating feature, but its discourse usage.

Table 6. Results of classification using discourse markers (DM) vs cue phrases (CP).

Classifier	Features used	No. features	P	R	F	Stat. sign.
Naïve Bayes	DM	100	0.682	0.662	0.653	
	CP	100	0.544	0.493	0.493	⇓
Decision Tree	DM	100	0.695	0.695	0.695	
	CP	100	0.560	0.559	0.549	⇓
Random Forest	DM	100	0.797	0.800	0.797	
	CP	100	0.606	0.603	0.595	⇓

4.4 Feature Combination

As Sects. 4.1 and 4.2 showed, regardless of the number of features, discourse features alone do not achieve the performance of BOW. However, since these features are complementary, we tried to combine them to see if discourse features could somehow improve the BOW model. We used the best performing

BOW approach (BOWtop2233) and augmented it with discourse relations only, discourse markers only and both discourse relations and markers.

As Table 7 shows, augmenting the BOW model with all discourse features (DM + DR) increases the F-measure of all classifiers significantly. For example, the F-measure of the Random Forest classifier increases from 0.796 to 0.878 with discourse features. Note that this increase in performance is significant and only requires the addition of 112 features (12 DR + 100 DM). The difference between the effect of discourse markers and discourse relations is not significant - however, because the two features measure essentially the same linguistic phenomena, the use of both features may not be necessary. The best performance was in fact achieved using a Random Forest classifier with the BOW model and only the addition of discourse markers (F = 0.884). Considering that discourse relations are made up of only 12 features, they might constitute a good choice to augment the standard BOW approach.

Table 7. Final results of the classification task

Classifier	Features	# Features	P	R	F	Stat. sign.
Naïve Bayes	BOWtop2233	2233	0.761	0.745	0.733	
	BOWtop2233+DR	2245	0.809	0.807	0.803	⇑
	BOWtop2233+DM	2333	0.806	0.804	0.801	⇑
	BOWtop2233+DM+DR	2345	0.806	0.804	0.801	⇑
Decision Tree	BOWtop2233	2233	0.757	0.746	0.748	
	BOWtop2233+DR	2245	0.809	0.810	0.809	⇑
	BOWtop2233+DM	2333	0.811	0.811	0.811	⇑
	BOWtop2233+DM+DR	2345	0.810	0.810	0.810	⇑
Random Forest	BOWtop2233	2233	0.816	0.798	**0.796**	
	BOWtop2233 +DR	2245	0.879	0.879	0.870	⇑
	BOWtop2233+DM	2333	0.884	0.884	0.884	⇑
	BOWtop2233+DM+DR	2345	0.878	0.878	**0.878**	⇑

5 Conclusion and Future Work

This paper aimed to measure the influence of textual genre on the usage of discourse relations and discourse markers. Specifically, we evaluated to what extend the use of discourse relations and discourse markers are correlated to textual genre and consequently can be used to predict textual genre. To do so, we have used a subset of the British National Corpus and compared a variety of discourse-level features on the task of genre classification.

The results show that individually, discourse relations and discourse markers do not outperform the standard bag-of-words approach even when the number of features is identical. However, discourse features do provide a significant increase in performance when they are used to augment the bag-of-words approach. Using discourse relations and discourse markers allowed us to increase the F-measure of the BOW approach from 0.796 to 0.878. This seems to show that discourse information models a linguistic phenomenon which the bag-of-words does not. The bag-of-words approach can model well textual topic, however for textual genre, this approach is not enough, and can be complemented by discourse information.

Our investigations also showed that although the BOW approach achieves a better performance, the actual words used as features need to be identified for each corpus and hence these features are tailored for the corpus at hand. On the other hand, discourse relations and discourse markers both constitute a small feature set (12 relations and 100 markers) and the features are fixed for all corpora. Using only these features, the Naïve Bayes and Decision Tree approaches achieve lower results than the BOW approach; however, with the Random Forest classifier, discourse markers alone achieve results that are statistically equivalent to the tailored BOW approach. If tailoring the feature set to the corpus is not an option, discourse markers constitute a very good alternative for genre classification.

Another interesting result is the fact that not all cue phrases are discriminating for genre classification. Our results show that using cue phrases (*since, and, because* ...) that actually mark a discourse relation produce a higher performance than using all cue phrases regardless of whether they signal a discourse relation or not. This is not surprising, as most cue phrases are grammatical words that have a low discriminating power. The fact that discourse markers are good indicators of textual genre may indicate that their usage to mark discourse relations is different across genres.

Our work has focused on the analysis of 4 genres of the British National Corpus. An obvious question to investigate is the validity of our results on other corpora. The Brown corpus [7], for example, also provides samples of different genres and could be used to validate our results.

Another very interesting question is to investigate how the conclusions drawn from our work can be used to improve discourse parsing. Our experiments seem to show that different genres have different discourse properties that are significant enough to be used as a basis for textual genre classification. Therefore, as Webber [26] noted, it might be the case that knowing the textual genre of a document could be an interesting feature to improve discourse parsing. Knowing, for example, that the text being parsed is an academic prose as opposed to a work of fiction might be useful to properly label a text span with a discourse relation as opposed to another.

Acknowledgement. The authors would like to thank the anonymous reviewers for their feedback on the paper. This work was financially supported by NSERC.

References

1. Bachand, F.-H., Davoodi, E., Kosseim, L.: An investigation on the influence of genres and textual organisation on the use of discourse relations. In: Gelbukh, A. (ed.) CICLing 2014. LNCS, vol. 8403, pp. 454–468. Springer, Heidelberg (2014). https://doi.org/10.1007/978-3-642-54906-9_37

2. BNC Consortium: The British National Corpus, version 3 (BNC XML Edition) (2007). http://www.natcorp.ox.ac.uk/

3. Carlson, L., Okurowski, M.E., Marcu, D.: RST Discourse Treebank. Linguistic Data Consortium, LDC2002T07, University of Pennsylvania (2002)

4. Davoodi, E., Kosseim, L.: On the influence of text complexity on discourse-level choices. Int. J. Comput. Linguist. Appl. **6**(1), 27–42 (2015)

5. Fang, C.A., Cao, J.: Text Genres and Registers: The Computation of Linguistic Features. Springer, Heidelberg (2015). https://doi.org/10.1007/978-3-662-45100-7

6. Finn, A., Kushmerick, N.: Learning to classify documents according to genre. J. Am. Soc. Inf. Sci. Technol. **57**(11), 1506–1518 (2006)

7. Francis, W.N.: A manual of information to accompany a standard sample of present-day edited American English, for use with digital computers. Department of Linguistics, Brown University (1971)

8. Freund, L., Clarke, C.L.A., Toms, E.G.: Towards genre classification for IR in the workplace. In: Proceedings of the 1st International Conference on Information Interaction in Context, New York, pp. 30–36 (2006)

9. Hall, M., Frank, E., Holmes, G., Pfahringer, B., Reutemann, P., Witten, I.H.: The WEKA data mining software: an update. ACM SIGKDD Explor. **11**(1), 10–18 (2009)

10. Karlgren, J.: Stylistic experiments in information retrieval. In: Strzalkowski, T. (ed.) Natural Language Information Retrieval. Text, Speech and Language Technology, vol. 7, pp. 147–166. Springer, Dordrecht (1999). https://doi.org/10.1007/978-94-017-2388-6_6

11. Karlgren, J., Cutting, D.: Recognizing text genres with simple metrics using discriminant analysis. In: Proceedings of the 15th Conference on Computational Linguistics (ACL), Las Cruces, vol. 2, pp. 1071–1075 (1994)

12. Kessler, B., Numberg, G., Schütze, H.: Automatic detection of text genre. In: Proceedings of the 35th Annual Meeting of the Association for Computational Linguistics and Eighth Conference of the European Chapter of the Association for Computational Linguistics (ACL/EACL), Madrid, pp. 32–38 (1997)

13. Knott, A.: A data-driven methodology for motivating a set of coherence relations. Ph.D. thesis (1996)

14. Laali, M., Davoodi, E., Kosseim, L.: The CLaC discourse parser at CoNLL-2015. In: CoNLL 2015, pp. 56–60, Beijing (2015)

15. Laali, M., Kosseim, L.: Inducing discourse connectives from parallel texts. In: Proceedings of the 25th International Conference on Computational Linguistics (COLING), Dublin, pp. 610–619 (2014)

16. Lee, D.Y.: Genres, registers, text types, domains and styles: clarifying the concepts and navigating a path through the BNC jungle. Technology **5**, 37–72 (2001)

17. Lin, Z., Ng, H.T., Kan, M.-Y.: A PDTB-styled end-to-end discourse parser. Nat. Lang. Eng. **1**, 1–34 (2012)

18. Pitler, E., Nenkova, A.: Revisiting readability: a unified framework for predicting text quality. In: Proceedings of the Conference on Empirical Methods in Natural Language Processing (EMNLP), Honolulu, pp. 186–195, October 2008

19. Prasad, R., Dinesh, N., Lee, A., Miltsakaki, E., Robaldo, L., Joshi, A.K., Webber, B.L.: The penn discourse TreeBank 2.0. In: Proceedings of the 6th International Conference on Language Resources and Evaluation (LREC), Marrakech, pp. 2961–2968 (2008)

20. Prasad, R., McRoy, S., Frid, N., Joshi, A., Hong, Y.: The biomedical discourse relation bank. BMC Bioinform. **12**, 188 (2011)

21. Rayson, P., Garside, R.: Comparing corpora using frequency profiling. In: Proceedings of the Workshop on Comparing Corpora, Hong Kong, pp. 1–6 (2000)

22. Stamatatos, E., Fakotakis, N., Kokkinakis, G.: Text genre detection using common word frequencies. In: Proceedings of the 18th Conference on Computational Linguistics (ACL), vol. 2, pp. 808–814 (2000)

23. Swales, J.: Genre Analysis: English in Academic and Research Settings. Cambridge University Press, Cambridge (1990)

24. Taboada, M., Anthony, C., Voll, K.: Methods for creating semantic orientation dictionaries. In: Proceedings of the 5th International Conference on Language Resources and Evaluation (LREC), Genova, pp. 427–432 (2006)

25. Taboada, M., Grieve, J.: Analyzing appraisal automatically. In: Proceedings of AAAI Spring Symposium on Exploring Attitude and Affect in Text, pp. 158–161. Stanford University (2004)

26. Webber, B.: Genre distinctions for discourse in the Penn TreeBank. In: Proceedings of the Joint Conference of the 47th Annual Meeting of the ACL and the 4th International Joint Conference on Natural Language Processing (ACL-AFNLP: Volume 2), Suntec, pp. 674–682, August 2009

Features for Discourse-New Referent Detection in Russian

Svetlana Toldova[1] and Max Ionov[2,3](✉)

[1] National Research University "Higher School of Economics", Moscow, Russia
toldova@yandex.ru
[2] Goethe University Frankfurt, Frankfurt, Germany
max.ionov@gmail.com
[3] Moscow State University, Moscow, Russia

Abstract. This paper concerns discourse-new mention detection in Russian. This might be helpful for different NLP applications such as coreference resolution, protagonist identification, summarization and different tasks of information extraction to detect the mention of an entity newly introduced into discourse. In our work, we are dealing with the Russian where there is no grammatical devices, like articles in English, for the overt marking a newly introduced referent. Our aim is to check the impact of various features on this task. The focus is on specific devices for introducing a new discourse prominent referent in Russian specified in theoretical studies. We conduct a pilot study of features impact and provide a series of experiments on detecting the first mention of a referent in a non-singleton coreference chain, drawing on linguistic insights about how a prominent entity introduced into discourse is affected by structural, morphological and lexical features.

Keywords: Coreference resolution · Discourse-new referent
Discourse processing · Natural language processing · Machine learning

1 Introduction

The task of tracking all mentions referring to the same entity in discourse, or coreference resolution task, is essential for many text-mining applications. This task has received much attention over the last years [15, 21] and nowadays it has achieved a rather high precision. Although current machine learning approaches to coreference resolution perform quite well, many studies have sought to improve the quality, not only via improving the basic algorithms but also via taking into consideration various theoretical assumptions from the discourse models of referential choice (cf. the hierarchy of referential accessibility, e.g. [1]). Among others, there are issues such as discourse-new recognition. The decision on whether an NP is introducing a new entity into discourse could improve the quality of coreference resolution [10, 20]. Moreover, a particular type of an introductory NP could be a clue to the discourse role of a referent as to whether it is an entity that is the main topic of a long discourse span or it is an occasional one.

© Springer International Publishing AG, part of Springer Nature 2018
A. Gelbukh (Ed.): CICLing 2016, LNCS 9623, pp. 648–662, 2018.
https://doi.org/10.1007/978-3-319-75477-2_47

There is a comprehensive analysis of different features for discourse new (DN) detection for English ([10, 16, 19, 28], etc.). However, there is much less investigation concerning the first mention detection for languages without articles such as Russian. While for the former the only task is to decide whether an NP introduces a new referent in spite of an overt definite marker, for the latter the task is to differentiate among nearly all the NPs (except anaphoric pronouns and NPs with demonstratives) as to whether they introduce a new referent or not. This task is complicated for less resourced languages as there is no freely available high quality syntactic parsers or rich semantic resources such as WordNet. Thus, one could hardly rely upon some complicated syntactic NP properties or semantic NP heads relatedness. However, there is some theoretical research (cf. [2, 26]) concerning the specific NP structure features and lexical features that serve as markers for introducing a new referent. Moreover, there are special markers for introducing prominent referents (referents in focus of attention, cf. [6]) into discourse. The question is whether these features could be helpful within a shallow machine learning approach for discourse-new mention detection.

In our work we conduct a pilot study of discourse-new detection in Russian. We take a subset of basic features for discourse-new mention detection for English and enrich it with features mentioned in the theoretical literature (mainly for Russian). The task is to check whether the theoretically assumed features really work. We suggest that the analysis of these features impact on the first mention vs. repeated mention classification given that the singletons (NPs referring to the entities mentioned only once) are filtered out in the previous stages of analysis.

The paper is structured as follows. In Sect. 2 we present the prerequisites for our experiment. In Sect. 2.1 the overview of general approaches to coreference resolution task is given. In Sect. 2.2 the approaches for the first mention detection in English are discussed. The Sect. 2.3 deals with various theoretical investigations of first-mention NP properties. Section 3 is devoted to the investigation of discourse-new detection features, their distribution in corpus and their correlation with the introductory NPs. In Sect. 4 we describe our experiments on DN mention detection and suggest the analysis of the features impact.

2   Background

2.1   Coreference Resolution Task and Discourse New Referent Detection. Task Settings

Coreference resolution is the task of grouping NPs referring to the same entity (mentions or referring expressions) into disjoint sets. An example is given in Example 1, where mentions from three sets are marked with brackets and corresponding indexes:

(1)   ...Probovali sravnivat' text [zapisnoj knižki]$_{a1}$ s rukopisjami [Nahimova]$_{a2}$.
<... >issledovatelej zainteresovala [ešjo odna [jego]$_{a2}$ zapisnaja knižka]$_{a3}$.

(they) tried to compare (the) text of (the) [notebook]$_{a1}$ with (the) [Nahimov]$_{a2}$'s manuscripts. The researchers were interested in [another ([his]$_{a2}$) notebook]$_{a3}$.

This example includes different types of referential expressions: the named entity (*Nahimov*), the possessive pronoun *jego* referring to this named entity, and NPs referring to two different entities of the same ontological class ('notebooks'). The NP_{a1} *zapisnaja knizhka* (notebook) has no overt markers as to whether it is the first mention of an entity, a repeated one or if it is a generic use of 'notebook' referring to a class of entities. Such ambiguity is a specific challenge for coreference resolution in languages without articles. Another difficulty is that NP_{a1} and NP_{a3} have a common part *zapisnaja knizhka* (notebook), though they refer to different entities. However, the latter has *esche odna* (one more) as a specific introductory modifier. Thus, there are cases where the DN detection might be helpful. Moreover, there are special lexical DN markers that serve as a signal for introducing the reference in discourse (cf. [17], see Sects. 2.3 and 3.4 for details).

The procedure for establishing coreference in the text such as that above can involve different types of information ranging from simple formal features such as token distance, NPs and NP heads equivalence, morphological congruence, syntactic features etc. ([15, 16] etc.) up to high level discourse and semantic features ([10, 13, 19, 21] and others). Moreover, it can include the task of separating singletons (NPs for entities mentioned only once in discourse) from coreferential NPs [24, 28].

Further knowledge that could influence the coreference resolution performance is discourse-new mention detection (see discussions in [10, 16, 19] etc.). This paper focuses on this task.

2.2 Overview of Discourse-New Detection Algorithms for English

The majority of previous works on discourse-new (DN) detection deal with English. One of the most detailed overview of various approaches to this task is given in [20]. The approaches suggested in [3, 19, 28] are among them. Though Ng and Cardie [16] reported that the incorporation of DN recognition into coreference resolution systems did not affect their general performance, further testing of the features from [16] as well as the development of other algorithms shows that the way in which DN detection is combined with the basic coreference resolution module also matters (see [10, 20] for details). In this paper we put the question of the DN utility aside and focus on the features that are used for DN detection.

The motivation for developing the algorithms for the DN detection task in English was the corpus study carried out by Poesio and Viera [22] where they reported that 52% of definite descriptions are discourse new. Thus, the main issue of DN recognition in English (apart from detecting indefinite descriptions) is to identify the class of definite descriptions. Vieira and Poesio in [29] suggested an algorithm for identifying five categories of definite descriptions that are licensed to occur as the first mention, on semantic or pragmatic grounds: (i) semantically functional descriptions [14] such as *the best* or *the first*; (ii) "descriptions serving as disguised proper names, such as The Federal Communications Commission"; (iii) predicative descriptions, including appositives and NPs in certain copular constructions, e.g. *Mr. Smith, the president of...,* etc.; (iv) "descriptions established (i.e., turned into functions in context) by restrictive modification [14] as in *The hotel where we stayed last night was pretty good*; (v) larger situation definite descriptions [8], i.e., definite descriptions like *the sun, the pope,* etc., which denote uniquely on the grounds of shared knowledge about a situation (Löbner's 'situational functions')".

Regarding the various features used for DN mention detection for English, a comprehensive set of features was analyzed and tested in [16]. They used 37 features to identify anaphoric and not-anaphoric NPs taking into consideration the information proposed in the previous works. Their set of features is based on grammatical features of an NP (whether an NP is a certain type of pronoun, a proper noun etc. or whether an NP has a modifier of a certain grammatical type), some string relation features (cf. 'lexical features') based on the identity of an NP or of its head to a preceding NP or the head of a preceding NP. Another class of features concerns some special constructions such as an NP with a relative clause, an NP with the superlative modifier, etc. These constructions require a definite marker in an NP, regardless of whether the NP was used anaphorically or not. This class of features is not relevant for Russian (see Sect. 2.3). Ng and Cardie also took into consideration semantic relatedness and NP sentence position. Another type of "constructional" features mentioned in the literature refers to the question of whether an NP is in an appositive or predicative construction [3]. However, these features need special syntactic analysis for Russian, which is beyond the scope of the present study.

The core aim of our experiment is to test the theoretical assumptions related to the referent first mention NP in Russian and find out whether they can be helpful for the task. Hence, even a subset of formal features is enough to set the baseline for this task. We took into consideration only string relations and grammatical features as the basis for our experiment (see Sect. 3.2 for details).

2.3 Discourse-New Referent Recognition in Russian: Theoretical Accounts

As has been shown in Sect. 2.1, the DN detection task for Russian (as a language without articles) differs significantly from that for English. In Russian, there are no overt grammatical markers of definiteness. Almost all the NPs (excluding anaphoric pronouns, refelexives and NPs with demonstratives) can refer to a newly introduced entity. Moreover, common NPs can have up to three interpretations (e.g. definite, indefinite or generic description, see Example 1 in Sect. 2.1).

One of the approaches for resolving ambiguous interpretations of an NP in article-less languages is to take into consideration the discourse status of the NP itself. This approach goes along with different cognitive-based coreference models and various typological findings. It has deep theoretical motivation, e.g. the hierarchy of referential accessibility suggested in Ariel [1] (see also [5,23] etc.). This hierarchy regulates the choice of anaphoric expressions: the more accessible a referent is, the less informative expressions can be used to refer to it; meanwhile the less accessible entities (e.g. newly-introduced into discourse) need more informative anaphoric expressions. The following hierarchy of NP types: zero < anaphoric pronouns < demonstratives < full NP corresponds to the referent accessibility hierarchy. The notion of accessibility varies through different discourse models, cf. 'topicality' as in [5], salience, activation [11,12], or whether an entity is in the focus of attention or not [6].

In [26], the theoretical account for the distribution of different full NP types in discourse is suggested. It is based on the notion of focus of attention (cf. [6]). The licensing of certain NP types for a referent depends on the basis of its discourse properties on whether it is in focus in a particular discourse unit or not.

Arutyunova [2] analyses the full NP structure with respect to the referent first-mention/non-first mention description. The main properties of the first-mention NPs specified by Arutyunova are as follows: (i) the NP length (they tend to be longer than the non-first ones); (ii) the number of adjectives (if it is higher than average); (iii) the semantics of adjectives (there is a tendency to use evaluative and qualitative adjectives). She also mentions special predicate types for referent introduction such as existential predications (c.f. features for discourse new descriptions detection suggested by Ng and Cardie [16]).

These observations are summed up in [4,26]. The latter presents a corpus analysis of introductory NPs in mass media texts of a special kind. The relevant features that we have taken into account from this work are special lexical clues for the introduction of a new referent in the focus of attention such as some types of indefinite pronouns, non-identity markers, novelty markers etc.

3   Features Analysis

Below we suggest an analysis of the features proposed in the theoretical literature for Russian (some of them coincide with the features discussed above). We rely only on a subset of structural features (see Sect. 2.1 for the discussion) as the basis for the DN mention classifier and explore the impact of the features mentioned in Sect. 2.3.

3.1   Data

Our experiments were conducted on RuCor, a Russian coreference corpus released as a part of the RU-EVAL campaign [27][1].

The corpus consists of short texts or fragments of texts in a variety of genres: news, scientific articles, blog posts and fiction. The whole corpus contains about 180 texts and 3 638 coreferential chains with 16 557 noun phrases in total. Each text in the corpus is tokenized, split into sentences and morphologically tagged using tools developed by Sharoff [25]. Noun phrases were obtained using a simple rule-based chunker [9]. The corpus was randomly split into train and test sets (70% and 30% respectively).

---

[1] The corpus may be downloaded on http://rucoref.maimbava.net.

Since anaphoric, reflexive, and relative pronouns cannot be used as first (non-anaphoric) mentions, it seemed more fair to ignore all the instances of these pronouns.

3.2   Basic Features

Firstly, we use string relation features. On the one hand, if an NP matches one of the preceding NPs or its head, it is highly probable that the former is a non-first mention. The number of occurrences of an NP or its head in the preceding text can be used as a feature (for English these features were used, for example, in [16]). Using these undoubtedly important features, we are setting the base level for the first mention detection.

Figure 1 shows a distribution of these features in the training set. In most cases there is a low number of preceding NP or head occurrences for the first mention NPs (c.f. more than 80% preceding NP match for non-first mentions).

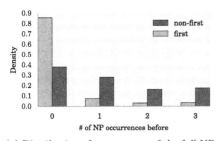
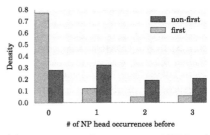

(a) Distribution of occurrences of the full NP   (b) Distribution of occurrences of NP head

Fig. 1. Distribution of string feature values

We also used some grammatical features of an NP as basic features (e.g. if the NP is a proper noun, has a demonstrative as a modifier etc.). For the full list see Table 4.

3.3   Structural Features of First Mention NPs

One of the introducing strategies is an extensive description of a new referent within the first-mention NP [2,4]. Consider Example 2 where the new Pope is introduced into a discourse. The introductory NP is composed of five tokens in length and has two adjectival premodifiers.

(2)   *76-letnij novyj glava milliarda katolikov.*
The 76 year-old new head of a billion of Catholics.

This assumption entails two classes of useful features: **NP length** (introductory NPs tend to be longer than non-first mentions) and **NP POS structure**

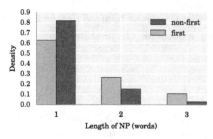

(a) Distribution of NP lengths

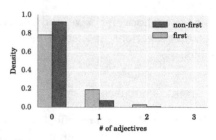

(b) Distribution of number of adjectives in NP

Fig. 2. Distribution of structural features

(introductory NPs have more adjectival modifiers). Distributions for those features are presented on Fig. 2. Though the one-token NPs are quite frequent in both classes, the longer NPs are predominately DN referent mentions.

The number of premodified adjectives is higher for introductory NPs relative to the average number of premodified adjectives.

3.4   Special Lexical Clues

Special attention for two more classes of features which are less discussed in the literature, is deserved. Usually the focus is on the special determiners that mark the identity of NP referents. However, there is a special class of so-called alterators (term used by Palek [17]). They mark the inequity of an NP referent to any preceding NP referent. Several classes of such alterators were suggested in [26]:

(a) indefinite markers, e.g. *odin* 'one', *nekij* 'a person';
(b) the inequity markers such as *drugoy, inoj* 'other' etc.;
(c) the similarity markers such as *takoj* 'such, *this* kind of', *podobnyj* 'analogous', *pohozchij* 'similar' etc.;
(d) markers for elements from a set *odin iz* 'one of the';
(e) *ostal'nue* 'the rest';
(f) the order of introduction: *pervyj is (nih)* 'the first', *vtoroj* 'the second', *poslednij* 'the last'.

Although the alterators do not occur very frequently in discourse, they can be reliable features for discourse-new descriptions detection.

Though Russian lacks articles there are still certain types of indefinite pronouns (specific indefinites, such as *kakoj-to* 'some of', *nekij* 'one of') that also occur only in first-mention NPs (cf. the feature 'indefinite pronouns' in [16]). The same is true for some quantifiers such as *mnogije* 'many of' or *bol'shinstvo* 'the majority' and others (cf. the feature 'quantifiers' in [16]).

There are also certain semantic classes of adjectives (see [4]), such as adjectives denoting the novelty of an entity *novyj* 'new', *nedavno otkrytyj* 'newly discovered' or evaluative adjectives such as 'mysterious', 'strange', 'curious', 'nice', 'modern', 'promising' and others. The corpus analysis provided in [4] has shown that these adjectives are typical for introductory contexts. The last class worth mentioning is a class of so-called classifiers whose nouns have very general semantics such as *chelovek* 'a man', *predmet* 'a thing', etc.

We take the lexical classes enumerated above as features for testing and compile the corresponding lists manually on the basis of lists suggested in the literature.

3.5   Automatically Extracted Lexical Features

The lexemes described above are not very frequently occurring features. In order to enlarge the list of adjectives that are specific for introductory NPs we performed an experiment that aimed to extract them automatically. Since this experiment in relies heavily on our main experiments, its design and results are presented hereafter in Sect. 4.3.

4   Experiments

4.1   Task

To check if the features proposed in the previous sections are adequate for distinguishing discourse-new mentions from recurring ones, we performed a series of experiments. For each experiment we built a classifier with a subset of features.

For our experiments we assumed that each noun phrase in a text belongs to one of the following three classes:

1. First mention (DN): the first mention of the coreference chain,
2. Recurring mention (Non-DN): the mention of an already introduced coreference chain,
3. Singleton: the mention of a referent that appears only once in the text.

Classifying mentions into these classes can be done in a number of ways. The most straightforward model for this task is multiclass classification with mentions as instances and 3 target classes. However, the distinction between the first two classes is different from the distinction between any of these classes and a singleton. In the first case there are two mentions (elements of some coreferential chain) with different discourse roles; however singletons do not usually count as mentions. Therefore, the task of identifying first mentions among all non-singleton mentions and identifying singleton mentions should be treated as two different tasks.

Since the focus of this paper is on the differences between discourse-new and referring coreferent mentions, in our experiments, singleton mentions were filtered out beforehand. Identifying singleton NPs for Russian is a matter of further research. Some approaches of filtering singletons in English are discussed in [24, 28].

4.2   Data

As has been mentioned in Sect. 3.1, we use RuCor for our experiment, taking 70% of it as a training corpus and 30% as a testing corpus. Since the dataset is unbalanced (instances of classes are in the ratio 1:4), we performed a balancing operation on a training set. Because of the relatively small size of the corpus we preferred oversampling to undersampling techniques. We chose Borderline-SMOTE1 method [7] over several popular oversampling algorithms since on average it yielded best results. Table 1 shows the number of instances before and after balancing.

Table 1. Number of instances before and after balancing (SMOTE ratio = 3.626)

	No balancing	With balancing
First	3 411	12 738
Non-first	12 367	9 868

Since there is no baseline for the DN-detection task for Russian, we created a simple heuristic baseline classifier with a set of heuristic rules. It works as follows:

- NP is a **first occurrence** if there are no other occurrences of this NP before
- NP is a **recurring occurrence** otherwise

4.3   Features

In our experiments we used several groups of features:

1. String relation features (class "string")
2. NP structural features ("struct")
3. Theoretically motivated lexical features ("lists")
4. Automatically extracted lexical features ("lists")

Below we describe these features.

String Relation Features. This group contains features that take into account the presence of the NP or its part in the preceding text. As we have seen in Sect. 3.2, most of the instances have low values of this feature. Given this, we may reformulate those features as the following binary features:

- **str_match_before**, number of occurrences of this NP in the text before: 0, less than 2, less than 3, 2 or more;
- **head_match_before**, number of occurrences of the head of this NP in the text before: 0, less than 2, less than 3, 2 or more;
- **is_proper_noun**, true if the whole NP is a proper noun
- **latin**, true if NP contains Latin letters;
- **uppercase**, true if NP is uppercase.

NP Structural Features. This group consists of two sets of features: those concerning the length on NP in tokens and those concerning the number of adjectives in the NP.

More specifically, we have extracted these binary features:

- **len_np**, the length of the NP in tokens: less than two words, more than two words;
- **n_adj**, the number of adjectives in the NP: 0, more than one, more than two;
- **conj**, true if there is a conjunction in the NP.

Pronominal Modifiers and Theoretically Motivated Lexical Features. We manually compiled several lists of words corresponding to theoretical expectations of discourse-new markers[2]. More precisely, we used the following lists:

- Demonstrative pronouns;
- Possessive pronouns;
- General class names: nouns that define a class (*building, manager,* etc.);
- Indefinite pronouns;
- New referent introductory adjectives (*contemporary, latest,* etc.);
- Non-identity and similarity markers: *another, similar,* etc.;
- Common knowledge markers (*famous, legendary,* etc.);
- Adjective markers of a discourse role in an NP (*main, small,* etc.);
- Subjective markers (*good, prestigious,* etc.)

Automatically Extracted Lexical Features. To extract the most important adjectives for the classification we performed a univariate feature selection operation using the χ^2 metric and 'bag-of-adjectives' as features: each feature meant the presence/absence of a unique adjective that we encountered in the training corpus. After this procedure we manually cleaned this list by removing the pronouns and words erroneously tagged as adjectives. From the cleaned list we extracted the 50 most important adjectives.

Top 10 adjectives from the list are given in Table 2.

4.4 Results

To implement the classifier, we used a Random Forest classifier from the scikit-learn Python library [18]. Since the test portion of our data set is unbalanced, overall classifier quality is not as important as the quality for the minority class. Results for this class are shown in Table 3. For each set experiment we state precision, recall and F1-measure.

In Table 3 we compare the performance of the classifier using structural features such as NP length and number of adjectives against the baseline and the classifier trained only on the basic feature set. The results show the increase in precision by more than two percent. However, the recall has decreased.

---

[2] Pronouns are a closed grammatical class therefore it may be treated as a list.

Table 2. Top 10 adjectives most valuable for classification

#	Adjective	Translation
1	novij	New
2	radioakivnij	Radioactive
3	russkij	Russian
4	pervij	First
5	sotsial'nij	Social
6	mestnij	Local
7	sobstvennij	Own
8	global'nij	Global
9	nebol'shoj	Not-big
10	regional'nij	Regional

Table 3. Classification results

	P	R	F1
Baseline	0.505	0.824	0.627
String	0.539	0.770	0.634
String + Struct	0.563	0.698	0.623
String + Struct + Lists	0.573	0.705	0.632

According to the table, the precision for all the features including the lexical lists increases by one percent. The recall has also increased slightly as compared to the classification without lists. Detailed analysis of the results shows the positive influence of those features on the performance. A minimal effect of structural and lexical features can be accounted for by some peculiarities of discourse structure (see Sect. 4.6) and the high sparseness of the list features. However, we find it a promising direction for further investigation of specific lexical features.

4.5   Analysis of Feature Contribution

To measure the importance of each feature we created a logistic regression classifier and trained it on our training data. Each coefficient of a trained classifier showed the impact of the corresponding feature. The results are shown in Table 4.

Our study shows that all the classes of features discussed above do matter for the DN mention detection task. For example, an NP length of more than one token has a positive correlation with DN class. The lexical features are not homogeneous. The non-identity and similarity alterators show a negative effect. However, a study of particular examples of this class has shown that this feature needs more precise analysis. The common-knowledge adjectives as well as indefinite pronouns and classifiers are more reliable features than subjectivity

Table 4. Feature importances for the DN detection task

Feature	Class	Coefficient
$str\_match\_before = 0$	String	1.0904
$str\_match\_before < 2$	String	−0.0704
$str\_match\_before < 3$	String	0.1171
$str\_match\_before > 2$	String	−0.4735
$head\_match\_before = 0$	String	1.2394
$head\_match\_before < 2$	String	0.3106
$head\_match\_before < 3$	String	−0.2663
$head\_match\_before > 2$	String	−0.0901
uppercase	String	−0.6143
latin	String	−0.4735
is_proper_noun	String	−0.9239
conj	Struct	−0.1072
$len\_np < 2$	Struct	−0.2964
$len\_np > 2$	Struct	0.3172
$n\_adj = 0$	Struct	−0.0161
$n\_adj > 1$	Struct	0.0012
$n\_adj > 2$	Struct	−0.5231
in_list_refer_to_CommKnowl	Lists	0.4179
in_list_adj-top50	Lists	0.9041
in_list_role_assess	Lists	−0.4250
in_list_NewRef	Lists	−0.5883
in_list_non-identity_sim	Lists	−0.4523
in_list_possessives	Lists	−2.3525
in_list_subjectivity	Lists	−1.0180
in_list_class	Lists	0.9906
in_list_demonstratives	Lists	−1.6133
in_list_indef	Lists	1.2672

adjectives. The automatically extracted lexical features (see Sect. 3.5) work better than manually built lists of subjectivity adjectives.

4.6   Discussion

Some of the most frequent types of precision mistakes are predicative NPs and appositive NPs that we did not take into consideration in our analysis. The fact is that these NP types tend to be long in tokens and include the vast description of an entity as in Example 3:

(3)   *On yavlyaetsya glavnym sponsorom kluba.*

The other class of cases when the longest NP for a referent is not the first one is a special discourse strategy of a referent introduction in fiction. The first mention is just a proper name without any details (cf. *Mashka* 'Mary') and the second mention is a detailed description of an entity (*Khuden'kaya bol'sheglazaya devochka* '(the) slim big-eyed girl').

The feature **in_list_non-identity_sim** has a negative coefficient. More detailed analysis of the examples shown that these are primarily NPs with the modifiers *takoj* 'such as', *podobnyj* 'similar to'. These cases are highly problematic for annotation for the majority of negative examples that concern abstract NPs with no referent. These are the cases of near-identity. Thus, these modifiers work well with concrete NPs, however, they occur more frequently with non-referential NPs.

The other type of mistakes (recall mistakes) comes from underestimation of one-token NPs. However, this type of DN NP is frequent for two-element chains, while the referents that are mentioned more than twice in discourse tend to be introduced with longer NPs. Though this hypothesis needs further investigation.

Thus, testing theoretically-accounted features and analysis of the cases contradicting with theory reveal new theoretically interesting phenomena and suggest new features for analysis.

5   Conclusions

In the present work we discussed various features used for the first mention detection classification task as a subtask of coreference resolution systems. We presented preliminary research on first mention detection with special emphasis on the Russian language.

The focus was on the analysis of the features used for discourse-new descriptions detection. We analyzed theoretically predicted features (based on typological and cognitive approaches to DN detection) and estimated their contribution. A set of additional features for Russian DN descriptions detection was suggested.

We tested classifiers that can distinguish between first mentions and recurring mentions. We also set a baseline for further experiments and tested special lexical features for referent novelty and inequality marking.

The analysis of the results of this first experiment for DN detection in Russian has shown that the lexical features are quite promising for this task and need further investigation and enhancing. The other promising direction of the research is the correlation of different features with the referent prominence (or the length of the coreference chain). In our future work we are planning to examine the contribution of more elaborated features to this task.

Acknowledgments. The authors would like to thank anonymous reviewers for their helpful comments, the Lomonosov Moscow University students who participated in the corpus markup, and Dmitrij Gorshkov for software support.

This research was supported by grant from Russian Foundation for Basic Research Fund (15-07-09306).

References

1. Ariel, M.: Accessing Noun-Phrase Antecedents. Routledge, London (1990)
2. Arutyunova, N.: Nomination, reference, meaning. [nominaciya, referenciya, znacheniye] (in Russian). In: Nomination: General Questions. [Nominaciya: obshie voprosi]. Nauka (1980)
3. Bean, D.L., Riloff, E.: Corpus-based identification of non-anaphoric noun phrases. In: Proceedings of the 37th Annual Meeting of the Association for Computational Linguistics on Computational Linguistics (ACL 1999), pp. 373–380. Association for Computational Linguistics, Stroudsburg (1999)
4. Bonch-Osmolovskaya, A., Toldova, S., Klintsov, V.: Introductory noun phrases: a case of mass media texts. [strategii introduktivnoj nominacii v teksrah smi] (in Russian) (2012)
5. Givón, T. (ed.): Topic Continuity in Discourse: A Quantitative Cross-Language Study. John Benjamins, Amsterdam (1983)
6. Grosz, B.J., Weinstein, S., Joshi, A.K.: Centering: a framework for modeling the local coherence of discourse. Comput. Linguist. **21**(2), 203–225 (1995)
7. Han, H., Wang, W.-Y., Mao, B.-H.: Borderline-SMOTE: a new over-sampling method in imbalanced data sets learning. In: Huang, D.-S., Zhang, X.-P., Huang, G.-B. (eds.) ICIC 2005. LNCS, vol. 3644, pp. 878–887. Springer, Heidelberg (2005). https://doi.org/10.1007/11538059_91
8. Hawkins, J.A.: Definiteness and Indefiniteness: A Study in Reference and Grammaticality Prediction. Croom Helm, London (1978)
9. Ionov, M., Kutuzov, A.: Influence of morphology processing quality on automated anaphora resolution for Russian. In: Proceedings of the International Conference Dialogue-2014. RGGU (2014)
10. Kabadjov, M.A.: A comprehensive evaluation of anaphora resolution and discourse-new classification. Ph.D. thesis. Citeseer (2007)
11. Kibrik, A., Linnik, A., Dobrov, G., Khudyakova, M.: Optimizacija modeli referencial'nogo vybora, osnovannoj na mashinnom obuchenii [Optimization of a model of referential choice, based on machine learning]. In: Computational Linguistics and Intellectual Technologies, vol. 11, pp. 237–246. RGGU, Moscow (2012)
12. Kibrik, A.A.: Reference in Discourse. Oxford University Press, Oxford (2011)
13. Lappin, S., Leass, H.J.: An algorithm for pronominal anaphora resolution. Comput. Linguist. **20**(4), 535–561 (1994)
14. Löbner, S.: Definites. J. Semant. **4**(4), 279–326 (1985)
15. Mitkov, R.: Anaphora resolution: the state of the art (1999)
16. Ng, V., Cardie, C.: Identifying anaphoric and non-anaphoric noun phrases to improve coreference resolution. In: Proceedings of the 19th International Conference on Computational Linguistics, vol. 1, pp. 1–7. Association for Computational Linguistics (2002)
17. Palek, B.: Cross-Reference a Study from Hyper-syntax. Universita Karlova, Prague (1968)
18. Pedregosa, F., Varoquaux, G., Gramfort, A., Michel, V., Thirion, B., Grisel, O., Blondel, M., Prettenhofer, P., Weiss, R., Dubourg, V., Vanderplas, J., Passos, A., Cournapeau, D., Brucher, M., Perrot, M., Duchesnay, E.: Scikit-learn: machine learning in Python. J. Mach. Learn. Res. **12**, 2825–2830 (2011)
19. Poesio, M., Kabadjov, M.A.: A general-purpose, off-the-shelf anaphora resolution module: implementation and preliminary evaluation. In: Proceeding of LREC, pp. 663–666 (2004)

20. Poesio, M., Kabadjov, M.A., Vieira, R., Goulart, R., Uryupina, O.: Does discourse-new detection help definite description resolution. In: Proceedings of the Sixth International Workshop on Computational Semantics, Tillburg (2005)
21. Poesio, M., Ponzetto, S.P., Versley, Y.: Computational models of anaphora resolution: a survey (2010)
22. Poesio, M., Vieira, R.: A corpus-based investigation of definite description use. Comput. Linguist. **24**(2), 183–216 (1998)
23. Prince, E.F.: The ZPG letter: subjects, definiteness, and information-status. In: Discourse Description: Diverse Analyses of a Fund Raising Text, pp. 295–325 (1992)
24. Recasens, M., de Marneffe, M.C., Potts, C.: The life and death of discourse entities: identifying singleton mentions. In: Human Language Technologies: The 2013 Annual Conference of the North American Chapter of the Association for Computational Linguistics, pp. 627–633. Association for Computational Linguistics, Stroudsburg, June 2013
25. Sharoff, S., Nivre, J.: The proper place of men and machines in language technology: processing Russian without any linguistic knowledge. In: Proceedings of Dialogue, Russian International Conference on Computational Linguistics, Bekasovo (2011)
26. Toldova, S.: Struktura diskursa i mehanizm fokusirovaniya kak vazhnie faktori vibora nominatsii ob'ekta v tekste (Discourse structure and the focusing mechanism as important factors of referential choice in text) (1994)
27. Toldova, S., Rojtberg, A., Ladygina, A., Vasilyeva, M., Azerkovich, I., Kurzukov, M., Ivanova, A., Nedoluzhko, A., Grishina, J.: RU-EVAL-2014: evaluating anaphora and coreference resolution for Russian. Comput. Linguist. Intell. Technol. **13**(20), 681–694 (2014)
28. Uryupina, O.: High-precision identification of discourse new and unique noun phrases. In: ACL Student Workshop, Sapporo (2003)
29. Vieira, R., Poesio, M.: An empirically based system for processing definite descriptions. Comput. Linguist. **26**(4), 539–593 (2000)

A Karaka Dependency Based Dialog Act Tagging for Telugu Using Combination of LMs and HMM

Suman Dowlagar[✉] and Radhika Mamidi

Kohli Center on Intelligent Systems (KCIS), International Institute of Information Technology, Hyderabad, Gachibowli, Hyderabad 500032, Telangana, India
{suman.d,radhika.mamidi}@research.iiit.ac.in

Abstract. The main goal of this paper is to perform the dialog act(DA) tagging for Telugu corpus. Annotation of utterances with dialog acts is necessary to recognize the intent of speaker in dialog systems. While English language follows a strict subject–verb–object(SVO) syntax, Telugu is a free word order language. The n-gram DA tagging methods proposed for English language will not work for free word order languages like Telugu. In this paper, we propose a method to perform DA tagging for Telugu corpus using advanced machine learning techniques combined with karaka dependency relation modifiers. In other words, we use syntactic features obtained from karaka dependencies and apply combination of language models(LMs) at utterance level with Hidden Markov Model(HMM) at context level for DA tagging. The use of karaka dependencies for free word order languages like Telugu helps in extracting the modifier-modified relationships between words or word clusters for an utterance. The modifier-modified relationships remain fixed even though the word order in an utterance changes. These extracted modifier-modified relationships appear similar to n-grams. Statistical machine learning methods such as combination of LMs and HMM are applied to predict DA for an utterance in a dialog. The proposed method is compared with several baseline tagging algorithms.

1 Introduction

In NLP, dialogs are nothing but the conversations between human and human or human and computer as shown in Fig. 1.

Dialogue acts are sentence-level units that represents the states of a dialog. According to [1], an utterance in a dialog is a kind of action performed by the speaker. To annotate the action of an utterance, dialog acts are needed. [2] said that, the dialog acts represent the meaning of an utterance in the context of a dialog. Tagging an utterance with its respective action is termed as dialog act (DA) tagging. DA tagging of an utterance is equivalent to understanding the utterance at a more general level which helps in recognizing the intent of that utterance.

© Springer International Publishing AG, part of Springer Nature 2018
A. Gelbukh (Ed.): CICLing 2016, LNCS 9623, pp. 663–674, 2018.
https://doi.org/10.1007/978-3-319-75477-2_48

Telugu	విద్యార్థి	:	నమస్కారం	సర్.		
Gloss	student	:	hello	sir		
Translation	Student	:	Hello sir.			
Telugu	లైబ్రేరియన్	:	నమస్కారం.			
Gloss	librarian	:	hello			
Translation	Librarian	:	Hello.			
Telugu	విద్యార్థి	:	నేను	ఈ	పుస్తకము	తీసుకోవాలి.
Gloss	student	:	I	this	book	take_should
Translation	Student	:	I want to check out this book.			

Fig. 1. A sample conversation written in the Telugu language with gloss and English Translation

A lot of research is being conducted regarding DA tagging on various languages. With the help of artificial intelligence and machine learning, several statistical approaches are developed for DA tagging. Out of all the statistical approaches, the use of n-grams for DA tagging gave better accuracy [3]. [4] observed that the dialog acts can be inferred from their constituent words using a very simple unigram model. [5] considers n-gram dialog act tagging method by taking n-grams of order 1 to 3. [6] extended the approach of [5] and proved that the n-grams of order 1 to 4 with certain threshold limits on n-grams not only decreases the complexity of data but also results in better accuracy. But [6] pointed out that the threshold concept works only if the corpus is large enough. [2] proved that using dependency features, DA tagging is possible. [7] lists several other approaches of DA tagging such as using memory based learning, Bayes network, Hidden Markov model etc.

The methods discussed above were developed by considering the dialog corpus in English which follows strict SVO syntax. These methods cannot be directly applied to free word order languages like Telugu. In Telugu, like most other South Asian languages, the word order of grammatical functions like subject and objects are largely free. Internal changes in the sentences or the phrases will not affect the grammatical functions of the nominals. Before we use statistical n-gram methods, a grammatical model must be implemented such that the model restricts the movement of the words or the word clusters by establishing some relationships between them [8].

The grammatical model used to establish the relationships between the words or the word clusters is based on Paninian framework [9]. Paninian framework treats a sentence as a series of modifier-modified elements. A sentence is supposed to have a primary modified, which is generally the main verb of the sentence. The elements modifying the verb participate in the action specified by the verb. The participant relations with the verb are called *karakas*. The appropriate mapping of the syntactic cues helps in identifying the appropriate *karakas* ('participants in an action'). *Karakas* are the grammatical relations that bind the modifier and the modified elements. Through Paninian framework we know that the relationship between words does not change even though the word order changes. Therefore this framework would be better suited for sentence analysis of Telugu, which is relatively a free word order language.

Why karaka based dependencies are chosen over Stanford dependency relations? According to [10] Indian Languages(ILs) are morphologically rich and have a relatively flexible word order. For such languages syntactic subject-object positions are not always be able to elegantly explain the varied linguistic phenomena. In fact, there is a debate in the literature whether the notions 'subject' and 'object' can at all be defined for ILs [11]. Behavioral properties are the only criteria based on which one can confidently identify grammatical functions in Telugu. It is difficult to exploit such properties computationally. Marking semantic properties such as thematic role as dependency relation is also problematic. Thematic roles are abstract notions and will require higher semantic features which are difficult to formulate and to extract as well. So, thematic roles are not marked at this juncture. On the other hand, the notion of karaka relations provides us a level which while being syntactically grounded also helps in capturing some semantics.

[12] describes DA tagging of Telugu using karaka dependencies to obtain n-grams. Dependency parsed data is converted to n_gram_karaka format and Katz's back-off is applied as language modeling at an intra utterance level. After that, memory based learning algorithm viz. KNN is applied at the utterance level. We extend the above approach as

1. Instead of considering naive language model(LM), we consider a combination of language models.
2. The context of previous utterances is considered for predicting the DA tag of the current utterance.

The combination of language models considered is back-off [13] and interpolation [14].

We know that a task oriented dialog is continuous flow of data. It can be broadly divided into 4 categories. First comes, conversation initiators. For example, *hello, hi, how are you?* etc. Followed by the request for start of the task, like, *I want to take this book.* Then comes the accomplishment of the task, like, *Now, you can take this book,* finally follows the conversation terminator, like, *thank you.* Due to this dialog flow, we can say that each utterance in a conversation depends on its previous utterances. The context of the previous utterances must be maintained to predict the tag of the current utterance.

In [12], dependency parsed data is extracted manually. In our method, we automate the extraction of karaka dependencies by employing a constraint based approach.

The paper is structured as follows. The 2nd section describes the corpus used. The 3rd section gives a detailed description of our technique. The 4th section describes results and 5th section concludes the paper.

2   Dialog Data

To perform DA tagging, the dialog data considered is the ASKLIB corpus obtained from [12]. This corpus focuses on the interactions that takes place between a student and a librarian. ASKLIB corpus consists of a total 225 task oriented library domain oriented dialogs.

The tagset that is considered for DA tagging of the ASKLIB corpus is DAMSL tagset developed by [15] along with some domain dependent tags. The DAMSL tagset is designed to be a domain independent tagset and is applicable for any type of corpus. In our tagset, a few domain dependent tags are included along with the DAMSL tagset. The reason is that, the dialogs in the ASKLIB corpus are divided into 4 categories. They are ISSUE, RETURN, REISSUE and ENQUIRY. To make the tagset more application specific, the above three categories i.e. ISSUE, RETURN, REISSUE are included as tags. The fourth category i.e. ENQUIRY is already a part of the DAMSL tagset which is present in the DAMSL tagset as INFO_REQUEST. The DA tagset used for tagging the ASKLIB corpus is obtained from [12]. A total of 20 tags are used to tag the ASKLIB corpus. Figure 2 gives DA tagged Telugu conversation between a student and a librarian with gloss and its English translation (Figs. 3 and 4).

Speaker		Dialog in Telugu with gloss and its English Translation					DA tag
విద్యార్థి	:	నమస్కారం	సర్.				
student	:	hello	sir				GREETINGS
Student	:	Hello sir.					
లైబ్రేరియన్	:	నమస్కారం.					
Librarian	:	hello					GREETINGS_REPLY
Librarian	:	Hello.					
విద్యార్థి	:	ఎన్ని	రోజులలో	ఈ	పుస్తకం	ఇచ్చేయాలి?	
student	:	how_many	days_in	this	book	give_should	INFO_REQUEST
Student	:	By when this book has to be returned?					
లైబ్రేరియన్	:	ఒక్క	నెలలో.				
Librarian	:	one	month_in				ANSWER
Librarian	:	In one month.					
విద్యార్థి	:	ధన్యవాదములు.					
student	:	thanks					GREETINGS_EOC
Student	:	Thank you.					

Fig. 2. A dialog taken from ASKLIB corpus with DA tags

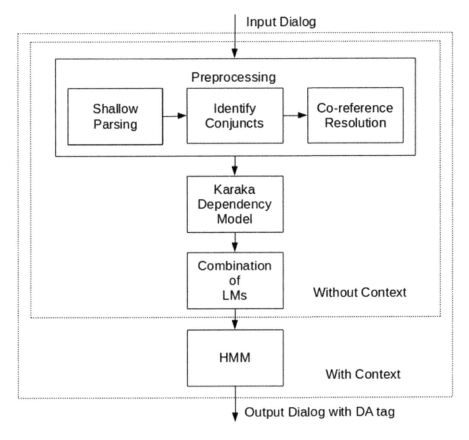

Fig. 3. Figure showing karaka dependency based approach for DA tagging using combination of LMs and HMM.

3   Our Approach

Our approach is based on the basic assumption that every word in an utterance must contribute to its DA tag. But using only unigrams to obtain the DA tag of an utterance yields accuracy below the baseline because the unigrams do not gather any local context. Instead of just using unigrams, if we capture dependency information, it might help in predicting the best DA tag for an utterance. This can be obtained with the help of karaka dependencies.

In our approach, a grammatical model which extracts the karaka dependencies between the words or the word clusters is implemented. The word clusters with karaka dependencies between them are given to the statistical language modeling unit at the utterance level and HMM is applied at the context level to perform the dialog act tagging.

3.1   Preprocessing

Before extracting a list of modifier_karaka_modified chunks for an utterance, the preprocessing of the utterance is a must. This includes shallow parsing, handling conjuncts and co-reference resolution.

Shallow parsing by [16] is done to establish intra chunk dependencies. Here, the Telugu utterances are given to the shallow parser(SP). The shallow parser has mainly 3 modules. They are morph module, POS module and chunker module. The morph module extracts the morph and suffix information. Then the POS module annotates each word in an utterance with its respective POS tags. Finally the chunker module establishes intra chunk dependencies.

After shallow parsing, by observing the output data, it is clear that there are certain words which are chunked into separate units by the shallow parser but a single meaning is obtained by joining them. These are known as conjuncts. The problem of conjuncts is solved by establishing constraints and grouping those conjuncts as a single unit.

The third step is the co-reference resolution. In the library domain, we come across utterances such as "give me the book". When it is said by a librarian, the utterance's intent might be to 'return the book', but when it is said by a student the utterance's intent might be 'to issue the book'. If the personal pronoun "me" is resolved with the respective anaphora i.e. either the librarian or a student, this further leads to better intention recognition of the utterance. The co-reference resolution is purely a set or rules. By observing the data, rules are written to resolve the personal pronouns.

3.2   Karaka Dependencies

After preprocessing, a constraint based karaka dependency module is implemented. By observing the ASKLIB corpus, rules are written by considering chunk, conjunct information and parts of speech tags. These rules act as syntactic cues and help in predicting the modifier-modified relationships between the chunks in the preprocessed utterance. Here, the preprocessed utterance is given as input to the constraint based karaka dependency module and the output obtained will be a list of modifier_karaka_modified chunks for the corresponding utterance.

Where "k1" is the karta, the one who carries out the action and "k2" is the karma or the object of the verb.

3.3   Language Modeling Combination

As n_gram_karakas can be treated likewise to n-grams, n-gram language modeling methods can be applied. The statistical n-gram LM technique used here is the language modeling combination of linear interpolation and back-off. We modified the simple linear interpolation method. Instead of obtaining the lambda values from the held out data, we replaced the lambda values using back-off weights. LM combination of interpolation and back-off is used because, it has been proven

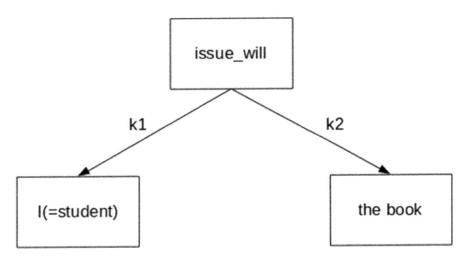

Fig. 4. Figure showing karaka dependencies for an example utterance *"I will issue the book"*.

that combination of LMs decreases the error rate when compared to the naive LM methods as the negative effects of interpolation and back-off cancel out. Another advantage of using back-off weights is that we can assign higher priority to the n_gram_karakas when compared to the non dependencies. The basic constraint of all the weights summing up to one is kept same. Along with LM combination, naive smoothing method is also applied.

Telugu language has a rich and productive morphology. The number of word forms might be infinite in such a case. The use of just n_gram_karakas is not enough while LM method is performed. Hence we go for the morphed form of n_gram_karakas, they are known as morphed n_gram_karakas. While identifying modifier-modified relationships, it is not necessary that all words in the utterance satisfy the modifier-modified criteria. In such a case, these words should not be left alone. Hence we reduce the n_gram_karakas to non karaka dependent chunks such as $chunk_1$ and $chunk_2$ as shown in Eq. 3. Similarly, morphed n_gram_karakas are further reduced to two morph chunks such as $morph\_chunk_1$ and $morph\_chunk_2$ as shown in Eq. 7.

Finally, in the interpolation method with back-off, the n_gram_karakas, the morphed n_gram_karakas, the non karaka dependent chunks and the morphed non karaka dependent chunks are interpolated by considering the back-off weights.

The equations that are used for the calculating probabilities over the DA tagset T using a combination of LMs are given below:

$$P_1 = \frac{C(n\_gram\_karaka, T)}{C(n\_gram\_karaka)} \tag{1}$$

$$P_2 = \frac{C(morph\_n\_gram\_karaka, T)}{C(morph\_n\_gram\_karaka)} \tag{2}$$

$$n\_gram\_karaka = chunk_1 + karaka + chunk_2 \tag{3}$$

$$P_{3_1} = \frac{C(chunk_1, T)}{C(chunk_1)} \tag{4}$$

$$P_{3_2} = \frac{C(chunk_2, T)}{C(chunk_2)} \tag{5}$$

$$P_3 = P_{3_1} + P_{3_2} \tag{6}$$

$$morph\_n\_gram\_karaka =$$
$$morph\_chunk_1 + karaka + morph\_chunk_2 \tag{7}$$

$$P_{4_1} = \frac{C(morph\_chunk_1, T)}{C(morph\_chunk_1)} \tag{8}$$

$$P_{4_2} = \frac{C(morph\_chunk_2, T)}{C(morph\_chunk_2)} \tag{9}$$

$$P_4 = P_{4_1} + P_{4_2} \tag{10}$$

$$P_{LI} = \sum_{i=1}^{4} \alpha_i * P_i \tag{11}$$

$$where \; \alpha_{i+1} = 0.4 * \alpha_i \; and \; \sum_{i=1}^{4} \alpha_i = 1 \tag{12}$$

where C refers to the count, T is the DA tagset used, P is the probability calculated for i^{th} tag T_i for the given chunk. Equation 11 gives the language model combination of interpolation with back-off. The condition given in Eq. 12 are taken from the stupid back-off by [13]. Stupid back-off is considered because of its simplicity and linearity in back-off weights which provides a good integration into the deleted linear interpolation method. The language model is applicable only at utterance level. For a better DA tagging, the context of the utterances must be considered. Hence, for context level we have used Hidden Markov Model (HMM).

3.4   Adding HMM for Context Handling

Now coming to the context preserving, we know that the task oriented dialogs have conversational flow which lead to accomplishment of the required task. Flow of conversation can be shown in the Table 2.

This flow can be captured by using the forward backward algorithms. We have used HMM as one of the forward backward algorithms to capture that flow. As HMM works on the principle of the Markov model, i.e. the present state of the sentence is predicted by observing its previous states. We modified it as, *the present DA tag of the utterance is predicted by observing the previous 'k' DA tags in the dialog.* The Viterbi algorithm is a dynamic programming algorithm and is the most common decoding algorithm used for HMMs, whether for part-of-speech tagging or for speech recognition. We used this for the DA tagging of the sentences.

Why context handling for ASKLIB corpus? In ASKLIB corpus there might be situations where the implicit meaning of the speaker is different from what the utterance literally conveys i.e. the words in an utterance might point to a particular DA tag based on the LM output but while considering the context they might point to an another DA tag. An example is shown in Table 1.

Table 1. An example which proves that context handling is necessary.

DA tagging without context handling	DA tag
Student: I would like to issue this book	ISSUE
Librarian: This book has already been issued to you. Would you like to reissue the book?	REISSUE_INFO_REQUEST
DA tagging with context handling	*DA tag*
Student: I would like to issue this book	REISSUE
Librarian: This book has already been issued to you. Would you like to reissue the book?	REISSUE_INFO_REQUEST

From the Table 1[1], after observing the 2 utterances, one can say that the student wants to reissue the book. On the other hand, if we use only the combination of LMs, the information conveyed by the librarian will be missed as LMs modeled here work at utterance level only. Hence without the context i.e. only at LM level, the DA tag given will be ISSUE. But with the help of context viz. with the librarian's utterance we can say that the appropriate DA tag is REISSUE.

$$\hat{Q} = argmax_Q \ P(Q)P(S|Q) \tag{13}$$

$$\hat{Q} = argmax_Q \prod_{i=1}^{n} P(Q_i|Q_{i-1})\frac{P(Q_i|S_i)}{P(Q_i)} \tag{14}$$

Equation 13 which is further simplified to Eq. 14. Equation 14 calculates the best tag for the given utterance considering its previous context.

Viterbi algorithm consists of the emission matrix and the transition matrix. The emission matrix values are obtained from the LM combination method mentioned above and the transition matrix values are obtained from conversation flow of the training data. The Viterbi algorithm is run two times. Firstly, the dialog is given to the Viterbi algorithm and the output with DA tag and its probabilities are taken. During the second run, the dialog is reversed and given to Viterbi algorithm and the corresponding DA tag for utterance and its probabilities are taken.

---

[1] The dialogs given here were originally written in Telugu but translated to English for reader's understanding.

Then the two probabilities obtained for each utterance are compared and the best tag which has the greater probability is given to the utterance.

This approach helps in complete coverage of the utterance i.e. this approach not only considers previous tags but indirectly it takes next tags to predict the output of the current utterance.

4   Results

We compared our approach with the various state of art tagging algorithms. The results are shown in Table 2

The ASKLIB data is divided equally into 4 sets. The concept of leave-one-out validation is applied here. Every time one of the set is used for testing and remaining are combined for training.

Table 2. Accuracy obtained by comparing various classification algorithms.

Classification algorithms	Result
Unigrams approach	58.17%
Bigrams approach	54.71%
Trigrams approach	47.79%
n_gram_karakas using katz's backoff as LM plus kNN	73.34%
Current approach with out context(i.e. without HMM)	75.89%
Current approach with context(i.e. with HMM)	78.67%

Firstly, from ASKLIB corpus n-grams of length 1 to 3 are extracted and Bayesian interpretation is used for DA tagging. From the results in Table 2, it is clear that when the n-gram length increases, there is a decrease in accuracy. The reason is, as Telugu is a free word order language, even though the utterance's intent is the same, there might be change in the position of words in both the training data and the test data. Hence, extracting only fixed order n-grams for DA tagging will not work for free word order languages like Telugu.

Next half of the Table 2 shows the method of karaka dependencies to extract n-grams. The current method is compared with various classification algorithms such as Katz's backoff and a combination of LMs.

From the results presented in Table 2 it is clear that our method has shown an improvement in accuracy. The improvement in the accuracy is due to the addition of the following factors.

1. **Efficient processing of data by considering the linguistic features of Telugu language.**
   Linguistic features such as shallow parsing and the use of conjuncts helped in the optimized grouping of word clusters. This in-turn helped in the better establishment of the modifier-modified relationships.

2. **Coreference resolution for speaker recognition.**
   Coreference resolution of speaker has differentiated the student and the librarian utterances which contributed effective DA tagging of the utterance.

3. **The use of combination of LMs instead of just Katz's back-off.**
   As the combination of the interpolation and back-off was used, high-order n-grams i.e. the n_gram_karakas which were sensitive, and had sparse counts were given high back-off weight. The reason was that n_gram_karakas have more detailed information for accurately predicting DA tag of the sentence. The low-order n-grams i.e. the non karaka dependent chunks (both inflected and morphed) having robust counts were given low back-off weight because the contribution of non karaka dependent chunks in predicting DA tag was low.
   Due to this LM combination the negative effects of both the naive interpolation and back-off on the data were reduced.

4. **Consideration of the context of the previous utterance while predicting the DA tag of the current utterance.**
   Context handling and backtracking by Viterbi algorithm helped in preserving the flow of the dialog, which in-turn helped in efficient DA tagging.

Note that the final accuracy is the multi-class accuracy that is obtained by calculating the accuracy of each and every DA tag, then taking the average of the normalized DA tagset accuracy as given in the Eq. 15

$$Accuracy = \sum_{i=1}^{T} \frac{Count\,(T_i)}{Count(T)} Accuracy(T_i) \tag{15}$$

where T stands for the tagset, T_i is the ith tag in the tagset, $Accuracy(T_i)$ gives the accuracy of the single tag T_i. $Accuracy(T_i)$ is calculated by comparing the tags obtained from the model w.r.t the manual tagged output.

5   Conclusion

In this paper, we present a DA tagging method which can tag the morphologically rich, free word order languages like Telugu. Our proposed method uses karaka dependencies in the modifier_karaka_modified format. The use of modifier-modified karaka dependency model converts the free word order to a fixed grammatical format. By careful observation, this format is similar to the n-grams in English. Hence, a statistical LM method is considered. Further, an improved version of statistical LM method which is a combination of interpolation and back-off is used. Contextual handling is done with the help of HMM using Viterbi algorithm.

References

1. Austin, J.L.: How to do Things with Words, vol. 367. Oxford University Press, Cambridge (1975)
2. Král, P., Cerisara, C.: Automatic dialogue act recognition with syntactic features. Lang. Resour. Eval. **48**, 419–441 (2014)
3. Ivanovic, E.: Dialogue act tagging for instant messaging chat sessions. In: Proceedings of the ACL Student Research Workshop, pp. 79–84. Association for Computational Linguistics (2005)
4. Garner, P.N., Browning, S.R., Moore, R.K., Russell, M.J.: A theory of word frequencies and its application to dialogue move recognition. In: ICSLP 1996 Proceedings of Fourth International Conference on Spoken Language, vol. 3, pp. 1880–1883. IEEE (1996)
5. Louwerse, M.M., Crossley, S.A.: Dialog act classification using n-gram algorithms. In: FLAIRS Conference, pp. 758–763 (2006)
6. Webb, N., Ferguson, M.: Automatic extraction of cue phrases for cross-corpus dialogue act classification. In: Proceedings of the 23rd International Conference on Computational Linguistics: Posters, pp. 1310–1317. Association for Computational Linguistics (2010)
7. Král, P., Cerisara, C.: Dialogue act recognition approaches. Comput. Inf. **29**, 227–250 (2012)
8. Bharati, A., Sangal, R., Sharma, D.M., Bai, L.: AnnCorra: annotating corpora guidelines for POS and chunk annotation for Indian languages. LTRC-TR31 (2006)
9. Bharati, A., Chaitanya, V., Sangal, R., Ramakrishnamacharyulu, K.: Natural Language Processing: A Paninian Perspective. Prentice-Hall of India, New Delhi (1995)
10. Begum, R., Husain, S., Dhwaj, A., Sharma, D.M., Bai, L., Sangal, R.: Dependency annotation scheme for Indian languages. In: IJCNLP, pp. 721–726. Citeseer (2008)
11. Mohanan, K.P.: Grammatical relations and clause structure in Malayalam. Ment. Represent. Gramm. Relat. **504**, 589 (1982)
12. Dowlagar, S., Mamidi, R.: A semi supervised dialog act tagging for Telugu. In: ICON 2015 : 12th International Conference on Natural Language Processing (2015)
13. Brants, T., Popat, A.C., Xu, P., Och, F.J., Dean, J.: Large language models in machine translation. In: Proceedings of the Joint Conference on Empirical Methods in Natural Language Processing and Computational Natural Language Learning. Citeseer (2007)
14. Jurafsky, D., Martin, J.H.: Speech & Language Processing. Pearson Education India, Noida (2000)
15. Core, M.G., Allen, J.: Coding dialogs with the DAMSL annotation scheme. In: AAAI Fall Symposium on Communicative Action in Humans and Machines, Boston, MA, pp. 28–35 (1997)
16. PVS, A., Karthik, G.: Part-of-speech tagging and chunking using conditional random fields and transformation based learning. Shallow Parsing South Asian Lang. **21** (2007)

Author Index

Printed in the United States
By Bookmasters